The Biology of Butterflies

Previous Symposia of the Royal Entomological Society

NO. 1. INSECT POLYMORPHISM edited by J. S. Kennedy
London: 1961

NO. 2. INSECT REPRODUCTION edited by K. C. Highnam
London: 1964

NO. 3. INSECT BEHAVIOUR edited by P. T. Haskell
London: 1966

NO. 4. INSECT ABUNDANCE edited by T. R. E. Southwood
Blackwell Scientific Publications, Oxford: 1968

NO. 5. INSECT ULTRASTRUCTURE edited by A. C. Neville
Blackwell Scientific Publications, Oxford: 1970

NO. 6. INSECT/PLANT RELATIONSHIPS edited by H. F. van Emden
Blackwell Scientific Publications, Oxford: 1973

NO. 7. INSECT FLIGHT edited by R. C. Rainey
Blackwell Scientific Publications, Oxford: 1976

NO. 8. INSECT DEVELOPMENT edited by P. A. Lawrence
Blackwell Scientific Publications, Oxford: 1976

NO. 9. DIVERSITY OF INSECT FAUNAS edited by L. A. Mound and N. Waloff
Blackwell Scientific Publications, Oxford: 1978

NO. 10. INSECT CYTOGENETICS edited by R. L. Blackman, G. M. Hewitt and M. Ashburner
Blackwell Scientific Publications, Oxford: 1980

NO. 11. THE BIOLOGY OF BUTTERFLIES edited by R. I. Vane-Wright and P. R. Ackery
Academic Press, London: 1984

Future titles in this series

NO. 12. INSECT COMMUNICATION edited by T. Lewis
Academic Press, London: 1984

E. B. Ford

The Biology of Butterflies

Symposium of the Royal Entomological Society of London
Number 11

Dedicated to E. B. Ford

Edited by
R. I. VANE-WRIGHT and P. R. ACKERY

Published for The Royal Entomological Society 41 Queen's Gate London
by

19  84

ACADEMIC PRESS

(Harcourt Brace Jovanovich, Publishers)

LONDON ORLANDO SAN DIEGO SAN FRANCISCO NEW YORK
TORONTO MONTREAL SYDNEY TOKYO SÃO PAULO

ACADEMIC PRESS INC. (LONDON) LTD
24/28 Oval Road
London NW1 7DX

United States Edition published by

ACADEMIC PRESS INC.
(Harcourt Brace Jovanovich, Inc.)
Orlando, Florida 32887

British Library Cataloguing in Publication Data
The Biology of butterflies—(Symposia of the
Royal Entomological Society of London,
ISSN 0080-4363;11)
1. Butterflies—Congresses
I. Vane-Wright, R. I. II. Ackery, P. R.
III. Series
595.78'9 QL452

ISBN 0-12-713750-5
LCCN 83 71684

Phototypeset by Dobbie Typesetting Service, Plymouth

Printed by Page Brothers, Norwich

Contributors

Ackery, Phillip R.
Department of Entomology, British Museum (Natural History), Cromwell Road, London SW7 5BD, UK

Baker, R. Robin
Department of Zoology, University of Manchester, Oxford Road, Manchester M13 9PL, UK

Boppré, Michael
Universität Regensburg, Zoologie 2 SFB 4/B6, Universitätstrasse 31, D-8400 Regensburg, Federal Republic of Germany

Brakefield, Paul M.
Department of Biological Sciences, Perry Road, University of Exeter, Exeter EX4 4QG, UK

Brower, Lincoln P.
Department of Zoology, University of Florida, Gainesville, Florida 32611, USA

Chew, Frances S.
Department of Biology, Tufts University, Medford, Massachusetts 02155, USA

Clarke, Sir Cyril A.
Department of Genetics, University of Liverpool, Brownlow Street, P.O. Box 147, Liverpool L69 3BX, UK

Courtney, Stephen P.
Department of Zoology, University of Liverpool, Brownlow Street, P.O. Box 147, Liverpool L69 3BX, UK

Dempster, Jack P.
Natural Environment Research Council, Institute of Terrestrial Ecology, Monks Wood Experimental Station, Abbots Ripton, Huntingdon PE17 2LS, UK

DeVries, Philip J.
Department of Zoology, University of Texas at Austin, Austin, Texas 78712, USA

Edgar, John A.
CSIRO, Division of Animal Health, Animal Health Research Laboratory, Private Bag No. 1, Parkville, Victoria 3052, Australia

Ehrlich, Paul R.
Department of Biological Sciences, Stanford University, Stanford, California 94305, USA

Evans, Fred. J.
Department of Pharmacognosy, School of Pharmacy, University of London, 29-39 Brunswick Square, London WC1N 1AX, UK

Gibson, Dianne O.
Department of Genetics, University of Liverpool, Brownlow Street, P.O. Box 147, Liverpool L69 3BX, UK

Gilbert, Lawrence E.
Department of Zoology, University of Texas at Austin, Austin, Texas 78712, USA

Gordon, Ian J.
Department of Biology, Oxford Polytechnic, Headington, Oxford OX3 0BP, UK

Harrison, S. J.
Department of Biological Sciences, University of Maryland Baltimore County, 5401 Wilkens Avenue, Catonsville, Maryland 21228, USA

Kitching, Ian J.
Department of Entomology, British Museum (Natural History), Cromwell Road, London SW7 5BD, UK

Lane, Richard P.
Department of Entomology, British Museum (Natural History), Cromwell Road, London SW7 5BD, UK

McLeod, Leonard
Quartier des Ecoles, 84330 St Pierre de Vassols, France

Marsh, Neville
Department of Physiology, Queen Elizabeth College, Campden Hill Road, London W8, UK

Morton, Ashley C.
Department of Biology, Building 44, The University, Southampton SO9 5NM, UK

Nakanishi, Akinori
Biological Laboratory, College of General Education, Kyushu University, Ropponmatsu, Fukuoka, 810 Japan

Parsons, Michael J.
Hurst Lodge, Hurst Lane, Egham, Surrey TW20 8QJ, UK

Pierce, Naomi E.
Museum of Comparative Zoology Laboratories, Harvard University, Cambridge, Massachusetts 02138, USA

Platt, Austin P.
Department of Biological Sciences, University of Maryland Baltimore County, 5401 Wilkens Avenue, Catonsville, Maryland 21228, USA

Pollard, Ernest
Natural Environment Research Council, Institute of Terrestrial Ecology, Monks Wood Experimental Station, Abbots Ripton, Huntingdon PE17 2LS, UK

Porter, Keith
Department of Biology, Oxford Polytechnic, Headington, Oxford OX3 0BP, UK

Pyle, Robert M.
Swede Park, Loop Road, Box 123, Gray's River, Washington 98621, USA

Robbins, Robert K.
Department of Entomology, NHB 127, National Museum of Natural History, Smithsonian Institute, Washington DC 20560, USA

Rothschild, Miriam
Ashton, Peterborough PE8 5L2, UK

Saigusa, Tohohei
Biological Laboratory, College of General Education, Kyushu University, Ropponmatsu, Fukuoka, 810 Japan

Shapiro, Arthur M.
Department of Zoology, University of California, Davis, California 95616, USA

Shima, Hiroshi
Biological Laboratory, College of General Education, Kyushu University, Ropponmatsu, Fukuoka, 810 Japan

Silberglied, Robert E.
Smithsonian Tropical Research Institute, Apartado 2072, Balboa, Republic of Panama

Singer, Michael C.
Department of Zoology, University of Texas at Austin, Austin, Texas 78712, USA

Smith, David A. S.
Department of Biology, Eton College, Windsor, Berkshire, SL4 6EW, UK

Suzuki, Yoshito
Department of Biophysics, Faculty of Science, Kyoto University, Kyoto, 606 Japan

Thomas, Jeremy A.
Institute of Terrestrial Ecology, Furzebrook Research Station, Wareham, Dorset DH20 5AS, UK

Turner, John R. G.
Department of Genetics, University of Leeds, Leeds LS2 9JT, UK

Vane-Wright, Richard I.
Department of Entomology, British Museum (Natural History), Cromwell Road, London SW7 5BD, UK

Williams, Thomas F.
Department of Biological Sciences, University of Maryland Baltimore County, 5401 Wilkens Avenue, Catonsville, Maryland 21228, USA

Yata, Osamu
Biological Laboratory, College of General Education, Kyushu University, Ropponmatsu, Fukuoka, 810 Japan

Yoshida, Akihiro
Life Sciences Institute, Sophia University, Tokyo 102 Japan

List of Participants

Ackery, P. R. *(UK)*
Adams, M. J. *(UK)*
Arnold, R. A. *(USA)*
Arora, R. *(UK)*
Bacon, J. *(UK)*
Balletto, E. *(Italy)*
Baker, R. R. *(UK)*
Betts, E. *(UK)*
Birch, M. C. *(UK)*
Boppré, M. *(W. Germany)*
Bowden, S. R. *(UK)*
Brakefield, P. *(Netherlands)*
Brower, L. P. *(USA)*
Bullini, L. *(Italy)*
Cain, R. *(UK)*
Cassidy, A. C. *(UK)*
Chew, F. *(USA)*
Cianchi, M. R. *(Italy)*
Clarke, C. A. *(UK)*
Clarke, F. *(UK)*
Clarke, S. A. *(UK)*
Classey, E. W. *(UK)*
Cook, L. M. *(UK)*
Cottrell, C. B. *(Zimbabwe)*
Courtney, S. *(UK)*
Crane, R. *(UK)*
Davis, R. H. *(UK)*
de Jong, R. *(Netherlands)*
Dempster, J. *(UK)*
Dening, R. C. *(UK)*
Dennis, R. L. H. *(UK)*
DeVries, P. J. *(USA)*
Dover, J. *(UK)*
Dowdeswell, W. H. *(UK)*
Duggan, A. *(UK)*
Duthie, D. *(UK)*
Dyte, C. E. *(UK)*
Edgar, J. A. *(Australia)*
Edwards, C. *(UK)*
Ehrlich, P. R. *(USA)*
Eldin, S. *(UK)*
Eliot, J. N. *(UK)*
Forsberg, J. *(Sweden)*
Frazer, J. F. D. *(UK)*
Gange, A. *(UK)*
Gardiner, B. O. C. *(UK)*
Gibson, D. O. *(UK)*
Gilbert, L. *(USA)*
Gill, A. C. L. *(UK)*
Gordon, I. J. *(Ghana)*
Greatorex-Davies, J. N. *(UK)*
Haeuser, C. *(W. Germany)*
Harborne, J. B. *(UK)*
Heath, J. *(UK)*
Hill, C. J. *(UK)*
Hobbs, R. N. *(UK)*
Holloway, J. D. *(UK)*
Huxley, J. *(UK)*
Jenniersten, O. *(Sweden)*
Kielland, J. *(Norway)*
Kitching, I. J. *(UK)*
Klÿnstra, J. W. *(Netherlands)*
Lane, R. P. *(UK)*
Larsen, T. B. *(UK)*
Lewis, T. *(UK)*
Longino, J. *(USA)*
Lyal, C. H. *(UK)*
Makings, P. *(UK)*
Malcolm, S. B. *(UK)*
Marks, L. S. *(USA)*
McCaffrey, J. *(USA)*
McLeod, L. *(France)*
Mikkola, K. *(Finland)*
Monteys, V. S. *(Spain)*
Morris, M. G. *(UK)*
Morton, A. C. *(UK)*
Mound, L. A. *(UK)*
Muggleton, J. *(UK)*
Nash, R. *(UK)*
Ng, S. M. *(UK)*
Owen, D. *(UK)*
Panchen, A. L. *(UK)*
Paxton, R. *(UK)*
Peachey, C. *(UK)*
Pellmyr, O. *(Sweden)*
Penrose, A. *(UK)*
Petersen, B. *(Sweden)*
Pierce, N. E. *(USA)*
Pitkin, L. M. *(UK)*
Platt, A. P. *(USA)*
Pollard, E. *(UK)*
Porter, K. *(UK)*
Pyle, R. M. *(USA)*
Racheli, T. *(Italy)*
Rainey, R. C. *(UK)*
Ratcliffe, N. M. *(UK)*
Remington, C. L. *(USA)*
Rivers, C. F. *(UK)*
Robbins, R. K. *(USA)*
Robinson, G. S. *(UK)*
Rothschild, M. *(UK)*
Rydon, A. H. B. *(UK)*
Ryrholm, N. *(Sweden)*
Sattler, K. S. *(UK)*
Sbordoni, V. *(Italy)*
Scheermeyer, E. *(Australia)*
Schneider, D. *(W. Germany)*
Shapiro, A. *(USA)*
Shreeve, T. G. *(UK)*
Silberglied, R. E. *(Panama)*
Simcox, D. *(UK)*
Singer, M. *(USA)*
Smiles, R. L. *(UK)*
Smith, A. G. *(UK)*
Smith, D. A. S. *(UK)*
Soberon, J. *(UK)*
Softly, R. *(UK)*
Solomon, M. E. *(UK)*
Southwood, T. R. E. *(UK)*
Sperling, F. A. H. *(Canada)*
Stubbs, A. E. *(UK)*
Thomas, J. *(UK)*
Tremewan, W. G. *(UK)*
Tubbs, R. *(UK)*
Turner, J. R. G. *(UK)*
Tweedie, M. W. R. *(UK)*
Usher, M. B. *(UK)*
Vane-Wright, R. I. *(UK)*
Van Loon, J. J. A. *(Netherlands)*
Waloff, N. *(UK)*
Warner, R. *(UK)*
Warren, M. S. *(UK)*
Watkins, D. *(UK)*
Watkins, G. G. *(UK)*
Watson, A. *(UK)*
Whalley, P. E. S. *(UK)*
Wiklund, C. *(Sweden)*
Wilson, A. *(UK)*
Winder, C. L. *(UK)*
Wooff, W. R. *(UK)*
Yata, O. *(Japan)*

Preface

Without doubt the Lepidoptera are not only one of the largest Orders of insects, but also the most popular. While it must be admitted that much of this interest borders on the merely philatelic, a vast amount of amateur energy and professional time has been expended on the biology of these beautiful animals. It is curious, then, that the only readily available introductory texts on their biology remain E. B. Ford's outstanding *Butterflies* (1945) and *Moths* (1955).

The idea for a symposium on butterfly biology arose from a series of five one-day workshops, organised by the Butterfly Research Association, with the help of the Royal Entomological Society of London (see *Antenna* **1**:20-1, **2**:47-8, **3**:40-1, **4**:22-3, **5**:34-5). In June 1980 we were invited by the Society to organise such a symposium, to take place in September 1981. During preparations for this exhausting, but wholly stimulating affair (*Antenna* **6**:179-81; *Yadoriga* **109, 110**:15-20), it was realised that the meeting would take place during Professor Ford's 80th year. If an excuse was ever needed to dedicate the symposium to E. B. Ford, this lucky chance provided it. Those who attended the meeting were honoured not only by his acceptance, but also by his presence throughout.

Butterflies have featured in a wide range of experimental, observational and evolutionary studies, involving important work on biochemistry, physiology, embryology and parasitology. However, our intention was to organise a meeting which addressed butterflies as butterflies—in other words, butterflies as whole organisms, communicating with each other, interacting with their environment and evolving within our biosphere. It is in this very area, so well-fostered by Professor Ford, that butterflies have come into their own as challenging, fascinating and instructive creatures to study those most "biological" of all biological disciplines: ecology, genetics and behaviour. We hope this volume will help re-double efforts in these pursuits, and stimulate a wider appreciation of the successes and failures of attempts to understand the biology of butterflies.

February 1984

Dick Vane-Wright, Phillip Ackery

Acknowledgements

This volume comprises 31 of the 33 papers read at the 11th Symposium of the Royal Entomological Society of London, held at the British Museum (Natural History), 23-26 September 1981. Additional papers published here are by M. Parsons (not read at the Symposium: in absentia) and P. R. Ackery (specially prepared for this volume).

The Society is grateful to Professor Ford for graciously accepting the dedication, and his generous response in helping to defray the cost of the meeting. The cost of the colour plates was met by the Cyril O. Hammond Bequest (Royal Entomological Society of London).

The Editors, who were also the Symposium convenors, gratefully record their thanks to the Staff of the Royal Entomological Society, the Staff of the Summer Accommodation Centre of Imperial College, and the warding staff of the British Museum (Natural History) for their help and consideration during the hectic days of the meeting itself. Special thanks are also due to our BMNH Butterfly Section colleagues, Ramnik Arora and Robert Smiles, for invaluable help both during the meeting and with the subsequent editing of the volume. Cyril Simsa, Cindy North and Philip DeVries have also freely given much valuable assistance. Pamela Forey undertook the difficult task of preparing the indexes.

Additionally, we wish to thank Professor Ford, G. G. Bentley, Sam Bhattacharyya, Lincoln Brower, Luciano Bullini, Robert Campbell, Sir Cyril Clarke, Kit Cottrell, Robin Crane, Paul Ehrlich, Ian Gordon, John Huxley, Ian Kitching, Kevin Murphy, Charles Remington, Rachel Hampshire, Miriam Rothschild, Elly Scheermeyer, Ken Smith, Valerio Sbordoni and Osamu Yata, together with all the contributors to this volume, for their helpful co-operation. We are also indebted to all the delegates to the meeting for much lively discussion, and to Academic Press for undertaking the task of publication.

Finally, we are happy to record our grateful thanks to the Royal Entomological Society of London for the double honour of being asked to organise the Symposium and to edit this volume. We have grown older—and perhaps a little wiser—in the process!

February 1984 *Dick Vane-Wright, Phillip Ackery*

Contents

Part I. Systematics

Part II. Populations and Communities

Part III. The Food of Butterflies

Part IV. Predation, Parasitization and Defence

Part VI. Sex and Communication

Part VII. Migration and Seasonal Variation

Part VIII. Conservation

C. B. Williams, David Gifford and R. E. Silberglied

Prior to the Symposium, one of the prospective speakers, David Gifford, died in Brazil. Since this sad loss, itself so close to the death of C. B. Williams, Bob Silberglied, an outstanding participant at the meeting, was killed in a tragic accident.

C. B. Williams must be regarded as the 'father' of all modern ideas and work on insect migration. His undisputed heir to this subject, Robin Baker, pays tribute to this great pioneer in Chapter 26. Accounts of Williams' life and work appear in *Atalanta* Würzburg **11**:237-54, **12**:313; *Antenna* **6**:2-3; and *Annals of Applied Biology* **100**;413-4.

David Gifford was a forest ecologist who had been working in Brasilia for several years prior to his sudden death in 1981. Some years before, when working in Africa, he produced a most original faunistic account, *A List of the Butterflies of Malawi* (1965). David had intended to present a paper at the Symposium on the problems and significance of refugial theory for the conservation of butterflies in South America. News of his death, although not entirely unexpected, came as a great shock: David Gifford was one of the most intelligent, honest and thoroughly entertaining characters in Lepidopterology.

Bob Silberglied captivated all who met him with his infectious enthusiasm and boundless energy. This was never more true than at the Symposium meeting, when he was in great form, buzzing with ideas, information and humour. His terrible death, in the Washington air disaster of 13th January 1982, not only robbed biology of a considerable talent but also took from us a delightful friend. He was only 35 years old. We believe that his final major paper, published here and finished just a few days before his death, is of outstanding importance and serves as its own memorial. Tributes from others appear in *Psyche*, *Cambridge* **88**:197; *Noticas de Galapagos* (**35**):28; and *News of the Lepidopterists' Society* **1982** (1):2.

R. E. Silberglied

February 1984

Dick Vane-Wright, Phillip Ackery

Dedication: Henry Ford and Butterflies

E. B. Ford is the author of the best book ever written about butterflies—a redoubtable achievement, for there is general agreement about this—and to weld entomologists into a coherent body must in itself be something of a *tour de force*. Apart from this notable contribution, Ford has used butterflies as tools with which to elucidate pigment chemistry, fluctuation in numbers, evolution, genetics, polymorphism, the structure of populations, sex-ratios, the evolution of dominance, mimicry and so forth. Twenty-eight masterly papers dealing with the Lepidoptera describe experiments and observations made both in the field and the laboratory. But my audience knows all this, and fully appreciates the thousand and one finer points of this impressive output, so I will pass on to a more personal appreciation. I had the good fortune to see E. B. Ford at work in Oxford, arriving there rather unexpectedly in 1956—an enthusiastic but uneducated amateur, without even a pass degree to my credit, but, possibly for just those reasons, more objective than the average zoologist. Working in the department one soon became aware of his all-pervading influence on Natural Science in the University. We all recognised it and I fancy this chorus of praise proved a trifle irritating to our opposite numbers in Cambridge, for I recall Professor Wigglesworth—as he then was—accusing the Oxford zoologists of being a dedicated mutual admiration department. I am not qualified to comment on a possible mutual element, but if Sir Vincent meant that we recognised the source of our original ideas and the not inconsiderable advances made at that time, he was right. I can sum up my own reactions by saying that Henry Ford's influence on scientific thought in Oxford at this time was like the combined effect of Juvenile Hormone and Ecdysone—anyhow that is how I described it to myself.

Curiously enough, up to the time I moved to Elsfield, the flea and butterfly lovers had not actually met—only corresponded. Soon after my arrival, with not a little hesitation, I asked if I might call and pay my respects. Professor Ford suggested I came along next day about 11 o'clock, and I duly did so, knocking on his door punctually as the clock struck. After a moment's silence there was rather a plaintive long drawn out cry: "Come in!" I opened the door and found an empty room. I looked round nervously—not a soul to be seen, but an almost frightening neatness pervaded everything. Each single object, from paper knife to *Medical Genetics* was in its right place. Each curtain hung in a predestined fold, and you felt that if a slight breeze or an unexpected earth tremor had disturbed one of them, it would have automatically resumed its rightful position. An unkind fate seems to have decreed that I share all my rooms with Typhoon Agnes: the sight of this distilled essence of neatness and order took my breath away. I stood there, probably with my mouth open, trying to reconcile this vacant room with that ghostly cry—had I dreamed it?—when suddenly Professor Ford appeared from underneath his desk like a graceful fakir emerging from a grave. Apparently he had been sitting crosslegged on the floor in the well of his writing table, lost in thought, but he held out his hand to me in a most affable manner. His explanation for this rather startling welcome was: "My *dear* Mrs Lane—I didn't know it was you." Dr Snow told me that another great butterfly lover, my uncle the second Lord Rothschild, who stood 6 feet 3 inches tall and weighed 22 stone stripped, received a nervous young entomologist at Tring, stark naked, swinging in a hammock. I'm sure Henry Ford won't mind me saying that the really distinguished butterfly people are usually a trifle eccentric, and you never have to ask a great man for an explanation. But only great men find time to sit (or swing) and *think*.

As a new girl at Oxford I soon became aware of another of Professor Ford's special attributes, and that was his rare skill in bringing enthusiastic bores together so that they cancelled each other out. At this period I suffered from insect smells on the brain. I was obsessed by the pungent odour of ladybirds, which I thought I detected everywhere, from stinging nettles to Monarch butterflies and froghoppers. At the same time a gifted chemist, Dr Courtenay Phillips, brought what Henry Ford called a "new engine" to Oxford, which occupied his mind—with far more justification—as much as the smell of ladybirds occupied mine. In those far off golden days the new engine was used exclusively for looking at samples of oil and petroleum, and

Courtenay Phillips decided it should be used for some biological purpose as well. E. B. Ford immediately saw a great opportunity here: "You must at once discuss this with Miriam Rothschild!" he exclaimed—and freed himself from the smell of Ladybirds* and the new engine for ever afterwards. In fact these were the first insects to be introduced to the gas chromatogram. Few entomologists realise that the initial halting experiments in this vast field-to-be were inspired by Ford in Oxford in the fifties.

During this period I had occasion to learn of an extremely kind and understanding aspect of Henry Ford's many-sided and gifted personality—which he tries to conceal behind a smoke screen of a characteristic and biting form of wit. A tragic event occurred in our new home at Elsfield a few months after my arrival—for the young woman who shared the house with me died suddenly, leaving me with six children under ten to rear solo. Suddenly the Professor appeared at Elsfield and without beating about the bush announced in a stern voice: "Mrs Lane you need a governess!" "A governess?" I echoed weakly, "But surely such people don't exist nowadays, do they?" "I will find one immediately," said Professor Ford. And he did. This sounds improbable, but within a week Mrs Brown, with a small, wan-looking boy in tow, was on the doorstep, and lived with us at Elsfield for the next few years. I was left marvelling at Henry's remorseless organising ability—for I had not expected it to extend into the field of domestic biology.

A little later Professor Ford and I became associated with Cyril Darlington, Julian Huxley and R. A. Fisher in preparing a scientific, genetically orientated report for the Wolfenden Committee, which was at the time considering possible changes in the laws relating to homosexuality. Our joint report was eventually well-received, and large chunks of it were incorporated almost verbatim in the Committee's final publication. Few people who have appreciated the beneficial changes which followed, are aware of the major role played by E. B. Ford in achieving this objective. Attitudes and opinions have changed so radically during the last few decades that it is difficult to believe how much courage was needed to produce a document of this sort in the fifties and sixties. I saw with the greatest admiration and respect how Professor Ford pursued our aims with relentless tenacity, courage and ultimate success. Moreover he retained his amusing sense of the ridiculous throughout. "What did Fisher think of our report?" I asked Henry one day, hoping for a few compliments from the great man. "Think of it?" said Henry, his eyebrows climbing, "My *dear* Mrs Lane, he said 'I will sign this report because Jesus Christ would have signed it'. I think the important point is he did so."

It is sad that the English language lacks the French adjective *genial*, for this word could have been coined to describe Henry's versatile mind and original ideas, coupled with his incisive attention to detail and persistent pursuit of a fruitful line of research. Furthermore he has attracted innumerable students—professionals as well as amateurs—into the butterfly field, recognising how the aesthetic appeal of these matchless insects, which caught our imagination as children, and are associated with the nostalgia of the golden age, can provide us, now we are older and possibly wiser, with one of the best biological tools ever invented. It is not only a rewarding field in itself, but one which carries with it a love of natural history and ageless delight. I will always be grateful to Henry Ford for suddenly endowing my fleas with wings.

Miriam Rothschild

*It is only this year that Dr B. Moore succeeded in identifying this smell, which is 2-methoxy-3-isopropylpyrazine.

Reflections in honour of E. B. Ford

Let me add here some recollections about my dear friend and mentor, E. B. Ford, known to us all as "Henry". We first met 26 years ago, beneath summer skies in the Rocky Mountains. I was at this time doing my PhD research, under Charles Remington, on the *Papilio glaucus* group, and Philip Sheppard and Henry joined us in Colorado for a few, gloriously inspiring days. The next time we met was in 1957, at Oxford, where I had the great good fortune of spending a postdoctoral year in Henry's laboratory, bubbling at that time with several fine British vintages, including Bryan Clarke, Lawrence Cook, David Jones, Kennedy McWhirter and, of course, Bernard Kettlewell—whose *explosive* research programme influenced us all. To this marvellous gathering occasional spice was added by visits from Philip Sheppard, Miriam Rothschild, and Bunny Dowdeswell.

Throughout my stay at Oxford, repeated in 1973, Henry was the most cordial of hosts, giving generously of his time in introducing the finest traditions of England. E. B. Ford is a teacher by example *par excellence* and I am indebted to him in many ways—but perhaps most of all for his gift of confidence in the value of scientific natural history. I remember him telling me, "My dear Lincoln, carefully controlled observations *are* experiments", and later, at 5 Apsley Road, or in the warmth of All Souls College, "You *do* know, Lincoln, that good science is an art; fortunately the reverse is not true."

It was indeed a pleasure to be back in England for this Royal Entomological Society Symposium on the Biology of Butterflies. Nothing could please me more than to know that our collective efforts are in honour of E. B. Ford, F.R.S.

Lincoln P. Brower

Introduction

R. I. Vane-Wright, P. R. Ackery and *P. J. DeVries

British Museum (Natural History), London and Department of Zoology, University of Texas, Austin, Texas

Interest in the descriptive aspects of butterfly natural history, particularly foodplants and 'transformations', began long ago (see Ford 1945; Ch.1). Eighteenth century pioneers in systematics, such as Johann Denis and Ignaz Schiffermüller, started the process of applying this information to Lepidoptera classification—a path littered with good intentions but still poorly trod. General reviews of butterfly biology are given in Hering (1926), Bourgogne (1951), Common (1970) and Richards & Davies (1977). However, serious interest in the dynamic aspects had to wait until the arrival of Henry Walter Bates, who, with his theory of mimicry among South American butterflies, first threw the spotlight of the Evolution Debate onto the Rhopalocera. Thereafter, the Lepidoptera, and often the butterflies in particular, have regularly been at the centre of new developments: insect chemical communication (Fritz Müller in the 1870s, 'Bombykol' in the 1950s), insect migration (taken up with such vigour by C. B. Williams), neoDarwinism (Ronald Fisher and Philip Sheppard on the evolution of mimicry), saltationism (Richard Goldschmidt), ecological genetics (E. B. Ford), insect reproductive behaviour (Niko Tinbergen), and insect/plant coevolution (C. T. Brues), to name only some of the more obvious examples.

Thus a huge body of data and theory concerning population biology has accumulated, involving real butterflies or fantasy models of them. An introduction can be found in Ford (1945), Owen (1971*a*) and Gilbert & Singer (1975). At the planning stage of this Symposium, 16 outstanding contributors to major areas of ecology, genetics and communication were sought who could, between them, review the whole current range of topics associated with the population biology of butterflies. The shorter, contributed 'discussion' papers were intended to amplify or highlight accepted dogma or difficulties. Finally, to give a practical systematic framework, we felt it necessary to add a review chapter drawing together the principal descriptive, taxonomic and faunistic works.

During the course of the meeting and subsequent editing, a considerable number of problems and questions have naturally arisen. A few such doubts and speculations, together with some of our own general thoughts, are given here by way of introduction, taking the various sections and chapters in turn.

Systematics

In Ch.1, Ackery shows how, despite a massive and unending effort, our systematic knowledge of the butterflies remains inadequate. However, even with its flaws and limitations, the systematic literature remains fundamental. Ackery brings together much of the basic taxonomic and faunistic literature; unfortunately, many of the works mentioned are difficult for field and laboratory workers to obtain. Faunistic treatments have profoundly influenced the research presented in this volume (where would the study of butterflies be without 'Seitz'?), and it is clear that more basic natural history and faunistic studies are needed. By appropriately classifying and cataloguing scattered information, systematists will continue to be of great assistance to other biologists by improving the basic framework from which to begin.

Populations and Communities

Part II introduces general aspects of population and community biology, with review chapters by Ehrlich (Ch.2) and Gilbert (Ch.3). Fundamental to all population studies is the ability to estimate effective population size. The reliability of 'MRR' methods (mark-release-recepture: Blower *et al.* 1981) often applied to butterflies for this purpose has recently been questioned (Singer & Wedlake 1981, Morton 1982, Lederhouse 1982). Perhaps

The Biology of Butterflies
0-12-713750-5

a more general problem concerns the very success of the Ehrlich and Gilbert 'schools'. There are perhaps dangers in utilizing two somewhat 'aberrant' colonial fritillaries, *Euphydryas* and *Heliconius*, to draw general conclusions on the population and community ecology of butterflies: a variety of species with contrasting biologies should now be investigated in similar depth—not an easy thing to do! Morton (Ch.4) describes some of his misgivings about MRR techniques, while Pollard (Ch.5) describes an alternative approach. Although the work on *Heliconius* and *Euphydryas* seems unlikely to have suffered significantly from these problems, it is now generally agreed that, in future, preliminary studies ought to be carried out to ascertain the most appropriate sampling technique for a particular species or population. Such methods as temporary anaesthesia may prove useful and should be investigated. According to Ehrlich, MRR techniques may ultimately prove more useful for studying butterfly population *structure* rather than as an ultimate indicator of absolute numbers; reliable population size estimates will probably come to depend on careful integration of MRR, grid and transect methods.

The Food of Butterflies

The great majority of the 200 000 or so species of Lepidoptera are herbivores that parasitize a wide range of angiosperms and other plants. Following the original work of Brues (1920, 1924), we have seen a major development in ecology and genetics—the emergence of the coevolution concept, seeded by the seminal paper of Ehrlich & Raven (1965) on butterflies and plants. However, difficulties remain with the notion that butterflies and plants have significantly coevolved (Benson *et al.* 1976, Vane-Wright 1978, Janzen 1980; but see also Berenbaum 1983 and Futuyma 1983), a caveat to be borne in mind when reading this section.

The first problem facing an embryonic butterfly is simply coping with the maternal 'choice' of oviposition site. Chew & Robbins (Ch.6) give a broad overview of oviposition behaviour, a fashionable aspect of current research, and their closing section returns anew to the problem of how the great diversity of plant families utilized by the butterflies has evolved. Singer (Ch.7) amplifies some of these themes, particularly with respect to the seemingly curious oviposition sites often chosen by females, and the complex factors affecting larval feeding efficiency and survival (see also Sims 1980, Wiklund *et al.* 1983). While indicating that so-called 'mistakes' may not be mistakes at all, Singer rightly attacks the Panglossian notion of optimality so often invoked by evolutionists and ecologists. Courtney (Ch.8) considers hostplant shifts at lower taxonomic levels to be plausibly as much related to habitat selection as to competition or coevolution with host species. The entire question of competition amongst phytophagous insects remains open (Simberloff 1982). Also, the problem of high tropical species diversity needs to be considered in this light—how many tropical butterflies besides *Heliconius* can be used to show habitat selection?

Edgar (Ch.9) considers a complex case of hostplant shift registered at high taxonomic level: the problem of pyrrolizidine alkaloid dependence in the biology of the Danainae and Ithomiinae, and the apparent shift of most of the latter group away from the Apocynaceae/Asclepiadaceae to the Solanaceae. Edgar's viewpoint has been challenged by Boppré (1978; see also Ch.25), but seems consistent with the cladistic studies of Ackery & Vane-Wright (in press *a*, *b*).

Some butterfly larvae, notably Lycaenidae, supplement their vegetable diets with insect meat—a few are entirely aphytophagous. This important dietary facet, and the relationships between lycaenids and ants, are reviewed by Cottrell (1984). (See also Atsatt (1981*b*), Henning (1983) and Chs 6, 19.) The significance of adult feeding, mentioned throughout this volume (e.g. Ch.3), including the use of spermatophores by females as nutrient sources (Ch.6), represents one of the least investigated areas—undoubtedly an aspect of butterfly biology of great importance for future research (Norris 1936, Gilbert & Singer 1975, Vane-Wright 1981*b*, Adler 1982, Gilbert 1983, DeVries 1983, and in prep.*c*).

Predation, Parasitization and Defence

Since Bates (1862) first published his theory of mimicry, butterflies have been at the forefront of disagreements over the significance of protective coloration and its evolution. As Turner (Ch.14) rightly comments, whatever explains mimicry can also explain the evolutionary process. Arguments have raged. Are there predators that eat butterflies? Even if there are, is the evolution of mimicry trivial because density regulating predation and disease only affect the early stages, not the adults (Nicholson 1927)? Are any butterflies really chemically protected to an effective degree?

Dempster (Ch.10) concludes that while there is plenty of evidence for attack on butterflies at all stages, density-dependent regulation of numbers through predation or parasitization of the early stages is far from well-documented, and may indeed be the exception rather than the rule. Dempster's point, that we still know very little about the real impact or biology of the natural enemies of butterflies, is demonstrated by Lane

(Ch.11), who provides many questions but tantalizingly few answers concerning ectoparasitic midges found on Rhopalocera.

The gamut of chemical defences so far discovered in butterflies is given encyclopaedic treatment by Brower (Ch.12), who goes to great pains to relate the problem to the essential physiological, pharmacological and psychological literature. Brower establishes an important distinction between 'Class I' defensive chemicals, potentially harmful to a predator, and 'Class II' compounds, which merely 'deceive' its perceptual system. Marsh, Rothschild & Evans (Ch.13) look at cytotoxins stored in larvae, and ponder the physiological mechanisms involved—and their possible application to cancer research. Both accounts indicate that even though most butterflies remain unknown quantities chemically, the few well-known groups (e.g. *Danaus*) have nonetheless generated vast amounts of work on theories of butterfly defences. It is curious that *Heliconius* and the acraeine and ithomiine butterflies, which play key roles in our concepts of the dynamics of mimicry, are still virtually unknown chemically (Nahrstedt & Davis 1981, 1983).

Turner (Ch.14) reviews the classic arguments concerning the evolution of Batesian and Müllerian mimicry, and the special difficulty of the evolution of aposematism itself—given that you may actually contain a 'Class I' chemical, how do you begin to advertise the fact without losing your life in the process? Having answered this problem, Turner relates ideas on mimicry to Goldschmidtian saltationism and its fashionable descendant—punctuationalism. Finally, Gibson (Ch.15) looks at the phenomenon of automimicry—how and why do some aposematic species (notably certain danaines) display such a wide palatability spectrum? This paper generated much discussion at the Symposium meeting—partly reflecting the difficulties of presenting a plausible theoretical model to a hall full of cynical field biologists! Two major problems with the model concern the need to invoke density-dependent predation (see Ch.10) and the role of pyrrolizidine alkaloids in the defence of danaines (Chs 9, 12, 25; see also Ackery & Vane-Wright in press *a*).

Genetic Variation and Speciation

This section addresses the existence and role of genetic variation in the microevolution and speciation of butterflies. Brakefield's major synthesis on *Maniola jurtina* surveys the work by E. B. Ford and his collaborators that established the field of 'ecological genetics'. Having always held strong selectionist views on the evolutionary process, it must be ironically satisfying to Professor Ford to read Brakefield's account of how the underside spot variation of *M. jurtina* and its inheritance can now be related to varations in seasonal and sexual behaviour. The extremely complex manifestation of this whole phenomenon had previously led Ford (1955) to an uncharacteristic conclusion: that individual selection was unlikely to be directly involved in this particular case. Brakefield's plea for the genetic investigation of behavioural traits in butterflies would, if answered, open up a vital area for understanding evolution and speciation in this group (cf. Part VI).

In Ch.17, Kitching gives a brief account of his studies on enzymic variation in the Danainae. While unearthing a tantalizing problem concerning heterozygosity levels and possible population history, he rightly points to the great unknown—what does enzyme work really represent in butterfly biology? Further work on the electrophoretic variation of well-known butterflies, like *Euphydryas* (Ehrlich *et al.* 1975) and *Heliconius*, would obviously have great potential—but the problem remains, even as the Drosophilologists have found, of relating enzyme variation to at least some notion of functional significance. More positively, Kitching suggests that enzyme work could contribute to historical aspects of butterfly biogeography and migration (cf. Ch.26).

The veiled and mysterious problem of speciation is touched upon by Gordon (Ch.18), who relates population dynamics and mimicry to the origin of two *Acraea* species, and Pierce (Ch.19), who invokes the notion of accidental hostplant shifts (Ch.6) caused through oviposition 'mistakes' (Ch.7) related to ant dependence, to explain the very high numbers of lycaenid species. Pierce gives a brief account of the remarkable effects of ant attendance on the survival of lycaenid early stages, and the biochemical and communication factors involved. For the latter aspects see also Henning (1980, 1983) and Cottrell (1984).

This section lacks any general account of the phenomenon of genetic polymorphism in butterflies. Reviews can be found in Ford (1953, 1975*a*) and Sheppard (1975), to which Bernardi (1974) and Vane-Wright (1975) add some systematic considerations, while works on enzyme variation are reviewed in Geiger (1981) and Kitching (1983). Many important references concerning pupal polymorphism in butterflies are included in Sims (1983). The possible relationship between butterfly dispersal, polytypic variation and speciation is discussed by Robbins & Small (1981).

Sex and Communication

The only sensory modalities for which we have significant information about communication between butterflies are vision and smell. Aural (Swihart 1967*b*, Kane 1982), chemotactile, and tactile modes are undoubtedly important, but are virtually unstudied.

The first two review papers, Silberglied (Ch.20) and Smith (Ch.21), both deal with the role of visual signals in sexual communication. However, there are significant differences of opinion between the two. Smith (in litt.) considers that Silberglied's chapter should have been entitled 'Visual communication and non-random mating among butterflies', because Silberglied did not differentiate 'true' sexual selection (i.e. preferential mating leading to an increase in mean fitness in progeny) from the other forms of non-random mating (which can lead 'merely' to a change of appearance or signal pattern value). Furthermore, Silberglied was of the opinion that work on sexual selection in butterflies has been too concerned with possible intersexual effects, intrasexual phenomena (male-male interactions in particular) having been largely ignored. Smith takes exactly the opposite view! Perhaps this contradiction can be explained by taxonomic scale: Ford (1964), so influential with the butterflies, subscribes to the view that female choice is sufficient to prevent the evolution of mimicry by males in such *Papilio* species as *glaucus*, *memnon*, *polytes* and *dardanus*. But on a broad entomological scale, Smith is probably right to say that the influence of Wallace (1889) and Huxley (1938*a*) and others has been to discount female choice. Ultimately, the differences of opinion relate to whether or not visual communication in butterflies is viewed as an object of individual selection, or as some form of social signal. As pointed out by Maynard Smith (1978, and see also Searcy 1982); the lack of additive fitness between parents and offspring does cause some special difficulties for purely neo-Darwinian interpretations of sexual selection. In the light of all this, we have not attempted to expunge the contradictions between Ch.20 and Ch.21.

Silberglied (Ch.20) thoroughly reviews visual communication in butterflies and reports on his experiments with painted *Colias* and *Anartia*. His principal conclusions are that visibly bright coloration is used primarily for long-distance communication, and has evolved through advantage to males in agonistic, male-male interactions. In Ch.21, Smith demonstrates the complexities of distinguishing the various forms of selection and non-random mating in detailed accounts of two species, *Hypolimnas misippus* and *Danaus chrysippus*. The results are interesting and important, but it might be unwise to form hasty general conclusions about sexual selection and polymorphism from just these two peculiar, ecologically related species.

In the case of *Papilio glaucus*, Platt, Harrison & Williams (Ch.22) found no evidence for sexual selection or preference affecting the colour polymorphism of this frequently studied American species. It may be relevant that several *Papilio* species, sharing patterns very similar to the *glaucus* male, co-exist with it over much of its range. Vane-Wright (Ch.23) picks up the Silberglied theme, but invokes both co-operative as well as agonistic behaviour to explain male pattern constancy. By also suggesting that something strange often seems to happen to butterfly signal patterns during speciation, he raises the saltationalist spectre again. Clarke (Ch.24) discusses the problem of sex-ratio control—and the potentially instructive deviations from the 1:1 rule, involving *Lymantria dispar* (more Goldschmidt!), and *Hypolimnas bolina*, a species intimately related to the equally curious *H. misippus* (Vane-Wright *et al.* 1977, Hiura 1983).

Boppré's challenging review of chemically-signalled interactions (Ch.25) raises many questions concerning the inadequacy of our understanding of butterfly pheromone biology, and demonstrates the need for a far wider range of hypotheses about potential pheromone functions. Boppré also revitalizes Lincoln Brower's (1963) theme: co-adaptation between male scent organs, aposematism and mimicry. Most appropriately, he finally reminds us that scent-communication is primarily a *biological*, not a chemical subject, adding that amateurs can still contribute much to this field through good observations and simple experiments.

Migration and Seasonal Variation

The challenging tone continues into part VII. Baker (Ch.26) chastises us for not studying butterfly migration assiduously or imaginatively enough. Certainly, with the exception of *Danaus plexippus*, there has been a strong tendency to reduce insect migration to the realms of pure chance, thereby robbing the subject of rigour and excusing us from the need to explain the phenomenon at all. Paralleling the MRR difficulties, Baker exposes the problem of the lifetime track: it seems extraordinary that we can be so uncertain of the day to day activity of even a cabbage white!

In a very different way, Shapiro (Ch.27) tries to revive interest in another 'forgotten' phenomenon, seasonal polyphenism, by placing it in the difficult yet exciting framework of Waddingtonian genetic assimilation. Shapiro reports on some of his heroic, but so far inconclusive attempts to unravel the genetics of polyphenism in a South American pierid. Porter (Ch.28) demonstrates how seasonal effects, through larval behaviour, can be translated into a change of adult sex-ratio, something usually considered under genetic control (Ch.24). McLeod (Ch.29), while disposing of the notion that seasonal morphs in the tropics, in *Precis* at least, are induced by rainfall or humidity changes, puts forward a number of interesting hypotheses, including a possible link with speciation. The application of Shapiro's genetics to McLeod's chosen animals might prove rewarding. Finally, Yata, Saigusa, Nakanishi, Shima, Suzuki & Yoshida (Ch.30) point to some rather surprising conclusions about the polyphenism of *Pieris rapae* (perhaps consistent with one of McLeod's hypotheses). Yata *et al.* indicate that this agricultural pest seems to be less specialised in its seasonal responses than its close relatives. Such work should be actively pursued, because an ultimate goal of butterfly biology must be to explain the success of species such as *P. rapae*, *Danaus plexippus*, *D. gilippus/chrysippus*, *Hypolimnas misippus*, *Phoebis argante*, *Cynthia cardui*, *Lampides boeticus* and so on, not merely in autecological terms, but by reference to a range of less invasive species and, in particular, by intimate knowledge of their 'less successful' close relatives.

Conservation

Such considerations as the 'success' or relative 'failure' of butterfly species to adapt (or become pests—Ch.2) lead naturally to the last section of the volume, conservation. Pyle (Ch.31) looks at the response of butterflies to natural large-scale habitat destruction by vulcanism. In Ch.32, Parsons considers the practical problems of species conservation when faced with a severely threatened insect with a very real price on its head (including 'wanted' posters!): *Ornithoptera alexandrae*, the world's largest butterfly. In the last chapter, Thomas (Ch.33) reviews a major body of knowledge and experience of attempts to conserve butterflies in the man-made environments of northern Europe and America.

All three chapters provide ample evidence that conservation of organisms without an understanding of their genetics, ecology and behaviour is impossible. From this it is also easy to appreciate that we cannot intelligently manage those organisms beneficial or inimical to us without continual access to many species often regarded as 'unimportant'. Conservation and pure research, as pointed out by Ehrlich (Ch.2) and Gilbert (Ch.3), are of vital interest to us all.

Part I
Systematics

1. Systematic and Faunistic Studies on Butterflies

Phillip R. Ackery

British Museum (Natural History), London

Most taxonomic studies on butterflies can be placed within two broad categories—faunistic, that is detailing the fauna of a particular area, or systematic, dealing with a particular group of butterflies over its entire range. To accommodate this, my review is divided into two distinct sections; firstly a summary of the systematic literature within a classificatory framework, and secondly, brief surveys of the major literature sources pertaining to each biogeographic region.

Although now much out-dated, the series of volumes edited by Adalbert Seitz (for details see faunistic section below) remains the only comprehensive treatment of the entire butterfly fauna. Such modern works as Lewis (1974), Watson & Whalley (1975) and Smart (1976), while valuable colour-guides, are not all-embracing. *Lepidopterorum Catalogus* (see references in systematic section below), although incomplete, is the most wide-ranging systematic catalogue and used in conjunction with the *Rhopalocera Directory* (Beattie 1976), compiled from the systematic index of *Zoological Record* up to 1971, provides ready access to much of the basic taxonomic literature on butterflies. In addition, the pioneering catalogue of Kirby (1871, 1877) is still of great value for early works.

With most aspects of biology covered elsewhere in this volume, there is no need for general references here. However, a number of serials deal exclusively or regularly with Lepidoptera and are worth consulting routinely, particularly *Alexanor*, *Atalanta* Würzburg, *Australian Entomological Magazine*, *Bulletin of the Allyn Museum*, *Entomologist* (now defunct), *Entomologist's Gazette*, *Entomologist's Record and Journal of Variation*, *Journal of the Lepidopterists' Society*, *Journal of Research on the Lepidoptera*, *Lambillionea*, *Lepidoptera Group 68 Newsletter* and *Papilio International*, *Malayan Nature Journal*, *Nota Lepidopterologica*, *Queensland Naturalist*, *Revista de la Sociedad Mexicana de Lepidopterológica*, *The Rhopalocerists' Magazine*, *Systematic Entomology* (and the former *Transactions* and *Proceedings of the Royal Entomological Society of London*), *Transactions of the Lepidopterological Society of Japan* and *Yadoriga*. All these journals commonly record systematic, experimental and field observations on butterflies, together with many anecdotes on butterfly behaviour.

The Systematics of Butterflies

It is perhaps ironic that while reviews of the biology of butterflies already fill a substantial volume, two centuries of effort by taxonomists still fail to provide a convincing systematic framework in which to place this wealth of data. Only Reuter (1896) has presented a detailed phylogenetic system for all butterflies and apart from the studies of Ehrlich (1958) and Kristensen (1976) there have been few significant advances in recent years. Ehrlich's analysis is phenetic and based almost entirely on adult morphology; Kristensen adopts a phylogenetic approach, largely re-analysing Ehrlich's original data. We can now at least feel re-assured that the butterfly superfamilies Hesperioidea (skippers) and Papilionoidea (true butterflies) together form a monophyletic group; Brock (1971) treats them as a single superfamily. Kristensen suggests the following characterization: ditrysian Lepidoptera with mesal fusion of the dorsal laminae of the secondary metafurcal arms; a 'butterfly type' brain shape with very large optic lobes and a small deutocerebrum; the mesothoracic aorta with a 'horizontal chamber'; the descending part of the aorta, morphologically anterior to the horizontal chamber, running adjacent and parallel to the ascending part of the chamber; the mesothoracic lateral dorsal muscle twisted. For an outline of the relationships between the Papilionoidea-Hesperioidea

The Biology of Butterflies
0-12-713750-5

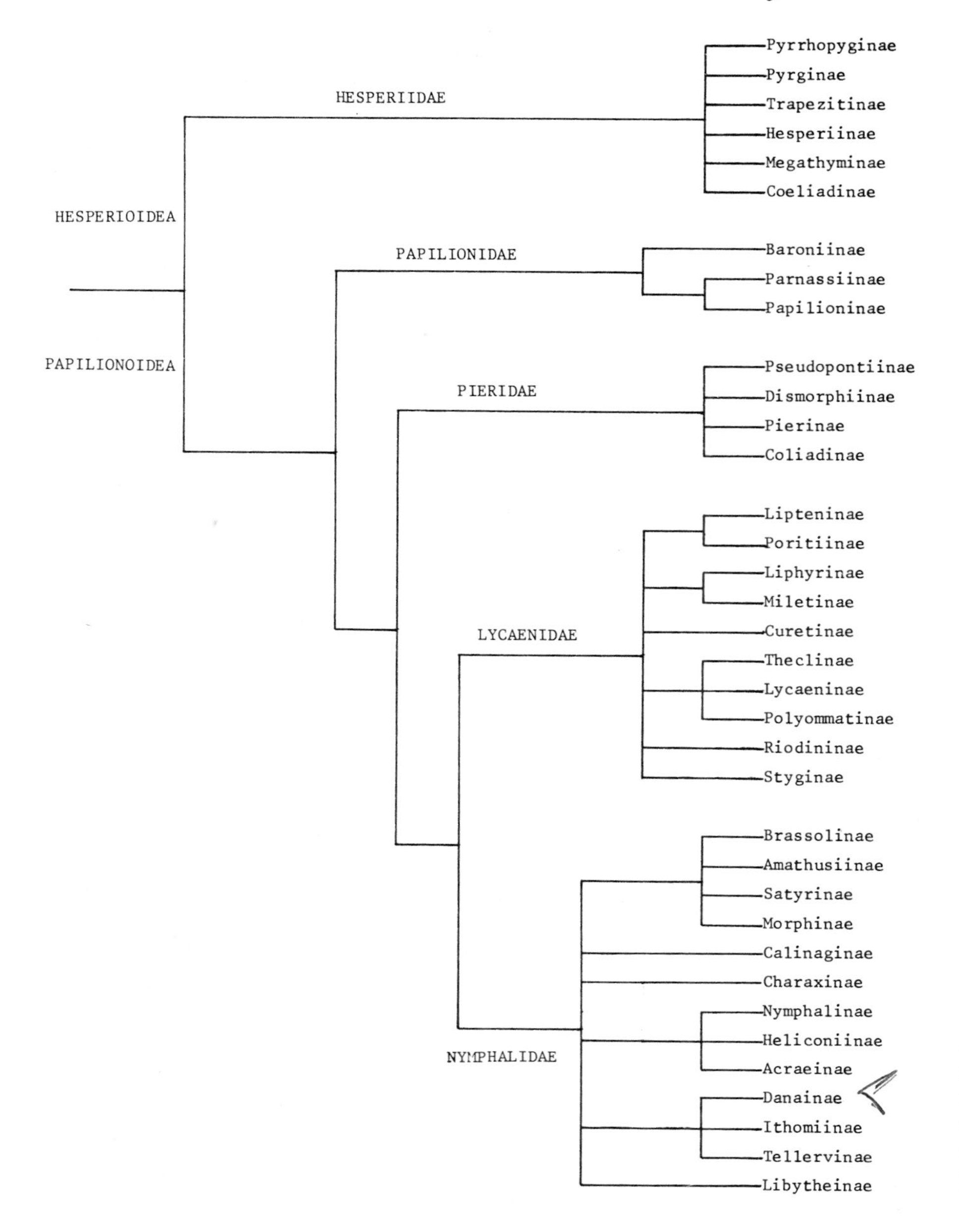

Figure 1.1. Diagram to show arrangement and classification of butterfly families and subfamilies adopted in this work. (Based mainly on the works of Evans, Ehrlich, Eliot and Kristensen—see text.)

and the rest of the Lepidoptera see Common (1970, 1974, 1975) and Brock (1971).

Within the Papilionoidea four families, namely the Papilionidae, Pieridae, Nymphalidae and Lycaenidae, may be recognized with confidence. At the subfamily level there is broad agreement within the Papilionidae (Munroe & Ehrlich 1960), Pieridae (Klots 1933, Ehrlich 1958) and the Lycaenidae (Ehrlich 1958, Eliot 1973). However, the Nymphalidae remain only unconvincingly resolved. Although this family without doubt contains many recognizable natural groups, their separation ultimately leads to an uncharacterizable, probably polyphyletic or paraphyletic residue. This is particularly unfortunate as so many recent advances in the study of butterfly biology have utilized such nymphalids as danaines, heliconiines and 'melitaeines'—groups whose affinities remain obscure. The Hesperioidea are monobasic. Few have dared to follow Evans into the mire created by his own studies; although he adopted different criteria in his various faunistic works, all his subgroupings appear to be assignable to six subfamilies.

The relatively conservative classification adopted here (Fig. 1.1.) and throughout this volume is based on a synthesis of the more recent literature, not a re-analysis of the available characters. The family-level characterizations and sequence are abstracted from Kristensen (1976). Subfamilies are based on Ehrlich (1958), with some more recent opinions incorporated (e.g. Miller 1968, Eliot 1973). Although such a hybrid classification is bound to contain weaknesses, it is unlikely that future research will result in any significant changes in the proposed family-level groupings other than perhaps in rank. With further advances almost certainly needing to take into account features of the early stages (see Fracker 1915, Mosher 1915, Hinton 1946, Döring 1955), brief outlines are given of the larvae and major hostplant families. More detailed information on these aspects of biology can be obtained by consulting references in the faunistic section.

Superfamily HESPERIOIDEA

Neither the Papilionoidea nor the Hesperioidea has a convincing array of autapomorphies. Kristensen (1976) gives the following for the Hesperioidea—forewing with no 'stalked' peripheral veins; forewing with CuP (Cu_2 auctt.) absent (it is also absent in most Papilionoidea but retained as a short vestige in Papilionidae); antennae with subapical thickenings. There is only one included family—the Hesperidae.

Family Hesperiidae

I follow Evans in differentiating six subfamilies—the members of the Heteropterinae (see Miller & Brown 1981, Higgins 1975) were included by Evans within the Hesperiinae while recognition of the Megathyminae as a distinct family (Freeman 1969) possibly renders the residual 'Hesperiidae' poly- or paraphyletic. Recent estimates suggest a global fauna of approximately 3500 species (Robbins 1982).

Major references. Lindsey *et al.* 1931, Evans 1937, 1949, 1951, 1952, 1953, 1955. In addition, see also the works of Atkins, A. F. [1973-on]; Berger, L. A. [1955-on]; Burns, J. M. [1960-on]; Eliot, J. N. [1950-on]; Freeman, H. A. [1941-on]; de Jong, R. [1972-on]; Mielke, O. H. H. [1966-on]; Miller, L. D. [1962-on].

Subfamily Pyrrhopyginae

Anntenal club stout and blunt, obtusely angled about its commencement, so that the entire club lies in the apiculus (i.e. reflexed part of the club) of 19-21 segments, all of which are denuded of scales anteriorly. Forewing cell long, longer than the dorsum and more than two-thirds the length of the costa; vein M_2 originating nearer M_3 than M_1 (Evans 1951). All known species are confined to the New World tropics.

The larvae are generally strikingly coloured, often reddish or purple, with complete or incomplete yellow segmental rings, and usually clothed with abundant fine, long hairs. The larvae construct shelters from the leaves of their hostplants.

Major hostplant families. Annonaceae, Clusiaceae, Flacourtiaceae, Malpighiaceae, Meliaceae, Myristicaceae, Myrtaceae.

Major references. Mabille 1912, Evans 1951.

Subfamily Pyrginae

Palpi variable: either erect with the second segment appressed to the face and the third segment not protruding in front of the second or entirely porrect, with the third segment protruding in front of the second. Mesotibiae without spines. Forewing vein M_2 not curved at the base and usually about intermediate between M_1 and M_3, or nearer M_1 at its origin; cell either long, approximately two-thirds the length of the costa, or short, less than two thirds the length of the costa (Evans 1952). Representatives occur in each major biogeographic region.

The larvae may be brightly coloured, or uniformly green, brown or whitish, and slender and pilose. The caterpillars form shelters by folding and joining the leaves of their hostplant.

Major hostplant families. Combretaceae, Fabaceae (Leguminosae), Lauraceae, Malvaceae, Myrtaceae, Rosaceae, Sterculiaceae, Verbenaceae.

Major references. Warren 1926 (as Hesperiinae), Lindsey *et al.* 1931, Shepard 1931-6, 1936, Evans 1937, 1949, 1952, 1953, Burns 1964.

Subfamily Trapezitinae

Body stout. Antennae about one-half the length of the costa; apical club often hooked or pointed. Hindwing cell distally truncate; median vein forked within the cell, the lower branch terminating at vein M_3 instead of near M_2; vein M_2 always tubular, arising nearer M_1 than M_3 (Common & Waterhouse 1981). The group is characteristic of the New Guinea region and Australia.

The larvae are variable. Commonly stout and brownish with a large, dark head; occasionally slender, translucent and greenish, often with longitudinal stripes—the head is large and roughened, often greenish or brownish and usually with dark bands; rarely the larvae are hairy and whitish. They construct tubes by joining together the leaves of their hostplants (Common & Waterhouse 1981).

Major hostplant families. Cyperaceae, Iridaceae, Xanthorrhoeaceae.

Major references. Shepard 1936, Evans 1949, Common & Waterhouse 1981.

Subfamily Hesperiinae

Male lacking hairpencils on the metatibiae. Forewing vein M_2 arising nearer M_3 than M_1; male forewing upperside with brands or a stigma but never with a costal fold. Wings held erect when at rest (Evans 1955). Distribution is worldwide, with representation in every major biogeographic region.

The larvae are slender, either smooth or finely pubescent, with a well-defined neck; commonly pale green, yellowish or whitish. The larvae often construct tunnels of grass stems woven together.

Major hostplant families. Agavaceae, Arecaceae, Combretaceae, Cyperaceae, Flagellariaceae, Heliconiaceae, Liliaceae, Marantaceae, Musaceae, Pandanaceae, Poaceae (Gramineae), Rubiaceae, Sapindaceae, Zingiberaceae.

Major references. Lindsey *et al.* 1931, Shepard 1937-9, Evans 1937, 1949, 1955.

Subfamily Megathyminae

Head small, half the width of the thorax. Antennae close together, no longer than half the length of the costa, with a well-formed club either with or without spicules, or compressed at the end and blunt. All tibiae spined; protibia with only a short epiphysis in the middle of the distal half; meso- and metatibiae with a single pair of terminal spurs, those on the metatibiae being concealed by hairs; terminal claws longer and wider apart than usual. Venation similar to the Hesperiinae—forewing cell shorter than the dorsum with vein M_2 'decurved' at its origin; hindwing vein M_2 well-marked, appearing as a slit on the hindwing upperside; wings rounded, hindwing costa equal to the dorsum. Female abdomen often very heavy (Evans 1955). All known species are restricted to the southern USA and Central America.

The larvae are pale creamy or yellow, clothed in short, light brownish hairs; head reddish. Stem and root borers.

Major hostplant families. Agavaceae.

Major references. McDunnough 1912, Evans 1955, Freeman 1969.

Subfamily Coeliadinae

Antennae short, less than half the length of the forewing costa, slender clubbed with a sharp-pointed, slender, curved apiculus. Labial palps with second segment stout and held erect, close to the face, and the third segment long and slender, projecting forwards. Hindwing distinctly lobed, sometimes produced at the tornus; M_2 tubular, arising nearer M_1 than M_3 (Common & Waterhouse 1981). All known species are either Austro-Oriental or Afrotropical.

The larvae are often brightly coloured and patterned with complete or incomplete hoops. Shelters are formed by folding and joining the hostplant leaves.

Major hostplant families. Araliaceae, Asclepiadaceae, Combretaceae, Euphorbiaceae, Fabaceae (Leguminosae), Malpighiaceae, Myristaceae, Myrsinaceae, Sabiaceae, Zingiberaceae.

Major references. Shepard 1933, Evans 1937, 1949.

Superfamily PAPILIONOIDEA

Kristensen (1976) suggests just two (very weak!) autapomorphies for the Papilionoidea—wing coupling amplexiform in both sexes (it is similar in the vast majority of heperioids, but the retinaculo-frenate condition occurs in male *Euschemon*); antennae with apical clubs (poorly developed in many species). There are four included families—Papilionidae, Pieridae, Nymphalidae and Lycaenidae.

Family Papilionidae

Larva with osmaterium. Pretarsal arolium and pulvilli reduced. Forewing vein 3A free to the margin. Mesothoracic portion of the dorsal vessel (aorta) without ostiae; mesothoracic aorta lacking a 'horizontal chamber'. Uppermost bundle of mesothoracic inner dorsal longitudinal muscle thick; metathoracic inner dorsal longitudinal muscle almost vertical. Ehrlich (1958) and Munroe (1961) agree in recognizing three subfamilies (see also Munroe & Ehrlich 1960) containing, in all, 700 species (Robbins 1982).

Major references. Munroe & Ehrlich 1960, Munroe 1961, d'Almeida 1966, Hancock 1978, 1983, Igarashi 1979.

Subfamily Baroniinae

Forewing median spur absent. Hindwing with two anal veins. Tarsal claws symmetrical. Spinasternum not produced laterally at the spina (Ehrlich 1958). The subfamily is represented by a single species *(Baronia brevicornis)* endemic to Mexico.

The larvae are smooth, green laterally and pale brown dorsally.

Major hostplant family. Fabaceae (Leguminosae).

Major references. Bryk 1923, 1934, Vazquez & Perez 1961.

Subfamily Parnassiinae

Forewing median spur absent or vestigial. Hindwing with one anal vein. Tarsal claws usually asymmetrical in males, sometimes also in females. Spinasternum produced laterally to some extent (from Ehrlich 1958). There are approaching 50 currently recognized species, with a largely Holarctic distribution.

The larvae are cylindrical and black, usually with red tubercles.

Major hostplant families. Aristolochiaceae, Crassulaceae, Fumariaceae, Saxifragaceae.

Major references. Bryk 1923, 1934, 1935, Ackery 1975*b*.

Subfamily Papilioninae

Forewing with median spur complete or (rarely)

vestigial, if vestigial then face extremely protuberant. Tarsal claws symmetrical. There are about 650 species (Ehrlich & Raven 1965) with an overall cosmopolitan distribution although greatest diversity is in the Old World tropics.

The larvae are variable. Leptocircini--mature larva green or greenish, usually smooth or with single pairs of conical spines on one or more thoracic segments and on the terminal abdominal segment. Papilionini—young larva bearing rows of spiny tubercles, typically dark with red or pale spots and a pale saddle on segments six and seven; older larvae sometimes retaining juvenile facies but more often with spines lost except for rudiments on the first and last segments, and with the pattern variously modified. Troidini—larva cylindrical with fleshy, usually red, segmental tubercles, and commonly with a white saddle on segments six and seven (Munroe 1961).

Major hostplant families. Annonaceae, Aristolochiaceae, Canellaceae, Hernandiaceae, Lauraceae, Magnoliaceae, Rutaceae, Winteraceae.

Major references. Rothschild 1895, Rothschild & Jordan 1906, Bryk 1923, 1929-30, 1934, Haugum & Low 1978-9, [1982], Tsukada & Nishiyama 1980.

Family Pieridae

Distinctly bifid pre-tarsal claws. Pronotum with medio-posterior membranous cleft. Lateral 'pre-spiracular bar' absent in the abdominal base. Profurco-laterocervical muscle absent. Origin of the mesothoracic tergo-coxal muscle partly behind that of the mesothoracic tergotrochanteralis. Wing scales containing pterine-type pigments. Klots (1933) recognizes three subfamilies; Ehrlich (1958) differs only in treating the Coliadinae as distinct from the Pierinae. There are about 1000 species (Robbins 1982).

Superficially the larvae of the Pieridae look remarkably uniform. A single general description probably suffices (although to date there are no records for the Pseudopontiinae)—larva cylindrical, slightly tapering posteriorly and also laterally in later instars; shortly setose or pubescent, but without any longer processes except for an occasional short pair on the twelfth segment; usually cryptic, green or brown with lighter longitudinal stripes.

Major references. Talbot 1932-5, Klots 1933, Yata 1981. Most revisionary studies of the Pieridae are limited to individual genera rather than presenting comprehensive treatments of subfamilies or tribes. See particularly the works of Berger, L. A. [1948-on]; Bernardi, G. [1945-on] and Talbot, G. [1928-1944]; Feltwell's monograph on *Pieris brassicae* (1982) gives a broad overview of much of the literature pertaining to the Pieridae.

Subfamily Pseudopontiinae

Forewing with M_2 stalked with R_{3+4+5}. Hindwing with $Sc+R_1$ secondarily fused with R_s before the middle of the wing; M_2 stalked with M_1 (Ehrlich 1958). A single monobasic genus (*Pseudopontia paradoxa*) from West Africa is the sole representative of this subfamily.

Subfamily Dismorphiinae

Forewing with M_2 arising from the end of the cell and with five radial veins, all stalked. Hindwing with $Sc+R_1$ not secondarily fused with R_s; M_2 arising from the cell (Ehrlich 1958). The subfamily is primarily Neotropical, with one Palaearctic genus, and contains approximately 100 species (Ehrlich & Raven 1965).

Major hostplant families. Fabaceae (Leguminosae).

Major references. Lamas 1979.

Subfamily Pierinae

Forewing with vein M_2 arising from the end of the cell; 3-5 radials present, at least one arising from the cell. Hindwing with $Sc+R_1$ not secondarily fused with R_s; M_2 arising from the cell; humeral vein usually long. Tegumen longer than the uncus. Patagia unsclerotized (from Ehrlich 1958). Although cosmopolitan, greatest diversity is in the tropics with approximately 700 species overall (Ehrlich & Raven 1965).

Major hostplant families. Brassicaceae (Cruciferae), Capparidaceae, Loranthaceae, Santalaceae.

Major references. Talbot 1928-37, Bernardi 1947, Field & Herrera 1977.

Subfamily Coliadinae

Forewing with vein M_2 arising from the end of the cell; 3-5 radials present, at least one arising from the cell. Hindwing with $Sc + R_1$ not secondarily fused with R_s; M_2 arising from the cell; humeral vein greatly reduced or absent. Tegumen considerably shorter than the uncus. Patagia sclerotized (Ehrlich 1958). Although cosmopolitan, greatest diversity is in the tropics; approximately 250 species are known (Ehrlich & Raven 1965).

Major hostplant families. Asteraceae (Compositae), Fabaceae (Leguminosae), Rhamnaceae, Zygophyllaceae.

Family Lycaenidae

Antennal bases adjacent to the margin of the eye, which they often indent. Frontoclypeus usually less arched than in other butterflies. Patagia entirely membranous (this condition is also found elsewhere but not as a postulated groundplan feature). Eliot (1973) includes eight subfamilies to which should be added the Styginae and Riodininae. Robbins (1982) estimates more than 6000 species with an overall cosmopolitan distribution although greatest diversity is in the tropics (Ehrlich & Raven 1965).

Major references. Clench 1955, Clark & Dickson 1971, Eliot 1973.

Subfamily Lipteninae

Proboscis with a few sensory hairs on the shaft; terminal papillae not strongly developed. Labial sclerite completely sclerotized. Second joint of palpi usually clothed with appressed scales but sometimes with bristly or hairy scales. Male prothoracic coxae not extending spine-like below the articulation of the trochanter. Male prothoracic legs more than half the length of the pterothoracic legs, with the tarsi fused to a single stubby-tipped segment; meso- and metatibiae without terminal spurs. Forewing usually with all 12 veins present, although R_4 is occasionally absent; androconia usually present on male forewing. Hindwing without tail or tornal lobe; precostal vein sometimes present; basal vein absent along the costa. All representatives of this subfamily occur in the Afrotropical region (Ehrlich 1958; Eliot 1973).

The larvae have both dorsal and lateral tufts of hairs. The head is comparatively broad and barely retractile.

Major hosts. Lichens and microscopic fungi.

Major references. Draesake 1936, Stempffer & Bennett 1953, Stempffer 1967.

Subfamily Poritiinae

Labial sclerite completely sclerotized. Palpi clothed with appressed scales. Male prothoracic coxae not extending spine-like below the articulation of the trochanter. Male prothoracic legs more than half the length of the pterothoracic legs, with the protarsus fused to a single stubby-tipped segment; meso- and metatibiae without terminal spurs. Forewing with 10, 11 or 12 veins; vein R_1 anastomosed with SC and sometimes R_2. Hindwing with a very short or indistinct precostal vein, no tail or tornal lobe and no basal vein along the costa; male usually with androconial brands. All known species are restricted to the Oriental region (Ehrlich 1958, Eliot 1973).

The larvae are long and thin, barely tapered at either end, and with a dorsal thick covering of hair becoming sparser laterally.

Major hostplant families. Fagaceae.

Major references. Eliot *in* Corbet & Pendlebury 1978.

Subfamily Liphyrinae

Proboscis usually wholly or partly atrophied, when normally developed with a regular series of fine sensory hairs on the shaft. Labial sclerite completely sclerotized. Palpi clothed with appressed scales; may be very small. Male prothoracic coxae not extending spine-like below the articulation of the trochanter. Male prothoracic legs more than half the length of the pterothoracic legs, with the protarsus segmented, clawed and fully functional; meso- and metatibiae without terminal spurs. Forewing with 12 veins, veins SC, R_1 and R_2 being free. Hindwing tailless, basally lacking a vein along the costa. Greatest diversity is in the Afrotropical region but there are a few Austro-Oriental representatives (Ehrlich 1958; Eliot 1973).

The larvae are smooth and elliptical, with a shirt-like carapace beneath which the legs, head and anal claspers are hidden.

Larval hosts. Aphytophagous on homopterans, ant regurgitations and probably ant brood; perhaps occasionally cannibalistic (Cottrell 1984).

Major references. Bethune-Baker 1925, Stempffer 1967.

Subfamily Miletinae

Proboscis, except when partially atrophied, with a series of fine sensory hairs on the shaft. Labial sclerite completely sclerotized. Palpi variable, sometimes asymmetrical. Male prothoracic coxae not extending spine-like below the articulation of the trochanter. Male prothoracic legs more than half the length of the pterothoracic legs, with the protarsus segmented or fused to a single segment; meso- and metatibiae without terminal spurs. Forewing with 11 veins, SC, R_1 and R_2 being free. Hindwing tailless and lobeless, basally lacking a vein along the coastal margin. Most species are Oriental or Afrotropical but there are a few Holarctic representatives (Ehrlich 1958, Eliot 1973).

The larvae are onisciform (widest and highest in the middle, with the dorsal surface convex).

Larval hosts. Probably wholly aphytophagous on living homopterans, homopteran secretions and the regurgitations of ants; *Lachnocnema* has been recorded on plant sap (Cottrell 1984).

Major references. Stempffer 1967, Eliot in prep.

Subfamily Curetinae

Proboscis with a prominent series of sensory hairs on the shaft; terminal papillae strongly developed. Labial sclerite completely sclerotized. Palpi normal, clothed with appressed scales. Male prothoracic coxae slightly extended spine-like below the articulation of the trochanter. Male prothoracic legs more than half the length of the pterothoracic legs, with the protarsus fused to a single segment and ending in a tapered, down-curved point; meso- and metatibiae with inconspicuous terminal spurs. Forewing with 11 veins, SC, R_1 and R_2 being free; R_5 ends on the termen. Hindwing tailless, without a basal vein along the coastal margin; male lacking alar androconia. Mostly exclusive to the Oriental region but a few representatives just extend into the Palaearctic (Ehrlich 1958, Eliot 1973).

The larvae are smooth, with permanently exserted cylindrical tubercles on the eleventh segment.

Major hostplant families. Fabaceae (Leguminosae).

Major references. Evans 1954, DeVries 1984.

Subfamily Theclinae

Proboscis often smooth-shafted. Labial sclerite completely sclerotized. Palpi variable. Male prothoracic coxae not extending spine-like below the articulation of the trochanter. Male prothoracic legs more than half the length of the pterothoracic legs, with the protarsus usually fused to a single segment, although segmented, clawed and fully functional in a few genera; meso- and metatibiae commonly with paired terminal spurs. Forewing with 10, 11 or 12 veins; vein R_5 usually ends on the costa or apex when only 10 or 11 veins are present; veins SC and R_1 nearly always free. Hindwing without a precostal vein or a vein along the basal part of the costal margin; sometimes tailless but usually with 1, 2, 3 or rarely 4 tails; tornal lobe usually well-developed. Alar androconia usually present in males. Represented in all major biogeographic regions (Ehrlich 1958, Eliot 1973).

The larvae are usually more or less onisciform, but sometimes prominently shouldered or waisted.

Major hostplant families. Fagaceae, Loranthaceae, Oleaceae, Santalaceae and diverse other families (See Ehrlich & Raven 1965)—additionally some species are known to be aphytophagous, at least in some instars, on homopterans and ant regurgitations; some may also be cannibalistic (Cottrell 1984).

Major references. Riley 1958, Evans 1957, Shirôzu & Yamamoto 1956, Stempffer & Bennett 1958, Cowan 1966, 1967, Stempffer 1967, Tite & Dickson 1973.

Subfamily Lycaeninae

Proboscis smooth-shafted. Labial sclerite completely sclerotized. Palpi clothed with hairy or bristly scales. Male prothoracic coxae not extending spine-like below the articulation of the trochanter. Male prothoracic legs more than half the length of the pterothoracic legs, with the protarsus fused to a single segment and ending in a sharp or rounded point; meso- and metatibiae spurred. Forewing with 11 veins, veins R_5 and M_1 usually narrowly separated, but sometimes connate or with a short stalk. Hindwing tailed at Cu_{1b} or tailless, tornus lobed or rounded; costal margin basally lacking a vein. Males lacking alar androconia. Most diverse in the Holarctic although there is weak representation in all other regions (Ehrlich 1958; Eliot 1973).

The larvae are onisciform.

Major hostplant families. Polygonaceae.

Major references. Stempffer 1967, Higgins 1975, Miller & Martin Brown 1981.

Subfamily Polyommatinae

Proboscis usually smooth-shafted with the terminal sensory area usually very weakly developed. Labial sclerite completely sclerotized. Palpi variable. Male prothoracic coxae not extended spine-like below the articulation of the trochanter. Male prothoracic legs more than one half the length of the pterothoracic legs, with the protarsus fused to a single segment and commonly ending in a tapered, down-curved point; meso- and metatibiae with terminal spurs. Forewing usually with 11 veins, veins R_5 and M_1 separate at their origins although sometimes only narrowly so. Hindwing tailless or with a filamentous tail at vein Cu_{1b} only; tornus rounded or occasionally with a vestigial lobe; basally lacking a vein along the costal margin. Represented in all major biogeographic areas (Ehrlich 1958, Eliot 1973).

The larvae are onisciform.

Major hostplant families. Crassulaceae, Fabaceae (Leguminosae), Lamiaceae, Myrsinaceae, Primulaceae, Rutaceae—additionally many species are known to be aphytophagous, at least in some instars, on homopterans, ant broods or ant regurgitations; others may be saprophagous or cannibalistic (Cottrell 1984).

Major references. Bethune-Baker 1910, Nabokov 1945, Tite 1963*a*, *b*, Stempffer 1967, Sibatani & Grund 1978, Eliot & Kawazoé 1983.

Subfamily Riodininae

Labial sclerite completely sclerotized. Male prothoracic coxae extending spine-like below the articulation of the trochanter. Male prothoracic legs less than one half the length of the pterothoracic legs, with the tarsi unsegmented and rarely bearly claws. Hindwing humeral vein usually present; hindwing with a vein along the base of the coastal margin (always present when the humeral vein is absent) (Ehrlich 1958).

The larvae are usually denisciform, broadest in the middle and tapering to both ends, with a clothing of short hairs.

Major hostplant families. Asteraceae (Compositae), Fabaceae (Leguminosae), Myrsinaceae, Myrtaceae — additionally recorded on many other diverse families (see Ehrlich & Raven 1965). The likely ant-associations of riodinines remain poorly documented (see Callaghan 1977, Robbins & Aiello 1982).

Major references. Stichel 1928, 1930–1.

Subfamily Styginae

Labial sclerite sclerotized principally behind (strongly) and between (lightly) the palpal sockets. Male prothoracic tarsi segmented and bearing claws (Ehrlich 1958). The early-stage biology of single genus and species (*Styx infernalis*) placed in this subfamily remains unknown; it occurs in Peru.

Family Nymphalidae

Male forelegs always reduced and clawless. Pupa not fastened by a silken girdle, usually suspended from cremaster at the abdominal tip. Most antennal segments with two ventral grooves. Very distinct separation of the upper laterocervico-tentorial muscles into two bundles. To the eight subfamilies recognized by Ehrlich (1958) should be added the Brassolinae (Miller 1968), Heliconiinae (Emsley 1963), the Libytheinae (Kristensen 1976), Tellervinae (Ackery & Vane-Wright in press *a*) and probably the Amathusiinae (Kirchberg 1942). It is likely that future studies will also reveal further subdivisions of the Nymphalinae as here characterized but at present their recognition would be premature. Robbins (1982) estimates some 6000 nymphalid species, approximately one-third the total number of butterflies. The nymphalids are rivalled only by the Lycaenidae in terms of species diversity.

Major references. Clark 1947, 1948, Ehrlich 1958.

Subfamily Brassolinae

Mesothoracic anepisternum present as a distinct sclerite; mesothoracic pre-episternum well-developed, varying in size; pre-episternal suture usually well-developed. Forewing veins not swollen basally; vein 3A not free at the base. Hindwing cell closed by a tubular vein; precostal cell present (adapted from Ehrlich 1958, Miller 1968). There are approximately 80 species, all restricted to the Neotropics. The curious S. American genus *Bia*, conventionally included in the Satyrinae, may prove to be a brassoline.

The larvae are smooth, with a bifid tail (except *Brassolis* and *Penetes*). The head capsule is generally armed but in *Brassolis* itself such ornamentation appears to be lacking.

Major hostplant families. Arecaceae, Bromeliaceae, Heliconiaceae, Musaceae, Poaceae (Gramineae).

Major references. Stichel 1909, 1932, Miller 1968.

Subfamily Amathusiinae

Mesothoracic anepisternum present as a distinct sclerite; mesothoracic pre-episternum well-developed, varying in size; pre-episternal suture usually well-developed. Forewing veins not swollen basally; vein 3A not free at the base. Hindwing cell open (some suggestion of closure in *Xanthotaenia*); precostal cell absent (adapted in part from Ehrlich 1958). There are about 80 species, all restricted to the Indo-Australian tropics.

The larvae are smooth but distinctly hairy. The tail is bifid and the head capsule armed although not apparently in *Discophora* species.

Major hostplant families. Arecaceae, Musaceae, Poaceae (Gramineae), Smilacaceae.

Major references. Stichel 1912, 1933, Kirchberg 1942, Aoki *et al.* 1982.

Subfamily Satyrinae

Mesothoracic-anepisternum present as a distinct sclerite; mesothoracic pre-episternum usually greatly reduced or separated from the katepisternum by a very weak pre-episternal suture. Forewing vein 3A not free basally (except in *Haetera* and its allies); usually with at least one vein basally swollen. Hindwing cell always closed by a tubular vein; precostal cell absent (except *Elymnias* and its allies) (adapted from Ehrlich 1958, Miller 1968). Ehrlich & Raven (1965) estimate up to 1500 recognized species distributed world-wide, including considerable extratropical diversity.

The larvae are usually smooth with a bifid tail; head capsule occasionally armed with horns.

Major hostplant families. Arecaceae, Cyperaceae, Poaceae (Gramineae).

Major references. Gaede 1931, Warren 1936, Avinoff & Sweander 1951, Miller 1968, Condamin 1973, Kudrna 1977, Aoki *et al.* 1982.

Subfamily Morphinae

Mesothoracic anepisternum present as a distinct sclerite; mesothoracic pre-episternum well-developed, varying in size; pre-episternal suture usually well-developed. Forewing veins not swollen basally; vein 3A not free at the base. Hindwing cell open; precostal cell absent (adapted in part from Ehrlich 1958). Currently a monobasic subfamily, limited to the Neotropics and according to Le Moult & Real (1962) containing approximately 70 species; most workers consider this figure excessive. While Ehrlich (1958) includes the Brassolinae and Amathusiinae within the Morphinae, Miller (1968) places the Brassolinae as a subfamily of the 'Satyridae'. Ehrlich's investigations suggest that the Satyrinae, Morphinae (*Morpho*), Brassolinae and Amathusiinae together form a monophyletic group but their mutual affinities remain obscure. Vane-Wright (1972a) suggests that the 'satyrine' genera *Antirrhea* and *Caerois* may be the closest relatives of *Morpho*, a view endorsed by DeVries *et al.* (in press) based on observations of the early stages.

The larvae are smooth with characteristic dorsal tufts of hairs. The tail is indistinctly bifid and the head capsule ornamentation rudimentary.

Major hostplant families. Fabaceae (Leguminosae), Poaceae (Gramineae).

Major references. Le Moult & Real 1962.

Subfamily Calinaginae

Mesothoracic anepisternum present as a distinct sclerite; mesothoracic pre-episternum well-developed; parapatagia lacking traces of sclerotization. Male with prominent superuncus. Female protarsus with small but perfect tarsal claws. Forewing veins not swollen basally; vein 3A free at the base. Hindwing cell closed by a weak tubular vein; precostal cell absent (Ehrlich 1958). Thorax clothed at least in part with orange or red hairs. This subfamily, represented by a single genus from the Sino-Himalayan region, lacks obvious affinities with any other nymphaloid grouping. Up to four species occur sympatrically in western China. Note In Ehrlich's diagram of relationships (1958) the names Calinaginae and Charaxinae are transposed.

The larvae are smooth and thinly pilose. The head capsule is armed with a pair of prominent horns and the tail weakly bifid.

Major hostplant family. Moraceae.

Major references. Stichel 1938, Ashizawa & Muroya 1967.

Subfamily Charaxinae

Mesothoracic anepisternum present as a distinct sclerite; mesothoracic pre-episternum well-developed; pre-episternal suture variable; parapatagia with at least some trace of sclerotization. Forewing veins not swollen basally. Hindwing cell not closed by a tubular vein (Ehrlich 1958). Ehrlich & Raven (1965) estimate between 3-400 species distributed throughout the tropics, only rarely extending into more temperate regions.

The larvae are smooth, generally with a bifid tail and an armed head-capsule although at least some *Anaea* species lack these adornments.

Major hostplant families. Convolvulaceae, Euphorbiaceae, Fabaceae (Leguminosae), Flacourtiaceae, Lauraceae, Piperaceae, Sapindaceae.

Major references. Rothschild & Jordan 1898-1900, Stichel 1939, Comstock 1961, Rydon 1971, Van Someren 1975, Smiles 1982.

Subfamily Nymphalinae

Mesothoracic anepisternum absent as a distinct sclerite; mesothoracic pre-episternum at its widest at least one half the width of the katepisternum; pre-episternal suture usually well-developed. Tarsal claws simple, symmetrical. Females lack odoriferous glands between the eighth and ninth abdominal segments. Gnathos usually well-developed. Forewing vein 3A usually not free at the base (adapted from Ehrlich 1958). Ehrlich & Raven (1965) estimate some 3000 species distributed almost world-wide.

It is doubtful if the Nymphalinae as here characterized form a natural assemblage. Clark (1948) includes within the Nymphalinae such groups as the Limenitini (see Chermock 1950; presumably the Neptini, sensu Eliot 1969, also belong here), Argynnini (see Warren 1944), Marpesiini, Nymphalini and Ergolini; the Apaturini (see Le Moult 1950) he placed with the charaxines as a separate family. Clark's approach was phenetic, based on larval features (Clark 1947, see below), and is only comprehensive for Nearctic genera. Although there are no doubt many identifiable natural groups within the Nymphalinae, their characterizations and relationships remain obscure. Note The familiar holarctic nymphalines including the tortoiseshells, painted ladies, red admirals, buckeyes, Peacock, Mourning Cloak etc. are often incorrectly referred to as the 'Vanessini', a synonym of Nymphalini.

Larva. These are usually cylindrical and spinose. Marpesiini—median dorsal spines only. Nymphalini—complete and well-developed row of median dorsal spines together with lateral spines. Ergolini—fully developed lateral spines with median dorsal series reduced or absent except towards the head and tail. Argynnini, Limenitini—lateral spines only, the median dorsal series being wholly absent (adapted from Clark 1947). Apaturini—not regularly spinose; head capsule usually armed and tail bifid.

Major hostplant families. Acanthaceae, Asteraceae (Compositae), Caprifoliaceae, Euphorbiaceae, Flacourtiaceae, Moraceae, Plantaginaceae, Sapindaceae, Scrophulariaceae, Ulmaceae, Urticaceae, Verbenaceae, Violaceae.

Major references. Stichel 1938, Warren 1944, Dillon 1948, Chermock 1950, Le Moult 1950, Eliot 1969, Field 1971, Higgins 1981.

Subfamily Heliconiinae

Mesothoracic anepisternum absent as a distinct sclerite; mesothoracic pre-episternum at its widest at least one-half the width of the katepisternum. Females with odoriferous glands between abdominal segments eight and nine. Tarsal claws simple, symmetrical. Forewing vein 3A not free at the base. Male hindwing upperside always with androconia present along two or more veins (Ehrlich 1958, Emsley 1963). Although included within the Nymphalinae by Ehrlich (1958), the heliconiines are now widely recognized as a separate subfamily (see Emsley 1963). There are approximately seventy species, all New World and mostly confined to the Neotropics although it has recently been suggested that the Old World passifloraceous-feeding nymphaline genera, *Cethosia* and *Vindula*, should more correctly be placed in the Heliconiinae (Brown 1981). The heliconiines will probably prove to represent a highly specialized subgroup of the Argynnini s.l.

The larvae are spinose, with six rows of longitudinal spines, each adorned with stiff bristles. The head is similarly armed with a pair of dorsal spines.

Major hostplant families. Passifloraceae.

Major references. Stichel & Riffarth 1905, Neustetter 1929, Stichel 1938, Michener 1942, Emsley 1963, Benson *et al.* 1976, Brown 1979, 1981.

Subfamily Acraeinae

Mesothoracic anepisternum absent as a distinct sclerite; mesothoracic pre-episternum at its widest at least one half the width of the katepisternum (except *Pardopsis*). Gnathos absent, or at most vestigial (*Pardopsis*). Tarsal claws usually toothed or asymmetrical, especially in males (normal in *Pardopsis*). Forewing vein 3A not free at the base. Hindwing cell closed by a well-developed tubular vein (Ehrlich 1958). There are approximately 250 species, largely tropical with the greatest diversity in Africa (Ehrlich & Raven 1965).

The larvae are spinose, with six longitudinal rows of spines, each adorned with stiff bristles. The head is hairy, lacking spines or horns, and the tail simple.

Major hostplant families. Asteraceae (Compositae), Passifloraceae, Tiliaceae, Urticaceae.

Major references. Eltringham & Jordan 1913, Carcasson 1961, Pierre 1983.

Subfamily Danainae

Mesothoracic episternum absent as a distinct sclerite; mesothoracic pre-episternum at its widest much less than one half the width of the katepisternum. Mesomeron with a prominent caudal bulge. Male with abdominal hairpencils. Female protarsus four-segmented, ankylosed and strongly clubbed. Forewing vein 3A free at the base. Antennae naked (Ehrlich 1958). There are approximately 150 currently recognized species (Ackery & Vane-Wright in press *a*); although a few are found in more temperate areas, most are essentially tropical with greatest diversity in the Orient and least in the New World.

The larvae are smooth and conspicuously patterned, lacking both head ornamentation and a bifid tail; dorsally with between one and nine segmental pairs of fleshy tubercles.

Major hostplant families. Apocynaceae, Asclepiadaceae, Moraceae.

Major references. Bryk 1937*a*, *b*, Stichel 1938, Forbes 1939, Morishita 1981, Ackery & Vane-Wright in press *a*.

Subfamily Ithomiinae

Mesothoracic anepisternum absent as a distinct sclerite; mesothoracic pre-episternum at its widest much less than one-half the width of the katepisternum. Mesomeron with a prominent caudal bulge. Female protarsus 4- or 5-segmented, not strongly clubbed. Forewing vein 3A free at the base. Male hindwing upperside with conspicuous costal hairs. Antennae scaled (Ehrlich 1958). There are more than 300 currently recognized species (Mielke & Brown 1979), all restricted to the Neotropics but poorly represented in the Antilles.

The larvae are smooth and variously patterned, lacking both head ornamentation and a bifid tail. Occasionally a single pair of dorsal fleshy tubercles occur on the metathorax and rarely lateral tubercles on segments 4–11.

Major hostplant families. Apocynaceae, Solanaceae.

Major references. Bryk 1937*b*, Fox 1940, 1956, 1960, 1967, Fox & Real 1971, Drummond 1976, Haber 1978, d'Almeida 1978, Mielke & Brown 1979.

Subfamily Tellervinae

Mesothoracic anepisternum absent as a distinct sclerite; mesothoracic pre-episternum at its widest much less than one-half the width of the katepisternum. Mesomeron with a prominent caudal bulge. Female protarsus 5-segmented, not strongly clubbed. Forewing vein 3A free at the base. Antennae scaled (Ehrlich 1958). The Tellervinae have often been included as a tribe of the Ithomiinae (Fox 1956, Ehrlich 1958) but Ackery & Vane-Wright (in press *a*) advocate subfamily status. Together, the Ithomiinae, Danainae and Tellervinae almost certainly form a monophyletic group with the Tellervinae only weakly characterized by the absence of the secondary sexual features of its sister subfamilies. *Tellervo*, the only included genus, is monobasic according to Fox (1956) although Hulstaert (1931) recognized five distinct species all restricted to the New Guinea subregion.

The larvae are dark and smooth, lacking both a bifid tail and head ornamentation. A single pair of dorsal fleshy tubercles occur on the metathorax.

Major hostplant families. Apocynaceae.

Major references. Hulstaert 1931, Bryk 1937*b*, Fox 1956.

Subfamily Libytheinae

Patagia not prominent or rounded and bearing only small lateral sclerotizations; metanotum essentially entirely below the mesoscutellum which covers it. Male prothoracic legs atrophied, in female only slightly so. The generally long labial palps, a 'traditional' character for the libytheines, are short enough in some species to overlap with the lengths found in some other nymphalids (Ehrlich 1958). There are only 10 species but their overall distribution is cosmopolitan (Ehrlich & Raven 1965).

The larvae are smooth, lacking both head capsule ornamentation and a bifid tail.

Major hostplant families. Ulmaceae.

Major references. Pagenstecher 1901, 1911, Aoki *et al.* 1982.

Faunistic Studies

The regions adopted in this section are broadly in accordance with Hollis (1980) except, because of its relative paucity, the Polynesian fauna comes within the Australasian survey. References are included to major faunal studies and catalogues, together with primary information sources on butterfly biology. The expansive nineteenth-century literature, although now out-dated, often contains useful data (see Pagenstecher 1909 for a comprehensive review).

The Palaearctic Region

There is no single modern catalogue to the butterflies of the Palaearctic region; taken together, the works of Higgins (1975) and Korshunov (1972) probably give the widest coverage. Bang-Haas' (1930) work is particularly useful, indexing by country and area the then published faunal lists of Palaearctic butterflies—this can be updated in part from the selection of faunal studies, together with the references therein, included in Higgins & Riley (1970). To date, Seitz (1907-9) remains the only comprehensive study of this region. Higgins and Riley's field guide covers only the western Palaearctic and the work to be edited by O. Kudrna (in prep.) will probably be similarly limited. However, there are many faunal studies on countries within the area, together giving a remarkably broad coverage. The most recent and important refer to the UK (Howarth 1973*b*), Scandinavia (Langer 1958), Spain (Manley & Allcard 1971), Germany (Forster & Wohlfahrt 1955), the Middle East (Larsen 1974, 1982*a*, Larsen & Larsen 1980), Afghanistan (Sakai 1981), the eastern USSR (Kurentsov 1970) and Japan (Fujioka 1975, Kawazoê & Wakabayashi 1977). Biological information included by Iwase (1954, 1964), Shirôzu & Hara (1960-2), Brooks & Knight (1982) and Hendriksen & Kreutzer (1982) supplements these studies.

The Afrotropical Region

Although Peters' catalogue (1952) remains useful, it is now very much outdated and likely to be largely replaced by Carcasson's forthcoming work (in prep.), itself the basis of D'Abrera's (1980) comprehensive colour guide and Carcasson's own handguide (1981). Otherwise only Seitz (1908-25) gives complete coverage of the entire region. Most faunistic studies of individual countries concentrate on southern and eastern Africa with just Liberia (Fox *et al.* 1965) and Zaire (Berger 1981) to the west having modern treatments; the Nigerian checklist (Cornes *et al.* 1973) is the only other major contribution. Pennington (1978), the latest in a succession of monographs on South African butterflies, covers the region south of Angola and the Zambesi, while Gifford (1965) gives a useful illustrated key to the Malawi fauna. The standard studies for Kenya and Uganda (Van Someren & Rogers 1925-31, Van Someren 1935-9) also contain a wealth of biological data, and are supplemented in part by Carcasson's works (1961, 1963, 1975) on the Danainae, Acraeinae and Papilionidae. Additional information on aspects of biology is to be found in many works, amongst which Lamborn (1914), Farquharson (1922), Haig (1936-8), Birket Smith (1960), Clark & Dickson (1971), Owen (1971*a*), Owen & Owen (1972, 1973), van Someren (1974), Sevastopulo (1975), Rosevear (1978), Pierre-Baltus (1978) and Fontaine (1981*a*, *b*, 1982) are outstanding.

The Oriental Region

The butterflies of various countries within the Oriental region are known from individual lists, particularly Thailand (Godfrey 1930), Hainan (Joicey & Talbot 1924-32) and Hong Kong (Hill *et al.* 1978); additionally Shirôzu's (1960) faunal study on Taiwan includes a checklist, but there is no single catalogue for the entire region. The area is covered within Seitz (1908-28), the 'Indo-Australian' fauna, and also by D'Abrera (1982) in the first part of a planned series of volumes on Oriental butterflies. Talbot (1939, 1949), although incomplete, is probably the most important faunistic study, covering India, Bangladesh, Pakistan, Sri Lanka and Burma (see also Woodhouse 1952 for Sri Lanka). Hong Kong (Johnston & Johnston 1980), China (Leech 1892-4), Korea (Kim 1976, Lee 1982) and Taiwan (Shirôzu 1960) have all received comprehensive treatment. Three major references sources, Pant & Chatterjee (1950), Varshney (1977) and Sevastopulo (1973), catalogue the described life-histories, local lists and larval hostplants of the Indian fauna; additional biological data on Oriental butterflies is to be found most importantly in Johnston & Johnston (1980), Hoffman *et al.* (1938) and Muroya *et al.* (1967*a*, *b*).

The Austro-Oriental Region

There is no single catalogue applicable to this region. Corbet & Pendlebury (1978) include a checklist of the Malaysian butterflies in their faunal study and although several island faunas west of New Guinea are similarly covered, often in now out-dated nineteenth century works, their details are outside the scope of this brief overview (see Ackery & Vane-Wright in press *a* for a detailed consideration of the faunistic literature, based on the Danainae). Otherwise only Java has received a comprehensive, reasonably modern treatment (see Piepers & Snellen 1909-18, updated by Roepke 1935-42); together with Corbet & Pendlebury (1978), these works are the primary reference sources for biological data on the butterflies of the region although Fountaine (1925-6) is also helpful. The butterfly faunas of the Philippines (see Semper 1886-92), Sulawesi, the Lesser Sunda Islands and the Moluccas are perhaps the least known within the South-East Asian islands, particularly with respect to their biology.

The Australasian Region

Moulds' (1977) bibliography of Australian butterflies is the primary reference source for this area—covering the years 1773-1973, it can be readily supplemented from the continuous bibliographies published regularly in the *Aust.ent.Mag.* In the absence of a comprehensive systematic catalogue, D'Abrera's (1977) study is probably the most useful single work, illustrating much of the region's butterfly fauna with the exception of the Hesperiidae; Common & Waterhouse (1981) only include species found in Australia itself. Among other areas covered by faunistic studies are Fiji (Robinson 1975), New Caledonia (Holloway & Peters 1976), Samoa (Hopkins 1927) and Oceania (Viette 1950). The entire fauna falls within Seitz (1908-28) 'Indo-Australian' region. Although most of these works contain biological data, Parson's forthcoming study on the biology of New Guinea butterflies (in prep.) will be a major additional contribution on one of the least known faunas in the Old-World tropics.

The Nearctic Region

The recent publication of Miller & Brown's (1981) catalogue of North American butterflies completes a basic quartet of reference works on this region—Howe (1975) is the most recent of several works covering the fauna of North America; Field *et al.* (1974) list by state and province the catalogues, checklists and faunal studies pertaining to Nearctic butterflies while Tietz (1972) indexes the described life-histories, early stages and hostplants published prior to 1950. Although this latter work is now rather out-dated, it can be supplemented in part by reference to Ehrlich & Ehrlich (1961) and Howe (1975). The Nearctic butterflies are covered within Seitz (1907-24), 'The American Rhopalocera'.

The Neotropical Region

The published regional lists of South American butterflies are catalogued by Lamas (1977); it is sufficient here just to mention the principal works —those for Argentina (Hayward 1973), Bolivia (Ureta 1941), Chile (Ureta 1963), Costa Rica (DeVries in prep.*b*), Guyana (Hall 1940) and Mexico (Hoffman 1940-1). The proposed work by Heppner (in prep.) should fulfill the need for a comprehensive catalogue covering the entire region. Although now much out-dated, Seitz (1907-24) gives the sole broad overview of the Neotropical butterfly fauna. Within continental South America, only Argentina (Hayward 1948-

67) is adequately, if incompletely, covered; Barcant (1970), Brown & Heineman (1972) and Riley (1975) cover the Antilles and Trinidad while the study by Godman & Salvin (1879-1901) is still the basis of our knowledge for Central America. Eventually the series of volumes started by D'Abrera (1981) will provide a useful colour-guide to the Neotropical region. Although these major faunal works generally contain some biological data, further information on ecology and life-histories is to be found in various studies (particularly Bates 1861, 1862, 1864-5, Müller 1886, Fountaine 1913, d'Almeida 1922, Miles Moss 1920, 1933, 1949, Comstock & Garcia 1961, D'Araújo e Silva *et al.* 1967-8, Hayward 1969, Ross 1975-7, DeVries in prep. *b*; see also the works of Allen Young [1971-on] and Alberto Muyshondt [1973-on]).

Postscript

This review serves two distinct purposes; firstly to provide a formal outline of the classification adopted throughout this volume, and secondly to emphasize the availability of an already expansive faunistic butterfly literature. Such works remain largely untapped as sources of data on many aspects of butterfly biology, particularly in relation to hostplant and habitat preferences, and life-histories. Much of this information, used in conjunction with current research, still has a part to play in investigations of the biology of butterflies. In conclusion, it should be emphasized that a great body of information exists in the extensive Japanese literature, data thus largely unavailable to many butterfly workers.

Part II
Populations and Communities

2. *The Structure and Dynamics of Butterfly Populations*

Paul R. Ehrlich

Department of Biological Sciences, Stanford University, Stanford, California

Explaining the arrangement of organisms in space and the fluctuations in their numbers through time is the basic task of population ecology. Butterflies are an almost ideal group in which to investigate these subjects, and more is now known about the structure and dynamics of their populations than about any comparable group of invertebrates—possibly more than any group of non-human animals. Yet so far this accumulation of knowledge has led to the induction of surprisingly few generalities even about butterflies, let alone about herbivorous insects, for which one might hope butterfly systems would serve at least as a partial model (see also Ch.3).

In this paper I deal with some general problems of population ecology and review what is now known about, first the structure, and second the dynamics of butterfly populations. After discussing generalizations that may be based on that research, I finally suggest some directions for future work that might increase the value of butterflies, both as model systems for ecological investigations and as 'indicator organisms', in a world in which the decay of organic diversity is becoming an increasingly serious problem (Ehrlich & Ehrlich 1981*a*).

A comprehensive treatment of work done prior to the early 1970s will not be attempted, since that period has been covered by two previous reviews (Ehrlich *et al.* 1975, Gilbert & Singer 1975). Taxonomic treatment will follow the general rule that obligatory categories (for example, genera and families, as opposed to subgenera and subfamilies) should be treated conservatively.

Population Ecology

In their seminal work, *The Distribution and Abundance of Animals*, Andrewartha & Birch (1954: 5) summarize the basic questions of the ecology of animal populations: 'Why does this animal inhabit so much and no more of the earth? Why is it abundant in some parts of its distribution and rare in others? Why is it sometimes abundant and sometimes rare?' Such questions are important not only because they are central to ecological and evolutionary theory, but because of their increasing significance in an overpopulated world with declining resources. By understanding the behaviour of natural populations, humanity can hope better to manipulate the populations of organisms of direct economic importance: encouraging those populations valued for harvest; discouraging those that endanger or compete with *Homo sapiens*.

Geographic Distribution

Unfortunately, it is usually difficult to determine the limits of ranges and the numbers of organisms in a given location. These are, of course, related problems. The edge of the range of a species is difficult even to define. If one defines the distribution as that area outside of which populations cannot be maintained, then the question immediately arises, 'maintained for how long?' Many Nearctic butterflies, for example, move northward or upward in elevation during the summer (Shapiro 1973*a*), establishing populations that may go through several generations before winter kills them. *Mestra amymone* in the Great Plains (Ehrlich & Ehrlich 1961), *Plebejus melissa* in Colorado (C. L. Remington, pers. comm.), and *Pieris rapae* and *Precis coenia* in the Sierra Nevada of California are examples. Similar patterns occur in northern Europe (e.g. Ekholm 1975).

This temporal variation in distribution reaches its extreme in migratory species (e.g. Baker 1978; Ch.26), but all distributions, even those of quite sedentary organisms, change at some rate over

The Biology of Butterflies
0-12-713750-5

time. Thus the question of species distribution, when examined closely, often turns out to be a set of questions about the structure and dynamics of individual peripheral populations.

Investigating Structure and Dynamics

Logistic difficulties appear, however, when one attempts to answer questions about the structure and dynamics of individual populations. The investigatory problems vary among taxonomic groups. Ideal organisms for the study of population ecology would be well understood taxonomically, and easily recognized and marked in the field, yet amenable to manipulation in the laboratory and have a short generation time (since full understanding of the ecology of a population must include its genetic attributes). Butterflies possess these characteristics: they are better known systematically than birds (Olson 1981), and many make quite reasonable laboratory animals, with generation times measured in weeks. A major drawback is that they have holokinetic chromosomes, making karyotyping difficult (Pearse & Ehrlich 1979), although there is some hope that new techniques may overcome this problem. Above all, butterflies share with birds a most important attribute. They, too, are studied by numerous and enthusiastic amateurs, and because of this a far larger body of information exists on butterfly distribution, foodplants and the like than could possibly have been assembled by professional biologists alone.

But, even with all these advantages, research on the biology of butterfly populations is fraught with problems (see also Ch.4). In order to get reliable data from mark-release-recapture (MRR) experiments (usually the most effective method for small, mobile animals) very high marking intensities are required (see Begon 1979 for overview). Furthermore, great care must be taken to record positions of capture and recapture so that demographic units can be defined (Ehrlich 1965, Brown & Ehrlich 1980). Failure to do so makes interpretation of changes in population size impossible. For example, if three demographic units are erroneously treated as a single unit, one can explode in size, one fluctuate, and one decline to extinction, and these events can be jointly misinterpreted as an increase in the single unit. This, for example, would have happened if three separate units (called C, G, and H) of *Euphydryas editha* had not been recognized in 1960–65 on Stanford University's Jasper Ridge Biological Preserve near San Francisco, California (Ehrlich *et al.* 1975).

Failure to define demographic units has been one factor leading to the development of models of population dynamics, such as that of Taylor & Taylor (1977, 1979), that are at best of limited or questionable applicability (Rankin & Singer in prep.), and at worst preposterous. For example, Taylor & Taylor (1977) show the Monarch, *Danaus plexippus*, aggregating at low density and flying 3200 km at high density. In fact its migration shows little relationship to density, and when travelling southward individuals move *toward* points of the highest density ever observed in large butterflies (Brower *et al.* 1977*a*). I am very sceptical about both Taylor & Taylor's assumptions and the biological reality of most of the patterns they have chosen to analyse.

A final problem is the difficulty of getting the reliable estimates of population size per generation (rather than per day) that are often the most useful for answering basic questions. Such estimates must ordinarily be made over many generations if the factors controlling dynamics are to be elucidated. In addition, even when high marking intensities are achieved, nagging questions about such things as recapture probabilities, long-distance movement, and sex ratio may persist (Singer & Wedlake 1981, Lederhouse 1982; Ch.4). For these reasons, in spite of the manifest advantages of butterflies as tools for understanding the biology of populations, their potential in this area has been developed only slowly.

Population Structure

The term 'population structure' is used here in the restricted sense of spatial structure: the patterns in which individuals move relative to each other and the environment, and their probabilities of reproducing in different places. Adult movement clearly controls butterfly population structure; larval movement is generally minimal, although it may be important in the dynamics of populations (e.g. Dethier 1959*b*, Holdren & Ehrlich 1982).

Patterns of Structure

Investigations of butterflies by MRR methods have revealed a great variety of population structure to exist both between butterfly populations (Gilbert & Singer 1975) and in the same population in different years (White & Levin 1981). Some butterflies, for example, show very 'tight' structure, occurring in clearly delimited demographic units with little exchange of individuals. Examples include *Maniola jurtina* on the Isles of Scilly (Dowdeswell *et al.* 1957), the three populations of *Euphydryas editha* at Jasper Ridge (Ehrlich 1965, Brussard *et al.* 1974), and populations of *Lycaena virgaureae* in southern

Sweden (Douwes 1970, 1975). Others have a 'loose' structure, being virtually ubiquitous over wide areas, such as *Erebia epipsodea* in montane Colorado (Brussard & Ehrlich 1970*a*, *c*), the skipper *Hylephila phylaeus* in the Sacramento Valley of California (I. D. Shapiro 1977), the pierid *Aporia crataegi* in the hills of Hokkaido (Watanabe 1978), and *Pieris rapae* (Emmel 1973*b*) and *Danaus plexippus* in many areas.

Such variety might be expected in almost 1000 genera and perhaps 15 000 species (Ehrlich & Raven 1965), even though the butterflies are a rather uniform group phenetically. Interestingly, while the taxonomic diversity of butterflies is largely tropical, the diversity of population structure has been largely revealed by studies in the temperate zones. Whether the more diverse tropics will show more patterns remains to be seen.

Resources and Structure

Many factors influence the structure of butterfly populations. One of the most important is the distribution and abundance of nutritional resources, primarily foodplants for larvae and liquid sustenance for adults. Consider first the case where these resources co-occur. The ultimate in such co-location is represented by the thecline *Hypaurotis crysalus*. The larvae attack *Quercus gambellii*, and, since nectar sources are absent from the habitat, the adults feed on sap oozing from the oak twigs (Scott 1974*a*). Scott suggests that the adults have evolved a flight period (in the season of afternoon rains; males patrol in light rain or cloudy weather, not in prolonged periods of sun) that minimizes desiccation, since the oak sap is rather viscous.

H. crysalus adults have shorter proboscides than do flower-feeding Theclini and will 'puddle' on the sandy edge of streams, although in most cases streams do not run near the oaks and the butterflies will not move far in search of water (whether they sip from raindrops is not known). The vast majority of individuals stayed in a 50 × 120m grove; no marked individuals were recaptured in another grove separated from the study site by only 100m of grassland. In another Colorado study, the only one of eleven subalpine butterflies in Colorado that appeared to have a tight population structure, *Plebejus saepiolus*, also used its patchily distributed larval host, *Trifolium*, as a preferred nectar source (Sharp & Parks 1973, Sharp *et al.* 1974).

In many cases the plant species providing nectar and those providing larval food are different, but grow together in the same habitat patches. For example, for *Euphydryas editha* populations at Jasper Ridge and nearby Edgewood Road, the key larval resource (Ehrlich *et al.* 1975) is the perennial *Orthocarpus densiflorus* (Scrophulariaceae), which in the San Francisco area is now abundant only on serpentine soil. A number of satisfactory nectar sources also occur on serpentine. In serpentine areas where larval resources are abundant, demographic units can be maintained and there is little tendency for individuals to disperse (Ehrlich 1961) as long as nectar resources are sufficient in quantity. When adult resources are not satisfactorily abundant, however, there is evidence that dispersal, especially of females, increases (Ehrlich *et al.* 1982).

When larval and adult resources are separated, adults naturally move between them. For example, in a population of the pierid *Leptidea sinapis* in central Sweden, adults take nectar in open woodlands and oviposit in an adjacent meadow (Wiklund 1977*a*); moving between the two even though they are weak fliers. *Euphydryas editha* in the Inner Coast Range of California (Del Puerto Canyon) feeds as larvae on *Pedicularis* distributed along ridge tops, while adult resources (especially *Eriodictyon californicum* and *Achillea millefolium*) are scattered throughout the habitat and may be in bloom only along the creek at the bottom of the canyon late in the flight season (Gilbert & Singer 1973). In these circumstances, individual *E. editha* move, on the average, much greater distances than they do at Jasper Ridge.

Varying Structure in *Euphydryas*

The determination of population structure by resource requirements is underlined by the varying structure found in *Euphydryas* populations. At Pioneer Resort in Gunnison County, Colorado, for example, patterns of movement in *E. editha* are more reminiscent of Del Puerto than of Jasper Ridge, even though adult and larval resources are co-located in a montane meadow (Ehrlich & White 1980). Here both sexes flew some 200m downslope from the meadow to probe moist soil at a seep, possibly to obtain sodium (Arms *et al.* 1974).

Indeed, probably the loosest population structure found in *E. editha* living in nonephemeral habitats is that shown by populations occupying montane meadows dominated by sagebrush (*Artemisia*), such as that at Pioneer Resort. In Gunnison County the preferred larval foodplant, *Castilleja linariifolia*, is common in such areas around 2500m to 3000m, and the butterflies occur (Ehrlich group, unpublished) in most places where that plant is sufficiently abundant and nectar sources are found. The highest concentrations occur in conjunction with the appropriate topography. The latter seems to consist of hillsides with a southern exposure where snow does not persist late enough to delay the development of post-diapause larvae, and bare patches of soil or rock outcrops are available as perches for males to await females. In fact, the presence of suitable male perching sites may be a key resource for these populations. We have evidence (observations, gene-frequency uniformity) suggesting that movements between areas of concentration are frequent.

Patterns of movement in *E. editha* can change within and between years in response to changing resource distributions (Gilbert & Singer 1973, Ehrlich *et al.* 1975, and unpublished). For example, at Lower Otay in a semi-desert area near San Diego, California, movement increases in dry years, especially in those that follow wet years when post-diapause larvae may totally devour the sparse crop of oviposition plants (White & Levin 1981).

Other Nearctic *Euphydryas* show a variety of resource-related patterns similar to those of *E. editha*. In Colorado the structure of alpine *E. anicia* populations resembles that of *E. editha* at Del Puerto (Cullenward *et al.* 1979, White 1980) as does the structure of *E. chalcedona* on Jasper Ridge (Brown & Ehrlich 1980). *Euphydryas phaeton*, the only *Euphydryas* in the eastern United States, occurs in small, scattered colonies in moist places where its scrophulariaceous foodplant, *Chelone glabra*, grows. At least in upper New York state, these colonies appear, however, to be united by a relatively high level of gene flow caused by dispersing females. These, unlike dispersants of *E. editha*, are not generally faced with rapidly drying host plants in the habitat patches into which they move (Brussard & Vawter 1975).

Euphydryas gillettii, the Nearctic representative of the predominantly European *maturna* group, shows a tight population structure reminiscent of that of *E. editha* at Jasper Ridge (Williams in press, Williams *et al.* in press). These butterflies show an extreme tendency to 'stay put' wherever the larval host, *Lonicera involucrata* (Caprifoliaceae), is found with suitable nectar sources and coniferous trees to serve as perches for males.

In a Colorado colony founded in 1977 by transplanted eggs and larvae from Wyoming (Holdren & Ehrlich 1981), all fourth-generation (1981) Colorado individuals captured were found within the roughly 100 × 150m area of the original transplantation. No egg mass was found outside this area until that fourth generation, when one was located about 300m from the centre of release in the direction of the most continuous combination of *Lonicera*, nectar sources and spruce trees (*Picea engelmannii*) (Ehrlich group, unpublished). This transplant experiment indicated that the previous absence of *E. gillettii* from Colorado was due to its inability to cross the Wyoming Basin lowland gap in the Rocky Mountains, and adds to the mass of evidence (Ehrlich & White 1980) that Nearctic *Euphydryas* do not readily cross substantial areas of unsuitable habitat.

In Europe, *Euphydryas* are probably also sedentary, occurring in isolated colonies as *E. aurinia* does in England (Ford & Ford 1930). For instance, like *E. gillettii*, *E. intermedia* (also *maturna* group) is reported to be associated with *Picea* (Higgins & Riley 1970), and I suspect it also crosses open country with reluctance.

Varying Structure in *Colias*

An array of variable population structures has also been observed in montane pierids in Colorado (Watt *et al.* 1977, 1979). Dispersal varied within and among four species of *Colias*: *C. eriphyle*, *C. alexandra*, *C. scudderi*, and *C. meadii* in various subalpine and alpine communities. The highest dispersal rate was seen in the subalpine *C. alexandra*. In 1970, large continuous areas containing intermixed larval foodplants and nectar sources were available. Drier 1971 made nectar resources patchy, and the *C. alexandra* were apparently more confined to the patches. Average dispersal radius of those that dispersed (moved between study sites) was 1.2–1.3km in both years, with moves of 5, 7 and 8km observed. *C. meadii*, flying in alpine and subalpine areas, showed population structure varying from a small isolate to a continuous occupancy of extensive swathes of suitable habitat. *C. scudderi*, the least intensively studied of the four, is confined to the vicinity of patches of willows, its larval foodplant.

Unlike the other three *Colias*, which are univoltine, *C. eriphyle* is bivoltine in the montane study area. It is sympatric but largely allochronic with *C. alexandra*, experiences a similar resource distribution (*C. eriphyle* larvae attack several legumes, including *Lathyrus leucanthus*, which is the major resource for *C. alexandra*), and showed, in two dry years, an average dispersal distance similar to *C. alexandra*. Interestingly, there was no sign of the long-range dispersal seen in the latter species. A complex interaction between reproductive behaviour and brood population density is thought to produce differences in proportion of dispersers by sex and brood.

Watt and his co-workers (1979) distinguish two models for dispersal in *Colias*. In one, the *activity level* model, various factors cause changes in the general patterns of movement (activity level) of individuals in the population. Individuals classified as 'dispersants' represent simply one tail of an activity distribution in which butterflies move far enough to transfer between two arbitrarily established study sites. The second *excited state* model sees individuals probabilistically being boosted into a dispersal phase by environmental factors. The population then has a bimodal activity distribution, containing stay-at-homes and dispersers. The former model would predict a positive correlation between the proportion of dispersers and the average distance moved by dispersers; the latter would not. Data gathered by the Watt group tend to support the excited state model. For example, in one location *Colias alexandra* showed a threefold decrease in the proportion of dispersers between 1970 and 1971, but there was no significant difference between those years in the distance the dispersers travelled.

The variation in movement patterns in *Colias* is reminiscent of that seen in *Euphydryas*. In general, however, for a given amount of movement, population structure is likely to be tighter in *Euphydryas* than in *Colias*. *Colias* court and mate repeatedly (Watt *et al.* 1977) and lack the genital plugging mechanism of *Euphydryas* (Labine 1964, Ehrlich & Ehrlich 1978). They also lay eggs singly, not in masses. Therefore *Colias* dispersers are relatively likely to reproduce after dispersing (Watt *et al.* 1977), while *Euphydryas* are much less likely to do so (Ehrlich *et al.* 1975).

Structure in Other Temperate-Zone Butterflies

In contrast to this sedentary behaviour, many nymphalid butterflies apparently have sufficiently widely distributed larval and adult resources to form virtually continuous populations, with little or no tendency to produce isolated demographic units. In montane Colorado this appears to be true for the nymphaline that is the closest sympatric relative of the local *Euphydryas*, *Chlosyne palla* (Schrier *et al.* 1976). The same is almost certainly the case for the very abundant *Phyciodes campestris* (Ehrlich, unpublished) and *Speyeria mormonia* (C. Boggs, unpublished). In contrast, A. M. Shapiro reports (pers. comm.) that in California's Central Valley, *P. campestris* forms discrete colonies but disperses widely, and that in one northern California locality *S. mormonia* has an extremely sedentary, low density colony.

In Colorado habitats the satyrines are the least colonial as a group. *Erebia epipsodea* is the most thoroughly studied of the four common subalpine satyrines. It occurs in vast, effectively panmictic populations, which may cover hundreds of square kilometers (Brussard & Ehrlich 1970*a*, *c*), and epideictic behavior may contribute to greater-than-random dispersion of males (Brussard & Ehrlich 1970*b*). (Interestingly, I. D. Shapiro, 1977, found many parallels to montane *E. epipsodea* in the population biology of the grass-feeding hesperiid *Hylephila phylaeus* in the hot Central Valley of California.) *Coenonympha tullia* and *Cercyonis oetus* seem as ubiquitous in Colorado as *E. epipsodea* (Sharp *et al.* 1974) and presumably have similar structures although MRR studies have not been done. *Oeneis chryxus* is also very widespread and abundant, but unlike the preceding two species, may require appropriate perching locations for males.

None of these satyrines makes heavy use of nectar, and unfortunately the degree of larval host specialization within the grasses and sedges is unknown. It is commonly assumed that in the temperate zones satyrine larvae have rather catholic tastes (Emmel 1975), although tropical species are sometimes quite host-specific and thus patchily distributed (Singer & Ehrlich in press, and unpublished). The distribution of some Colorado satyrines (*e.g. Erebia theano*, *E. magdalena*, *Oeneis jutta*, *Cercyonis pegala*) suggests that patchy distribution of some resource, possibly a specific larval host, and therefore a more colonial population structure.

The same is suggested by the very localized occurrence of *Erebia epipsodea* within the La Sal Mountains of Utah. Especially interesting is the case of *Oeneis uhleri*, which is superficially almost indistinguishable from *O. chryxus*, flies with it, behaves similarly, and yet has a much patchier distribution. My guess is that *O. uhleri* has much more specialized larvae than does *O. chryxus*. In Europe the resource-use patterns underlying the complex distributions of the numerous *Erebia* species, especially of the *tyndarus* group (Lorkovìc 1957), badly need investigation.

Structure in Tropical Butterflies

Food-resource dominated population structure seems to be the rule in the few tropical butterflies thoroughly studied. For example, Blau (1980) found that *Papilio polyxenes* in Costa Rica dispersed continuously to take advantage of the period of suitability of ephemeral habitat patches. It feeds on a successional umbellifer, *Spananthe paniculata*, and has high survival rates only when eggs are deposited as the first flowers are produced. Emerging females lay some eggs in the old patch (which has a life span of about six months) and then disperse. Thus one or two additional broods may be produced in the deteriorating patch, but each female also has an opportunity to find an optimal patch.

In contrast to this successional *Papilio*, the nymphaline *Heliconius ethilla* in a montane forest in Trinidad showed no discernible tendency to disperse (Ehrlich & Gilbert 1973). Movement patterns, in the presence of relatively continuous and abundant larval resources, are controlled by the distribution of three species of plants from which trap-lining adults gather pollen (from which they extract amino acids: Gilbert 1972) and nectar. Removal of just two individual flowering vines altered the spatial pattern of a demographic unit and the frequency with which it exchanged individuals with an adjacent unit.

Thus, not surprisingly, the distribution of nutritional resources is probably the major factor controlling the structure of non-migrating butterfly populations in both temperate and tropical areas. Indeed, it probably holds the key to migration behaviour as well (see below).

The relative importance of larval and adult resources needs more attention. At first glance it would appear that larval resources, where requirements tend to be highly specific, should dominate. Some butterfly populations may survive without nectar for a generation or more (Ehrlich & Murphy 1981), and adults are mostly generalists when it comes to choosing nectar sources. But it is also clear from the works cited above that in many situations suitable flowers are scarcer than larval food, and the distribution of nectar and pollen sources may exert great influence on the population structure, as may the occurrence of suitable puddling sites.

Mate Location

Mate-locating behavior in butterflies (Scott 1974*b*), which is influenced by the distribution of nutritional resources, also affects population structure. Two kinds of behaviour predominate: patrolling and perching. Patrolling males fly continuously in search of females but in perching populations they take up stations and dash out to examine flying objects, and court them if they prove to be females. Both behaviours are widely distributed taxonomically in temperate-zone butterflies, and casual observation makes it seem likely that both occur commonly in the tropics as well. From the standpoint of population structure, perching is the more interesting behaviour. Males select characteristic perching sites, which then become a resource for the population. The distribution of such a resource will influence the population structure, as it does in the Colorado *Euphydryas editha* and *E. gillettii* populations mentioned above. Different perching sites have, in at least one case, been shown to separate closely related species that have otherwise virtually identical behavior (Scott 1973*e*).

More research is needed on the factors that lead to perching *vs* patrolling behaviour. The issue is discussed by Scott (1974*b*). I suspect that the dominant factor controlling its evolution is the ease with which individuals can detect areas likely to be occupied by the opposite sex. This in turn would seem to be a function of the distribution of nutritional resources, topography, and population density. When the distribution is highly patchy with abundant co-located adult and larval resources, protandry and patrolling by males produces a high probability of rapid mate location. Both are the rule in Jasper Ridge *Euphydryas editha*, although in area H there is some perching by males in a bare area on a hilltop.

Protandry makes adaptive sense, especially in species where mated females are plugged (Labine 1964, Ehrlich & Ehrlich 1978) or unreceptive. It increases a male's chances of successfully mating, and minimizes the females' pre-reproductive energy use and risk of predation. The advantages of protandry (Scott 1977, Wiklund & Fagerström 1977) may be one reason why male butterflies are often smaller than females (Singer 1982*b*)—completing development earlier may more than balance any advantages of larger size. More work on protandry is needed, however. In *Speyeria mormonia*, for example, Carol Boggs (pers. comm.) finds that the gap between male and female emergence apparently varied with the weather (see also Ch.28), and may be so extreme that many males do not survive to mate.

When nutritional resources are no longer co-located, as with Del Puerto Canyon *E. editha*, or when their distribution is more or less continuous, as in montane Colorado *E. editha*, then perching sites become a cue for females in search of males. Low population density also puts a premium on establishing a prominent station to which females can proceed with a high probability of being inseminated. For butterflies, hilltops often serve this function (Shields 1968, Scott 1970, 1974*b*). Rare animals dare not count on random wandering to produce mating encounters, as has been demonstrated by that most unbutterfly-like creature, the Indian Rhino (Hutchinson & Ripley 1954).

Overall, it is clear that ecological factors, not taxonomic affinity, are the chief determinants of mate-locating behaviour (note convergent behaviour in a satyrine and a skipper living in the same habitat and using partially overlapping resources: Scott, 1973*a*). That such behaviour often appears constant throughout a species is, I suspect, either because most or all populations are in ecologically similar situations or because populations in different situations have not been investigated. Dennis (1982) has recently shown that both mate-locating strategies are employed by a British satyrine.

Plasticity of Flight and Dispersal Behaviour

Indeed, the degree to which general flight-dispersal behaviour in butterflies is phenotypically plastic remains an open question. Scott (1975*a*, *c*), on examining flight patterns of 11 species of butterflies and skippers, concluded there was a strong genetic programming of flight distances, based on the similarities of distances in two closely related species pairs and the dissimilarity of a pair of distantly related species occupying the same area. Genetic programming for flight distance is also suggested by the taxonomic clustering of migratory behaviour, as found in the Nymphalini: *Nymphalis antiopa* (Roer 1970), *Nymphalis californica* (Ehrlich group, unpublished), *Nymphalis urticae* (Roer 1961*a*, 1962, 1965*a*, *b*, 1968), *Nymphalis io* (Roer 1961*a*, 1969), *Vanessa atalanta* (Roer 1961*b*), and *Vanessa cardui* (Williams 1970) all migrate. And, of course, sexual differences in flight behaviour (*e.g.* Roer 1959, Shapiro 1970, Scott 1975*a*, Ehrlich *et al.* 1984) also imply genetic involvement.

The most interesting question is not, however, whether there is some genetic influence on flight behaviour. The answer to that must be 'yes' from first principles—after all, the morphological differences between *Pseudopontia paradoxa* and *Danaus plexippus* are clearly largely genetically determined, and effectively debar *Pseudopontia* from migrating thousands of kilometres as does the Monarch. The question is whether divergent flight behaviour has evolved in populations confronting different environments, or whether that behaviour is plastic and can be greatly modified by the individual in response to rapid environmental change. The best way to answer this question is with transplant experiments, but even these may be difficult to interpret in the absence of thorough knowledge of the cues controlling behaviour in the home environment.

To date, transplant experiments have yielded ambiguous results. Gilbert & Singer (1973) released larvae of *E. editha* from Del Puerto Canyon on *Pedicularis* at Jasper Ridge. These gave rise to adults about one month after the Jasper Ridge population had completed its flight. The behaviour of the introduced butterflies in some ways more resembled behaviour at Del Puerto than at Jasper Ridge: they routinely flew over bushes and trees and were not restricted to grassland areas. Gilbert & Singer interpreted the result as indicating a genetic basis for observed differences in flight behaviour between the two localities.

However, it has since become clear that when nectar resources are scarce, Jasper Ridge butterflies will disperse out of the normal areas occupied by demographic units much more readily, flying over the chaparral (Ehrlich *et al.* 1984). A key control experiment, delaying the emergence of some Jasper Ridge adults until a month beyond the normal flight season and then observing *their* behaviour in the then nectar-poor environment, would be logistically difficult, but will be attempted by our group if the Jasper Ridge populations rebound from current low levels.

That movement patterns within the same population of *E. editha* can change dramatically with environmental conditions has been shown for the Lower Otay population near San Diego (White & Levin 1981, Ehrlich group, unpublished). Dispersal in this and other semi-desert populations is vastly greater in very dry years than in wet ones. A transplant experiment, moving adults from a population near Jasper Ridge to Lower Otay, resulted in much greater dispersal at the latter locality than at the former. The possibility, however, could not be excluded that the increase was simply a reaction to displacement to a totally strange environment (White & Levin 1981). Clearly, more experiments are required. Of course, transplant experiments with *Euphydryas* and in other systems under long-term scrutiny must be carried out with great care to avoid gene flow or other effects that might disturb their natural behaviour.

Dispersal and Migration

Perhaps the most basic question about dispersal (movement beyond the home range) is 'why does it occur at all?' Everything else being equal, the First Law of Dispersal should be for each individual to remain where it matured, for its own success is an indicator of suitable habitat. How then can dispersal behaviour in general, and migratory behaviour (directed long-distance seasonal movement) in particular, evolve? (See also Ch.26.)

This question can be analysed as follows: assume there is genetic variance in a population for dispersal distance. For simplicity assume two kinds of genotypes: stay-at-homes and dispersers. Obviously, there is no way for disperser genes to accumulate within a demographic unit, if they are continually being carried away by emigrants more rapidly than they are being introduced by immigrants. In demographic units occupying discrete patches of habitat where resources remain abundant (no significant density effects), such as those of *Euphydryas editha* at Jasper Ridge (see below), selection will strongly favour the stay-at-home genotypes unless the patches are so close that virtually all dispersers survive the trip and reproduce in their new homes. When patch quality always tends to deteriorate, though, or when the resources required by adults tend to be more abundant far from those utilized by larvae ('diffuse habitat' of Baker 1969), genotypes that move readily will be favoured. Individuals that immediately disperse on eclosion will be more likely to reproduce successfully than those that stay. The demographic unit, in this case, will be much larger, since individual movements will be much longer. Baker (1968*a*, *b*) argues that such butterflies moving in search of resources would evolve mechanisms ensuring the same areas are not searched repeatedly; he gives evidence for one mechanism, flight at a constant angle to the sun.

Of course, there is a spectrum between the two extremes 'patch almost always remains satisfactory' and 'patch deteriorating'. Many butterflies are denizens of successional habitats, and many are also capable of increasing their population density to the point where they can threaten the quality and quantity of resources. This can lead to the evolution of what Baker (1969) defined as 'evolutionary displacement': departure from a still-satisfactory area. Dispersal has been shown to increase with density in, for example, *Chlosyne harrisii* (Dethier & MacArthur 1964) and in *Pieris protodice* (Shapiro 1970). In the latter case, harassment of mated females by males leads to the emigration of the females.

Other butterflies seem to be genetically 'programmed' to move on without density cues; for example, *Papilio polyxenes*, studied by Blau (1980) in Costa Rica, and *Pieris rapae* and *P. brassicae* (Baker 1968a). However, the *Pieris*, unlike the *Papilio*, show a peak geographical flight direction —i.e. rather than being simply dispersers, they are migrants as defined by Baker (1969). Baker (1968*a*, 1969) adduces a variety of evidence that migratory patterns in butterflies reflect selective responses to environmental gradients, particularly temperature. It appears that species with very widespread resources, such as the two *Pieris*, may be able to increase their fitness by moving in directions that lead to higher fecundities or more broods per year.

Migrations such as occur in *Pieris*, unlike those of the Monarch, *Danaus plexippus* (Brower 1977), do not appear to have a complete return flight, there being a steady attrition of *Pieris* individuals in the northern fringe of their range (Baker 1969). A similar situation may exist with migrations northward and toward higher altitudes in temperate-zone butterflies (*e.g.* Shapiro 1973*a*). The population genetics of Baker's model are complex (Gilbert & Singer 1975) and clearly need further investigation. So does the evolution of migratory behaviour in such species as *Danaus plexippus*, since the advantages of the long trip from the eastern United States to overwintering sites in Mexico (as opposed to an *in situ* diapause) remain obscure. Progress in unravelling the phenomenon of butterfly migration, first brought to prominence by C. B. Williams (1930), has remained depressingly slow.

Population Dynamics

We turn now to the numbers of butterflies that make up their structured populations and how and why these numbers change through time. This section will deal with four topics: the magnitude of butterfly numbers, fluctuations in butterfly population size, the causes of fluctuations, and extinction.

Numbers

Interestingly, it is the migratory Monarch that has generated the largest number of butterflies censused in a single place. Brower *et al.* (1977*a*) estimated conservatively that there were 14.25 million *Danaus plexippus* at Mexican overwintering site Alpha in early 1977. We have no way of determining how many demographic units were represented by that great agglomeration, although Eanes & Koehn (1979) considered the eastern USA Monarch to be panmictic (on an annual basis). There is a literature on comparative butterfly abundance through time obtained by counts along transects and in gardens (e.g. Ekholm 1975, Moore 1975, Owen 1975, Pollard *et al.* 1975, Pollard 1977 and Ch.5), but it does not provide much insight into population sizes.

Where demographic units have been defined, numbers are much smaller—sometimes less than a hundred, generally a few hundred to a few thousand (Arnold 1980*a*, *b*, 1981, 1983, Brown & Ehrlich 1980, Cook *et al.* 1976, Davis *et al.* 1958, Dempster 1971*b*, Dowdeswell *et al.* 1940, 1949, Ehrlich & Gilbert 1973, Ehrlich *et al.* 1975, Scott 1974*a*, *c*, *d*, Scott & Opler 1975, Turner 1963, Watt *et al.* 1977, Vasconcellos Neto 1980, Warren 1981, Watanabe 1979*c*, Young & Thomason 1974). The largest population sizes recorded for well-defined demographic units were 50 000–100 000 each for both a high alpine population of *Euphydryas anicia* in 1976 (Cullenward *et al.* 1979) and the Edgewood Road colony of *E. editha* in 1981 (Ehrlich group, unpublished). In instances where demographic units have not been well-defined, censuses of relatively limited areas have generated numbers that tend to remain in the lower range (e.g. Abbott 1959, Cook *et al.* 1971, Dethier & MacArthur 1964, Fosdick 1973, Parr *et al.* 1968, Watanabe 1979*c*).

As any collector can testify, extremely dense butterfly populations are the exception rather than the rule. Pest species such as *Colias eriphyle* may reach densities of hundreds per hectare (Tabashnik 1980), and I have seen *Lycaena epixanthe* in densities of tens of thousands per hectare in cranberry bogs in the pine barrens of New Jersey.

Demographic units restricted to relatively small habitat patches, such as Jasper Ridge *Euphydryas editha* (Ehrlich 1965) and montane Colorado *E. anicia* (Cullenward *et al.* 1979), can reach densities of a thousand per hectare or more, although a few hundred is more typical (Ehrlich *et al.* 1975 and Ehrlich group, unpublished). For 'common' butterflies without such habitat restriction, densities are usually below a hundred per hectare (ha) (Brussard & Ehrlich 1970*a*, Watt *et al.* 1977, 1979).

Male butterflies are commonly observed to outnumber females (Ehrlich & Gilbert 1973, Scott 1973*a*, *c*, *e*, 1975*a*, Schreier *et al.* 1976, Watt *et al.* 1977, 1979, Cullenward *et al.* 1979). For example, only 16.9% of 4491 *Erebia epipsodea* marked by Brussard & Ehrlich (1970*a*), and only 21.6% of 4730 *Colias eriphyle* individuals captured two years by Tabashnik (1980), were females. Several hypotheses have been put forward to explain these deviations from the 1:1 sex ratio normally found in laboratory-reared butterflies (Shields 1968, table 19; Ch.24). Brussard & Ehrlich could not rule out differential mortality of larval and pupal stages in the field (see also Ch.28), but leaned toward the conclusion that behavioural differences made females less susceptible to capture. Tabashnik was able to show that males were both more abundant *and* more catchable. Apparently, female *C. eriphyle* develop more slowly than males and are more subject to pre-adult mortality; they also fly less and are thus less likely to be seen and captured. In contrast, Bernstein (1980) suggested that differential emigration of females accounted for skewed sex ratios in *Colias lesbia* in alfalfa fields in Argentina, but his data fit well to Tabashnik's conclusion of lower female catchability in uncut fields.

Ehrlich *et al.* (1984) did an extremely intensive MRR study, one objective of which was to determine the explanation for male-biased sex-ratios in *Euphydryas editha*. Having excluded 'catchability' differences, they found that males were almost exactly twice as abundant as females in Jasper Ridge area H in 1981. They also found evidence that differential emigration of females in response to a scarcity of nectar resources was at least partially responsible for the deviation from 1:1. Whether differential pre-adult mortality had a significant influence on the sex-ratio of this protandrous population could not be determined, but at least some affect probably occurred. It seems likely that sex-ratios in butterfly populations will often change from generation to generation with ecological conditions. (See also Chs 21 and 24, where Smith and Clarke discuss the few well-documented, genetically controlled exceptions to the 1:1 sex-ratio.)

Dynamics

Unfortunately, size changes have been followed in detail over significant periods of time in relatively few butterfly populations. The dynamics of three demographic units of *Euphydryas editha* have now been tracked closely for twenty-two generations, 1960-1981 (Ehrlich *et al.* 1975, and unpublished). Numbers in the two larger units, C and H, have fluctuated from a hundred or so each to a few thousand, with the changes quite asynchronous in the early years of the study and nearly synchronous in the latter. The smallest unit, G, went extinct in 1964, was re-established in 1966, and went extinct again in 1974. Its peak population size was about 200.

This pattern of wide fluctuation in the size of demographic units is now well-established for *Euphydryas* (Lenz 1929, Ford & Ford 1930, Ehrlich *et al.* 1975, Cullenward *et al.* 1979, Ehrlich & White 1980, Brown & Ehrlich 1980, Ehrlich *et al.* 1980, Murphy & Ehrlich 1980), and also seems to be the common pattern in other temperate-zone butterfly populations judging from the few studies available (Watt *et al.* 1977, 1979, Shapiro 1979*c*, Hayes 1981), the comparative abundance studies previously cited above, and the casual lore of collectors.

Scanty as detailed data are on population size changes in most temperate-zone species, there is even less information on butterflies of the wet and seasonal tropics. Ehrlich & Gilbert (1973) followed two demographic units of *Heliconius ethilla* for more than two years in a mature second-growth montane forest in Trinidad. Population size was virtually constant throughout the period, showing no significant fluctuations with the passage of wet and dry seasons. Benson & Emmel (1972) censused gregariously roosting *Marpesia berania* over a period of six months and found a similar picture of near constant population size, as did Owen & Chanter (1972) working with *Charaxes* in Sierra Leone.

In contrast to this picture of constancy, Vasconcellos Neto (1980) found considerable fluctuation in the size of populations of five ithomiines in a subtropical forest reserve in São Paulo State, Brazil. Three species, *Mechanitis lysimnia, M. polymnia* and *Hypothyris ninonia* are brightly coloured members of the same Müllerian mimicry complex. Their populations increase in size during the rainy season but, in the dry winter months, population growth is interrupted, while the three species aggregate and reach high local densities. *Dircenna dero* does not aggregate, and population growth occurs mainly in fall and winter. *Mcclungia salonina* builds its population during summer and early fall and then plunges to low levels during the winter and spring.

Similar fluctuations have been observed in members of a rain-forest butterfly community in Panama (Emmel & Leck 1970), in *Acraea encedon* in Uganda (Owen 1971*a*), and in Neotropical acraeines (Gilbert & Singer 1975). The Ithomiinae and Acraeinae tend to play similar ecological roles (small to medium sized, slow-flying, aposematic, largely restricted to a single family of larval foodplants), which may lead to similar dynamic properties for their populations. But many more investigations of the dynamics of tropical butterfly populations are required before generalities can be induced.

Causes of Dynamic Trends: *Euphydryas editha*

Ehrlich *et al.* (1974), modifying the basic scheme of Andrewartha & Birch (1954), divided the environment of an animal into four components: medium (weather), resources, other nonresource organisms, and hazards. The first three of these have all be shown to affect death rates in butterfly populations significantly, and at least the first two can affect migration rates.

Weather and resources interact complexly in the population dynamics of *Euphydryas editha* at Jasper Ridge and Woodside, California. The size of these demographic units is controlled primarily by mortality of pre-diapause larvae caused by senescence of the larval foodplant, *Plantago erecta* (Singer 1972, Ehrlich *et al.* 1975, Singer & Ehrlich 1979). Thus there is an annual resource shortage for *E. editha* larvae, caused by weather, not by consumption of the resource by competitors of the same or different species. The drier the year, the earlier the senescence and the greater the *Plantago* shortage, everything else being equal. But everything else is not equal, and there are complex factors involving the phase relationship of the butterfly and the *Plantago* that tend to lessen the impact of isolated dry years and increase the impact of two consecutive dry years (Singer & Ehrlich 1979).

Although in moist years or in favorable areas, pre-diapause development can be completed on *Plantago*, the major route to successful diapause in the Jasper Ridge and Woodside populations is through successful transfer of pre-dispause larvae from senescing *Plantago* to still-edible flowers of the scrophulariaceous annual *Orthocarpus densiflorus*, a hemiparasite (Atsatt 1965, 1970, Atsatt & Strong 1970) on which development to diapause size is completed (Singer 1972). Indeed, the abundance of *Orthocarpus* in a given year is a good predictor of the abundance of *E. editha* the next year (Ehrlich *et al.* 1975, and unpublished). This, of course, raises the question of the determinants of abundance of the *Orthocarpus* itself, which is now under investigation by our group. We hope we will not discover that *Orthocarpus* density depends primarily on the condition of *its* hosts, thus removing a major cause of *Euphydryas* fluctuations one step further!

During the 22 years of study of *E. editha*, it has been clear that population size changes in the serpentine-*Plantago* ecotype have been density-independent and that migration plays no immediate role in changes in population size (i.e. the large increase in area H in the early 1960s was not due to a flux of immigrants, and extinction in area G was not due to a mass exodus). Recently, however, it has also become clear that, even within the ecotype, the factors affecting dynamics can vary dramatically. For example, the demographic unit at Edgewood Road, only 10km away, differs from those at Jasper Ridge in patterns of phenology, oviposition, and adult and larval behaviour (Murphy *et al.* 1983), but their exact effect on dynamic patterns remains obscure.

In *E. editha* populations of the serpentine *Plantago* ecotype, adult resources seem to play a most interesting role in population dynamics. Although a female *E. editha* can produce about two full-sized egg masses without access to nectar, egg production is greatly enhanced by nectaring (Murphy *et al.* 1983). In years of low rainfall, nectar availability would be of minimum importance, since only larvae from the first egg mass or so generally reach diapause size while suitable plant food is available. In moist years, however, time constraints on development are relaxed, and larvae from later masses may successfully reach diapause size. Later masses are those increased by the availability of nectar.

Murphy has concluded that the primary role of nectar in this density-independently regulated ecotype is probably to aid long-term population survival by increasing population size in good years. This would permit the population to fluctuate at a higher level and thus have a greater buffer of numbers to avert extinction in bad years, provided the amplitude of fluctuations does not show too great an increase (Ehrlich & Murphy 1981, Murphy *et al.* 1983). The demographic unit in area G at Jasper Ridge has gone extinct twice. It has always been relatively depauperate in nectar sources, and, if the 'Murphy hypothesis' is correct, this may have prevented it from building a population size adequate to avoid frequent extinction.

Factors controlling population size in other ecotypes of *E. editha* are as diverse as those ecotypes. In many, weather and density effects interact. For example, in semi-desert populations near San Diego, California, prediapause mortality is high in dry years when food is sparse, and the shortage of nectar and oviposition plants leads to large movements of adults. In wet years, prediapause mortality is generally low, and adults are sedentary in the presence of abundant nectar sources and oviposition plants. If there are several consecutive wet years, the population size increases greatly and the resulting competition among post-diapause larvae is so intense that the habitat is denunded of *Plantago* (Murphy & White 1983). In such years, many of the post-diapause larvae starve, some probably return to diapause, but some become adults that disperse, since no oviposition sites are available. Persistence of the population is probably due to larvae that return to diapause, although this has not been fully demonstrated.

In the laboratory, post-diapause *E. editha* larvae are very sensitive to food quantity and quality and will re-enter diapause at the slightest 'excuse' (Singer & Ehrlich 1979). Inner Coast Range populations of *E. editha* in northern California also often undergo density-dependent mortality as pre-diapause and post-diapause larvae devour the limited supply of the perennial foodplant, *Pedicularis densiflora* (White 1974).

Causes of Dynamic Trends: *Euphydryas chalcedona*

Euphydryas chalcedona at Jasper Bridge also seems to be partially food-limited. The population there uses two larval hosts, *Diplacus aurantiacus* and *Scrophularia californica* (both Scrophulariaceae). While both are perennials, development on them is time-constrained (Ehrlich & Murphy 1981), and the condition of the plants is influenced by rainfall (Brown & Ehrlich 1980).

Due to human disturbance a large increase in the amount of *Scrophularia* in one area caused the biggest increase in *E. chalcedona* population size observed in 10 years of study. This in turn resulted in substantial impact on the *Scrophularia* itself (Brown & Ehrlich 1980). The latter plant, well-defended biochemically, is generally a less satisfactory host (Lincoln 1980, Lincoln *et al.* 1982, Mooney *et al.* 1980, 1981, Williams 1983, Williams *et al.* 1983*a*, *b*). One possible cause of continued oviposition on *Diplacus* is that periodic over-exploitation of the *Scrophularia* occurs. Near San Diego, *Scrophularia* and *Diplacus* grow intermixed, and the former is preferred by both ovipositing females and larvae. When the *Scrophularia* is defoliated, however, *Diplacus* is then used (D. D. Murphy, pers. comm.). Another possible advantage of oviposition on *Diplacus* at Jasper Ridge may be its greater proximity to a prime nectar source, *Eriodictyon californicum* (a hypothesis soon to be tested).

Overall, quality and quantity of prediapause larval resources (rather than availability of nectar) seem to be the crucial factors in the dynamics of *Euphydryas* populations in western North America. Young larvae are generally in a race to reach satisfactory dispause size before the foodplants dry up, are devoured by others, or decline in quality. Indeed, where several nutritionally equivalent plant species are available, the phenological characteristics of the larval foodplants tend to determine the choice of species for oviposition (Holdren & Ehrlich 1982). Similarly, abundance of their *Aster* larval foodplants is the major factor in the abundance of *Chlosyne harrisii* (a close relative of *Euphydryas*) in the eastern United States (Dethier 1959*b*, Dethier & MacArthur 1964). As succession reduces *Aster* populations, the butterflies decline to extinction.

Causes of Dynamic Trends: *Heliconius ethilla*

In contrast to the key role played by the quantity and quality of larval resources in temperate *Euphydryas*, the availability of larval foodplants seems to have little influence on the tropical *Heliconius ethilla* (Ehrlich & Gilbert 1973). *Passiflora cyanea* vines are very abundant in the study area, where they are scoured by predacious ants and wasps and searched by parasitoids. There being no sign of larval food shortage, Ehrlich & Gilbert concluded that *H. ethilla* had evolved a life-history strategy that minimized time spent in the dangerous larval stage, a deduction supported by Young (1978*b*). Essentially no resources are sequestered by the larvae for egg production; instead, the long-lived (up to six months) adults obtain the necessary resources by gathering pollen from *Anguria* and *Gurania* flowers (Gilbert 1972). The availability of this adult resource was limiting, placing an upper limit on egg production within the population. Other *Heliconius* that are also long-lived, such as *H. erato* (Turner 1971*a*), probably have similar ecological strategies.

The observed constancy of population size in this species would seem to be due to a relatively constant limiting flow of the pollen resource combined with great adult longevity, which would tend to damp out fluctuations caused by variable pre-adult mortality. In *H. ethilla* an increase in adult recruitment cannot lead to a great increase in egg production because of the pollen limit; a decrease in adult recruitment will mean more pollen for the adults already in the population. *Euphydryas*, in contrast, acquire the resources to mature eggs as larvae and emerge with several hundred 'ready to go'. Changes in pre-adult mortality are thus quickly translated into changes in egg production in the population.

Causes of Dynamic Trends: *Colias alexandra*

In between the *Euphydryas* case, where females dump masses of eggs as rapidly as possible, and that of *Heliconius ethilla*, which dribbles out a few per day for months, are organisms like *Colias alexandra* where females lay several hundred eggs singly over a life-span of a week or so. Hayes (1981) has made detailed life tables and a key factor analysis (Varley & Gradwell 1960) of this species in montane Colorado. She found that, in spite of heavy pre-adult mortality, the major controlling variable in the dynamics of the population was the ability of females to get the eggs out—an ability curtailed not only by mortality but also by unsuitable weather and inadequate nectar resources. Similarly, Warren (1981) found the impact of temperature on fecundity to be a key factor in the dynamics of English *Leptidea sinapis*.

Causes of Dynamic Trends: Weather

The degree to which weather can directly produce mortality in butterfly populations, rather than by influencing resource availability, fecundity or other factors, is unknown. Clearly, events such as hail or wind storms and temperature extremes are capable of causing mortality, but their roles in population dynamics have not been elucidated. The extermination of a montane population of *Glaucopsyche lygdamus* by an unseasonable snowstorm (Ehrlich *et al.* 1972) almost certainly was caused by destruction of the lupine inflorescences upon which the larvae depended, rather than from thermal effects. Indeed, flowering at a time when it is vulnerable to frost may be an evolved defence strategy of the perennial lupine. By sacrificing one season's seed production, the plant gained a decade of virtual freedom from its major seed predator (Ehrlich group, unpublished).

Still, humid weather was found to be a major factor causing mortality in *Lycaena dispar* populations (Bink 1972), but it probably operated by creating ideal conditions for fungal attack, not by directly killing larvae. Dempster (1971*b*) concluded, however, that lack of sunny weather in which females could oviposit was the direct factor causing a decline of *Lycaena phlaeas* in the Monks Wood area in the mid 1960s, and on Jasper Ridge weather is the primary determinant of butterfly activity (pers. obs.). Rausher (1979*b*) found that egg and larval survival of three *Aristolochia* swallowtails was higher in shady than sunny habitats (possibly due to differential predation by ants and spiders, or to the greater exposure to torrential rain away from the protection of trees).

Causes of Dynamic Trends: Predation and Parasitism

Similarly, the roles of predation and parasitism in the dynamics of butterfly populations remain largely speculative (see Ch.10). Predators, parasitoids, and disease take a high toll of eggs, larvae and pupae of Japanese *Papilio xuthus*, especially of eggs and early larval instars (Tsubaki 1973, 1977, Watanabe 1976, 1979*b*). In two consecutive generations a hymenopterous egg parasite killed 52 and 29% of the eggs of a Japanese lycaenid *Artopoetes pryeri* (Watanabe & Omata 1978). A braconid killed a large proportion of the larvae in a Swedish population of *Anthocharis cardamines* (Wiklund & Ahrberg 1978), and *Nymphalis urticae* is often heavily attacked by tachinids (Pyörnilä 1976). Mites and mirid bugs wiped out entire egg masses and entire groups of hatchlings from masses of *Euphydryas gillettii* in Colorado (Ehrlich group, unpublished).

Adult butterflies may be taken in large numbers by lizards in the tropics (Ehrlich & Ehrlich 1982); birds also may eat numerous adults in some circumstances (Brown & Vasconcellos Neto 1976, Pough & Brower 1977, I. L. Brown & M. D. Bowers unpublished); and small mammals, spiders and predatory insects also take them. But to date neither predation nor parasitism has been shown to be the cause of observed changes in butterfly population size, even though in theory they may be important controlling factors.

Natural Extinctions

Evidence is accumulating that extinction may be a regular feature of the dynamics of many butterfly populations. For instance, work by our group with *Euphydryas* in the western United States has made it apparent that extinction of demographic units is a regular occurrence. For the last few hundred years, *Euphydryas editha* has existed in restricted patches of serpentine grassland in the San Francisco Bay area. The demographic units in these patches are differentially affected by drought and, presumably, by other environmental stresses, and frequently die out (Ehrlich *et al.* 1980, Murphy & Ehrlich 1980). Empty patches are recolonized eventually, perhaps after 50-100 years, by these relatively sedentary insects.

A 'mosaic' pattern of population regulation (Ehrlich & Birch 1967, Ricklefs 1973) similar to that seen in *Euphydryas* has been described by Shapiro (1979*d*) for *Pieris protodice* in California's Central Valley. Shapiro believes that the mosaic (extinction-recolonization) model of population dynamics may be very widely applicable in butterflies—so much so that, 'if time scale for population turnover is taken into account, the mosaic model could prove so robust in the face of differing life styles that it is reduced to a truism'. Shapiro notes that many butterfly populations must have turnover rates much lower than those observed in *Euphydryas editha* and *Pieris protodice*. Like our group (Ehrlich *et al.* 1975: 227), he sees the need for pressing on with the time-consuming and logistically difficult task of determining patterns of population regulation in a much wider sample of butterflies (to say nothing of other organisms!).

Extinctions Caused by Humanity

The role of extinctions in the 'normal' dynamics of butterfly populations is, sadly, just one aspect of butterfly extinction today. The other is a unidirectional and virtually global trend toward the loss of butterfly populations and species as a result of the expansion of the populations and activities of *Homo sapiens* (Arnold 1980*a*, *b*, 1981, 1983, Bielewicz 1967, Brown 1970, Kloppers 1976, Pyle 1976*b*, Emets 1977, Pyle *et al.* 1981, Ehrlich *et al.* 1980, Ehrlich & Ehrlich 1981*a*). Because of the wide amateur interest in butterflies, they make an excellent index of the loss of populations and species of 'obscure' organisms, a loss that threatens the 'public service' functions of ecosystems that are so crucial to the maintenance of our society (Ehrlich & Ehrlich 1981*a*).

Disappearance of well-documented populations has been depressingly frequent. In the San Francisco area *Cercyonis sthenele*, *Glaucopsyche xerces*, and key demographic units of *Euphydryas editha* (Murphy & Ehrlich 1980) have gone extinct due to urbanization, and *Callophrys mossii* and local populations of *Speyeria callippe*, *Callophrys viridis*, *Plebejus icarioides*, and *Apodemia mormo* (Arnold 1980*a*, 1981, 1982) are threatened by human activities. England has suffered the loss of *Lycaena dispar* to habitat destruction and overcollecting (Duffey 1968); of *Papilio machaon* at Wicken Fen to habitat change (Dempster *et al.* 1976); and of *Maculinea arion* to changes in grazing that decimated their ant allies (Muggleton & Benham 1975, Thomas 1980*a*).

British scientists have contributed much to our knowledge of the habitat requirements of butterflies as they have attempted to understand extinctions and, often, to re-establish populations. Duffey (1968, 1977), for example, showed that subtle specialized needs in size and situation of their *Rumex hydrolapathum* foodplants made individuals from Dutch populations of *Lyceana dispar* less than perfectly adapted to the British marshland where reintroduction was being attempted. Dempster and his co-workers (1976) developed fascinating morphometric evidence that, as the habitat of *Papilio machaon* was reduced in the Wicken Fen area, it evolved a low-mobility phenotype.

But more depressing than the disappearance of prominent populations is the steady, undocumented attrition occurring everywhere, so obvious to those who do extensive field work. Every collector has stories of favourite localities where the butterflies no longer occur. I well remember Wilfred Dowdeswell taking me to see one of his *Maniola jurtina* localities near Winchester in 1964. We arrived just in time to see the last of the field ploughed under.

Our research group has recently undertaken an extensive project on the island biogeography of butterflies and their plant resources, in the mountains of the basin and range province of the western United States. The research is designed to provide information on the design of reserves for insects and plants. Familiar as we are with habitat destruction in the American West, we were unprepared for the impact of overgrazing by cattle and sheep in those mountains. For example, in June of 1981 Dennis Murphy and I went to sample an isolated *Euphydryas chalcedona* population in the Pine Nut Mountains of Nevada. When we reached the locality, we discovered that sheep had moved through. Not only were the butterflies gone, but not a scrap of their *Orthocarpus* foodplant could be found. Some overgrazed smaller mountain ranges in the Great Basin are now primarily inhabited by tree-, grass- and weed-feeding species such as *Papilio rutulus*, *Cercyonis oetus*, and *Nymphalis milberti* (Ehrlich group, unpublished). Similarly, the shared *Penstemon*-palatability spectrum of *Poladryas minuta* and stock animals seems to have led to the extinction of populations of *Poladryas* over most of its previous range in Texas (Kendall & Kendall 1971, Scott 1974*d*).

While the cause of such extinction is obvious, in others, declines in population size may or may not be connected with human activities, and these cry out for thorough investigation. For example, in the upper East River drainage of Gunnison County, Colorado, *Erebia epipsodea* is now quite scarce; between 1960 and 1970 it was abundant. The major obvious change in its habitat is an increase in dust from increased vehicular traffic. This could account for the decline of *E. epipsodea*, but if so it must be a differential effect. Some other butterflies have disappeared from the area (e.g. *Parnassius phoebus*, possibly because *Sedum lanceolatum* became too sparse in the patch of roadside habitat it once occupied; see Scott 1973*c*) or are much less common than previously (*Oeneis chryxus*), but others (*Speyeria mormonia*, *Papilio rutulus*) are as abundant as ever.

Perhaps the saddest threat faced by butterfly populations is overcollecting. Market collecting probably played a role in the extermination of the British ecotype of *Lycaena dispar* (Duffey 1968), and Hayes (1981) has recently shown that in some circumstances collecting adults from a population can affect the population size. I regret to report that market hunting is not a thing of the past and that it may now threaten the newly-discovered *Boloria acrocnema* in the Colorado Rockies (Gall 1981, T. Coborn, pers. comm.). Much more education is still needed to counter the 'stamp collector' mentality that remains common among those interested in butterflies.

Discussion and Conclusions

Butterflies are an extremely successful group of insects, having occupied every significant land mass except Antarctica. Occurring in a vast array of biotopes, from rainforest canopy to stark desert and frigid tundra, it is not surprising that they have evolved a great diversity of ecological strategies. Limited as our detailed knowledge of these strategies is, some points stand out.

Diversity within Species

Remarkable differences in population biology have evolved within taxonomic species. Ecologically, *Euphydryas editha* from Del Puerto Canyon in California's Inner Coast Range is more like *E. chalcedona* on Jasper Ridge than it is like *E. editha* on Jasper Ridge. Work with butterflies demonstrates clearly the bankruptcy of the 'endangered species' approach to the conservation of biological diversity. Butterfly populations are not interchangeable, ecologically or genetically, even when they are grouped under the same specific name. Dutch *Lycaena dispar* is not ecologically equivalent to British *Lycaena dispar*, as Duffey has so elegantly shown. Moreover, there is every reason to believe that this conclusion is robust not just for butterflies but for most organisms (Ehrlich & Mooney 1983). It calls for a conservation strategy aimed at preserving *all* remaining diversity, with the main tactic being comprehensive habitat protection (Ehrlich 1980, Ehrlich & Ehrlich 1981*a*).

Dynamics and Dispersal

Although a great many dynamic patterns have been observed in butterfly populations, in the vast majority there is little sign that dispersal is the proximate cause of observed fluctuations in numbers. It certainly *is not* in organisms such as *Euphydryas editha*, *Colias alexandra* or *Heliconius ethilla*. What the situation is in insects such as *Pieris rapae*, or Costa Rican *Papilio polyxenes* remains to be investigated. Those investigations, however, will require studies of individual movement and careful definition of demographic units.

Limits on Butterfly Adaptation

In spite of the great ecological adaptability shown by butterflies, one is continually astonished at what they cannot or do not do. Why does *E. editha* at Jasper Ridge persist in laying its eggs on *Plantago erecta* when the best hope of larval survival is on *Orthocarpus densiflorus*? At nearby Edgewood Road, most of the eggs are deposited on *Orthocarpus*, and in many populations *E. editha* shows a high degree of discrimination among potential hosts. Why do migratory butterflies that invade northward and upward in temperate zones, only to be killed by winter's cold (Ekholm 1975), not evolve a diapause that permits them to overwinter? Butterflies have evolved diapause at every stage of the life cycle (Scott 1981), and often those that do not diapause have congeners that do. The selective advantage for the summer invaders that evolved the ability to diapause would obviously be enormous.

Indeed, the general question of why butterfly range limits occur where they do has received little attention; it is mainly addressed in connection with an observed shrinkage or expansion. Thus Pollard (1979*b*) has been able effectively to explain the expansion of *Limenitis camilla* in Britain; a combination of changed forestry practices and unusually favourable weather led to a northward spread. Warm Junes apparently shorten time spent as late larvae and pupae vulnerable to bird predation. Conversely, Chew (1981) has shown that the disappearance of *Pieris napi oleracea* from areas where it was once common in southern New England was due to changes in land use that extirpated its favourite cruciferous foodplants.

In these cases, however, one might ask 'why hasn't' questions. Why hasn't *L. camilla* switched to the willow, birch, or other foodplants eaten by its Nearctic congeners and adopted the trick (whatever it is!) that permits *L. weidemeyerii* to thrive at 10 000 feet in the Rocky Mountains in a temperature regime considerably more rigorous than that of central England? Similarly, why hasn't *Pararge aegeria* been able to match the rapid development of its more widespread congener, *P. megera* (Lees 1962*b*)? Why hasn't *P. napi* evolved the ability to thrive on various crucifers (and other plants containing mustard-oil glycosides) that are happily gobbled by *P. rapae*?

The mechanics for making hostplant shifts clearly persist in many (all?) butterflies (Wiklund 1981, Singer 1982*a*, Tabashnik *et al.* 1981).

Evolution of Pest Status

Similarly, little is understood of how and why a few butterfly species have been able to extend their range by entering anthropogenic monocultures and becoming agricultural pests. A large literature has developed around the population biology of what is perhaps the most famous butterfly pest, *Pieris rapae* (e.g. Baker 1970, Dempster 1967, 1968, Gossard & Jones 1977, Harcourt 1966, Ives 1978, Jones 1977*a*, *b*, Jones & Ives 1979, Jones *et al.* 1980, Kobayashi & Takano 1978, Osada & Itô 1974, Parker 1970, Richards 1940, Takata & Isheda 1957). Only recently, however, have studies begun that bear on the crucial question of why *P. rapae* became an important pest, while *P. napi*, its close relative, has not. Yamamoto & Ohtani (1979) show that *P. rapae* has fecundity two to three times as high as that of *P. napi*, and *P. rapae* prefers more open areas while *P. napi* shows a preference for shaded situations (Ohsaki 1979, Yamamoto 1981).

These findings indicate why *P. rapae* is a pest while *P. napi as is*, is not. But it doesn't tell why *P. napi* did not evolve pest status in response to the availability of abundant food in cabbage fields. Exclusion by competition with a better pre-adapted *P. rapae* is one obvious answer. But Yamamoto's results (1981) cast some doubt on this facile notion, since, when both are confronted with a novel situation (both invading an imported weed, *Rorippa sylvestris*, in northern Japan), *P. napi* shows signs of being a better competitor (high r, less susceptible to density-dependent mortality). In California both species also attack the introduced crucifer, *Nasturtium officinale* (= *Rorippa nasturtium-aquaticum*) (Shapiro 1975*b*), but their relative competitive ability has yet to be assessed.

From recent studies by Tabashnik (1980, 1983), we know that knowledge of nonpest populations may be a poor indicator of pest potential. He found that pest populations of *Colias eriphyle* in Colorado alfalfa fields show a considerably different population ecology from their wild relatives in Colorado grasslands. Similarly, Shapiro & Masuda (1980) found that *Papilio zelicaon* had become a pest of *Citrus* twice in California, once in the south and more recently in the north. *Citrus* is in all respects an inferior host to *Foeniculum vulgare*, the usual foodplant, but *Foeniculum* is not found around orange groves. As Shapiro & Masuda say, 'there is nothing wrong with using a suboptimal host if that is the only host around.' But, then, why hasn't *Pieris napi* moved on to suboptimal hosts like radish and cabbage when its prime foodplants are not available?

Evolution and Optimization

The answers to the 'why hasn't' questions are obviously among the most difficult to solve in biology. One reason is that generally we have a very poor understanding of the developmental or integrative constraints under which selection operates. It may be a mistake to seek a selective 'story' to explain the basic colour pattern that is repeated (with variation) on the underside of polyommatine butterflies or the limits to a butterfly's geographic or hostplant range. It may simply be that the evolutionary costs of changing certain characteristics may be too high—that achieving, say, pupal diapause may be too disruptive to the integrated whole that is an individual of species A, but not too disruptive in its congenor B. This issue is discussed in detail in a provocative article by Gould & Lewontin (1979); suffice it to say that we should keep in mind that organisms are inevitably bundles of compromises, and that one should not expect every aspect of a butterfly's ecological behaviour to be optimized.

Desiderata

Finally, let me emphasize that, although the population ecology of butterflies is probably better known than that of any other group of invertebrates, there is still an enormous amount to learn. An effort should be made to concentrate attention on long-term detailed studies of a sample of butterflies stratified taxonomically and geographically. Considering that butterflies are a tropical group, information from that part of the world is sadly lacking. The same can be said for tundra species, where the problem of time-constrained development should be especially fascinating.

Butterflies and Plants

Exploration of the role of adult resources in determining structure and dynamics has hardly begun. The first such investigation (Clench 1967) inferred that they were responsible for the phenology of a skipper community. What is already clear, however, is that nectar (and, in *Heliconius*, pollen) sources can play an important role in habitat selection (works cited above; Douwes 1978, Jennersten 1980).

I have found only one detailed investigation of the dynamics of a plant population initiated with the intention of understanding the dynamics of a butterfly that attacks it—Watanabe's (1979*a*) investigation of *Zanthoxylum ailanthoides*, a host of *Papilio xuthus*. Much more work in the area of joint plant-herbivore dynamics needs to be done, following the growing literature that focuses primarily on the impact of butterfly herbivory on foodplants (*e.g.* Ehrlich & Raven 1965, Breedlove & Ehrlich 1968, 1972, Dolinger *et al.* 1973, Rausher 1980).

Genetics and Dynamics

Although knowledge of their population ecology has made butterflies an extremely important system for investigating genetic problems such as the 'neutrality' controversy (Ehrlich & Murphy 1981, Watt 1983), the role of genetics in butterfly population dynamics and structure remains largely unknown. Data are being accumulated on changes in gene frequencies at allozyme loci in *Euphydryas* populations over numerous generations (Ehrlich group, unpublished), but at the moment lack of adequate theory makes their interpretation very difficult.

Voltinism

A major area of butterfly population ecology requiring further investigation is the diversity of patterns of voltinism found in the group. Time constraints obviously again play a major role here, and the importance of the availability and condition of larval foodplants, emphasized by Slansky (1974), is apparent in many cases. But other factors are also clearly involved (Shapiro 1975*b*; Ch.27), and detailed analyses of the selective advantages and disadvantages of multiple broods in different ecological situations are badly needed. Similarly, more work is needed on the role of male 'territoriality' in population structure and dynamics.

Taxonomic Sampling

Work on butterfly population ecology has concentrated on the Nymphalidae and Pieridae, probably because they present fewer logistic problems than papilionids or lycaenids. But what work has been done with the latter groups is extremely interesting, especially that on the flight behaviour of the swallowtails (Dempster *et al.* 1976), the intricate ecological requirements of hairstreaks (Thomas 1974), and the fascinating extra 'resource' for the blues—appropriate ants (Malicky 1969) that have recently been shown to provide protection against parasitoids (Pierce & Mead, 1981; Ch.19). The taxonomic sample of butterflies under study clearly needs broadening.

The kinds of investigations most needed require time and patience, but little equipment. Butterflies are so well known largely because they have been studied so intensively by dedicated amateurs (indeed most professionals working with butterflies today were once amateur butterfly collectors!). It is to be hoped that the energies of amateurs can be channeled increasingly into careful studies of population structure and dynamics, and away from the silly nomenclatural exercises that so often masquerade as 'science' in the lepidopterological literature (Ehrlich & Murphy 1982).

Acknowledgements

I thank A. H. Ehrlich, C. Boggs, F. S. Chew, J. P. Dempster, L. E. Gilbert, R. W. Holm, N. D. Johnson, D. D. Murphy, J. Roughgarden, A. M. Shapiro, M. C. Singer, B. Tabashnik and W. B. Watt for criticizing the manuscript. This work was supported in part by a series of grants from the National Science Foundation, the most recent of which are DEB 78-22413 and DEB 78-02067, and a grant from the Koret Foundation of San Francisco.

3. *The Biology of Butterfly Communities*

Lawrence E. Gilbert

Department of Zoology, The University of Texas, Austin, Texas

The great variety of form, behaviour, and life history in the butterflies has come about through evolution within a community of species. Yet, for the most part, the focus of butterfly biology has been upon the traits of individual species rather than the traits of communities.

Robert MacArthur (1972) wrote 'ecologists primarily interested in separate species have never made any progress in unravelling community patterns.' The converse statement should not be true. Indeed, it is safe to assume that an improved perspective on how a species fits into its community will help explain many things about it, especially those attributes whereby a species conspicuously departs from its close relatives. The purpose of this paper is to summarize our knowledge of communities as viewed from the perspective of a butterfly biologist, and to point to ways in which community analysis may further the understanding of all aspects of butterfly biology. It will conclude with a discussion of how the study of these insects may contribute to general theories of community organization.

Community Biology

No strict definition of community has emerged in the biological literature, but the term is almost always used to refer to a 'many species population' (Pielou 1969: 236) or 'coexisting, interdependent populations' (Price 1975: 3). Much of modern community ecology restricts itself to 'all organisms in a chosen area that belong to the taxonomic group that the ecologist is studying' (Pielou 1969: 203).

The philosophical and practical problems involved in selecting a meaningful community for study are compounded versions of the problems faced in the study of single populations. An arbitrarily selected area may not contain a representative sample of the species of the community to be analysed. Likewise, taxonomically delimited communities (warblers, ithomiine butterflies, etc.) usually lack taxa of potentially important competitors, parasites or mutualists of the selected community which may greatly influence its structure.

Community structure is 'the pattern of resource allocation among the species of a community, and the pattern of their spatial and temporal abundance...' (Cody & Diamond 1975: 7). An understanding of such structure is one aim of community biology, a field I define as the study of attributes of life which can be strongly influenced by interspecific interactions. These attributes can be found at different levels of organization. While community biology tends to focus on such community level patterns as species richness, limiting similarities, and mean niche breadths, etc., there exist traits of populations such as spatial structuring, genetic diversity (e.g. mimetic polymorphisms), average density, and numerical dynamics which demand some consideration of interspecific effects. Likewise traits of individuals such as colour pattern (especially in mimetic systems), host preference, habitat restriction and timing of activity are potentially shaped by interactions between species of a community.

Interaction within a local assemblage of species is not the only means by which a particular attribute of the community or of one of its components might have arisen. The species richness or host partitioning within a local area may be determined not by interactions observable locally, but on a scale commensurate with the population structure of interacting species, which in the case of migrants, would be regional in scale. Traits of individuals which appear to be coevolved within a community may be characteristic of an old, widespread, taxon (Shapiro 1980*c*) and thus unlikely to be the results of evolution within a local community.

The Biology of Butterflies
0-12-713750-5

It is also possible that strictly random events could produce what appears to be organized pattern in communities of non-interacting species. Null hypotheses generated by random models of such communities can be rejected if observed patterns are significantly non-random (Strong *et al.* 1979). Although potentially powerful as an analytical tool in community biology, neutral models are sensitive to biological details and can lead to incorrect, yet statistically significant, conclusions if not sufficiently precise and realistic (Cole 1981, Colwell & Winkler 1981).

Community Biology of Butterflies

Compared with birds, there is no large body of research on the community biology of butterflies (e.g. Cody 1974). My own review (*in* Gilbert & Singer 1975) was largely synthesized from systematic, faunistic, biogeographic and autecological studies. In other words, much of the data from which a picture of community patterns in butterflies could be built were not collected with community-level hypotheses or theories in mind. It is therefore not surprising that the study of butterflies has played little role in the development of general theories of community organization (see also Ch.2).

Gilbert & Singer (1975) provide a reasonably complete summary of the literature on butterfly community ecology up to the mid 1970s. The basic framework of the field remains much as it was then. This paper therefore focusses largely on those aspects that can be re-examined in the light of recent research (some previously overlooked work is also included).

The central question of butterfly community ecology is fundamental to biology: why are there so many species? Lepidopterists tend to find this one of the most aesthetically pleasing problems of science. It is also one of the more difficult to deal with empirically. Aside from the problem of analysing complex networks of species, one must also separate the question of diversity into an evolutionary component (why have so many species evolved in a region?) and into an ecological component (what limits the number of species which coexist in a given area?). Two one-hectare plots may contain different numbers of species, not because of local ecological differences between the areas, but because the plots lie in larger regions which themselves differ in diversity for reasons unrelated to small scale ecological patterns. Thus major historical and biogeographic patterns must be recognized and factored out (Gilbert & Smiley 1978).

Patterns of Diversity on Islands

One of the landmarks of community biology was the publication of the equilibrium theory of island biogeography by MacArthur & Wilson (1967). Intended to explain the pattern of species diversity on islands, it has since been applied to the analysis of diversity gradients across habitat patches, and

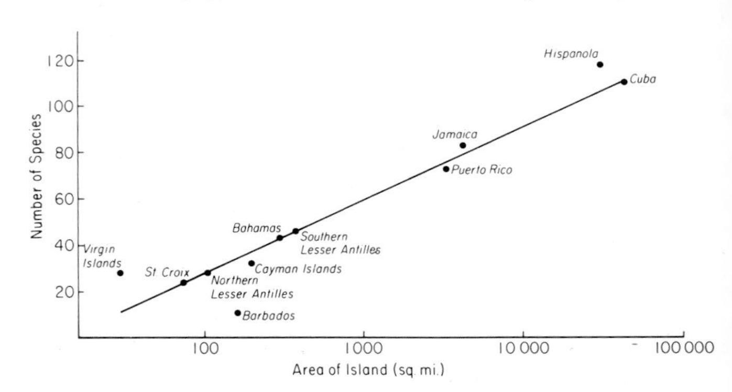

Fig. 3.1. Eugene Munroe's little known discovery of species area relationships in Caribbean butterflies. Graph redrawn from his 1948 dissertation.

between hostplants. However, the equilibrium theory of insular biogeography was first developed, in modern form, by Munroe (1948: 117–118) in his thesis on the distribution of Caribbean butterflies (Fig. 3.1):

> It is seen . . . that the size of the fauna is closely correlated with the area of the island inhabited, and that this correlation far overshadows any there may be with the other factors. . . . If the two quantities are plotted on semi-logarithmic paper, it becomes evident that, except in a few cases, the size of the fauna appears to be linearly related to the logarithm of the area of the island. The exceptions to the rule are: the Virgin Islands and Hispaniola, in which the faunas are too large, and Barbados, in which it is too small; as we shall see, it is possible in each case to find an hypothesis to account for the departure.
>
> A correlation of this kind is as interesting as it is unexpected, for it suggests the existence of an equilibrium value for the number of species in a given island, a value which acts as a limit to the size of the fauna. The processes which determine the equilibrium value for an island of given size must be, on the other hand, the extinction of new species, and, on the one hand, the formation of new species within the island, and the immigration of new species from outside it. Now, in each of the three islands or groups of islands which departs from the general series, one of these last two factors has an exceptional value. In Barbados, a small island, situated a considerable distance to the windward of the nearest sources of immigration, themselves small islands, the immigration rate must be unusually small. In the Virgin Islands, on the other hand, an abnormally high immigration rate may be expected, owing to the proximity of the relatively

large fauna of Puerto Rico. In Hispaniola there is no reason to expect an abnormal immigration rate, but here the rate of speciation has been higher than elsewhere, due to the expansion of the genus *Calisto*, which . . . has taken place largely within the island; and in fact, if all but one species of *Calisto* are subtracted from the total, the corrected value for the size of the fauna falls much closer to the semi-logarithmic line.

It is clear that this semi-logarithmic relationship is dependent on the rate of extinction of species, which must in turn be governed in some way by the area which they occupy: i.e., by their population numbers. Restriction of the population number, more severe the smaller the island, must increase the hazard of extinction for each species, whether through accidental reduction of the already small population, through contest with other species for the essentials of life, or through deterioration of the population genotype, due to the Sewall Wright effect. We can, then, think of an equilibrium value for the number of species in a given island fauna, depending on the one hand on the probability of extinction, correlated inversely with island size, and on the other hand on the probability of reinforcement, correlated directly with proximity to an area of richer fauna.

The clarity and insight of Munroe's discovery are obvious from this passage, yet it was apparently an idea out of context or ahead of its time, even for its author. I believe this case beautifully illustrates one of the reasons suggested by Gilbert & Singer (1975) for the slow progress of butterfly ecology: it has often been an afterthought of systematic or genetic studies. Interest in using butterflies for studying general ecological problems, now rapidly increasing, might have begun 20 years earlier had Munroe published the ecological parts of his work.

Whether the principles of island biogeography as applied to habitat patches or hostplant 'islands' provide a predictive theory for butterfly community diversity is uncertain. Recent work in Britain (Shreeve & Mason 1980) has shown that 50% of the variance in species number between isolated woodland patches is due to area. However, there is considerable uncertainty in this study concerning the number of actual breeding resident species in an area, so that the biological meaning of the species/area correlation is not yet clear.

Analyses of host-specialist communities such as heliconiines and ithomiines do not support Janzen's 'hostplant as an island' extension of equilibrium theory (Gilbert & Smiley 1978). Under this theory, a habitat containing many geographically widespread hostplants should maintain more species specializing on those hosts than a habitat with an equal diversity of endemic host species.

On the other hand, for some groups of butterflies, local diversity is a strong correlate of host species richness (Gilbert & Smiley 1978), and such relationships have been found to hold well

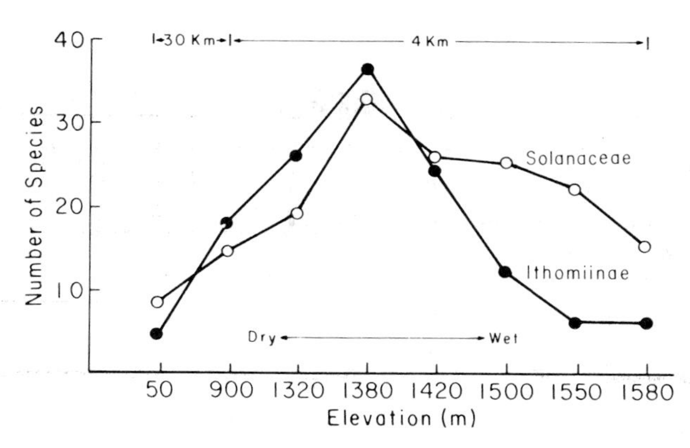

Figure 3.2. Patterns of ithomiine and Solanaceae species richness over moisture and elevation gradients in Costa Rica. Modified from Haber (1978).

over considerable elevation gradients in groups such as ithomiine butterflies (Fig. 3.2).

Niche Segregation Within Communities

If butterfly species diversity were always strongly correlated with hostplant diversity in local communities it would be tempting to argue that a precise understanding of hostplant partitioning would be sufficient to explain observed patterns of diversity. Unfortunately for this simple view, the local diversities of some groups such as *Euptychia* and related satyrines do not correlate with larval hostplant diversity (Singer & Ehrlich in press). Even for those groups which do exhibit positive correlations of insect and plant diversities, there are reasons to suspect that the primary dimension partitioned between two species may be microhabitat, with hostplants differentially used only as an indirect result of habitat choice (Gilbert & Smiley 1978).

It therefore seems important to discuss all the possible ways in which similar species are thought to differ and therefore coexist locally. Ultimately, the species richness and organization of butterfly communities is an outcome of adaptive radiation on the one hand, and coevolution of coexisting species on the other. This highly deterministic view of community is not generally accepted for all organisms, but appears to be particularly valid for host specialized, taxonomically well-studied groups such as butterflies.

Following Gilbert & Singer (1975), recent evidence concerning niche segregation in butterflies is discussed under the six following major niche dimensions: larval resources, by taxa; larval resources, by plant part; adult resources; microhabitat; time; and predator escape. These categories are useful for organizing discussions of butterfly community biology, but other ways of organizing the same information may be equally valid. I will return to this problem later.

Larval Resources: Taxonomic Partitioning

As a group, butterflies are relatively specialized on plant taxa used for larval food. This specialization involves host recognition by adults, gustatory responses and digestive tolerance by larvae. Therefore, most butterfly species in an area interact on larval resources with only a small fraction of sympatric butterfly species.

From this perspective there is no basis for thinking that, say, the list of butterflies from 4sq.km of Veracruz, Mexico, constitutes a community for which meaningful theories of organization and diversity can be developed. Since broad patterns of segregation onto particular hostplant taxa are worldwide and ancient (Ehrlich & Raven 1965) such segregation in a local community obviously does not result solely from local interactions.

In contrast, the set of species which utilizes a taxonomically small and chemically related subset of hostplants in an area constitutes a community whose patterns may result from ecological and microevolutionary interactions among the component species. The careful identification of such component communities (Gilbert 1980) is critical to the analysis of insect herbivore communities in general (Gilbert 1979).

It is also important to keep in mind that several orders of insects may share the larval hosts of a particular butterfly group and these cannot be ignored if host limitation for species of one order is caused by the activities of another. However, aside from one documented case of Monarchs affecting *Oncopeltus* bugs on milkweed (Blakley & Dingle 1978), it appears that such cases are rare. In situations where host limitation is not observed, it is legitimate to factor out just the butterfly assemblage sharing a hostplant taxon as a 'community' containing the bulk of information required to explain its own patterns.

This simplifying assumption is based on the logic that if hostplants are not directly limiting, predator/parasitoid escape and adult resources become the most likely niche components affecting species coexistence (see below). It also depends on the fact that such factors appear to be highly distinct for beetles, bugs, orthopterans, moths, and butterflies. Thus butterfly species which utilize a closely related set of hostplant species constitute a community comparable to MacArthur's warblers and fit Root's idea of 'guild', whether or not other insect orders use the same hosts. Only to the extent that the foregoing assumptions are valid is *butterfly* community biology an interesting special pursuit of ecology.

The general view of host partitioning remains basically that described by Gilbert & Singer (1975) and Gilbert & Smiley (1978). Butterfly species are frequently found to utilize a more restricted subset of locally available host species than would be predicted from larval host choice and feeding experiments (Smiley 1978*a*) or from summaries of hosts used over the entire range of species (Benson *et al.* 1976, Fox & Morrow 1981).

At first glance host partitioning within component communities would appear to result from competition. However, similar host restriction is known to occur where several suitable hosts are available yet not used, even in the total absence of interspecific competition. In *Euphydryas editha*, host specificity results from a complex interplay of factors which differentially affect the suitability of different host species, each of which support larval development under controlled conditions (Singer 1982*a*).

Another pattern inconsistent with a simple view of competition driving host separation is exemplified by Shapiro's (1975*a*) observation of two *Pieris* species specializing on the same species of *Lepidium* in an area with other potential hosts. Because both species prefer buds, flowers, and fruit, and because females appear to be extremely selective about oviposition sites, Shapiro (1975*a*) is not prepared to rule out competition, even in the absence of apparent host limitation. Based on the non-random distribution of eggs and larvae, he suggests that optimal oviposition sites are indeed limited and that emigration of females results when these sites are taken. Presumably, coexistence locally is a trivial result of one species, *Pieris protodice*, existing as much more vagile and widespread populations, so that microevolution of host partitioning is not necessarily expected where it overlaps with a more sedentary and geographically restricted competitor. Other studies of *Pieris* guilds (Ohsaki 1979) show that species utilizing the same larval hosts differ in other important respects which would reduce the likelihood of competition.

Benson's (1978) *Heliconius* work provided the first geographical comparison of host partitioning within a hostplant guild. These passion vine butterflies utilize the new shoots of *Passiflora* as oviposition sites. Benson assumed that *Heliconius* populations are limited by the amount of *Passiflora* new growth and therefore that they must compete for that resource. Comparing more equatorial, less seasonal areas (Trinidad, Costa Rica) with a highly seasonal locality (south Brazil), Benson found a higher degree of host specialization in communities occupying less harsh environments (*ca* 1.25 important hosts per species) than in south Brazil (*ca* 2.5 important hosts per species). This difference

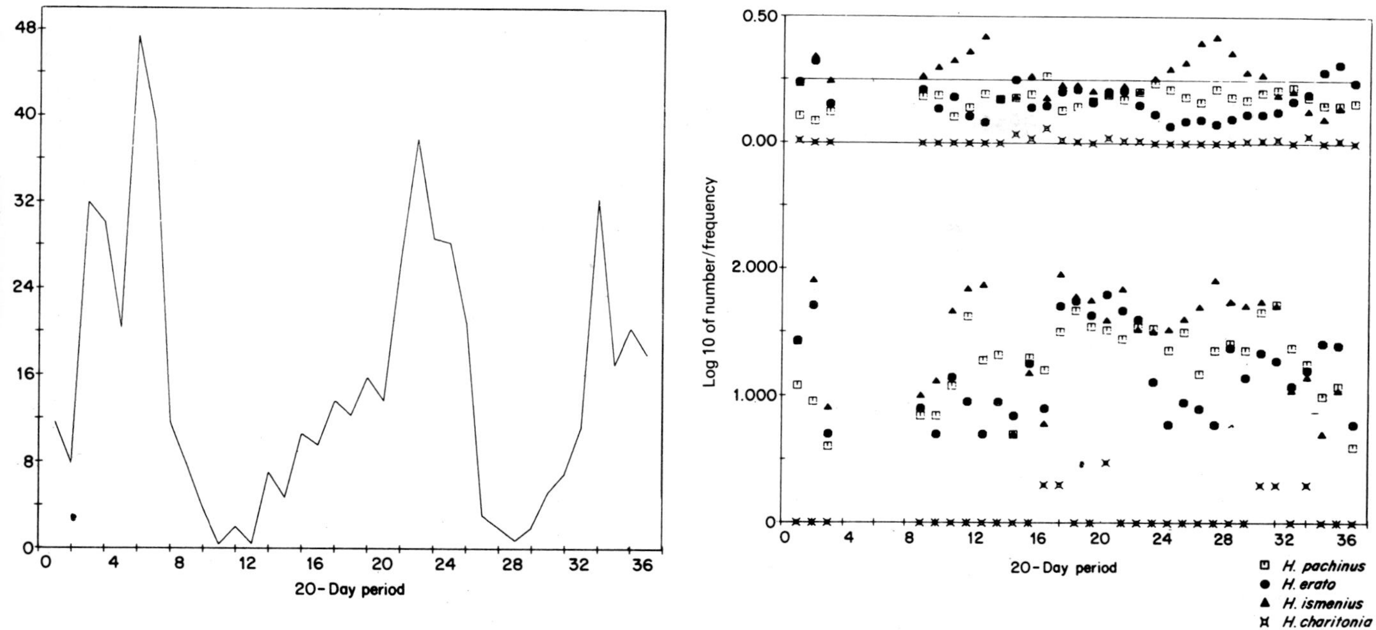

Fig. 3.3 Patterns of rainfall and *Heliconius* community changes at Sirena, Parque Corcovado, Costa Rica, from July, 1979 (period 1) to July, 1981 (period 37). Left: average daily rainfall for each 20-day period; driest seasons occur each year during February and March. Right: population and community data for four representative *Heliconius* spp. in a 30-ha area near Sirena. Top right: pattern of relative abundance for each sp. as measured by the total sample of *Heliconius* males for a period divided into the total sample of males for each sp. in that period. Bottom right: Manly-Parr estimates for male populations of each species of the four.

results from new growth production (by the various *Passiflora*) being seasonal in south Brazil and temporally staggered, forcing a wider host utilization by permanent residents of the area. Benson then argued that the absence of *H. melpomene* in the seasonal site is a case of competitive exclusion resulting from other species utilizing its host in that area. He suggested that each community seems to be 'saturated' (see also Gilbert & Smiley 1978).

The general patterns of community structure in *Heliconius* have been independently studied by Smiley (1978*b*). This study and my own population studies of the genus support the general picture of heliconiine community structure proposed by Benson. However, there remains room for questioning the role of competition for larval hosts in creating the observed patterns.

One problem with Benson's competition theory is the assumption that larval hosts limit populations of *Heliconius*. Such a conclusion was not obvious from the only published long-term study (Ehrlich & Gilbert 1973). More recently my students and I have gathered extensive data on eight species of *Heliconius* in Parque Corcovado. Costa Rica (Colour plate 1A). One species, *H. charitonia*, is always rare in the study area (Fig. 3.3) in spite of the apparent availability of an uncontested and underexploited larval hostplant (*Tetrastylis lobata*). In contrast, another consistently low-density species, *H. melpomene*, appears to face a hostplant limit judging from the low frequency of unoccupied shoots (Gilbert *et al.*, unpublished).

Benson asserts that in the seasonal site where most *Heliconius* have an expanded host range, competition for hosts selects for strong habitat separation. Unfortunately, this structural difference between communities is not quantified. But even assuming that it is a real trend, we cannot be sure what it signifies, since microhabitat preference might result from non-competitive interactions or competition for other resources. Only experimental population studies, some of which are now in progress, will resolve the role of competition for larval hosts in structuring *Heliconius* communities.

The work of Drummond (1976) and Haber (1978) on ithomiines suggests similar degrees of host partitioning. Direct host limitation and therefore competition for host leaf material does not appear to be sufficient to drive observed levels of specialization. Yet, since hostplant abundance seems roughly proportional to abundance of the butterflies using them, hostplant may limit butterfly populations in some indirect way (Gilbert & Smiley 1978), as will be discussed below.

Lawton & Strong (1981) have recently attacked what they believe to be a widespread belief that competition for hostplants structures herbivorous insect communities. Since the butterfly literature, including Gilbert & Singer (1975), was used to construct part of this particular straw man, it is appropriate to repeat the view we actually expressed (Gilbert & Singer 1975: 380): 'In none of the cases presented so far can we be sure of the ultimate basis for within-guild partitioning of host plants'. Throughout the 1975 review this was maintained for all described instances of ecological segregation, and the need for experimental studies was emphasized.

Larval Resources: Partitioning Parts of Hostplant

Specialization on particular leaf stages by related species sharing a hostplant appears to be one way of partitioning larval food. Benson, Brown & Gilbert (1976) and Benson (1978) show that heliconiine species can be divided into categories of new shoot versus older leaf specialists. *Heliconius* is characterized by new shoot specialists (Denno & Donnelly 1981) and appears to have recently radiated on this relatively scattered, ephemeral resource. Typically one finds a given *Passiflora* species utilized by one or two *Heliconius* and by one or two old-leaf specialists belonging to other genera of the tribe.

Has competition caused or is it now maintaining these niche differences? One approach to answering this question would involve obtaining answers to the following more specific questions: if all *Heliconius* were removed would *Dione* or *Eueides* begin to exploit new shoots? Second, if the old-leaf feeders were removed would *Heliconius* oviposition become less specific to new shoots?

Such questions seek to demonstrate competition indirectly by predicting niche expansion in its absence. For the heliconiines and for butterflies generally no such evidence exists. The behavioural, chemical and physical differences between these categories of leaves may require such specialization that the observed 'partitioning' would evolve in the absence of competitive influences. It may be that *Heliconius* were able to invade and radiate on to unexploited new growth because of the reproductive longevity provided by pollen feeding (Gilbert 1979).

Nevertheless, it is likely that defoliations by old-leaf larvae might influence the growth rate and phenology of new shoots, or that new shoot feeders might divert the energy and materials otherwise destined for older leaves. In a long term study of the community of five pierid species which utilize the tree *Capparis frondosa* in Tamaulipas, Mexico, Jordan (1981) has observed details of butterfly population changes and host leaf phenology which suggest that conditions may occasionally lead to competitive exclusion of young-leaf specialists by old-leaf specialists. It appears that by choosing forest understory for oviposition search, and by long residence time in the area, *Itaballia viardi* successfully exploits the very newest shoots of the tree even though the appearance of new growth is so widely spaced in time that few adults remain in the population when it reappears. Jordan noted that slight habitat changes (tree falls, etc.) which alter the rate of discovery of the plant by *Appias drusilla* and *Ascia monuste* may have the effect of increasing the interval between bursts of new growth, and this effect may be sufficient to eliminate *Itaballia* from a local community. It is reasonable to suggest that such interactions between specialists on different leaf stages might have forced the evolution of preference for forest habitats by female *I. viardi*.

Adult Resources

Discussions of possible competition within communities of phytophagous insects (e.g. Benson 1978, Lawton & Strong 1981), have typically only considered competition for larval hosts. But for insects like butterflies with complex life histories, it seems unwise to ignore the role of adult resources in shaping community patterns. Work on *Heliconius* illustrates this point.

Heliconius adults feed on nectar and pollen, the pollen accounting for about 80% of egg production and extended longevity (Dunlap-Pianka *et al.* 1977). Pollen resources appear to have greater influence on adults movement patterns than do larval hostplants (Ehrlich & Gilbert 1973). Females gather more pollen than males, older individuals gather more than younger, species in a community often show differences in quantity as well as type of pollen collected, and greenhouse experiments show that species may differ in ability to compete for pollen of a particular size (Boggs *et al.* 1981). Recent field studies on adult foraging in Parque Corcovado, Costa Rica (Murawski & Gilbert, unpublished) suggest that each patch of flowers has a carrying capacity for regular visitors and that an individual *Heliconius* becomes a regular visitor if pollen and nectar are found to be available during initial visits. Clumps of flowers saturated with experienced *Heliconius* (*i.e.* individuals that visit many times per day for weeks) are not likely to 'reward' newly arriving individuals. The working hypothesis is that when all flower clumps in an area are saturated with experienced butterflies, new adults entering the population may be forced to emigrate. Local population ceilings in *Heliconius* (Figs 3.3, 3.4) may thus be determined by dispersal; a mechanism thought to be common in many animal populations (Lomnicki 1980). Ehrlich (Ch.2) reviews the recent work relating adult resources to butterfly population dynamics.

What features of this *Heliconius* community may be explained by interactions around adult resources? One possibility is the rarity of *H. charitonia* in spite of the availability of unexploited larval hosts. This species, common less than 200km away in the dry forests of Guanacaste and in large open areas within Parque Corcovado

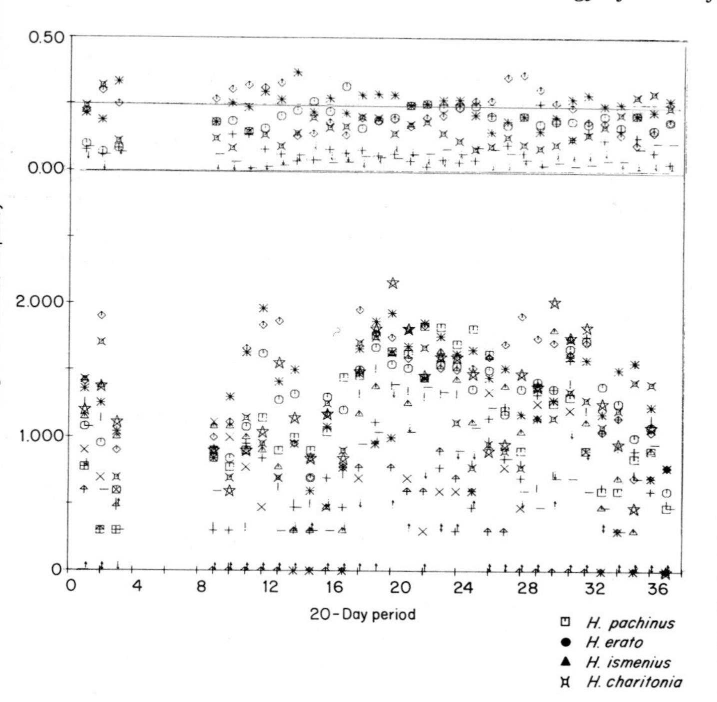

Fig. 3.4. Patterns of relative abundance and population estimates of seven resident breeding species of *Heliconius* at Sirena, Parque Corcovado, Costa Rica (*H. doris* is excluded as it is poorly sampled). Relative abundances shown above are for both sexes combined. Patterns seen for males only in Fig. 3.3 are still present but modified (*H. ismenius*, *H. erato*, for example). Below, 14 estimates of population size are shown for each period, one each for males and females of the seven species. Remarkably, in 448 separate estimates (14 estimates/period × 32 periods having estimates) only two, both female *H. hecale* rise above 100 individuals in the 30-ha area. Thus six to eight adults per ha appears to be the upper limit to population size for a *Heliconius* species in this area, i.e. three to four of each sex. Another general pattern emerging from these data is the fact that some species are characteristically rare while others are always more abundant in the area.

itself, may not be able to compete for adult resources with wet-forest *Heliconius* around our study site. Limited adult resources may also explain the general failure of any of the seven species to achieve estimated populations above 6–8 adults per hectare (ha) during a two-year period, despite abundant unexploited *Passiflora* in the area (Fig. 3.4).

Heliconius are unique in their utilization of pollen, but they are probably representative of many butterflies with respect to the general importance of adult resources to population and community structure.

Habitat Partitioning

Congeners possessing great ecological similarity are sometimes stratified along mountain slopes and vertically within a forest, or segregated among successional stages within an area (Gilbert & Singer 1975). Cases where hostplant distributions do not account for the butterflies' habitat preferences are of particular interest, and lead us to investigate other factors which may correlate with the apparent habitat partitioning, since space *per se* is not likely to be a resource for most butterflies. However, little quantitative work has been done.

In the case of elevational segregation, such as that described for Andean pronophilines by Adams & Bernard (1977), it is difficult to imagine many factors which would be limited to a 100 metre elevational band, as at least one of these satyrids appears to be. Observations of coexisting pronophilines indicate that all share adult resources on the forest floor, yet separate vertically for other activities. In the absence of striking ecological gradients and competition for adult resources, it would appear that selection for reproductive isolation may be the most likely factor driving habitat partitioning in this group.

The elevational segregation of other species may involve competition. However, any interpretation of biotic interaction must take into account the thermal environment and the constraints of the butterflies' thermoregulatory mechanisms. Thus, Kingsolver (1983*a*) has compared the thermal physiology of low and high elevation species of *Colias*, and has shown that, under the same conditions, differences in pigmentation and pubescence result in significant differences in time available for daily flight activity.

I hypothesize that in zones of potential overlap between 'low' and 'high' *Colias*, or other genera for which nectar is an important adult resource, the degree of divergence in thermoregulatory strategies would greatly influence the outcome of competition. Species able to be active earlier and longer at a particular elevation should have a major advantage even if their advancement and extension of flight time is only slight. Assuming that nectar availability influences adult habitat choice, the boundary between low and high adapted species would become more discrete as floral resources become more limited and more monopolized by first visitors. Thus, competition can sharpen elevation boundaries between two species, even if it had nothing to do with driving the prior evolution of their thermoregulatory differences.

Kingsolver (1983*b*) suggests that flight activity time is a limiting resource for *Colias*—but the situation is complex. Elevation translates into temperature, temperature into flight time, flight time into resource exploitation. The results of species interactions in zones of overlap are most likely to be perceived as fine adjustments in elevational boundaries, rather than as the partitioning of time of day when flight occurs.

Intrahabitat partitioning by height or by vegetation type is equally complex. Papageorgis (1975) made a notable attempt to document and explain the concurrence in Peru of different mimicry complexes possessing highly distinctive colour patterns. By observation she obtained flying-height distributions for each complex. Not surprisingly, the more clear-cut separation between 'blue', 'red', 'tiger', and 'transparent' pattern groups was found in tall primary forest. Papageorgis suggested that changing light conditions favour different types of signals at different levels of the forest and ruled out thermal gradients or predator stratification as possible reasons for the observed patterns.

Papageorgis' study has encouraged other workers, more involved with hostplant relationships, to examine microhabitat separation more carefully and in a more quantitative fashion. It appears that specialization on larval resource leads to increasing microhabitat restriction: this in turn may result in the evolution of co-mimicry with other species in the same microhabitat (Benson 1978). Although it is impossible to be certain of the sequence of these specializations, it is clear that they should influence one another. For example, both Benson (1978) and Smiley (1978*a*) note that '*melpomene* group' species are 'generalists' on *Passiflora* when they live as forest interior populations, but are highly specialized on larval host where populations live in secondary succession. Haber (1978), in his detailed study of ithomiine communities, notes a number of life history tactics, including adult size, egg number and egg size which, along with colour pattern, correlate roughly with adult microhabitat.

It is not clear how these studies can be harmonized with the observations of Papageorgis. It is likely that the local relation between vegetation structure and hostplant distribution is highly important. For example, in Parque Corcovado, forest plants such as *Passiflora pittieri* may be found on the edge of pastures and in early second growth vegetation as an artifact of recent human disturbance. Under these conditions, the typically forest interior species *Heliconius hewitsonii* can be found in relatively open, sunny habitats.

Central American workers have failed to detect clear vertical partitioning in heliconiines and ithomiines, but this could be due to the methods employed to measure habitat-specific activity. For example, I have made extensive observations at one particular *Anguria* vine, the pollen and nectar source for *Heliconius* at Parque Corcovado. By virtue of its position on the edge of a treefall, this plant produces flowers from 1-18m above ground level. Virtually all local *Heliconius* species, regardless of general microhabitat affiliation or colour pattern (Colour plate 1A), have been seen visiting flowers at any level. Nevertheless, as Smiley's (1978*b*) work emphasizes, microhabitat separation is more clearly seen during oviposition activity.

Perhaps the most similar pattern to Papageorgis' vertical stratification has been noted by J. Mallet (unpublished) in his study of roosting groups of *Heliconius*. His data show a clear-cut stratification such that pairs of mimetic species usually roost within the same height interval, and often occur together in mixed species roosts. Likewise, in a study of fruit feeding nymphalids trapped at baits placed in the canopy and in the understorey in Costa Rica, P. DeVries (in prep. *a*) found vertical stratification in several genera and species.

Thus microhabitat partitioning has a vertical and horizontal component, it is based on a complex of interacting factors, and the patterns perceived within a community depend heavily on what behavioural activities are observed. There is no direct evidence that competition and/or predation have caused the major patterns of microhabitat separation (see also Ch.2). Indeed, since the most closely related subsets of *Heliconius* and ithomiines appear to be distributed locally among different microhabitats and colour pattern groups (Smiley 1978*b*), reproductive isolation may be the raison d'être of microhabitat separation. Frequent interspecific hybrids in greenhouse cultures of *Heliconius* (see Colour plate 1B) support this suggestion. However, selection by predators leading to habitat-specific mimicry associations in allopatry may have allowed subsequent coexistence as 'species' by populations differentiated by little other than colour pattern and habitat.

Patterns in Time

Daily and seasonal patterns of activity, occurrence and abundance of species in butterfly communities suggest that coexistence by partitioning time may be possible for sets of closely related species (Gilbert & Singer 1975). This possibility has been explored in depth by Shapiro (1975*b*) for temperate zone butterflies. Both reviews stress that interactions with climate and hostplants are sufficient to explain most of the observed patterns.

But it would be a mistake to ignore the possibility that competitive interactions may be involved in shaping the daily patterns of flight activity in certain groups like neotropical skippers, where some species are crepuscular or nocturnal. Especially where exploitative competition is less important than direct interference, temporal separation of congeners could evolve. Plants whose nectar production straddles major light intensity gradients (early afternoon to midnight) provide the opportunity for competitive interactions to segregate species in time, especially in the tropics where night temperatures allow butterfly flight, but no such cases are clearly documented at this time.

Predator Escape: Vertebrate Predation on Adults

It has been argued that modes of predator avoidance, such as particular warning colour patterns, may constitute one kind of limit on local species diversity in mimetic species. Both theory (e.g. Charlesworth & Charlesworth 1975) and empirical observations (e.g. Brown & Benson 1974) deal with the maintenance of genetic diversity within populations, rather than species diversity.

Within a tropical forest habitat, such as those in Costa Rica with which I am familiar, a number of basic colour pattern systems coexist (Colour plate 1C). The species richness of these associations varies from dozens of species ('tiger stripe' complex) to three (black and yellow '*pachinus*' complex). A particular aposematic signal may be 'occupied' by few mimic species because it is neither frequent, predictable, nor sufficiently avoided by predators (i.e. if it signals only mild unpalatability or if a subset of the model species is actually edible). Alternatively, aposematic patterns may vary in the degree of developmental alteration necessary for new mimics to evolutionarily 'colonize' a pattern system. Thus, low density, mildly unpalatable species having novel shape, pattern and colour combinations should not attract many Batesian parasites, and may involve small groups of closely related species having similar developmental pathways. This seems to apply to *Heliconius*. Indeed, when *Heliconius* belong to large mimicry complexes there are good reasons to think that they do so as Batesian mimics of ithomiines. They are often polymorphic (Brown & Benson 1974) and trials with captive Rufous Tailed Jacamar, an important butterfly predator in Costa Rica and elsewhere, indicate that ithomiines are much less acceptable than *Heliconius* (Peng Chai, pers. comm.).

Thus we can view different warning signals in the way that we view different hostplant phenotypes. Whether one of these 'resource units' contributes to community diversification through maintaining populations of specialist species-parasites or mutualists depends on how easily it can be 'colonized' by specialists, and how well populations persist through time when specialized on the resource. Colonization depends upon target size (population density of hostplant or model species), the competitive environment (whether other species are already there), the extent of genetic steps required (as in the case of hostplants with novel chemistry or model species with patterns which depart radically from standard lepidopteran schemes), and the population structure of the colonizing species (this is discussed with reference to hostplant shifts by Gilbert (1979) and parallel arguments can be applied to shifts into new mimicry systems).

Putting aside the evolutionary origins of niche diversification, the question of local diversity in host-specialist butterflies reduces ultimately to the number of resources predictably available in sufficient abundance in an area, and to the distinctiveness of alternative resources in morphology, chemistry, behaviour and microhabitat. Gilbert & Smiley (1978) and Gilbert (1979) have discussed the diversity of *Heliconius* communities in light of *Passiflora* diversification (see also Gentry 1981). Here I wish to extend the same reasoning to the concept of model phenotypes as resources. The number of warning patterns which coexist locally, and their degree of distinctiveness amount to limits on the species diversity of mimicking groups.

In community ecology we generally start with the number and relative abundance of resources (host species, model patterns, etc.) as given, then attempt to account for the community structure of a guild as an outcome of resource pattern. But ultimate answers to community structure must deal with questions like why so many *Passiflora* species are available, or why so many aposematic patterns are used in one area. As discussed above, Papageorgis (1975) initiated the attempt to deal with the latter question. Inevitably, the interesting answers about butterfly communities must involve the study of broader systems of interacting species (Gilbert & Singer 1975).

A legitimate criticism of such speculations has been that the colour pattern evolved by a population may have little bearing upon its numerical dynamics. This conclusion was based largely on theoretical work by Nicholson, Haldane, Sheppard and others (see Vane-Wright 1981*b*). However, recent field studies suggest how mimicry association may be significant for the persistence of a species in a local community. Jordan (1981) noted that although phenology and dynamics of shoot production by hostplants explain the essential features of abundance and dynamics of *Itaballia viardi* populations, their persistence may frequently depend on the lives of two or three sedentary females bridging the gap between episodes of shoot availability. Were those females not mimics of *Heliconius charitonia*, I suspect that the local extinction rate of *I. viardi* would be much higher. Note that any climatic or biotic perturbation increasing the interval between periods of shoot growth would increase the value of the model's 'protective umbrella'. Mimicry probably does not account for any aspects of population dynamics except persistence during bad times. Remington (1963) anticipated this conclusion.

Some population data collected at Parque Corcovado illustrate this possibility, in the mimicry association between two *Heliconius* species. Figure 3.5 illustrates the estimated numbers of *H. erato* and *H. melpomene* in a 30ha sub-area over 160 days (July 1980–January 1981). Both species are relatively abundant during wet months, when birds presumably have abundant insect prey. Rather than interacting mutualistically at that time, both species may actually compete weakly on adult pollen sources (note reduced pollen scores). When numbers are high, predation low,

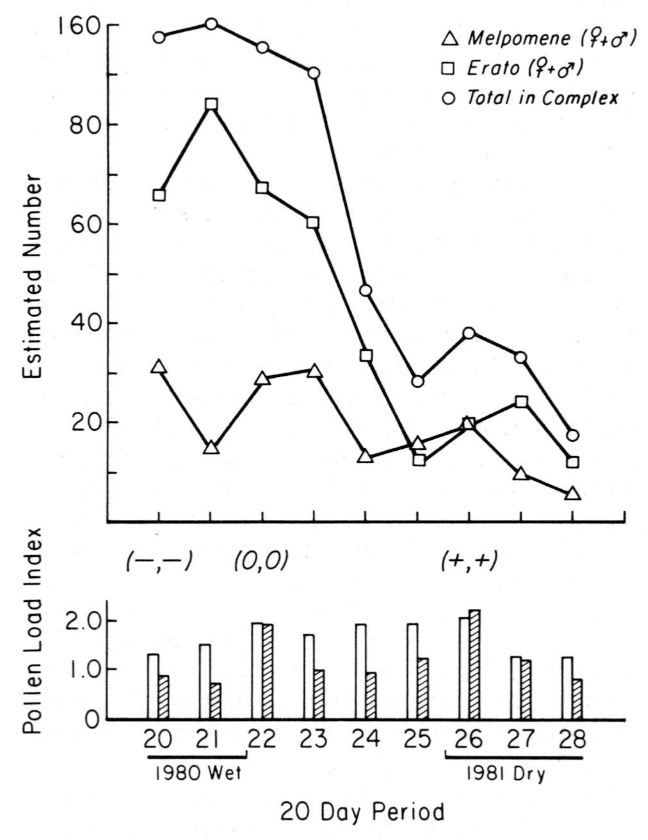

Figure 3.5. Population dynamics of a mimetic pair of *Heliconius* at Sirena. A 160-day interval, ending with the beginning of the 1981 dry season, is shown. Periods are comparable to those of Fig. 3.3. Pollen load index is obtained by summing average of male and female pollen scores for each period (see Boggs *et al.* 1981 for methods). Open bars represent *H. erato* pollen scores, hatched bars represent those of *H. melpomene*. The net interaction coefficients shown between the two graphs are seen to reflect the coincidence of various levels of adult food availability, with various densities of the butterflies, and hypothesized seasonal shifts in predator response to *Heliconius*.

and adult resources not limiting, the net interaction may be neutral since larval hosts are not shared. However, during early dry season (periods 26–28) both species decline to a point where the sum of both populations may only equal the number of jacamars in the area, i.e. about one per hectare (estimated by Peng Chai, pers. comm.). At that point, if these birds are faced with declining amounts of favoured insects, one or both *Heliconius* could undergo local extinction of adults, merely on the basis of increased jacamar 'sampling'. At this point, *H. erato* and *H. melpomene* are clearly mutualists, since an increase in the adult numbers of one will reduce the probability that the other will be eliminated. Such bottlenecks in population size correlating with peaks in selection for mimicry may account for the remarkable resemblances of the parallel races of *erato* and *melpomene* group species (see Ch. 14 and Gilbert 1983).

Predator Escape: Arthropods on Early Stages

A growing number of workers are interested in the role of insect predators (especially ants) and parasitoids (primarily Hymenoptera) in the population and community biology of butterflies. Although few new generalities can be added since Gilbert & Singer's (1975) summary, there is now solid experimental evidence for the role of ants in providing protection from parasitoids (Pierce & Mead 1981) and additional documentation of ant-dependent oviposition (Atsatt 1981*a*).

Vane-Wright (1978) suggests that mutualism with ants should be treated as a separate category from 'predator escape' in organizing data about butterfly niche diversification. I choose not to do this only for the sake of argument. Although ants occasionally become the food of butterflies, their positive benefit to butterflies typically involves their role in providing enemy-free space, and since 'enemy' here implies arthropod enemies, ant protection is but one of the ways used by butterflies to avoid such enemies. But, to the extent that communities of ant-dependent butterflies such as riodinids are specializing upon, as well as partitioning ant species at the oviposition stage (while being rather generalized on hostplants), ant mutualism should be considered as a separate niche dimension. If, on the other hand, individual females are simply choosing among individuals of the hostplant species on the basis of which individual plants are occupied by favoured ants, then I see no reason to separate ants from other factors which signal enemy-free parts of hostplant biomass. Investigations of such issues are in an early stage (see Ch. 19).

Ants, as predators or mutualists, and parasitoids appear to exert an indirect yet powerful influence on how larval hosts are utilized and partitioned among species of a butterfly community, and butterflies appear to be excellent candidates for attacking the problems of multitrophic level interactions (Gilbert & Singer 1975, Gilbert 1980, Price *et al.* 1980).

Ants may reduce the extent of safe oviposition sites within a hostplant population to the extent that there is effective host limitation (Benson 1978). Different assemblages of ants consistently associated with otherwise similar hosts may lead to selection for more narrow oviposition preferences than would be predicted from larval growth on other available host species (Smiley 1978*b*).

Parasitoids, by virtue of their greater specificity and more rapid population response to changing densities of particular butterflies, exert a less conspicuous yet potentially far more potent effect on butterfly populations and communities. While an ant colony may recruit to a bush covered with butterfly larvae, bushes elsewhere in the area controlled by less dangerous ant species may be relatively safe. In contrast, parasitoids have the capacity to rapidly affect an entire host population of one or more butterfly species.

One potential consequence of shared parasitoids is local extinction of the rarer of two species should they be too ecologically similar. This phenomenon has occasionally occurred in my greenhouse cultures of *Heliconius*, which usually contain an egg parasitoid, *Trichogramma* sp. If mortality reaches 99%, a common species laying 1000 eggs per day expects to hatch 10 eggs, while a rare species laying 50 eggs in the same area can expect only 0.5 eggs to become first instar larvae each day. In other words, the rare species is eliminated by 'spill-over' of parasites generated by the dense species.

Haber (1978) suggested that a similar phenomenon might account for host specificity in ithomiine butterflies. Should egg parasites tend to concentrate searching activity in areas of high egg density, rare species sharing hosts with other species would tend to be differentially eliminated by shared parasitoids. Selection, Haber argued, should then favour shifts to alternative hosts, or at least a more careful search by the rare species for isolated plants.

Haber may have recognized an important cause of within-guild partitioning of microhabitat, phenology (Gilbert & Singer 1975) and, in particular, hostplant species. However, a more realistic model of parasitoid-forced host partitioning should begin with a community of 'generalist' butterflies all using the full range of available species of a hostplant family. For example, if one begins with a local assemblage of *Heliconius*, all generalists on *Passiflora*, several assumptions about hosts, parasitoids, and how the butterflies forage for hosts must be made in order for shared parasites to select for host partitioning. These assumptions are first, that host species are patchy and differ in relative abundance; second, that hosts are encountered at random by ovipositing females which subsequently revisit previously discovered host individuals and patches more frequently than expected; third, that coexisting *Heliconius* vary in relative abundance and that some are typically common and others rare in an area; fourth, that parasitoids respond to high prey density by increasing searching time in patches with high density, and they use plant odours for host location.

Given these assumptions, most of which are based on observations in real communities, we would expect both rare and common species to leave offspring more often when they placed eggs on patches of a rare host, but that selection would be much more intense on the rare species to confine its eggs to the rare *Passiflora*. There are at least two reasons for this. First, it is possible to envisage periods during which a very rare butterfly species would by chance, place all of its eggs on the common host. Second, while the rare species would do equally well as the common species when each placed eggs on isolated and different hosts, on shared hosts crowded with eggs the rare species might generally fail to produce offspring. The common species would also have a high rate of success on the rare, low plants, but would produce many more offspring in absolute terms on the crowded plants. Conversely, the rare species obtains both relative and absolute advantage in offspring production if it evolves a mechanism for non-random discovery of the rare *Passiflora*. That mechanism might be the use of a chemical cue specific to the rare host.

One can therefore predict that in habitats dense with parasitoids and possessing a variety of common and rare *Passiflora* species, the *Heliconius* more generally at low densities should become more strongly specialized than the common species. Furthermore, in habitats lacking parasites and with all *Passiflora* of equal abundance, *Heliconius* will be generalist (assuming that eggs are laid singly).

These crude predictions are in general agreement with host relations found by Smiley (1978*a*) in Costa Rican Atlantic lowland forest. In the forest, where hosts are rare and parasites lacking (Smiley 1978*b*), one finds the guild's most generalized feeder, *H. cydno*. In secondary succession where *Passiflora* are abundant, most *Heliconius* species are host specialists with the rare species, i.e. *H. melpomene*, the most strict in its oviposition preference. Smiley (1978*b*) accounts for such patterns in terms of foraging theory combined with higher predation in non-forest habitats. He argues that where hosts are rare and predation is relatively light (in the forest interior) a generalist strategy should evolve, except in the case of cluster laying species which can afford to specialize (e.g. *H. sapho*, *H. hewitsoni*), since only one suitable host need be discovered each day or so. In habitats (successional patches) with abundant *Passiflora*, Smiley argues, the problem is not finding hosts, but finding safe hosts. In such habitats the butterflies can afford to specialize if necessary. Smiley's theory thus complements both that of Haber (1978) and the variation of Haber's theme presented above.

Patterns generated by such interactions in the *Heliconius* system should be temporally and geographically dynamic. Thus, *H. melpomene*, where it occupies forest (fewer parasites, no abundant *Passiflora*) is a generalist (Benson *et al.* 1976, Benson 1978). Moreover these differences between species appear to be microevolutionary since it is possible to cross *H. cydno* and *H. melpomene* (Colour plate 1B). Thus a development of less anecdotal and more rigorous theory pertaining to host partitioning and specialization should be testable by the geographical and genetic analysis of the hostplant ecology within this and other butterfly guilds.

Predator Escape:
Congeneric Cannibalism, Pupal Plunder

Cannibalism of eggs and small larvae by larger larvae is known for many phytophagous insects. In butterflies generally, and in the genus *Heliconius* in particular, most species which lay single eggs are cannibals to a degree. *H. erato*, one of the most aggressive *Heliconius* species in the larval stage, when compared to a less aggressive heliconiine, exhibits a much slower rate of population increase and rarely defoliates all hosts in the greenhouse because only one caterpillar survives the 'war' on each growing branch. Other species that share a resource with *H. erato* tend to place eggs away from the meristem area where *erato* eggs are placed (Benson 1978). Therefore, it seems reasonable to assume that other *Heliconius* species might evolve mechanisms to avoid hostplant species dominated by cannibalistic species such as *H. erato*.

One way to avoid losing offspring to cannibals involves recognition and avoidance of previously laid eggs. One consequence of this is the evolution of egg mimicry by some *Passiflora* (Williams & Gilbert 1981). Similar structures in Brassicaceae are regarded by Shapiro (1981*c*, *d*, 1981*b*) as indirect evidence for competition. Aside from providing one more anecdotal example of competition, this phenomenon illustrates the kind of coevolutionary step which may promote community diversity by increasing the differences between related hostplants, thereby increasing the likelihood of specialization and host partitioning among the insects (Gilbert 1979, Gilbert & Smiley 1978).

Another form of interspecific interference which may cause adult mortality is the harassment of emerging adults by congeneric males. *Heliconius erato*, *H. hewitsonii* and allied species (Brown 1981) share the habit of male assembly on pupae prior to emergence. In some species (*H. erato*) mating occurs as females emerge (Gilbert 1976) while in others (*H. charitonia*, *H. hewitsonii*) males actually penetrate the pupa prior to eclosion and couple with the female as she drops from the pupal case.

In the insectary I have frequently observed pupal maters sitting on the pupae of other species (Colour plate 1D) including those of non-pupal mating groups. As this phenomenon occurs in crowded cultures when female chrysalids of pupal maters are available and possibly releasing pheromones, I had concluded that this behaviour, which usually leads to the death of the non-pupal-maters, was an artifact of culture conditions.

Do such insectary observations predict potential natural events? In the field, *H. hewitsonii* pupates in groups on or around hostplants where 'excess males' are frequently seen sitting on male pupae and would undoubtedly sit on other species were they nearby (Longino & Gilbert, unpublished). Further, males searching the area investigate all vertical objects resembling pupae. The immediate vicinity of the hostplant of a pupal-mating *Heliconius* may therefore resemble the artificial culture for another species pupating there. That is to say, interspecific sexual harassment may be much more probable in hostplant patches dense with pupae.

There is therefore a variety of potential disadvantages to sharing hostplant with another *Heliconius*, ranging from the increased effective density for shared parasitoids, larval cannibalism, and mortality of tenerals if a pupal-mating species is involved. From a population perspective, these phenomena amount to competitive interactions which would select for hostplant and/or microhabitat separation within the community, even without host defoliation. Thus, one-dimensional discussions of whether or not phytophagous insect communities are structured by hostplant competition are somewhat artificial (*e.g.* Lawton & Strong 1981).

Approaches and Prospects

The study of communities faces both practical and theoretical difficulties. Easily measured properties of multi-species systems, such as relative abundance estimated from spot samples, are of questionable significance (Wiens 1981). The correspondence between such measures and underlying populations dynamics is problematic. In the *Heliconius* data shown in Fig. 3.3, the relative abundances (above) are less chaotic than the population estimates made at the same periods. A species may increase in relative abundance in the community, while increasing, decreasing or keeping the same absolute numbers over a given time interval, depending upon the status of other species.

Further, trying to make sense of things by correlation analysis of various community, resource and habitat parameters provides no sure guide to underlying causes for observed community patterns. Thus, although there is a strong correspondence between *Heliconius* and *Passiflora* species richness across habitats, there is no single factor to account for the correlation (Gilbert & Smiley 1978). Furthermore, general classifications of habitat characteristics (e.g. predictable *vs* unpredictable), frequently used in testing ideas about communities, may not apply accurately to a particular group of species (Gilbert 1979).

Another approach is to make mathematical models of a community by variously combining single population models into an interacting system. In theory, experimental manipulations of the community would be able to eliminate resulting hypotheses and improve assumptions of the model. Unfortunately, models of this type (May 1976; Ch.2) become chaotic and/or predict many equally likely outcomes of perturbation well before achieving the complexity inherent in even a simple community of *Heliconius*.

In spite of these difficulties, I am confident that focus on the community will be an increasingly important area of butterfly ecology. I also see immense potential in the interplay of population and community biology for answering questions at each level. For example, in our study of *Heliconius*, questions about the regulation of individual populations of the community are approachable by comparative and experimental means. Thus, this community provides various replicated sets of species whose population traits, if different, suggest manipulative experiments which automatically have built-in controls (the other species). Further, by carefully selecting the species, one can reduce the number of variables which must be considered. For example, *H. erato* and *H. melpomene* share phenotype, microhabitat and climatic regime. Differences between them must involve remaining factors which differ between them (e.g. hostplant). Thus, the community provides an experimental arena for population studies.

Population studies also contribute to understanding community organization. For example, important niche dimensions such as pollen feeding were discovered during demographic studies of a single species when measured lifespan and egg production could not be accounted for without external input by adult feeding. There seems no way to interpret emergent community patterns without knowing something of the important interactions within and between populations of an entire food web. On the other hand, studies of single species systems provide insight concerning which patterns of specialization need not involve interspecific competition.

Butterflies are uniquely suited to work at the population community interface (Ch.2). They can be observed, visually identified, caught and marked, much as birds can. On the other hand, more generations of butterflies may be observed within a human lifetime and butterfly population studies in the field can be co-ordinated with studies of laboratory cultures. If the population basis for community pattern is to be understood, it is likely that butterflies will play an important role in bridging this important gap.

Because it has become obvious that 'general' theories based on groups such as birds or ants do not necessarily apply to phytophagous insects, butterflies are now becoming an important tool for the development of general ideas about the earth's most important ecological interactions—those between insects and plants. Of course, because of their pivotal ecological position between plants, parasitoids and predators, butterflies offer the opportunity to study many ecological questions within a single system. Because they are typically host specific and require complex adult resources, butterfly communities provide the best rapid indication of habitat quality (Singer & Gilbert 1978). We can also anticipate that studies of butterfly communities will contribute to an understanding of how complex ecological systems are organized and thus how best to apply biology to the management of the remaining natural diversity of this planet (Gilbert 1980).

Acknowledgements

My thanks go to many colleagues and students, especially Bill Haber, Craig Jordan and John Smiley for making available information in the form of reprints, manuscripts and dissertations. Paul Feeny came across the Munroe dissertation at Cornell and pointed it out to me. Phil DeVries, Donald Harvey, Duncan Mackay and Michael Singer all read and made constructive comments concerning drafts of the paper. The *Heliconius* population and community data presented represents the joint efforts of those working on my NSF DEB-7906033 project in Parque Corcovado, especially Jack Longino, Jim Mallet, Darlyene Murawski and Annie Simpson. The effort of Sharon Bramblett in typing and editing the manuscript is greatly appreciated.

4. *The Effects of Marking and Handling on Recapture Frequencies of Butterflies*

Ashley C. Morton

Department of Biology, The University, Southampton

A number of entomologists who employ mark-release-recapture (MRR) methods to estimate population parameters accept the view that 'the marks were not conspicuous and thus were unlikely to affect either the survival of marked individuals or their probability of being recaptured.' The more cautious generally recognize that handling and incarceration may also have an effect, and reject data with low recapture rates. A much smaller minority take the trouble to compare observed numbers of successive recaptures with those expected from a Poisson distribution. I hope to demonstrate that not only are the effects of handling and using marks of different colours and sizes rather unpredictable, but also that high recapture frequencies and good fits to Poisson distributions are unreliable indicators of the suitability of a MRR technique.

The first study concerns a woodland population of *Melanargia galathea*. On the first day, 225 males and 225 females were assigned to nine different treatments (A-I, Table 4.1) using random number tables. The 25 males and 25 females in each treatment were marked with a unique number using fine-tipped marker pens, on the upper forewing and the underside hindwing, and released immediately. The treatments investigated the effects of colour of mark (red, black or green), size of mark and the effect of further disturbance (Table 4.1). Specimens marked with large numbers (10-12mm × 1-5mm) did not need to be netted to record their details. Thus these could be divided into those which were left undisturbed (designated 'not handled') and those which were netted anyway (designated 'handled').

Table 4.1. Details of marking and capture techniques.

Size of mark	Handled (+) or not handled (-)	Colour of mark Black	Red	Green
large	-	A	B	C
large	+	D	E	F
small	+	G	H	I

Marked specimens were allowed two days to mix with the remaining population and thereafter sites were sampled daily, following the same route each time. In order to ensure that the specimens designated 'not handled' were not accidentally disturbed, all specimens discovered were examined at rest and none were taken in flight. Specimens encountered several times during the same day were counted once only.

The differences in recapture frequencies between sexes were examined by 2 × 3 table *G*-tests. Only treatment B showed a significant difference between the sexes (G_2 = 7.82, $P < 0.05$), so data for both sexes were pooled to provide larger classes in subsequent analyses.

If the probability of an insect being captured on any sampling occasion is constant, the number of successive recaptures regardless of previous captures should describe the Poisson distribution for a particular sample size and mean. A significant deviation from the expected values may indicate a tendency for marked individuals to be recaptured more or less often than expected.

The observed and expected frequencies of recapture are shown for each treatment in Table 4.2. Four of the nine treatments show significant differences from the Poisson distribution, which may indicate that these treatments affect the behaviour of the marked specimens. This can be examined by *G*-tests between different groups of treatments.

The influence of different coloured marks may be examined in three ways: *G*-tests may be performed on the 'not handled' specimens with large marks of

The Biology of Butterflies
0-12-713750-5

Table 4.2. Tests of goodness-of-fit of data obtained by pooling sexes to the Poisson distribution.

Treatment	Recapture frequency	Observed	Expected	G-test (all one df)
A	0	22	15.06	
	1	13	18.07	
	2+	15	16.86	4.59*
B	0	19	20.33	
	1	20	18.30	
	2+	11	11.37	0.25
C	0	25	19.15	
	1	13	18.38	
	2+	12	12.47	3.41
D	0	31	30.94	
	1	15	14.85	
	2+	4	4.20	0.01
E	0	30	25.33	
	1	10	17.23	
	2+	10	7.44	5.18*
F	0	31	24.34	
	1	7	17.53	
	2+	12	6.31	17.58***
G	0	29	25.33	
	1	11	17.23	
	2+	10	7.44	3.89*
H	0	26	23.39	
	1	14	17.77	
	2+	10	8.84	1.29
I	0	26	21.16	
	1	12	18.20	
	2+	12	10.64	3.60

df: degrees of freedom
$^{*}P < 0.05$
$^{***}P < 0.001$

each colour (treatments A, B, C); on the 'handled' specimens with similarly large marks (D, E, F) and on those specimens with small marks of each colour (G, H, I). None of these comparisons show significant differences between the treatments, nor are any shown by pooling the data for each colour regardless of other aspects of treatment (Table 4.3). We are therefore left with the surprising finding that adding coloured marks to the cryptic underside of this species seems to make little difference to probability of recapture.

The effect of using easily applied large marks rather than small (1-4mm × 1-2mm) inconspicuous ones may be examined by using data from all the handled specimens, i.e. treatments D, E and F *vs* G, H and I. Again, no significant differences in recapture frequencies are found (Table 4.4).

The remaining effect is that of handling. All the specimens in this study had to be handled to mark them; we would therefore expect no differences between the two treatments at the zero recapture class. At first recapture some specimens are handled and some are not. Thus, there should be no difference in the number of specimens caught

Table 4.3. Effect of colour of mark on recapture frequencies (data for each colour pooled, with expected values in parentheses).

Frequency of recapture	Black	Red	Green		Total	G-value
0	82	75	82	(79.67)	239	0.41
1	39	44	32	(38.33)	115	1.92
2	15	24	19	(19.33)	58	2.10
3	9	3	12	(8.00)	24	5.97
4+	5	4	5	(4.67)	14	0.15
TOTALS	150	150	150		450	

TOTAL $G_8 = 10.55$

Table 4.4. Effect of size of mark on recapture frequencies (expected values in parentheses).

Frequency of recapture	Large marks		Small marks	Total	G-value
0	92	(86.5)	81	173	0.70
1	32	(34.5)	37	69	0.36
2	20	(20.0)	20	40	0.00
3+	6	(9.0)	12	18	2.04
TOTALS	150		150	300	

TOTAL $G_3 = 3.10$

once only, but the differences between handled and not handled specimens should then increase with subsequent recapture. Using pooled data for handled (treatments D, E, F, G, H, I) and non-handled specimens (A, B, C), this is exactly what is found (Table 4.5). It is clear that the 'handled' specimens are recaptured less frequently than the 'not handled' ones.

Returning to the comparisons with the Poisson distributions, it is interesting to note the lack of association between significant differences from the Poisson and significant differences between treatments, as shown by the *G*-tests. For example, treatment D shows a remarkably good fit to the

Table 4.5. The effect of handling on recapture frequencies (expected values in parentheses).

Frequency of recapture		Not handled (A-C)		Handled (D-I)		
1 or less	(118)	112	(236)	242	354	0.46
2	(19.33)	18	(38.67)	40	58	0.14
3	(8.00)	11	(16.00)	13	24	1.61
4+	(4.67)	9	(9.33)	3	14	5.57*
TOTALS		150		300	450	

TOTAL $G_2 = 7.32^{*}$

$^{*}P < 0.05$

Poisson distribution, but consists of 'handled' specimens which seem to be affected by this treatment. I would therefore suggest that the information given by the comparison to a Poisson distribution could be very misleading if taken as an indication of the suitability of an MRR technique.

The other measure of suitability, that of eliminating data with low recapture rates, also fails to discriminate reliably between treatments since even the lowest percentage of specimens recaptured on at least one occasion (38%) is high by comparison to other studies.

Given that the effect of handling is real, we are left to question its extent and significance in terms of population parameters. Work with other species since 1977 seems to confirm that the effect is widespread but unpredictable. *Maniola jurtina*, *Colias eurytheme* and *Thymelicus sylvestris* clearly show the effect of handling, whilst the mobile *Clossiana selene* does not (Table 4.6). Each of these species was investigated using the method described above, which fails to separate the effects of marking from those of handling, since it is difficult to mark the specimens without first catching them.

However, Singer & Wedlake (1981) managed to do just this with *Graphium sarpedon* and demonstrated a very significant difference in recapture frequencies between specimens captured prior to marking and those marked without capture. A similar method may be used with some European lycaenid butterflies, since they tend to roost together in the evening and become quite torpid. A number of specimens may then be marked, using marker pens as before, in a short space of time, but only on the underside of the wings. On subsequent evenings the specimens may be scored without the need to capture them. The results for *Polyommatus icarus*, *Lysandra coridon* and *Cupido minimus* are shown in Table 4.7. Only

Table 4.6. The effect of handling on recapture frequencies (expected values in parentheses).

Frequency of recapture	Handled		Not handled	Total	G-value
Maniola jurtina					
0	68	(67.5)	67	135	0.00
1	51	(51.5)	52	103	0.00
2	24	(18.5)	13	57	3.32
3+	7	(12.5)	18	25	5.51*
TOTAL	150		150	300	
TOTAL $G_1 = 8.83$**					
Thymelicus sylvestris					
0	53	(52.0)	51	104	0.04
1	44	(44.5)	45	89	0.01
2	43	(32.5)	22	65	6.91**
3	6	(14.0)	22	28	9.73**
4+	4	(7.0)	10	14	2.66
TOTAL	150		150	300	
TOTAL $G_2 = 19.30$***					
Colias eurytheme					
0	37	(36.5)	36	73	0.01
1	29	(29.5)	30	59	0.02
2	29	(24.0)	19	48	2.10
3+	5	(10.0)	15	20	5.23*
TOTAL	100		100	200	
TOTAL $G_1 = 7.33$**					
Clossiana selene					
0	18	(18.0)	18	36	0.00
1	14	(13.0)	12	26	0.15
2	12	(10.5)	9	21	0.43
3+	6	(8.5)	11	17	1.49
TOTAL	50		50	100	
TOTAL $G_1 = 1.92$					

*$P < 0.05$
**$P < 0.01$
***$P < 0.001$

Table 4.7. Comparison of recapture frequencies of specimens captured prior to marking with specimens marked without capture (expected values in parentheses).

Frequency of recapture	Captured prior to marking		Marked without capture	Total	G-value
Polyommatus icarus					
0	34	(25.0)	16	50	6.63**
1	6	(9.0)	12	18	2.04
2	4	(8.0)	12	16	4.19*
3+	6	(8.0)	10	16	1.01
TOTAL	50		50	100	
TOTAL $G_3 = 13.87$**					
Lysandra coridon					
0	33	(25.5)	18	51	4.48*
1	9	(12.0)	15	24	1.52
2	4	(5.5)	7	11	0.83
3+	4	(7.0)	10	14	2.66
TOTAL	50		50	100	
TOTAL $G_3 = 9.49$*					
Cupido minimus					
0	28	(24.0)	20	48	1.34
1	11	(10.5)	10	21	0.05
2	5	(7.0)	9	14	1.16
3+	6	(8.5)	11	17	1.49
TOTAL	50		50	100	
TOTAL $G_3 = 4.04$					

*$P < 0.05$
**$P < 0.01$
***$P < 0.001$

C. minimus fails to show a difference between handled and unhandled specimens.

Taken together with those of Singer & Wedlake, I feel these findings demonstrate the need to investigate the effects of size and colour of marks and, in particular, the effect of handling before choosing a technique for MRR study of a population. Moreover, there is even the possibility that less costly methods of estimating population density, such as the transect methods of Douwes (1970), Moore (1975) and Pollard (1977, Ch.5), although apparently less precise than MRR, may yield more accurate estimates due to avoiding disturbance of the study insects.

5. *Synoptic Studies of Butterfly Abundance*

Ernest Pollard

Natural Environment Research Council, Institute of Terrestrial Ecology, Monks Wood Experimental Station, Cambridgeshire

A national scheme for monitoring the abundance of butterflies commenced in Britain during 1976. At a few sites in eastern England recording has been in progress since 1973 or 1974. Results obtained provide a new perspective on butterfly population ecology. Data for two species of grass-feeding satyrid butterflies, the Meadow Brown (*Maniola jurtina*) and the Ringlet (*Aphantopus hyperantus*) are presented in this paper. Both occupy similar grassland habitats or wide grassy tracks in woodland. In many places, they can be found flying together, although the Meadow Brown is characteristic of more open areas. Both are 'sedentary' and normally fly within or close to their breeding areas. Data from Monks Wood National Nature Reserve are compared with those from other sites in eastern England.

Methods

The recording technique is based on transect counts (Pollard 1977), which do not involve marking (cf Ch.4). A regular walk is made at each site around a fixed route, divided into sections which coincide with changes in the habitat. All butterflies encountered within fixed limits are noted, with recording conditions standardized as far as possible. Weather conditions must meet prescribed minima, and the time of recording is restricted to a period around the middle of the day. The season is divided into 'recording weeks' and lasts from April to September inclusive. From the weekly counts an index of abundance is calculated for each brood.

The use of the method depends on the assumption that changes in index values are closely related to changes in the actual abundance of butterflies. For some species, there is independent evidence that this is so (Pollard 1979*b*, Warren 1981). For the Meadow Brown and Ringlet there is no direct evidence, but the results themselves leave no room for doubt that the changes in index values are associated with changes in abundance (Fig. 5.1). The butterflies fly at the same time and the counts are likely to be similarly affected by weather conditions, yet the Meadow Brown index increased by a factor of 10 between 1973 and 1980, while the Ringlet index decreased by a similar factor over the same period. The pattern of weekly counts, showing a relatively smooth rise to a peak as butterflies emerge and then decline as they die, is also convincing. Major effects of weather conditions on counts would result in erratic fluctuations.

Results

The data provide information on phenology, and comparative studies can be made with other sites and other years. Information is also obtained on the distribution of counts around the transect routes and these can often be related to habitat characteristics. In this account, however, the main focus of interest is on abundance.

The Monks Wood site is just one of 80 at present in the national scheme in which the methods described here are used. The number of sites in eastern England increased from 7 in 1974 to 13 in 1980, with the main scheme starting in 1976. The data from these sites (excluding Monks Wood) are collated using the method of ratio estimates (Cochran 1963) to show trends over the period, and these trends can be used for comparison with the Monks Wood data (Fig. 5.2). For both species the patterns of annual fluctuations at Monks Wood are very similar to the collated regional fluctuations. This is an important point. Annual changes which occur over a wide area are almost certainly attributable, either directly or indirectly, to weather. Data for a number of other species show

The Biology of Butterflies
0-12-713750-5

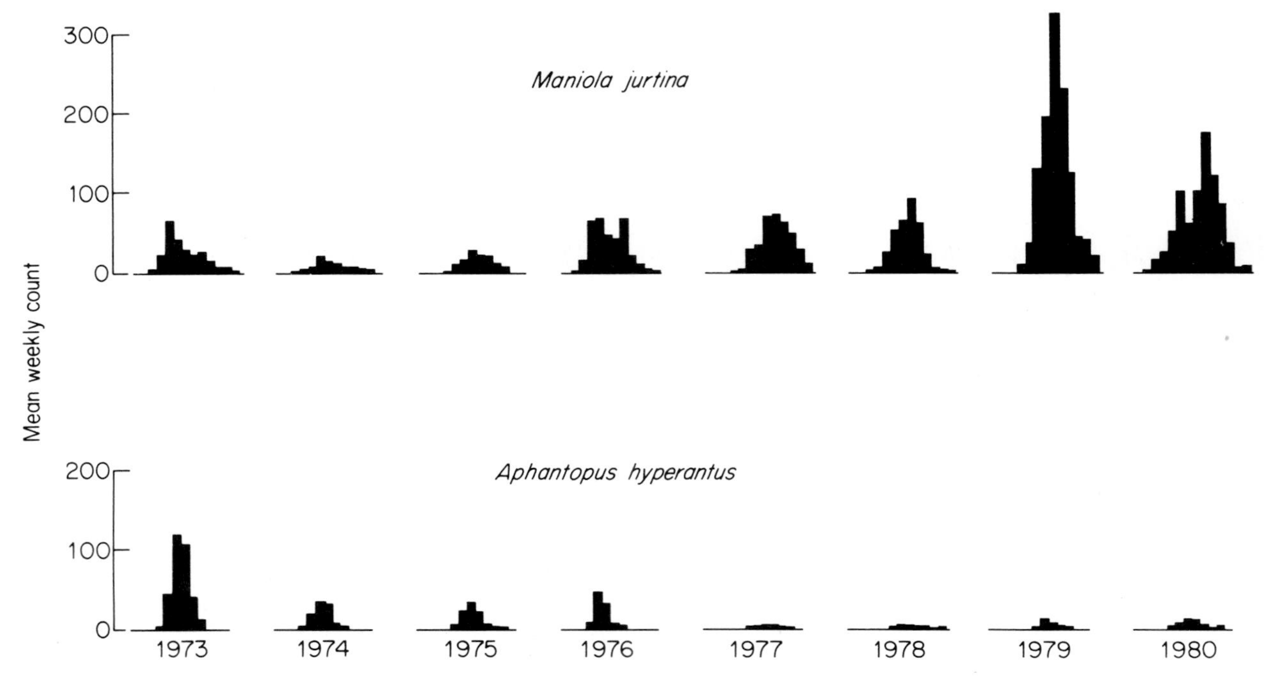

Figure 5.1. Mean weekly counts for the Meadow Brown *Maniola jurtina* and the Ringlet *Aphantopus hyparantus* at Monks Wood 1973-80.

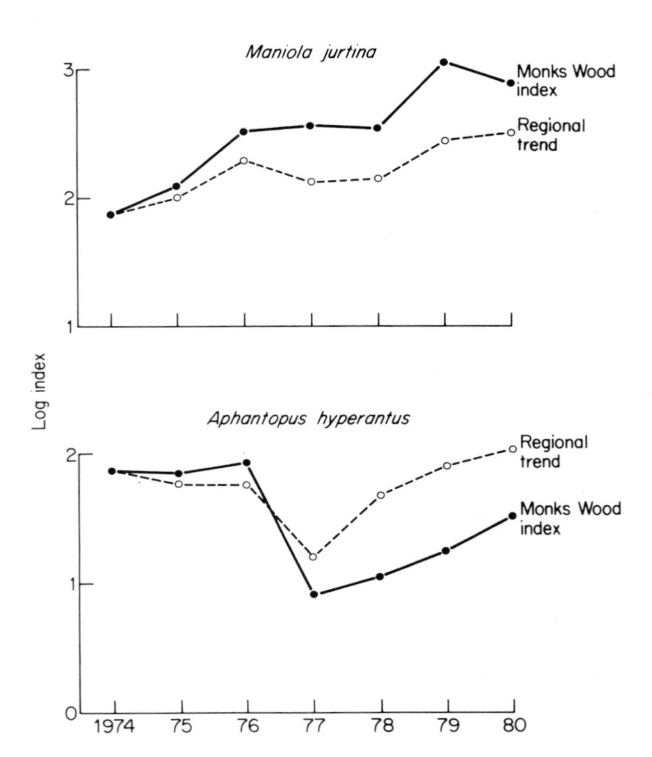

Fig. 5.2. Comparison of the index value for *Maniola jurtina* and *Aphantopus hyperantus* at Monks Wood with the collated results from eastern England.

similar close agreement over the region and this suggests that for many, probably most, butterflies in this area the annual fluctuations are caused by variable weather conditions. The details of the relationship with weather are not known and will require longer series of data before analysis will be possible. Several species, including the Ringlet, declined sharply in 1977, after the severe drought of 1976.

If the collated trend is taken to be representative of the region as a whole, the effect of habitat change at an individual site is shown by departures from the regional trend. The nature of the annual fluctuations remains the same but the population level may change because of changes in the carrying capacity of the habitat. In Fig. 5.2 the index values of the Meadow Brown can be seen to increase steadily in relation to the regional trend, while those of the Ringlet fell much more sharply in Monks Wood between 1976 and 1977 and remained much lower.

The division of the transect route into sections makes it possible to examine changes within the transect. Just one example is given here. Within Monks Wood are two fields, of 4.2 hectares (ha) and 1.7ha. The transect route runs through the larger of these fields. In the early 1970s, the fields were becoming overgrown with scrub, but a programme of winter scrub-cutting has effectively returned the area to open grassland. The recorded relative increase of the Meadow Brown in the field (Fig. 5.3) is in accord with its known association with open areas. On similar grounds the Ringlet would be expected to decline, but its extinction in the field in 1976-7 was surprising. In spite of the presence of colonies in nearby woodland rides, it has not subsequently recolonized. Mark and recapture data indicated that in 1975 the field

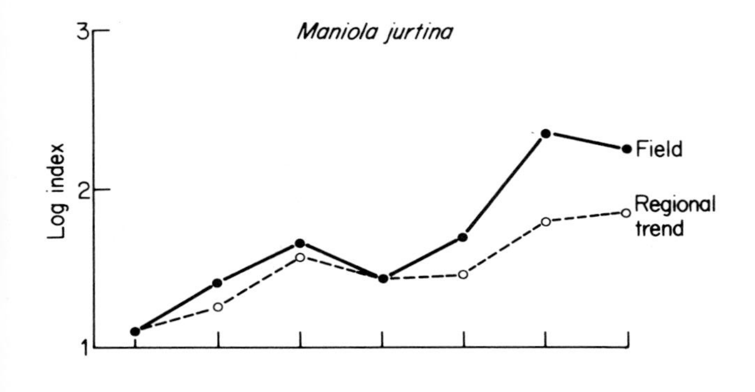

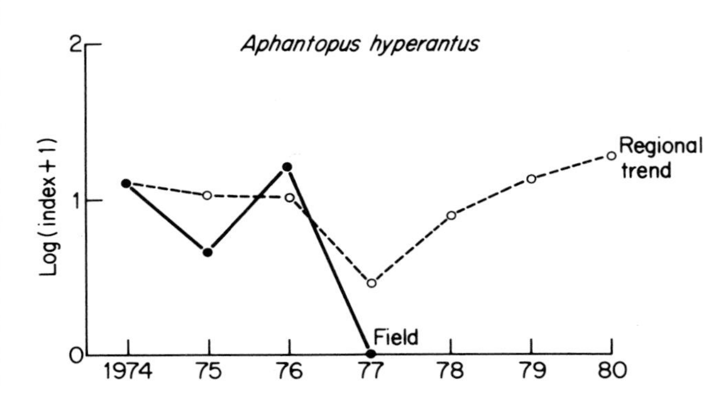

Fig. 5.3. Comparison of the index value for *Maniola jurtina* and *Aphantopus hyperantus* in the field at Monks Wood with the collated results from eastern England.

population was of the order of 500–1000 individuals (Pollard 1977). The loss of the field population of Ringlets may explain the sudden relative fall shown in 1976-7 in the site data as a whole and the subsequent failure to recover. The field may be able to support a population of Ringlets but has yet to be recolonized.

Discussion

The results from this study suggest that the population data for many butterfly species in Britain can be conveniently considered as made up of two components. The annual fluctuations, common to a large area, are almost certainly caused by direct or indirect effects of weather. The longer term changes in 'mean carrying capacity' at individual sites may be the result of long term management effects or of successional changes.

The example of the Ringlet, which was lost from the field in a year in which this species declined generally, shows the importance of annual fluctuations. In most population studies these annual fluctuations, conveniently analysed by *k* factor analysis of life table data (Varley & Gradwell 1960), are the main focus of interest. Sometimes these annual changes are the result of short term changes in carrying capacity (Dempster & Pollard 1981), and this was probably true of the effect of 1976 drought on the Ringlet and other species. Longer term trends in carrying capacity have, however, frequently been neglected in population studies, probably because the annual fluctuations are often so pronounced. Where the objective of a study is to assess the long term prospects of survival of a species, as in most conservation orientated studies, changes in long term carrying capacity are just as important as annual fluctuations.

These synoptic studies were begun with the intention of simple monitoring of population change. The data obtained suggests that they will also make a useful contribution to the understanding of the processes involved in the population ecology of butterflies.

Acknowledgements

I would like to thank the Nature Conservancy Council for financial support of the monitoring scheme, M. L. Hall and T. J. Bibby, who provided assistance in many ways, and the many recorders who have made the scheme possible.

Part III
The Food of Butterflies

6. *Egg-laying in Butterflies*

Frances S. Chew

Department of Biology, Tufts University, Medford, Massachusetts

and

Robert K. Robbins

Department of Entomology, Smithsonian Institution, Washington

When a female butterfly lays an egg, thereby abandoning her offspring, her behaviour is the result of physiological and ecological factors as well as evolutionary history. In this paper we present a broad overview of oviposition in butterflies with the goal of integrating into an evolutionary framework those factors that influence egg-laying. This is a massive subject; of necessity, we have covered only a few areas in depth. We emphasize multidisciplinary methodologies as the approach yielding the most complete understanding of oviposition. We refer readers to related reviews emphasizing optimal foraging methods (Rausher 1982*a*, Stanton 1982*b*, Miller & Strickler, in press; but see the criticisms of Gould & Lewontin 1979 to this approach) or focusing on the chemical interactions mediating oviposition (Feeny *et al.* 1983). We focus our discussion on four major questions: What factors influence (1) egg production and oviposition rate? (2) specificity of egg-laying site? (3) patterns of egg dispersion? and (4) evolution of oviposition specificity?

Patterns of Egg Production

The gross anatomy of female butterfly reproductive systems is similar to that of other Lepidoptera with two genital openings (Ditrysia): a ventro-posterior copulatory opening for mating and receiving the spermatophore (the bursa copulatrix), and a posterior ovipore for laying eggs (Stern & Smith 1960, Richards & Davies 1977, Hinton 1981). There are two ovaries, each composed of four ovarioles. Eight oviduct branches connect the ovarioles to a common oviduct that leads to the ovipore. The common oviduct contains openings to the spermatheca, where sperm are stored after they migrate via the ductus seminalis from the bursa copulatrix, and to the colleterial glands, which secrete the cement that glues eggs to the oviposition substrate, and which in some Lepidoptera may cover the eggs with a polysaccharide-rich layer called spumaline. Arnold & Fischer (1977) and de Jong (1978) discuss aspects of the functional anatomy of copulation and oviposition.

Male lepidopterans produce both ‘normal’ nucleated (eupyrene) sperm and anucleated (apyrene) sperm (Meves 1903). These sperm types undergo different spermatogenetic development and activation (Shepherd 1974*a*,*b*, Friedlander *et al.* 1981 and references therein). Apyrene and eupyrene sperm migrate from the corpus bursae of the female to separate spermathecal sacks (Holt & North 1970, Katsuno 1977) although occasionally the ‘apyrene sack’ is absent (J. G. Shepherd, pers. comm.). Apyrene sperm disintegrate in the spermatheca (Holt & North 1970, Riemann & Gassner 1973), and do not fertilize eggs (Friedlander & Gitay 1972). Although apyrene sperm occur in all lepidopterans tested (butterflies include *Pieris* (*Artogeia*) *rapae* and *Papilio polyxenes*, J. G. Shepherd, pers. comm.) as well as snails and rotifers (Friedlander 1975), their function is unclear. Physiological hypotheses are that they facilitate eupyrene migration to the spermatheca (Iriki 1941, Friedlander & Gitay

The Biology of Butterflies
0-12-713750-5

1972), and that they provide male contributed nutrition for the female (Riemann & Gassner 1973). A proposed evolutionary explanation for apyrene sperm is that they inactivate eupyrene sperm from previous matings (R. E. Silberglied & J. Shepherd, pers. comm.).

The following brief description of oogenesis in butterflies is summarized from a number of sources (Stern & Smith 1960, Telfer 1965, Ehrlich & Ehrlich 1978, Dunlap-Pianka 1979, Hinton 1981, Herman & Dallman 1981). Each ovariole contains a string of progressively enlarging and maturing follicles. Each follicle comprises a primary oocyte and seven nurse cells enclosed in a follicular envelope. As in other insect eggs, lepidopteran yolk protein appears to be synthesized as vitellogenin (probably a glycoprotein—Engelmann 1979, Aldrich *et al.* 1981) by the fat body, and taken up in the oocytes by pinocytosis from the female's haemolymph. As vitellogenesis progresses, nurse cells reach their maximal volume, then transfer all their cytoplasm to the oocyte. After completion of oocyte growth, the follicle cells secrete the chorion, and the mature oocyte ovulates from the follicular envelope into one of the eight branches of the oviduct. From the oviduct branch, the mature chorionated egg passes into the common oviduct. After fertilization, the egg then passes down the common oviduct, past the opening to the colleterial glands and through the ovipore. Recent descriptions of the external morphology of insect eggs include portfolios of scanning electron micrographs of chorion surfaces (Downey & Allyn 1980, 1981 and references therein, and Hinton 1981). Kafatos (1981) reviews developmental and genetic aspects of chorion protein synthesis.

Ovary maturation, oogenesis and vitellogenesis are markedly enhanced by juvenile hormone (Herman 1975, Pan & Wyatt 1971, 1976, Deb & Chakrovorty 1981, Herman & Dallman 1981; cf. Nicholson 1976). Other interactive factors include photoperiod (Herman 1975), temperature (Barker & Herman 1973), diapause (Edwards 1973, Herman 1981), mating (Herman & Barker 1977), and nutrition. Interactions between endocrinological and nutritional factors in butterflies are not well understood (Tojo *et al.* 1981). The nutritional maintenance regimes utilized by some endocrinologists do not seem sufficiently controlled to invite speculation (e.g. Herman & Dallman 1981).

Nutritional Correlates of Fecundity

Realized fecundity depends upon potential fecundity (the number of primary oocyte cells—e.g. Dunlap-Pianka *et al.* 1977) and various ecological factors, particularly those related to nutrition. The purpose of this section is to discuss ecological factors that affect actual fecundities. We first note, however, that measures of fecundity in the literature vary. Contents of ovarioles of field-caught females (Ehrlich & Ehrlich 1978), tabulations based on several methods (Hinton 1981) and comparison of various methods for one species (Yamanaka *et al.* 1978) are not equivalent.

Specialization of larvae for assimilation and growth enables them to accumulate limiting nutrients that may be used in the formation of eggs. Herbivorous larvae often assimilate nitrogen preferentially (in the form of amino acids or proteins—Scriber & Slansky 1981). Correlation between adult size and fecundity (Baker 1968*a*, references in Hinton 1981) suggests that larval nutrition may generally determine reserves available for oogenesis. Female adult insects tend to be heavier than males and often contain greater caloric (fat) content (Scriber & Slansky 1981).

Adult females may supplement larval nutrient reserves. Nutrients and water assimilated by adult females support metabolism for egg maturation (e.g. *Colias eurytheme*, Stern & Smith 1960). Some butterflies discriminate among potential nectar sources, and preferentially visit flowers that produce nectars rich in water, monosaccharides, and nitrogen-rich amino acids (e.g. *Colias* species, Watt *et al.* 1974; DeVries, *in litt.* reports that Costa Rican butterflies feed on parts of rotting fruits that are rich in amino acids). Females of some ithomiines follow army ants, and feed on ant-bird droppings, which are probably rich sources of nitrogen (Ray & Andrews 1980). Amino acids from pollen contribute to adult longevity and oogenesis in *Heliconius* (Gilbert 1975, Dunlap-Pianka *et al.* 1977, Boggs *et al.* 1981; see also Ch.3). Availability of sugars in the adult diet markedly enhances female longevity and number of egg clusters in *Euphydryas editha* (Ch.2); availability of amino acids, however, extends production of maximum-sized eggs in later clutches, in contrast to the otherwise approximate 10% decrease in egg size (Murphy 1981). Murphy's experiments, together with those of Telfer & Rutberg (1960), suggest that age-dependent decreases in egg size in cecropia moths, *Pararge aegeria* and *Pieris rapae* (Jones *et al.* 1982, Wiklund & Persson 1983, see below) reflect depletion of nitrogen stores. For butterflies whose adults eat nitrogen-rich foods, resources for egg production may not be limited by larval reserves (Ch.3).

Adult males may also provide females with nitrogen for vitellogenesis and egg maturation. Spermatophores contain protein that may be absorbed by females and contribute to maintenance and oogenesis (Boggs & Gilbert 1979, Boggs & Watt 1981). Subsequent matings may contribute to the fitness of later-mating males (because of sperm precedence—Labine 1964, Boggs & Watt 1981), and females, which may obtain a greater supply of protein. Boggs & Watt (1981) and Boggs & Gilbert (1979) suggest that male spermatophores may represent an important nutritional resource for butterflies which do not manipulate pollen sources as supplements of larval nitrogen reserves (e.g. *Colias*, *Danaus plexippus*).

Realized fecundity may be uncorrelated with nutritional status in temperate areas where cold weather unsuitable for flight activity occurs frequently (Douwes 1976*a*, Gossard & Jones 1977, Courtney 1980, Wiklund & Persson 1983). For example, Wiklund & Persson (1983) found that individual fecundities for *Pararge aegeria* were uncorrelated with adult size or number of spermatophores received, but were strongly correlated with the amount of time a female was able to fly and oviposit.

Egg Size

Female age may affect weight or size of eggs produced. Telfer & Rutberg (1960), Murphy (1981) and Jones *et al.* (1982) noted that eggs produced by individuals become smaller as females age. Wiklund & Persson (1983) observed a similar progression in *Pararge aegeria*, but interestingly, demonstrated that egg weight is not correlated with hatchability, mortality, larval growth rate, or eventual pupal weight. For *Euphydryas editha*, however, there appears to be an egg weight threshold below which development is abnormal (D. D. Murphy, pers. comm.).

Egg size varies substantially among butterfly species with comparable adult body sizes. Labine (1968) found a twenty-fold range in egg volume among four species. Egg weights or volumes of congeners, however, apparently differ by less than a factor of two (*Heliconius*, Dunlap-Pianka 1979; *Parides*, *Papilio*, *Siproeta* [as *Victorina*], Young 1972). More approximate estimates, extrapolated from measurements of single dimensions (e.g. egg height) are consistent with this trend. (Such egg measurements are widely scattered in the literature; see e.g. Scudder 1889, Chew 1981.)

Species that produce small eggs can be more fecund, but often produce a smaller total volume or weight of eggs than do species that produce larger eggs (Labine 1968, Young 1972). Because adults of fecund species may be of comparable size to those of less fecund species, smaller total egg volumes or weights may reflect proportionally smaller energy allocation to reproductive effort. However, because measures of reproductive effort must account for possible supplementation of larval reserves by adult nutrient intake (e.g. Boggs 1981*a*), accurate estimates of reproductive effort will require additional data on adult feeding habits and energy expenditure.

Egg size may be related to number of larval instars and adult size. Dyar (1890) noted that larval head capsule widths of some 28 lepidopteran species increase 1.2 to 1.4 times between instars; this increase approximates a doubling of volume in each instar. The significance of this finding is that doubling egg size theoretically allows a species to eliminate one instar while producing the same sized adult. To the extent that Dyar's law is valid (see Richards & Davies 1977 for exceptions, Richards 1949 and Wright & Clarke 1981 for factors affecting size changes between instars), this theoretical deduction may provide a convenient 'rule of thumb' for investigating the relations among egg size, number of instars, and adult size. Species with variable numbers of instars, such as some brassolids (Aiello & Silberglied 1979), would be particularly appropriate experimental animals for such a study. To our knowledge, these relationships have not been explored, but the foregoing discussion on nutritional influences suggests that the relationships may be complex.

Rate of Oviposition

Reproductive patterns involve consideration not only of the number and size of eggs, but also egg distribution over time and space (onset of oviposition, oviposition rate, cluster size—Labine 1968). We consider here factors that influence egg distribution over time and defer to a later section discussion of spatial aspects of egg distribution. Gossard & Jones (1977) distinguish between factors affecting the rate of oogenesis and those affecting the rate at which eggs are laid. Butterflies with different egg-laying rates may require about the same amount of time to mature single eggs, but have different numbers of simultaneously maturing oocytes (Dunlap-Pianka 1979).

Available data on rate of oviposition (records of eggs produced by individual females per unit time under various conditions—e.g. Yamanaka *et al.* 1978, Yamamoto & Ohtani 1979) during adult lifespans appears to form three general categories. We describe them as distinct patterns but they may well be modes of a continuum. First, females may eclose with large numbers of mature eggs in their oviducts, and begin laying eggs very soon after mating (e.g. *Euphydryas editha*, Labine 1968). The presence of a very large egg load in the oviducts at the time of eclosion would probably preclude much pre-oviposition dispersal by females (Labine 1968). Second, females may eclose with few mature eggs, but oocytes quickly mature after the female mates and begins to nectar, e.g. *Pieris rapae*, *Colias eurytheme*, *Agraulis vanillae* and *Anartia fatima* (respectively, Gossard & Jones 1977, Yamamoto & Ohtani 1979, Stern & Smith 1960, Dunlap-Pianka *et al.* 1977; R. E. Silberglied, pers. comm.). These butterflies, which are often widely-dispersing, produce relatively few eggs during the first few hours or days of adult life, but oviposition rates quickly rise to a peak and then sharply decline after a few days. The third pattern is characterized by oviposition rates that remain constant for comparatively long periods. These butterflies, e.g. *Heliconius*, produce fewer, usually somewhat larger eggs during their lives than those in the first two patterns, and tend to have longer active lives.

These generalized oviposition patterns may vary among congeners. For example, while *Pieris rapae* exemplifies the second grouping, *P. napi nesis* produces eggs in a pattern that more closely resembles that of *H. charitonius* in the third grouping (Dunlap-Pianka 1979, Yamamoto & Ohtani 1979).

Scales on Eggs

Females of some lycaenid and hesperiid species have tufts of scales (at the posterior end of the abdomen) that adhere to eggs as they are laid. Lycaenid records include *Chaetoprocta odata*, *Nordmania acaciae*, *N. myrtale*, *Crudaria leroma*, *Phasis wallengrenii* and *P. argyrophaga* (de Nicéville 1890, Clark & Dickson 1971, Nakamura 1976). Females of *Pseudaletis* have large tufts of anal scales that presumably adhere to eggs (Stempffer 1967). Scales and dust accumulate on the eggs of *Japonica lutea* and *J. saepestriata* (Shirôzu & Hara 1960–1962, Nakamura 1976), but the females lack conspicuous tufts of abdominal scales. Hesperiid records include *Daimio tethys*, *Tagiades trebellius*, *T. litigiosus*, and *Matapa aria* (Shirôzu & Hara 1960–1962, Johnston & Johnston 1980). These species represent several thecline tribes and two hesperiid subfamilies, indicating repeated evolution of the phenomenon. Hinton (1981) summarizes records among moths.

Proposed explanations for the occurrence of adherent scales on butterfly eggs include protective and nutritive functions. The scales might shield the egg like armour, camouflage it, or emit a chemical that deters parasitoids (de Nicéville 1890, Nakamura 1976, Downey & Allyn 1981). Alternately, these scales may be nutritionally important to newly hatched larvae. Downey & Allyn (1981) state that larvae of lycaenid species with adherent scales generally consume the entire eggshell after hatching, in contrast to the usual lycaenid behaviour of eating only a small exit hole. Larvae of *C. leroma*, however, do not eat their eggshells (Clark & Dickson 1971). As mentioned by Downey & Allyn (1981), removal and addition experiments are required to determine the functions of these specialized scales.

Specificity of Oviposition Site

The idea that taxonomic specificity of butterfly-angiosperm associations is mediated by plant secondary compounds (e.g. Fraenkel 1959) has strongly influenced studies of determinants of oviposition specificity. Because many studies on larvae implicate plant secondary compounds as important determinants of larval taxonomic specificity, for physiological (e.g. Erickson & Feeny 1974) as well as behavioural reasons (e.g. Waldbauer 1968), the conspicuous, 'characteristic' compounds associated with individual plant families have been the focus of many investigations of oviposition cues. However, determinants of larval foodplant suitability and determinants of oviposition specificity need not be the same (see e.g. Courtney 1981, 1982*a*, who suggests that foodplant suitability of the crucifer *Hesperis matronalis* (Brassicaceae) for *Anthocharis cardamines* larvae is determined primarily by mechanical defences while ovipositing females are attracted to visually conspicuous, blooming flowers). Ovipositing females perceive plants using many sensory modalities. Any stimulus, if correlated with first instar larval success in locating suitable foodplants, may provide a cue for oviposition sites.

Oviposition sites are not equivalent to larval foodplants because some apparently 'mistaken' choices are normal behaviour, even if they are imprecise assessments of what is suitable for larval development. Examples of apparent 'mistakes' include oviposition on the following: non-plant substrates (Dethier 1959*a* and references therein; Singer *et al.* 1971); withered, or otherwise unsuitable plant parts (Tutt 1899, Wiklund 1977*b*); plants that do not support larval development and that are taxonomically and phytochemically unrelated to suitable larval hostplants (Dethier 1959*a*, Singer 1971, Chew 1977); plants that do not support larval development, but which are taxonomically and phytochemically related to suitable larval hostplants (Hefley 1937, Straatman 1962, Sevastopulo 1964, Smiley 1978*a*, Rodman & Chew 1980, Berenbaum 1981); failure to discriminate among plants suitable for larvae, but whose ingestion results in significantly variable developmental times or adult weights (Dolinger *et al.* 1973, Chew 1975, 1977); suitable plants of insufficient size to support complete larval development (Dethier 1959*a*, White 1974, Chew 1977); ignoring specific habitats in which suitable larval hostplants grow (Petersen 1954, Owen 1959 and references therein, Shapiro & Cardé 1970, Singer 1971, Benson 1978 and references therein, Smiley 1978*a*, Ohsaki 1979, Courtney 1980, Chew 1981, Jordan 1981); ignoring plant species suitable for larval growth that occur in habitats where adults fly (Emmel & Emmel 1969, Gilbert & Singer 1975 and references therein, Neck 1973, 1977). But many 'mistakes' do not necessarily have dire consequences, and many represent normal oviposition behaviour for the specific butterfly species involved (see Tutt 1899; Ch.7). They do, however, provide a tool for probing oviposition processes in much the same way that mutations are routinely used to reveal genetic processes. They also emphasize the necessity of studying other life-history stages to assess the role of adult choice in the nutrition of larval offspring.

Measuring Preferences

To study oviposition specificity, we must be able to assess butterfly preference among potential oviposition sites. Methods for detecting and quantifying preference are extensively discussed in literature on foraging behaviour (e.g. Pyke *et al.* 1977). These methods usually compare the frequency with which potential resources are chosen (Berube 1972, Chew 1977, Jaenike 1980) or observe how encounters with particular resources affect subsequent searching behaviour (Jones 1977*b*, Stanton 1982*a*). We refer readers to discussions of specific methodologies (e.g. Stanton 1982*a*, Singer 1982*a*).

Sampling procedures may bias measures of oviposition preferences. Variation in the plant species chosen by individuals has been documented in laboratory or cage tests (Singer 1971, Wiklund 1981, Tabashnik *et al.* 1981) and in the field (Singer 1982*a*, Chew 1977, Rausher 1978). These observations are based on obtaining a limited sample of behaviour from many 'randomly chosen' individuals rather than obtaining a large sample of behaviour from each chosen individual (cf. Douwes 1968; Tabashnik *et al.* 1981, Wiklund 1981). This procedure may underestimate the variance in an individual's behaviour and correspondingly overestimate the variance among individuals. For example, the number of crucifer species chosen by ovipositing *Pieris* increases with the length of the observed oviposition sequence; differences among individuals are probably artifacts of the small number of oviposition choices recorded for each individual (Chew 1977).

Observed preferences may vary with the motivational state of females, season, or geographic location. Females that carry heavy egg loads or have not recently oviposited, may more readily accept a variety of oviposition substrates (Singer 1971, 1982*a*, Jones 1977*b*, Wiklund 1981). Oviposition preferences may vary within a season or between seasons (Rausher 1978, 1981*a*, Stanton 1980, 1982a). Geographic variation in flight behaviour and oviposition choice has been documented in *Euphydryas editha* (Singer 1971, Gilbert & Singer 1973, White & Singer 1974). This variation suggests that care is necessary in generalizing results from studies of single populations.

Secondary Plant Compounds as Determinants of Specificity

Although the clear roles prematurely ascribed to secondary plant compounds in herbivore-plant interactions by early work (e.g. Verschaffelt 1911) have now been rendered more complex by subsequent findings, they nonetheless provided impetus for examining butterfly phytochemical specificity. Many papers written since 1950 on foodplant choice by butterflies and other herbivores are steeped in corollaries of insect-plant coevolution and spiced with the diverse flavours and aromas that angiosperms produce (Dethier 1970, Feeny 1975). As a result, we know more about butterfly response to secondary plant compounds than about other determinants of specificity. However, although some butterflies oviposit in response to single compounds (or classes of compounds, e.g. *Pieris brassicae* in response to glucosinolates, David & Gardiner 1962), such unequivocal results probably represent isolated cases.

Perception of secondary compounds

In adults and caterpillars, peripheral chemoreceptors may respond to specific compounds or to a variety of stimuli. Specialized receptors appear to be restricted to a few types, which vary among species and which are correlated with the known feeding habits of particular species. For example, *P. brassicae* has specialized receptors for glucosinolates (Ma & Schoonhoven 1973; Dethier 1978, 1980*a* gives other examples).

Most contact chemoreceptors and antennal olfactory receptors, however, respond to a wide variety of stimuli. Individuals possess many generalized receptors that differ in their sensitivities and electrophysiological responses to given stimuli. The greater number of these different receptor types in herbivorous caterpillars and their adults enables these insects to perceive a large diversity of compounds, so giving them an expanded sensitivity when compared, for example, with the carnivorous dipteran *Phormia* (Dethier 1980*a*)—whose limited receptors make it 'blind' to many stimuli. A similar positive correlation exists between number of receptor types and diet breadth in grasshoppers (Chapman & Blaney 1979 and references therein).

Females may perceive foodplants chemically before alighting (Minnich 1924). Various lepidopterans respond differentially to potential larval foodplants before contact (Hovanitz & Chang 1964, Douwes 1968, Vaidya 1969*a*,*b*, Berube 1970, Rothschild & Schoonhoven 1977, Mitchell 1978); the chemical basis for differential response, however, is little explored (see further below). Female *P. brassicae* have olfactory sensilla on most distal flagellar segments of the antennae (Behan & Schoonhoven 1978; Den Otter *et al.* 1980).

Once a female alights, contact chemoreceptors respond to plant compounds that may stimulate or inhibit oviposition (e.g. Lundgren 1975). Females may tap or 'drum' the plant surface with their foreleg tarsi (Ilse 1956, Fox 1966, Myers 1969, Vaidya 1969*b*, Calvert 1974; see also Calvert & Hanson 1983). Fox (1966) suggested that combinations of leaf-abrading spines and chemosensory hairs on foretarsi permit females to detect plant compounds. Both 'B-type' chemosensory hairs of the Pieridae (Ma & Schoonhoven 1973) and clustered trichoid sensilla found in the Nymphalidae, which Calvert (1974) believes are chemosensory, are paired with spines on proximally adjacent tarsomeres. A female may also probe plant surfaces using her proboscis (Platt 1979, DeVries, 1983). Trichoid sensilla have been described on the proboscis of lepidopterans (Frings & Frings 1949).

Homologies between tarsal sensory receptors in different species are unclear. Sensory receptors in *Chlosyne lacinia* include clustered trichoid sensilla, approximately 13 pairs of other structures that show chemoreceptor-like activity as determined by electrophysiological responses, and 'pit setae', which point towards the terminus of the foretarsus (Calvert 1974). In *P. brassicae*, 'A-type' tactile bristles are oriented distally, and may be homologous to Calvert's 'pit setae'. 'B-type' chemosensory receptors in *P. brassicae* are similar in size to the 13 paired 'chemosensory' hairs in *C. lacinia* (Ma & Schoonhoven 1973).

Prior to oviposition, some butterflies curl their abdomens (Saxena & Goyal 1978) and/or dip their antennae towards the plant (Calvert 1974). Contact receptors occur on antennal tips and on ovipositors of pyralids (e.g. *Chilo partellus*, Chadha & Roome 1980). Functionally similar sensilla on the ovipositor valves of *P. brassicae* respond electrophysiologically to oviposition deterrent pheromone (Klijnstra 1982).

Compared with the large number of phytochemical compounds to which female butterflies are potentially exposed, the actual number of chemoreceptor sensilla is relatively small. This limited number involves redundancy, however, because removal of antennae, fore-, meso-, or metathoracic tarsi (but not all tarsi simultaneously) does not inhibit oviposition in *P. brassicae* (Ma & Schoonhoven 1973); Myers (1969) obtained similar results for tarsi of *Danaus gilippus berenice*. Whether such removal compromises the butterfly's ability to differentiate among potential oviposition substrates is not clear.

Deciphering the gustatory and olfactory 'coding' of complex response profiles of lepidopterans has proved difficult. Specialized cells, for example the G (glucosinolate) cells in tarsal chemoreceptors of *P. brassicae*, response to increasing concentrations of glucosinolates by increasing their firing rates (Ma & Schoonhoven 1973). However, mixtures stimulate responses that are difficult to interpret. Plants that are behaviourally rejected or accepted (by lepidopteran larvae) induce no standard electrophysiological responses correlated with acceptance or rejection. Further, plants accepted by two caterpillar species stimulate different electrophysiological responses in each (Dethier 1973). Failure to find universal modalities for rejection (or acceptance) is not surprising given the variety of reasons for which a plant might be rejected (e.g. presence of deterrent, lack of attractant—Dethier 1973). When responses to chemically-similar compounds are studied, electroantennograms (which represent summations of responses of many antennal olfactory cells) may be useful in comparing receptor sensitivities (Behan & Schoonhoven 1978, Den Otter *et al.* 1980). However, the variation among responses of individuals (Ma & Schoonhoven 1973, Schoonhoven 1977) does not encourage us that this 'coding' in peripheral receptor output will be easily solved. On the other hand, to the extent that the wide variation observed in electrophysiological responses correlates with variation in individuals' behaviours, these inconsistencies may reflect ecologically and evolutionarily significant differences among individuals.

Isolating biologically active compounds

Correctly identifying the chemical group responsible for behavioural variation can be a problem in research on secondary compounds. Although early work with *Pieris brassicae* suggested that single compounds could attract butterflies and elicit oviposition (David & Gardiner 1962), compounds that stimulate oviposition need not be the same ones that attract females (Rothschild & Fairbairn 1980, Feeny *et al.* 1983). For example, the Apiaceae and families related to these umbels contain essential oils, some of which are attractive to larval *Papilio* (Dethier 1941). However, attempts to isolate and identify the oviposition stimulant in foodplants of *Papilio* species suggests that active fractions do not contain essential oils, although essential oils may initially attract females to the plants (Feeny *et al.* 1983; see also Rothschild & Fairbairn 1980). Feeny *et al.* (1983) suggest that the active components of biologically active fractions may be compounds that are widely distributed botanically, rather than unique secondary compounds (see also Jones *et al.* 1970, who suggest this is the case for the moth *Heliothis zea*). The oviposition 'mistakes' observed by Berenbaum (1981) suggest that coumarins or biosynthetically related compounds, rather than the somewhat more conspicuous essential oils, may be responsible for variation in *Papilio* responses to these plants. At issue are problems of separating correlative from causative relationships, and testing possible synergistic interactions among compounds. These observations warn us against uncritically assuming, without appropriate evidence, that specific, conspicuous compounds are involved in choice behaviour by particular butterflies.

Response to variation within chemical classes

Of primary interest is the hypothesis that variation among compounds belonging to a single class (i.e. sharing the same chemical functional group, with many R-group substituents) are the bases for differentiation of potential foodplants (e.g. Chew & Rodman 1979). Evidence for this hypothesis rests on a relatively small number of cases in which both behavioural and chemical aspects of oviposition choice have been examined.

The most extensive evidence that butterflies respond to variation among members of a single class of secondary plant compounds concerns pierids and their cruciferous foodplants. Hovanitz & Chang (1963) provided early evidence that in laboratory tests, *Pieris rapae* females respond differentially to extracts of different Brassicaceae. These observations suggested that *P. rapae* females detect variation in both glucosinolates (mustard oil glucosides) and isothiocyanates (mustard oils, one of the aglycone products of enzymatic degradation of glucosinolates) in extracts of crushed plants. However, Ma & Schoonhoven (1973) detected electrophysiological responses in the tarsal chemoreceptors to individual glucosinolates, but not to isothiocyanates. These electrophysiological observations do not exclude the possibility that olfactory receptors (Behan & Schoonhoven 1978) respond differentially to R-group substituents of isothiocyanates, but to our knowledge, the possibility has not been pursued. Chew and Rodman (Chew 1975, 1977, 1980, Rodman & Chew 1980) have provided correlational evidence that *Pieris occidentalis* and *Pieris napi macdunnoughii* respond to variation among glucosinolates in the field.

Another set of data on responses to secondary plant metabolites comes from work on pierids and lycaenids that use leguminous foodplants. Stanton (1979) tested the behavioural responses of *Colias* butterflies in the laboratory to extracts of various legumes. The butterflies distinguish among legumes and legume extracts at contact range (i.e. stimulation was gustatory rather than olfactory). Females of *Glaucopsyche lygdamus* distinguish among perennial *Lupinus* species (Breedlove & Ehrlich 1968, 1972, Dolinger *et al.* 1973) in localities where plant species are characterized by invariant species-specific alkaloid profiles. However, where plant populations are characterized by considerable intraspecific variation, butterflies seem unable to distinguish those species that support larval survival from those that do not. It is not clear whether these butterflies sometimes are unable to detect differences among R-group substitutents of the alkaloids, or whether their perceptual ability to distinguish these compounds sometimes was rendered useless by the low correlation between alkaloid profile and larval foodplant suitability.

Other Cues as Determinants of Specificity

Because ovipositing butterflies possess a number of sensory modalities, they may utilize different plant characteristics as oviposition cues at different distances from the plant, or during successive parts of an oviposition sequence (Courtney 1980, Stanton 1980). Many authors (e.g. Ives 1978, Wolfson 1980, and Jordan 1981 and references therein) show that ovipositing pierids discriminate among plants and plant parts on the basis of factors such as water content of leaves (important for larval growth efficiency—Scriber & Slansky 1981). The presence of ants may stimulate myrmecophilous species to oviposit (Atsatt 1981*a*). Colour and shape of plants, particularly leaves, are conspicuous distinguishing features. Ilse (1937) provided early evidence that ovipositing *Pieris brassicae* are attracted to certain shades of green, a finding corroborated for *Papilio demoleus* (Vaidya 1969*a*,*b*, Saxena & Goyal 1978).

Recent work has established the importance of leaf shape as an oviposition cue. In *Heliconius*, ovipositing butterflies select larval foodplants for their offspring by leaf shape, which varies substantially among *Passiflora* species in each community. The butterflies thus exploit, and possibly engender, the diversity of leaf shapes that characterize *Passiflora* (Gilbert 1975, Benson 1978).

In *Battus philenor*, ovipositing females utilize leaf shape to distinguish *Aristolochia* species, whose suitability for larvae varies seasonally (Rausher 1978, 1979*a,b*, 1980). Females that consistently oviposited on *Aristolochia* plants, whose leaves were a particular shape, more frequently approached leaves (of any plant species) of similar shape. Upon alighting on visually chosen plants, however, females may reject the plant, presumably after assessing phytochemical cues.

Colias butterflies similarly utilize leaf shape to distinguish potential larval foodplants (Stanton 1980, 1982*a*). Females sometimes were confused by legumes of similar leaf shape (but of differing suitability for larval development) during approaches to plants. However, gustatory and olfactory stimuli at close range are probably responsible for differences in acceptance after alighting (Stanton 1979).

Behavioural responses to leaf shape may vary with vegetational characteristics of the habitat, and may depend on the degree of size, shape, colour, or textural contrast between foodplant species and the surrounding vegetation (Rausher 1982*a*, Stanton 1982*b*). Ovipositing *Euphydryas editha* display different behaviours in adjacent habitats with different densities and species of potential foodplants. In one habitat, searching behaviour resembles that of *B. philenor* and *C. alexandra*: females encounter preferred plant species more frequently than expected on a random basis by flying to areas where the preferred species is more abundant. In the other, females encounter potential foodplant species in proportion to their relative abundance (Mackay, *in litt.* 1982).

That the preferences of females for plants with different leaf shapes can change within seasons (Rausher 1978, 1981*a*) or between seasons (Stanton 1982*a*) or localities (Stanton 1980), suggests that female butterflies learn to associate leaf shape with other plant characteristics. They may then use these conspicuous visual cues to increase searching efficiency in diverse natural communities. Such associative learning (Traynier 1979) may enable butterflies to locate larval foodplants more efficiently, while retaining ability to later reassess the suitability of potential foodplants in a community. Visual cues (involving leaf shape) appear to enhance the rates at which *Battus philenor* and *Colias alexandra* encounter potential foodplants (Rausher 1978, Stanton 1982*a*); similar results involving flowers were obtained for *Anthocharis cardamines* (Courtney 1980). Similar responses have been observed in nectaring butterflies, which exhibit flower constancy, and for which visual cues of floral colour and morphology are strongly associated with nectar of given composition (see e.g. Watt *et al.* 1974, Levin & Berube 1972).

Further evidence for specificity based on visual cues involves female recognition of previously laid conspecific eggs, and plant 'egg-mimics'. Pierids (Rothschild & Schoonhoven 1977, Shapiro 1981*c,d*), heliconiines (Gilbert 1975), and papilionids (Rausher 1979*a*) avoid ovipositing on plants carrying confamilial or conspecific eggs (see also Ch.3). Some of this response is chemically mediated (Behan & Schoonhoven 1978, Schoonhoven *et al.* 1981), but the presence and efficacy of similarly shaped and coloured 'egg-mimics' in preventing oviposition on plants (Shapiro 1981*c, d*, Williams & Gilbert 1981) indicates the importance of visual stimuli.

Female preference may be influenced by the presence of plant parts on which larvae preferentially feed. Plant parts include leaves of a specific age (Jordan 1981), apical meristems (Benson *et al.* 1976) and reproductive organs (Downey 1962*a*, Breedlove & Ehrlich 1968). An unusual example is two lycaenids, *Azanus jesous* (Clark & Dickson 1971 and *Celastrina lucia* (Robbins unpublished), that occasionally oviposit on galls on their normal foodplants; the larvae feed on the galls.

The predilection of butterflies for ovipositing in specific habitats has been widely observed and, in some cases, is attributable to specific causes. Among these are adequate density of larval foodplants (Courtney 1981, Smiley 1978*a*), avoidance of predators (Singer 1971 and references in Gilbert & Singer 1975), and adult preference for specific environmental conditions (Petersen 1954, Singer 1971, Cromartie 1975*b*, Williams 1981). In other cases, however, the sighting of butterflies is not necessarily reliable information on where eggs are deposited (Chew 1981). Habitat selection (Ch.8), rather than partitioning of potential larval foodplants, appears to be responsible for maintaining reduced potential competition between closely related butterfly species (Gilbert & Singer 1975, Shapiro & Cardé 1970, Chew 1981). In other cases, however (e.g. Emmel & Emmel 1969 on *Papilio* species), butterflies that fly in the same habitats partition larval foodplant species even though larvae of both species develop normally on all potential foodplants. Larvae compete for the same foodplants only in years when environmental deterioration renders a large proportion of the foodplants unsuitable for larval growth.

Oviposition Cues and Larval Polyphagy

Although conspicuous plant secondary compounds stimulate oviposition behaviour in some butterflies, these chemicals may lack a stimulatory role in species whose larvae exhibit extreme taxonomic polyphagy. In some cases where insects have restricted feeding habits, such as leaf mining or boring into buds, plant anatomy rather than plant phytochemistry or phylogenetic relationships may primarily determine which plant tissues are acceptable as food (van Emden 1978, Powell 1980). In other cases, chemicals present in virtually all plants may stimulate oviposition (Jones *et al.* 1970), but deterrents distributed more or less at random among plant taxa determine the range of foodplants that are used (Thorsteinson 1960). Perception of deterrent compounds by larvae (and presumably by ovipositing females) is postulated to occur in central processing, not in peripheral sensory receptors (Dethier 1980*a*). As a result, restriction of the range of foodplants might evolve more quickly than if specific receptors were needed to perceive each potential deterrent.

Butterfly oviposition and larval feeding on plants in more than three or four taxonomically disparate families apparently only occurs with regularity among the species-rich (over 1000 species) primarily Neotropical eumaeine hairstreaks (Lycaenidae). Many eumaeines are strikingly more polyphagous than other butterflies (Robbins & Aiello 1982; an extensive list of records is in preparation) with the exception of some true nymphalids (Ehrlich & Raven 1965). This taxonomic polyphagy is correlated with oviposition on and larval boring into plant reproductive tissues (especially flower-buds and flowers) rather than feeding on foliage. It is not correlated with the presence of ants; obligate myrmecophily is unrecorded among eumaeines (Robbins unpublished).

Lycaenids that feed on buds, flowers, developing seeds and fruits eat food with a greater ratio of amino acids and protein to alkaloids or other secondary compounds than if they fed on foliage. Developing ovules and reproductive structure generally contain higher concentrations of proteins and amino acids than do leaves (McNeill & Southwood 1978, Mattson 1980). The higher assimilation efficiencies documented for insects feeding on plants with high nitrogen levels (e.g. McNeill & Southwood 1978) suggests that lycaenid larvae feeding on such tissue might require smaller quantities, and complete development with less exposure to alkaloids in their diets. Also, meristematic tissues often contain no or small vacuoles rather than large, well-organized vacuoles often found in foliage (Esau 1965, Matile 1976). Very young cells that have not yet become vacuolized contain small concentrations of compounds such as alkaloids than do more mature tissues (Robinson 1979). We consider it likely that a broad-range stimulant (such as the shape of buds or flowers), or a chemical or chemical composition common to young reproductive tissues (cf. Feeny *et al.* 1982, Jones *et al.* 1970), combined with an absence of deterrent compounds that might be sequestered in well-organized vacuoles, make these reproductive plant parts attractive to ovipositing lycaenid females.

Spatial Distribution of Eggs

Dispersion of Eggs

Distribution of eggs on foodplants may be classified as clumped, random or regular based upon the ratio of the variance to the mean in numbers of eggs per plant, branch or quadrat (see Myers 1978 for measures of dispersion). These patterns may result from female assessments of foodplant quality or of 'egg load', or may be a statistical result of female movements and plant spatial distributions. Some of the most detailed knowledge of oviposition patterns comes from work on *Pieris rapae* in agricultural plots (Harcourt 1961, Kobayashi 1965, 1966, Jones 1977*b*, Ives 1978, Jones & Ives 1979). Egg distributions in agricultural plots are clumped, and can be fitted rather well to negative binomial distributions (but see Myers 1978). Jones (1977*b*) successfully modelled oviposition behaviour of *P. rapae*. Her model shows that egg distributions, including 'edge effects' (more eggs on peripheral plants) can be explained for this species in agricultural plots as a statistical result of female movement patterns.

Eggs of *Pieris napi microstriata*, *Euphydryas editha*, and *Euptychia libye* also have clumped distributions (Shapiro 1980*c*, Rausher *et al.* 1981, Singer & Mandraccia 1982, respectively). These distributions are the result, in part, of a preponderance of eggs being laid on peripheral and isolated plants. Several explanations may account for this phenomenon (see also Ch.3). When a female butterfly picks the nearest plant from a random point (her location), this choice process results in isolated plants being chosen more frequently than would be expected on a random basis (Pielou 1969, Mackay & Singer 1982). Isolated plants may also be easier to see from a distance (e.g. Cromartie 1975*a*,*b*). Additionally, the preponderance of eggs on peripheral or isolated plants may result from female movements as they cross from nectaring to oviposition areas, or from one habitat to another (Kobayashi 1965, 1966, Gilbert & Singer 1973, Wiklund 1977*a*, Courtney & Courtney 1982).

Eggs or other immature stages on a particular plant inhibit oviposition of some species (Urquhart 1960, Gilbert 1975, Rothschild *et al.* 1975*a*). This assessment behaviour usually results in a regular distribution of eggs. Gilbert (1975) reported that some heliconiines do not oviposit if they see an egg on a potential foodplant. Rothschild & Schoonhoven (1977) supplied laboratory evidence that visual cues, olfactory stimuli, and tactile chemosensory input from previously laid clusters inhibit female *Pieris brassicae* from ovipositing. They found similar patterns for *Danaus plexippus* and *P. rapae*, but egg distributions of *P. rapae* on other continents and *P. napi* in North America show no evidence of egg-load assessment (Jones & Ives 1979, Chew, unpublished data). Larval frass and macerated larval foodplant (cabbage) are reported to deter oviposition in some lepidopteran pests (e.g. *Trichoplusia ni*, Renwick & Radke 1980, 1981). Shapiro (1980*c*, 1981*c,d*) has shown that some pierids assess egg loads visually, and that the resultant egg distributions are statistically regular. He experimentally demonstrated that those species that lay visually conspicuous red eggs, or whose eggs turn red a few hours after oviposition, are inhibited from ovipositing on plants or branchs with red eggs. Dispersion of eggs may ensure adequate larval food supply and prevent cannibalism.

Some plants have evolved 'egg-mimics' to deter females from ovipositing (see Gilbert 1975, Shapiro 1981*d*, Williams & Gilbert 1981, and Ch.3).

Cluster Laying

Females of many butterfly species lay their eggs in clusters rather than singly. Stamp (1980) lists records for Papilioninae (one species), Pierinae, Nymphalinae, Danainae (one species), Acraeinae and Lycaeninae. Representative genera in other subfamilies (and other Papilioninae) that lay clusters of more than ten eggs include the following: Papilioninae—*Papilio*, *Battus* (Miles Moss 1920); Parnassiinae—*Luehdorfia* (Shirôzu & Hara 1960-1962); Ithomiinae—*Mechanitis* (Fox 1967; Satyrinae—*Neope* (Shirôzu & Hara 1960-1962); Brassolinae—*Brassolis* (Bondar 1940), *Caligo* (Malo & Willis 1961); Amathusiinae—*Faunis* Johnston & Johnston 1980) Morphinae—*Morpho* (refs. in Young 1973); Riodininae—*Euselasia* (Hoffmann 1931, Kendall 1976), *Hades* (Robbins unpublished), *Audre* (Bourquin 1953); Coeliadinae (Hesperiidae)—*Bibasis* (Johnston & Johnston 1980); Hesperiinae—*Thymelicus* (Shirôzu & Hara 1960-1962). The widespread occurrence of this life-history trait is evident. Clusters may contain the eggs of more than one female (Mallet & Jackson 1980) or be laid next to previously deposited clusters (DeVries 1977, Stamp 1981*b*). We include larval gregariousness in our discussion below because it is so closely correlated with cluster-laying.

In his original theoretical discourse on the evolution of distastefulness, Fisher (1930) predicted that cluster-laying species would tend to be distasteful and conspicuously coloured. Unpalatability was unlikely to evolve by selection on individuals since 'any individual tasted would seem almost bound to perish . . .' However, in species with gregarious larvae, a tasted larva would afford protection to its siblings. (This discussion was the original formulation of the kin selection concept: Hamilton 1963.) Once evolved, distastefulness and conspicuous coloration would reinforce the gregarious habit by enhancing aposematic advertisement. Thus, aposematic insects would tend to be cluster-layers even if their unpalatability did not evolve by kin selection. In some instances gregarious behaviour would be 'lost' when its advantages were outweighed by a shortage of suitable food. (See also Ch.14.)

Evidence supports Fisher's proposed association between cluster-laying and aposematic coloration. Cluster-layers shown to be relatively unpalatable include adults or larvae of various species of Heliconiinae, *Euphydryas*, Acraeinae, Ithomiinae, *Battus*, *Mylothris*, *Delias*, and *Pieris brassicae* (references in Brower & Brower 1964, Bowers 1980, 1981). Cluster-layers whose larvae appear aposematic include the lycaenid *Eumaeus* (with bright red larvae; DeVries 1977) and the riodinine *Hades*. Stamp (1980) discusses examples of insects that lay clusters of toxic, aposematic eggs. However, many butterfly eggs, whether clustered or single, are relatively unpalatable to an array of invertebrate predators (Swynnerton 1915*a,b*; but see Baker 1970).

Aposematically coloured butterflies that oviposit singly indicate that other factors influence cluster-laying. Benson (1971) noted the poor correlation between cluster size and palatability of heliconiines, but later (Benson 1978) showed that cluster size is partly a response to competition for limited larval food resources. Troidines are among the most unpalatable, conspicuously coloured butterflies, but females of most genera (e.g. *Parides*, *Atrophaneura*) oviposit singly or in small clusters (Straatman & Nieuwenhuis 1961, Young 1972). Danaines are also exceptions (except *Amauris albimaculata*, a cluster-laying species, Van Son 1955).

Gregarious lycaenid and riodinid larvae are more likely to have obligate myrmecophilous relationships than solitary larvae. Ants may protect these larvae in return for nutritious 'honeydew' (Ross 1966, Pierce & Mead 1981). Groups of larvae may produce more honeydew or other chemical attractants and should be more dependably tended by ants. Further, since gregarious larvae are not cannibalistic (unlike solitary lycaenid larvae—Downey 1962*a*), females can oviposit near larvae, and secure protection for their eggs through the presence of ants.

Available evidence suggests that cluster-laying lycaenids are usually ant-attended. All Australian lycaenids that lay clusters of ten or more eggs are obligate myrmecophiles ($n=11$) while only about a third of the species that oviposit singly ($n=26$) have an obligate relationship with ants (Kitching 1981). Females of *Ogyris amaryllis* lay larger egg-clusters after tactile contact with ants (Atsatt 1981*a*). There is a paucity of data on riodinines, but species of *Audre* oviposit small to moderate sized clusters, and apparently are obligate myrmceophiles (Bruch 1926, Bourquin 1953, Robbins & Aiello 1982). A striking counter-example, however, is the lycaenid cluster-layer *Eumaeus atala*, which is not tended by ants (J. Weintraub, pers. comm.) but whose bright red larvae appear aposematic. DeVries (pers. comm.) reports similar findings for *E. minyas* in Costa Rica; however, the larval aggregations are attended by langurid beetles.

A female may arrange her eggs in any of a variety of designs. The simplest arrangement consists of non-contiguous eggs. (The pattern is probably statistically regular.) Examples include some heliconiines (Benson *et al.* 1976), ithomiines (Fox 1967), and the parnassiine *Luehdorfia* (Shirôzu & Hara 1960–1962).

A more complex design is a single layer of eggs touching each other. Some hesperiids, satyrids and brassolids lay a single row of eggs on blades of grass (Shirôzu & Hara 1960–1962, Malo & Willis 1961). Other species (*Audre*, *Nymphalis*) lay eggs in a compact hexagonal pattern (Bourquin 1953, Shirôzu & Hara 1960–1962) resembling the arrangement of cells in honey-comb. Still others (*Melitaea*) lay eggs partially in a compact hexagonal pattern and partly loosely arranged, but usually contiguous (Shirôzu & Hara 1960–1962).

The most complex arrangement consists of contiguous eggs more than one layer deep. Two variations differ in manner of deposition. First, some species (e.g. *Aporia*, *Aglais*, *Chlosyne*, *Papilio anchisiades*) lay masses of eggs up to three or four layers deep (Shirôzu & Hara 1960–1962, Robbins unpublished). Females of *Chlosyne* lay clusters one layer at a time. We think it likely that other species behave similarly. Second, some species (e.g. *Hamadryas amphinome*, *Polygonia*, *Araschnia*, *Papilio liomedon*) lay strands or stacks of eggs (Edwards 1870, 1882, Bell 1911, Shirôzu & Hara 1960–1962, Robbins unpublished), usually several strands to a cluster. Females of *Hamadryas* oviposit one strand at a time rather than one layer at a time (Robbins unpublished).

Available substrate space may influence egg deposition. Hexagonal arrangements and multi-layered clusters permit females to lay more eggs on a smaller surface, but falling eggs constrain the number of layers. 'String clusters' of *Hamadryas amphinome* (up to 15 eggs long) break during afternoon thunderstorms (Robbins, unpublished). Females of *Chlosyne lacinia* lay up to three layers on the undersides of leaves. However, in an aberrant cluster of five layers, the outermost layers fell prematurely (Robbins unpublished).

Some egg arrangements protect eggs in the centre from parasitoids. Wasps may preferentially parasitize eggs at the edge of a single layer cluster (e.g. Van Den Berg 1971) or attack the outer layers of egg masses (Howard & Fiske 1911, Stamp 1981*b*). The innermost eggs in clusters of Panamanian *C. lacinia* are protected, even when an excess of wasps is experimentally placed on an egg cluster. The wasps cannot squeeze their abdomens between closely packed eggs (Robbins, unpublished). (However, in Texas populations, all eggs in a cluster of *C. lacinia* may be successfully parasitized: Drummond *et al.* 1970). As clutch size increases, the proportion of gypsy moth eggs parasitized decreases with the ratio of surface area to volume (Dowden 1961, Doane 1968).

Females of *Hypolimnas antilope* stand over their egg clusters and prediapause larvae (Rothschild 1979). Although this behaviour does not afford absolute protection for their eggs (peripheral eggs may be preyed upon or parasitized), we presume that guarded clusters have lower mortality rates than unguarded ones. This result was shown in an analogous case with a pentatomid bug (Eberhard 1975).

Gregarious larvae may suffer lower rates of parasitism in groups than as solitary individuals. *Euphydryas phaeton* larvae in groups that are larger or smaller than those occurring naturally suffer greater rates of parasitism (Stamp 1981*a*). Gregarious larvae may show stereotypic behavioural displays that presumably deter predators or parasitoids (see Stamp 1980).

Gregarious larvae may be better able to overcome physical plant defences, such as tough leaves and trichomes, than solitary larvae (Ghent 1960 for sawfly larvae). We know of no unequivocal evidence for butterflies. Gilbert (1971) proposed that trichomes on the leaves of *Passiflora adenopoda* pierce and kill larvae of some *Heliconius* species. Rathcke & Poole (1975) suggested that gregarious larvae of *Mechanitis polymnia isthmia* spin a web as an adaptation for crawling over trichomes without being pierced. However, larvae of some heliconiine species, which may lay single eggs, successfully feed on *P. adenopoda* in the wild (Benson *et al.* 1976, K. S. Brown, pers. comm.). Further, Panamanian larvae of *Mechanitis p. isthmia* (second or third instar) are not adversely affected by placing them individually on their solanaceous foodplants, even though the leaves have sharp recurved spines (Robbins unpublished).

Gregarious larvae feed more efficiently in groups. Long (1953, 1955) reared larvae of *Pieris brassicae* singly and in groups of 60, and found that larvae in groups developed faster; individuals apparently stimulated their neighbours to feed. Larvae of *Chlosyne lacinia* hatching under normal group conditions spin a web and then feed. Solitary larvae, however, walk until they starve (80% mortality within two days, n = 31), or as occurred in 20% of the cases, spin a small web and then feed (Robbins, unpublished). Group feeding, however, does not facilitate feeding efficiency in *Euphydryas phaeton* (Stamp 1981*a*).

That normally gregarious larvae feed more efficiently in groups or have reduced rates of parasitism in groups is not surprising; it is invalid, however, to interpret these results as advantages of egg clustering (e.g. Stamp 1980). As gregarious feeding evolves, for whatever reason, behavioural adaptations for group feeding also evolve. We visualize 'normal behaviour' as an adaptive peak (Wright 1967) steepened by modifying behaviours. Comparing group feeding efficiency, for example, with abnormal solitary feeding supports the adaptive peak concept, but cannot be used to infer the general advantages of gregariousness. Even if the original factors favouring gregarious behaviour should change, the gregarious habit may well be retained because of these modifying behaviours and their secondary advantages.

Other advantages of clustering eggs have been suggested. Tutt (1899) proposed that strings of geometrid eggs resemble tendrils of their foodplant. Stamp (1980) suggested myriad other hypothetical possibilities; the reader is referred to her paper. Many of these hypotheses, and the one discussed above, could be tested by manipulating arrangements of eggs.

Systematics may be one of the most useful techniques for determining why life-history traits, such as egg-clustering, evolve. The combination of ecological methods and systematics has produced some remarkably original papers (e.g. Ehrlich & Raven 1965, Benson *et al.* 1976). We believe that systematics, particularly cladistics, may contribute much to evolutionary analysis (Mitter & Brooks 1983). Arnold (1981) presents a mostly jargon-free introduction to the methods used by cladists in reconstructing phylogenies. We present below a representative example of how cladistics might profitably be used.

The relationship between egg-clustering and type of larval foodplant (annual or perennial) is presently unclear. Stamp (1980) found that some cluster-laying Nearctic nymphalids use annual larval foodplants while others use perennials, and concluded that these factors are not related. An alternate explanation is that cluster-laying evolved once in a species that used perennial foodplants and was ancestral to extant nymphalid cluster-layers. Further, the cluster-laying trait persisted through changes to annual foodplants because gene complexes adapted for gregariousness, such as those determining larval behaviours, presented too steep an adaptive peak for reversion to single egg-laying.

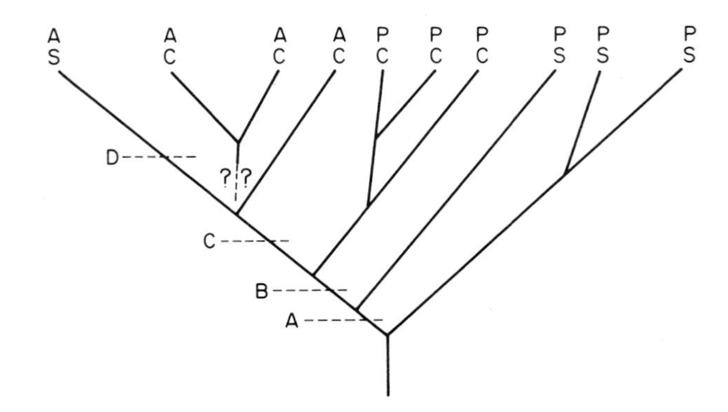

Fig. 6.1. Hypothetical cladogram of ten species with type of larval foodplant (annual [A] or perennial [P]) and and manner of egg deposition (single [S] or clusters [C]) noted.

A cladogram can be used to test these hypotheses. We present an hypothetical cladogram of ten species, annotated for manner of egg deposition and type of larval foodplant (Fig. 6.1). We would infer that at point (A), the ancestral species laid single eggs on perennials, at point (B) cluster-laying evolved, at point (C) there was a change to annual foodplants, and at point (D) egg deposition reverted to single eggs. This cladogram would support the alternative hypothesis above whereas one in which cluster-laying evolved a number of times, sometimes when species were feeding on annuals and sometimes on perennials, would support Stamp's no-correlation hypothesis.

Cladistic analysis presents certain difficulties. First, cladograms are based on parsimony methods not readily amenable to statistical testing. There is no objective way to measure how well a cladogram represents phylogenetic relationships. Second, reasoning may be circular if the trait being examined is used to construct a cladogram. For instance, if manner of egg deposition had been used to construct Fig. 6.1, then the analysis would be biased against changes in egg deposition because fewer changes are more parsimonious than more changes. Despite these difficulties, cladistic analysis potentially provides a window for viewing evolutionary history that is not otherwise available.

Evolution of Oviposition Specificity

We discuss here mechanisms promoting changes in specificity to novel foodplants, considering first cases in which butterflies change to a related foodplant, and then instances in which they switch to unrelated foodplants. Undoubtedly most changes involve a switch to a plant that is phylogenetically and phytochemically closely related to the old foodplant, such as from one cruciferous species to another. Work on variation in oviposition preferences, conditioning of preferences and the heritability of specificity, although scanty, provides a framework for our discussion (in the first part of this section) of such changes. Theoretical cost-benefit analysis of exploiting new foodplants offers a different approach to this problem (Levins & MacArthur 1969, Jaenike 1978).

We believe that large 'jumps' in specificity are responsible, at least in part, for the wide range of plants that are used by butterflies as a group. For example, the use of gymnosperms, monocots, and a diverse group of dicot families (Loranthaceae, Ericaceae, Asteraceae, Verbenaceae, etc.) by the 'subgenera' of *Callophrys* (Howe 1975, Robbins & Aiello 1982) provides a representative example in which large changes in specificity must have occurred rapidly on an evolutionary time scale. It is unlikely that long-term association during phylogenetic branching (see Ehrlich & Raven 1965, Edgar *et al.* 1974, Benson *et al.* 1976, Edgar 1982 for proposed examples), or accumulated small changes in specificity between related plants, could account for this diversity of taxonomically unrelated larval foodplants. We propose mechanisms in the second part of this section that might account for large changes in specificity.

Foodplant Changes among Related Plants

Conditioning larvae to specific foodplants has been suggested as a mechanism for generating variation and for changing preferences in oviposition behaviour within population. The Hopkins Host Selection Principle (Hopkins 1917) proposed that adult female insects prefer to oviposit on those hosts that they ate as larvae. By this means, an individual larva that eats a novel foodplant could introduce that plant species to its population's resources; oviposition specificity for the plant would be induced.

We know of no unequivocal evidence that larval conditioning persists through metamorphosis to the imago. Larval preferences may be influenced by experiences as young larvae (e.g. David & Gardiner 1966, Hanson 1976 and references therein, Copp & Davenport 1978*a*; but see also Wiklund 1975*b*, Copp & Davenport 1978*a*, and Chew 1980 for negative results). Adult oviposition preferences may be similarly influenced by experience as adults (e.g. Traynier 1979, Rausher 1978, Stanton 1980; but see Tabashnik *et al.* 1981 for negative results). However, attempts to link larval experience with adult preference have either confused selection with conditioning effects (Hovanitz & Chang 1963) or provided negative evidence (Takata 1961*a,b*, Wiklund 1974*b*, 1975*b*, Copp & Davenport 1978*b*, Tabashnik *et al.* 1981). Claridge & Wilson (1978) and Jaenike (1982) provide negative results in Homoptera and *Drosophila* respectively, and review evidence in other insects.

Extensive behavioural observations of individuals document variation of oviposition preferences among individual females (Stanton 1979, Tabashnik *et al.* 1981, Singer 1982*a*); this variation is a prerequisite to changes in oviposition specificity. Some of this variation is attributable to the motivational state of females; females accept 'suboptimal' oviposition substrates when preferred ones are rare or absent (Singer 1971, Stanton 1982*a*, Wiklund 1981) or when they are strongly motivated to oviposit by heavy egg loads (Singer 1971, Jones 1977*b*). Some evidence suggests that females in a population share similar 'generalist' behaviour (each female oviposits on a similar range of plants, Chew 1977) while other evidence shows that a population may appear to contain a diversity of 'specialists' that exhibit consistent individual preferences (Tabashnik *et al.* 1981, Singer 1982*a*). Wiklund (1981) suggests that female *Papilio machaon* accept potential foodplant species in the same hierarchical order, but some individuals retain preference for the 'optimal' larval foodplant longer than others.

Evidence that variation in oviposition preference is heritable includes correlation of parent-offspring preferences (Tabashnik 1981) and artificial selection experiments (Hovanitz & Chang 1963; see also Wasserman & Futuyma 1981 for work on a bruchid beetle). To the extent that oviposition preferences are highly correlated with the fitness of a female, e.g. as might be the case if young larvae were very sedentary, we would expect heritabilities to be low (Falconer 1960).

Heritable variation in oviposition preference may not, however, be a particularly important factor in mediating changes in specificity. The detailed studies of Tabashnik (1981) on host shifts from native to crop plants in *Colias* suggest that although behavioural differences among females may have a genetic basis, genetic changes are not necessarily a prerequisite to foodplant shifts among related plants. If phenotypic plasticity produces sufficient ability to exploit newly encountered resources, selection may favour increased phenotypic plasticity rather than genetically fixed predisposition to exploit specific (novel) resources (Mayr 1963).

Foodplant Changes to Unrelated Plants

In this section, we discuss two general mechanisms which we believe account for many large 'jumps' in butterfly foodplant specificity. The first mechanism occurs when old and new foodplants chronically grow in physical proximity to each other. The second involves adults that oviposit on flowers and are chronically exposed to other flowers on which they nectar. Both mechanisms require long-term association between the butterfly and its new foodplant, thus allowing time to select for improved ability to exploit the new host. It is hard to believe that large changes in specificity to phylogenetically and phytochemically unrelated plants occur frequently without such contact.

Large changes in specificity may occur when old and new foodplants grow in physical proximity to each other. Larvae are exposed to taxonomically unrelated potential foodplants for many generations; selection favours those larvae that can advantageously supplement their diets with the new foodplant (see King 1971 and Chew 1975 for further discussion of the consequences at the population level of supplementing diet). Actual change in oviposition behaviour, which is independent of larval digestive capabilities (see previous section), might occur in various ways as outlined in the following examples.

Females of some species normally oviposit in the vicinity of their foodplant, not on it, and may thus oviposit on unrelated plants that larvae may eventually utilize. *Euptychia* species generally use grasses as larval foodplants, but frequently oviposit on plants near grasses. Singer *et al.* (1971) found a Panamanian *Euptychia* species that feeds on lycopsids growing in the vicinity of grasses. They reasoned that habitual oviposition on lycopsids, coupled with the advantages that would accrue to larvae able to complete development on lycopsids, might explain this switch in foodplant specificity.

Females of many lycaenids and riodinids preferentially oviposit in the vicinity of ants (e.g. Atsatt 1981*b*). Larvae are thus chronically exposed to those homopterans that ants tend, to ant brood, and to lichens and fungi that grow in 'ant runs' and ant nests. The cannibalistic tendencies of lycaenid larvae, coupled with long-term exposure to ant-tended homopterans and ant brood, has undoubtedly resulted in the repeated independent evolution of carnivorous feeding on these animals (Ehrlich & Raven 1965). Significantly, lycaenid larvae are not known to feed on insects other than those tended by ants. The switch to lichens, which frequently grow in 'ant runs', by the African liptenines (over 500 species; Ehrlich & Raven 1965, van Someren 1974) may have a similar explanation (see Atsatt 1981*b*), but may also have occurred as described in the next mechanism.

Females may chronically oviposit near to or on epiphytes (or parasites) that commonly grow on a butterfly's foodplant. This behaviour may mediate large 'jumps' in specificity from ancestral foodplants to epiphytes. Additionally, epiphytes also may act as a conduit in foodplant specificity over evolutionary time if species switch from the original foodplant to the epiphyte, and thence to another unrelated plant species on which the epiphyte grows.

A possible result of oviposition near parasitic plants is the repeated changes of specificity to mistletoes (Loranthaceae) among pierids, lycaenids, and nymphalids; these changes are not easily explained by similarity of secondary compounds (Ehrlich & Raven 1965). This is a difficult hypothesis to test because in most cases the switch to Loranthaceae occurred too long ago to determine the ancestral foodplant. Further, subsequent changes in foodplant specificity from mistletoes growing on the old foodplant to those growing on other plants may have occurred. The only valid test is to examine genera whose species feed on a well-defined set of plants (i.e., the proposed ancestral foodplant) except for one or two species feeding on Loranthaceae. An example would be species of *Euthalia* (Nymphalidae). These feed on trees of the family Ancardiaceae, except for one species that feeds on mistletoes which grow on these trees (L. Young 1907). Another possible example is the Nearctic genus *Mitoura* (Lycaenidae). Most species feed on Cupressaceae, but two species switched (probably before they speciated) to mistletoes that grow only on gymnosperm trees (Howe 1975).

A second mechanism for large changes in specificity occurs when butterflies which oviposit on flowers are chronically exposed to other flowers on which they nectar. If tactile cues or absence of chemical deterrents stimulate oviposition (see above), then females may occasionally oviposit on flowers which they visit for nectar. Further, larvae may occasionally complete development on these flowers because they might be exposed to relatively low concentrations of secondary compounds (as compared to larvae which find themselves on *leaves* of a 'strange' plant). Selection would then favour those individuals that oviposit (and whose larvae can develop) on flowers of the new plant, particularly in years or seasons when the usual foodplants produce few or no flowers. This mechanism would cause an increase in polyphagy which might be followed by specialization, if favourable conditions for it arise, on either the new or the old foodplants. This mechanism may account for the taxonomic polyphagy of individual lycaenid species (e.g. *Strymon melinus*) and for the wide range of plants used by the flower-feeding lycaenids as a group (Robbins & Aiello 1982).

This mechanism predicts the testable hypothesis that the plants which a flower-feeding species uses as larval foodplants are a subset of those that it uses as sources of nectar. Available evidence is scanty, but there is some 'anecdotal' evidence supporting this hypothesis among Neotropical eumaeine lycaenids. Males and females of *Strymon 'basilides'* are the only lycaenids in Panama that nectar on *Heliconia* (Moraceae) flowers (Robbins unpublished); they are also the only lycaenids in Panama known to oviposit on these flowers (Robbins & Aiello 1982). *Cordia* (Boraginaceae) and *Mangifera* (Anarcardiaceae) flowers attract more species of nectaring lycaenids than any other Panamanian plants; a phylogenetically wide range of lycaenid species has been reared from flowers of each of these plants (d'Araújo e Silva *et al.* 1967–68, Robbins unpublished). This evidence suggests that the mechanism is plausible and worthy of testing in the future.

Acknowledgements

For helping with various aspects of this paper, including providing access to unpublished data, reading early drafts and giving us benefit of stimulating discussion, we thank Annette Aiello, Deane Bowers, Phil DeVries, Paul Feeny, Ira Heller, Charles Mitter, Dennis Murphy, Jan Pechenik, Miriam Rothschild, Julian Shepherd, Robert Silberglied, Michael Singer, Maureen Stanton, and Margaret Thayer. For financial support of some of our research described in this paper, we thank the National Science Foundation (DEB 7805960 to FSC), the US Department of Agriculture (CRGO 7900435 to FSC), and the Smithsonian Institution (to RKR).

7. *Butterfly-Hostplant Relationships: Host Quality, Adult Choice and Larval Success*

Michael C. Singer

Department of Zoology, University of Texas, Austin, Texas

This chapter discusses effects of host suitability on butterfly larvae. It considers the likelihood that a newly-hatched larva will find itself on a very suitable, marginal or unsuitable host as a result of its mother's oviposition behaviour. The relationships between adult choice and host suitability for larvae are compared for different butterfly species and for populations differing in breadth of host use. The ways in which larvae may respond to variation in host quality are also described.

Oviposition Mistakes

Many authors have reported that ovipositing butterflies make mistakes, with the result that the first plant encountered by the newly-hatched larva is unsuitable. As Chew & Robbins (Ch.6) remark, some of these examples are misinterpreted. Dethier (1959*a*) quotes the 'British Argynnid' as occasionally ovipositing on tree trunks, though its larvae feed on violets. South (1941) indicates that such behaviour is normal in *A. paphia*—the larvae diapause in bark crevices without feeding, then descend to find food in the spring. After quoting other examples of butterflies repeatedly ovipositing on non-host species or on individual plants too small to support larvae to maturity, Dethier remarks that 'errors made by adults in the selection of sites may contribute appreciably to losses . . . lack of precision in egg laying by species whose larvae are restricted feeders represents behaviour of questionable adaptive value. Compensation of the deficit rests with the larvae.'

As in the case of the 'British Argynnid', many of these observed behaviour patterns are likely to be characteristic of the species concerned. Most of Dethier's records of oviposition on soil, sticks or non-hosts were made in the tropics. My own

Table 7.1. Oviposition and habitat of euptychiine satyrids (described in *Euptychia*).

	Habitats			
Species	Open, near trees	Partly shaded	Shady second growth	Light gaps in mature forest
Species that oviposit on their host grasses				
arnaea			X	X
metaleuca			X	X
sp. n.		(X)	X	X
libye	X	X	X	X
alcinoe	X	X	X	?
hesione		X	X	
B. Species that oviposit off their hosts				
hermes	X	X	(X)	(X)
usitata	X	X		
labe	X	X		
renata	X	X	X	

observations in tropical habitats indicate that oviposition away from the host is frequent in disturbed habitats but less common among forest species (cf. Ch.32). This relationship between habitat and oviposition behaviour is shown in Table 7.1 for neotropical euptychiines. However, the euptychiines that oviposit away from their host grasses do not distribute their eggs at random. For example, Trinidadian *Euptychia (Hermeuptychia) hermes* alight on various plant species when searching for oviposition sites, often 'drumming' them with their foretarsi. After alighting on non-hosts they fly on, usually without great change of direction, and test other plants. After testing a host, an insect may either reject it by flying away or accept it by flying slowly for a few inches, often in a small circle, then alighting and laying an egg. The

The Biology of Butterflies
0-12-713750-5

egg may be laid on a dead leaf or stick—but sometimes it is placed on a green leaf belonging to any low-growing species. Occasionally this may be a host, though when this happens it appears to be accidental. These observations suggest that *E. hermes* is sensitive to host quality, since a high proportion of tested hosts (>60%) are rejected. In order to ask whether this species is selecting the best sites for its eggs, it would be necessary to measure the suitabilities of accepted and rejected hosts for larvae and compare the fates of eggs laid close to the chosen host with fates of eggs laid on the host itself. Comparisons of populations with different behaviour patterns could be useful. In Costa Rica, *E. hermes* itself typically oviposits on the host in some areas and off the host in others. However, comparison of the fates of eggs laid on and off the host would be time-consuming, since eggs laid off the host can only be found by following searching females, while those laid on hosts are much easier to find. If this difference in 'findability' applies to non-human predators or parasites, the explanation of off-host oviposition may be uncomplicated (Benson *et al.* 1976).

The examples given above show that oviposition behaviour must be studied in some detail before observations of egg-laying on non-hosts can be interpreted as 'mistakes'. With this reservation, the sections below discuss the relationships between adult host choice behaviour and host suitability for larvae.

Discrimination Between Individual Plants of the Same Species

Much of the within-species discrimination practised by ovipositing butterflies is obscure. If one observes an insect rejecting most individuals of a host species and accepting only a few it seems likely that discrimination is occurring, although the possibility remains that the insect is accepting at random only a few of the many hosts it encounters. If the discrimination is based on plant size, dispersion, or the presence of eggs or larvae, then this will become evident from comparison of the characteristics of accepted and rejected plants. However, discrimination that is not correlated with differences in measurable characters will be hard to demonstrate unless an insect encounters the same individual plants several times and gives consistent responses to them. Rausher *et al.* (1981) were able to show discrimination by *Euphydryas editha* females testing individual *Pedicularis semibarbata* plants. Butterflies that had been observed to reject one plant and accept another (of the same species) were captured before the first egg was laid and replaced by hand on each plant in turn. They showed a significant tendency to duplicate their previous responses.

Most butterfly species would not be amenable to the experimental manipulations performed by Rausher *et al.* Consequently, observations of insects in the field rejecting some hosts and accepting others could not be classified as showing discrimination between hosts in different categories unless consistent differences were found between accepted and rejected plants. Nonetheless, the behaviour patterns observed are often highly suggestive of conspecific discrimination. For example, the Costa Rican pierid *Perrhybris pyrrha* feeds on an understory tree in the genus *Capparis* (Capparidaceae). During its search, it usually rejects many (>10) such trees before selecting one. Then it spends up to 30 minutes apparently comparing leaves on the same tree. It flies to a leaf, alights on the tip, runs up the leaf to the base, then takes off and repeats the procedure on another leaf. Eventually, the number of leaves receiving this treatment declines from perhaps a dozen to two or three, and the insect finally vacillates between these before settling down to oviposit. If disturbed while laying eggs, it flies into the canopy but subsequently returns, relocates the site, and completes its egg-mass. Such complex and time-consuming behaviour suggests sophisticated discrimination among conspecific hosts and individual leaves. Detailed investigation of such phenomena, especially chemical comparisons of accepted and rejected leaves or plants, would be interesting and relevant to both ecological and evolutionary aspects of insect-plant relationships.

Studies Showing Conspecific Discrimination

Rausher & Papaj (1983) found that *Battus philenor* females did discriminate in favour of the *Aristolochia* plants on which larvae could achieve greater size before migration. They demonstrated this experimentally by placing newly-hatched larvae on plants that had been accepted or rejected by searching butterflies, then following the fates of these larvae. Rausher (1979*a*) had already shown that a large larva forced to migrate has a better chance of surviving than a small individual because of its better host-locating ability.

Flying insects may show preferences for hosts growing in particular spatial arrangements (e.g. Jones 1977*a*), but such preferences are hard to demonstrate unequivocally. For example, Mackay & Singer (1982) exposed clumped and isolated plants to captive *Euptychia libye* and found that isolated hosts were attacked more heavily than those in groups. However, if they defined preference as a systematic departure from random behaviour, they found that the insects actually preferred plants in clumps. Mackay (1982) worked with post-alighting preference and found that

Euphydryas editha were more likely to oviposit after alighting in dense stands of *Collinsia torreyi* than when they landed on a *C. torreyi* in a low density area. He found that survival of larvae to diapause was significantly lower on plants classified as acceptable (to adults) than on those classified as rejected.

Butterflies may discriminate between micro-habitats and hence between hosts growing in them. Rausher (1979*b*) studied three *Aristolochia*-feeding swallowtail species and found that one laid more eggs in shady than in sunny habitats, while the other two showed the reverse preference. Larval survival was higher in the shade in all three species.

Several species are known to be deterred by the presence of eggs on plants (see Chs 3, 6). Rausher (1979*a*) has performed field experiments and shown that, at least in the *Battus* population he studied, this discrimination increases larval survival.

Failure to Show Conspecific Discrimination

Dethier (1959*a*) remarked that several species fail to discriminate between plants of different sizes, the result being that small larvae must often leave the initial host and migrate in search of additional food. At such an early stage their ability to locate food is poor. He observed this phenomenon in detail in one species, *Chlosyne harrisii* (Dethier 1959*b*, as *Melitaea*). In other species there may be nutritional advantages in selecting small plants that outweigh the disadvantages of early migration, but I know of no documented case in butterflies.

In my own work I have found two examples in which insects fail to show discrimination that would increase larval survival. The first is the failure of two neotropical grass-feeding euptychiines to respond to conspecific eggs although their hosts are so small and sparse that this results in early forced migration when several larvae feed on the same plant (Singer & Mandracchia 1982). The second is the failure of *Euphydryas editha* at Jasper Ridge, near Stanford, to discriminate between hosts (*Plantago erecta*) that would senesce before egg hatch and those that would remain edible. This resulted in larval mortality in the order of 80% in the years of study (Singer 1972). In contrast, *E. editha* in populations that feed on *Collinsia tinctoria* do perform such discrimination, and thereby successfully avoid oviposition on plants that will be dry by the time the eggs hatch (Singer 1971).

Discrimination Among Plant Species

Since no butterflies are complete generalists, all of them discriminate between the various plant species in their habitats. However, they do so to differing extents, with resulting differences in diet breadth. Adaptive explanations of this diversity of diet breadth can be tested by investigating the consequences of adding or deleting host species. Wiklund (1982) favours working with oligophagous insects (see Ch.8), for which a relatively small number of plant species can be classified as potential hosts if newly-emerged larvae placed upon them can survive. Other authors classify as potential hosts species in particular taxonomic groups or possessing key secondary compounds. After identifying potential hosts in a habitat, one can then ascertain by experiment the consequences of oviposition on both actual and potential hosts. These consequences should ideally include effects on the following fitness components: survival and fecundity of adults searching among different plant species; egg survival and duration; larval growth rate, development rate, and survival; and adult time of emergence and fecundity. In some cases there may be effects associated with proximity to the host of suitable pupation sites, larval diapause sites, and nectar sources. In practice, most studies have concentrated on larval growth and survival. The results have been very diverse. Some butterflies fail to use apparently suitable potential host species, while others lay many or even most of their eggs on relatively unsuitable hosts (cf. Ch.32). Perhaps Dethier was correct in his estimation that the oviposition behaviour of many butterflies is rather far from optimal. Below, I summarize some case histories, including results of a detailed ongoing study of *E. editha* populations with different diet breadths. The same topic is addressed with respect to plant-feeding insects in general by Rausher (1982*b*).

Chew (1977) and Courtney (1981, 1982*a*,*b*), both working with pierids, found poor correspondence between adult oviposition and larval suitability. Chew found ovipositions on toxic hosts that also stimulate larval feeding. These plants are introduced, and are biochemically similar to suitable native hosts (Rodman & Chew 1980). Courtney (1982*a*) found that the host species that received most eggs in proportion to its abundance was the least suitable species, although survival on it was possible for some larvae. This host had been established in the habitat for over 125 years and has a longstanding association with the insect in other areas. Courtney felt that the apparently maladaptive behaviour could not be explained on the basis of exposure of the insects to recently-introduced plants. Effects of gene flow between populations exposed to different selection pressures were considered unlikely because the butterfly population was isolated and sedentary, and because the relative suitabilities of the host species were the same in all populations tested.

Smiley (1978*a*) studied two monophagous and one oligophagous species of *Heliconius*. He raised captive larvae on several potential hosts. The oligophagous species, *H. cydno*, developed well on all *Passiflora* species. One of the monophagous butterflies, *H. erato*, grew faster on its normal host than on any of the others. However, the other monophage, *H. melpomene*, gave results similar to those obtained from *H. cydno*. Among this small group of butterflies, oviposition specificity was not well correlated with digestive specialization of larvae. Smiley gave circumstantial evidence that the survival of *H. melpomene* larvae on unused potential hosts would be lower than that on the actual host species because of greater risk of parasitism. However, he was not able to test this in the field; his growth rate experiments were carried out in a controlled environment chamber, using potted plants. This procedure has the advantage that small differences in larval performance can be detected. Smiley expressed the opinion that some of his findings, such as the significantly faster growth of *H. erato* on its natural host, were quantitatively so small that they would have been almost impossible to detect in the field. Unfortunately, the relative quality of plant species may change when they are brought into cultivation. Thus, field trials should be performed in order to test further the hypothesis that *H. melpomene* avoids, for ecological reasons, plants that are just as suitable digestively as its actual host.

While Smiley classified any *Passiflora* species as a potential host of *Heliconius*, Holdren & Ehrlich (1982) were able to utilize the interpopulation variation in host use of *E. editha*, and to classify as potential hosts members of genera that are used by other populations of the same insect species. By this means they classified *Castilleja chromosa* and *Penstemon strictus* as potential hosts in a monophagous population of *E. editha* feeding on *Castilleja linariifolia*. Like Smiley, they found that captive larvae grew well on both actual and potential hosts. They then gave circumstantial evidence (based on plant density, phenology and predictability) that larval survival in the field would be highest on the species used, but they did not test this directly. Rausher (1982*b*), also working with *E. editha*, found interpopulation differences in larval specialization. He compared the population at Del Puerto Canyon (coded DP; see Ehrlich *et al.* 1975 for map) with that at Indian Flat (IF). The host at DP is *Pedicularis densiflora*, and that at IF is *Collinsia tinctoria*. In laboratory trials, *C. tinctoria* was more suitable than *P. densiflora* for larvae from both populations. However, the performance of DP larvae on *P. densiflora* was much better than that of IF larvae. Blau (1981) has obtained identical results in a comparison of *Papilio polyxenes* from New York and Costa Rica. Larvae from both areas grew faster on *Spananthe paniculata* (the Costa Rican host) than on *Daucus carota* from New York. However, the difference in growth rates was much less for New York larvae.

Where possible, larval success on potential hosts should be assessed both in the field and under controlled conditions. Newly-emerged larvae of *Anthocharis cardamines* have been placed on plants in the field by Wiklund (1982) and their survival measured. The fates of naturally-laid eggs of the same species have been followed by Courtney (1981) at several sites in Northern England (see above). Singer, Mackay and Moore (unpublished) have used a variety of techniques with *E. editha*. They have compared two monophagous and two oligophagous populations of this species, and have asked how diet breadth is related to larval growth and survival on actual and potential hosts. The first comparison is between a monophagous population feeding on *Collinsia tinctoria* (population IF) and an oligophagous population (Schneider Meadow, coded SN) in which most insects prefer *Collinsia parviflora*, but a few prefer *Plantago lanceolata*, an imported European weed. Eggs are also laid on *Penstemon heterodoxus* at SN, though no insects have been found to prefer this species.

We have been able to cause SN insects to oviposit on the three hosts at this site, but at IF we were unable to do this, and used dental floss to attach *Collinsia* sprigs bearing egg masses to the test plants. Survival of larvae to mid-first instar and second instar was recorded for each group. These results are summarized in Table 7.2. Table 7.2A shows survival of local (IF) and transplanted (SN) egg masses on all potential hosts at IF, and Table 7.2B shows equivalent data from SN. Survival of IF larvae at IF is significantly different on any pair of plants compared. The most suitable plant is the one actually used by this monophagous population, and the order of suitability (*Collinsia* > *Orthocarpus* > *Castilleja* > *Penstemon*) corresponds to the order of preference of those insects that have yielded complete data. One plant classified as a potential host, *Penstemon breviflorus*, appears not to be. All the larvae emerging on it attempted to feed, consumed very small amounts of leaves or flowers, and then died. This plant is host at this site to a closely-related butterfly (*E. chalcedona*) but it is apparently toxic to *E. editha*.

Different results were obtained at SN, the oligophagous population. The plants here were less variable in their suitability, and the species preferred by most butterflies (*C. parviflora*) was the least suitable of the three. However, the local *Penstemon*, *P. heterodoxus*, did support larval growth. Why do larvae survive so much better on SN *Penstemon* than on IF *Penstemon*? IF *Penstemon*

Table 7.2. Survival of *Euphydryas editha* larvae on actual and potential hosts at two sites.

Origin of eggs	Host genus	Number of eggs	Proportion of individuals surviving to: mid-first instar	second instar
A. Egg masses placed on the sole host (*Collinsia tinctoria*) and on potential hosts at Indian Flat (IF)				
IF	*Collinsia*	945	0.259	0.163
IF	*Orthocarpus*	725	0.131	0.063
IF	*Castilleja*	1301	0.040	0.022
IF	*Penstemon*	825	0.002	0.000
SN	*Collinsia*	261	0.080	0.054
SN	*Castilleja*	546	0.097	0.075
SN	*Penstemon*	765	0.003	0.003
B. Egg masses laid or placed on the three host species at Schneider Meadow (SN)				
SN	*Plantago*	2764	0.341	0.242
SN	*Penstemon*	1074	0.211	0.128
SN	*Collinsia*	1810	0.142	0.037
IF	*Plantago*	638	0.080	0.008
IF	*Penstemon*	1268	0.110	0.042
IF	*Collinsia*	463	0.134	0.058

Mean egg mass sizes were about 30
Significance of differences:
IF larvae on IF plants: Any two pairs of plants differ with at least $P < 0.05$
SN larvae on SN plants: Survival on *Plantago* greater than on either of the other hosts with $P < 0.01$; *Collinsia* and *Penstemon* not significantly different.
Transplanted larvae: If survival of transplanted larvae on a particular host is compared with that of local larvae on the same host, only one comparison is significant: IF larvae on SN *Plantago* did more poorly than SN on SN *Plantago*, with $P < 0.01$.
Test: Mann-Whitney, using the proportion surviving in each group as data points.

may be unsuitable for *E. editha* in general, or IF *E. editha* may be more specialized in larval digestive traits, just as they are more host-specific than SN butterflies in their oviposition behaviour (Singer 1982*a*). The fates of larvae from transplanted egg masses (Table 7.2) allow us to distinguish between these alternatives. SN larvae died on IF *Penstemon* and survived on IF *Collinsia* and *Castilleja*, while the survival of IF larvae on SN *Penstemon* was not significantly different than that of the local insects. Thus, the difference between survival of SN larvae on SN *Penstemon* and that of IF larvae on IF *Penstemon* is due more to inter-site differences between the plants than to inter-population variation of *E. editha*. However, there is one effect that confirms Rausher's (1982*b*) finding that *E. editha* larvae do show interpopulation variation in larval digestive traits; IF larvae on SN *Plantago* fared significantly less well than did the local SN insects.

The second comparison, between populations DP and GH, has yielded similar results: the host preferred by most insects in the oligophagous population is not the species on which larval survival is highest, while in the monophagous population there are potential hosts that could support larval survival, but the actual host is the most suitable species available. Further work along these lines with this and other butterfly species would help to evaluate the various assumptions that lie behind attempts to predict how diet breadth should vary with ecological conditions.

Host-associated Fitness Components Other than Larval Success

Although I have quibbled with Dethier's specific examples, the discussion above clearly supports his conclusion that butterflies often fail to select the optimum hosts for their larvae. There are three possible explanations: the host choice behaviour may really be maladaptive, the success of larvae may not have been measured adequately, or the behaviour patterns that maximize the fitness of the parent may not involve preference for the most suitable larval hosts. I shall not discuss the first possibility in detail, since the reasons why optimality is not always to be expected have been adequately discussed by other authors (Gould & Lewontin 1979). Optimality is rightly less fashionable than it used to be. P. Whittaker (pers. comm.) has remarked: 'I don't believe in optimality, I believe in pessimality—the organism is no better than the environment forces it to be.'

The second possibility is that not all the important components of larval success have been measured. Attempts to measure additional components have not yet provided adaptive explanations of instances in which adult preference has been poorly correlated with larval survival. Courtney (1981) found that fitness components were positively correlated: plant species on which survival was low also supported slow development to produce small pupae that were more likely to suffer bacterial diseases. Nonetheless, they received more eggs per plant than the more suitable species. Singer, Mackay and Moore (unpublished) have also failed to account for preference of *E. editha* by investigating additional fitness components. At the GH (Generals' Highway) population, most butterflies prefer *Pedicularis semibarbata* and most eggs are laid on this species, but larval survival has been higher on *Collinsia torreyi* in each of the three years 1979-1981. Further investigation has shown that rates of parasitism are also different between the two hosts, and that additional advantages of feeding on *Collinsia* stem from predation rates on searching adults and on egg masses, growth rates of larvae and timing of emergence in the following generation. Larvae growing on *Collinsia* produce adults more than a week earlier than those on *Pedicularis*. These early adults lay eggs with higher survival, no matter which host they are on. If these eggs are laid on *Collinsia*, their earliness reduces the risk that the host will senesce before they are large enough to diapause. On *Pedicularis* they suffer less from intraspecific competition than do later larvae.

The third possibility listed above, that optimal oviposition behaviour does not always involve selection of the hosts on which larval growth and survival are best, could come about in several ways. Examples given by Rausher (1979*b*) and Courtney (1982*b*) invoke effects of host preference on adult fecundity. Rausher suggested that two papilionid species avoided plants growing in shady habitats and selected sunny sites where larval survival was lower because searching in the sun would probably increase the rate at which adults could find oviposition sites. Courtney & Duggan (1983) have calculated that the *A. cardamines* they studied only laid 15% of their eggs because of shortage of time when the weather was suitable for insect activity. Kingsolver (pers. comm.) has found that *Colias* spp. in Colorado are also limited in their fecundities by the numbers of oviposition sites that they can find. Courtney (1982*b*) points out that this situation entails selection pressure for expansion of the host range of the population. An insect that alights on a host, even a relatively poor quality host, should oviposit, since to do so reduces future oviposition only to the extent that the time taken to oviposit reduces time available for future searching. Based on this idea and on his field data, Courtney calculates that selection does indeed favour oligophagy over monophagy in his study populations. However, he assumes that handling time is negligible, and that an *A. cardamines* individual that alighted on four host species would not alight on species (A) less often than it would if it were monophagous, and alighted only on species (A). This biases his model in favour of finding advantages of oligophagy over monophagy. The situation he describes in *A. cardamines* is curious in that such low realized fecundities should entail strong selection for laying larger eggs or egg clusters, even though the larvae are cannibals. However, while the situation persists it certainly should give rise to increased diet breadth.

Rausher (1982*b*) makes the general point that flight search behaviour should evolve to maximize rates of encounter with hosts, while post-alighting preference should reflect the relative suitabilities of hosts for larvae. Thus, pre-alighting preference ranks are not necessarily closely related to host suitability for larvae. Courtney's work could fit this prediction, but he does not give data on post-alighting discrimination between host species. However, he shows that there may be post-alighting discrimination between conspecific hosts of different qualities—only about 12% of encounters with *Alliaria petiolata* are followed by oviposition regardless of the age of the plant. This high rejection rate is puzzling, since survival on *A. petiolata* is five times as great as on *Hesperis matronalis*, a host that receives more eggs per plant (Courtney 1981). Perhaps the observed rejections were responses to eggs already present.

Shortage of time for oviposition should affect pre-alighting and post-alighting components of preference in different ways. Species that are not short of time (e.g. cluster layers) have less selection for pre-alighting discrimination but greater selection for post-alighting accuracy, since oviposition on poor hosts does reduce the number of eggs that could be laid on good hosts in future. My own observations show various degrees of pre-alighting discrimination. *Laxita teneta* and *Chlosyne janais* (Singer and McKey unpublished) show no pre-alighting discrimination at all, just like the *Leptidea sinapis* observed by Wiklund. *Euptychia insolata*, feeding on an epiphytic moss (*Neckeropsis undulata*) merely restricts its search to green tree trunks. Those *Euptychia* species that feed on *Selaginella* seem to discriminate the host genus while in flight and then discriminate between species after alighting (but more data are required to be sure of this). *Siproeta stelenes* alights only on very small dicotyledonous plants, while *Perrhybris pyrrha* alights on all plants except ferns while searching for its Capparidaceous hosts. It is interesting that *P. pyrrha* does alight on ferns when investigating the environs of a host it has found. *E. editha*, characteristically, shows different degrees of pre-alighting discrimination in different populations.

In summary, Courtney considers that shortage of time leads to increased diet breadth, and that cluster-layers and tropical butterflies that have more time should be more restricted in their diets. Wiklund (1982) compares three Scandinavian species and argues that diet breadth increases in response to lower host predictability in space and time. Both Rausher and Wiklund separate preference into pre- and post-alighting components; Wiklund notes that pre-alighting discrimination should be relatively poor in species whose hosts are abundant or (to the insect) cryptic.

Larval Responses to Host Quality

I have discussed how oviposition behaviour is related to host quality. The outcome of this relationship is that the newly hatched larva finds itself on a plant that may be a good host, a poor host, or not a host at all. How does it respond? Wiklund (1975*b*) argued that the searching abilities of larvae are poor relative to those of the adults and, in consequence, the range of plants acceptable to larvae should be broader than that acceptable to adults. In other words, an adult that finds a plant of marginal quality may search elsewhere without ovipositing, but a larva that hatches on a marginal host should attempt to feed, since its chances of finding another host are low. Although Chew (1977, 1980) considered that the searching abilities of newly hatched *Pieris* larvae were good enough to reduce selection for adult ovipositional accuracy, she did confirm Wiklund's prediction. The range of plants accepted by larvae was often greater than that normally accepted by adults. Smiley's experiments, described above, depended on the fact that *H. melpomene* larvae would accept a wide range of *Passiflora* species. In this case the adults are so strictly host-specific that oviposition mistakes seem unlikely to have provided the selection pressure responsible for the breadth of diet acceptable to larvae. The gregarious larvae of *E. editha* seem unable to maintain their group integrity if they are forced to move more than a few inches in search of a host when newly hatched. Rather than move, they will attempt to feed on a wider range of plants than will support development (e.g. the toxic *Penstemon* at IF: Table 7.2).

At some point in its life, a larva may leave the plant on which it has been feeding in search of a new host. A host may become less acceptable to a larva as a result of a change in either host quality or larval preference. Changes in plant quality may result from maturation (Lincoln *et al.* 1982) or induced defences (Haukioja & Neimalaa 1977). A poor host may be rejected by a larva that has learned to avoid it after feeding on it and becoming 'ill' (Dethier 1980*b*). However, changes in larval preference often involve conditioning to the host—larvae come to prefer the species on which they have been feeding (Jermy *et al.* 1968). This is likely to be advantageous if the insect becomes digestively adapted to its host through the induction of appropriate enzymes. These phenomena have been documented for moths, not butterflies, but they are likely to be widespread, and circumstantial evidence is readily obtained. For example, the behaviour of a larva that cuts circular trenches in leaves, then eats the enclosed tissue (*Melinaea* spp.), may be indicative of induced plant defences. Larvae of *Danaus plexippus* cut the main leaf veins on the host-plant and then feed on the drooping leaf. DeVries (unpublished) reports that this habit occurs in several neotropical butterflies.

The necessity for larval migration often results from the selection of very young plants by adult insects. Captive *D. plexippus* prefer very young plants (Rothschild, pers. comm.), and DeVries reports that it is characteristic of most non-lycaenoid species in Costa Rica. *Euptychia (Cissia) confusa* at La Selva, Costa Rica, feeds principally on the seedlings of the palm *Euterpe macrospadix* (though at other sites its hosts are grasses). Although the palm seedlings vary in size, the butterflies select the very youngest and it is impossible for a larva to mature without having to migrate. This can be deduced from the correlation between size of the larva and defoliation of the plant; single larvae in third instar can be found on plants they have completely defoliated; conversely, large fourth (last) instar larvae can be found on plants with very little damage. In contrast, immature larvae of *D. plexippus* can be found migrating away from *Asclepias* plants that still bear partially-eaten leaves of all ages. The dispersal pattern of leaf damage suggests the existence of plant defences induced by larval feeding.

The search behaviour of migrating larvae has been studied by Jones (1977*a*) who found that starved *Pieris rapae* larvae moved in straighter lines than those only recently deprived of food. Her model of larval search showed that the behaviour of starved larvae was appropriate for locating a 'patch' of hosts, while well-fed larvae would be efficient at finding individual plants within a 'patch'.

Larval Growth and Metamorphosis

Scriber & Slansky (1981) have reviewed the data on growth rates and efficiencies of larvae on plants or artificial diets of varying qualities. This kind of information has been used by other authors (e.g. Futuyma & Wasserman 1981) to compare the growth of host generalist and specialist species and test the hypothesis that the advantage of specialization lies in more efficient exploitation of the host. No generalities have been forthcoming. One reason is likely to be variation among species in the extent of larval storage of reserves used by adults for reproduction (Boggs 1981*a*). Lederhouse *et al.* (1982) have recently shown that male *Papilio polyxenes* are more efficient than females at converting food into larval biomass. However, female pupae contain higher proportions of fat and protein than male pupae. The apparent inefficiency of female larvae, then, is probably due to their consuming large quantities of host material in order to extract compounds that are present in the plant in relatively low concentration. In general, a larva that has reached a size at which it could become an adult may either pupate or continue to store reserves for reproduction. *P. polyxenes* males become adults sooner while females store more reserves; this affects their apparent relative efficiencies as larvae measured in the last instar; probably earlier instars would differ less or not at all. Singer (1982*b*) suggested that holometabolous insects can trade time as large larvae against time as adults, and that the result of this trade-off 'should depend on the relative qualities of larval and adult habitats, especially such factors as relative predation and relative accessibility and quality of resources'. Thus, if host quality is high, a larva should be more likely to store reserves gleaned from the host and emerge as a (female) with mature eggs and potentially high fecundity early in adult life. I have observed that, among six species of neotropical grass-feeding euptychiines, slow larval growth and high food intake seem to be correlated with high adult fecundity early in life. The data are meagre, and I cannot yet assert that larval storage of reserves is generally correlated with slow growth. However, if this does occur, and if storage is a response to high host quality, then the effects of variation in host quality on larval growth rates and apparent efficiencies will be masked by this interaction between larval and adult stages. If generalists are indeed less efficient than specialists at utilizing their food, they may respond by allocating a greater proportion of reproductive reserve gathering to the adult stage, thereby becoming adult sooner and increasing their apparent larval digestive efficiency. Perhaps this problem could be overcome by measuring efficiencies of young larvae rather than those in their final larval instar, and by analyzing data for males and females separately, since the sexes are likely to respond differently to variation in host quality (Singer 1982*b*).

Host quality may also affect the length of the life cycle. A larva that finds itself on a poor host may either lengthen its generation time and become an adult of normal size or conserve its larval duration and become a smaller adult or one with fewer reserves. Where generations are discrete rather than overlapping the latter strategy seems more likely. However, the evidence from *E. editha*, which has only one generation per year, suggests that both effects occur. The weight at emergence of female insects shows a significant decline during the flight season (Singer and Moore, unpublished). The most likely explanation of this is that larvae growing in adverse conditions produce adults that are both small *and* late. Variation in larval environment masks, and indeed reverses, the expected positive correlation between adult size and larval development time.

This chapter has discussed the relationships between host quality and behaviour of both adult and larval butterflies and summarized the behavioural evidence that female butterflies discriminate between plants of the same or different species. The nature of this discrimination is related to the suitabilities of actual and potential hosts for larvae of monophagous and oligophagous butterfly species. Responses of larvae to host quality may involve compromises between size of adult and time of emergence, or between gathering of resources for reproduction by larval and adult stages. These relationships have consequences for interpretation of apparent larval efficiencies on different hosts.

Acknowledgements

I thank S. Davies, D, Harvey, J. Longino, D. A. Mackay, J. L. B. Mallet, R. A. Moore, D. Ng, P. J. DeVries and C. Wiklund for their help.

8. *Habitat versus Foodplant Selection*

Steven P. Courtney

University of Liverpool, Liverpool Department of Zoology,

Since the seminal work of Ehrlich & Raven (1965), a dominant theme in studies of butterfly-hostplant relationships has been plant characteristics, especially defence chemicals. Here I wish to consider another factor, widely discussed in optimal foraging theory, the time taken to find a host.

Time Constraints

Wiklund (1974*a*, 1975*b*) noted that, in addition to truly monophagous (M) or polyphagous (P) species, oligophagous butterflies might be classified into two groups—those species which utilize only one plant species in any one population ('monophagous type'—OM), and those which use all available foodplants ('polyphagous type'—OP). Examples are *Papilio machaon* (OM) and *Pieris (Artogeia) napi* (OP). The Orange Tip, *Anthocharis cardamines*, is an OP species which has been studied intensively in Sweden and Britain, with the aim of understanding how the OP habit may have evolved. Survival of immature stages is seen to be very poor on several regularly used hostplants. Why, then, are these species included in the diet? Earlier studies suggest that environmental heterogeneity in space (Chew 1977) or time (Wiklund & Ahrberg 1978) might be responsible, but neither idea is supported by work on British populations. More satisfactory is an explanation based around time shortage for searching females: frequently it will be advantageous to oviposit on poor hostplants rather than die with unlaid eggs. Key factor analysis suggests that other Holarctic species may be under similar constraints (Dempster, 1983). A model combining the selective effects of foodplant quality and availability, and of time constraints, leads to several predictions and tests.

Predictions and Conclusions

Female butterflies from populations where time constraints are important should include in the diet/oviposition range as many species of foodplant as are tolerable. Plant characteristics, especially toxins, will determine whether a butterfly is P or OP. Selection strongly favours catholicism in both larval tastes and female oviposition range. Populations of some butterflies not under time constraints (for example, in the tropics where adult lifespan may be long) should be more commonly monophagous (M or OM), and more selective as to where eggs are laid. Scriber (1973) has described how foodplant specialization by Papilionidae increases towards the equator.

Temperate region butterflies can evade time constraints if they lay batches of eggs, when the number of hosts to be found is drastically reduced. Species should then be of M or OM type. Species which employ a searching image (Rausher 1978) may be of OM type when specialization on a particular foodplant type *reduces* searching time. This may underlie the OM strategy of *Papilio machaon*.

Butterfly species with similar hostplant affinities should coexist by habitat segregation, *not* by segregation of foodplant types or species. Specializing on habitat types reduces the amount of time spent searching for hosts (MacArthur & Pianka 1966). For example, microhabitat selection by *Euchloe belemia* in south Morocco may greatly increase searching efficiency, reducing the time spent and increasing the number of eggs laid. Females search predominantly in thorn patches, rarely inspecting foodplants (Brassicaceae) outside. This results in a high contact rate with hosts, which occur at much higher densities within thorns.

Habitat and foodplant selection was further investigated in a guild of three Moroccan Euchloeini, all of the OP type, and all feeding on Brassicaceae. A total of 48 study sites were grouped on habitat types, and on the crucifers which occurred in each. Habitat type resulted in good

The Biology of Butterflies
0-12-713750-5

segregation of records for the three species, whilst grouping on foodplant did not. Specialization on a habitat type may produce an apparently monophagous relationship when only one foodplant species is present. A well-known example is *Pieris (Artogeia) virginiensis* (Hovanitz 1963). Similarly, *Anthocharis belia*, now known as an OP species, was previously thought to be of M type, due to its marked preference for rocky areas where its host, *Biscutellata*, is a predominant crucifer.

From this it follows that at lower taxonomic levels we should not expect foodplant shifts to be as important as at higher levels, in the manner first formulated by Ehrlich & Raven (1965). Especially at specific and intraspecific levels, habitat shifts should be more important in allowing coexistence. Some British populations of *Papilio machaon* (Dempster *et al.* 1976), *Pieris napi* and *Anthocharis cardamines* exhibit narrower habitat tolerances than other populations. In continental European *P. napi* at least, such habitat segregation is associated with speciation. Hence it may be argued that species richness of a genus may be related *not* to the degree of foodplant specialization shown by its species, but to the degree of habitat specialization. This appears to be the case for butterflies described by Higgins & Riley (1975) and Ehrlich & Ehrlich (1961), in the Western Palaearctic and America north of Mexico respectively.

Habitat selection is neither an easily studied nor attractive subject when compared to foodplant selection, although a few workers have now started investigations (e.g. Ohsaki 1979). Butterflies, however, make an ideal study group for this problem, and an open discussion on appropriate methods would currently be useful.

Acknowledgements

Christer Wiklund has contributed much to the development of the ideas presented above. Fieldwork in Morocco was supported by grants from NERC, and from the National Geographic Society (to F. S. Chew). This work was carried out during tenure of a NERC Research Fellowship.

9. *Parsonsieae: Ancestral Larval Foodplants of the Danainae and Ithomiinae*

John A. Edgar

CSIRO, Division of Animal Health, Animal Health Research Laboratory, Parkville, Victoria, Australia

A variety of factors determine the foodplants chosen by butterflies and their larvae (Chs 6–8), amongst which plant secondary chemistry is especially important (Fraenkel 1959, Ehrlich & Raven 1965). It is generally believed that many secondary chemicals evolved as defence agents, protecting plants from a variety of herbivores, pathogens and competitors. However, certain insects can metabolize or otherwise deal with particular classes of defensive chemicals, and specialize on plants containing these substances, in some cases storing the chemicals for their own defence (Rothschild 1972*c*; Ch.12).

As well as evolving increasingly effective means of detoxifying or tolerating hostplant secondary chemicals, specialist feeders are likely to have evolved sensory systems for locating chemically appropriate plants, and behavioural responses to ensure oviposition takes place upon them. The plants, by focusing encounters between the more chemically motivated insects, may also act as centres for mate location. Thus the advantages derived from utilizing particular hostplants could more rapidly and effectively accrue to the offspring of females which only became receptive to males in the vicinity of the plants, and this characteristic could be selected for.

If this is so, then it is conceivable that chemosensory systems first evolved for hostplant location may ultimately be used by the insects for inter- and intra-specific communication (see also Ch.25). This process would involve developing the ability to store and release plant chemicals. Selection would mainly occur at the behavioural level and could be considered an extension of the previously evolved activities centred around and triggered by foodplant volatiles (Edgar 1975). Release by males of a chemical signal indicating association with a particular class of advantageous (e.g. protective) secondary chemicals could form a basis for mate detection or selection remote from the foodplant (see Conner 1979). Such a process might be extended even further and give rise to territorial and other social signals.

Plants under attack from such chemically-dependent specialist feeders are likely to respond by evolving a different class of defensive secondary chemicals, still affording protection from general herbivores but also making them less 'apparent' to and protected from the specialists (Feeny 1976). Some insects may have the capacity to adapt to a change in the secondary chemistry of their hostplants and so retain links with them as they evolve. Other lines, however, may survive on unrelated but perhaps chemically similar plants, and still others may remain feeding on plants which, because they have been subjected to less feeding pressure, retain the original secondary chemistry. Within a group of closely related insects, plant associations may therefore exist reflecting steps in an evolutionary sequence, including some which display the characteristics of ancestral associations. Such an assessment of the possible evolutionary changes in the larval foodplant associations of two very closely related butterfly subfamilies, the Danainae and Ithomiinae, and the secondary chemistry of their ancestral larval foodplants is presented here.

Larval Foodplants of the Danainae and Ithomiinae

Danaine larvae feed mainly on the closely related plant families Apocynaceae and Asclepiadaceae, with some species favouring the Moraceae. The Ithomiinae feed almost exclusively on Solanaceae, the exceptions being a few which utilize Apocynaceae (Table 9.1). This suggests that ancestral Ithomiinae-Danainae fed on Apocynaceae,

The Biology of Butterflies
0-12-713750-5

Table 9.1. Larval foodplant families of the Danainae and Ithomiinae.

Danainae[1,2] (56 species)			
Asclepiadaceae	Apocynaceae	Moraceae	Other
42	17	15	11
Ithomiinae[2,3] (63 species)			
Solanaceae	Apocynaceae		
56	7		

Data from [1]C. C. J. Culvenor and author, [2]P. R. Ackery pers. comm. and [3]Drummond 1976. Some species of Danainae use foodplants in more than one of the plant families listed. The numbers in all cases represent the number of butterfly species, within the given subfamily, known to feed on the plant family indicated.

and have since diversified to, or coevolved with the Solanaceae, Asclepiadaceae and Moraceae.

This view is supported by the observation that the genera of Ithomiinae (e.g. *Aeria*, *Tithorea*) which feed on Apocynaceae have been judged to be morphologically among the more primitive (Fox 1956). The larvae of these genera have danaine characteristics, notably paired, flexible filaments (Gilbert & Ehrlich 1970, Drummond 1976, Young 1978*a*). Another example is *Tellervo*, found in northeastern Australia, New Guinea and adjacent islands, which thus occurs remote from all remaining ithomiines, otherwise exclusively inhabitants of Central and South America (Fox 1956). *Tellervo* larvae feed on *Parsonsia* (Apocynaceae) and, like *Aeria* and *Tithorea*, have a pair of flexible filaments on their dorsal surface (Common & Waterhouse 1981). Chromosome data suggest that *Tellervo zoilus* is primitive within the Ithomiinae (Emmel *et al.* 1974). *Tellervo* has thus apparently retained the primitive hostplant association and larval morphology, and has probably evolved in isolation from the Neotropical Ithomiinae, at least since the final break-up of Gondwanaland during the Eocene, when the last links between Australia and South America were severed.

Pyrrolizidine Alkaloid Requirement

Adult danaine and ithomiine butterflies are strongly attracted to and feed on plants containing 1,2-dehydropyrrolizidine alkaloids (Edgar *et al.* 1973, Edgar 1975, Pliske 1975*a*). These PA alkaloids and their metabolites are utilized for defence, courtship and territorial behaviour (see Edgar 1982; Ch.25). The shared requirement for PAs presumably pre-dates their division into separate subfamilies.

The danaine and ithomiine involvement with 1,2-dehydropyrrolizidine alkaloids would be unexceptional if the alkaloids were widely available in the larval foodplants: the PA requirement would be a 'normal' example of insects evolving the chemosensory mechanisms essential for locating their larval foodplants, as well as a storage capacity for defence, and subsequently exploiting both for inter- and intra-specific communication. However, PAs are not found in the great majority of danaine and ithomiine hostplants. My colleagues and I have therefore suggested that the PA requirement possibly stems from an ancestral association involving hostplants which did contain PAs (Edgar *et al.* 1974; an alternative hypothesis is discussed by Boppré 1978—see also Ch.25). In 1971, when this idea was conceived, 1,2-dehydropyrrolizidine alkaloids were unknown in any of the larval foodplant families of the Danainae and Ithomiinae. The Apocynaceae are noted for indole alkaloids, aminosteroids and cardenolides. Cardenolides also occur in the Asclepiadaceae and Moraceae, while the section of the Solanaceae most favoured by the Ithomiinae contain steroidal alkaloids. Our search for larval foodplants containing 1,2-dehydropyrrolizidines has now revealed these alkaloids in the Parsonsieae tribe of the Apocynaceae (Edgar & Culvenor 1975, Edgar *et al.* 1980)—precisely the section of the Apocynaceae in which the larval foodplants of the 'primitive' Ithomiinae are to be found (Table 9.2). Parsonsieae, notably species of the pyrrolizidine containing genus *Parsonsia*, are also used as larval foodplants by several Danainae

Table 9.2. Larval foodplants in the tribe Parsonsieae (Apocynaceae).

Species	Foodplant
Ithomiinae	
Tithorea harmonia salvadoris	*Fernaldia pandurata*[1]
Tithorea tarracina duenna	*Prestonia guatemalensis*[1]
Aeria eurimedia agna	*Prestonia* sp.[2]
Aeria olena	*Prestonia coalita*[2]
Tellervo zoilus zoilus	*Parsonsia velutina*[3]
Danainae	
Euploea treitschkei aenea	*Parsonsia spiralis*[4]
Euploea crameri bremeri	*Parsonsia helicandra*[5]
Euploea core corinna	*Parsonsia straminea*[6]
Tirumala hamata hamata	*Parsonsia* sp.[3]
Tirumala septentrionis	*Vallaris* sp.[7]
Ideopsis similis	*Parsonsia* sp.[7]
Idea hypermnestra linteata	*Parsonsia helicandra*[5]
Idea leuconoe	*Parsonsia* sp.[8]

Pyrrolizidines have been found in *Parsonsia* spp. (Edgar & Culvenor 1975, Edgar *et al.* 1980, Edgar, unpublished), *Fernaldia pandurata* (= *Urechites karwinsky*) (Borges del Castillo *et al.* 1970, see Edgar 1975) and *Prestonia guatemalensis* (Harvey & Edgar, unpublished). Foodplant records are from [1]Muyshondt *et al.* 1976, [2]Young 1978*a*, [3]Common & Waterhouse 1981, [4]Edgar 1982; [5]Kirton *et al.* 1982, [6]Sankowsky 1975, [7]P. R. Ackery, pers. comm. and [8]Muroya *et al.* 1967*a*.

(Table 9.2). The prediction and subsequent discovery of PAs in the shared apocyne hostplants of the two subfamilies supports the view that the Apocynaceae, and the tribe Parsonsieae in particular, represent the ancestral larval foodplants of these butterflies.

Genera containing cardenolides and aminosteroids are also found in the Parsonsieae (Hegnauer 1964). A change in the secondary chemistry of the ancestral larval hosts from PAs to cardenolides and aminosteroids, suggested by this coincidence of different secondary chemicals, may help to explain the present-day preference of the Danainae and Ithomiinae for foodplants containing one or other of these classes of natural product.

In the ancestral association, PAs were probably acquired by the larvae and carried through to the adult stage for storage and metabolic conversion to semiochemicals: the present-day acquisition of alkaloids by the adults is almost certainly a secondary development (Edgar 1982). Larval acquisition has so far only been demonstrated unequivocally in one instance—a single specimen of *Euploea treitschkei aenea* from the Solomon Islands reared on its natural hostplant, *Parsonsia spiralis* (Edgar 1982). It seems certain, however, that all species which feed as larvae on PA-containing plants will eventually be shown to sequester these alkaloids and retain them through to the adult stage. Wild-caught *Tellervo z. zoilus*, for example, contain a specific PA also found in the larval foodplant, *Parsonsia velutina*. Evidence that the mechanisms for larval acquisition of alkaloids still exist even in the more highly evolved species has been demonstrated by adding PAs to the diet of *Danaus plexippus* larvae, and finding them in the newly-emerged adults (Rothschild & Edgar 1978). However, the specificity of the transfer process from larva to adult has not been established.

Alkaloids of the Ancestral Larval Foodplants

1,2-Dehydropyrrolizidine alkaloids are found in the Boraginaceae, together with sections of the Asteraceae and Papilionaceae, and now the Apocynaceae. Several types can be defined. The macrocyclic diester alkaloids, such as trichodesmine (Fig. 9.1a) and senecionine (Fig. 9.1b) are typical of *Crotalaria* (Papilionaceae) and the tribe Senecionae of the Asteraceae, while alkaloids which are esters of the α-isopropylbutyric acids, such as indicine (Fig. 9.1c), are found in the Boraginaceae and the tribe Eupatorieae of the Asteraceae (Culvenor 1978). These alkaloids have a common aminoalcohol moiety but differ in the type of aliphatic branched-chain acids which form ester links with the alcohol groups (Fig. 9.1d). Both macrocyclic and α-isopropylbutyric ester alkaloids are stored by Danainae (Edgar 1982, Edgar *et al.* 1979, Edgar *et al.* 1976*a*) and are suitable for conversion into danaine dihydropyrrolizine semiochemicals (Fig. 9.1 e-g). However, only the α-isopropylbutyric type can provide the acids (Fig. 9.1h) needed for production of certain semiochemicals (Fig. 9.1i,j) secreted by several species of Danainae (Edgar 1982) and Ithomiinae (Edgar *et al.* 1976*b*). The specificity of this requirement, and the absence of semiochemicals derived from the esterifying acids of alkaloids such as those shown in Fig. 9.1 a and b, indicates that the ancestral larval foodplants probably synthesized alkaloids similar to indicine (Fig. 9.1c). The alkaloids found in the Parsonsieae are of this type, adding further support to the view that the Parsonsieae do indeed represent the ancestral hostplants of these butterflies.

(e) R = CHO, R' = H
(f) R = CHO, R' = OH
(g) R = CH_3, R' = O

(h) R = H
(i) R = CH_3

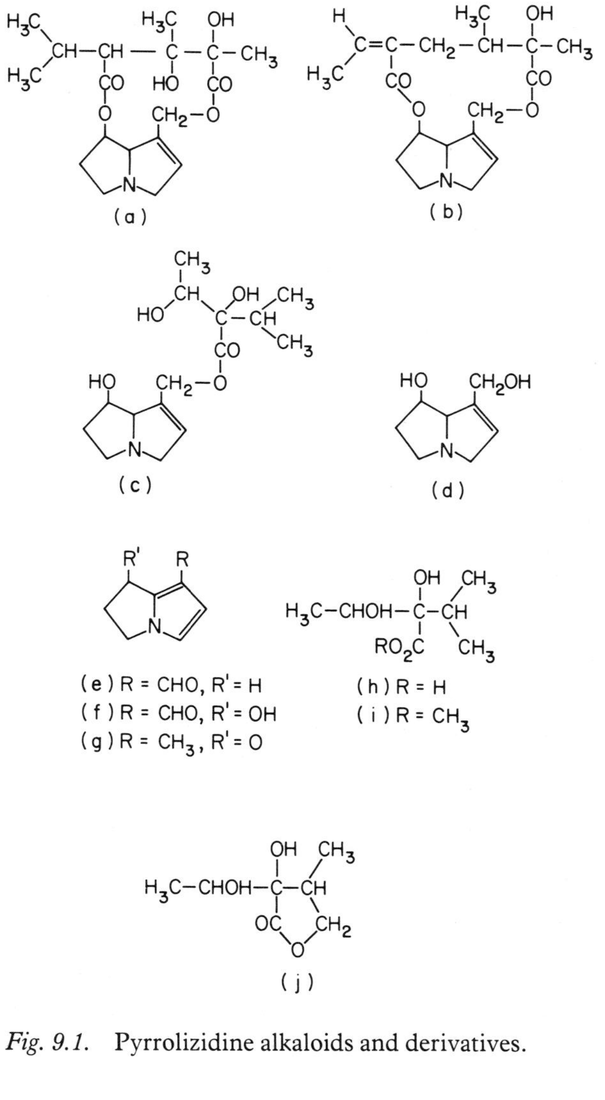

Fig. 9.1. Pyrrolizidine alkaloids and derivatives.

Part IV
Predation, Parasitization and Defence

10. *The Natural Enemies of Butterflies*

Jack P. Dempster

NERC Institute of Terrestrial Ecology, Monks Wood Experimental Station, Cambridgeshire

One has only to look at the huge array of defence techniques employed by butterflies to appreciate the impact of natural enemies on their evolution. Butterflies show an amazing range of morphological and chemical adaptations for predator-avoidance, such as camouflage, warning coloration and mimicry, distastefulness, protective silk webs and body hairs (Eltringham 1923, Hering 1926, Cott 1940, Ford 1945, Edmunds 1974*a*; Chs 12, 14). They also show many behavioural traits which appear to give protection against predation, such as the spacing out of camouflaged species and the gregarious behaviour of warningly coloured ones, the restriction of larval feeding to times when predators are least active, the elaborate interaction of some species with ants, and the defensive reactions by individual larvae when disturbed (e.g. the jerking of the body, erection of osmateria, vomiting, etc.). This range of predator-avoidance techniques gives the impression that predators have played a more important role in the evolution of butterflies than either parasitoids or pathogens. However, this could be misleading since many of the defences evolved against the latter (phagocytes, antibodies, encapsulation, etc.) are far less obvious to the casual observer, and some of those against predators may also give protection against parasitoids.

A number of papers in this volume discuss the evolutionary interaction between butterflies and their natural enemies. This paper considers the effects of these enemies on butterfly abundance. Of course, the two aspects of population dynamics and evolution are interlinked, but it is important to realize that high selection pressures imposed by natural enemies do not necessarily imply that they are having a large impact on butterfly numbers. Selection will simply determine which individuals survive, and in theory there can be marked genetical changes in response to predation within a stable population.

Considering the length of time that entomologists have been studying the ecology of butterflies, and noting attacks made by predators, parasites and diseases, there is surprisingly little quantitative evidence of their impact on butterfly populations. The literature is full of anecdotal comment, but there are few hard data from which to assess their effect. This will not be surprising to anyone who has tried to assess the effects of natural enemies in the field, since this is extremely difficult. Predation usually leads to the complete disappearance of the butterfly prey, frequently at night, so that identification of the cause of the disappearance by direct observation is often a matter of chance. Various indirect methods of assessing predation have been devised, but all have their limitations (Kiritani & Dempster 1973).

Assessment of the rates of attack by parasitoids and disease organisms appears easier at first sight, but the most commonly used method of study is by rearing individuals which have been brought into captivity from the field, and this may give misleading results. By removing individuals from a field population possible interactions with other mortality factors are ignored. For example, parasitized or diseased individuals may be more prone to predation. Added to this, rearing in the laboratory may well stress the individual, so increasing its chance of succumbing to disease.

Even if a reasonably accurate estimate of the number of prey killed by an enemy can be obtained, assessment of the impact that it is having on the prey population can be made only in relation to the other factors affecting population size. The destruction of a large proportion of the prey population does not necessarily affect the net survival of the prey, since it may simply replace some other cause of mortality without materially altering the number surviving. Similarly, a high, but constant, mortality caused by natural enemies

The Biology of Butterflies
0-12-713750-5

will have little effect on variations in abundance between generations.

The assessment of the impact of natural enemies is not an easy matter, and most of the available data must be treated with some caution, since most are crude, often biased, estimates. Nonetheless, there is a growing body of information which appears to be applicable to many butterfly species and also appears to fit theoretical considerations. In attempting to review this information, each of the major groups of natural enemy (predators, parasitoids and diseases) is considered before trying to generalize about their likely effects. Rather more space is devoted to reviewing what is known about predators than the other two categories, since reviews of the general biology of the latter can be found elsewhere (Shaw & Askew 1976, Rivers 1976).

Predators

Butterfly populations are attacked at all stages in their life cycle by a wide range of predacious animals, but especially by arthropods and vertebrates. Most of the predator-avoidance techniques, mentioned earlier, appear to be concerned with vertebrate predators which hunt by sight, but the early stages of many Lepidoptera are also eaten by a range of invertebrates, many of which feed at night. Against these, camouflage and warning coloration appear to give little defence. For example, the young caterpillars of the Cinnabar moth (*Tyria jacobaeae*) are attacked by mites, spiders, phalangids and beetles, which regularly take 30–90% of the first two instars, whilst vertebrates tend to leave the warningly coloured, poisonous larvae of this moth alone (Dempster 1971*a*, Dempster & Lakhani 1979).

Many workers have suggested that invertebrate predators are the main cause of death to the early stages of butterflies, particularly small caterpillars, although frequently the actual mortality attributable to invertebrate predators has not been assessed. Such studies include the following on *Pieris rapae* (Itô *et al.* 1960, Pimentel 1961, Dempster 1967, 1968, Parker 1970), *Papilio xuthus* (Tsubaki 1973, Watanabe 1976, 1979*b*); *Papilio machaon* (Dempster *et al.* 1976), *Papilio polyxenes* (Blau 1980); *Lycaena phlaeas* (Dempster 1971*b*), *Artopoetes pryeri* (Watanabe & Omata 1978), *Thecla betulae* (Thomas 1974), *Ladoga camilla* (Pollard 1979*b*), *Mechanitis isthmia* (Young & Moffett 1979) and *Euphydryas aurinia* (Porter 1981).

The best quantitative evidence of the effects of invertebrate predators is for *Pieris rapae* (Dempster 1967). In this study, arthropod predation, together with unknown factors, accounted for 52.5% and 63.4% of the early stages in two separate years. Using the serological technique known as the precipitin test, *Pieris* material was identified in the guts of a dozen species of arthropod, of which *Harpalus rufipes* (Carabidae) and *Phalangium opilio* (Phalangida) were numerically the most important. The bulk of arthropod predation occurred at night and was concentrated on the youngest instars (Fig. 10.1). Assuming that one caterpillar was eaten for every positive reaction obtained in the precipitin tests, a rough estimate of the minimum

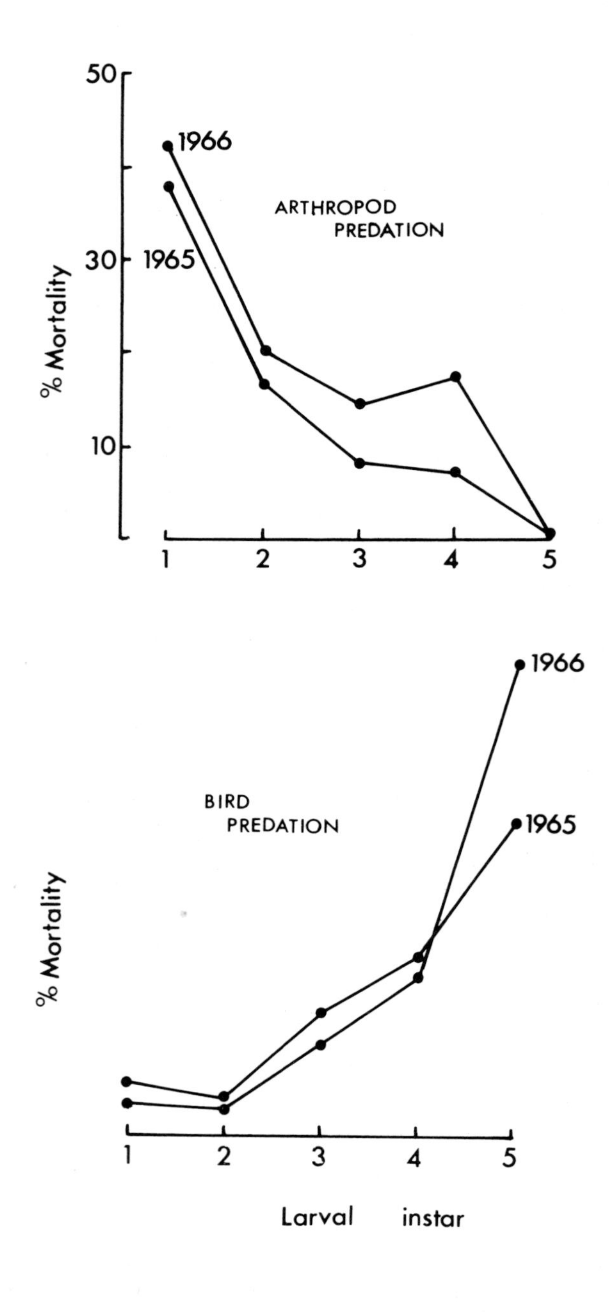

Fig. 10.1. Larval mortality of *Pieris rapae* caused by arthropod and bird predators in two years, 1965 and 1966 (Dempster 1967).

number eaten could be obtained from estimates of the number of each predator present, the percentage of guts giving a positive reaction, and the length of time that a meal remained detectable. This gave an estimate of 12 697 compared with a figure of 14 912 which were unaccounted for by other known mortalities, suggesting that arthropod predators accounted for the bulk of these losses. Further evidence of the importance of these predators was obtained by showing that survival of *P. rapae* caterpillars was significantly better after spraying with DDT. This eliminated many predators in the crop (Dempster 1968).

Although many authors have suggested that arthropod predators are the main cause of high mortalities amongst young caterpillars, some have specifically discounted this factor. Harcourt (1966) considered heavy rainfall to be the main cause of death in young caterpillars of *P. rapae* in Canada. His evidence rested largely on a correlation between mortality and rainfall. As with any correlation, this does not necessarily imply a causal relationship, and the possibility cannot be ruled out that rainfall was acting through its effect on predators, many nocturnal arthropods being less active in dry weather. However, the suggestion that arthropod predators play such an important role in determining the survival of the young stages of butterflies is based upon evidence from a very small number of studies. Equally, although it is probably true that most arthropod predation occurs on the young larvae, some arthropods, such as the wasps *Polistes* and *Vespa*, attack mainly older caterpillars of butterflies (e.g. *Papilio xuthus*: Tsubaki 1973, Watanabe 1976).

The predatory fauna may be markedly affected by the structure (height and density) of the vegetation in which the butterfly foodplants grow. In general, the denser and taller the vegetation, the larger is the number of predators present. Predation on *P. rapae* was greatly increased by the presence of weeds within the crop (Dempster 1969), whilst predation by spiders caused heavy mortalities of the young caterpillars of *Papilio machaon* when the foodplants were growing in tall, dense vegetation (Dempster *et al.* 1976). A similar effect was shown for the Cinnabar moth (Dempster 1971*c*), where rabbit grazing was shown to reduce the predatory fauna attacking young caterpillars.

In all cases where invertebrates have been shown to feed on the caterpillars of butterflies, they have been found to be polyphagous, i.e. feeding on many other prey species. Estimates of invertebrate predation covering more than a very small number of generations of a butterfly have yet to be obtained. It is therefore difficult to predict whether attack by invertebrates is likely to be density-dependent. It is theoretically possible that polyphagous predators may concentrate on particularly abundant prey, either by switching between different prey species, or by aggregation onto areas of high prey density. This will be discussed more fully below.

Of the vertebrates, birds appear to be the most important predators of butterflies (Moss 1933, Richards 1940, Carpenter 1941, Heslop 1955), but again the amount of quantitative evidence is very limited. In recent years, estimates of bird predation by direct observation (Baker 1970) and the use of exclusion cages (Dempster 1967, Duffey 1968, Pollard 1979*b*) suggest that birds will feed on all stages of the life cycle, from eggs to adults, but that they take mainly older larvae (Fig. 10.1) and pupae.

Baker's (1970) study of bird predation, by continuous observation in a small garden, is particularly interesting. He found that sparrows (*Passer domesticus*) took mainly eggs and young larvae of *Pieris rapae*, whereas older larvae and pupae were eaten mainly by tits (*Parus major* and *P. caeruleus*). Larger larvae were taken by a range of ground-feeding birds at the time of leaving the foodplants in search of pupation sites. Baker also found that bird predation was the main cause of larval mortality of *Pieris rapae* in the garden, although replaced in importance by arthropod predation in a field crop. This is in agreement with the findings of Dempster (1967) working on a field crop of brassicas. In two years he estimated that birds accounted for 22.8% and 21.9% of the eggs and larvae of *P. rapae*, compared with 52.5% and 63.4% taken by arthropods. The combined effects of arthropod and bird predation accounted for the bulk of larval mortality in this study.

The main evidence of bird predation on adult butterflies comes from research into the evolutionary advantages of mimicry. The now famous debate about the use of beak-marks to assess the frequency of bird predation (Carpenter 1941, 1942, Wheeler 1939) has since been put on a firmer basis by field experiments, such as those of Brower *et al.* (1967*b*), Cook *et al.* (1969) and Benson (1972) (see also Chs 2, 22). There can be little doubt that bird predation on adult butterflies has been a considerable selective force in the evolution of wing patterns. Even so, what evidence there is suggests that birds frequently take far higher proportions of caterpillars and pupae than of adults.

All terrestrial vertebrate predators (birds, mammals, reptiles, amphibians) are polyphagous, and probably switch onto particular prey organisms as they become numerous enough to be exploitable. Studies by Gibb (1958, 1966) showed that tits (*Parus caeruleus* and *P. ater*) concentrate onto areas of high prey density when feeding on the young stages of the moth, *Ernarmonia conicolana* (Tortricidae), in pine cones. Below a certain density, the tits ignored the moth. Similar, but less convincing data were obtained with *Papilio machaon* in Norfolk (Dempster *et al.* 1976, Dempster & Hall 1980)—the caterpillars were preyed upon more heavily in years of high density than in years of scarcity (Fig. 10.2). Although the same bird species were present at Wicken Fen, no predation occurred on an introduced population of the butterfly, except for one year when numbers were particularly high. Even then, distribution of predation was very patchy, and gave the impression of a small number of birds having learnt that the caterpillars were a good source of food. After developing a specific 'searching image' (Tinbergen 1960), they preyed heavily on the butterfly locally.

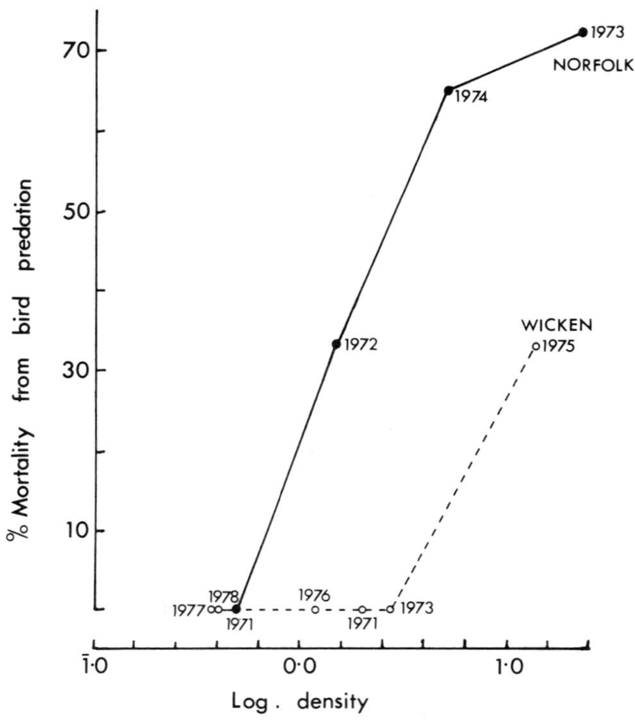

Fig. 10.2. The effect of larval density of *Papilio machaon* on the rate of predation by birds (Dempster *et al.* 1976, Dempster & Hall 1980).

There is little quantitative evidence of the effects of other vertebrates on butterfly numbers. Small mammals are possibly important predators of the pupae of some species (e.g. *Papilio machaon*, Dempster *et al.* 1976; *Thecla betulae*, Thomas 1974) and lizards take adult butterflies in the tropics (Ch.2), but there are no field studies of their impact.

The ability to aggregate onto areas of high prey density suggests that vertebrate predation may be density-dependent, at least over a range of densities (see below). However, it must be stressed that the field evidence for this is extremely sketchy for all types of prey, let alone butterflies.

Parasitoids

All stages of butterflies, except the adults, are attacked by parasitoids (Hymenoptera and Diptera; Shaw & Askew 1976). The tendency for lepidopterists to obtain parasitoids when rearing butterflies for their collections has led to a wealth of casual observations on rate of attack. Sometimes exceedingly high percentages of caterpillars have been found to contain parasitoids, but it is not always easy to interpret data of this sort, since the butterfly hosts have often not been collected systematically. For example, parasitized individuals often developed more slowly than healthy caterpillars, thus apparently very high rates of attack can be due to sampling late in the season. Nevertheless, there can be no doubt that some parasitoids are capable of killing very high percentages of their hosts in some years—Ford & Ford (1930) recorded 90–95% of *Euphydryas aurinia* parasitized by *Apanteles bignelli* in years following high numbers of the butterfly.

Whilst all predators feeding on butterflies are polyphagous, and are therefore not dependent solely on any one species of prey, parasitoids range from being polyphagous to totally specific. Frequently, a parasitoid species will attack a limited range of closely related hosts: *Apanteles rubecula* attacks both *Pieris rapae* and *P. napi*, whilst *Apanteles glomeratus* will attack all three British species of *Pieris*, *Aporia crataegi* and *Pontia daplidice* (Nixon 1974, Ford 1976), although *Pieris brassicae* is probably its main host.

When a parasitoid is specific, or nearly specific, its numbers are likely to depend upon the abundance of its host. An example of this is seen in Fig. 10.3, which shows the fluctuations in the number of *Apanteles rubecula* larvae leaving their hosts in each of ten generations. Its numbers closely followed those of *Pieris rapae*, with a fairly constant percentage of the hosts attacked each year.

Interdependence between parasitoid and host abundance can lead to classical intergeneration oscillation between the two, resulting in delayed density-dependence. In multivoltine species, the length of the generation times for host and parasitoid will greatly affect the end result and in many temperate butterflies, high rates of parasitism are usually found only at the end of the summer. Examples of this can be seen in *Apanteles medicaginis* attacking *Colias eurytheme* (Michelbacher & Smith 1943) and *Apanteles glomeratus* attacking *Pieris rapae* in Japan (Itô *et al.* 1960).

When the parasitoid has several generations to one of its host, it has a greater chance of adjusting its attack to the host's density, thus making more efficient use of the host population. For example, the attack of *Trichogramma* spp. (mainly *T. papilionis*)

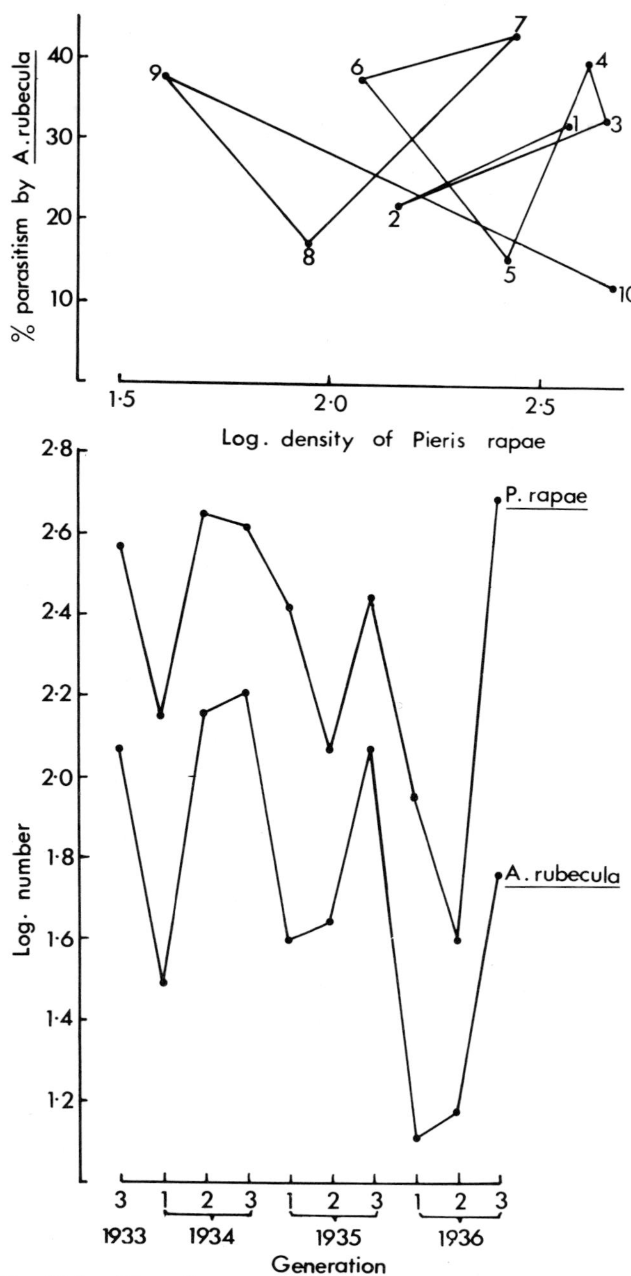

Fig. 10.3. The effects of larval density of *Pieris rapae* on the numbers and percentage attack of *Apanteles rubecula* (Richards 1940).

on the eggs of *Papilio xuthus* appears to be density-dependent, as a result of producing several generations in one generation of its host (Fig. 10.4; Hirose *et al.* 1980), although the *Trichogramma* are not specific in their attack on *Papilio*.

As with any two or more species exploiting the same resource, the presence of more than one species of parasitoid in one individual host frequently leads to interspecific competition and the death of one of the parasitoids. Richards (1940) showed that in *Pieris rapae* larvae, *Apanteles rubecula* was superior to the tachinids *Phryxe vulgaris*, *Epicampocera succincta* and *Compsilura concinnata* when in the same host. Again, this

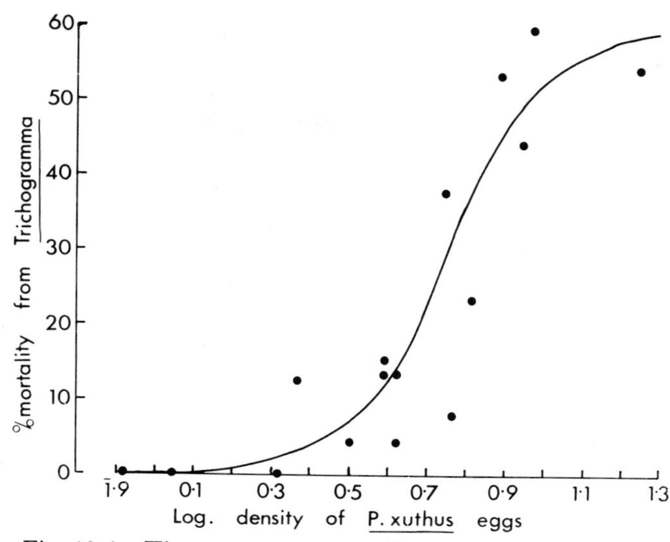

Fig. 10.4. The percentage mortality caused by *Trichogramma* at different densities of *Papilio xuthus* eggs (Hirose *et al.* 1980).

interaction between parasitoids can make interpretation of casual observations of parasitism more difficult.

The importance of parasitoids in the population ecology of butterflies is then difficult to assess. In the small number of detailed population studies which have been made on butterflies, parasitoids appear to have played a very minor role. In terms of numbers killed, they appear to have a smaller impact than do predators, and most appear to be density-independent in their action. However, once again, it must be stressed that the number of studies in which the rates of attack by parasitoids on butterflies have been measured over a reasonable number of generations is very small.

Parasites and Diseases

In common with all other animals, butterflies are subject to attack by a wide range of parasites and pathogens, such as fungi, bacteria, viruses, protozoa, nematodes and mites (Rivers 1976). The identification of these organisms usually requires specialist knowledge, thus few studies of butterflies have included any detailed assessment of their importance. Although diseases are frequently fatal, they can cause chronic infections in which the hosts do complete their life cycle. However, resulting adults often have a reduced fecundity as a result of the infection, as has also been found with some parasites, such as nematodes (Welch 1963).

Viruses are probably the most important pathogens of butterflies, causing what is often referred to as 'wilt disease'. A number of different types of virus disease have been described, differing in their microscopic structure, e.g. nuclear polyhedrosis viruses, granulosis viruses, and cytoplasmic polyhedrosis viruses. These tend to differ in their specificity, with cytoplasmic polyhedrosis viruses having a fairly wide range of hosts and granulosis viruses being most specific in their attack (Tanada 1965). Viruses attack all stages in the life cycle of butterflies, but they are least common in adults. However, adults can transmit some viruses via the eggs to their offspring.

The effects of virus diseases tend to be density-dependent, since they are more easily spread at high density. 'Stress' of one sort or another (crowding, poor nutrition, inclement weather, insecticides, etc.) can also cause outbreaks of virus disease, but whether this is due to activation of a chronic or latent infection is difficult to establish. In *Pieris rapae*, high death rates from virus have been associated with high densities (Harcourt 1966), wet weather and low temperatures (Dempster 1967), and DDT applications (Dempster 1968). Similarly, Michelbacher & Smith (1943) associated deaths from virus in *Colias eurytheme* with high densities and poor weather at the end of the summer.

Bacterial infections are commonly reported to cause the deaths of insects, including butterflies. Many species of bacteria are found in the guts of healthy insects, but if these are allowed to pass through the gut wall, sometimes as a result of damage from some other cause, such as another parasite, they can multiply in the haemocoel, causing septicaemia. The body contents darken and become viscous, giving off an unpleasant smell—a characteristic which distinguishes bacterial from viral diseases. There is no quantitative evidence of the effect of bacterial diseases in butterflies.

Because of their conspicuous mycelia and fruiting bodies, fungi are the most easily recognized pathogens of insects. They are probably most commonly found attacking hibernating pupae, but they will attack all stages in the life cycle. Many fungi are saprophytes, feeding on dead organisms, and this has resulted in some being mistakenly described as pathogenic. Nevertheless, fungi such as *Cordyceps*, *Metarrhizium* and *Beauvaria* are important pathogens of insects, including butterflies. None of these entomophagous fungi is specific in its attack and all depend upon the right weather conditions for their spread. Although they are commonly found attacking Lepidoptera in culture, they are probably only of minor importance in the field (but see also Ch.32).

Protozoa make up the fourth important group of pathogenic micro-organism. In the Lepidoptera, members of the Microsporidia appear to be most frequently encountered: *Nosema bombycis* caused heavy losses in the silkworm industry at the end of the nineteenth century. Like some of the viruses, the microsporidians can be passed from generation to generation through the eggs. From the limited information available, this group of diseases is not common in butterflies.

Parasitoids and predators may carry pathogens from one prey to another, and, as we have seen, some pathogens can be spread from females to their offspring. However, the principal source of infection is the release of pathogens by sick or dead individuals. Infected individuals regurgitate and defecate pathogens onto vegetation before they die, and the rotting corpses liberate even more. These are picked up by healthy individuals, usually by feeding on contaminated vegetation. Most pathogens are dependent upon the right weather conditions for survival outside the body of their hosts, making outbreaks of disease very dependent upon the weather.

The interplay between 'stress' and disease has scarcely been studied, but there is circumstantial evidence to suggest that the physiological state of the host plays an important role. Most data on the incidence of disease in butterflies have been obtained by rearing field-caught individuals. Artificial conditions are likely to cause 'stress' and so influence the results obtained, and this makes the high death rates from disease, reported by some workers, of doubtful significance.

Pathogens usually have very high rates of reproduction and short generation times, and so can respond rapidly to increases in prey abundance. On the other hand, they vary considerably in their virulence and often appear to require the right environmental conditions for adequate spread, so that any density-dependence in mortality tends to be imprecise. Far more information is required on the interaction between pathogens and their host populations.

Some Theoretical Considerations

Population ecologists try to explain two features of the populations which they study. First, they wish to know the causes of the fluctuations in abundance from generation to generation, and secondly, they are interested in what determines the extent of these fluctuations. The first involves the identification of what has come to be known as the 'key factors' determining population changes, and the second involves the recognition of regulatory, density-dependent factors which might be capable of limiting the extent of these changes.

Any factor which has a large and variable impact on the reproduction, survival, or dispersal of butterflies could theoretically act as a key factor, and there are some indications in the literature to suggest that natural enemies can act in this way.

Of the small number of more detailed butterfly population studies, those by Harcourt (1966, virus in *Pieris rapae*), Thomas (1974, predators of *Thecla betulae* pupae), Watanabe (1979*b*, arthropod predators of *Papilio xuthus*), and Pollard (1979*b*, bird predators of *Ladoga camilla*) suggest that natural enemies are key factors determining fluctuations between generations. In contrast, factors other than natural enemies have been shown to be the key factors for other species of butterfly (e.g. Courtney 1980, Hayes 1981). This evidence must be treated with some caution, since the methods of analysis involved are not directly comparable. Also, although bird predation was the key factor in the case of *Ladoga*, this was dependent upon summer temperatures which determined the period of availability of older larvae and pupae to the birds.

The role of natural enemies in the natural regulation of butterfly populations is far more difficult to assess, and here we are much more dependent upon theoretical considerations than on direct evidence. Over the last fifty years, there has been a huge number of laboratory and theoretical studies of the response of natural enemies to changes in the density of their prey, and these give us some insight into their possible roles in regulating populations of insects, such as butterflies. This is not an appropriate place to review such studies, but certain points are particularly relevant to this paper.

In theory, all natural enemies (predators, parasitoids, or pathogens) may respond numerically to changes in prey density as a result of improved survival, or reproduction, when prey numbers are high. This will be more likely to occur with specific enemies than polyphagous ones, since the performance of the former will be more closely dependent on the one species of prey. The effect of such an intergeneration response (Hassell 1966) on the prey population will depend upon the relative rates of reproduction of the enemy and prey. This will be affected not only by the enemy's innate powers of reproduction, but also by its generation time. A short generation time compared with that of the prey, as with some pathogens and parasitoids, can theoretically result in an immediate density-dependent effect on the prey population. An example of this was seen with *Trichogramma* attacking the eggs of *Papilio xuthus* (Fig. 10.4). *Trichogramma* has several generations to one of *Papilio*, but even so, there is an indication that its ability to respond in a density-dependent way breaks down at low prey densities.

When the generation time of the natural enemy is as long as, or longer than, that of its prey, the effect of the intergeneration response is likely to be delayed density-dependent, or not related to the prey density at all. This probably holds for most of the enemies of butterflies.

Those natural enemies which actively seek out their prey, such as predators and most parasitoids, may also respond to changes in prey density by modifying their behaviour. For example, they may catch or consume prey more rapidly at high prey densities, they may aggregate onto areas of high prey density, or they may switch from one prey organism to another in response to their relative abundances. Of course, such behavioural responses

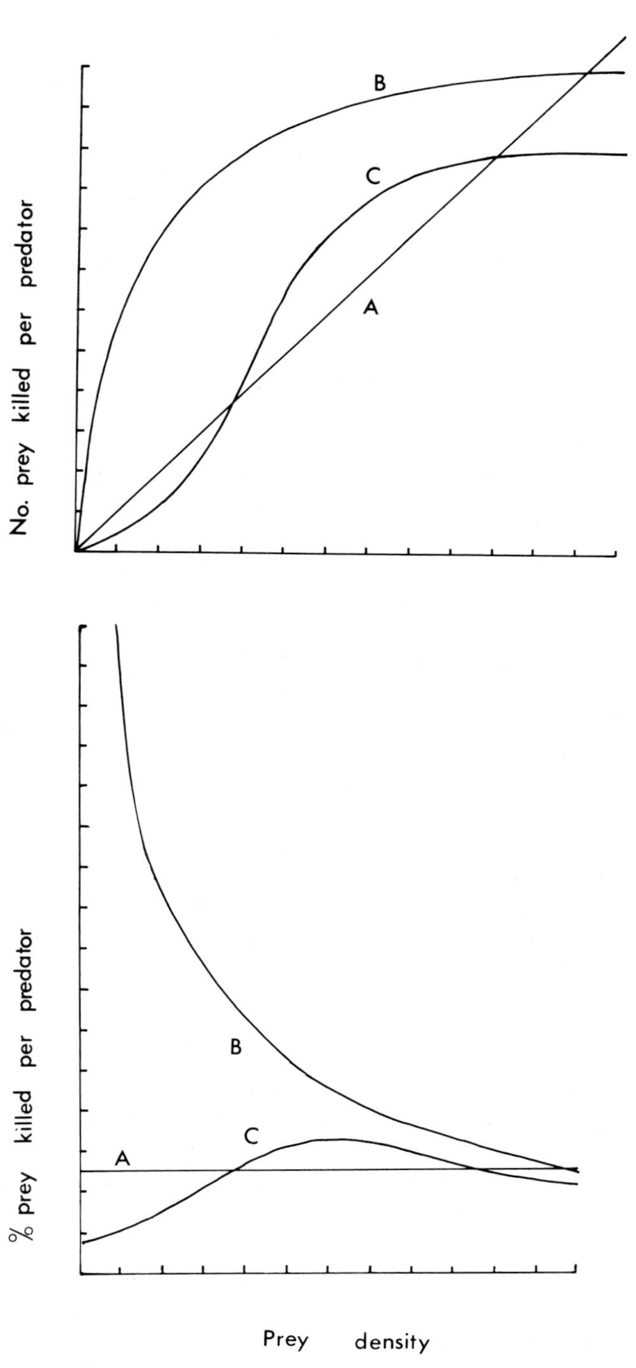

Fig. 10.5. The theoretical effects of behavioural responses by predators to changes in prey abundance.

(Hassell 1966) cannot occur for those organisms which are transmitted passively, such as most parasites and pathogens.

Changes in behaviour in response to changes in prey abundance are likely to lead to an increased number of prey killed with increasing prey density. Whether or not they will result in an increased percentage killed, i.e. density-dependent mortality, depends however on the shape of the response curve. Holling (1959) described from laboratory studies three basic types of response curve (Fig. 10.5) but only type C of these is likely to produce a density-dependent response, and then only over a limited range of prey densities.

Polyphagous predators, such as birds, may switch from one prey to another, so maintaining a higher density than more specific predators. They are also capable of aggregating in areas of high prey density, and so are most likely to react with a type-C response curve. At low prey densities their effect will be very small and density-independent. As prey becomes sufficiently abundant to be worth exploiting, their response is likely to be density-dependent, but limitations in their ability to deal with really large numbers of prey may result in a density-independent or even an inverse density-dependent effect at very high densities (Fig. 10.5).

The overall effect of an actively hunting natural enemy on a prey population depends on the combined effect of its behavioural and inter-generation responses, and clearly this may, or may not, be related to prey density. Added to this, proof of density-dependence does not show that the factor is capable of regulating the population. This will depend upon the slope of the density response.

Conclusions

There can be no doubt that natural enemies are amongst the most important causes of mortality of butterflies. As a group, they frequently cause very high death-rates and may sometimes be the key factors determining fluctuations in numbers. However, there is little evidence to suggest that the mortality caused by natural enemies is sufficiently density-dependent for them to play an important role in population regulation. Those with rapid rates of reproduction and very short generation times (pathogens and a few parasitoids) could theoretically respond to changes in prey numbers sufficiently quickly to cause density-dependent mortality, but there is little evidence to suggest that they regularly do so. Of the other forms of natural enemy, only the polyphagous predators are likely to act in a density-dependent way, as a result of their behavioural response to increases in prey density. However, the range of densities over which this response is likely to lead to density-dependent mortality will probably be limited.

Perhaps the most important point to be made is to draw attention to the scarcity of quantitative field data covering a reasonable number of butterfly generations, from which to assess the role of natural enemies. In spite of the long interest that entomologists have had in butterflies, we still know very little about their natural enemies.

11. *Host Specificity of Ectoparasitic Midges on Butterflies*

Richard P. Lane

British Museum (Natural History), London

Females of the Ceratopogonidae, a family of small biting-midges, feed on the blood of other animals, ranging from insects to birds and mammals. The insectivorous ceratopogonids can be divided into two general groups: those approximately equal in size to their hosts (mainly the insectivorous members of the Ceratopogoninae), and those mainly belonging to the Forcipomyiinae feeding on very much larger insects. Downes (1978) gives a detailed review of the biology of the insectivorous Ceratopogoninae.

The subfamily Forcipomyiinae comprises two genera, *Forcipomyia* and *Atrichopogon*. Both contain species ectoparasitic on a wide range of hosts, including harvestmen, dragonflies, damselflies, stick-insects, grasshoppers, alderflies, lacewings, beetles, butterflies, moths, crane-flies, mosquitoes, sawflies and coreid bugs. Undoubtedly, the host range (127 species) and the number of parasite species (about 50) are very incompletely known.

Both larval and adult Lepidoptera are attacked by midges. Larvae host the ubiquitous *Forcipomyia fuliginosa* and allied species. The poorly understood taxonomy of this species limits conclusions concerning host-specificity; *F. fuliginosa* may represent a species complex attacking a wide range of eruciform larvae, including Tenthredinoidea, Rhopalocera and Heterocera. Their potential use as vectors in biological control has promoted interest in their biology (Wirth 1972).

All midges attacking adult Lepidoptera belong to the subgenus *Trichohelea* of *Forcipomyia*. Although this subgenus is found worldwide, host records of Lepidoptera are more restricted geographically—probably due to lack of observations. Adult midges are usually found attached to the wing veins, where they draw haemolymph, becoming increasingly engorged until fully distended. Unlike *Forcipomyia* parasitic on other insects, there are few records of the Lepidoptera parasites at alternative sites of penetration, such as the thorax.

Lepidoptera midges can be recognized with reasonable confidence, even in the absence of host data, by morphological modifications apparently connected with attachment to and feeding from scaly wings. In contrast to the slender mandibles found in most *Forcipomyia*, those of the Lepidoptera feeders are broadly rounded, with blunt teeth. The maxillae are also unusual, each having two broad, leaf-like expansions apically, with parallel strengthening struts. Such expansions may be found in other *Trichohelea*, but the long, blunt shape is typical of the butterfly and moth ectoparasites. The tarsal claws are also modified, being either very long, slender and strongly recurved, or shorter and more open-angled. The latter type have basal spines or combs on the inner surface. Both claw types probably represent adaptations for attachment to scale-covered hosts (Lane 1977). However, claw modification shows little correlation with host specificity—a South-East Asian moth parasite, *F. pectinunguis*, has a comb on the claw, as does an African *Charaxes* parasite, *F. collinsi*. The combs may interlock with scale ridges (Lane 1977) and so possibly restrict these midges to hosts with particular scale characteristics, rather than taxonomic affinities.

Host Specificity from Published Data

Parasite species for which several records have been published (*F. aeronautica*, *F. pectinunguis*) are not confined to any host taxonomic category lower than the rank of 'suborder', e.g. Rhopalocera and 'Heterocera' (Table 11.1). All three recorded hosts of *F. baueri* are Lycaenidae, and both hosts of *F. monoplectron* are *Charaxes*. The late René Lichy

The Biology of Butterflies
0-12-713750-5

Table 11.1. Summary of all previously published records of ectoparasitic midges on adult Lepidoptera.

Ectoparasite	Host	Host family or subfamily	Country of origin	Location of midge on host
F. aeronautica Macfie	*Catoblepia berecynthia*[1]	Brassolinae	Brazil, Pará	'on wing'
	Catoblepia xanthus[2]	Brassolinae	Guyana	'on wing'
	Morpho menelaus[3]	Morphinae	Cayenne	'on wing'
	Morpho patroclus agamedes[3]	Morphinae	Venezuela	'on wing'
F. danaisi Floch & Abonnenc	*Danaus eresimus*[4]	Danainae	Venezuela	'on wing'
	Eurema nise[4]	Coliadinae	Venezuela	'on wing'
F. mexicana Wirth	*Pyrrhogyra otolais neis*[3]	Nymphalinae	Mexico	dorsal mid-hindwing
F. baueri Wirth	*Mitoura siva*[3]	Theclinae	USA, Arizona	ventral wings
	Philotes enoptera dammersi[3]	Polyommatinae	USA, Arizona	ventral wings
	Celastrina argiolus[5]	Polyommatinae	USA, Arizona	ventral hindwings
F. papilionivora Edwards	*Pieris napi*[6]	Pierinae	Britain	ventral forewings
F. auronitens Kieffer	*Phalera bucephala*[3]	Notodontidae	—	—
	Ectropis crepuscularis[7]	Geometridae	Switzerland	dorsal forewings
	Ectropis repandata[7]	Geometridae	Switzerland	dorsal forewings
	Lymantria monacha[7]	Lymantridae	Switzerland	dorsal forewings
F. ?auronitens Kieffer	*Perizoma didymata*[7]	Geometridae	Denmark	'on wing'
F. monoplectron Lane	*Charaxes z. zoolina*[8]	Charaxinae	Tanzania	ventral hindwing
	Ch. castor flavifasciatus[8]	Charaxinae	Tanzania	ventral hindwing
F. collinsi Lane	*Ch. n. numenes*[8]	Charaxinae	Ethiopia	ventral hindwing
F. tsutsumii (Tokanaga)	*Apatele rumicis oriens*[9]	Noctuidae	Japan, Honshu	'sucking blood'
F. pectinunguis (Meijere)	*Miltocrista cruciata*[10]	Arctiidae	Sumatra	'on wing'
	Simplicia marginata[10]	Noctuidae	Sumatra	'on wing'
	Heterocera[11]	—	Sumatra	'sucking blood'
	*sphingid moth[12]	Sphingidae	Caroline Islands	—
	**Chromis erotus eras*[13]	Sphingidae	Caroline Islands	base forewings, dorsal thorax and base abdomen

Specimens from the record for *Perizoma didymata* have been lost subsequently, so that accurate identification is impossible.
*Two records refer to the same observation.
[1]Specimen in United States National Museum, Washington DC; [2]Macfie 1935; [3]Wirth 1956; [4]Floch & Abonnenc 1950; [5]Ehrlich 1962; [6]Edwards 1923; [7]Edwards 1925; [8]Lane 1977; [9]Tokanaga 1960; [10]de Meijere 1923; [11]Macfie 1934; [12]Tokanaga 1940; [13]Esaki 1940.

(1946) gave a list of 23 Rhopalocera species and one ctenuchine on which he found ectoparasitic midges in Venezuela. Unfortunately, only midges from two host species were sent for identification (they now comprise the type-series of *F. danaisi*). It cannot be assumed, however, that all the records refer to this parasite, as several lepidopteran species listed by Lichy are recorded here (Table 11.1 and below) as hosts of *F. aeronautica* in the Brazilian Amazon. Furthermore, several new species of ectoparasitic *Trichohelea* from northern South America await description (Lane, unpublished). From this survey of published records it is clear that far more data are required before any conclusions on host-specificity can be reached—and such data should be obtained by systematic sampling methods rather than accumulating sporadic or casual observations.

A Field Observation

In March 1980 I was able to make observations on host-specificity in Belém, Pará State, Brazil. An attempt was made to sample all butterflies occurring at a single location, on a single day, for the presence of midge ectoparasites. The study area, a 100m length of track some 2-3m wide, was situated in dense wet lowland forest, at Utinga, near Belém. All butterflies entering the area, except a few small *Euptychia* which escaped, were collected during the period 08.30-12.00h. Each butterfly was inspected whilst still inside the net as engorged midges may leave their host immediately on its capture or death. This behaviour probably accounts for the scarcity of parasite records from butterflies. By noon most butterflies had ceased flying.

During the 3½ hour sampling period, 45 butterflies were examined (22 species in 14 genera), from 12 of which 27 midges were collected. Only one midge species was present, *F. aeronautica*. Details of the butterflies collected, their systematic position, dominant colour and wing length are presented—Table 11.2 for those with parasites, Table 11.3 for those without.

Table 11.2. Butterflies found with ectoparasitic midges at Belém, Pará, Brazil.

Butterfly species	Subfamily	Dominant colour		Wing length (mm)	Position on host	Number of midges
Morpho menelaus	Morphinae	dark brown/blue	(i)	85	dorsal fore-wing	4
Morpho achilles	Morphinae	dark brown with blue stripe	(i)	60	dorsal hind wing on cubital vein	5
			(ii)	70	dorsal hind wing and in net	5
			(iii)	64	dorsal hind wing on cubital vein	4
Taygetis andromeda	Satyrinae	dark brown	(i)	36	in net	2
			(ii)	39	dorsal fore wing	1
Haetera piera	Satyrinae	translucent and dusty brown	(i)	32	in net	1
			(ii)	33	in net	1
Euptychia hesione	Satyrinae	brown and white	(i)	20	in net	1
Pierella hyalinus	Satyrinae	blue and white	(i)	—	dorsal hind wing	1
Eurybia halimede	Riodininae	dark brown with orange spots	(i)	26	in net	1
Parides lysander	Papilioninae	black with pink patches	(i)	45	dorsal thorax ? feeding	1

Table 11.3. Butterflies found without ectoparasitic midges at Belém, Pará, Brazil.

Butterfly species	Subfamily	Dominant colour	Wing length (mm)	Number examined
Taygetis andromeda	Satyrinae	dark brown	35	1
Euptychia hermes	Satyrinae	dark brown	16-17	2
Euptychia hermes group	Satyrinae	dark brown	16-18	4
Euptychia mollina	Satyrinae	white with brown stripes	18-19	4
Euptychia ? myncea	Satyrinae	dark brown	19-20	3
Euptychia penelope	Satyrinae	dark brown	19-20	3
Euptychia renata	Satyrinae	dark brown	19-24	2
Anartia j. jatrophae	Nymphalinae	light brown and white	32	2
Heliconius e. erato	Heliconiinae	dark brown, orange and yellow	38	1
Mechanitis polymnia	Ithomiinae	black, yellow, and orange	31-38	5
Eurema venusta	Coliadinae	yellow and brown	17-19	2
Helicopsis acis	Riodininae	orange, white and brown	25	1
Stalachtis phlegia	Riodininae	black and orange	26	1
Leucochimona hyphae	Riodininae	brown with white bands	13	1
Hyalothyris leucomelas	Pyrginae	dark brown and white	20	1

Discussion

The results of the Belém observation produce more questions than answers. Tables 11.2 and 11.3 show that *F. aeronautica* is not specific to any one butterfly group. Whether this infers only limited, if any, host-specificity beyond 'Rhopalocera', or that midges are selecting hosts according to other criteria or purely opportunistically, is difficult to determine. The great majority of hosts are relatively large (over 20mm wing length), predominantly dark in colour, slow-flying (when not disturbed) and more prevalent in the darker areas of the forest. The midges may therefore be selecting, or best able to infest hosts of a particular ecological or behavioural type. The daytime activities of the sexes of butterflies are often very different, but both males and females were parasitized. Nocturnal roosting, communal feeding, or other activities indulged in equally by both sexes may be important for location by midges.

None of the six Ithomiinae or *Heliconius* were found to have been attacked (Table 11.3), and midges were not present on dozens examined subsequently in the area, perhaps suggesting that *F. aeronautica* avoids these butterflies. However, although both butterfly groups concerned are well known for their apparent unpalatability to vertebrates, they may not be unpalatable to midges: the closely related *F. danaisi* feeds on the chemically-protected Danainae. Furthermore, L. Gilbert (pers. comm.) reports roosting *Heliconius* in Venezuela to be attacked by midges, and P. DeVries (pers. comm.) has made observations of parasitization on both *Heliconius* and ithomiines in Costa Rica.

Most midges found on the host were on the dorsum, most frequently affecting the hindwings. This might appear to be a precarious position, but a boundary layer of still air trapped during the upward 'clap' of the wings in flight is presumably sufficient to prevent midges from either being damaged or dislodged. Although relatively few individuals of most butterfly species were found to have parasites, the presence of midges on all four *Morpho*, three of the four *Taygetis andromeda*, and both *Haetera piera* specimens, suggests that these species are particularly attractive to midges.

A complicating factor is that the number of parasites per host appears to be related to host size, ranging from one midge on small species, up to four or five on *Morpho*. However, although this may be true for *F. aeronautica*, it is not necessarily a general phenomenon. Edwards (1923) found nine midges on an individual of *Pieris napi*, and single midges have been found on large butterflies (e.g. *Charaxes*, *Morpho*) on several occasions. With respect to maximum recorded numbers, it is interesting to note that the powerful *Charaxes numenes* is still able to fly when parasitized by up to 30 midges (Lane 1977), with a similar number also recorded from a single moth (Tokanaga 1960).

Many questions remain completely unanswered. How does a midge locate its host? Is the host attacked whilst in flight or at rest? Do midges select hosts, and if so, is this on the basis of taxonomy, chemistry, behaviour or ecology? What effect, if any, is there on the host? How do midges avoid being dislodged? As this phenomenon of midge ectoparasitization becomes more widely known to lepidopterists, it is to be hoped that answers to some of these questions will be forthcoming.

12. Chemical Defence in Butterflies

Lincoln P. Brower

Department of Zoology, University of Florida, Gainesville, Florida

Our knowledge of chemical defence in plants and animals is based on a rich background of nineteenth and twentieth century natural history, and in the past 25 years has been greatly advanced by interdisciplinary research in evolutionary ecology, biochemistry, animal behaviour, pharmacology, and toxicology. Historically, we are indebted to H. W. Bates and F. Müller for their theories of mimicry, which postulated that model insects are avoided by predators because they contain substances which are unpalatable. The search for these chemicals, their identification, how they are deployed by the prey, and their effects upon predators have been subjects of intense research. Of all organisms, butterflies and their avian predators have provided the greatest general insights into the complexities of chemical defence. It is therefore singularly appropriate to write upon this subject in honour of E. B. Ford.

My review consists of three parts. The first presents an overview of the ecology and behaviour of the enemies of butterflies. Here I develop a new definition of chemical defence based on the properties of what I have called Class I and Class II chemicals as they relate to the physiological sensitivities and behavioural repertoires of predators. I also discuss common misconceptions about the nature and functioning of chemical defence, and the possibility of chemical mimicry. In the second part, I review the kinds and pharmacological effects of chemicals that are known or thought to be of importance in chemical defence. The third part surveys the many chemical defence mechanisms in the egg, larval, pupal and adult life history stages of representative butterfly taxa. I conclude by comparing and contrasting the purported efficacies of pyrrolizidine alkaloids and cardenolides as defensive chemicals in butterflies.

Chemical Defence in the Context of Predator and Parasitoid Behaviour

Predators and Parasitoids of Butterflies

Chemical defence operates in all stages of the life history of butterflies. Because more eggs or larvae are killed than adults does not lessen the importance of adult mortality; indeed, each additional day that an adult survives increases its reproductive potential (Nicholson 1927, Fisher 1930). Eggs and young larvae are subject to extensive arthropod predation, especially by mites, harvestmen, spiders, ants, wasps, bugs, and beetles, whereas later larval instars and chrysalids are more subject to being eaten by vertebrates including lizards, birds, shrews, mice, and primates (Clausen 1940, Imms 1951, Doutt 1959, Wigglesworth 1964, Dempster 1967, Buckner 1971). Adults share many of these predators and are also eaten by fish, frogs, toads, snakes, mantids, dragonflies and robberflies. The early feeding experiments of Poulton (1887) and Pocock (1911) in England, of Marshall & Poulton (1902), and of Swynnerton (1915*c*, *d*, 1919) in Africa, of Finn (1895-1897) in India, of Jones (1932, 1934, 1937) in North America, and of Darlington (1938) in Cuba involved numerous species of predators. Recent discussions of predators of butterflies at various stages in their life history include Rothschild 1967, 1972*c*, Reichstein *et al.* 1968, Baker 1970, Rettenmeyer 1970, Owen 1971*a*, Pasteur 1972, Marsh & Rothschild 1974, Edmunds 1974*a*, Enders 1975, Nielsen 1977, Turner 1977*a*, Gilbert & Singer 1975, Stamp 1980, Rausher 1980, 1981*a*, Wright 1981, West & Hazel 1982, and Dempster 1967, and Ch.10. Finally, judging from the following quotation from Ford (1955: 103), human predators need also to be considered: 'I personally

The Biology of Butterflies
0-12-713750-5

have made a habit, which I recommend to other naturalists, of eating specimens of every species which I study'.

Vertebrates are long-lived and can utilize learning to avoid their predators (but see Bolles 1970). In contrast, because of their small brain sizes and short life-spans (Wigglesworth 1972, Wilson 1971: 219), insects have little opportunity to learn to cope with their enemies. As a consequence, the antipredator and antiparasitoid behaviours of insects are largely genetically determined, stereotyped responses which enhance the insects' morphological and chemical defences.

Chemical defence reduces the risk of predation and is a system of behaviour control of obvious benefit to the prey. At the same time, learning to avoid chemically defended prey is an important fitness component in the behavioural repertoire of predators. Thus prey which contain noxious substances may result in a predator's loss of time, lowered ability to care for offspring, and greater exposure of the predator to its own enemies. If a prey contains an emetic substance, a predator can lose previously eaten food and be incapacitated for up to 30 min (Brower *et al.* 1968). This could well be dangerous in the natural environment. For example, Gibb (1954, 1960) found that the English Goldcrest spends virtually 100% of its waking time feeding during winter, and that individual birds require an average of one food item every 2.5s to survive.

Predators of insects exhibit wide ecological partitioning which differentially affects the risk to various life history stages. Foliage gleaning birds will most frequently find caterpillars, aerial hunters adults, and ground feeders as well as trunk gleaners will more often encounter pupae (Morse 1971, West & Hazel 1982). The relative development of senses in predators is also important. Thus among lizards, anolids hunt mainly by vision in arboreal habitats, whereas lacertids use both vision and olfaction in searching for ground dwelling prey (Johki & Hidaka 1979).

Behavioural rigidity of predators can also determine which life history stages serve as prey. Mantids, frogs and toads strike only moving objects and therefore probably never eat eggs or pupae. Likewise, the anthophilous habitat of many assassin bugs results in their capturing mostly adults. Large, generalized 'sit and wait' predators such as orb web spiders frequently capture adults, but probably capture larvae rarely, and would probably never encounter pupae. In contrast, a small species of spider spins webs at the base of *Aristolochia* plants and preys heavily on young *Battus* swallowtail larvae (Rausher 1981*a*).

In certain circumstances even predators which are not specialized insect eaters can become major enemies (see also Ch.10). For example, Black-headed Grosbeaks, *Pheucticus melanocephalus*, have become one of two major predators of Monarch butterflies in Mexico (Calvert *et al.* 1979; Fink & Brower 1981, Fink *et al.* in press) and in mixed flocks with Black-backed Orioles, *Icterus abeillei*, were estimated to kill from 3500–35 000 adults per day (Fig. 12.1; Brower & Calvert in prep.).

Nocturnal (see Ch.10) and diurnal partitioning of the environment is another major factor which has shaped defensive adaptations in butterflies. Adults of most species must enjoy relative freedom from

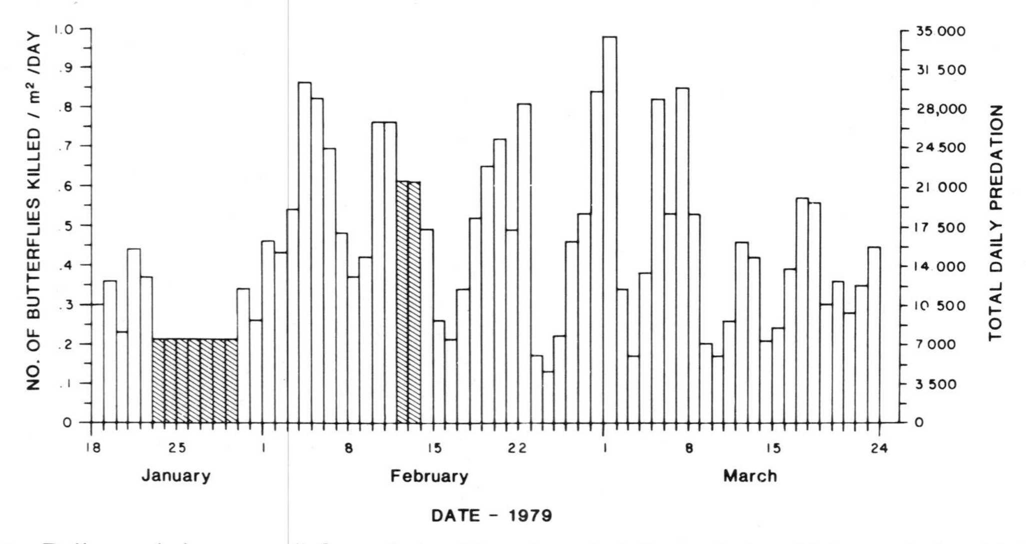

Figure 12.1. Daily predation rates (left vertical axis) and total daily predation (right vertical axis) for Monarch butterflies at Site Alpha, Michoacan, Mexico during the 1978-79 overwintering season. Large flocks of Black-headed Grosbeaks and Black-backed Orioles attacked the roosting Monarchs during morning and afternoon feeding forays. The colony area was *ca* 35 000m^2 and total mortality for the overwintering season was conservatively estimated to be 1.808 million butterflies (from Brower & Calvert, in prep.).

nocturnal predators because they are highly dispersed and totally inactive during the night. Moreover, they are largely free from attack by parasitoids. As a consequence, birds, and to a lesser extent lizards, must be the principal biotic agents driving the evolution of defensive adaptations in adult butterflies.

Strictly nocturnal predators cannot employ colour vision, and probably most insectivorous mammals have foregone visual orientation in favour of acoustical, olfactory, tactile, and/or gustatory cues (Dunning & Roeder 1965, Dunning 1968, Braveman 1977), although more mammals have colour vision than once thought (Jacobs 1981). Some nocturnal predators may be able to associate chemicals in butterfly foodplants with the presence of immature stages, but they would probably not be able to use this strategy in locating adults which in most species roost independently of their foodplants. On the other hand, if eggs were to have the specific odour of the leaves they were laid on, then scent oriented parasites and predators might have greater difficulty in finding them. Finally, it is well to remember that generalizations from one predator or prey species to others may not be warranted, and what applies to one life history stage may not apply to another.

How do Chemicals Defend Prey?

Terminological multiplicity

Since Bates (1862) originally introduced the term 'unpalatable' to describe butterflies in Amazonia which he reasoned were not attacked by predators, many words have been used to describe purportedly defended insects. Vane-Wright (1976, 1981*a*) has pointed out that *ad hoc* terminology has impeded our understanding of mimicry. Uncritical and synonymous usage has likewise led to semantic obfuscation of both the evolutionary ecology and underlying mechanisms of chemical defence. A non-exhaustive list among literally many dozens of descriptive terms found in the literature includes distasteful, nauseous, bad, vile, disagreeable, repulsive, nasty, repellent, offensive, inedible, deterrent, unacceptable, noxious, irritating, poisonous and toxic.

Class I and Class II defensive chemicals

There are in fact two distinct classes of defensive chemicals which can deter enemies. Those in *Class I are noxious* by virtue of their capacity to irritate, hurt, poison, and/or drug an individual predator or parasitoid; they may or may not stimulate the olfactory or gustatory receptors. Those in *Class II are innocuous* chemicals which harmlessly stimulate the predators' olfactory and/or gustatory receptors. The seeming paradox that innocuous chemicals can be defensive is at the heart of much confusion.

In terms of classical conditioning theory, Class I chemicals are unconditioned negative stimuli which stimulate external and/or internal neural receptors either directly or indirectly and produce a variety of noxious effects. When their effect is immediate (= active phanerotoxins, see Russell 1971), prey possessing them can avert predators directly and often survive being attacked. When Class I chemicals have delayed effects (= passive cryptotoxins), rejection by predators is possible only through learning trials in which the prey are likely to be killed and eaten. This difference is of profound importance for the evolution of chemical defence.

There are three possible ways in which predators can learn to avoid prey containing Class I chemicals. First, Class I chemicals can condition the prey's visual appearance, acoustical properties, and tactile characteristics as aposematic stimuli. Secondly, the noxious chemical may also itself possess specific odour and/or taste qualities. In this case, the chemical's noxious effect may condition avoidance of its accompanying odour and/or taste properties. Thirdly, even if odourless and flavourless, Class I chemicals may condition any other innocuous odorous and/or flavoured molecules present in or on the prey to become Class II chemical aposematic stimuli. Class II chemicals also can become aposematic stimuli in prey which are physically noxious because they are extremely tough, possess sharp spines, or are able to bite, *etc.*

The solution to the seeming paradox of the innocuousness of defensive chemicals is as follows: prey which contain only Class II chemicals can be chemically aposematic without being chemically noxious. It also seems likely that Class I chemicals may occur in prey at concentrations which are too low to be noxious, but still can be sensed. In this case they are effectively reduced to the Class II category.

Failure to recognize the distinction between the Class I and Class II properties of defensive chemicals is seen in two recent statements. In discussing strychnine and amygdalin (which on hydrolysis yields HCN), Bate-Smith (1972: 53) said the fact that they '. . . happen to be extremely toxic is incidental—it is their bitterness which is repellent rather than the fact that they are deadly.' Rhoades (1979: 40) stated the exact opposite. 'Animals avoid certain foods because they possess toxicity or some other noxious quality, not because they have a bad taste or smell'.

Class I compounds which are lethal do exist but ecologically are far less efficient than those which can condition the avoidance of entire populations of predators (Williams 1966: 228, Brower 1971, Smith 1975, 1977, 1978, Greene & McDiarmid 1981; Huheey 1980*a*, and in press).

Proximal and distal avoidance of chemically defended prey

Defensive chemicals (as well as physical noxiousness) produce two types of conditioned avoidance. The first is conditioned *proximal avoidance* in that the experienced predator must come into actual contact with the prey in order to taste (and perhaps also smell) or feel it before remembering to reject it. The second is conditioned *distal avoidance* in which the predator perceives the prey from a distance and by remembering its odour, sound and/or visual appearance avoids physical contact. Marshall & Poulton (1902), Swynnerton (1919) and Rothschild (1961) recognized the importance of distal aposematic cues, which Huxley (1938*b*) and Cott (1954) had termed allaesthetic characters, and which recent literature, in the context of information theory, calls predator-selected-signals (Matthews 1977, Dawkins & Krebs 1978, Jacobs 1981). A useful index to conditioning signals for various species is in Barker *et al.* (1977).

Chemicals which facilitate distal avoidance *via* animals' telereceptors are of advantage to both predator and prey: the predator need not waste time and energy in pursuit, capture, handling and possible rejection of the prey, and the prey benefits in being actively avoided and completely unharmed by the predator (Brower 1969, Garcia *et al.* 1974, Bobisud & Potratz 1976). In contrast, chemicals which condition only proximal avoidance require the predator physically to contact and possibly damage the prey before rejecting it. The difference between proximal and distal avoidance is thus of immense ecological and evolutionary importance (Ford 1975*a*).

Conditioned distal aversion of mammals by nocturnal moths which lack the ability to produce warning sounds requires an odorous compound, whereas conditioned distal aversion of birds by butterflies does not, because butterflies are diurnal and can utilize visual signals. This is probably why odours in butterflies are less common than in moths. Thus learned visual avoidance in birds can be based on a single noxious chemical. If Eisner & Grant (1981) are correct that no known insect odours are Class I deterrents, then learned odour avoidance by mammals requires at least two chemicals, one that is noxious and a separate one that is odorous. It is also interesting to speculate that the association of painful stimuli with odour at night may be more efficient than associating the delayed effects of ingested noxious substances with odour. This might explain why more moths than butterflies utilize chemically irritating spines in their defensive adaptations. Finally, some tropical nymphalids are active at dusk including the Malaysian *Amathusia phidippus* (Vaughan 1982) and neotropical owl butterflies (*Caligo*, Emmel 1976), and might be subject to mammalian predation which could be related to their possession of complex scent glands.

A definition of chemical defence

Based on the above considerations, I propose the following generalized definition: *chemical defence exists in a species when individual prey organisms contain one or more noxious chemical substances which facilitate proximal and/or distal rejection by predators or parasites; rejection can occur after a predator partially to completely ingests one or more prey individuals, or after the predator (or parasite) simply smells or tastes the prey.* Prey which contain only noxious Class I chemicals or contain both noxious Class I and innocuous Class II chemicals, as well as those prey which are physically noxious, are considered *unpalatable* to vertebrate predators. Those containing only Class II chemicals are not chemically unpalatable, but they may be chemically aposematic, either because they are physically unpalatable, or because they are palatable chemomimics. By this definition, Batesian mimics which deceive by visual, auditory and/or tactile stimuli are not chemically defended; however, both Batesian and Müllerian chemical mimicry are possible (see p.118).

I also propose that the terms *noxious* or *unpalatable*, and *innocuous* or *palatable* should take precedence. Other terms must be defined explicitly in discussions of chemical defence, and the precise *modus operandi* of purported defensive chemicals as Class I or Class II compounds should be clearly articulated. My choice of 'unpalatable' is in deference to Bates (1862) who introduced the term in his classical mimicry paper.

Stimulus Potentiation of Distal by Proximal Stimuli

Recent research on avoidance learning by Garcia and others revealed the phenomenon of stimulus potentiation, increasing the importance of the relationship between Class I and Class II chemicals. These researchers found in several species of mammals, including *Peromyscus*, that delayed sickness resulted in conditioned avoidance of the flavour and/or the odour of a chemical solution (Hankins *et al.* 1973, Robbins 1978, 1979, Garcia & Rusiniak 1979). Remarkably, when both flavour and odour cues were simultaneously associated with illness, the taste stimulus *potentiated* the odour stimulus, i.e. odour became extremely effective in eliciting later distal avoidance (Best *et al.* 1976, Rusiniak *et al.* 1979). Moreover, rats had great difficulty in learning to avoid substances based on their visual appearance (Garcia & Koelling 1966, Garcia *et al.* 1974, Best *et al.* 1977, see also Braveman 1977). In contrast, quail, pigeons and red-tailed hawks readily associated visual colour cues of solutions and/or food after lithium-caused emesis (Wilcoxin *et al.* 1971) and the presence of a salty or bitter flavour greatly potentiated both the pigeons' visual avoidance of blue-dyed water (Clarke *et al.* 1979) and the hawks' visual avoidance of mice (Brett *et al.* 1976).

Thus, when the ingestion of food was followed by sickness, proximal gustatory stimuli potentiated strong distal avoidance of visual cues in birds and of odour cues in mammals, i.e. potentiation promoted learned distal avoidance in both diurnal and nocturnal vertebrates. The great importance of potentiation for chemical defence is that it is a behavioural mechanism which reduces the number of predator-prey encounters necessary to facilitate long-lasting distal avoidance in both birds and mammals.

The apparent absence of potentiatable odour avoidance in birds is consistent with their poorly developed olfactory epithelia, few neural receptors, and weakly developed neuroanatomical pathways linking the visual and olfactory brain centres. Even species such as geese which can perceive odour (Welty 1975) have difficulty in associating visual with odorous cues, and species of thrush, robin, tit and pigeon are unable to do so (Neuhaus 1963). Most experiments purporting to show olfactory conditioning in birds have confused it with visual conditioning based on minute visual cues (e.g. Zahn 1935, as discussed by Neuhaus 1963) or failed to rule out non-olfactory responses to high concentrations of chemical compounds by the common (general) chemical sense. This is an ancient sensory system in all animals which permits avoidance of dangerous non-biological concentrations of dissolved or vaporized noxious chemicals, and which is frequently confused with odour and taste perception (Prosser & Brown 1961, Moncrieff 1967).

Neophobic Rejection of Prey

Many animals are conservative in their feeding and initially reject novel, unfamiliar foods. This neophobia appears to be based on both innate and learned components (Morgan 1896, Hogan 1977, Matthews 1977, Wilcoxin 1977) and is a behavioral strategy to avoid the high probability of poisoning or physical harm when an animal encounters food with which it has not had previous experience (Richter 1950, Rubinoff & Kropach 1970, Freeland & Janzen 1974, Domjan 1977, Rozin 1976, 1977, Laycock 1978). The stimulus bases for the initial rejection include visual cues in birds (Mostler 1934, Mühlmann 1934, Hogan 1965, 1966, Reiskind 1965, Rabinowitch 1968, Coppinger 1969, 1970, Shettleworth 1972*a*), gustatory and olfactory cues in mammals (Nichol 1938, Krueger, in Laycock 1978, Domjan 1977, Rozin 1977, Zahorik 1977, Leon 1974, see also papers by Garcia *et al.*), and gustatory and olfactory cues in lizards and snakes (Loop & Scoville 1972, Boyden 1976, Johki & Hidaka 1979).

Many species of Lepidoptera have exploited visual neophobia through behaviours and variable colour patterns which decrease predictability or surprise predators. These have been called anomaly by Sargent (1973), and aspect diversity by Rand (1967) and Ricklefs & O'Rourke (1975). Visual neophobia is a graded response, with the least tendency to reject unfamiliar cryptic prey and the greatest to reject prey which possess deimatic (frightening) coloration—which is usually combined with erratic, startling, or protean behaviour (Humphries & Driver 1967, 1970, Sargent 1981). Examples of frightening coloration include resemblances to vertebrate eyes (Blest 1957, Owen 1980), monkey faces (Hinton 1974), snakes (Poulton, in Marshall & Poulton 1902: 397, Edmunds 1974*a*, Smith 1975, 1977, 1978), and crocodilians (Poulton 1924*a*, Cott 1940; Brower 1968, 1971). Neophobic rejection of aposematic prey appears to be intermediate between these two extremes. Turner (1975*a*) has suggested that selection produced by neophobia has favoured aposematic patterns which are as different from familiar cryptic prey as possible, and recent studies indicate that avian predators learn to avoid artificial unpalatable prey faster if they are brightly coloured than if they are cryptic (Goodale & Sneddon 1977, Gittleman *et al.* 1980, Gittleman & Harvey 1980).

I agree with Schuler (1982) that most insectivorous birds have an initial tendency to avoid warningly coloured prey but that aposematism ultimately depends on chemical or physical punishment. Furthermore, Batesian mimicry and frightening (deimatic) displays will fail when the deceptive prey become too common (Blest 1957, Curio 1976). Nevertheless, as Brower *et al.* (1970) and Pough *et al.* (1973) have calculated, Batesian mimicry can sustain a selective advantage even when the proportion of highly unpalatable models in the population is extremely small.

Neophobia can lead to misinterpretation of predators' behaviour. For example, captured wild birds may initially refuse experimental insects and when finally induced to attack them may still refuse to eat them, e.g. the peck, mangle, kill, etc. categories in the experiments of J. Brower (1958 *a*, *b*, *c*), Brower *et al.* (1963), Rothschild & Kellett (1982), Eisner *et al.* (1978), Bowers (1980), and Jarvi *et al.* (1981*a*). Final refusal to swallow prey could be because they taste unfamiliar, regardless of whether or not they contain noxious substances.

Cultural Transmission of Learned Behaviour: Bias and Empathic Learning

Neophobia is undoubtedly reinforced by culturally transmitted feeding behaviour, and both positive and negative biases are possible. For example, if parents had never themselves attacked aposematic prey, or if they had learned not to after suffering the consequences, it seems unlikely that they would offer aposematic prey to their young or attack them while hunting and feeding in the presence of their young. Each new generation of predators would thus be biased towards attacking familiar prey and simultaneously biased against attacking aposematic prey (Manders 1911, Alcock 1973). Evidence for such parentally induced bias exists for birds (E. Turner 1965, Curio *et al.* 1978), and rats (Bronstein & Crockett 1979, Leon 1979). It is conceivable that the orioles' and grosbeaks' extensive feeding upon Monarch butterflies in Mexico (Fig. 12.1) results from cultural tradition.

Bias may also arise through empathic learning, a type of behaviour in which individual predators learn to avoid prey by observing active rejection by other individuals. Considerable discussion and some evidence for this exists for birds (Swynnerton 1915*d*, Klopfer 1957, Rothschild & Lane 1960, Gans 1964, E. Turner 1965, Rothschild & Ford 1968, Alcock 1969*a*, *b*, Curio 1976, Mason & Reidinger 1981) and possibly also rats (Galef 1977). Of particular interest is a report by Mason & Reidinger (1982) of emesis-based observational learning in Red-winged Blackbirds (*Agelaius phoeniceus*). (See also Coombes *et al.* 1980).

The Evolution of Chemical Defence and Warning Coloration

The paradox

The evolution of warning coloration poses a paradox which was initially recognized by Wallace (1867*a*) and Darwin (1871: 326), and discussed at length by Poulton (1887; see also Ch. 14). The question is how natural selection can provide an advantage to an initial variant that shifts from a cryptic to a conspicuous way of life. Such an individual would surely have an increased likelihood of being discovered and killed by predators. Fisher (1927, 1930, 1958) recognized that the evolution of defence by slow acting Class I chemicals posed a similar problem, whereas defence by rapidly acting ones did not. Thus a genetic change producing a noxious Class I chemical secretion on or in the body surface, of sufficient potency to facilitate immediate rejection by the predator, would allow the prey to survive. We now know that many such chemicals can originate through micromutations affecting primary metabolic pathways (Swain 1976*a*, Whittaker & Feeny 1971, Rodriguez & Levin 1976, Duffey 1977, 1980, Blum 1981).

In contrast, noxious Class I chemicals such as emetic poisons which are slow acting would appear to provide their initial bearers with no individual advantage. Fisher (1927, 1930, 1958) formulated the theory of kin selection to provide a mechanism explaining how such chemicals as well as warning coloration might be preserved by selection. Understanding chemical defence has thus played a major role in the development of evolutionary theory and continues to be debated (Wright 1945, Hamilton 1964*a*, *b*, Williams 1966, Benson 1971, Brower 1971, Eshel 1972, Rothschild *et al.* 1972*b*, Marsh & Rothschild 1974, Turner 1971*a*, *b*, 1975*b*, 1981, Wilson 1975, Dawkins 1976, Bertram 1978, Harvey & Greenwood 1978, Roughgarden 1979, Garcia & Rusiniak 1979, Jarvi *et al.* 1981*a*, Harvey & Paxton 1981, Harvey *et al.* 1982, and Huheey, in press).

Ecological chemistry of noxious chemicals

Ecological chemistry as defined by Brower & Brower (1964) and Brower (1969) refers to the biological sequestration of secondary plant compounds which the sequesterers utilize for various purposes, especially defence. (For a recent review of the many plant chemicals which might be utilized, see Levin 1976.) Uptake of plant poisons by caterpillars was originally proposed by Distant (1877) and Slater (1877), elaborated by Haase (1896), and finally proven for milkweed cardenolides and Monarch butterflies in pharmacological experiments by Parsons (1965), in feeding experiments by Brower *et al.* (1967*a*) and in chemical analyses by Reichstein (1967) and von Euw *et al.* (1967).

Although the principle of ecological chemistry is widely accepted, it has never been fully appreciated that its underlying processes can obviate kin selection as a requirement for the evolution of passive unpalatability in butterflies. This is because the development of the ability to colonize poisonous plants as a result of ecological pressures (Brues 1946, Ehrlich & Raven 1965) which are initially independent of chemical defence could automatically result in the larvae of entire populations fortuitously containing defensive chemicals in their guts, thus constituting what Gould & Vrba (1982) term an 'exaptation', permitting the rapid evolution of further refinements (p.121).

Conclusion

Although the evolution of passive chemical defence and aposematism is still not resolved, neophobia, cultural bias, and potentiated distal rejection can all contribute to reducing the number of individual prey which must be killed in the learning process. As a consequence of these behavioural phenomena, as well as the phenomenon of ecological chemistry, the selective force required for the evolution of unpalatability and aposematism may not be as great as originally thought (see also Ch.14).

Caveats in the Study of Chemical Defence

Chemical defence can never be absolute

Many authors (e.g. Poulton 1887, Mostler 1934, Cott 1940, Ford 1945, Wigglesworth 1964, Rettenmeyer 1970, Rothschild & Ford 1970, Rothschild 1971, 1981*a*, Dawkins & Krebs 1978, 1979, Huheey, in press) have emphasized that no matter how elegant a defence may be, total protection can never be achieved because of the positive feed-back nature of the coevolutionary arms race (Roughgarden 1979). Roeder (1967) referred to this as an ongoing contest, or game, in which the interaction is never one-sided and leads to ever-increasing sophistication in the defensive adaptations of the prey and in the sensory, behavioural, and physiological capacities of the predators, including even their memory spans (Orians 1981).

Seven reasons of special interest why protection is not perfect in butterflies are listed here. First, seasonal or local decrease in availability of alternative palatable prey leads to eventual acceptance of less palatable species which predators had previously learned to reject (Holling 1965, Sexton 1960, Alcock 1970, Garcia *et al.* 1974, Schuler 1974, Estabrook & Jesperson 1974, Dill 1975, Atsatt & O'Dowd 1976, Boyden 1976, Matthews 1977). Secondly, naive gourmands in each new generation may eventually overcome their neophobia or be forced through lack of familiar prey to sample aposematic species. The evolution of Batesian (palatable) mimics will favour occasional testing by predators (Pough *et al.* 1973, Dill 1975). Periodic forgetting is a general property of learned behaviour (Thorpe 1956, Gonzalez *et al.* 1967) and is an evolved component of optimal foraging strategies (Orians 1981). Fifth, predators can behaviourally counter chemical defence adaptations (Eisner *et al.* 1963, Eisner 1970). For example, Great Tits (*Parus major*) peck off the heads of caterpillars and pull out and discard the guts with their contained plant toxins (Vince 1964). In Mexico, Black-backed Orioles, have apparently adapted their fruit pulp-stripping behaviour (Bent 1965) in eating Monarch butterflies (Calvert *et al.* 1979) which can result in their avoiding more than 75% of the cardenolide in each butterfly (Fink & Brower 1981, Colour plate 2B, Brower *et al.* in prep.). Similarly, to avoid possible noxious chemicals in the exoskeleton a neotropical tanager (*Pipraeidea melanonota*) squeezes out and eats the abdominal contents of ithomiid and other aposematic butterflies (Brown & Vasconcellos N. 1976). Sixth, predators may belong to phylogenetic lines which possess metabolic pathways that are partially to completely unaffected by various defensive chemicals (Heim de Balsac 1928, Kear 1968). Examples of known butterfly predators in Mexico which are insensitive to the emetic effects of cardenolides include the Black-headed Grosbeak, Rufous-sided Towhee (Fink & Brower 1981, Fink *et al.* in press), and mice of the genera *Microtus* and *Peromyscus* (Brower *et al.* in prep.). Lastly, some predators appear to undergo developmental changes in their response to prey. Thus the Cuckoo (*Cuculus canorus*) during its fledgling stage is said to find aposematic prey distasteful and indigestible, but later in life eats the same prey with impunity (Rothschild 1981*a*).

Problems with injection bioassays

Injection bioassays of purportedly Class I chemicals which have been isolated in either pure or unpure form from insects or their foodplants have yielded useful toxicological and pharmacological information. However, the same compounds can be totally without effect if administered orally. For example, Dempster (1967) found that extracts of cabbage butterfly larvae (*Pieris rapae*) were extremely toxic if injected into rabbits, although the living caterpillars are freely eaten by numerous species of birds in the wild, with no evidence of adverse effects. As Duffey (1977:336) pointed out, '. . . the relevance of wholesale injection of unpurified extracts containing a multitude of contaminants is questionable'.

It is imperative that appropriate feeding bioassays be carried out together with injection assays if the latter are to be of ecological and evolutionary significance. Another problem is in using bioassay animals which are primarily herbivorous; they may have evolved tolerance or insensitivity to toxic compounds in their natural foods. For further caveats on this methodology, see Marsh & Rothschild (1974, and Ch.13), and Blum (1981).

The interdependence of laboratory and field research

Our knowledge of chemical defence is based on natural history observations, deductions from the general body of biological knowledge, feeding experiments, and chemical and pharmacological investigations. However appealing the evidence may seem, the ultimate test is that chemical defence can be shown to operate in the natural environment (Duffey 1977). The best illustration of the difficulty of obtaining field verification is the case of the Monarch butterfly. Although its cardenolide-based defence has been thoroughly documented in the laboratory, it has yet to be proven effective against any predators in the wild (see also Petersen 1964).

Defensive Chemicals and Their Physiological Effects

Odorous and Non-odorous Volatiles

Historical perspective

Linnaeus' (1756) classification of odours included aromatic, fragrant, repulsive, and nauseous and it is within this framework that naturalists have described the scents produced by butterflies. Pleasant butterfly odours were considered sexual signals, whereas malodorous secretions were assumed defensive (Bates 1862, Trimen 1869, Müller 1877*a*, *b*, *c*, 1878*a*, Haase 1896, Swynnerton 1915*c*, *d*, 1919, Carpenter 1921, Clark 1926, 1927, Carpenter & Ford 1933, Cott 1940, Alexander 1961*b*, Brower *et al.* 1965). Defining the functions of volatile chemicals based on odour qualities as judged by humans is tautological, fails to distinguish between Class I and Class II volatiles, incorrectly excludes compounds which are odourless to humans (Ford 1955), and often assumes that odours have only one function when they may have several (Rothschild *et al.* 1973, Gilbert 1976). For example, odourless volatile pyrrolizidine alkaloid derivatives occur in the hairpencils of danaines and ithomiines (Edgar *et al.* 1976*b*) and are required for successful courtship in *Danaus gilippus* (Myers & Brower 1969, Pliske & Eisner 1969, Schneider & Seibt 1969, Myers 1972). The reverse situation is exemplified by an unidentified hairpencil component in several danaines which is strikingly odorous but elicits no or very weak responses from the butterflies (Schneider & Seibt 1969, Rothschild *et al.* 1973, Schneider 1975—see also Ch.25). Further, the presence, absence, concentrations, isomers, and ratios of components must all be considered in understanding how chemicals operate in defence (Claesson & Silverstein 1976, Silberglied 1977, Roelofs 1980).

Volatile secretions in adult butterflies

Modern instrumentation, including gas chromatography and mass spectrophotometry, have facilitated the identification of compounds within the molecular weight range of approximately 50–300, the effective limit of volatility for biologically active compounds (Wilson *et al.* 1969, Blum 1981). In addition to the secretions of larval osmeteria, examples of purportedly defensive odorous chemicals in adult butterflies include those secreted by the abdominal glands of Neotropical heliconiines, yellow fluid exudates (and froths) produced by some African danaines and acraeines (Poulton, in Marshall & Poulton 1902: 413–414, Owen 1971*a*, 1980) and also a sticky brown odorous exudate of the aposematic Neotropical lycaenid, *Eumaeus minyas* (DeVries 1976). The exudate of *Amauris* is said to be highly repugnant and that of *Acraea* to contain chemical irritants which cause human eyes to water (Rothschild *et al.* 1972*a*), and also to contain HCN. Rothschild (1961) also stated that the odours of *Amauris echeria* and *Acraea horta* when inhaled cause a sensation close to pain. Other *Danaus*, *Euploea*, *Amauris*, and *Acraea* butterflies exude odorous fluids from wing veins, leg joints and antennal tips (Carpenter & Ford 1933), which Cott (1940) referred to as nauseous.

Recently identified volatiles in adults include alcohols, aldehydes, terpenes, ketones, and pyrrolizidine alkaloids. Since several of these occur in defensive glands of various insects, e.g. pinenes and myrcene in termites, bugs, and/or ants (Blum 1981), they may play a defensive role in butterflies as Class I and/or Class II chemicals. The fact that odours often occur only in, or are more concentrated in, males of many butterfly species (Brower 1963, Myers 1972, Seibt *et al.* 1972) argues that they are sex pheromones, but they may also be used in defence (Ch.25). Seven volatiles occur in the male hairpencil secretion of *Amauris ochlea*, including those with citrus, rose-orange, wintergreen and clove-oil scents (Petty *et al.*, 1977). Males of the strongly smelling Japanese papilionine *Atrophaneura alcinous* produce a medley of volatile substances in numerous hair-like scales in a fold on the inner margin of the hindwings. These include linalool (an intense floral odour), four aldehydes including *n*-heptanal (a harsh odour), benzaldehyde (a bitter-almonds scent), phenylacetaldehyde (a 'greenish' odour), 2-phenylpropanal (an odour of hyacinth), and methylheptanone (a lemon-scented ketone, Honda 1980). Defined odours in other species include the lemon-scented limonene and citral (= gerenial and neral), linalool, and the pine-scented pinenes, myrcene, and *p*-cymene from *Pieris* (Bergström & Lundgren 1973, Hayashi *et al.* 1978). A European blue (*Lycaeides argyrognomon*) secretes nonanal, which has a strong fatty-rose-orange scent, the camphor-scented cadinol, and hexadecanol which is odourless (Lundgren & Bergström 1975). These odour descriptions are from the perfume industry's Givaudan Index (Anon. 1961), Moncrieff (1967), and Robinson (1975).

Ecological chemistry of butterfly odours

Most volatile secretions of butterflies are *de novo* products (Blum 1981) but some may be derived from larval foodplants and pyrrolizidine alkaloids definitely are obtained during adult feeding. McCann (1952) maintained that adults of *Tirumala limniace* acquire the odour of *Crotalaria* plants and Marsh & Rothschild (1974) speculated that odour components derived from *Brassica* plants caused a magpie to reject *Pieris brassicae* pupae. Referring back to Haase (1896) and Trimen (1887), Rothschild *et al.* (1970) and Rothschild (1972*c*) agreed that danaine and troidine larvae sequester volatiles from *Asclepias* and *Aristolochia* foodplants that were assumed to be aposematic deterrents to both vertebrate and invertebrate herbivores. Surprisingly, research on food selection by large herbivorous mammals has largely failed to consider the possible aposematic function of odours (Arnold & Hill 1972; Cronin *et al.* 1978).

Odours as Class I or Class II chemicals

Marshall & Poulton (1902: 356) proposed that defensive odours must be linked with other poisonous or unwholesome qualities to be effective, and Swynnerton (1919: 224), based on experiments with *Acraea*, *Amauris* and *Danaus* butterflies with caged European Rollers (*Coracias garrulus*), concluded that 'smells and tastes, like appearance, are of use in averting the eating of their possessor mainly when they have become associated in the enemy's mind with unpleasantness of a more fundamental character . . . probably, in most cases, indigestibility'. Rothschild (1961: 106) reemphasized this in stating

> My own belief is that . . . aposematic insects contain a poisonous or unwholesome element which causes discomfort, nausea, or pain *after* they have been swallowed, and that both the bright colour pattern and scent pattern are part of a mechanism which directs the attention of different types of predators to these dangerous qualities.

Rothschild & Ford (1970) further speculated that the scent of Monarch butterflies is an *aide-memoire* to birds. Eisner & Grant (1981: 476) reaffirmed the Marshall-Swynnerton-Rothschild hypothesis.

> There is no evidence to indicate that odors themselves, at their natural concentrations, are intrinsically repellant to predators or play any direct role in chemical defense.

Moreover, Stoddart (1979) maintained that most experiments purporting to show innate avoidance of odours (i.e. that they can be Class I defensive chemicals) lack adequate controls. Thus Feather's report (in Carpenter 1921: 95) that acraeine and danaine odours deterred two *Cercopithecus* monkeys can be interpreted as innate avoidance, neophobic rejection, or prior learned avoidance rejection of any of several distal stimuli.

The discovery that taste potentiates odour aversion is definitive evidence for a Class II function of odours (Müller-Schwarze & Mozell 1976, Henessy & Owings 1978, Stoddart 1979) and reaffirms the view that butterfly odours can serve a Class II aposematic function towards mammals, and possibly also some lizards (Johki & Hidaka 1979), snakes (Burghardt *et al.* 1973), and invertebrates. The possibility that odour can deter insectivorous birds seems nil.

In view of all that has been written on the subject, it is astounding that no definitive experiment has ever been carried out on odorous butterfly chemicals with respect to any vertebrate predator.

Gustatory Stimulants

General considerations

Gustatory response to dissolved chemicals appears to be of major importance in chemical defence. The sense of taste in numerous species throughout the animal kingdom is remarkably uniform in its responsiveness to four basic modalities: saltiness, sweetness, sourness and bitterness. This indicates that gustation is a generalized response of ancient phylogenetic origin (Moncrieff 1967, Bate-Smith 1972, Garcia & Hankins 1974, Brower & Glazier 1975, Beidler 1976, Swain 1977, Chapman & Blaney 1979, Garcia & Rusiniak 1979). The ability to sense saltiness serves principally to maintain ionic balance and is unimportant in chemical defence. Sweetness generally elicits feeding responses and characterizes natural organic substances which are calorically nutritious and usually non-toxic. Sourness generally elicits aversion, and bitterness virtually always elicits aversion. Vertebrates, with few exceptions (Swain 1976*b*, 1977), respond to bitter substances at concentrations two orders of magnitude lower than to sour substances, their next most sensitive taste modality (Moncrieff 1967). Experiments with many species of insectivorous birds have proven their negative response to prey or food to which bitter compounds have been added (Brower & Glazier 1975, Eisner *et al.* 1978). Some non-insectivores such as the pigeon (Prosser & Brown, 1961) may prove to be exceptions which need explanation.

Aversion to bitter substances is a basic adaptive response. Virtually all compounds of biological origin which are toxic are also bitter-tasting to humans (Allport 1944, Richter 1950, Brower & Brower 1964, Beidler 1966, Bull *et al.* 1968, Conn 1969, Fischer 1971, Robinson 1975). Examples of known bitter toxic substances include saponins (Applebaum & Birk 1979), HCN, all alkaloids (Bate-Smith 1972, Laycock 1978, Vickery & Vickery 1981), aristolochic acid (Henry 1949), tannins (Whittaker 1970), terpenes and polyphenolics (Gardner 1978), and most glycosides (Kingsbury 1964, Shallenberger & Acree 1971, Hegnauer 1964–1973), including both cyanogenic (Bate-Smith 1972) and most cardiac glycosides (other references in Brower & Glazier 1975). The fact that most birds eat only ripe fruits (Turcek 1963, Corner 1964, van der Pijl 1966, 1969, Snow 1971) reflects the ecological significance of the four taste modalities: unripe fruits are generally cryptic, sour, bitter, astringent and/or sickening, but as they ripen major biochemical changes occur: bright colours develop, sweet sugars accumulate, sour acids and bitter alkaloids disappear, and astringent tannins polymerize to innocuous higher molecular weight compounds (McKey 1974, Robinson 1975, Janzen 1978, Rhoades 1979).

The generalization that all natural toxic substances are bitter does not necessarily imply the reverse: some naturally occurring bitter compounds may not be toxic. This raises the additional possibility that chemical mimicry exists (p.118).

Gustatory stimulants as Class I and Class II chemicals

A major source of confusion has been the assumption that the chemical identification of extracts or pharmacological proof of their toxicity is *prima facie* evidence that they actually are Class I deterrents in the butterflies. In fact, they may occur at such low concentrations that they can operate only in a Class II capacity. Thus we must carefully examine the evidence that cardenolides, pyrrolizidine alkaloids, mustard oil glucosides, aristolochic acids, cyanogenic glycosides, and other compounds occur in sufficiently high concentrations and amounts to act as Class I aversive chemicals, or whether they, as appears to be the case for odours, function mainly as Class II aposematic chemicals.

The research of Mostler (1934) and Liepelt (1963) showed that the venom *per se* in the stinging apparatus of Hymenoptera caused gustatory rejection by several insectivorous birds. Windecker (1939) similarly proved that the haemolymph of Cinnabar moth adults as well as substances in their larval skin deterred numerous species of birds and mammals. As discussed below (p.121), laboratory feeding experiments have repeatedly found that birds taste-reject butterflies. This supports the contention that some chemicals in butterflies have a sufficiently noxious effect upon the gustatory receptors of predators that they are active Class I deterrents (see also Boyden 1976, Turner 1981, Jarvi *et al.* 1981*a*).

Chemical Mimicry

Theory

A predator which had learned to associate the odour component of a secretion with the effects of a noxious component present in the same species might avoid individuals in another species possessing the odour component alone. This would be deceptive Batesian odour mimicry. If two species of prey had the same odour, and the same or different noxious components, then the situation would exemplify Müllerian odour mimicry. Analogously, Batesian taste mimicry could also exist. For example, the model could be bitter and noxious, and the Batesian mimic bitter but not noxious. Müllerian taste mimicry likewise could occur and may in fact be widespread because apparently all naturally occurring toxic substances are bitter. Chemical automimicry could also occur if both noxious and innocuous individuals within a species shared the same odour and/or taste modalities.

Taste and odour modifiers

Chemicals, such as one found in artichokes, which enhance, selectively eliminate or switch taste responses (Kurihara & Beidler 1968, Eisner & Halpern 1971, Kurihara 1971, Bartoshuk *et al.* 1972, Rohan 1972) could play an additional role in taste mimicry and chemical defence. Analogously, the enhancement of defensive odours by carotenoid pigments was suggested by Rothschild (1971) and Rothschild *et al.* (1972*b*). Many species of insects sequester carotenoids from plants (Feltwell & Rothschild 1974, Rothschild *et al.* 1977*b*) which then appear in the haemolymph and defensive secretions. However, extensive research failed to substantiate that carotenoids enhance sensory perception (Rothschild 1978).

Examples

Chemical mimicry as a form of aggressive interspecific exploitation definitely exists. Examples include the attracting of moths by spiders (Eberhard 1977), the luring of insect pollinators by various species of orchids (Wickler 1968, Bergström 1978, Dressler 1981), and the symbiotic relationships of various insects including butterflies with ants. Proponents of the hypothesis that chemical mimicry is effective against vertebrates include Rothschild (1961, 1963, 1964*a*), Holling (1965), Brower (1969), Rhoades (1979), and Eisner & Grant (1981). However, experimental evidence is lacking (Rothschild 1964*b*, Edmunds 1974*a*, Garcia & Hankins 1975, Ford 1975*a*, Burtt 1981, Cloudsley-Thompson 1981, Huheey, in press).

Emetic Substances

Nature and control of vomiting

The following summary is from Borison & Wang (1953), Wang (1965), Miller *et al.* (1977), Bowman & Rand (1980), and *Webster's Dictionary* (Anon. 1950). Vomiting is the forceful expulsion of the contents of the stomach and/or duodenum through the mouth and is regarded as one of the most primitive protective behaviours with which animals are endowed. The various components of vomiting behaviour are coordinated by a specific vomiting centre in the medulla of the brain. Retching represents vomiting movements without the expulsion of material, while nausea is the human sensation associated with the anticipation of retching or vomiting.

Vomiting can be caused directly by chemical irritation of the gastrointestinal mucosa and indirectly by the excitation of several neural pathways leading from various parts of the body to the vomiting centre. These include nerves from gut chemoreceptors, from visceral pain receptors in the stomach, intestine, liver, heart, urinary and genital tracts, from the CNS chemoreceptor trigger zone which responds to digested chemicals circulating in the blood, and from the cerebral cortex which mediates vomiting as a learned response. This last pathway can activate emesis solely on the sight, taste, or odour of previously experienced emetic food. Such conditioned emesis (and retching) occurs in pigeons (Riddle & Burns 1931), mammals (Lieb & Mulinos 1934), and Blue Jays (Brower 1969, 1971).

As we have seen above, compounds which cause internal irritation and pain can cause nausea and vomiting (Bowman & Rand 1980) and it is therefore possible that histamines, mustard oils, acetylcholine, cyanogenic glycosides, or aristolochic acids may be involved as emetic deterrents. However, their ability to avert predators as Class I or as Class II chemicals remains hypothetical.

Other Chemicals

Rothschild *et al.* (1979*a*) found that larvae of Tobacco Hornworm (*Manduca sexta*, Sphingidae) which they fed upon the solanaceous plant *Atropa belladona*, incorporated atropine which the pupae stored. Atropine is a parasympathetic depressant and might influence insect palatability by drying up the predators' saliva (Bowman & Rand 1980). Tannins, which are present in the leaves of most plants and therefore inevitably end up in caterpillar guts (Perrins 1976), are well known both for their bitterness and for their astringency which results from their complexing of saliva proteins (Bate-Smith 1972). Some tannins also reduce protein digestibility (Whittaker 1970, Feeny 1970, 1976, Swain 1979) and Perrins (1976) reported slowed growth of nestling Blue Tits (*Parus caeruleus*) which ate larvae with tannins in their guts. Tannins, with some exceptions (Bernays 1981, Bernays & Woodhead 1982), are considered powerful feeding deterrents to most herbivores (Swain 1978, 1979). To date neither they nor atropine are proven defensive chemicals in butterflies.

The recent finding of Marsh *et al.* (1977) of an unknown cardioactive agent in *Hypolimnas bolina* that may not be emetic supports the hypothesis that noxious effects on the heart *per se* may be an important chemical defence. Classical cardenolides, in addition to causing nausea, retching, and/or vomiting, produce cardiac arrhythmias, anorexia, diarrhoea, visual disturbances, muscular spasms, delirium, *etc.* (Bowman & Rand 1980, Fink 1980). Cardenolides, therefore, have the possibility of being Class I deterrents, with a plethora of noxious effects.

The *modus operandi* of certain toxic chemicals is paradoxical in that their Class I noxious effects involve long term lethality yet the compounds are said to fail to deter herbivores (Zahorik & Houpt 1977, 1981). For example, when sheep ingest cyclopamine in *Veratrum californicum* (Liliaceae), gross cyclopean deformities in fetal development result (Evans *et al.*, 1966, Keeler & Binns 1968). Likewise, pyrrolizidine alkaloids, which are sequestered by danaines ithomiines and possibly some acraeines, cause liver deterioration and death from a few days to several months after they are ingested (Mattocks 1972, Hooper 1978*a*). However, the significance of PAs as lethal poisons may be exaggerated: range mammals usually avoid plants containing PAs and most poisoning occurs due to crowding, overgrazing, poor range management, or by the inadvertent harvesting of poisonous plants in hay or fodder (Bull *et al.* 1968: 115, Kingsbury 1975, 1978, James 1978, Laycock 1978).

It is possible that pyrrolizidine alkaloids may condition avoidance *via* pain pathways from the liver before lethal amounts are accumulated. However, whether PAs can act as Class I chemicals has yet to be established and Robinson (1979: 439) has even stated that 'sweeping generalizations about the protective functions of alkaloids are groundless'.

Various secondary plant chemicals affect cerebral activity and might be used in defence. For example, Rothschild *et al.* (1977*a*) discovered the psychoactive component (delta^1 tetrahydrocannabinol = THC) in the larval cuticle of the arctiid moth *Arctia caja* which the larvae had sequestered from *Cannabis sativa*. Such compounds might increase the perceived intensity of Class I chemicals and, as Rothschild (1967) speculated, might also result in longer memory of aposematic tastes and odours. Tranquilizers raise the concentrations of compounds necessary to maintain learned aversive responses (Gollub & Brady 1965) and might lead a predator to ignore the bitter component of a noxious chemical and receive a more severe negative reinforcement. Likewise, Eisner & Halpern (1971) suggested that hallucinogens and psychotomimetic agents might disorient animals which had ingested them. Stimulants such as amphetamines directly increase the rate of avoidance learning and also reduce the desire to eat. The evolution of insect defence using such compounds would seem to require kin selection.

Chemical Defence in the Life History Stages of Butterflies

Biochemical Possibilities and a Caveat

The amount of defensive chemicals which larvae obtain from feeding on plants will be a function of concentrations of the substances in the plants and rates of ingestion, digestion, absorption, metabolic degradation, excretion, and defecation by the larvae. Some species may degrade the chemicals so rapidly that they are not protected at all, others may be poisonous only while actively feeding, and still others may be poisonous throughout their lives. Metabolic degradation or modification of stored chemicals is also possible during the pupal and/or adult stages (Roeske *et al.* 1976, Rothschild *et al.* 1977*a*, Duffey 1977, 1980, Dixon *et al.* 1978, Rothschild & Marsh 1978, Seiber *et al.* 1980, Nelson *et al.* 1981, Brower *et al.* 1982).

For species that store secondary plant compounds,

we can expect to find various levels of sophistication of chemical defence. One possible sequence in their evolutionary history has been discussed by Brower & Brower (1964), Eisner (1970), Duffey (1980), and Blum (1981) and is as follows: if the molecules sequestered by larvae pass through the gut wall into the haemolymph, without severely damaging the insect (Brower & Glazier 1975, Cohen & Brower 1982), absorption-excretion equilibria could become established leading to tissue noxiousness. The molecules could then bind to or be taken up by certain tissues or organs and this would set the stage for their conservation through metamorphosis and storage into the pupal and adult stages, as well as passage through the female reproductive organs into eggs of the next generation. Evidence for this exists in the Monarch butterfly. Brower *et al.* (in Thomashow 1975) established by thin layer chromatography that all stages in the life cycle of the Monarch reared on *A. curassavica* contained the same array of cardenolides (Colour plate 2A). It is also possible that a gland which evolved in relation to non-defensive functions could fortuitously concentrate noxious components when an insect species altered its feeding preferences and moved onto toxic hostplants (Ehrlich & Raven 1965, 1967) thereby invading a new adaptive zone (Simpson 1953, Gould & Vbra 1982). Another possibility, probably exemplified by the grasshopper *Poekilocerus bufonius* (Reichstein 1967, von Euw *et al.* 1967), is for a species to add components to an already existing defensive gland after moving onto a poisonous plant.

There is no *a priori* guarantee that feeding on poisonous plants proves that a species is unpalatable, nor that its palatability is constant either in different individuals or in different populations. Palatability must be studied on a populational basis and the potential chemical defence in each life history stage must be determined by separate chemical analyses, feeding experiments, and appropriate bioassays. This caveat also applies to the life history stages of species which, *de novo*, secrete their own purportedly defensive chemicals. The suggestion put forward by Rothschild & Marsh (1978) that all diurnal Lepidoptera are to some degree chemically protected does not seem a useful working hypothesis for understanding the intricacies of chemical defence in either evolutionary time or ecological space.

Defensive Chemicals in Eggs

Butterflies which lay brightly coloured eggs in clusters and have larvae and adults characterized by gregarious behaviour or other aposematic features, are presumed to have antiparasitoid and antipredator chemicals in their eggs (Stamp 1980; see also Ch.6). Examples include some tropical American heliconiines (Beebe *et al.* 1960, Alexander 1961*a*,*b*, Mallet & Jackson 1980), African acraeines (Owen 1971*a*), and New World troidine swallowtails of the genus *Battus* (Young 1979*a*, Hazel & West 1979), but not the related *Parides* which lay eggs singly and have cannibalistic larvae (Young 1977, Brown *et al.* 1981). Other aposematic egg clusterers are the North American nymphalines *Euphydryas phaeton* (Edwards 1884) and *Chlosyne lacinia* (Drummond *et al.* 1970), the neotropical ithomiine *Mechanitis* (Guppy 1894, Emmel 1976, Gilbert, in Silberglied 1977), and the neotropical lycaenid, *Eumaeus minyas* (DeVries 1976, 1977). Another nymphaline, *Hypolimnas antilope* from the Philippines, lays up to 200 eggs per cluster which the female stands astride until the larvae hatch. This behaviour did not prevent ant or parasitoid attacks (but see also Ch.6) but may deter birds (Rothschild 1979). Unpalatability is implied since *H. bolina* appears to contain a cardioactive toxin. Joint egg-clustering occurs by two or more females in some *Heliconius* (Turner 1971*b*, Mallet & Jackson 1980). Finally, chemically mediated defence of egg clusters occurs in several myrmecophilous lycaenids and riodinids.

That birds do eat butterfly eggs has been established by Baker (1970) who compared the mortality of singly laid eggs of *Pieris rapae* with clusters laid by *P. brassicae*. House Sparrows (*Passer domesticus*) and Garden Warblers (*Sylvia borin*) ate 20% of the *P. rapae* eggs, and one bird ate all eggs in a single *P. brassicae* cluster. More of the conspicuous *P. brassicae* clusters may not have been attacked because they contained mustard oil glycosides (Aplin *et al.* 1975).

Toxins need not be restricted to egg clustering species and in fact the only butterfly for which there is proof of poisons in its eggs is *Danaus plexippus* (Reichstein *et al.* 1968, Thomashow 1975), which lays its eggs singly (Brower 1961*a*). Utilizing a stock reared on *Asclepias curassavica*, Thomashow found that the mean dry weight of Monarch eggs was 171μg and that each contained on the average 0.97μg of cardenolide, or approximately 0.6% by dry weight. For comparison, 40 adults contained an average of 693μg, approximately 0.4% by dry body weight. (The lesser amount of cardenolids (0.24μg) per egg reported by Reichstein *et al.* 1968, appears due to differences in methodology).

How cardenolides may protect eggs is of interest. By combining Thomashow's (1975) 0.97 μg per egg with the emetic dose of 71.4μg of cardenolide per 85g Blue Jay (Roeske *et al.* 1976), a jay would have to eat only 74 Monarch eggs to sicken. In view of Baker's data (see above), eggs could lead to sickening of predators, particularly when species oviposit in clusters. Why, then, does only one known danaid species (Ackery & Vane-Wright, in press *a*) lay eggs in clusters? The answer is probably twofold: many foodplants lack cardenolides so that a female's eggs may lack toxins (furthermore, only certain danaids appear to sequester heart poisons), and many are too small to provide for more than one or a few larvae, each of which consumes substantial food (Roeske *et al.* 1976). These factors have apparently outweighed the advantage of clustering and an adaptation has evolved which prevents even fortuitous clustering: first instar larvae of both the Monarch and Queen (*D. gilippus berenice*) are egg cannibals (Brower 1961*b*). Similar behaviour also exists in some *Heliconius* (Williams & Gilbert 1981), and in *Parides* as discussed above. For a discussion of various oviposition strategies in butterflies, see Gilbert (1979 and Ch.6), and for a general review of the adaptive significance of cannibalism, see Polis (1981).

Sequential defoliation by *Battus philenor* larvae of as many as 25 *Aristolochia* plants occurs even though the probability of dying is high when a larva leaves one plant to search for another (Rausher 1980, 1981*a*, *b*). In these circumstances it is curious that egg clustering and early larval gregariousness in *Battus* retains a selective advantage. One explanation may be that *Aristolochia* plants vary in size so that some species can support many larvae. In fact, the oviposition behaviour of *Battus* appears labile, such that females search and avoid ovipositing on small plants which already have a cluster on them (Rothschild & Schoonhoven 1977, Dixon *et al.* 1978, Rauscher 1979*a*, Gilbert 1982).

Presumed toxins in the eggs of aposematic species may function in ways other than as predator or parasite deterrents (Marsh & Rothschild 1974, Swain 1977). Cardenolides (Rothschild *et al.* 1972*b*) as well as aristolochic acids (von Euw *et al.* 1968) have been suggested to protect eggs from infection, but this was not supported by the data of Frings *et al.* (1948).

Defensive Chemicals in Larvae

Tissue unpalatability

Brower *et al.* (1967*a*) selected for a strain of Monarch butterflies which could eat cabbage. Two Blue Jays were each offered 10 cabbage-reared larvae, and collectively ate 19 of them without signs of illness, whereas each bird sickened and vomited profusely after eating 2–3 larvae rearing on the cardenolide-rich *Asclepias curassavica*. Larvae which had evacuated their guts prior to forming pupae also elicited vomiting in another jay, thereby proving that the poisons had been sequestered and stored in the larval tissues, and ruling out the alternative explanation (Brower & Brower 1964, Eisner & Meinwald 1966) that the poisons were simply present in the undigested gut contents.

Brower & Brower (1961) offered four *Peromyscus leucopus* 14 fifth instar Monarch larvae, which they ate without ill effect. Since these larvae had been collected on *A. incarnata*, a milkweed which contains insufficient cardenolide to produce emetic Monarchs (Roeske *et al.* 1976), the potential for cardenolide-based unpalatability of Monarch larvae to mice is unresolved. While the experiment did suggest the absence of other Class I or Class II deterrents in the larvae, wild captured *Peromyscus* and *Microtus* mice in Mexico freely ate adult Monarchs containing substantial amounts of cardenolide (Brower *et al.* in prep.).

Larvae of *Euphydryas phaeton* (Bowers 1980) also derive emetic compounds from their foodplants. Larvae elicited vomiting in one Blue Jay, and after eating two larvae, a second jay rejected the next 18 on sight. Together with her more extensive results on adults, Bowers established that *E. phaeton* larvae are defended by Class I chemicals.

According to Aplin *et al.* (1975), the Pieridae are supposedly a well-protected group, generally aposematic in the adult stage, and in some species throughout their life cycle. The whites (Pierinae) feed principally on the Brassicaceae and Capparidaceae (Ehrlich & Raven 1965, 1967) which contain irritant glucosinolates (Chew 1979). Aplin *et al.* (1975) found one of these, allylisothiocyanate, in the eggs, pupae and adults of *Pieris brassicae*, and in the pupae of *P. rapae*. Chrysalids of *P. brassicae*, but not *P. rapae*, also contained sinigrin, suggesting greater chemical protection of the former. Because of the egg clustering and larval aposematism of *P. brassicae* compared to the laying of single eggs and larval crypsis of *P. rapae*, together with the fact that later instar larvae of *P. rapae* are more freely eaten by birds in the wild (Dempster 1967), Aplin *et al.* (1975) speculated that *P. brassicae* is the less palatable of the two species. Earlier observations (Lane & Rothschild, in Lane 1957) supported this: a tame *Kittacincla* flycatcher ate both *P. rapae* larvae and pupae, but rejected those of *P. brassicae*, and a captive crow (*Corvus corone*) found *P. brassicae* larvae unpalatable and rejected them for nine months (Rothschild 1964*a*). However, Baker's (1970) data are contradictory: *P. brassicae* larvae in their first three instars were avidly eaten by wild sparrows (*Passer domesticus*) and tits (*Parus* spp.), and, in later instars, by thrushes (*Turdus philomelos*).

Anal discharge and oral regurgitation

Little research has been done on anal discharges in Lepidoptera even though defecation on disturbance occurs in many species (Blum 1981). Monarch faeces contain substantial cardenolide (Roeske *et al.* 1976) which could contribute to toxicity, and larvae discharge anal fluid when severely disturbed. I have observed accumulations of faecal pellets on milkweed leaves below feeding larvae in Massachusetts, Florida and California, and suggest that this may serve as visual or chemical aposematic displays towards diurnal or nocturnal predators.

Oral discharge of various sequestered terpenoids, resin acids, oils, dyes, *etc.* (Blum 1981) by sawfly (Hymenoptera) and pyralid moth larvae has been extensively investigated by Eisner and his colleagues. These discharges appear highly aversive to ants, spiders, Swainson's Thrushes (*Hylocichla ustulata*), deer mice (*Peromyscus leucopus*), and a grasshopper mouse (*Onychomys leucogaster*) (Eisner 1970, Eisner *et al.* 1974, Morrow *et al.* 1976, Eisner 1980).

The importance of oral regurgitation as an active chemical defence in butterfly larvae has been carefully documented by my student Rauch (1977), who determined that all larval instars of the Monarch butterfly can regurgitate, except during their moulting periods. Urquhart (1960) noted that molested larvae drop to the ground and curl into a tight 'C' pattern, a behaviour which he termed 'playing possum', and to which he attributed no survival function. Rauch found that curling and dropping occur after mild tactile stimulation whereas if they are pinched they first regurgitate fluid and then curl. When larvae were held posteriorly with blunt forceps, they arched their heads backward, and in several whip-like motions repeatedly regurgitated fluid and spread it over most of the body. Rauch and I hypothesized that the yellow, white and black coloration of the larva in its curled position provides an ocellar target (Blest 1957) which directs the bird to attack what in effect is an oral effluent bullseye.

Dissection indicated no diverticulae as found in sawflies and it is probable that the effluent is stored in the crop. The regurgitant is a dark green homogeneous fluid lacking leaf particles, which indicates the presence of an active sorting system. Rauch determined that the mean wet weight of the stored effluent of fifth instar larvae of *D. gilippus* was about 73mg, or 9.1% of the larva's wet weight. The mean concentrations of cardenolide in the effluent of Monarchs and Queens reared on *A. curassavica* were 1658 and 914μg/0.1g dry weight, respectively. The effluent of the Monarch therefore contained about 2.4 times the concentration of cardenolide found in the larval tissues (Thomashow 1975) and 4.8 times that in *A. curassavica* leaves. Based on a dry/wet ratio of Monarch effluent of 7.2% (larvae reared by J. A. Cohen on *A. humistrata*), the effluent of Rauch's Monarchs contained an average of 87μg of cardenolide, an amount in excess of an emetic dose for a Blue Jay.

Rauch also tested the reaction of Blue Jays to mealworms dabbed with effluent obtained from Monarchs raised on *A. curassavica*, and with control effluent from larvae reared on *Gonolobus rostratus*, a milkweed lacking cardenolides (Brower *et al.* 1967*a*). Three birds manipulated the control mealworms for an average of 1.6s compared to an average of 31s for the experimentals. Four additional birds given a sequence of 10–12 choices of uncoated and coated mealworms ate 37 of 43 controls but only 10 of the 43 experimentals. Rauch applied approximately 100μl of oral fluid to each mealworm, an amount far less than the total effluent present in one larva (about 1/700). It thus seems likely that cardenolides can also serve as Class II gustatory deterrents to Blue Jays in amounts far below those which are emetic.

Rauch also determined that cardenolides in the effluent were absent after 12 hours of food deprivation, whereas those in the midgut were retained for at least 36h. From this we can conclude that a Monarch larva can lose its regurgitatory cardenolide defence if it overeats its milkweed food supply and is forced to wander in search of a new plant.

Osmeterial secretions of Papilio *larvae*

Swallowtail larvae are characterized by a gaudy gland behind the head capsule in a prothoracic pouch (Crossley & Waterhouse 1969) which they can evert and display. The gland is forked, brilliantly coloured yellow, orange, or red, glistens with a pungently odoriferous secretion, and is lashed with accuracy towards any point of attack by sidewards or backwards thrusts of the caterpillar's anterior. Eisner & Meinwald (1965) determined that the secretion of the osmeterium of the Palaearctic *Papilio machaon* consists mainly of a mixture of two aliphatic compounds, isobutyric and 2-methylbutyric acids. These also comprised most of the osmeterial secretion of *Papilio demodocus* from West Africa, five papilionine species from southeastern USA (*P. glaucus*, *P. troilus*, *P. palamedes*, *P. cresphontes* and *P. polyxenes*), three species from Trinidad (*P. cresphontes*, *P. thoas* and *P. anchisiades*) (Eisner *et al.* 1970, Lopez & Quesnel 1970) and several Japanese *Papilio* (Oshima *et al.* 1975). Small additional amounts of acetic, propionic, and *n*-butyric acids were found in *P. anchisiades*, which Lopez & Quesnel speculated might account for the more pungent odour of this species.

Eisner *et al.* (1970) reported that the primitive baroniine swallowtail (*Baronia brevicornis*) from Mexico and a graphiine, *Eurytides marcellus*, contained the same compounds. However, Eisner *et al.* (1971) then found that the troidine *Battus polydamas* from Florida differed from the above three taxa in containing two aromatic, non-steroidal sesquiterpenes. Similarly, Suzuki *et al.* (1979) examined two parnassiines (*Luehdorfia japonica* and *L. puziloi*) and discovered myrcene, a monoterpene. However, their technique involved puncturing, and myrcene may have been a haemolymph contaminant.

Prior to the discovery of these terpenes in the parnassiines and troidines, the most parsimonious explanation for the uniform occurrence of the aliphatic acids in the primitive baroniines and the more advanced papilionines (classification scheme according to Munroe & Ehrlich 1960) was that the osmeterial secretions are biosynthetic products of ancient monophyletic origin. The chemistry of this situation now merits further study (Blum 1981).

Earlier authors suggested that osmeterial secretions might have been sequestered from larval hostplants (Bourgogne 1951), and possibly were essential oils (Wigglesworth 1964). No evidence supports sequestering and the study of Seligman & Doy (1973) indicates biosynthesis from L-valine and acetate or leucine. However, no recent study has analysed residual material which may contain plant-derived chemicals and, as Oshima *et al.* (1975) point out, myrcene is present in many plant species.

Most Papilionini larvae are cryptically coloured and consequently thought to be palatable to vertebrates. As is true of many insects (Kettlewell 1959), these larvae employ several lines of defence. Their primary defence is to avoid detection by searching predators (Robinson 1969) and in the North American *Papilio troilus*, this includes a leaf shelter, cryptic coloration, and countershading. Their secondary defence involves deterring attack once discovered, and in *P. troilus* this is also multiple: at close range realistic false eyes are displayed and if the larva is touched the osmeteria are everted and disseminate the pungent odour. These combined stimuli probably constitute a protean display.

Recent publications on osmeteria have shed little light on their actual function in natural situations. Eisner & Meinwald (1965) showed that butyric acid repels ants and it may also deter oviposition by ichneumonid and tachinid parasitoids (Owen 1971*a*), as well as spiders. Honda (1983) also considers repellency to ants a likely function, and in this context it is interesting to note that Melkert (1983) reports an ant-repellent substance produced by larvae of the nymphalid *Issoria*. Seligman & Doy (1972, 1973) found different aliphatic acids in the osmeteria of mature and early instars of *Papilio aegeus*. This raises the possibility that different enemies are targeted during different developmental stages. Jarvi *et al.* (1981*a*), using 15 Great Tits (*Parus major*) as predators, attempted to compare the palatability of normal *Papilio machaon* larvae to those with their anteriors (including the osmeteria) removed. The birds were extremely hesitant to peck larvae either with or without their osmeteria, immediately released the few they did attack, and harmed none. The study provided no evidence either for unpalatability or effectiveness of the osmeterial display and the authors concluded that the coloration of the larvae induced the birds to reject them. Gustatory neophobia may also have been involved in rejecting the ones they pecked, if tasteable substances occur in the larval exoskeleton.

The question of how the osmeteria protect swallowtails remains moot but can be solved through integrated behavioural and chemical research. Is their utility against vertebrates only as part of a protean display, or are they chemically repellent? Do the acidic properties of the secretions qualify as Class I noxious components which irritate mucus membranes? What other components are in the secretions? Are their odours adapted to distal conditioning of nocturnal predators? Is chemical mimicry involved? Do these organs deter both vertebrates and invertebrates?

Chemomimetic manipulation of ants

Many riodinid and lycaenid butterflies associate with ants and numerous species are tolerated or protected by ants in one or more stages of their life histories (Vane-Wright 1978, Kitching 1981, Atsatt 1981*b*, DeVries 1983, Henning 1983). Some species live among ants as normal phytophages, others are carnivorous upon ant-tended aphids, and still others directly exploit ants by entering their colony and feasting upon the developing brood (Clark 1925, Hinton 1951, Schremmer 1978, Cottrell 1984).

This complex of interactions is a sophisticated manifestation of chemical defence which involves the chemical control of the ants' behaviour by the butterflies (Henning 1983). By analogy with other insect studies, these butterflies probably produce chemomimetic pheromones which are similar or chemically identical to those of the ants in eliciting dispersal, appeasement or attack responses (Hölldobler 1971, Howard *et al.* 1980). Thus *Ogyris* females in Australia are able to oviposit in the midst of ants, apparently by producing a dispersal pheromone (Atsatt 1981*a*, *b*), and freshly eclosed *Maculinea arion* may utilize appeasement pheromones during their ascent from the subterranean colonies to the outside world (Ford 1945). Larval and pupal glandular exudates probably contain appeasement and dispersal pheromones and definitely contain nutrients of value to ants (Ch.19). On attack by parasitoids or predators, lycaenid larvae may also release alarm pheromones to elicit the ants' social defences. Although defence against parasitoids was considered unlikely by Malicky (1970), Pierce & Mead (1981) proved that *Formica* ants do protect *Glaucopsyche lygdamus* larvae against braconids and tachinids. Pheromonal-mediated protection by the ants against other vertebrate and invertebrate predators, including other species of ants (Malicky 1969, 1970) is undoubtedly important (see Atsatt 1981*b*, for literature review).

Defensive Chemicals in Pupae

Why are so few pupae aposematic?

The fact that natural selection has generally enforced elaborate crypsis on the pupal stage of even the most toxic of butterfly species is of considerable interest. Poulton (1924*b*) and Cott (1940) speculated that the partial to total immobility of butterfly pupae would generally result in aposematic failure, because a chrysalis can neither flee nor behaviourally display. Moreover, no pupae are known which produce active secretions, so that the only way a predator could be exposed to defensive chemicals would be to eat or lethally penetrate the insect. (Trimen 1869 maintained that a disagreeable odour characterizes all stages of *Acraea horta*, but I have been unable to confirm this.)

The overriding importance of camouflage in the pupal stage is emphasized by the fact that species in most families have complex genetic control mechanisms in the larvae which interact with environmental cues, anticipate the most cryptic alternative in regard to the future background coloration, and produce either green or brown chrysalids (Owen 1971*b*, Edmunds 1974*a*, Wiklund 1975*a*, Hazel & West 1979). The apparent ability of *Danaus chrysippus* (Rothschild *et al.* 1978*a*) and also *D. gilippus* (Brower, pers. obs.) pupae to match their background colour, whereas Monarch pupae are always blue-green is unexplained.

The absence of pupal warning coloration in most noxious species is correlated with the tendency of those species which are gregarious as larvae to disperse and pupate singly. This includes *Nymphalis antiopa* of North America, in which the larvae are mechanically protected by spines, the purportedly unpalatable acraeines in Africa (Owen 1971*a*), and the highly unpalatable *Battus philenor* in eastern North America (Hazel & West 1979). A remarkable exception is *Heliconius doris*, a relatively unpalatable heliconiine from Trinidad (Brower *et al.* 1963). Females lay their eggs in clusters and development is synchronized such that the larvae hatch, moult and pupate together, and virtually all adults emerge on the same day about three weeks after the initial oviposition (Alexander 1961*a*, *b*, Brower & P. M. Sheppard, pers. obs.). Why *H. doris* has not evolved aposematic pupae is enigmatic.

Probably most butterflies of the American nymphaline tribe Melitaeini are unpalatable in all life history stages, and at least two species apparently have warningly coloured chrysalids. The best known is *Euphydryas phaeton* from eastern North America, which is gregarious in all but its pupal stage and is aposematic in all stages. (For superb coloured photographs, see Emmel 1976). Bowers (1981) found that *E. phaeton* pupae were initially killed and then rejected on sight by two Blue Jays, and she attributed their unpalatability to the same emetic substances which are in *E. phaeton* larvae and adults. In Mexico, another nymphaline (*Chlosyne ehrenbergi*) forms conspicuously coloured pupae in exposed positions (Brower, pers. obs.), as does the African *Acraea zetes* (Carpenter, in Carpenter & Ford 1933: 78). Another striking exception is found in the Neotropical lycaenid *Eumaeus minyas* which is highly conspicuous in all stages. The females oviposit in clusters on cycads (*Zamia*) which contain gastro-intestinal and liver toxins (Hooper 1978*b*), and the larvae feed and pupate gregariously (DeVries 1976, 1977).

Neither Bowers nor Stamp (pers. comm.) agrees with my opinion that *Euphydryas* pupae are aposematic. In contrast, West & Hazel (1982) maintain that overwintering chrysalids of *Battus philenor* on tree trunks are conspicuous, whereas the fact that the species exhibits pupal colour dimorphism argues that they are cryptic. These differing interpretations underscore the need for more research on pupae.

Noise production and chemical defence

Pupae of more than 150 lycaenid and riodinid species are capable of producing audible or ultrasonic sounds (Hinton 1948, Downey & Allyn 1973). Because ants hear by perceiving groundborne vibrations through acoustical organs located in their legs (Wilson 1971), it is likely that sound emission by the pupae combines with their chemomimetic pheromone release and mediates chemical and aggressive defences of the ants against attacking parasitoids (Hinton 1955). It is also conceivable that birds and mammals can become distally conditioned to avoid noisy pupae following one or more noxious encounters with ants which were aroused by the chrysalids.

Experimental studies

Chemical defence of pupae has been studied both in the natural environment and in the laboratory. Thomashow (1975) determined that the average cardenolide content of Monarch pupae reared on *Asclepias curassavica* was 978μg ($n = 10$), an amount equivalent to about 13.7 Blue Jay emetic units and considerably more than found in adults reared on this plant. Miller (1976) verified this: four jays vomited after eating one or more, whereas six control jays ate several which had been reared on the cardenolide-free *Gonolobus rostratus*, without sickness. Furthermore, the array of cardenolides as determined by thin layer chromatography was similar to that found in all the other stages (Colour plate 2A).

Wiklund, (1975*a*) reported that *Papilio machaon* pupae are heavily preyed upon by mammals and birds in Scandinavia. West & Hazel (1982) found that mice (*Peromyscus*) and shrews (*Blarina*) in Virginia prey extensively on overwintering chrysalids of both *Papilio glaucus* and *Battus philenor*, the adults of which are, respectively, palatable and unpalatable to birds. Wintering birds seemed to favour *P. glaucus* over *B. philenor*, in conformity with the known adult palatabilities. Brower & Brower (1961) offered three pupae of the Monarch reared on the relatively non-toxic *A. incarnata* to *Peromyscus leucopus*, which ate them completely. Baker's (1970) study of *Pieris rapae* and *P. brassicae* in England found that wild birds favoured pupae of the former. However, the birds' selection did not implicate palatability differences, but rather the fact that *P. rapae* pupae could more easily be removed from their substrates. The evidence for unpalatability of the early stages of these pierine species is thus substantially lacking, even though *P. brassicae* has aposematic pupae.

Birds as Enemies of Adult Butterflies: the Importance of Beakmarks

Many antipredator adaptations of adult butterflies have undoubtedly evolved in relation to selection exerted by avian predators. The early naturalists were fully aware of this but, as a perusal of the publications of the Entomological Society of London from 1862 onwards indicates, the question of whether birds really are significant predators of butterflies was hotly debated. Evidence involved direct observation of attacks and the discovery that numerous butterflies bear the marks of birds' beaks upon their wings. Poulton (in Marshall & Poulton 1902: 366) reported that the supposedly unpalatable acraeines and danaines more frequently bore beakmarks than did those of purportedly palatable species. He interpreted the data to indicate that birds catch, taste, and release individuals of the unpalatable species, whereas they catch, taste, and eat the palatable ones. Consequently, in a random sample of wild butterflies, more individuals of the unpalatable species should bear beakmarks. Carpenter (1941, 1942) analysed large numbers of beak-marked butterflies which supported this correlation. The value of this method for measuring relative predation pressures on various species as well as on the different forms within a single polymorphic species is now well established (Sheppard 1959, Sargent 1973, Edmunds 1974*a*,*b*, Shapiro 1974, Smith 1979, Robbins 1981).

An extension of the beak-marked butterfly methodology was made by Pough & Brower (1977) and Bowers & Wiernasz (1979), who distinguished beak-torn from beak-marked individuals in population samples. Whereas beak-marked butterflies are interpreted as being actively taste-rejected by the birds, beak-torn individuals are held to be butterflies which were caught and would have been eaten but escaped by breaking away from the bills of the birds. Thus a species of butterfly which has a higher proportion of beak-marked than beak-torn individuals in a population sample is judged less palatable than a species in which these frequencies are reversed. One caveat in comparing species is to control for wing fragility, which itself may be an antipredator adaptation.

Comparative Palatability Studies of Adult Butterflies

A study comparing the palatability of ten species of tropical butterflies to Blue Jays was done by Brower & Brower (1964) in Trinidad, W.I. On ten consecutive days each bird was simultaneously offered one individual of each of the ten butterfly species and after two hours the numbers not touched, pecked, 'killed' (i.e. damaged sufficiently to be incapable of reproducing), and eaten were tallied. We later repeated the experiment using two locally

Table 12.1. Reactions of six individually caged Blue Jays, seven Silverbeak Tanagers and six Parson's Tanagers to 10 individuals each of 10 species of wild captured butterflies in Trinidad, W.I. (From Brower & Brower 1964, and unpublished data.)

	BUTTERFLY SPECIES														
	Parides anchises or *neophilus*			*Danaus plexippus*			*Ithomia drymo*			*Heliconius sara*			*Lycorea cleobaea*		
% of Birds' Responses	BJ	SB	PT	BJ	SB	PT	BJ	SB	PT	BJ	SB	PT	BJ	SB	PT
Not Touch	17	49	67	20	80	78	23	52	71	15	48	75	18	59	85
Peck	48	16	15	27	13	3	37	11	9	27	9	13	3	4	7
Kill	35	31	18	51	7	18	23	17	10	27	27	12	30	24	8
Eat	0	4	0	2	0	0	17	20	10	32	16	0	48	13	0
	Biblis hyperia			*Euptychia* spp.			*Agraulis vanillae*			*Anartia amathea*			*Siproeta stelenes*		
	BJ	SB	PT	BJ	SB	PT	BJ	SB	PT	BJ	SB	PT	BJ	SB	PT
Not Touch	13	11	42	15	0	33	13	3	42	7	1	28	10	0	38
Peck	10	3	12	10	0	2	3	3	8	12	0	7	5	0	3
Kill	10	7	17	2	0	2	10	34	32	7	7	7	8	4	15
Eat	67	79	30	73	100	63	73	60	18	75	92	58	77	96	43

Each bird was simultaneously offered 10 dead butterflies on a tray for 2 h for each of 10 consecutive days.

collected Neotropical birds, the Silverbeak Tanager (*Rhamphocelus carbo magnirostratus*) and Parson's Tanager (*Tachyphonus rufus*) and the previously unpublished results are compared in Table 12.1. The order of acceptability of the butterflies for these three taxonomically and ecologically distinct bird species (Table 12.2) was nearly identical, upholding their classically predicted palatability (Brower & Brower 1964). This concordance diminishes both Benson's (1971) criticism of using wild and possibly experienced birds and Duffey's (1977) criticism of drawing general conclusions based only on the Blue Jay as a standard bioassay predator.

Table 12.2. Orders of preference of 10 species of Neotropical butterflies to the Blue Jay, Neotropical Silverbeak Tanager and Parsons Tanager. (From Brower & Brower 1964, and unpublished data.)

	Ranks by percentages of butterflies eaten[1]		
Butterfly Species	6 Blue Jays	7 Silverbeaks	6 Parsons
Siproeta stelenes	1	2	3
Anartia jatrophae	2	3	2
Agraulis vanillae	3	5	5
Euptychia spp.	4	1	1
Biblis hyperia	5	4	4
Lycorea cleobaea	6	8	7[1]
Heliconius sara	7	7	9[1]
Ithomia drymo	8	6	6
Danaus plexippus	9	10	8[1]
Parides anchises or *neophilus*	10	9	10[1]

[1]Parsons Tanager ate none of four species, so rank of data[1] is based on the percentage not touched. (1 = most preferred, 10 = least preferred.)

Palatability of the Satyrinae and Pieridae

Satyrids of the genus *Euptychia* are palatable to Blue Jays, Silverbeaks and Parson's Tanagers (Tables 12.1, 12.2). Bowers & Wiernasz (1979) provided three lines of evidence for the palatability of *Cercyonis pegala*. First, five jays ate an average of 26 out of 30 butterflies; secondly, the birds' handling times were short (mean 5.3–10.3s); thirdly, of the 9% wing damaged (58 out of 639) wild butterflies collected in Massachusetts, only four of the 58 (7%) were beakmarked, whereas 93% were beak-torn. This is evidence that the wild birds taste-reject very few *Cercyonis*. In comparison, Pough & Brower (1977) found that 160 (23%) of 697 *Ascia monuste* (Pieridae) individuals had been attacked by birds in south Florida of which 107 (67%) were beak-marked and only 33% were beak-torn, i.e. wild birds taste-rejected *Ascia* ten times more frequently than *Cercyonis*

Table 12.3 Analysis of beakmarked and beaktorn individuals in wild samples of 639 *Cercyonis pegala* (Satyrinae) from Massachusetts and 697 *Ascia monuste* (Pieridae) from Florida. (Data from Bowers & Wiernasz 1979, Pough & Brower 1977.)

Treatments	Butterflies	
	Cercyonis	*Ascia*
Beak-marked[1]	4 (7%)	107 (67%)
Beak-torn	54 (93%)	52 (33%)
Total	58	159
$X^2_1 = 63.5 \quad P < 0.001$		

[1]Given as 'symmetrical' in Pough & Brower.

(Table 12.3). The two satyrids are thus highly palatable butterflies whereas the pierid is of intermediate palatability. Pough & Brower (1977) independently reached this conclusion in feeding experiments with *A. monuste* and jays.

Finn's (1895, 1897*a,b*), Swynnerton's (1915*c,d*, 1919) and Carpenter's (1921) early observations and experiments with wild birds and monkeys suggested that pierids vary in palatability. This was deduced from two *Cercopithecus* monkeys' and several species of birds' slow picking apart, mouthing, and long eating times for several pierid species. Others (*e.g. Belenois, Eurema*) appeared completely palatable, whereas *Delias* appeared completely unpalatable, as Wallace (1867*b*) had predicted. Lane & Rothschild (Lane, 1957) offered a tame flycatcher (*Kittacincla malabarica*) various British pierids and stated that it found *Gonepteryx rhamni* palatable, *P. rapae* unpalatable but edible, and *P. brassicae, Anthocharis cardamines* and *P. napi* the most unpalatable. In contrast, *Archonias tereas* from the Osa Peninsula in Costa Rica was completely acceptable to two Blue Jays (Brower, unpublished) suggesting that this black, red and white pierid is a classical Batesian mimic of *Parides*. Interestingly, Beddard (1892) suggested that Bates' classical *Leptalis Dismorphia*) mimics may not be palatable, but this has never been tested.

The behaviour of Swynnerton's birds and Carpenter's monkeys towards various pierines, the behaviour of Blue Jays towards *Ascia*, the contradictory interpretations of the palatability in the various life history stages of *Pieris brassicae* and *P. rapae* (this chapter; also Jones 1932, Frazer & Rothschild 1960), and the jays' complete acceptance of *Archonias* all support a range of palatabilities among the Pieridae, as concluded earlier by Brower & Brower (1964). In contrast, satyrids are probably uniformly palatable, although Lane & Rothschild (Lane 1957) reported that *Kittacincla* was reluctant to eat female *Maniola jurtina*.

Palatability of the Genus *Papilio*

Jane Brower (1958*a*,*b*,*c*) offered more than 1000 wild captured *Papilio glaucus*, *P. palamedes* and *P. troilus* to nine individually caged Scrub Jays, *Cyanocitta coerulescens*, and the birds ate virtually all of them without hesitation. Four birds also ate 43 out of 44 *P. polyxenes*. In contrast, four birds attacked but did not eat a single Monarch out of 336 offered, and three of the same birds ate only two out of 207 *Battus philenor*. Eight Blue Jays similarly treated two of these species: they ate 75 of 80 *P. glaucus*, compared to only five of 80 Monarchs (Platt *et al.* 1971, Platt & Brower, in prep.). The finding of noxious chemicals in *Papilio antimachus*, the largest swallowtail in Africa (Owen 1971*a*), was therefore surprising: Rothschild *et al.* (1970) reported large amounts of cardenolides in a 70-year-old museum specimen. Rothschild's (1971) report that further individuals lacked cardenolides suggests an error in the original study and unfortunately was not noted in subsequent reviews (Pavan & Dazzini 1976, Blum 1981; but see Rothschild 1972*b*).

Possible Odour Mimicry in *Atrophaneura*

The Japanese swallowtail *Atrophaneura alcinous* secretes benzaldehyde and a variety of other volatiles in its wings (p.116) and may be an odour mimic. From the presence of this substance in the defensive glands of some millepedes, beetles, bees and ants (Blum 1981), it is deduced to be generally defensive. In millepedes, benzaldehyde is often found together with HCN because both compounds are derived from the same precursor, mandelonitrile (Duffey *et al.* 1974, Eisner *et al.* 1975, Huheey, in press). Interestingly, both chemicals smell of bitter almonds, although benzaldehyde is more aromatic and HCN is more pungent (Conn 1979). In *Atrophaneura*, benzaldehyde (and HCN if it also occurs) could effect Müllerian odour mimicry of other HCN-containing species. On the other hand, if benzaldehyde proves to be innocuous, it could effect Batesian odour mimicry of species containing HCN.

Palatability of *Troides* and *Battus*

All *Aristolochia*-feeding troidines so far tested proved highly unacceptable to birds, although Brown & Vasconcellos (1976) reported the tanager *Pipraeidea melanonota* eating *Parides bunichus* in southeastern Brazil. J. Brower (1958*b*) offered 68, 76, and 63 individual *Battus philenor*, respectively, to three experimental Scrub Jays. Two birds each pecked apart ('killed') six butterflies in their first twenty trials, ate only two, and then sight-rejected the majority of those subsequently offered. The third jay's pattern was similar except that it only pecked the butterflies and ate none. Platt *et al.* (1971) offered 10 *B. philenor* reared on *Aristolochia californica* to each of 23 Blue Jays. Most of these birds also taste-rejected *B. philenor* (Platt & Brower, in prep.).

Tables 12.1 and 12.2 compare the treatment by Blue Jays, Silverbeaks and Parson's Tanagers of two Neotropical *Aristolochia* feeders, *Parides anchises* and *P. neophilus*. The three species of birds treated *Parides* uniformly: out of 70 butterflies offered, seven Silverbeaks ate only three, and the other two bird species ate none of 60 offered. All birds 'killed' and/or pecked a significant proportion of the *Parides*, thus again exhibiting gustatory rejection of the butterflies. Rothschild *et al.* (1972*a*) found that Starlings (*Sturnus vulgaris*) initially ate but then rejected *Troides aeacus*.

Based on the classical literature as summarized in Brower & Brower (1964), the *Aristolochia*-feeding members of the tribe Troidini, including *Battus*, *Parides* and the Malaysian birdwings of the genus *Troides*, were predicted to sequester alkaloids from their foodplants which could serve as the basis of their unpalatability to birds. In a search for likely candidates, Rothschild and colleagues analysed several species of *Troides*, *Battus* and *Ornithoptera*. *Troides aeacus* was found to lack aristolochic acids but did contain about 10μg per 0.1g of a potentially noxious acetylcholine-like substance (Rothschild *et al.* 1970). Von Euw *et al.* (1968), Rothschild *et al.* (1970, 1972*a*) and Rothschild (1972*b*) found as much as 100μg of aristolochic acid-1 (and derivatives) in the adults of several other troidine species, including the New Guinean *Ornithoptera priamus*, the North American *Battus philenor* and *B. polydamas*, and the Ceylonese *Pachliopta aristolochiae*. The same chemical was also in two *Pachliopta* foodplants, *Aristolochia clematis* and *A. rotunda*. Rothschild *et al.* (1972*a*) also discovered two distinct aristolochic acids in the parnassiine *Zerynthia polyxena* reared on *A. clematis*. Three species of *Troides* proved negative for aristolochic acids, as did *P. hector*.

To test for possible noxious effects of aristolochic acids, von Euw *et al.* (1968) administered by stomach tube the equivalent of two crushed butterflies to quails (*Coturnix sylvatica*). The birds did not sicken and showed no negative symptoms, possibly because they are generally insensitive to many secondary plant compounds. Intravenous injection of the extracts of two to three crushed butterflies into guinea pigs produced sedation for about an hour without further consequences, whereas injection of pure aristolochic acid-I into two rabbits and one guinea pig resulted in their death after 2–14 days. However, as already noted, such administration of toxins is without obvious ecological significance, and the ability of these chemicals in butterflies to avert natural predators remains hypothetical.

Von Euw *et al.* (1968), in agreement with Ford (1955), concluded that the rejection of troidine butterflies as recorded in the early literature is based on taste. This hypothesis is supported by our jays' and tanagers' oral rejection of *Battus* and *Parides*. It seems likely that as yet unidentified Class I noxious chemicals exist in troidines which are sufficiently severe to invoke oral rejection without ingestion. As von Euw *et al.* (1968) noted, whether aristolochic acid-1 is one of these, has not been established.

Palatability of the Ithomiinae

Neotropical ithomiines feed mainly on the Solanaceae from which they were predicted to sequester alkaloids (reviews in Brower & Brower 1964, Drummond 1981). No chemical evidence for solanaceous alkaloids in ithomiines exists, although as adults the males of many and perhaps all species are highly attracted to pyrrolizidine alkaloid-producing plants and most species analysed contain PAs (Pliske *et al.* 1976, Edgar *et al.*, 1976*b*). Feeding experiments with three species of birds indicated that *Ithomia drymo* collected by PA-baiting in Trinidad consistently ranked as unpalatable, but are not as unpalatable as danaines or troidines (Tables 12.1 and 12.2). Alcock (1965) offered ithomiines of three genera to four species of North American birds and found them less acceptable than control *Cystineura* butterflies (Table 12.4). Brower (unpublished data) found that one experimental silverbeak ate 18 of 20 *Mechanitis lysimnia* in Trinidad, and a second bird, which would not attack most on sight, finally ate the last two offered. Unpalatability of the exoskeleton of three additional ithomiines was suggested by Brown & Vasconcellos (1976). Ithomiines probably range across a broad palatability spectrum for the reasons discussed by Pough *et al.* (1973) and Huheey (1976).

Palatability of the Heliconiinae and Acraeinae

Brower *et al.* (1963) experimentally offered 1177 individual heliconiine butterflies of seven species to 62 Silverbeak Tanagers in Trinidad. The species were predicted to exhibit a range of palatability, and the birds ate 25% of those considered most primitive (*Agraulis vanillae* and *Dryas iulia*) compared to less than 1% of the specialized and highly aposematic *Heliconius numata*, *H. melpomene*, *H. erato* and *H. sara*. Overall, the birds sight-rejected 76%, taste-rejected 15%, and ate 9% of the butterflies. Similar results were found when *H. erato* was offered to Forktailed Flycatchers, *Muscivora tyrannus* (Brower *et al.* 1971).

In a well-designed experiment, Boyden (1976) offered individual *Heliconius melpomene*, *H. erato* and control *Anartia fatima* butterflies to a small wild population of *Ameiva* lizards in Panama. Inexperienced lizards initially killed and taste-rejected most *Heliconius*, but rapidly learned to reject them both visually and gustatorily. Rubbing off the scales or painting the butterflies led the experienced individuals to renew their attacks. The lizards also found *Anartia fatima* palatable, and the overall results are similar to our tanager and flycatcher experiments in Trinidad.

Brower & Brower (1964) predicted, on the basis of the natural history and known plant chemistry that heliconiine butterflies would be found to sequester cyanogenic compounds from their *Passiflora* foodplants. Rothschild (1971) extended the prediction to those species of the Acraeinae which feed on *Passiflora*, and Owen (1970*a*, 1971*a*) stated that the yellow, foamy thoracic exudate of *Acraea encedon* does in fact contain HCN. Nahrstedt & Davis (1981), in important pioneering chemical analyses, have now confirmed the presence of variable amounts of the cyanogenic glycosides linamarin and lotaustralin in several African acraeines, including *A. caldarena*, *A. eponina* and *A. encedon*, and also in a pooled sample of various Neotropical heliconiines, including *H. erato* and *H. doris*. The authors incline to the view that these cyanogenic glycosides are produced through *de novo* synthesis in the butterflies rather than sequestered from their foodplants, but this is by no means conclusive.

Table 12.4. Reactions (% eaten) of four North American birds to three species of Neotropical Ithomiine butterflies. (From Alcock 1965.)

Butterflies	Catbirds[1] ($n=10$)	Myrtle Warblers[2] ($n=7,5,5$)	Chickadees[3] ($n=10,7,6$)	Whitethroated Sparrows[4] ($n=8$)
Hypoleria ocalea	40%	0%	44%	—
Hypothyris euclea	—	4%	71%	40%
Ithomia drymo	—	16%	43%	—

Each bird ate five single palatable *Cystineura arana* (Nymphalidae) paired with five individual Ithomiines in a randomized pair sequence.
[1]*Dumetella carolinensis* (Mimidae)
[2]*Dendroica coronata* (Parulidae)
[3]*Parus atricapillus* (Paridae)
[4]*Zonotrichia albicollis* (Fringillidae)

The feeding experiments carried out in Africa by Marshall (in Marshall & Poulton, 1902) and by Swynnerton (1915*c*,*d*, 1919) provided strong qualitative evidence that acraeines are unpalatable to several species of wild birds and monkeys. How HCN may be involved is uncertain.

Palatability of the Nymphalinae

Nymphalines appear to range the palatability spectrum. Platt *et al.* (1971) found that *Limenitis arthemis-astyanax* are highly palatable to Blue Jays, and experiments with various bird predators established that *Siproeta stelenes, Anartia jatrophae, A. amathea* and *Cystineura cana* are highly palatable, whereas *Biblis* is only moderately palatable (Tables 12.1, 12.2, and Brower *et al.* 1971). *Anartia fatima* in Panama was also palatable to lizards (Boyden 1976).

At the opposite extreme, Bowers (1979, 1980) found that the larvae, pupae, and adults of the North American *Euphydryas phaeton*, when reared on their normal scrophulariacious foodplant *Chelone glabra*, are emetic to Blue Jays but when reared on plantain (*Plantago*, Plantaginaceae) they are not (Table 12.5).

Table 12.5. Summary of blue jay responses to individually offered wild *Cercyonis* and reared *Euphydryas* butterflies from two foodplants. (From Bowers 1980.)

Jay Responses	To *Cercyonis*	To *Euphydryas* reared on *Chelone*[1,2]	To *Euphydryas* reared on *Plantago*[3]
Not Touch	4%	64%	38%
Peck	6%	3%	0%
Kill	3%	9%	7%
Eat	87%	23%	55%
TOTAL butterflies	225	225	29
TOTAL birds	10	10	3

[1]Three birds refused all *Euphydryas* on sight.
[2]Four of seven birds which ate two to five *Chelone*-reared *Euphydryas* vomited.
[3]None of these birds showed signs of sickness.

The cardenolide assay of both the butterflies and *Chelone* proved negative and Bowers (1981) suggested that the emetic substances may be iridoid glycosides sequestered from *Chelone*. Bowers also determined that 10 of 14 jays taste-rejected 37 out of 84 of the *Euphydryas chalcedon* and *E. editha* butterflies, and did not eat a single individual. In other words, 70% of the jays rejected these butterflies without being previously conditioned during captivity by their emetic properties.

The Asiatic *Hypolimnas bolina* has been assumed to be a palatable Batesian mimic of euploeines (Clarke & Sheppard 1975). Marsh *et al.* (1977) found that extracts of both ground-up adults reared on the poisonous *Ipomoea batatas* and of the plant material had cardioactive effects when injected into rats. Minimal responses occurred from similar injections when the butterflies were reared on *Asystasia*, and injections of the plant extracts were negative. Similar extracts of *Ipomoea* plant material and butterflies reared thereon when force-fed to pigeons, did not produce emesis. Reichstein (reported in Dixon *et al.* 1978) found cardenolides lacking in *Ipomoea*, and the authors concluded that *Hypolimnas* both stores a cardioactive non-cardenolide from *Ipomoea*, and also secretes *de novo* some type of cardioactive substance. The first of these conclusions appears unjustified because a test for cardenolides was not reported for the butterflies. The second conclusion, weakly supported by the data at best, is subject to alternative interpretations because of the problems with injection bioassays. Furthermore, since the authors did not assay the amount of toxins which were force-fed, the negative emesis tests with pigeons are inconclusive. The occurrence of Class I chemicals in *Hypolimnas bolina* thus appears debatable.

J. Brower (1958*a*,*c*) determined that the northern and southern races of the Viceroy (*Limenitis archippus*) are effective mimics of Monarch and Queen butterflies, respectively. In her experiments, three out of five Scrub Jays were initially hesitant to eat Viceroys and taste-rejected 6–15 of them. Platt *et al.* (1971) likewise found that 15 Blue Jays showed variable treatment in eating *L. archippus* compared to *L. arthemis* and *L. astyanax*. This raises the possibility that the classically assumed Batesian mimicry status of the Viceroy is incorrect; its mimicry may be Müllerian. Given the variability in emetic noxiousness of Monarchs, it is not inconceivable that the traditional assumed roles of Viceroy and Monarch might even be reversed in some populations.

Palatability of the Danainae

History

Danaine butterflies became the keystone of theory in the study of chemical defence because of Swynnerton's (1915*d*: 43) discovery of emesis-based unpalatability in East Africa. In one series of experiments with three Wood Hoopoes (*Irrisor erythrorhynchus* = *Phoeniculus purpureus*, Phoeniculidae), he found that one bird vomited after eating a single *Amauris echeria*, another after eating a single *Danaus chrysippus*, and a third after eating one *D. chrysippus* and one *Acraea encedon*. Interestingly, Swynnerton (1915*d*, 1919) also found that four other species of birds did not vomit even after eating up to 50 danaines in a single feeding session. These included three Rollers (*Coracias garrulus*, Coraciidae), two Yellow Billed Hornbills (*Lophocerus leucomelas* = *Tockus flavirostrus*, Bucerotidae), one Crowned Hornbill (*L. melanocephalis*

= *T. alboterminatus*), and one Babbler (*Crateropus kirki* = *Turdoides jardinei*, Turdoidae).

Vomiting in Swynnerton's hoopoes was probably caused by cardenolides and its absence in the other four bird species may have been due to cardenolide insensitivity. Rothschild (1971, 1972*b*), and Rothschild & Kellett (1972) have remarked on the historical priorities in the discovery of cardenolide-caused emesis in butterflies. In reviewing the original papers, I conclude that what is certain is that Swynnerton discovered emetic butterflies; that the emesis he observed was caused by cardenolides is debatable. This is because Swynnerton offered several wild individuals and different species in single experiments and we now know that not only is there a high proportion of cardenolide-free danaine butterflies in wild African populations of *Danaus chrysippus*, but the species is also a less efficient sequesterer than the Monarch (Rothschild *et al.* 1973, 1975*b*, Brower *et al.* 1975, 1978).

Proof that cardenolides are Class I defensive chemicals

Various lines of support for the existence of cardenolides in *Danaus plexippus* were provided by pharmacological experiments (Parsons 1965), feeding experiments (Brower *et al.* 1967*a*), and chemical analyses (Reichstein 1967, von Euw *et al.* 1967). Chemical proof that cardenolides *per se* are causative agents of vomiting was established by Brower *et al.* (1982) in experiments in which the emetic dose fifty of labriformidin isolated from the California *Asclepias eriocarpa* was determined to be 57μg per 85g Blue Jay. Labriformidin is ingested by Monarchs when fed on this plant and chemically reduced within the larva to desglucosyrioside which is then stored within the adult. Powdered butterfly and whole plant material gave ED_{50} values similar to those of pure labriformidin. Seiber & Lee (1982) have recently reviewed the chemistry of these and other *Asclepias* cardenolides.

Populational variation in cardenolide content and emetic potency

Variability in the cardenolide content and emetic potencies of Monarch butterflies has been investigated by Brower, Seiber, Nelson and their colleagues (Brower *et al.* 1972, 1982, Roeske *et al.* 1976, Nelson *et al.* 1981, Fink & Brower 1981). Wild larvae and pupae collected on seven species of *Asclepias* in California produced adults with unmeasurable to 1279μg of cardenolide per butterfly and their emetic potencies ranged from 0 to 26 Blue Jay emetic units (Brower *et al.* in prep). Wild butterflies from various migrating and overwintering North American populations ranged from 0 to 850μg and 0 to 9 emetic units (Fink & Brower 1981).

Variability in the butterflies results both from a mirroring of the differences among cardenolide profiles present in the various *Asclepias* species which the larvae ate, and from quantitative intraspecific variation in the gross concentrations in individual plants. Importantly, little of the variation appears to result from variability in the arrays of cardenolides within each plant species which in fact appear remarkably constant. Likewise, and contrary to the speculation of Rothschild & Marsh (1978) and Dixon *et al.* (1978), little if any evidence exists for metabolic modification of the cardenolides between the pupal and adult stages, at least in Monarchs reared on *Asclepias curassavica* (Colour plate 2A); see also Seiber *et al.* 1980). This constancy of pattern allows adult Monarchs to be 'fingerprinted' to their respective foodplants (Brower *et al.* 1982), contrary to the skepticism expressed by Blum (1981). Distinctive TLC patterns for adult male and female Monarchs reared on *Asclepias curassavica* and *A. syriaca* are compared in Colour plate 2C,D. Note how little variation there is by sex, by body section, or in the internal organs compared to the exoskeleton.

Gustatory rejection by birds

J. Brower's (1958*a*,*c*) experiments were the first controlled demonstration of gustatory rejection of butterflies. Four Scrub Jays refused to eat numerous Monarchs after either pecking or beginning to eat them and no bird actually ingested a Monarch. Four additional birds when offered more than 200 Queen butterflies (*Danaus gilippus*) sight- or taste-rejected most, ate only eight individuals, and none vomited. Six Blue Jays in Trinidad similarly rejected Monarchs, as did Silverbeaks and Parson's Tanagers (Table 12.1), and Platt *et al.* (1971, and unpublished data) found that eight Blue Jays taste-rejected 28 out of 75 Monarchs. Whether or not gustatory rejection in these birds was based on prior conditioning to emetic Monarchs in the wild, on Class I noxious gustatory stimulation caused by the bitterness of the cardenolides or of other substances in the butterflies, or on neophobia remains unresolved.

Are cardenolides overemphasized as defensive chemicals?

It has been suggested that cardenolides are over-emphasized in danaine research and that other sequestered secondary chemicals, as well as *de novo* synthesis products, may be of equal or perhaps even greater importance in defence (Rothschild & Kellett 1972, Rothschild *et al.* 1975*b*, Duffey 1977, Marsh *et al.* 1977, Boppré 1978, Edgar *et al.* 1979, Blum 1981). Indeed Monarch investigators have been referred to as 'hypnotized by the discovery of cardenolide storage' (Dixon *et al.* 1978: 462).

In searching for other substances, Marsh *et al.* (1977), Rothschild *et al.* (1978*b*), Rothschild & Marsh (1978) and Dixon *et al.* (1978) adduced evidence through rat heart perfusion experiments that both the Monarch and *Euploea core* contain cardioactive substances which may not be cardenolides. Rothschild *et al.* (1975*b*) and Dixon *et al.* (1978) also pointed out that African *Amauris* butterflies generally specialize on genera of asclepiads which lack cardenolide (Rothschild *et al.* 1970), but that dried specimens were very distasteful to grackles. The isolation and characterization of these purported cardioactive substances has yet to be done, and as already noted, injection assays need critical re-evaluation.

I certainly agree that broader studies of the chemical ecology of danaines should be carried out, particularly because of the many reports in the natural history literature of birds eating both danaines and euploeines (*e.g.* Manders 1911). Now that the systematics of the Danainae has been completely revised (Ackery & Vane-Wright, in press *a*), a comparative study of chemical defence in the rich assemblage of Danainae in Southeast Asia would be extremely valuable. However, the fact remains that cardenolides are the *only* proven Class I chemical deterrents of vertebrate predators that are known to occur in butterflies. Furthermore, cardenolides are the only known defensive chemicals that have been studied comparatively and quantitatively in natural populations.

Pyrrolizidine alkaloids

Other possible Class I and/or Class II deterrents found in danaines, acraeines and ithomiines are the pyrrolizidine alkaloids (PAs). When PAs were initially reported in the aposematic Cinnabar moth (*Tyria jacobaeae*, Arctiidae), Aplin *et al.* (1968) considered it unlikely that they could be actively repellent to vertebrates because of their slow acting effects. Rothschild *et al.* (1973) reversed this position, apparently based both on the potential or established co-occurrence of cardenolides and PAs in ctenuchid and arctiid moths together with the discoveries of PAs in new and old world danaines, including *Lycorea cleobaea*, *Danaus gilippus*, *D. chrysippus*, *D. plexippus* and several species of *Amauris* (Meinwald *et al.* 1966, 1969, 1971, 1974, Edgar *et al.* 1976*a*).

An important difference between PAs in moths and butterflies is that the moth larvae often feed on plants which contain both PAs and cardenolides, whereas ithomiine and danaine larvae, with the possible exception of the primitive ithomiine genus *Tithorea* (Pliske 1975*a*) and *Tirumala limniace* (McCann, 1952 —but see also Ch.9), feed on PA-negative foodplants and obtain PAs only as adults by sucking them up in fluids of decomposing vegetation or in nectar from various species of the Boraginaceae, Asteraceae and Fabaceae (Edgar & Culvenor 1974, Edgar *et al.* 1974, 1976*b*, 1979, Pliske 1975*a*, Edgar 1975, Schneider 1975, Bernays *et al.* 1977, Boppré 1977, 1978, 1981, Dixon *et al.* 1978, Rothschild & Edgar 1978). While it was previously thought that only adult males of ithomiine and danaine species were attracted to PA sources, Pliske (1975*a*) reported a few observations of females at PA plants and Edgar *et al.* (1979) later found PAs in females of several danaines.

Citing unpublished oral communications from Eisner's laboratory (see also Eisner 1980) purporting to show that PAs are distasteful to various invertebrate predators and to birds, these authors have arrived at the consensus that a dual PA-cardenolide based defence exists in danaine butterflies and that, overall, PAs may be more important than cardenolides (Rothschild & Marsh 1978).

Pyrrolizidine alkaloids versus *cardenolides*

The action of cardenolides in defence is clear: they are highly effective Class I chemicals with both noxious emetic and bitter properties. In contrast, PAs are analogous to certain other poisonous plant alkaloids in that their Class I noxious effects involve long-term lethality yet they fail to deter herbivores. Moreover, because the larvae sequester cardenolides, resulting pupae and adults can automatically contain them. In contrast, both larvae and pupae are unprotected by PAs as are the adults during both their crucial emergence period and the time it takes them to locate a PA source. Cardenolides would appear to have yet another advantage: because high concentrations are transferred during larval and pupal development into the exoskeleton (Colour plate 2C,D; Brower & Glazier 1975, Blum 1981, Brower, Seiber, Nelson and Bond, in prep.), they provide adult butterflies with gustatory aposematism. In contrast, if PAs are stored only in the internal body tissues, their effectiveness as gustatory aposematic cues is diminished.

It is possible that PAs can be released as chemical deterrents in male hairpencil displays of danaines (Brower *et al.* 1965, Rothschild *et al.* 1973, Boppré 1977, Ch.25) and of ithomiines (Pliske 1975*b*, Edgar *et al.* 1976*b*), or that they synergistically interact with cardenolides or other Class I chemicals. However, until further research is done, the efficacy of pyrrolizidine alkaloids in all aspects of chemical defence, and whether they operate as Class I or Class II chemicals must be considered hypothetical.

Conclusions and Summary

Chemical defence in butterflies involves complex adaptations which are held to be the outcome of natural selection resulting from coevolutionary arms races between butterflies, their foodplants, their predators, and their parasitoids.

Our knowledge of chemical defence against vertebrate predators and especially birds has been greatly advanced by the ecological and evolutionary synthesis of information from natural history, physiology, behaviour, pharmacology and chemistry. In contrast, our meagre understanding of chemical defence against invertebrate predators and parasitoids is based largely on the identification of chemicals and hypothetical inferences about their effects. An important exception to this is the emerging field of the chemomimetic manipulation of ants. Chemical defence produces chemically mediated behaviour control of vertebrate predators in ongoing dynamic relationships. This results in the maintenance of educated predator populations and is generally of advantage to both the individual prey and the individual predators. Synthesis of knowledge about defensive chemicals and predator behaviour indicates that the evolution of chemical defence and aposematism may not be as difficult as originally envisaged.

Two classes of chemicals are involved in defence and their different modes of operation have generally been confused in the literature. Chemicals in Class I are noxious by virtue of their capacity to irritate, hurt, poison, and/or drug individual enemies, and they may or may not stimulate the predators' olfactory and/or gustatory receptors. Chemicals in Class II are innocuous chemoreceptor stimulants. Class I chemicals can result in averting enemies in several ways, including the conditioning of Class II chemicals as aposematic cues which results in subsequent gustatory or olfactory rejection by predators. Thus prey which contain only Class II chemicals can be chemically aposematic without being chemically noxious, and chemical mimicry involving various predators is possible, although not proven. Class I chemicals which occur at low concentrations probably operate as Class II chemicals.

Chemical defence can mediate learned proximal avoidance of prey in which a predator must attack and feel or taste the prey before remembering to reject it. This is inefficient and can be dangerous for both prey and predator compared to learned distal avoidance in which the rejection is based on visual, olfactory or auditory cues. Differences in the kinds and deployment of defensive chemicals by butterflies and moths appear to reflect the fact that visual cues cannot generally be used at night.

The conservative feeding behaviour seen in the tendencies of many predators to avoid unfamiliar visual and chemical stimuli has undoubtedly evolved as a result of the severity of the effects of Class I chemicals. The importance to predators of avoiding chemically defended prey appears great. Chemical or pharmacological evidence that Class I chemicals occur in butterflies neither proves that the butterflies are chemically defended nor explains how the chemicals mediate defence. Future studies must combine chemical identifications with ecologically relevant bioassays, and experimental tests must be carried out to confirm that laboratory-based deductions are manifest in nature.

Students of butterflies can regale in the fact that the last 125 years of research have borne out Henry Walter Bates' (1863) famous statement that 'As the laws of nature must be the same for all beings, the conclusions furnished by this group of insects must be applicable to the whole organic world; therefore the study of butterflies—creatures selected as the types of airiness and frivolity—instead of being despised, will someday be valued as one of the most important branches of Biological Science'.

Acknowledgements

This review has had the benefit of several years of collaboration with my University of California (Davis) colleagues James N. Seiber and Carolyn Nelson. I also wish to thank my senior honors students at Amherst College who, together with Susan C. Glazier, Lorna L. Coppinger, Lee D. Hedrick, Christine M. Moffitt, Julia Frey, and Marjorie M. Holland, pursued numerous facets of ecological chemistry with me between 1965 and 1981. I owe much to Linda S. Fink and James A. Cohen of the University of Florida for critically reviewing several drafts of the manuscript, to Norris Williams for helping with the section on odours, to Esta Belcher and Paloma Ibarra for artwork, and to Martha Hoggard for helping me see the product through to completion. I am extremely grateful to The National Science Foundation for providing continued support for our research programme in ecological chemistry.

13. A New Look at Lepidoptera Toxins

Neville Marsh

Department of Physiology, Queen Elizabeth College, London

Miriam Rothschild

Ashton, Peterborough

and

Fred Evans

Department of Pharmacognosy, School of Pharmacy, London

Do Lepidoptera (the Spurge Hawk and Garden Tiger have been temporarily raised to the rank of butterflies in honour of E. B. Ford) store cytotoxic substances, and do they sometimes secrete them? The classical aposematic coloration of the larva of the Spurge Hawk moth (*Hyles euphorbiae*) strongly suggests that it sequesters and stores poisonous secondary plant substances. Moreover it is well known that many species of *Euphorbia* on which it feeds contain in their tissues the so-called co-carcinogens, highly inflammatory tumour-promoting diterpenes (ingenol ester, Fig. 13.1; Hegnauer 1966; Hecker 1970) while two species have been shown to produce, in addition, tumour-inhibiting substances (ingol ester, Fig. 13.2) which may be regarded as potential cancer drugs (Evans & Kinghorn 1977, Evans & Soper 1978). It also seemed possible that the larva could secrete some chemical substance which protects it from the effects of the *Euphorbia* toxins, and was therefore of interest from a medical standpoint. Hence we investigated the effect of aqueous extracts of both larval and pupal stages on cell growth rate of cultures of cervical carcinoma cells (HeLa cell), normal human lung fibroblasts (MRC5) cells, and the growth of Lewis Lung primary carcinoma cells inoculated into mice. We also examined extracts of various stages of development by thin layer chromatography (TLC), intraperitoneal injection into the laboratory mouse

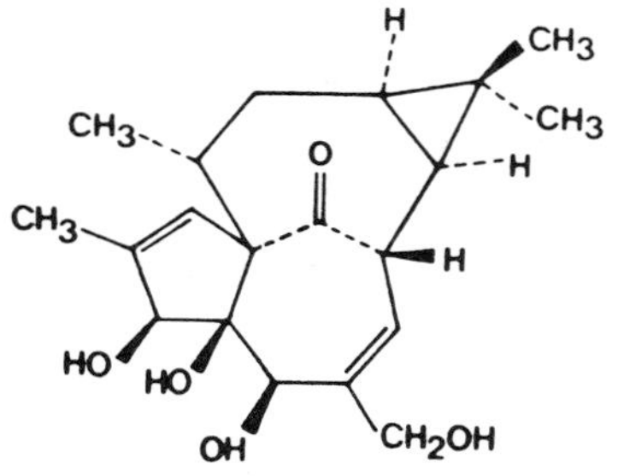

Fig. 13.1. Ingenol ester.

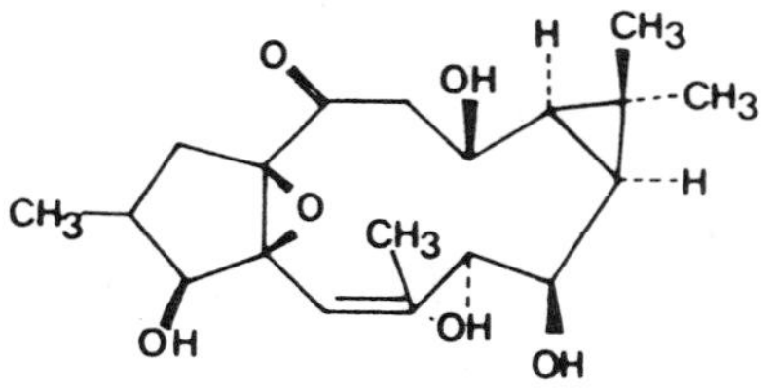

Fig. 13.2. Ingol ester.

(tables 1 and 2 in Marsh & Rothschild 1974), and introduction into the isolated rat heart (the Langendorff preparation—Marsh & Whaler 1980).

Since no spurge-eating cryptic species or artificial diet-reared specimens were available as controls, we used three known toxic Lepidoptera (tables 1 and 2 in Rothschild 1972*b*) for comparison. Detailed descriptions of our results and techniques will be

The Biology of Butterflies
0-12-713750-5

published elsewhere (see also Marsh *et al.*, in prep.). Here we pick out points of special interest.

Experiments With *in vitro* Cell Cultures and *in vivo* Tumour-bearing Mice

We reared *Hyles euphorbiae* larvae (which included both black and green 'races' of the caterpillar—Rothschild & Jordan 1903) on four different species of European spurge, *Euphorbia peplus, E. characias, E. cyparissias* and *E. polychroma*, all known to contain ingenanes (Evans & Soper 1978). Aqueous extracts of both the foliage of these plants and the larvae totally inhibited the growth of malignant and normal cell cultures. Chemical changes must occur during the pupal stage, depending on the species of food plant, for those pupae we tested reared on *E. characias* and *E. polychroma* produced different effects on the cultures of malignant and normal cells.

The growth of the malignant cells during the initial five days was stimulated by extract of *E. characias*, but by the eighth day this growth had diminished to zero. Normal cells were unaffected for the first five days, but their growth also diminished to zero by the eighth day. In the case of *polychroma*-reared pupae (two trials only) the growth of malignant cells was totally inhibited, while that of normal cells *was considerably increased* (Table 13.1).

Table 13.1. Effect of *Euphorbia polychroma* extract, and extract of *Hyles* pupae fed on *E. polychroma*, on cell cultures (^{3}H-thymidine uptake after two days—% change).

	MRC5	HeLa
E. polychroma seeds	− 92	− 26
E. polychroma foliage	− 55	− 83
Hyles larvae fed on *E. polychroma*	−100	−100
Hyles pupae ditto (two experiments)	+ 50 (+94, +6)	− 75 (−75, −75)

Extracts of both sexes of adult Large White butterflies (*Pieris brassicae*) reared on cabbage and adult male Garden Tiger moths (*Arctia caja*) reared on lettuce, killed both types of cells, but extracts from the female Tiger abdomen (reared on cabbage) were more selective, killing the malignant cell cultures but having little or no effect on normal cells. Extracts of another Arctiid, *Amphicallia bellatrix* (reared on *Crotalaria*) had no influence on the growth of either type of cell culture. (The main pyrrolizidine alkaloid stored by this species was not among those investigated for anti-tumour activity by Culvenor 1968.)

Figure 13.3 shows the effect of extracts of pupae reared on *E. characias* and *E. marginata* seeds on the growth of Lewis Lung primary tumour. There was a significant reduction in tumour growth in animals treated with these extracts. The number of secondary lung tumours at the end of the experiment (day 23) was reduced from a mean control value of 18.9 to 10.0 in animals treated with pupal extract.

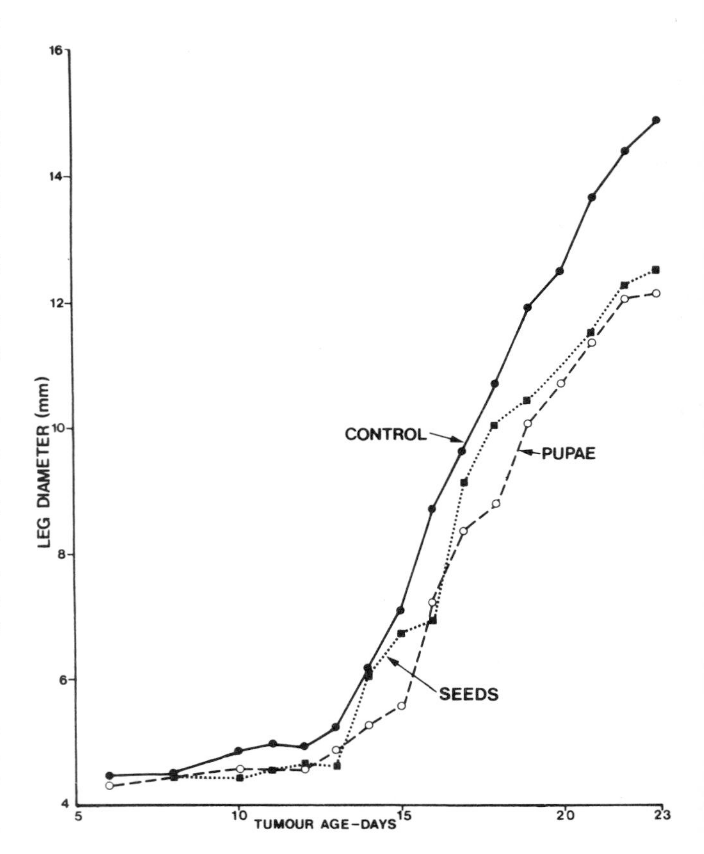

Fig. 13.3. Effect of *E. marginata* seed extract and *Hyles euphorbiae* (*E. characias*-reared) pupal extract on the growth of Lewis Lung primary tumour in mice. Measurement of the leg diameter assesses the size of the tumour growing within the thigh. The tumour is first noticeable around day 13, when the growth curves become significantly steeper.

Effect on the Isolated Rat Heart (Langendorff Preparation)

Aqueous extracts of abdomen of fresh female *A. caja* produced a dramatic irreversible paralysis of the heart (Fig. 13.4A), accompanied by a marked contracture of the whole organ. The action was reminiscent of that of another powerful cardiotoxin—bee venom containing melittin (Fig. 13.4B—Marsh & Whaler 1980). Fresh male extract and dried extract of both sexes had little effect on the rat heart. Consequently the causative agent of the arrest and contracture cannot be the protein described by Hsiao *et al.* (1980) which is also present in preserved material, but might be due to the effect of the low molecular weight substance extracted from live material by Rothschild

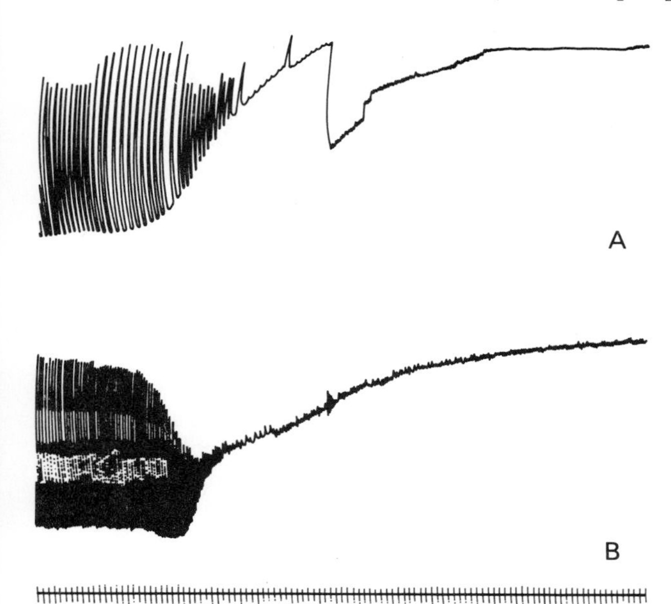

Fig. 13.4. Extract-effects on isolated rat heart. A: extract from the female Garden Tiger moth. B: bee venom.

et al. (1979*c*) from the ovaries and eggs, which they named 'cajin'.

The effect of extract of the Large White is most unusual. Immediately after application there is an uncoupling of atrioventricular conduction. The atria contract at about 20% of their resting rate whilst the ventricles continue at their own inherent rhythm, receiving no excitation from the partially paralysed atria. The phenomenon is remarkable because it is completely reversible (Fig. 13.5).

Extract of *Hyles euphorbiae* reared on *E. characias* and *E. polychroma*, apart from a transient and trivial slowing of the rate, had no effect on the isolated rat heart (Fig. 13.6B,C). It is a striking feature of the Langendorff preparation that various toxins such as the pyrrolizidine alkaloids also show only insignificant changes on the relevant tracings (see Fig. 13.6A).

Chemical Analysis by Thin Layer Chromatography

P. brassicae is known to sequester and store mustard oil glycosides from its food plant (Aplin *et al.* 1975) and *A. caja* cardiac glycosides from *Digitalis*, and both it and *A. bellatrix* sequester and store different types of pyrrolizidine alkaloids (Rothschild *et al.* 1979*b*). The spurge feeders, however, have not hitherto been analysed from the point of view of the secondary plant substances which they may ingest. Using the method described by Abo & Evans (1981*a*,*b*) we found small amounts of ingol macrocyclic diterpene esters (known cytotoxic but non-tumour promoting agents) in *H. euphorbiae* larvae, pupae and adults reared on *E. characias* and *E. polychroma* and in pupae reared on *E. cyparissias*. Further material (only three specimens in each group

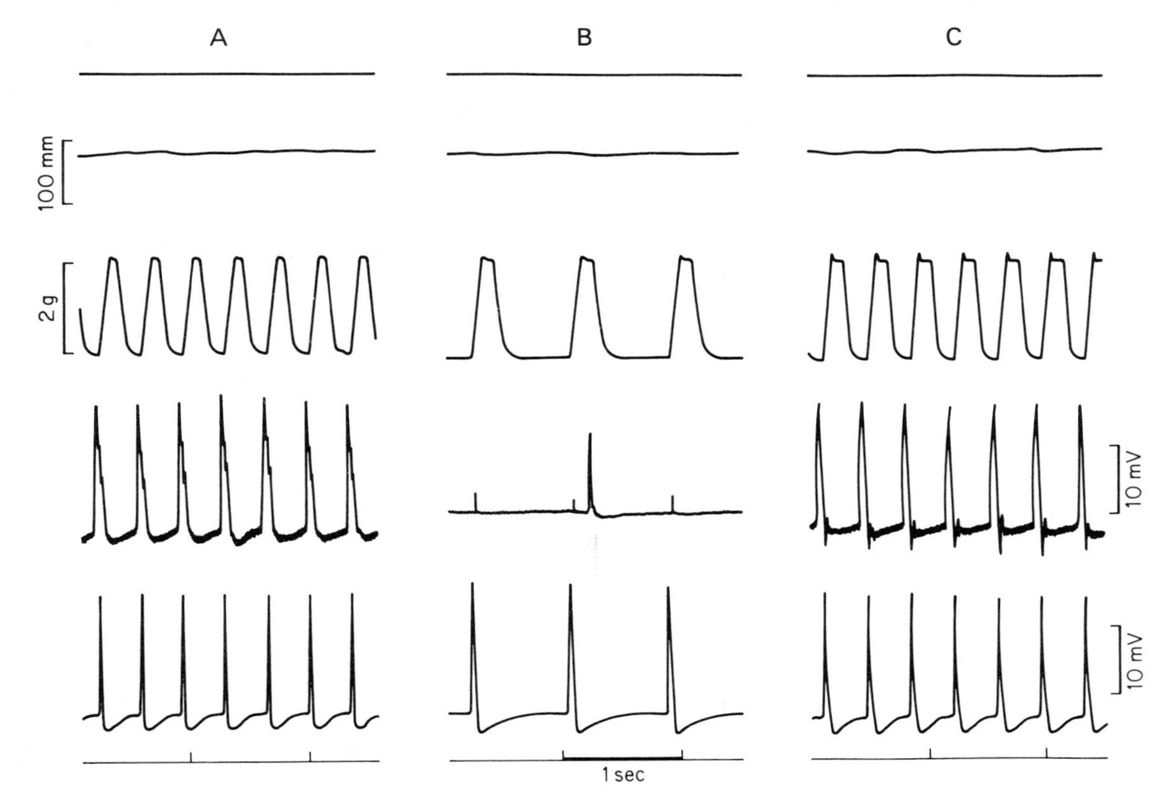

Fig. 13.5. The effect of extract from the Large White butterfly on the isolated rat heart. Tracings, from the top: perfusion pressure, ventricular contraction, atrial action potential, ventricular action potential. Record A: control; record B: after 15 seconds; record C: after 25s.

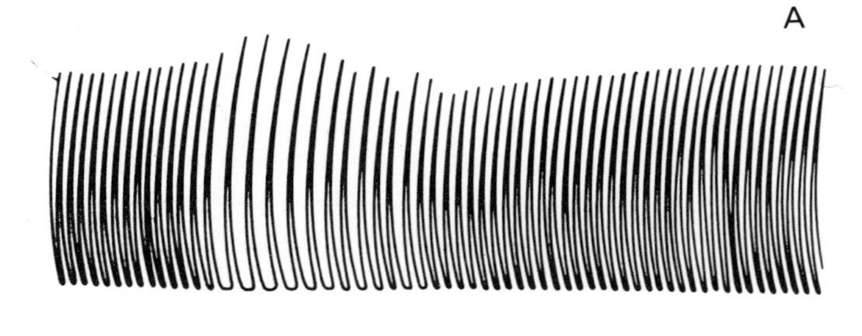

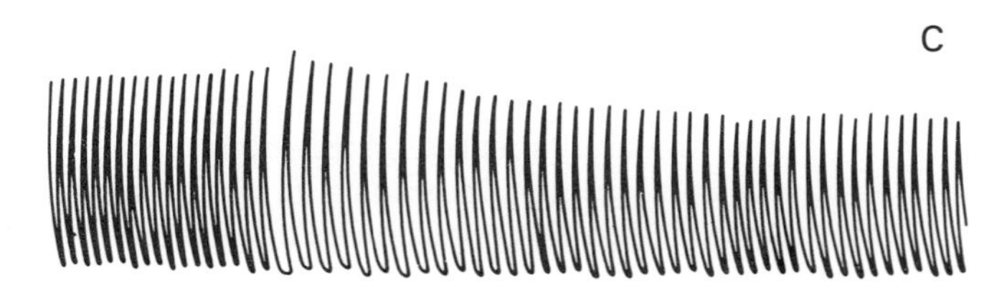

Fig. 13.6. Extract-effects on isolated rat heart. A: *Senecio vulgaris* extract (this plant contains high concentrations of senecionine, seneciphilline and integerramine). B: *Hyles euphorbiae* pupal extract (*Euphorbia polychroma*-reared). C: *H. euphorbiae* pupal extract (*E. characias*-reared).

were available for analysis) is required for positive identification. In extracts from nine larvae and two pupae reared on *E. peplus*, material which chromatographed with a marker of ingenol tetracyclic ester (a known tumour-promoting agent, Fig. 13.1) was more clearly visible.

Discussion

In view of its aposematic coloration and lateral row of defensive glands, instead of the super-crypsis characteristic of most hawk moth larvae, the storage of toxic material from the foodplant was almost a foregone conclusion. Nevertheless it is surprising that a caterpillar can tolerate within its body tissues such highly inflammatory substances as the ingenols. The contrast between the effect of pupal extracts on cell cultures derived from larvae reared on different species of *Euphorbia* is less remarkable, since it is well known that the chemistry of this enormous plant genus (6000 species) is very varied (Ourisson *et al.* 1979). It has been shown previously that the different ingol esters exert different cytotoxic and tumour-inhibiting effects (Abo & Evans 1981*a*). It seems likely from our studies that these chemicals are more widely distributed in the genus than the two previous records suggested. But they have never before been found in an animal.

During the pupal stage (and possibly at metamorphosis) chemical changes appear to take place by which, according to the species of foodplant, the tumour promoters are frequently metabolized or destroyed while the tumour depressants are retained. Similarly in the pupal stage of *Euploea core* no trace of the cardenolides sequestered by the larvae from *Nerium oleander* can be found (Malcolm & Rothschild 1983) while, conversely, those present in the late pupal stage of *Danaus plexippus* reared on *Asclepias curassavica* not infrequently become more actively emetic (Dixon *et al.* 1978).

The female Garden Tiger poses a special problem. The extracts which selectively inhibited the growth of HeLa cell cultures were from specimens reared on cabbage. *A. caja* does not store mustard oil glycosides and it is difficult to believe that any other cytotoxic substances were sequestered from this foodplant; it is more probable that they were self-secreted. Among its favoured foodplants in nature is *Senecio vulgaris*, from which the moth sequesters and stores pyrrolizidine alkaloids, chiefly senecionine —a substance shown by Culvenor (1968) to be a tumour depressant. Is this an analogous case to that of *D. plexippus* which sequesters cardioactive substances from its foodplant, but also secretes one itself? (Rothschild *et al.* 1978*b*).

The diversity and complexity of the biochemical defence mechanisms of even the few species mentioned here is positively awesome: the amines, the proteins, the alkaloids, the cannabinols, glycosides, ingenanes, pyrazines and so forth (Ch. 12). It is the tip of an enormous iceberg which can keep us inquiring and incredulous for a long time to come. Aposematic insects can be used as indicators of plants which contain active substances, potentially of medicinal importance, but it may also be possible to use such insects—particularly the pupal stages of aposematic species—as biochemists for our own purposes. The spurge feeders, among which we can include a number of arctiids, nymphalids and a few pierids, should prove one of the most interesting and rewarding groups to investigate.

Postscript

We have recently seen a note (Pettit *et al.* 1968) reporting on arthropod anti-neoplastic agents. Among the small number of species which gave 'encouraging results. . . . is the butterfly *Melanitis leda determinata*'.

Anti-neoplastic activity was also detected in extracts from the following species: *Catopsilia crocale*, *Ixias pyrene insignis*, *Pieris rapae cruavore* (sic), *Prioneris thestylis* and *Yoma sabina vasuki*. He also mentions the beetle *Allomyrina dichotomus* and Acrididae sp. 'producing substances which inhibit growth of Walker 256 carcinoma'. It is worth noting that in his investigation of the stag beetle, Pettit used 15kg of beetles—over 7000 specimens.

We subsequently used Walker 256 carcinosarcoma to screen some of our extracts and found that *H. euphorbiae* pupae reared on *E. polychroma* caused a significant reduction in tumour size (control tumour weight mean ± SEM: 9.86 ± 0.37 g; test tumour weight 2.24 ± 0.91 g).

14. *Mimicry: The Palatability Spectrum and its Consequences*

John R. G. Turner

Department of Genetics, University of Leeds, Leeds

This paper is dedicated also to the memory of Professor Ford's most distinguished graduate student, Philip Sheppard, who not only inspired most of the work recounted here, but did much of it as well.

Why Do Some Things Taste Nasty?

'Palatability spectrum' is ecojargon for the fact that all things are not equally nice to eat. The phenomenon is the product of the coevolution which occurs between all organisms, dictating their food preferences as a result of variations in nutrience, availability, competition, ease of capture and so on. The aspect of the palatability spectrum of most interest to the student of butterfly mimicry is outright nastiness: the production of some substance unpleasant or even poisonous to predators which would otherwise find the prey perfectly acceptable. As the work of Rothschild, Brower and many others (e.g. Brower & Brower 1964, Rothschild 1972*b*) has shown, the story generally starts with the production of chemical defences by plants (but see also Ch.12), to deter either herbivores or phytophagous insects. Sooner or later some insect overcomes the plant defence by detoxifying or sequestering the chemical. Provided the plant is an evolutionary success the insect has a new set of niches available, multiplying in evolutionary time, so that a whole group of insects, largely untroubled by competition, may diversify on the radiating plant group. Thus we find danaines on Asclepiadaceae, heliconiines on Passifloraceae (Benson *et al.* 1976) and ithomiines on Solanaceae.

But whether a success is in terms of multiplication of species or not, the insect has the additional option of using the defence substance of the plant for its own protection. Whether they simply use the plant chemicals direct (*Danaus plexippus*), or convert them into related substances, or have developed the ability to synthesize compounds independently of any input from the plant (*Zygaena*), the result is that some insects are not nearly no nice, or so safe, to eat as others. The stage is set for the evolution, first, of warning coloration, and then of mimicry.

Why Are Warning Colours Bright?

Some species develop chemical defence and leave it at that: the Buff-tip moth (*Phalera bucephala*) is distasteful, but like the majority of palatable species, it is cryptically coloured, being an excellent mimic of a birch twig (Fisher 1930). Probably a great many more develop the simple and conspicuous patterns of red, black, yellow and white stripes and dots which, as Philip Sheppard was fond of pointing out, being the same colour as road signs, stand out well against vegetation, and which we recognize as warning colours. Why do they do this? Surely it is better not to be seen at all, than to be somewhat mangled before being dropped from the beak as too hot to handle? That there is indeed a disadvantage in becoming conspicuous is shown by what appears to be a correlation between diurnal flight and warning colour: most fully diurnal moths are warningly coloured (one thinks in Britain of the only truly diurnal geometrid, the Chimney Sweeper moth (*Odezia atrata*) which is jet black, compared with the vast, cryptically coloured nocturnal family to which it belongs); and Batesian mimicry, which involves a similar adoption of conspicuous colouring, is found frequently among butterflies but very rarely among caterpillars. It is quite uncertain in the case of moths whether warning colour allows them to come into the open and operate on solar power, flying by day being much more economical of energy than flying by night (e.g. Wasserthal 1975, Vielmetter 1958, Douwes 1976*a*, Douglas 1979) or whether being diurnal imposes enormous evolutionary pressure to become aposematic (Rothschild 1972*a*). Butterflies

The Biology of Butterflies
0-12-713750-5

as a group are diurnal, but many are not aposematic: many give the impression of avoiding predators by being thin and gristly which their dependence on solar rather than metabolic heat allows, and rather agile; on the whole, they are just not worth the energy required to catch them. However this may be, it looks as though warning colour evolves more readily when the organism is already conspicuous because it is flying around in full view: *Zygaena* larvae are toxic, slow moving and cryptic; the diurnal adults also toxic, are warningly coloured.

But even putting to one side the awkward thought that there are many aposematic caterpillars, this still does not explain the adoption of the 'road-sign' coloration. Two explanations, both of which are probably correct, have been put forward. The first is that there is an inbuilt tendency within the vertebrate nervous system to learn rapidly to associate these colours with a nasty experience. Certainly vertebrates do show asymmetries of this kind in response to colour. Goldfish being trained by electric shocks to stay in one half of a tank, or go to the other, learned more rapidly when a red light meant 'go' and a green light 'stay', than they did when the lights had the conventional meaning of traffic lights (stay on red, go on green) (Bisping *et al.* 1974). More recently Gibson (1980) showed that finches learned to avoid the negative experience of having prey disappear through a trap door when the prey was red than when it was cryptically coloured (blue) to match the background. It is not certain here whether it is the red colour, or the contrast with the background which induces the faster learning, but subjective experience certainly suggests that there is something unrestful about red and yellow colours: diazo slides are much pleasanter to look at if they are white on blue (the preferred colour with most lecturers) than if they are white on red. Red for us has strong connotations of warning and of decoration, as in 'The masque of the red death' or the use of the same root in Russian to mean both 'red' and 'beautiful'.

In addition, or alternatively, it may be that aposematic colouring has evolved to take warning patterns away, beyond any possibility of confusion, from the green and brown colours of the palatable, cryptic prey for which predators are constantly forming search images (Turner 1975*a*). The idea was succinctly put by Fisher (1930), who pointed out that it was essentially similar to the tendency of models to evolve away from their Batesian mimics: 'to be recognised as unpalatable is equivalent to avoiding confusion with palatable species.'

I believe that there is a further, overlooked factor, in the learning rather than searching behaviour of predators (see also Ch. 12). Many aposematic species are poisonous, perhaps without being particularly distasteful (Rothschild 1971, 1972*b*, 1976, Rothschild *et al.* 1972*b*). In such cases there is a time-delay between eating the insect and receiving a nasty experience. Vertebrates are in fact capable of associating the symptoms of poisoning with a food item taken the appropriate time *before* the symptoms develop (e.g. Garcia *et al.* 1966, 1972), but if at that time they were consuming a series of insects, as for instance flycatchers and jacamars do, they may have difficulty knowing *which* insect caused the poisoning. Any insect conspicuously different from others will be much more readily associated with the unpleasantness.

A fourth possibility is that the conspicuous coloration causes a faster rate of learning simply because it induces a faster rate of attack. A predator attacks say ten conspicuous prey in a much shorter time interval than ten cryptic prey, and as learning and forgetting are time dependent, builds up its avoidance of the more frequently encountered prey much more quickly (see also Ch. 12). (This is analogous to the greater protection given to a common warningly coloured form than to a rare one (below), the difference in abundance in this case being replaced with a difference in apparency.) Gittleman & Harvey (1980, also Gittleman *et al.* 1980) found that chicks more rapidly attacked, and then sooner avoided, unpalatable crumbs which contrasted with the background than those which matched it, although in this case a fifth possibility, which is in fact the preferred interpretation of the authors is not ruled out: that contrast by itself induces faster learning. That it was not simply colour was shown by reciprocal experiments in which either blue or green crumbs contrasted with, or matched the background. This of course is not quite the same as the theory that 'road-sign' colours are more easily learned, in themselves.

It remains to be seen which of the mechanisms, innately fast learning of bright colours, innately fast learning of contrasting colours, fast learning of prey that is different, greater initial rate of attack on conspicuous colours, and avoidance of mimicry of cryptic colours, is the significant mechanism in the origin of aposematic colouring. Although some experiments will be easy to devise (to choose between faster initial attack rate and rapid learning of contrast it would only be necessary to make the matching prey so common that they were attacked at the same rate as the contrasting prey), the coevolution of the vertebrate nervous system with the insects may have made it impossible for us finally to disentangle the problem.

Although there are thus at least five plausible explanations for the use of 'road-sign' colours to advertise distastefulness, there is a well known and much discussed difficulty in evolving the warning colour in the first place. Being distasteful can be individually advantageous if the predator drops you relatively unharmed, but there is no advantage to the individual in attracting a predator by making itself conspicuous. This is most obviously the case if the 'advantage' of the bright colour is simply to generate faster learning through a faster rate of attack. The same applies to being poisonous: it profiteth one nothing to make the bird that ate one sick, as one cannot be delivered up whole, like Jonah out of the belly of the fish. The question was originally tackled by Fisher (1930), apparently the first to enunciate the principle of kin selection (see also Ch.12), put firmly on the map by Hamilton (1964*a*,*b*): if the death of an aposematic individual protects a relative, rather than an unrelated conspecific, then the gene causing the bright coloration, or toxicity, may be at an advantage. Harvey *et al.* (1981, 1982) have recently examined this situation mathematically, and have delimited the conditions under which aposematic coloration will evolve. The distasteful species must live, as do gregarious caterpillars, in family groups, which are rather widely scattered so that not very many families occur within the home range of any one predator. When a predator has eaten a few of the prey, it must learn to avoid them, and it must learn to avoid warningly coloured ones more quickly than similarly distasteful but cryptic prey. Finally, the warning pattern, while making the prey more memorable, must not make it excessively conspicuous. It seems that these conditions have been met in many aposematic species: aposematic larvae and bugs are frequently gregarious; and at least some warningly coloured butterflies have restricted home ranges and roost gregariously; the evidence for faster learning of bright colours has already been quoted; and a number of observers have noted that warningly coloured insects, although blatant in close up, are relatively camouflaged at a distance (review by Endler 1978).

I am uncertain about the statements that warningly coloured *Heliconius* are partly camouflaged in flight by a flicker effect (Papageorgis 1975), as at least one field observer believes the contrary—their patterns are adapted to make them *more* conspicuous against their normal background (Benson 1982). Hinton (1977) speculated that certain gregarious membracid bugs were black when young so as to form a black mass that would stand out against the background, whereas the adult colours, yellow spots on black, would tend to blend into grey on the young bugs. The contrast/camouflage story about bright colours can clearly be told both ways! But at least there are good grounds for believing that aposematic butterflies and dragonflies are not very much more conspicuous than cryptic butterflies when flying around in mid-air. It is surely significant that many aposematic butterflies have much duller colours on the underside, the surface exposed while at rest.

Järvi *et al.* (1981*a*,*b*) have recently challenged the view that kin selection is necessary for the evolution of aposematism, basing their view on experiments which show that Swallowtail (*Papilio machaon*) larvae are released, unharmed, by Great Tits. Here there is clearly an individual advantage in distastefulness, as there is in *Heliconius* and *Danaus*, where concentrations of the active compounds in the wings tend to facilitate their safe release (Boyden 1976, Brower & Glazier 1975). We need here to distinguish three aspects of aposematism: distastefulness, toxicity, and bright coloration (Harvey & Paxton 1981). Distastefulness, if produced by excreting the unpleasant substance on the outside of the prey, clearly has a strong individual advantage. Toxicity, without accompanying distastefulness and depending on delayed action within the predator's stomach, is of no individual advantage and can be selected only in kin groups (see also Ch.12). Bright colouring is the intermediate case: so long as the bright colouring is more easily remembered than cryptic colouring, a bright distasteful caterpillar is better protected than his cryptic brother in subsequent encounters *with the same individual predator*, but under most circumstances it seems unlikely that this form of selection will be strong enough to outweigh the disadvantage of attracting yet more predators by the bright colouring. In general we would expect family selection to have been influential, if not critical, in the evolution of bright colouring (but see Ch.12 for a somewhat different view).

Attempts to confirm many of the above hypotheses are likely to be hampered by further coevolution between predator and prey. The fact that warningly coloured butterfly larvae (Harvey *et al.* 1982) and even butterflies (Brown 1981, Turner 1981) are often gregarious, may not show that the gregarious habit was a precondition for the warning colour. First, in many circumstances it is advantageous for a predator that has encountered a distasteful prey, or an unrewarding 'patch' in its environment, to move on a bit before feeding again (Arnold 1978). Both this kind of predator behaviour and the effect of clumping the prey within the home ranges of a smaller number of predators (Turner 1975*b*), make it advantageous for distasteful prey to be gregarious. Although gregariousness can be individually advantageous for palatable prey, because of the 'selfish herd' effect, whereby a solitary individual is more likely to be eaten than a member of a clump (Hamilton 1971), this must often not be the case with camouflaged prey that are much smaller than their predator, who on discovering one will tend to wipe out the whole lot. Second, once bright coloration becomes the badge of toxic prey, there will be selection on predators to have an innate tendency to avoid such colours or to learn rapid avoidance of them. Schuler (1982) has recently presented evidence that there is just such an innate avoidance of 'wasp' patterns by starlings.

Any innate avoidance by the predators will produce individual selection for bright coloration. One could think of this as a very generalized kind of Müllerian mimicry. Third, there is an advantage to predators in finding toxic prey distasteful. Hence toxic prey will tend over time to become distasteful (again see also Ch.12).

The evolution of aposematism therefore seems likely to be a nested set of coevolutionary cycles between the prey and the predator, depending to some extent on kin selection but generating individual selective advantages at later stages. The whole scenario is roughly toxicity-clumping-distastefulness-blatant colour-gregariousness, with a slow trend from kin to individual selection as predators evolve in response to the evolution of their prey.

Are Batesian and Müllerian Mimicry Different?

Mimicry of the kind I am discussing (there are of course many other kinds—Wickler 1968, Vane-Wright 1976) is conventionally placed into two classes: Batesian mimicry, the resemblance between a palatable species and an unpalatable one and Müllerian mimicry, a mutual resemblance between two or more unpalatable species. Conventional wisdom states that in Batesian mimicry the model loses what the mimic gains in protection from predators, whereas Müllerian mimics all gain from the resemblance. Given that palatability is a complete spectrum, from the very nice to the very nasty, does this division make any sense? Are Batesian and Müllerian mimicry the extreme, limiting cases of a continuous range of phenomena (Huheey 1976, 1980*b*, Rothschild 1981*b*), or is it useful to divide the spectrum into two halves, say at the actual point of neutrality (Benson 1977)?

To answer this question we need to devise a system for predator behaviour (I shall write 'system' not 'model' for obvious reasons), and the answer we get will depend on how we think predators behave. Chiefly from the experiments of J. V. Z. and L. P. Brower, we know that after a number of encounters with unpalatable prey, a predator will make a 'decision' not to attack anything it 'recognizes' as this type of prey. At intervals it may make a 'mistake' and attack this kind of prey again (reviews Brower 1963, Rettenmeyer 1970, Turner 1977*a*). As it is clear that 'mistakes' of this kind allow predators to sample and adapt to a changing world, we might think of them as deliberate reversals of policy rather than as failures of memory. The opposite of course will happen to palatable prey: the predator learns to attack it after a number of pleasurable encounters. As unfamiliar prey is not automatically attacked, but is approached with some caution (e.g. Coppinger 1969, 1970, Shettleworth 1972*a*,*b*, Morrell & Turner 1970) we can think of neutral palatability *either* as the point in the middle of the spectrum where tasting the prey results in neither an increase nor a decrease in the rate of attack, *or* as the watershed between prey which provoke an 'attack' decision and those which provoke avoidance.

The simplest way of imitating (modelling) this system is by Monte Carlo simulation: imagine a jacamar sitting on a branch, catching insects. The general supply of flies and beetles keeps it moderately satisfied. The much lower density of passing butterflies (say one within reach every five minutes) is too small a part of the diet to affect its level of hunger. An unfamiliar butterfly has a 50% chance of being attacked. If it turns out to be nasty, the

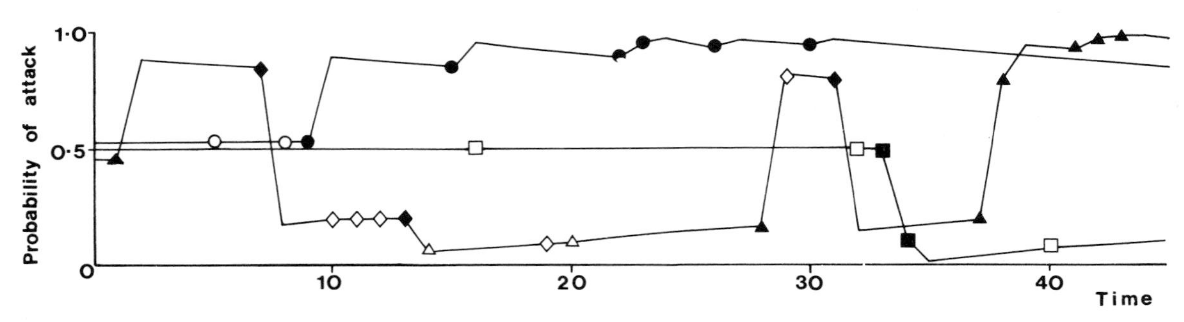

Fig. 14.1. Part of a computer simulation of a predator preying on a random sequence of four butterflies: nice (circles), nasty (squares), model (diamonds) and mimic (triangles). Graph shows current probability of the bird attacking the four types (model = mimic, as they cannot be distinguished on sight). Open symbols —prey not attacked; closed symbols—prey attacked. For example, the predator fails to attack the first two 'nasties' that it sees, then attacks the third and fourth, after which it has 'learned' to avoid them. Conversely, after ignoring the first two 'nice', it discovers that they are palatable and attacks them regularly. The model and mimic cause the predator repeatedly to 'change its mind': it attacks the first mimic, and the following model, avoids three models, makes a 'mistake' and attacks a model, avoids a model *and* two mimics, attacks a mimic, ignores one model but attacks the next, after which a 'mistake' leads it into attacking a series of mimics which appear in a cluster with no models to protect them. Original data, from simulation on a programmable calculator.

probability of subsequent attack is reduced, say to $0.5 \times 0.2 = 0.1$. Conversely, a pleasant butterfly increases the probability of attack to, say, $1 - (0.5 \times 0.2) = 0.9$. The bird also 'forgets': every five minutes the attack probabilities are reduced (or increased) to 98% of their current difference from the base line of 0.5. Figure 14.1 shows part of a computer run in which the bird encounters at equal frequency four kinds of butterfly: a nice one, a nasty one, and an indistinguishable pair of mimics. The mimic is as nice as the nice species, and the model is as nasty as the nasty species. The system, as can be seen, imitates the behaviour of a predator whose experiences build up to 'decisions' to attack or not attack a particular kind of prey, these decisions being reversed after a variable length of time.

The results of a series of runs with this system are shown in Fig. 14.2. The unpleasantness of the nasty and model butterflies is constant throughout, but the palatability of the 'nice' one and the mimic is varied right across the palatability spectrum, from inducing

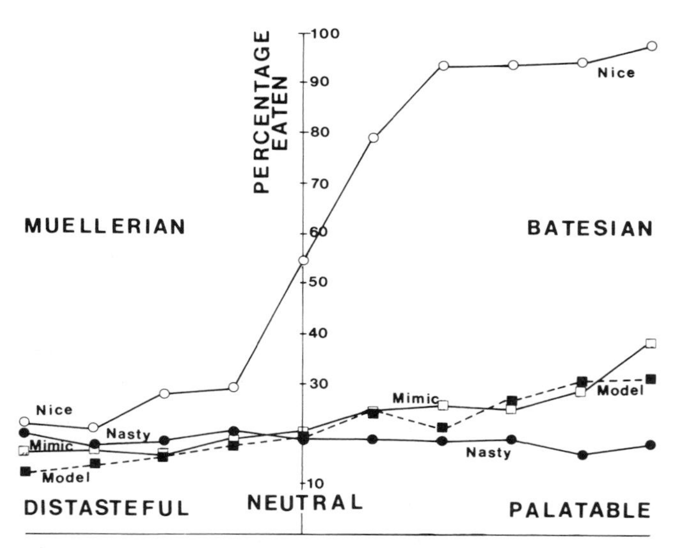

Fig. 14.2. Mimicry and the palatability spectrum. Attack rate on four types of prey, by a predator operating on the stochastic learning system shown in Fig. 14.1. The 'nasty' prey (closed circles) and the model (closed squares) are unpalatable throughout the experiment, reducing subsequent attack probability to 20% of its current value if they are eaten. The palatability of the 'nice' prey (open circles) and the 'mimic' (open squares), which is indistinguishable in pattern from the model, varies according to the values shown on the horizontal axis. To the left of the neutral point they are unpalatable, palatable to the right; at the extreme left all four prey are equally unpalatable. Note that to the left of neutrality both model and mimic are protected; to the right the mimic is better protected than the nice prey, which acts as a control, but the model is predated more than the nasty prey. Original runs on a programmable calculator, in part carried out by Ms E. Kearney as an honours project in the Department of Genetics, University of Leeds.

a 100% decision to attack, to being as unpleasant as the nasty and the model. The result is the one anticipated by conventional wisdom: when the mimic is palatable, the model-mimic pair is eaten much less than the nice butterfly, and a little more than the nasty; in the unpalatable half of the spectrum, all four butterflies are somewhat protected, but it is the model-mimic pair that receive the greatest protection (as predicted by Sheppard & Turner 1977).

The point of neutrality thus divides the palatability spectrum into two qualitatively different zones, as Benson (1977) postulated. On one side both species benefit; on the other the mimic benefits at the expense of the model. Müllerian mimicry and Batesian mimicry seem to me to be good names for the zones as the evolutionary consequences of being on one side or the other are, as I will show, rather different; although in the real, fluctuating world, a species whose palatability is in the region of neutrality will be neither quite one nor the other.

How much is this division of the spectrum into Müllerian and Batesian zones a result of the particular system which I have used? A system in which the predator decides firmly, rather than probabalistically, not to eat or to eat a particular pattern, and then reverses that decision after a particular time (the nastier or nicer the experience the longer the time) produces a result which is fundamentally the same except that the Batesian mimic and its model tend to be eaten at the same rate no matter how nice the mimic is. (The Monte Carlo system, although it works on a different computer algorithm, is basically the same system except that it allows the decision period to vary about its own mean.) If the predator makes a decision to avoid unpleasant prey not for a particular time, but until it has seen a particular *number* of these prey, then as was pointed out by Huheey (1976), who solved this system explicitly, there is no Müllerian mimicry as conventionally understood, as in a pair of mimics the nastier always suffers from the resemblance, even if its mimic is itself also distasteful. But on the whole I think it very unlikely that a predator who has decided to give Monarch butterflies a miss, starts counting the number it sees and then attacks, say, the eleventh one! Time-dependent reversal or forgetting seems much more likely to be the general rule, and therefore I believe that Fig. 14.2 represents the usual situation in the real world.

The other empirical fact about predator learning which must be understood is that the conditioned stimulus is 'generalized': that is to say the predator will treat as 'the same' not only an identical insect, but one that is somewhat like it. The probability that something is treated as 'the same' declines as its appearance departs more and more from the conditioned signal, is greater for really nasty experiences than for moderately nasty ones (Duncan

& Sheppard 1965, Goodale & Sneddon 1977), and is also heavily context-dependent: a pattern treated as a mimic of the model when a particularly attractive alternative food is available may be attacked a large per cent of the time if the alternative food is less attractive (Schuler 1974). It is important to note that generalization seems to be qualitatively the same when the conditioned stimulus is an unpleasant taste (Goodale & Sneddon 1977) as when it is an electric shock (Duncan & Sheppard 1965), so allowing some valid conclusions about mimicry to be drawn from shock experiments (Fig. 14.3).

For the purposes of this discussion, I am going to imagine the generalization of warning patterns to be constant for any one population of predators, and that predator behaviour with respect to model-mimic systems is like that shown in Fig. 14.2. This amounts to saying that I am ignoring some of the variance put into the system by such complexities of the real, fluctuating world as varying population sizes, changing availability of alternative prey, and the clumping of model populations.

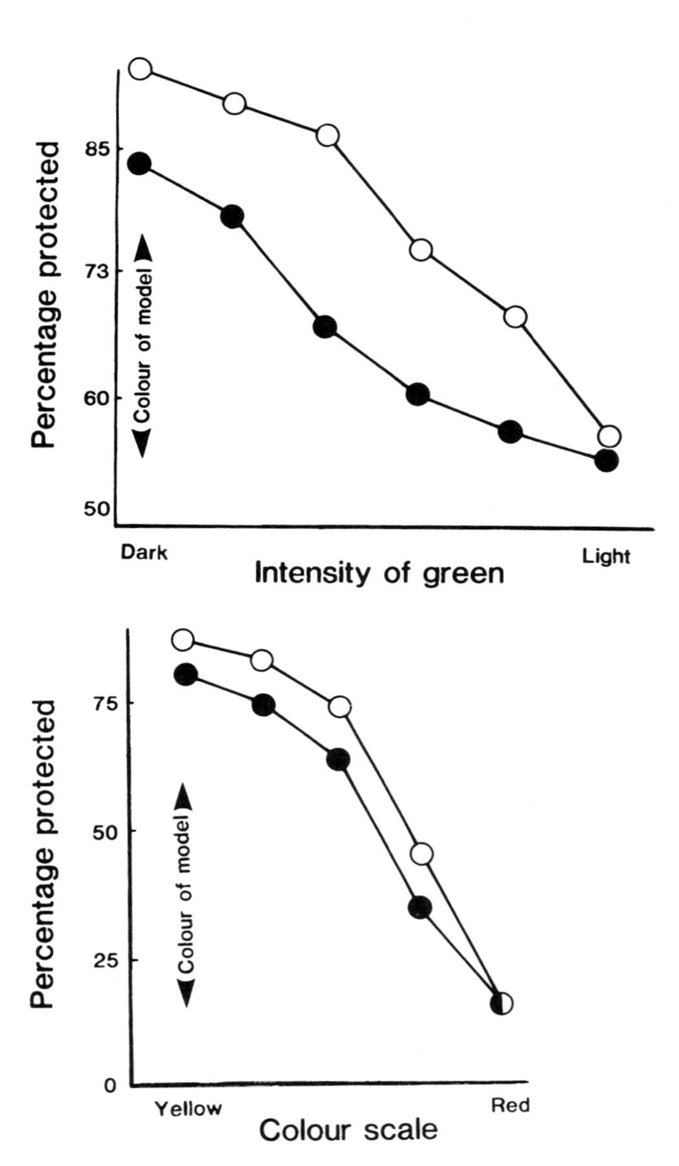

Fig. 14.3. Generalization curves produced in experiments by Duncan & Sheppard (1965) (upper) and Goodale & Sneddon (1977) (lower). The graphs show the predation rate on mimics as they depart in colour from the colour of the model. Open symbols: very unpleasant model; closed symbols: less unpleasant model.

Müllerian Mimicry—Gradual Convergence

The two systems of predator behaviour are now combined to show the way in which close Müllerian mimicry might evolve between two rather similar warningly coloured species. Both butterflies are imagined to have patterns which vary in the number of white spots on the wings, but they are not perfect mimics. Species A ranges from having no spots up to four spots, whereas the range in B is from one to five (Fig. 14.4). Densities and unpalatability, both equal in the two species, are the same as at the extreme left of Fig. 14.2, and generalization is taken into account by reducing the current attack probability for the 1 spot and 3 spot classes to 40% of its current value every time a 2 spot butterfly is eaten (the 2 spot attack probability is reduced to 20% of its current value), and similarly for all other spot classes; the flanking classes are thus somewhat protected by an encounter with the class in between them. (In real life, generalization would probably be much wider than this: the simplification, used in order to fit the algorithm into a programmable calculator, should not make any qualitative difference to the results.)

The protection afforded to the two species is shown by the heavy line in Fig. 14.4 (left). The crucial finding is that in species A the classes with fewer spots are on the whole less protected than those with more spots, whereas in species B it is the higher spot numbers that are less protected. The net result is to select for increase of spots in A, decrease in B, and, in short, for convergence of the two species onto the same spot distribution. Given that there is at least some genetic influence on spot number, the butterflies will, in the fullness of time, become very accurate Müllerian mimics.

For reasons which will become clear later, I shall not now give an actual example of this process in a wild species. However, the evolution of any mimicry which, as with Müllerian mimicry, does not involve an increase in conspicuousness, probably proceeds in small steps in this way. It is not difficult to imagine the gradual perfection of a leaf mimic from a simply-camouflaged green insect. Mimicry of snakes, almost the only kind of Batesian mimicry known in caterpillars (significantly always of cryptically coloured snakes) also probably evolves in this way. It is certainly possible to build up an

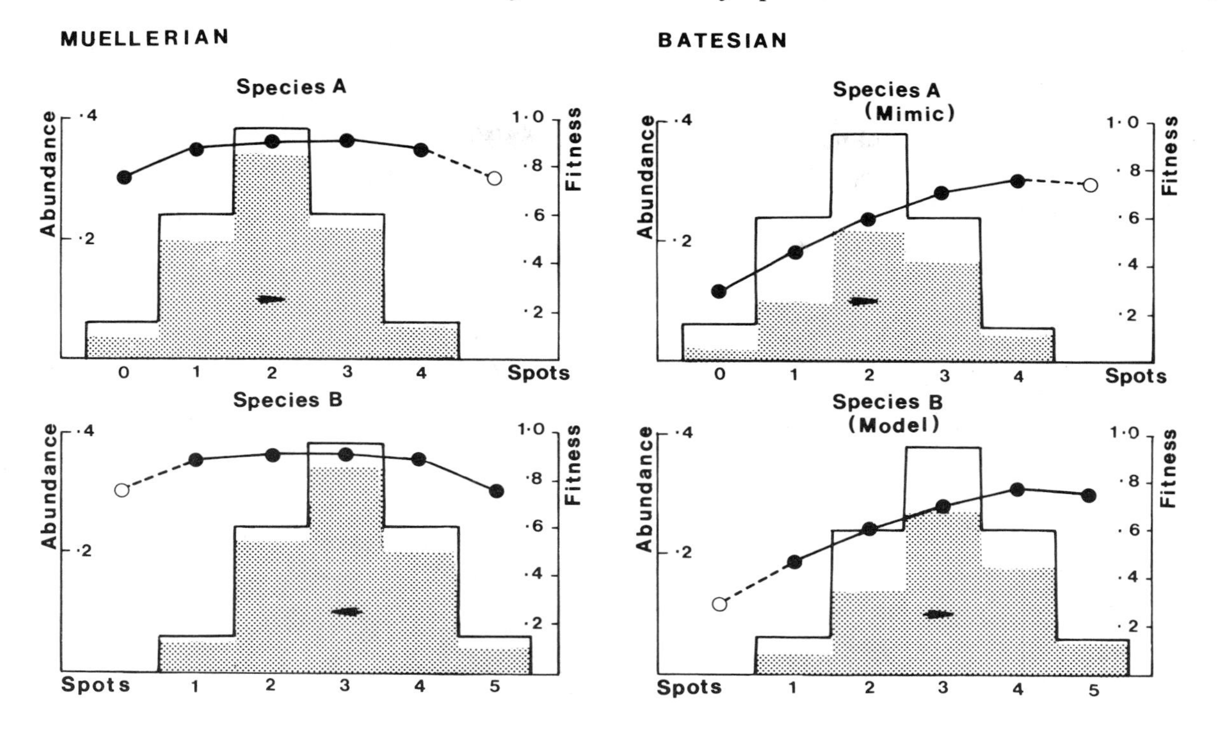

Fig. 14.4 left. Gradual evolution of Müllerian mimicry. Two species, both alike in density and unpalatability, have overlapping but not identical variation in the number of spots on the wing (outlined histograms). A predator, acting on the same stochastic learning system as in Figs 14.1 and 14.2, cannot distinguish them, but does generalize to the extent of avoiding adjacent spot-classes at 75% of the avoidance generated by the spot-class it has just encountered. The heavy line shows the 'fitness' (1 *minus* predation rate) of the various spot numbers, which is naturally the same for both species. The shaded histograms, showing the spot distribution after one round of such predation, show that the spot-distributions of the two species are converging toward each other (arrows, not to scale, show direction). One prey encountered in every time interval; other parameters the same as at the extreme left of Fig. 14.2.
Fig. 14.4 right. Gradual 'advergence' of a Batesian mimic to its model. The situation is the same as at the left, except that species A is now highly palatable (reducing probability of subsequent *avoidance* by 80% whereas the model, species B, reduces the probability of subsequent *attack* by 80%). In contrast to the Müllerian situation, both species are now selected for *increased* spot number (arrows, not to scale), but as shown by the shaded histograms (survivors after selection) the mimic is selected for much more rapid increase than the model, so that eventually both the species will have the same pattern distribution. All figures are original simulations of 5000 bird-butterfly encounters.

acceptable cryptic snake-like pattern from a number of small mutations. Take the caterpillar in Fig. 14.5a, a snake mimic not unlike the Elephant Hawk moth caterpillar, with frontally placed eye spots, and an intricate cryptic pink and brown pattern with short diagonal lines along the back such as are seen in many noctuids. The surprise is that this animal is nothing more than a domestic silkworm, but carrying four rather unusual mutations (Fig. 14.5): *Moricaud*, which comes from the wild ancestor (*Bombyx mandarina*) gives it the general, brown ground colour and the basis for the eye spots; the details of the pattern have been added by three mutations which have occurred within the domestic strain, *Zebra*, *Multilunar* and *quail*. Individually these add a little to the pattern, and it is instructive to see how few genetic changes are needed to turn a fairly plain brown caterpillar into a quite intricately patterned cryptic mimic. Aposematic colouring can also be achieved quite easily: a worm carrying *Multilunar* and *Striped* (an allele of *Moricaud*) is black with bright orange dots on each segment (Fig. 14.5g).

It is important to note, as there has been some genuine confusion on this point (Rothschild 1981*b*) that the individual genes may be identifiable (i.e. be 'major' genes) even when evolution has been gradual: whether a gene is a 'major' gene or a 'polygene' simply depends on our skill in genetic analysis; whether it produces 'minor' or 'major' changes in the colour pattern depends on its mode of action and the amount it changes the pattern relative to the amount of generalization by the predators.

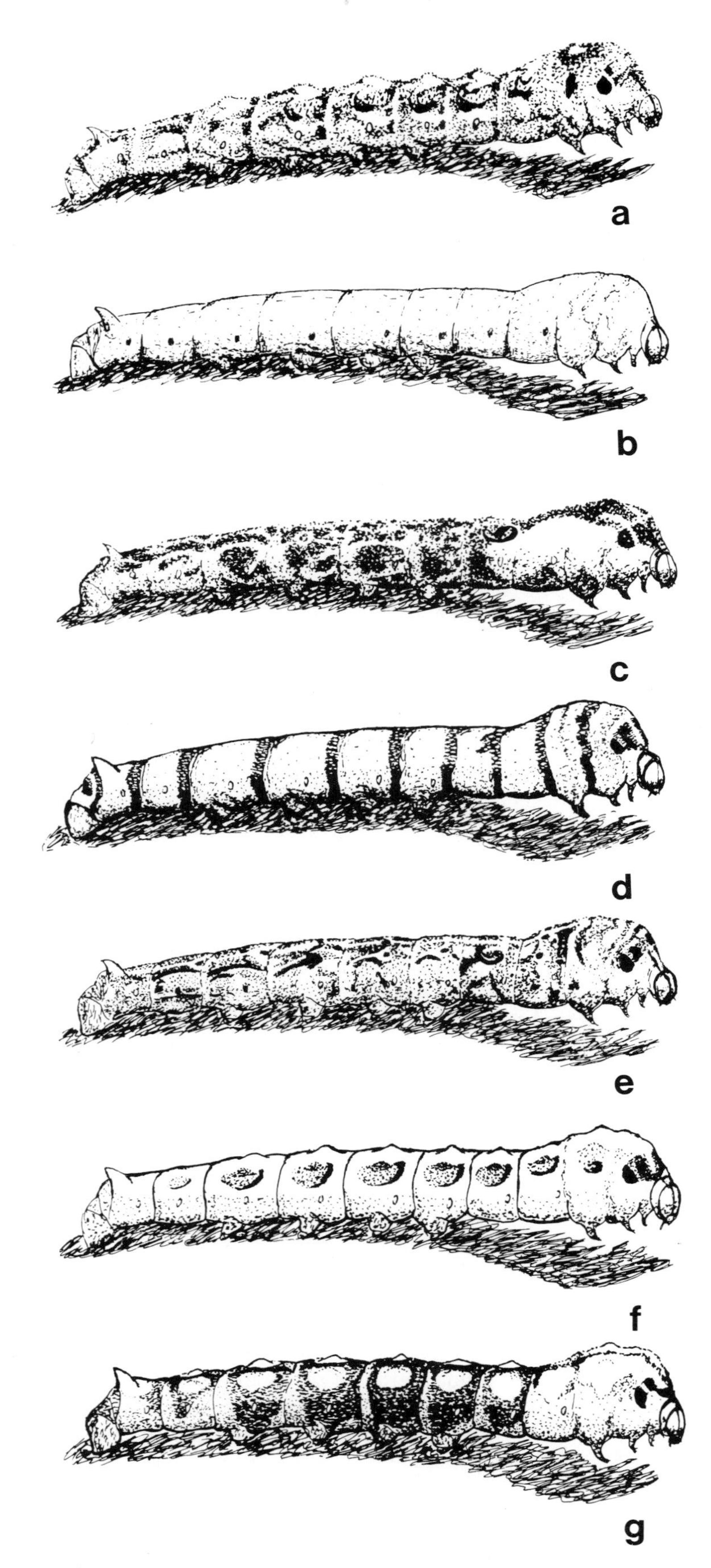

Fig. 14.5. Genetics of a caterpillar pattern which mimics a snake (a). The pattern can be built up on a white ground (b) by the four mutants *Moricaud* (c), *Zebra* (d), *quail* (e), and *Multilunar* (f). This last mutant, when combined with the mutant *Striped* (but not with the others), produces an aposematic pattern (g).

Batesian Mimicry — Two Phase Evolution

Whatever its causes, the phenotypic 'distance' between warningly coloured species and more or less cryptic, palatable prey presents a serious problem for the evolution of Batesian mimics. Or rather, it presents a serious problem for evolutionary theory: the problems of Batesian mimics appear to be less than insuperable, as there are plenty of them in the world! That there is a real 'problem' for the insects is shown by the fact, already mentioned, that Batesian mimicry of aposematic models is very rare in caterpillars, indicating that the increase in conspicuousness exerts a heavier price than the gain from resembling the model. An analogous resistance seems to occur in butterflies, in which the original non-warning colour may function as a sexual or other social signal. Silberglied (Ch.20) presents evidence that in *Colias* the male colour is used in signalling to other males (specifically the UV flash repels rivals, Silberglied & Taylor 1978) and hence may be an important component of male mating success. Female colour is known from many studies (Chs 20, 21, 23—but see also Ch.22) to affect male choice, but the single mating of the sperm storing females compared with the multiple matings of the males will make sexual selection considerably stronger, perhaps by one or two orders of magnitude, in males than in females. Sexual selection on the behaviour of the two sexes, as in most animals, will render the males much less discriminating than the females (Ch.20, Turner 1978—but see also Ch.23). Female colour is therefore expected to be subject only to weak stabilizing selection, whereas male colour, whether by selection for communication with other males, as both Silberglied (Ch.20) and Vane-Wright (1980) suggest, or alternatively because of discrimination by the females (Ch.21), is likely to be highly resistant to change. Hence the limitation of mimicry to females

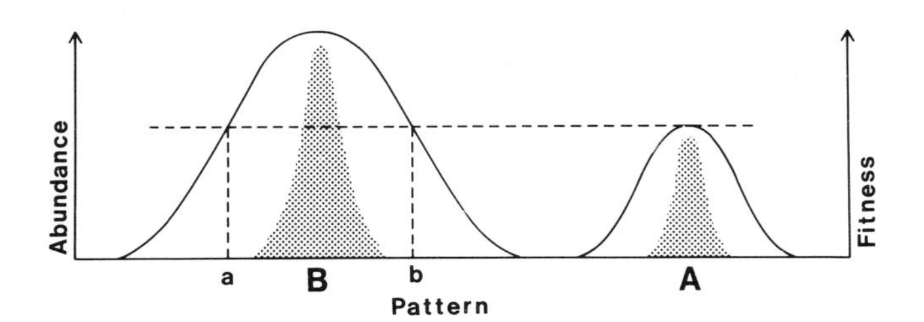

Fig. 14.6. Mimicry evolving by a single large mutation. Two species, whose phenotypes are shown by the shaded distributions, are rather dissimilar, but mimicry will evolve if species A, whose pattern has its own fitness (heavy curve) generated by whatever function (e.g. social or cryptic) it serves, can produce a mutation which need not perfectly resemble the model (B) (curve of protection generated by distastefulness and warning colour), but which has only to lie in the phenotype range *ab*. From Sheppard (1962) and Sheppard *et al.* (1984).

in a great many butterfly species provides good, although indirect, evidence that there is evolutionary resistance to the development of Batesian mimicry.

The 'problem' can be illustrated as in Fig. 14.6 (Sheppard 1962). The existing pattern has some function, which produces a peak of fitness in that part of the adaptive space. The generalization of the predators produces a peak of fitness round the pattern of the potential model. But between these is a region of the phenotype space where neither advantage can be experienced: a move into this by the potential mimic loses it the cryptic, or sexual, or perhaps thermoregulatory function of its original pattern, but fools not a single predator! The 'solution', as we all now know, is the occurrence of a mutation of comparatively large effect, great enough to place the pattern of the mimic somewhere in the region (*ab* in Fig. 14.6) where the protection given by the model confers greater fitness than that conferred by the original pattern.

Punnett (1915), working long before generalization by predators was understood, supposed that the mutation had to land the mimetic pattern bang on target, with very high quality mimicry. Goldschmidt (1945) also supposed that this happened, largely because it was consonant with other evolutionary theories which he held. Fisher (1927, 1930) on the other hand, again because the view fitted better with his other evolutionary theories, supposed that mimicry evolved gradually by the accumulation of many genes each of very small effect. It was Nicholson (1927) who proposed the compromise theory, later adopted by Sheppard (1962) and by Ford (1964), that mimicry evolves in two stages, and that once the first mutation, which may be of quite major effect, is established, the rather inaccurate resemblance it achieves can be considerably enhanced by the action of further 'modifier' genes. The only requirement is that the necessarily inaccurate mimicry produced by the first mutation confers greater fitness than whatever adaptation was served by the original pattern. When we remember that the original pattern must have been refined by thousands of generations of natural selection, it is not surprising that mimicry does not arise very often and that most species are non-mimetic.

Although the best studied mimics from a genetic point of view are the three polymorphic *Papilios* (*dardanus*, *memnon* and *polytes*), the considerable genetic complexity which arises from the polymorphism itself (see below) makes them rather poor exemplars of the major-gene/modifier system. The clearest demonstrations of the initial major mutation in Batesian mimics come from experiments with monomorphic species, of necessity conducted by rather difficult inter-species crosses. Thus Clarke & Sheppard (1955*a*) showed that mimicry of *Battus philenor* by the black swallowtail *Papilio polyxenes* is largely produced by a single gene converting the butterfly to black from the yellow colour of its relatives. The mimicry is refined by several further loci which control the details of the yellow spotting. Significantly, the black but non-mimetic species *Papilio brevicauda* carries the major black mutation but not the modifiers (Clarke & Sheppard 1955*a*, Sheppard 1961*b*). Similarly, Platt (1975) has shown that mimetic *Limenitis* differ from their non-mimetic relatives by a major mutation which wipes out the 'white admiral' bars which are the most prominent feature of the non-mimics.

Although in the polymorphic *Papilios* the original major mutation cannot be unambiguously identified, they do provide excellent evidence for the existence of modifying genes which specifically improve the mimetic pattern, and that pattern only. Thus the Ethiopian population of *P. dardanus* is polymorphic for mimetic forms and for a yellow non-mimic. The models are tail-less, but all the local forms of *dardanus* have tails, which presumably serve an aerodynamic function and must be favoured by selection in the males and the yellow females. By an elegant study, Clarke & Sheppard (1962*b*) showed that the population contained genes which specifically shortened the tails of the mimetic forms, but not of the non-mimetic yellow females. They were later able to demonstrate an analogous set of modifying genes in *Papilio polytes*, this time specifically *increasing* the length of the tails in the mimics, which copy a distasteful *tailed* swallowtail (Clarke & Sheppard 1972). Similar *specific modifiers* (Charlesworth & Charlesworth 1976) were found altering the colour pattern of a form of *P. dardanus* in parallel with the geographical cline of its model, *Amauris niavius*, for the amount of black on its wings, and the colour pattern of a form of *P. polytes* in response to geographical variation in the model, *Pachliopta aristolochiae* (Clarke & Sheppard 1960*b*, 1972).

It may now seem *passé* to say, yet again, that these results do not accord with Goldschmidt's theory that mimicry would arise by systemic mutations which immediately produced high grade mimicry. But the revival of interest in Goldschmidt's ideas (e.g. Gould 1980) makes it important not to ridicule him (that particular sport being long out of fashion), but to show where his ideas were wrong, in order to sort out those parts of his work which can be profitably pursued. On mimicry, Nicholson (1927) was the one who was closest to the mark, and Goldschmidt even cites Nicholson in support of his own theory. Batesian mimicry usually evolves in two phases: the establishment of a major mutation which produces only rather poor mimicry, followed by the selection of further genes which refine the resemblance of the mimic to the model. I have emphasized this point again, because in thinking about problems connected with the evolution of mimicry it is most important to keep these two processes separate in one's mind. The two phases involve very different phenomena.

Müllerian Mimicry— Two Phase Evolution Too?

We have now arrived at what, after the seminal work of Clarke and Sheppard, has been seen as the conventional picture of the evolution of mimicry: Batesian mimicry evolves by the two phase process; Müllerian mimicry can evolve by gradual convergence (Fig. 14.4). It was Fisher (1930) who produced an argument to show that Müllerian mimicry can arise in this way, as part of his attempt to defeat the saltational theories of Punnett. Fisher's presentation gives the impression even after quite careful reading that he has proved that Müllerian mimicry does not evolve by major mutations, although in fact he never says as much. Thus Sbordoni *et al.* (1979) noting that the moth *Zygaena ephialtes*, which they had shown to be distasteful, had used major gene mutations to produce its mimetic pattern, described it as a mimic with a mixture of Batesian and Müllerian properties.

Consider a rainforest, with a great diversity of warningly coloured butterflies, presenting many different patterns. Those resembling each other closely enough will converge, by the process described in Fig. 14.4, until they are good Müllerian mimics. In this way, like planets forming from a cloud of gas, clusters of mimetic species will arise, and form what we call mimicry 'rings'. Species occupying the spaces in between the rings will be pulled into them, but sooner or later these focal patterns, having absorbed all the available species, will stabilize. If they differ too much from each other they will not be able to converge, for birds will never mistake one for the other (Sheppard *et al.* 1984). Hence, in the South American rainforests there are no fewer than five different mimicry rings, all very distinct to the human eye, and probably to the avian eye as well (Papageorgis 1975) (Fig. 14.7). Likewise in the West African rainforest there are two butterfly rings (Owen 1974*b*) and among European bumblebees two rather distinct patterns, each comprising several species (Plowright & Owen 1980). Whether the little scenario I gave for the origin of these rings is correct is not very important—what matters is that the presence of all five of the South American butterfly rings throughout the American tropics (with some geographical variation), argues that they are persistent and relatively stable in evolutionary time.

Once distinct rings of this kind form, then further Müllerian mimicry will arise in the same way as Batesian mimicry. Fig. 14.6 could equally represent two protected, warningly coloured species. If the less protected of them (A) can produce an approximate

Fig. 14.7. The five mimicry rings to which most of the long winged aposematic butterflies of the South American rain forest belong, represented here by one species from each ring, as they appear in Trinidad. From Sheppard *et al.* (1984).

resemblance to the better protected one (B), then the two will become Müllerian mimics by a two stage process which, as with Batesian mimicry, will involve the substitution first of a major mutation, and second of an indeterminate number of modifiers, some of which may of course be individually detectable mutations themselves, which refine the resemblance.

A rather clear example of the first step was discovered by Keith Brown and Woodruff Benson, in company with Philip Sheppard, in the population of *Heliconius hermathena* near Faro on the Amazonas (Fig. 14.8). In most of its scattered colonies in the Amazon basin this butterfly is not a Müllerian mimic. It has its own stable pattern of yellow bars and red splotches, and is apparently so different from the other sympatric *Heliconius* that no single mutation can initiate mimicry. But at Faro it flies with a local population of the polytypic *Heliconius melpomene* (see Fig. 14.9) which differs from normal *hermathena* only in lacking the yellow bars. In this one population a single gene which largely removes the bars has risen to a high frequency, making *hermathena* and *melpomene* into quite good Müllerian mimics (Brown & Benson 1977).

However, research in *Heliconius* has to yield precedence to studies on the European moth *Zygaena ephialtes*, most recently reviewed by Sbordoni *et al.* (1979). We are fortunate in having excellent studies not only of the genetics of this species (Bovey 1941 and later, Dryja 1959), but also of its ecology and behaviour, and of the reactions of predators (Bullini *et al.* 1969, Sbordoni & Bullini 1971).

In northern Europe the moth is a comimic of other distasteful members of its own genus; in much of southern Europe it has departed from this certainly ancestral pattern and become a Müllerian mimic of the sympatric moth *Amata phegea* (Ctenuchidae). Crosses between the northern ancestral populations and the southern *Amata*-mimics show that most of the difference resides in two genes, one converting the red colour to yellow, and the other (unlinked) increasing the black (or dark green) markings and converting most of the spots on the wings to white. (Dryja (1959) believed, as all subsequent commentators have noted, that this 'gene' was a complex of two loci, and that some aberrant moths appearing in his broods were the result of crossing over between them. Dryja did not publish the data for this conclusion in full and as his material was destroyed by military action, it is unfortunately impossible to verify it.) There are then two possibilities (if we exclude the rather improbable eventuality that near-perfect mimicry was produced by both mutations occurring at the same time): either the pattern or the colour must have changed first. Thanks to experimental and field work by Sbordoni, Bullini and their colleagues, we know as certainly as one can know anything in evolutionary biology that the first change was in the pattern. Changing the colour from red to yellow produces a bright

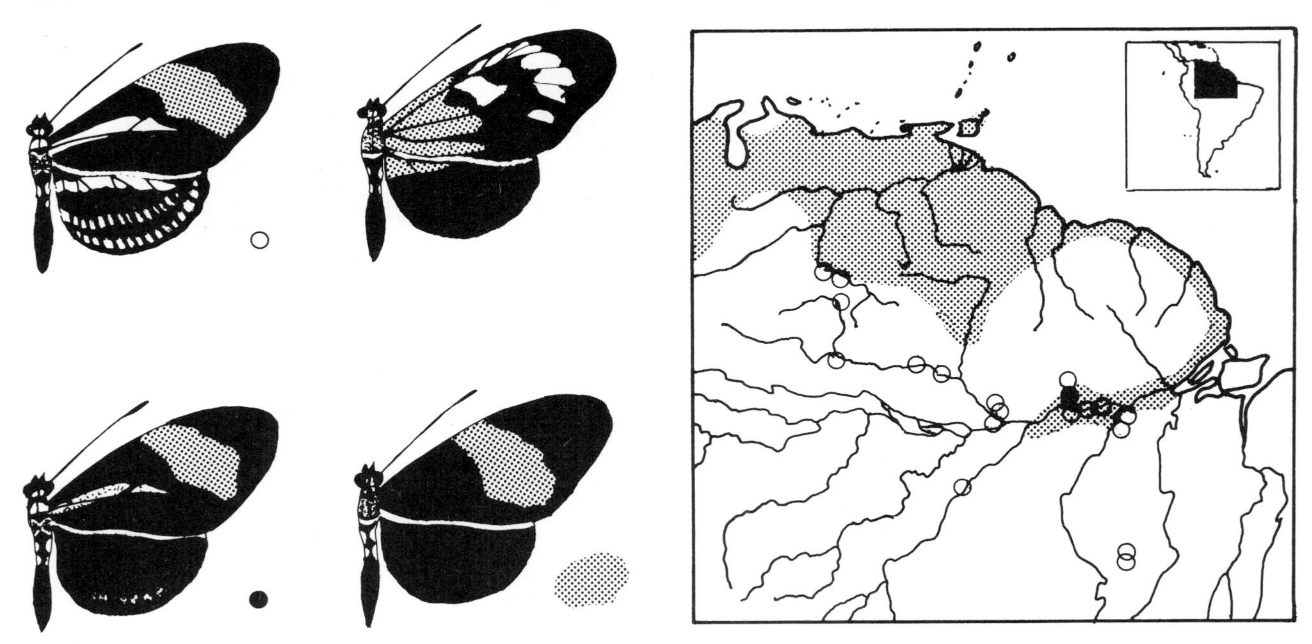

Fig. 14.8. Evolution of approximate mimicry by a single mutational step in *Heliconius hermathena*. The most widespread form of this localized Amazonian species (top left) bears no resemblance to the related species, such as *H. melpomene* and *H. erato* (top right) with which it flies. But at one locality (Faro) where it flies with *melpomene* and *erato* races (bottom right) from which it differs only in having yellow bars, a single mutation has largely removed the yellow bars from a majority of the population of *hermathena* (bottom left).

yellow *Zygaena* moth that does not look remotely like *Amata phegea*. On the other hand, changing the pattern without changing the colour produces a moth that bears a passable resemblance to *phegea*, except for some small marks near the thorax and the stripe on the abdomen being the wrong colour. Now this is known to be an adequate mimic of *phegea*, not only from experiments with caged birds, but from the fact that in an unusual population consisting entirely of this form, the moths adjust their posture to correspond with whichever of the other comimics (*Zygaena* spp. or *A. phegea*) is in flight at the time (Sbordoni & Bullini 1971). This is very good evidence that wild birds treat this form as a mimic both of *A. phegea* and of the other *Zygaena*.

Thus in terms of the system for the evolution of Müllerian mimicry in Fig. 14.6, one must imagine that mimicry was initiated by the pattern mutation, which carried *ephialtes* into the region of the phenotype space which was protected by *phegea*, without in this case totally losing the protection of the *Zygaena* mimicry ring: presumably the *Zygaena* and *Amata* protection curves overlap slightly. In most southern populations, this mimicry has been further refined by changing the colour from red to yellow, a change a geneticist would think of as a major gene, but which, as only a very few marks on the moth are now coloured, is in terms of the perceptual generalization of the predators, a rather minor change.

It could be that both *Z. ephialtes* and *H. hermathena* are unusual in using a rather large mutation to initiate mimicry, and that most other Müllerian mimics evolved by convergence of the gradual kind from patterns that were already rather similar. But in two other *Heliconius* species at least, changes in the mimetic pattern have involved genes which appear to be major to us, and probably also to their predators. *Heliconius melpomene* and *Heliconius erato*, almost always strict parallel mimics of one another, have diverged to an astounding degree *within* each species, so that their score of geographical races have only relatively recently been correctly assigned to just the two species (Fig. 14.9) (Emsley 1965, Turner 1965). Our genetic studies of six races of *melpomene* and eight of *erato* show that a relatively small number of major changes are involved in differentiating each race (Table 14.1), but, as predicted by the two stage theory, there are further minor genes which alter the expression of the major mutations. For example, the red rays of the Amazonian races of *melpomene* (produced by a single dominant allele) become much wider when placed on the genetic background of the East Brasilian race, which lacks these rays; in such a backcross hybrid, the rays are rather like the markings of two close relatives *Heliconius cydno* and *H. ethilla*. Thus the Amazonian races carry genes whose specific function is to make the rays narrow, in mimicry of *H. erato* and of several other *Heliconius* species.

As Dixey (1909) pointed out, there are two factors that determine which of a pair of distasteful species will be the better protected, and hence act as the model: distastefulness and abundance, a common

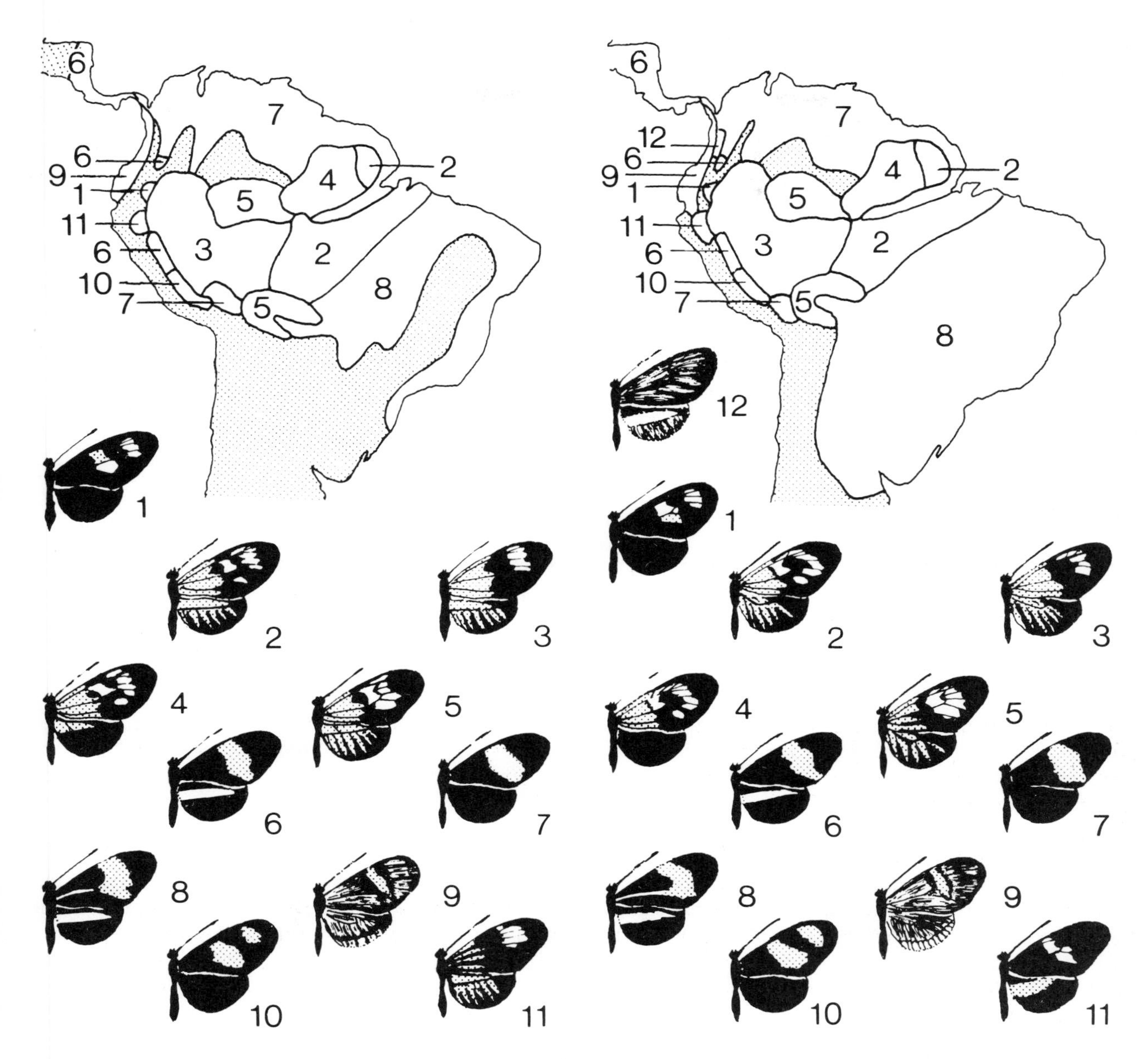

Fig. 14.9. Parallel Müllerian mimicry of the races of *Heliconius melpomene* and *H. erato*. From Turner (1981) courtesy of Annual Reviews Inc.

species being much better protected than a rare one. Sbordoni *et al.* (1979, also Bullini *et al.* 1969) have argued convincingly that although *Zygaena ephialtes* may in fact be nastier than *Amata phegea*, the much greater abundance of *Amata* (compared with *Zygaena*) which persists in undisturbed biotopes in Italy, and may have been general in southern Europe at the end of the glaciation, has caused *ephialtes* to converge to the pattern of *phegea*. Similarly we have postulated (Sheppard *et al.* 1984, also Turner 1977*b*, 1982) that the divergence of races within each of our two *Heliconius* species has been driven largely by marked spatial alterations in the abundance of the various warningly coloured butterflies of South America when the rain forest became fragmented into more or less isolated refuges during the latest cold dry period of the Pleistocene (Fig. 14.13), with *melpomene* and *erato* tending to mimic whichever was locally the most abundant and protected mimicry ring. (For reviews of Quaternary climate and vegetation in South America, see Prance 1982.)

Müllerian and Batesian Mimicry— Convergence or Advergence?

It is now important to bear in mind the distinction between the two phases of evolution. Although Batesian and Müllerian mimicry can and do evolve by the same process in the first phase, with the palatable or less protected species converging by a single mutation onto the better protected model (it

Table 14.1. Genetics of six races of *H. melpomene* and eight races of *H. erato*.

H. melpomene Race	1 East Ecuador	2 Lower Amazon	3 Upper Amazon	4 Guiana/ Manaus	7 Venezuela/ Trinidad	8 East Brasil	Gene
	+	0	0	0	+	+	*D*
	0	0	+	+	+	+	*R*
	0	+	+	+	0	0	*B*
	0	+	+	+	0	0	*N*
	0	0	0	0	0	+	*Yb*
	+	+	0	+	+	0	*C*
	0	+	+	0	0	0	*O*
	0	0	0	0	+	0	*F*
	0	+	+	+	+	+	*Rr*
	0	+	+	+	+	+	*S*
	+	0	0	0	0	0	*T*
	+	0	0	0	0	0	*Wh*

H. erato Race	1 East Ecuador	2 Lower Amazon	3 Upper Amazon	4 Guiana/ Manaus	5 Mato Grosso	6 Central America	7 Venezuela/ Trinidad	8 East Brasil	Gene
	+	0	0	0	0	+	+	+	*D*
	+	0	0	+	0	+	+	+	*R*
	0	+	+	+	+	0	0	0	*Y*
	0	0	0	0	0	0	0	+	*Yl*
	0	+	+	+	+	+	+	+	*St*
	+	+	0	+	+	+	+	+	*Sd*
	+	0	0	0	+	+	+	+	*Ly*
	0	0	0	0	0	0	0	+	*Cr*
	0	0	0	0	0	+	0	0	*P*
	0	+	+	0	0	0	0	0	*Or*
	+	0	0	0	0	0	0	0	*Ur*
	+	0	0	0	0	0	0	0	*Wh*
	0	+	+	+	+	+	+	+	*Ro*

+ indicates that this race has the recessive allele at this locus; 0 that it carries the dominant. The patterns produced by these alleles are shown in Fig. 14.9, which also indicates the race numbers. From Sheppard *et al.* (1984); also in Turner (1981, 1983*a*).

is fairly obvious that even if both species are distasteful, a mutation of the better protected species that resembles the less protected is at no advantage), the events in the second, 'modification' phase will be different. Müllerian mimics will converge mutually on some intermediate pattern; the model of a Batesian mimic, as has been frequently pointed out, is placed at a disadvantage by the mimicry, and should evolve away from the mimetic pattern. Brower & Brower (1972) coined the term 'advergence' for this process whereby mimic and model are involved in an 'arms race' (Dawkins & Krebs 1979) which the mimic somehow wins by evolving towards the model faster than the model can evolve away.

To the question 'how does the mimic win the race?' there are I believe two valid answers, both well expounded by Nur (1970), and dependent on considering the two distinct phases in the evolution of mimicry. Take first the gradual 'modifier' phase in which the mimic will evolve slowly toward the model. A hint as to why the model evolves at a slower rate is provided by Fig. 14.2: the advantage of being a mimic (compare predation on the nice butterfly with predation on the mimic) is considerably greater than the disadvantage of being a model (compare predation on the model with predation on the nasty butterfly). To show that this effect does indeed result in the mimic catching up with the model I have again simulated the situation of two species with varying numbers of white spots: all the conditions are the same, except that the species with the smaller number of spots is now a palatable Batesian mimic. Fig. 14.4 (right) shows that the rate of predation on each spot class is now skewed in such a way that both species are subject to selection for greater spot numbers, but that the mimic is much more strongly selected than the model. In fact for the particular numerical values used in this simulation, the mean spot number of the mimics will move during one generation of selection from 2.00 to 2.22, whereas the model mean moves only from 3.00 to 3.12, so that the 'gap' between the species has narrowed by 0.10 of a spot.

Given that spot number is inherited, both species will slowly increase their spot number, but stabilize as the mimic catches up with the model. It is not difficult to see that this result is a general one: the effect of the mimic is always to make phenotypes 'to the right' fitter, but as the model's phenotype is nearer to the right than is the mimic's, selection on the model will always be the weaker.

The model of course could escape from the mimic by producing a major mutation which carried its phenotype right away from the present pattern. But it cannot do this successfully, as the new mutant would be rare, would not be recognized by predators, and would be sampled by them: in warning coloration, nothing succeeds like being common (see the experiment by Benson 1972 on the increased predation rate on *Heliconius* with altered patterns). Thus the first phase of evolution, a large mutational 'jump', can be used by the mimic to initiate mimicry, but not by the model to escape (Nur 1970, Sheppard 1975). The only chance for the model is to make a 'jump' into a different, well-protected mimicry ring.

One further difference between Batesian and Müllerian mimicry is brought out by the two computer simulations: the final rate of convergence. Remember that all the parameters, except the palatability of the 'mimic', are the same in Fig. 14.4. In the Batesian case, as has been said, one generation of predation narrows the difference between the species by 0.10 of a spot. In the Müllerian case one species increases by 0.028 of a spot, and as both are equally unpalatable, the other declines by the same amount; thus the difference narrows by only 0.056 of a spot. The long-term response to these selection pressures will be determined by the genetic architecture of spot number, but again assuming that this is the same in the two cases, it is clear that the Batesian pair are converging nearly twice as quickly as the Müllerian pair. Marshall (1908) and many later students of mimicry have suggested that Batesian mimicry is expected to be the more accurate and it is said (although it would be very difficult to quantify) that Batesian mimics often show astoundingly close resemblance, whereas Müllerian mimics often show only a general similarity (e.g. Fisher 1930).

Even with equal heritabilities, the convergence of a Müllerian pair will not in general be equal: the less abundant or more palatable species will evolve faster and farther than the commoner or more distasteful one (Fisher 1930). In the limit, when one species is very rare (or of course, when it is of neutral palatability), this one will do all the evolving and the 'model' will remain unaltered. Whether this has happened in any *Heliconius* species we do not yet know, but on taxonomic grounds it is likely that few if any changes have occurred in *Amata phegea* as a result of its association with the normally much rarer *Zygaena ephialtes* (Sbordoni *et al.* 1979). It is sometimes suggested that the nastier of a pair of Müllerian mimics will evolve 'away' from the association, like the model in a Batesian system: Rothschild (1981*b*) for instance says that local and temporal variation in toxicity 'will mitigate against [*sic*] the tendency for the more poisonous species to, willy nilly, evolve away from the less poisonous and thus be forced out of the desirable companionship of its co-mimics.' It should now be clear from the two limiting cases of Müllerian mimicry (equal nastiness, or one mimic neutral) discussed that, given the system of predator behaviour postulated here, there is no such tendency for one of the Müllerian mimics to 'escape'. Unequal nastiness simply results in unequal but nonetheless mutual convergence.

Batesian Mimicry — Preadaptation

The frequency-dependent effect on aposematic species just mentioned is inevitably reversed for Batesian mimics: the commoner they become the less benefit accrues from the mimicry. The result is well known: mutations producing mimicry of another model are at an advantage when rare, so that the species tends to become polymorphic (Ford 1971). In a study of the first importance for our understanding of mimicry, Charlesworth & Charlesworth (1976) have investigated what happens to such mutations if they arise in a species which is already polymorphic for a mimetic and a non-mimetic form. If the new mutant is unlinked, or only loosely linked to the major gene which originally produced the mimicry, then only two outcomes are possible: either the new mimic is so advantageous that both the major allele and the new modifier spread to fixation, so converting the species to monomorphic mimicry of the new model, or the new mimic is not advantageous enough, and the new mutation does not spread. *Papilio dardanus* provides an example of this system: in subSaharan Africa both mimicry and loss of tails (unlinked genes) have reached fixation; i.e. all *dardanus* are tail-less and mimetic. In Ethiopia there is a polymorphism with non-mimics at 60–80% of the population, and here the *tail-less* allele has been unable to spread: both mimics and non-mimics have tails on the wings.

On the other hand, if the new mutation happens to be rather closely linked to the original major gene, then both loci remain polymorphic, and the butterfly becomes a polymorphic mimic of two different models, an extensive polymorphism builds up, in which the forms are controlled by clusters of tightly models, an extensive poly-morphism builds up, in which the forms are controlled by clusters of tightly linked loci which at first sight appear to be multiple alleles at one locus. Thus in the highly polymorphic

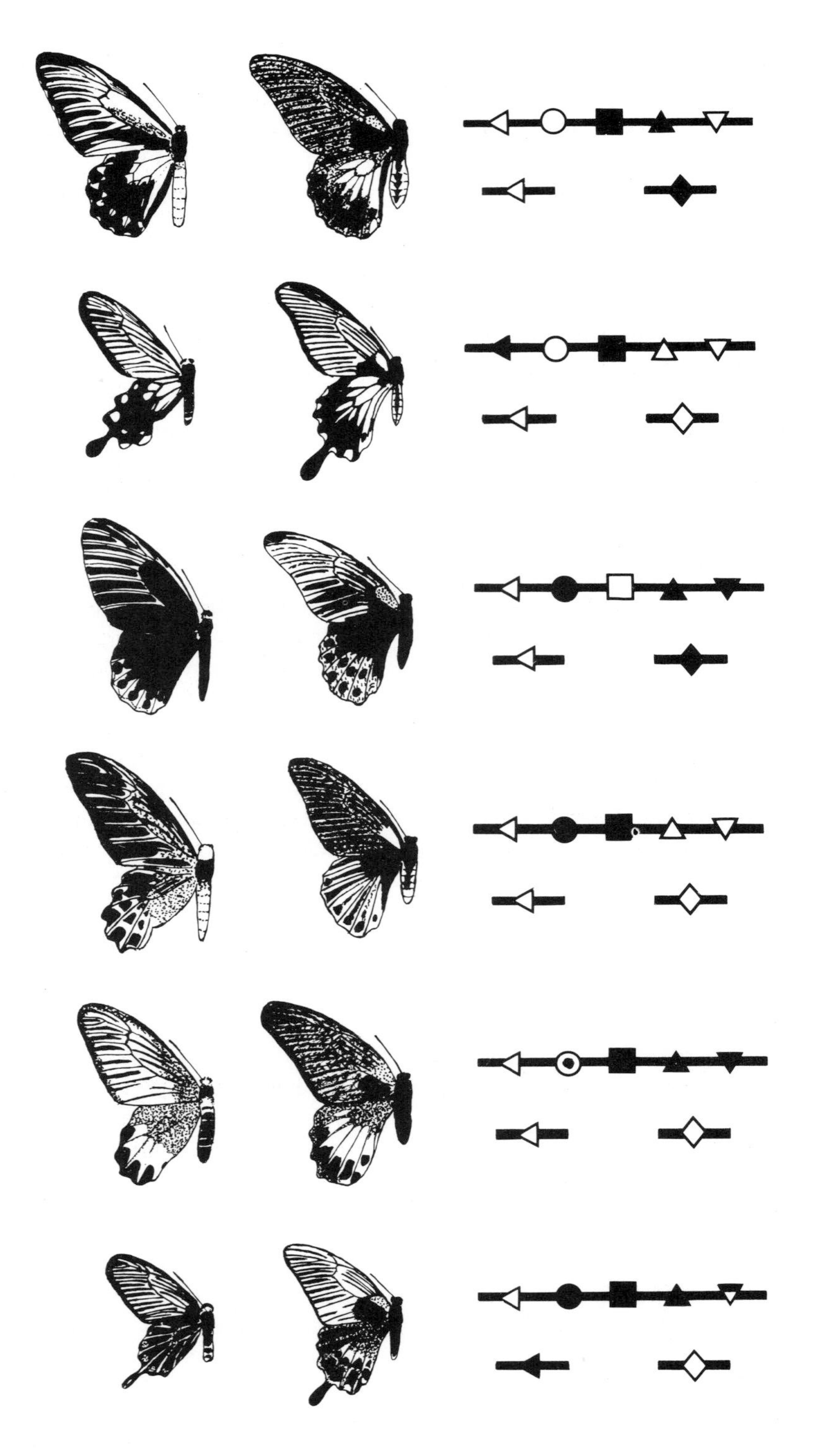

Fig. 14.10. Six of the forms of the batesian mimic *Papilio memnon* (centre) with their models (left). The diagrams (right) represent the way the forms are built up from combinations of a limited number of genes: ◁ tails (closed symbol, present; open, absent); ○ hindwing pattern (closed, dark; open, with extensive white marks; dot, with less extensive white); □ forewing (closed, dark; open, with white apex); △ 'shoulder' flash (mimics head or thorax colour of model—closed, red; open, white); ▽ abdomen (closed, black; open, yellow; particoloured, black with yellow tip); ◇ yellow suffusion of hindwing (closed, present; open, absent). Black bars represent chromosomes. Five of the genes (or six if one takes into account data showing that the 'gene' for hindwing pattern is at least two genes) are very tightly linked, and behave as if they were multiple alleles of a single locus. The remaining two are unlinked. Data from Clarke *et al.* (1968) and Clarke & Sheppard (1971).

Papilio memnon the occurrence of rare crossovers reveals that the cluster contains at least six loci (a 'supergene') affecting the colour and morphology of different parts of the wings and body (Fig. 14.10) (Clarke *et al.* 1968, Clarke & Sheppard 1971). In these circumstances, as Fisher first pointed out, selection will favour tightening of linkage between the loci. Evidence from the silkworm indicates that Lepidoptera have factors which exert very strong control over recombination in their own chromosome (Turner 1979, Ebinuma & Yoshitake 1982) and hence this tightening of linkage may be most effective (in males only of course; there is no recombination in females—Turner & Sheppard 1975). In *Papilio dardanus* and *Papilio polytes* the linkage has become so tight that the nature of the individual loci can no longer be discovered, given the limits on the number of offspring obtainable in butterfly breeding experiments (Clarke & Sheppard 1960*d*, 1972).

The curious and interesting thing about these findings is that the Fisherian theory of modification does not entirely account for the supergene: the loci involved must be fairly closely linked from the start, and the alternative view that selection will bring together loosely linked or unlinked loci to form the supergene is no longer tenable (see also Rothschild 1981*b*). Many people will find it improbable, if not smacking of special creation, that the loci should just happen to be appropriately linked. It is indeed improbable, and that is why there are so few spectacularly polymorphic mimics among butterflies: only a few species happen to have clusters of loci controlling wing pattern functions. These are the ones that attract our attention by becoming polymorphic: in a way we are performing a biased experiment. Even so, the 'successes' do not have things one hundred per cent in their favour; whereas *P. memnon* happens to have a tail locus linked to the colour pattern genes, *P. dardanus* and *P. polytes* do not, and consequently adjust less accurately to the presence of tailed and tail-less models.

The origin of supergenes in mimetic butterflies is in short an example of something I like to call a 'sieve': an evolutionary mechanism that picks out those species, or genes, which happen to have the required properties. The simplest example is provided by the dominance of most of the genes for industrial melanism in moths: although both dominant and recessive melanic mutations occur, it is the dominant ones that increase faster under natural selection and hence become the predominant industrial melanics (Haldane 1924, Sheppard 1975).

It is clear that a sieve operates also on the evolution of the colour patterns in mimetic butterflies. For

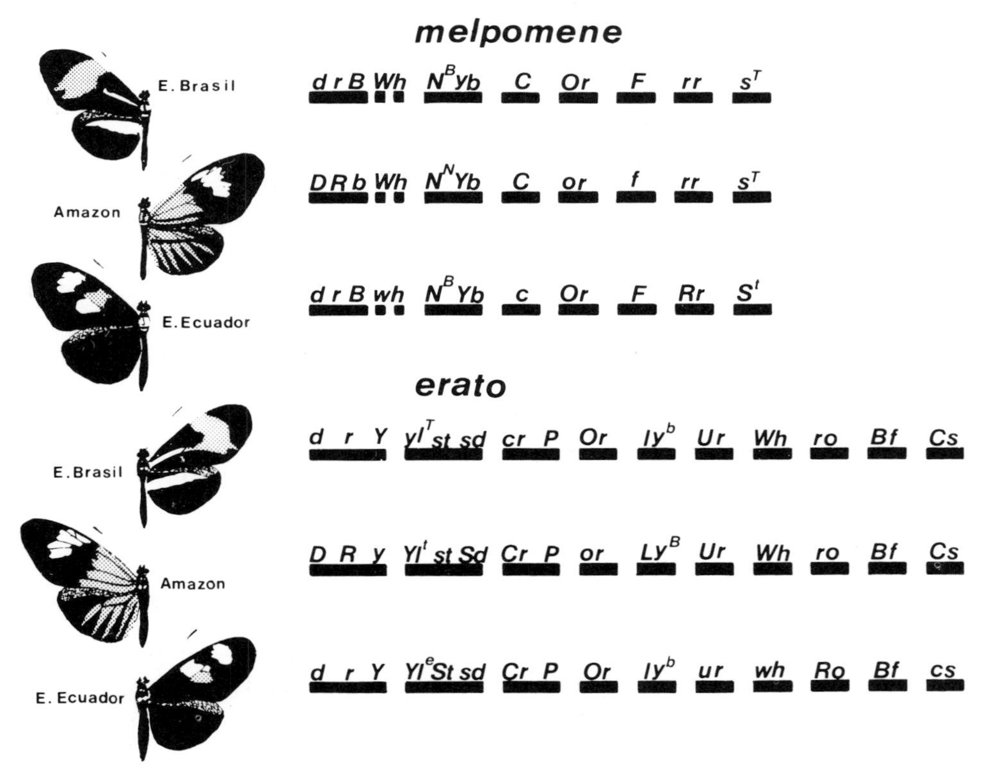

Fig. 14.11. The genetics of three parallel races of *Heliconius melpomene* and *Heliconius erato*. The way in which the alleles shown here combine to produce other races is shown in Table 14.1, where the dominant alleles are indicated by 0 and the recessive by +; in this figure dominance is indicated by a capital letter. Heavy bars are chromosomes. Data from Sheppard (1963), Turner (1972) and Sheppard *et al.* (1984).

example, when the three forms of *Papilio dardanus* which mimic *Amauris niavius*, *Amauris echeria* and *Amauris echeria septentrionis* are crossed into races where the forms do not occur, the genes producing the forms cease to be distinguishable: they produce, all three, the same pattern (Clarke & Sheppard 1960*c*, 1962*a*,*b*, 1963). Now, suppose that the *A. niavius* mimic was the first one. If the mutant converting this to *A. echeria* had occurred on *this* genetic background it would not have spread, for it would not have produced a new pattern, and likewise for the mutant changing this form to mimic *A. e. septentrionis*. In short, for these new forms to appear, some of their distinctive 'modifiers' must have existed in the population *before* the mutation occurred. The evolution of mimicry involves also a certain amount of preadaptation, or luck.

The occurrence of the fortunate and fortuitous linkage required for the evolution of supergenes is seen in *Heliconius*, in which a few of the genes controlling the mimetic patterns are linked, some of them rather closely (Fig. 14.11) (Sheppard *et al.* 1984). There is no balanced polymorphism in these species, and thus no prolonged period in which selection could produce, or substantially increase, the linkage between loci. A possible explanation for the linkage is that we expect a rather weak sieving effect which causes linked rather than unlinked modifiers to spread in the population if the mutations arise during the period when the major gene is spreading (Turner 1977*a*). However, as the distribution of the genes on the chromosomes is an excellent fit to a Poisson distribution (Table 14.2), there is no need to suppose that the linkage is anything other than random.

Table 14.2. Fit of distribution of loci controlling colour patterns between chromosomes in *H. melpomene* and *H. erato* to a random (Poisson) distribution.

	No. of loci per chromosome				
	0	1	2	3	TOTAL
Observed cases	24	12	3	3	42
Expected cases	22.0	14.2	4.6	1.1	39.9

I have said that what we need to do with the theories of Punnett and Goldschmidt is not to ridicule them, but to determine what parts of them might be correct. The theory that mimicry arises perfect from the start is shown in general not to be true: both Müllerian and Batesian mimics give evidence for the occurrence of subsequently selected modifying genes. But the theory that homologous genes could produce the same pattern in model and mimic, suggested in that simple form by Punnett (1915) and in the form of homologous developmental pathways by Goldschmidt (1945), is to some extent confirmed by a comparison of the genetics of *Heliconius melpomene* and *H. erato*. It seems too much of a coincidence that these two butterflies, which although not sister-species are still quite closely related, should both have a single linkage group which controls the yellow colour of the forewing band, the 'Dennis' marks on the forewing, and the rays on the hindwing, and probably in each case comprising three loci (*B*, *D*, *R* in *melpomene*, *Y*, *D*, *R* in *erato*), and another linkage group which in both species controls the yellow hindwing bar and white hindwing margin (Fig. 14.11). This is good evidence that homologous genes are being used in both butterflies. But the whole of the resemblance between the two species is not produced in this way. The superficially similar forewing bands of the East Ecuadorian races (Fig. 14.11) are produced by genes whose mode of action is quite different: there is for instance no analogue in *melpomene* of the *Ro* gene which rounds out the tip of the band in *erato*.

The problem with the homology hypothesis is not that it is wrong, but that Goldschmidt tried to make it explain too much. The less closely related are a pair of mimics, the less likely it will be that they will use homologous developmental pathways (Nicholson 1927). To some extent I have cheated with the silkworm mimic of the snake (Fig. 14.5); the ease with which this intricate pattern is produced is no doubt in part due to the fact that, on the background produced by the *Moricaud* gene introduced from the wild ancestor *Bombyx mandarina*, the effect of the mutations in *B. mori* may be to evoke developmental pathways which existed in the wild ancestor. The delicate reticulate pattern added by the *quail* mutation is not produced simply by that mutation, but by the interaction of *quail* with many other genes in the genome. The cryptic snake pattern is therefore not produced merely by the four mutations shown in Fig. 14.5, but by a considerable array of genes controlling developmental pathways, some of which may have been modified during the history of wild *Bombyx* to produce refined cryptic patterns. In fact, as even domestic *mori* retain the display behaviour appropriate to a snake mimic (which of course looks utterly meaningless when performed by a white silkworm), there is a strong presumption that the wild ancestor *was* a snake mimic.

Thus, although homologous developmental pathways cannot entirely explain mimicry (who would suppose that snakes and *Bombyx* caterpillars share a significant number of pathways?), there are two important principles implicit in Goldschmidt's thinking which, in the excitement which followed on disproof of his saltational theory, have been overlooked. First, a single mutation may be able to produce a refined and well-adapted pattern if it switches on again a hidden developmental pathway that has been refined by the selection of modifiers

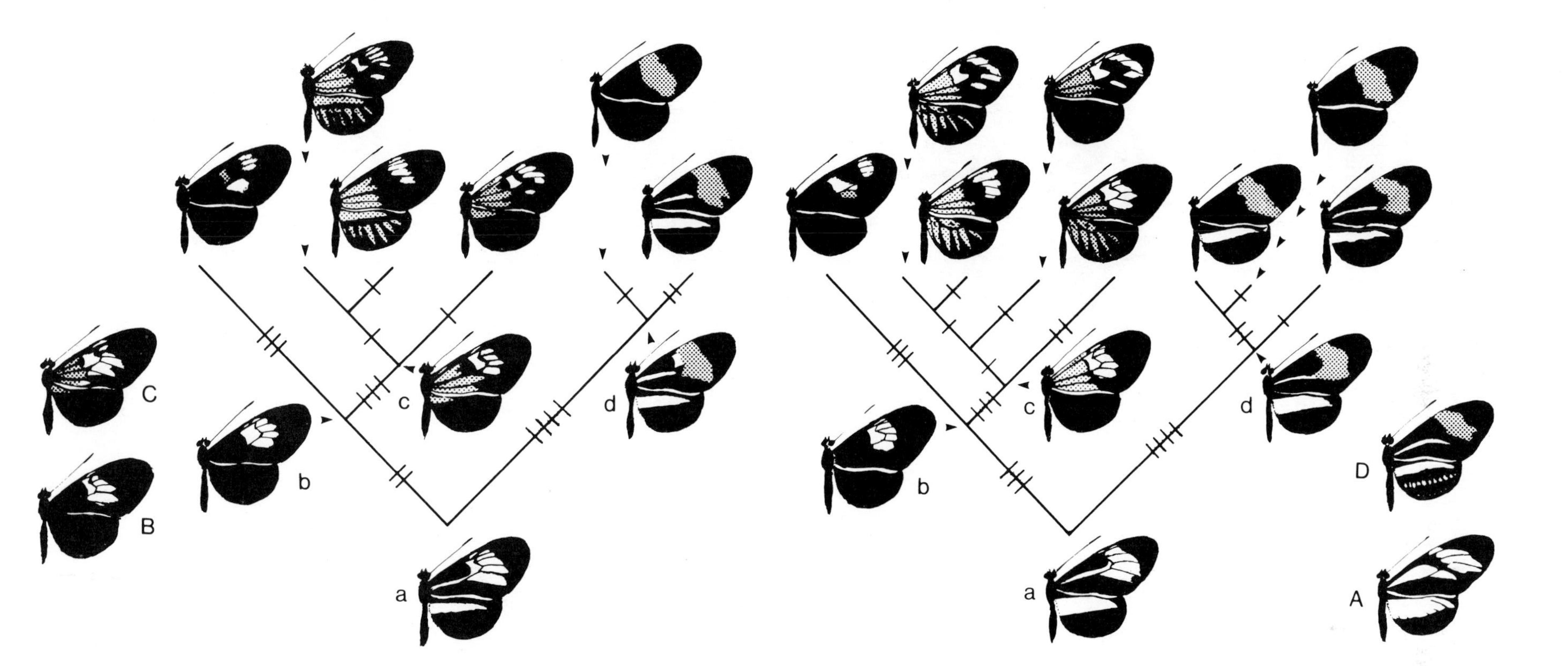

Fig. 14.12. Cladogeny of some races of *H. melpomene* (left) and *H. erato* (right). Minimum trees have been constructed using the genetic data in Table 14.1, by applying the dominance-sieve theorem and the weighted invariant step method (Farris *et al.* 1970). Of several equal length trees obtained, the most geographically probable have been selected for illustration. Reconstructed ancestors (a, b, c, d) may have been Müllerian mimics of patterns which are still extant in the relict species A, B, C, D (*H. nattereri*, *H. timareta*, *H. elevatus*, *H. hermathena*). Cross bars, placed conventionally in the centre of the branch, denote the substitution of a major gene mutation.

at some time in the past and then switched off. The selection in domestication of the *plain* alleles which remove all colour from the silkworm must have left largely intact, though subject to mutational damage, the developmental pathways, controlled by other gene loci, that produced the cryptic coloration of the wild ancestor. Second, more generally, the effect of a particular mutation, and therefore its success or failure in producing a new adaptation, depends critically upon the background genome in which it occurs. A mutation that produced tolerable mimicry on one genetic background might be hopelessly maladaptive on another. The resulting selection of only those mutations which 'suit' the genetic background, results in cases like the *Amauris*-mimics of *P. dardanus*, in which the genome appears to be preadapted to mimicry.

Müllerian Mimicry—Evolution by Jerks

As this last idea is somewhat akin, although by no means identical to some of the notions of Goldschmidt (1945—the chief difference being that whereas he maintained that the genome was set up in such a way as to produce immediate, high grade mimicry, I am making the more modest suggestion that it has to be set up in such a way that approximate mimicry is possible), it will be interesting to round off this discussion by looking at another theory in which Goldschmidt's idea of bang-on target systemic mutations have been invoked: the punctuational model of evolution (Gould 1980, Stanley 1979). According to this theory, evolution takes place by an alternation of long periods of stasis (no change) with short periods of very rapid evolution in which both speciation and the origin of new adaptive phenotypes take place.

Our genetic studies on *Heliconius melpomene* and *H. erato* have now reached the stage where we can use cladistic methods to reconstruct the evolution of the races. The technique is quite simple: the 'dominance sieve' mentioned above in relation to industrial melanism tells us that of any pair of alleles the recessive one is likely to be ancestral, and given that theorem the minimum tree can be estimated readily by the weighted invariant step method (Farris *et al.* 1970). The trees obtained (Fig. 14.12 shows those which correspond best with the current geographical distribution of the races) suggest that these two species have been mutual Müllerian mimics during the whole of their adaptive radiation, having both started as black butterflies with a yellow barred pattern not unlike the existing species *H. charitonia* and *H. nattereri*; perhaps they both inherited it from a common ancestor (Sheppard *et al.* 1984; Turner 1981). This is encouraging as it is far more likely that the beautiful parallel variation we see today would have been produced by parallel evolution than that all the races had converged, in pairs, from very disparate ancestors. Whether the trees and reconstructed ancestors are 'right' of course we cannot say: I nurse a suspicion that the common ancestors are 'wrong' in one feature, but am unable to prove this within the rules of cladistics. What is important is that, whatever the exact form of the tree, they must, like the ones figured, include the substitution of a few major genes in each branch (shown as cross bars).

Fig. 14.13. Approximate location of South American rain forests at the peak of the last glaciation, *ca* 18 000 years BP, deduced from a combination of biogeographical and palaeoecological data. After K. S. Brown, in Prance (1982).

Now if race formation took place in glacial refuges (Fig. 14.13), these trees span the 30 000 years of the most recent glacial cycle, at least. Complete substitution of a gene under selection of only 1% takes less than 4000 generations, or 400 years in *Heliconius* (Turner 1982). Thus if we could find fossil butterfly patterns, what we would see in these species would be the alternation of periods of stasis with rapid periods of change. In short, punctuated equilibria. However there is nothing in the process which is not perfectly describable in terms of population genetics and neoDarwinism. There are no systemic mutations, or catastrophic speciation events; merely the substitution of new dominant genes under natural selection.

I believe that we may here be seeing an example of an important evolutionary process: the appearance of new phenotypes, taking place quite rapidly in evolutionary time, when a species comes to occupy a new ecological niche. I have argued here that

Heliconius take up new patterns (an analogy for ecological niches) when some other pattern becomes well protected; niches of course are a sort of photo-negative of this: they tend to be occupied when they are 'empty', whereas mimicry rings become increasingly occupied when they are 'full'. I have elsewhere (Turner 1977*b*, 1981, 1982, also Sheppard *et al.* 1984) advanced the view that changes in pattern abundance, and hence protection, tended to be marked, persistent and therefore influential when the rainforest, fragmented by cool dry conditions during glacial maxima, underwent progressive extinction of parts of its fauna and flora (Fig. 14.13). Extinction empties ecological niches, and remaining species may evolve to occupy them. In ecologically stable periods, when all niches are occupied, there will be little evolution. Rapid changes occur in the comparatively rare circumstance that a new, imperfect adaptation, not yet improved by the long selection of 'modifier' genes, confers higher fitness than the old, refined adaptation. The removal of competition by the extinction of species will provide precisely this condition.

How often such changes involve major mutations remains to be seen. Although morphological changes seem more generally to be produced by a number of genes of individually small effect, single major genes are certainly involved not only in mimicry but in the evolution of host-resistance and parasite virulence (Sidhu 1975).

It is here, rather than in any new theory of systemic speciation, that we should seek for the causes of evolution by jerks.

Acknowledgements

My research on silkworms is funded by a grant from the Science and Engineering Research Council. I am most grateful to the Laboratory of Sericology, Kyushu University (H. Chikushi, H. Doira), and to the Department of Morphological Genetics, National Institute of Genetics, Mishima (Y. Tazima) for the supply of mutant stocks of *Bombyx mori*.

15. How is Automimicry Maintained?

Dianne O. Gibson

Department of Genetics, University of Liverpool, Liverpool

Danaus chrysippus, the African Queen or Plain Tiger, has a vast tropical and subtropical distribution, including Africa, Asia and Australia. The larvae feed mainly on milkweeds (Asclepiadaceae), many of which have secondary chemicals bitter or emetic to browsing vertebrates. These defensive compounds include the heart poisons, or cardenolides (Ch.12).

Larvae of *D. chrysippus* feeding on cardenolide-containing milkweeds can sequester these secondary plant compounds and retain them in both the pupal and aposematic adult stages, which become unacceptable to many predatory birds. However, adult *chrysippus* reared from milkweeds lacking heart-poisons are nonetheless identical in appearance to their chemically protected brethren, and thus act as intraspecific Batesian mimics—or *automimics* (Brower *et al.* 1967*a*; Ch.12).

Assuming that larval foodplants are not a limiting resource, why do some female *D. chrysippus* lay eggs on non-toxic plants? A population may contain over 80% adults non-toxic to birds (Brower *et al.* 1975). Naïve birds sampling such a population will have a high probability of first encountering a palatable individual, and so form a searching image for this conspicuous species. This may be re-inforced several times before an emetic butterfly is eventually eaten.

Despite theoretical predictions that 30% models will protect 70% mimics (Holling 1965), the demonstration of some counter selection against individuals feeding on toxic plants would make the maintenance of high mimic to model ratios more plausible or understandable. It has been suggested that sequestering cardenolides reduces larval growth rate (Brower *et al.* 1972, Brower & Glazier 1975). However, this idea has been countered by other findings indicating, at the very least, that the physiological cost is trivial (Dixon *et al.* 1978), or even that the presence of cardenolides actually enhances larval growth rate (Smith 1978).

A New Model for Automimicry

The explanations so far suggested for the maintenance of automimicry only take account of the direct effect on the fitness of individual adult butterflies of feeding on a particular milkweed. Here I wish to propose a new theoretical model, in which counter selection pressure comes from parasitoids differentially attacking the larvae on more toxic hostplants. The model incorporates three hypotheses. First, female butterflies prefer to oviposit on the plant species which they experienced as larvae (but see also Ch.6 concerning the Hopkins host selection principle). Second, birds distinguish between externally identical larvae by the shape of the plant on which a larva is feeding. Thirdly, female parasitoids select, for oviposition, larvae feeding on plants toxic to vertebrate herbivores. These three hypotheses have been tested and the results support the model.

Figure 15.1 outlines the model, in which *Apanteles chrysippi* has been taken as the parasitoid. This species can be successfully reared on *D. chrysippus* larvae with or without cardenolides. The cycle starts with the adults of both sub-populations at the point (*) where models are frequent enough to protect the automimics from bird predation. Larvae originating from eggs laid by these adults on non-toxic plants are subject to predation by birds, but few are killed by parasitoids. In contrast, larvae of this generation feeding on toxic plants suffer heavy losses through parasitoids, but are effectively free from bird attack. Parasitoids are known to be more effective population regulators than birds (Varley *et al.* 1973; but see also Ch.10). As a result, in the next generation the relative frequency of toxic to non-toxic butterflies will be reduced (Fig. 15.1 area 1). The population remains in this part of the cycle for a number of generations despite the steady reduction in the frequency of models. Eventually the ratio of models to mimics is

The Biology of Butterflies
0-12-713750-5

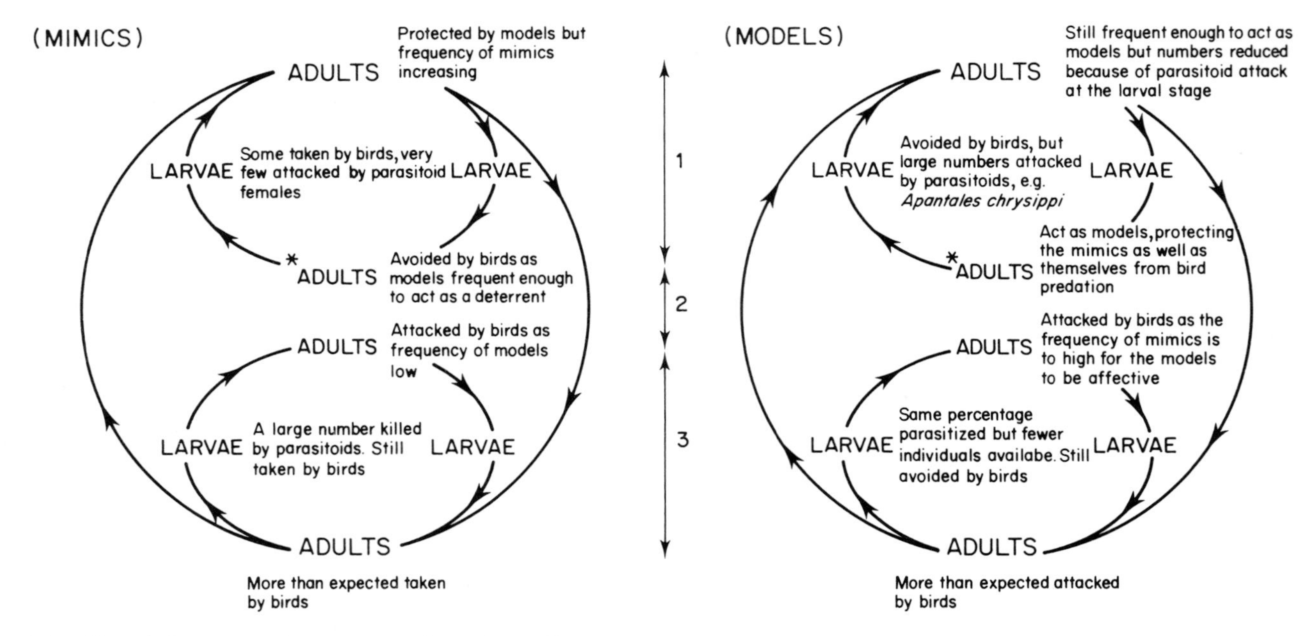

Fig. 15.1. Diagrammatic representation of the effects of parasitoid attack and bird predation on a population of danaine butterflies which contains both toxic and non-toxic members. 1: in this area, the frequency of the two groups is in balance; 2: in the transition generation, female parasitoids are unable to find enough of the preferred (toxic) hosts and begin to attack non-toxic larvae; 3: while the two groups are out of balance parasitoid attack continues until the balance-frequencies are restored. Left: Sub-population not containing cardenolide. Right: Sub-population containing cardenolide.

so low that birds attack rather than avoid all *D. chrysippus* adults. In the next generation so few eggs are laid on the toxic plants that the parasitoids cannot always find the preferred type of host, the one toxic to birds, and so start to oviposit in non-toxic larvae. Palatable larvae are now not only subject to bird predation but are also killed by increasing numbers of parasitoids. The butterfly population remains in this lower part of the cycle (area 3) until the relative frequency of palatable to toxic individuals is reduced to a level at which the models are again effective. The cycle then moves through the transition area 2 (Fig. 15.1) and the first part of the cycle begins once more.

Simulation, Field Data and Conclusion

A computer simulation has been made based on the three hypotheses mentioned earlier. Biological assumptions involved include non-overlapping *D. chrysippus* generations; a generation time for the bird predator much greater than that of the butterfly; milkweeds available to the butterfly larvae being of only two sorts, either with or without cardenolides; and that a small proportion of female butterflies do not oviposit on the expected plant species. Bird predation on the palatable larvae has not been assumed to be frequency dependent, although such a facility is available in the program. The program has been run with a number of different levels of predation and parasitoid attack, the extreme values of which are illustrated (Fig. 15.2), with a 0.05 probability that the females would choose a plant other than the one experienced as a larva.

There is a limited amount of suitable data on the population dynamics of *D. chrysippus*, and information has been found for only two elements in the model—the frequency of parasitoid attacks on larvae feeding on toxic milkweeds at one site, and the ratio of toxic to non-toxic adults in another population. Data on the effect of parasitoids on the toxic larvae of the African Queen came from a nine month study made in 1972/3 by Edmunds (1976) in Ghana. He found that *Apanteles chrysippi* was the most frequent of the parasitoids on toxic larvae, with 85% of the December 1972 sample infected. In the same year, a sample of 50 adult *D. chrysippus* was collected from Ghana by Edmunds (Brower *et al.* 1975) over a period of seven months. This site contained a range of milkweed species, some toxic and some palatable. Unfortunately, the site was 20km from the one where the parasitoids had been studied. One problem in interpreting these data in relation to a model dealing with cyclical changes is that only seven samples were taken, with as few as three insects per sample. In the December 1972 sample, 21% were toxic ($n=14$). Given this information, the results of the computer simulation would predict that a female parasitoid would totally reject, as oviposition sites, the first five non-toxic larvae encountered and would probably parasitize the next.

All the parameters of the model but one can be kept constant. If this one changing parameter is either the probability of bird predation on the non-toxic

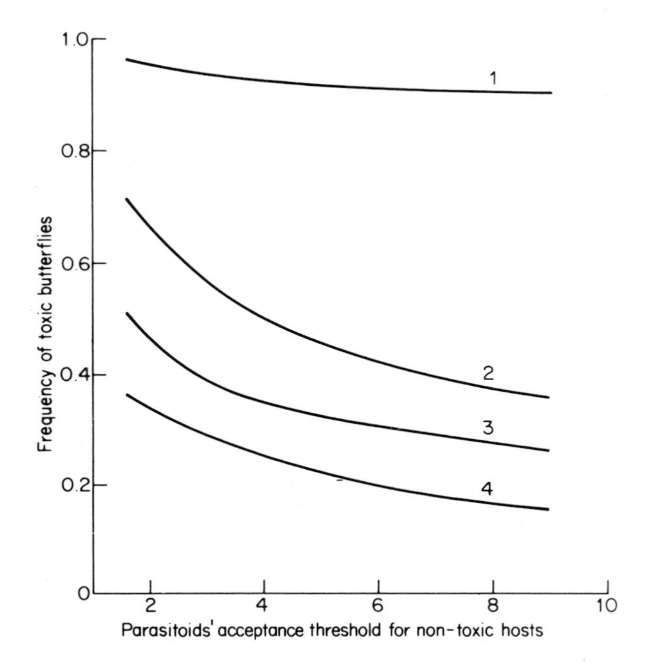

Fig. 15.2. Variation of the stable frequency of toxic butterflies, for different selective values of parasitoid and bird attack, plotted against the parasitoids' acceptance threshold for non-toxic larvae. The values are as follows. 1: fP = 0.45, fL = 0.75, fA = 0.25–0.75; 2: fP = 0.85, fL = 0.75, fA = 0.25–0.75; 3: fP = 0.45, fL = 0.25, fA = 0.25–0.75; 4: fP = 0.85, fL = 0.25, fA = 0.25–0.75; where fP = fraction of all larvae parasitized, fL = fraction of non-toxic larvae killed by birds, and fA = fraction of non-toxic adults killed if totally unprotected by resembling the toxic form. The number of larvae that survives parasitoid attack each generation is calculated as follows: *toxic larvae*, toxic population [1 − fraction parasitized (1 − density of non-toxic larvae)]; *non-toxic larvae*, non-toxic population [1 − fraction parasitized (1 − density of toxic larvae)].

larvae or parasitoid attack on all larvae, the expected relative frequency of the two toxic groups in the population is altered dramatically. If the probability of bird predation on the non-toxic *adults* is the one parameter to be changed then there is little or no effect on the relative frequencies of the two toxicity groups (Fig. 15.2). This would suggest that automimicry is not a primary cause of palatable butterflies being maintained at such high frequencies as 80% in a *D. chrysippus* population. Rather it is the behaviour of parasitoids combined with predation by birds on the larvae.

Postscript

The examination of automimicry is now being extended to include the effect of adult feeding on pyrrolizidine alkaloids (PAs). These plant alkaloids, like cardenolides, are bitter-tasting (Boppré 1978; Ch.12). Adult danaids without cardenolides are considered to be mimics, but after they have fed on sufficient PAs they probably become taste, as well as colour, co-mimics of the cardenolide containing models. Alcock (1970) has shown that merely being unpalatable can be an effective defence. Therefore birds naïve to the effects of cardenolides will be deterred from eating *D. chrysippus* adults even if those butterflies only contain PAs. However, although PAs may be toxic, the effect is delayed, so a bird is unlikely to associate the effects with the source (but see also Ch.12).

Acknowledgements

I am grateful to Dr G. S. Mani of the Department of Physics, Schuster Laboratory, at the University of Manchester for his assistance with the computer programming. This work was carried out during a Mr and Mrs John Jaffé Research Fellowship with the Royal Society. My thanks are also due to Professor D. A. Ritchie in whose department the research was carried out, and Miss D. O'Leary.

Part V
Genetic Variation and Speciation

16. The Ecological Genetics of Quantitative Characters of Maniola jurtina *and Other Butterflies*

Paul M. Brakefield*

Department of Population and Evolutionary Biology, University of Utrecht, Utrecht, The Netherlands

Ecological geneticists combine field and laboratory studies to investigate the dynamics of evolutionary processes. Their approach rests on the assertion that the causes of genetic variation cannot be understood without a knowledge of the ecology of the organisms concerned (Ford 1975*a*).

In genetics the term character is applied to any property of an organism for which individual variation, especially when of a heritable nature, can be recognized. A quantitative character normally varies in a continuous manner and its study depends on measurement. It is usually jointly determined by the interacting effects of a number of minor genes or polygenes. The same phenotype can be determined by different combinations of these polygenes. Selection on quantitative characters is usually recorded in terms of the population mean and/or variance since it cannot be represented in terms of gene frequency changes, as is possible for Mendelian characters controlled by major genes. However, the distinction between polygenes and major genes is not absolute (see also Ch.14). The relationship between these genes and the different types as recognized by molecular geneticists remains unclear (e.g. Mather & Jinks 1971, Cavalier-Smith 1978, Ayala & McDonald 1980).

Polygenic systems provide the basis for smooth adaptive change. Over 30 years ago Dowdeswell *et al.* (1949) chose to use variation in the number of small hindwing spots in the univoltine butterfly *Maniola jurtina* as an index of the fine adjustment and adaptation of populations. The field data now accumulated on this particular species represent the most extensive available on the evolution of quantitative characters in animal populations (spot-number variation reflects an underlying character whose variation is truly continuous). Indeed, spot patterns provide the most frequently studied examples of quantitative variation in butterflies. Unfortunately the genetic basis of such variation has only rarely been rigorously examined (see Robinson 1971). This must always be an initial aim in ecogenetical investigations. Furthermore, although differences among populations have sometimes been adequately quantified this has seldom led to the development of hypotheses regarding the specific nature of those factors influencing the observed variation. An important purpose of this contribution is to describe recent research on *M. jurtina* which seeks to expand on the questions of how the spot phenotypes are determined, and how their relative frequency within populations may be influenced by natural selection.

Spot Patterns

Nijhout (1978) reviews wing pattern formation in the Lepidoptera and develops a model for wing pattern determination based on the observation that the pattern in each wing cell is developed in a definite relation to a central focus. Experimental evidence for such a focus has been obtained in *Precis coenia* (Nijhout 1980*a*). Eyespots represent the simplest condition in which the pattern is laid down as a system of concentric circles around a focus. Modifications of this are envisaged by Nijhout as resulting from the interpretation process of the distribution of some form of gradient in positional values radiating from the focus. The position of a focus and hence of a spot may shift laterally along the cell midline. The pattern may be expressed to a different degree in each wing cell.

Schwanwitsch (1924, 1948; see also Süffert 1927, (1929) analysed the wing patterns of nine groups of Palaearctic Satyrinae. From each he selected a

*Present address: Department of Biological Sciences, Perry Road, University of Exeter, Exeter EX4 4QG.

The Biology of Butterflies
0-12-713750-5

number of representative forms, and by combining all their pattern elements he constructed a prototype wing pattern. The group which includes the genus *Maniola* showed the presence of a submarginal series of 5 forewing and 6 hindwing spots.

Dowdeswell & McWhirter (1967) studied the geographic variation in spot-number of *M. jurtina* throughout the species' distribution. They also undertook a preliminary analysis of two other species of *Maniola* in west Asia. All three showed a similar pattern of individual variation in spot-number. They suggested that the genes were trans-specific, trans-generic and trans-familial and therefore of great antiquity ('paleogenes'). Frazer & Willcox (1975) extended this study of *Maniola* and also examined species of the closely allied genus *Pyronia*. Six out of seven species showed considerable individual and geographic variation in spot-number, often on both forewing and hindwing. Examples of intraspecific variation in spot pattern have been recorded in most families of butterflies.

In *M. jurtina* the black hindwing spots lie within a band of lighter pigmentation on the ventral wing surface. Electron micrographs of individual spots show that changes in the morphology and the orientation of wing scales occur between the spot and the surrounding wing surface (Brakefield 1979*a*; see also Caspari 1941 for review of gene effects on scale structure).

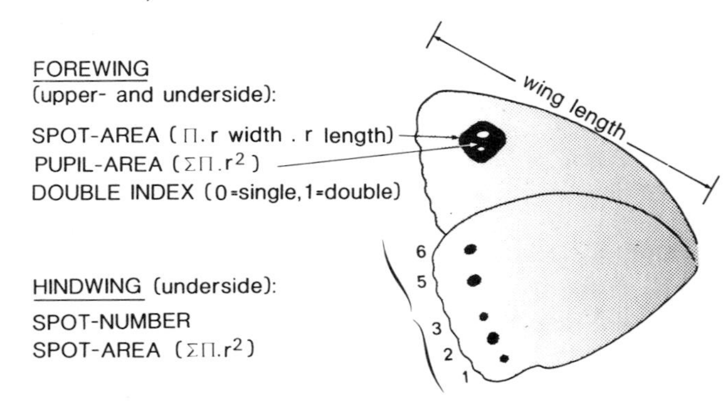

Fig. 16.1. Left wings of *M. jurtina* indicating the characters used in the spot pattern analysis. (Position numbers of hindwing spots are shown.)

The hindwing spots in *M. jurtina* occur at positions 1 to 6 of Schwanwitsch (1948) except that the spot at position 4 is rare or absent (Fig. 16.1; by convention only the left wings are scored). At each position the spots may be present or absent. Thus the range in spot-number is 0–5. Spots are usually encountered in only 13 out of 32 possible combinations (Table 16.1). I shall refer to these as spot types. McWhirter & Creed (1971) adopted a costality index to measure the spot-placing variation in populations. It is the percentage of the costal (positions 5 & 6) and anal (1 & 2) spots which are costally positioned. The neutral or centrally placed spots at position 3 and individuals with all 5 spots are not included in the calculation. In other satyrine species which show a variable spot pattern the spots also only occur in certain combinations, e.g. *Coenonympha tullia* (Turner 1963, Dennis 1972*a*), *Aphantopus hyperantus* (Seppänen 1981) and *Pyronia tithonus* (Brakefield 1979*a*). Furthermore, individual spots usually only develop in the presence of a particular spot or group of spots.

Table 16.1. The spot-combination types of the hindwing spots of *M. jurtina*. (Modified from McWhirter & Creed 1971.)

Name		Spot position					Notation
		costal		median	anal		
		6	5	3	2	1	
Nought		—	—	—	—	—	0
Costal	1	—	●	—	—	—	C1
Anal	1	—	—	—	●	—	A1
Costal	2	●	●	—	—	—	C2
Splay	2	—	●	—	●	—	S2
Anal	2	—	—	—	●	●	A2
Costal	3	●	●	—	●	—	C3
Median	3	—	●	●	●	—	M3
Anal	3	—	●	—	●	●	A3
Costal	4	●	●	●	●	—	C4
Splay	4	●	●	—	●	●	S4
Anal	4	—	●	●	●	●	A4
All	5	●	●	●	●	●	all 5

The name and notation refer to both spot-placing and spot-number. ●, spot present; —, spot absent

The forewing spot pattern of *M. jurtina* is concentrated on the single spot at position 5. This forms an apical black eyespot (with a white pupil) on both the dorsal and ventral surfaces. It may be large enough to cover the area of position 4, when it usually becomes 'double' with an additional pupil at the centre of position 4 (Fig. 16.1; the form nomenclature of *M. jurtina* is described by Thomson 1973).

Morphometrics of Spot Pattern

Samples of *M. jurtina* were obtained from 13 populations (Fig. 16.2; details in Brakefield 1979*a*). Measurements of the characters shown in Fig. 16.2 were all made on left wings, using a binocular microscope fitted with a micrometer. Bipupillation of the forewing spot was not analysed for males (only 2.7% showed a double spot, cf. Frazer 1961).

The matrix of Pearson correlation coefficients was calculated for each sample. The common (or weighted mean) correlation coefficients for these population values are given in Table 16.2. All values ($n=57$) are positive and all except four are significantly different from zero. Furthermore, only three show a significant heterogeneity amongst the corresponding population values. They can

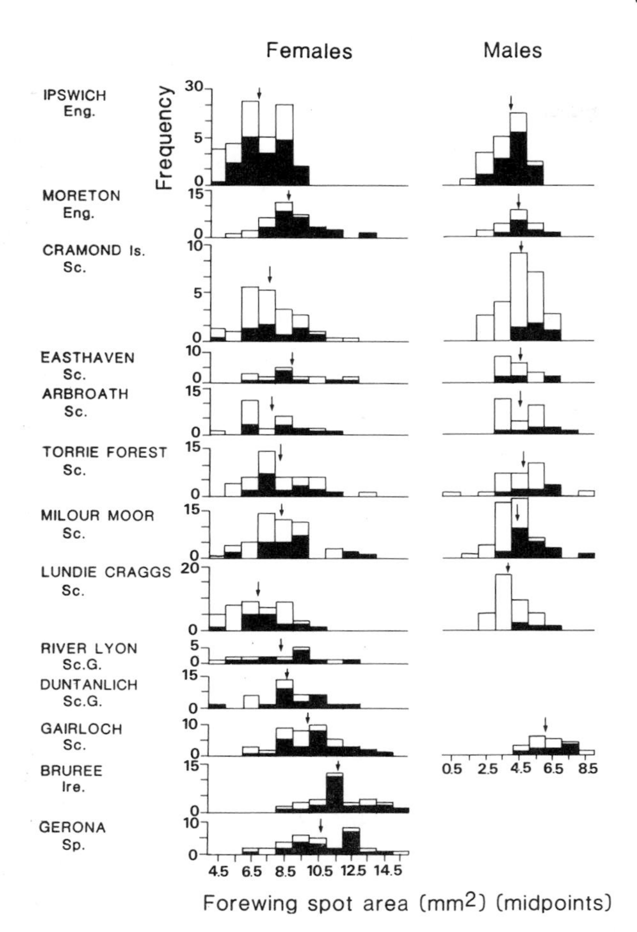

Fig. 16.2. Variation in underside forewing spot-size of *M. jurtina* in the populations indicated. (Unshaded areas of histograms show 0 spot females or 0–2 spot males. Shaded areas show all others. Arrows indicate mean spot areas. Abbreviations: Eng., England; Sc., Scotland; Sc.G., Scotland, Grampian Mountains; Ire., Ireland; Sp., Spain.)

therefore be considered to be good estimates for the species.

Table 16.2 groups the spot pattern characters by wing and additionally for the forewing by under- and upperside. The correlations for characters within these groups are consistently higher than those between the groups. There is a highly significant correlation between both the hindwing spot-number and spot-area and the forewing spot-area. In each population there is a more or less pronounced tendency for butterflies with relatively small forewing spots to be 0 spot females or low spotted (0–2) males (Fig. 16.2). Conversely, those with relatively large forewing spots tend to be spotted females or higher spotted males. Turner (1963) reported a 'good' correlation between the spot-number on the fore- and hindwings in a population of *Coenonympha tullia*. Using his data for the pooled sample the correlation amounts to +0.875 ($n=73$). Similar relationships may be a general feature in variable species of Satyrinae (Frazer & Willcox 1975).

Ehrlich & Mason (1966) and Mason *et al.* (1967, 1968) used morphometric techniques to study a large number of spot pattern characters in *Euphydryas editha*. The matrix of correlations suggested that the spots could be considered as being in several anterior-posterior columns affecting both wings. These results are consistent with what is known of the development of wing pattern in some Lepidoptera (see Sondhi 1963). Mason *et al.* (1968) suggested that changes in selected spot characters from each wing column in *E. editha* (which they called trend characters) could be considered to represent the temporal variation in the wing pattern as a whole. In *M. jurtina*, hindwing spot-number can be conveniently taken as a trend character because its expression is in terms of whole numbers. In contrast, spot dimensions or areas, which exhibit truly continuous variation, are more laborious to score. Furthermore, hindwing spot-area does not, in general, yield higher correlations with the other characters than does spot-number (Table 16.2).

The forewing eyespot of *M. jurtina* shows a wide variation in size (Fig. 16.2). The mean spot-area in females is larger by a factor of about 1.9. Significant heterogeneity of the population-means occurs in both sexes (Fig. 16.2; females—$F = 26.54$, d.f. 12 and 509; males—$F = 9.19$, d.f. 8 and 324, $P < 0.001$ for each value).

High positive correlations are found between the population-means of the sexes for each of the spot-number (cf. McWhirter 1957), hindwing spot-area and forewing spot-area characters (Brakefield 1979*a*). Furthermore, a female population with a high mean spot-number tends also to show a high mean for forewing spot-area ($r = 0.93$, d.f. = 11, $P < 0.001$). In males, for which fewer populations were analysed, the correlation is positive but not significant ($r = 0.17$, d.f. = 7). Each character shows a higher coefficient of variation for the population-means in females than males (Brakefield 1979*a*). This corresponds to the greater variability in spot frequency between female populations (see Ford 1975*a*).

Figure 16.3 illustrates the differences detected in the spot pattern both within and between the sexes. Individuals of two of the more abundant spot types in each sex which differ in spot-number are shown. In females there is a marked emphasis on the forewing spot pattern whilst males show a more even spot distribution between the wings. McWhirter & Creed's (1971) study of spot-placing showed that female populations tend towards a high costality index of about 65–85% whilst in males a value slightly lower than 50% is typical. Thus in females the hindwing spots tend to be more heavily expressed in the costal area of the wing which is closest to the apical forewing spot. My morphometric data show that the size of the individual hindwing spots

Table 16.2. The matrix of the common correlation coefficients and their corresponding χ^2 heterogeneity values for the wing-span and spot pattern characters in *M. jurtina* (correlation values are given above χ^2 values. Comparisons for males are given to the top right, and for females to the bottom left).

	Wing-span	Hindwing		Forewing—und.			Forewing—upp.	
		spot-number	spot area	spot area	pupil area	double index	spot area	pupil area
Wing -span	—	0.077 6.53	0.141 6.86	0.237 6.44	0.191 2.32	— —	0.284 4.97	0.215 3.24
Hindwing								
spot-number	0.099 7.39	— —	0.706 3.19	0.466 2.49	0.359 8.92	— —	0.397 5.01	0.319 8.77
spot area	0.089 12.76	0.828 35.73**	— —	0.514 4.20	0.363 13.46	— —	0.423 1.05	0.273 2.96
FW und.								
spot area	0.365 7.18	0.307 11.12	0.322 13.41	— —	0.655 7.80	— —	0.760 5.09	0.486 1.96
pupil area	0.193 12.81	0.175 16.12	0.139 17.53	0.468 24.76*	— —	— —	0.552 7.34	0.576 5.23
double index	0.044 9.33	0.173 14.43	0.126 13.54	0.288 6.01	0.441 17.07	— —	— —	— —
FW upp.								
spot area	0.238 2.09	0.168 4.36	0.186 6.14	0.698 4.28	0.351 17.50	0.241 2.89	— —	0.600 7.61
pupil area	0.135 4.72	0.180 9.89	0.183 3.11	0.488 3.09	0.687 34.53**	0.327 2.79	0.550 3.66	— —
double index	0.157 2.96	0.099 5.92	0.096 8.24	0.196 5.71	0.293 9.71	0.607 5.10	0.198 5.25	0.360 11.53

Total sample size (n) for the comparisons involving upperside forewing characters are male = 239 and female = 320, for all others they are male = 333 and female = 522. Correlation coefficients, *except* those underlined, are significant ($P < 0.05$) when degrees of freedom of $n-2$ is considered. Degrees of freedom for the χ^2 values of the comparisons involving upperside forewing characters are male = 5 and female = 7, for all others they are male = 8 and female = 12.
For χ^2 values: *, $P < 0.05$; **, $P < 0.001$.

increases with spot-number and with the relative frequency of the individual spots. Spots 2 and 5 are the most frequent and usually the largest.

Heritability of Spot Pattern Characters

It is assumed that variation in a quantitative character results from a combination of genetic and environmental differences (Falconer 1981). The initial aim of a genetic investigation is to divide the total or phenotypic variance (V_P) into its components, the additive genetic variance (V_A) and the environmental variance (V_E). Heritability (h^2) is a parameter indicating the proportion of the total variance which is additive:

$$h^2 = V_A/V_P$$

The heritability can be estimated from the slope of the regression line of offspring on mid-parent value. It is a property not only of a character but also of the population and of the environmental conditions experienced by individuals. Therefore, a value for h^2 refers to a particular population under particular conditions. The principal use of h^2 is to predict response to directional selection. This is possible because h^2 gives the expected similarity between relatives.

The only available estimates of heritability for a quantitative character in butterflies are those obtained by McWhirter (1969) for spot-number in *M. jurtina*. McWhirter raised four broods of the Isles of Scilly race under temperature conditions fluctuating around 15°C. The brood sizes were 8, 9, 19 and 53. This limited material was analysed by linear regression of all individual offspring on mid-parent values (usually mean offspring values are used). The estimates for h^2 were 0.14 in males and 0.63 ± 0.14 in females. McWhirter also analysed the data by an analysis of variance of spot-number between and within broods. This yielded an estimate of h^2 based on double the full-sib contribution of 0.83 in females. McWhirter suggested that the latter estimate was more reliable because of the small broods, the difference in environment under which the parents (some collected *in copula*) and progeny developed,

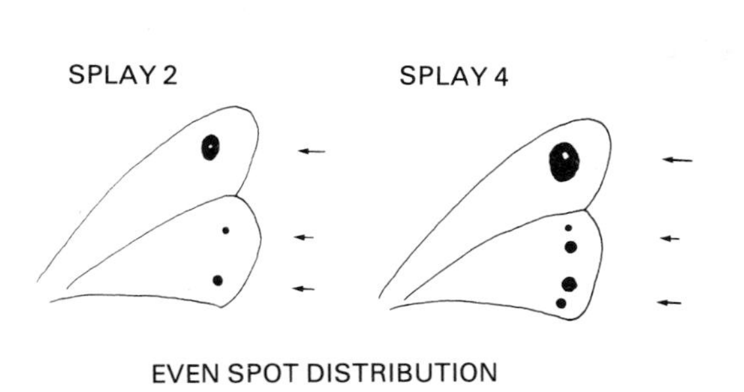

Fig. 16.3. A diagrammatic representation of the variation in the underside spot pattern of *M. jurtina*. (Arrows indicate the relative emphasis on different elements of the spot pattern. Full explanation in text.)

and the different estimates obtained for the sexes.

This suggestion of a difference in heritability between the sexes indicated the necessity of obtaining more complete breeding data (see Tudor & Parkin 1979). Furthermore, McWhirter's data provide no information about the inheritance of other spot pattern characters. In 1979, I obtained a sample ($n=30$) of female *M. jurtina* from a population at Oude Mirdum in Friesland, N. Holland. The population is exceptionally high spotted (spot averages: 1979, female = 2.77, n = 35; male = 3.875, n = 62. 1980, female = 2.67, n = 9; male = 3.60, n = 35). The costality is similar to populations in S. England (costality index: female = 65.22%, male = 53.03%; cf. McWhirter & Creed 1971). In the early-mid flight period of 1980 a Lincoln Index estimate of daily population size for males of 55 ± 14 was obtained. This suggests a total population size in the order of several hundred insects (Brakefield 1982). More than 300 adults were raised from eggs, and crosses then set up between selected adults to provide material for the estimation of heritability. The regression of offspring on parents is not affected by the selection of parents (Falconer 1981).

Parents and offspring were reared in similar conditions. *M. jurtina* can easily be paired in net cages and females will lay readily in small plastic boxes covered with cotton net. Young larvae were raised on seedling grasses (from a lawn grass seed mix) sown in 20cm diameter pots. Mid to late instar larvae were fed on grass (mainly *Poa annua*) transplanted from outside into 45cm square boxes. Broods were kept in an unheated laboratory with dampened temperature fluctuations in comparison to outdoors. During the pupation period for the broods (59 days) the daily maximum and minimum temperatures were 20.5 ± 0.6°C (range 15–29.5°C) and 14.8 ± 0.4°C (9–21.5°C) respectively. In addition to conventional strip lighting (*ca* natural day length) the larvae were raised under 'gro-lux' lamps which emit UV light. Percentage mortality within broods was 42.65 ± 3.24% (measured from first/second instar larvae to adults). Eggs from English mainland stocks have proved difficult to raise because great mortality occurs from the third instar due to a bacterial pathogen (McWhirter 1965). Misyalyunene (1978) carried out experiments with *Pieris brassicae* which showed that irradiation of a bacterial pathogen with sunlight prior to inoculation of larvae reduced subsequent mortality to low levels in comparison to controls. Thus the 'gro-lux' lamps possibly act as an artificial bactericide. However, McWhirter & Scali (1966) found that larvae of *M. jurtina* were strongly selective as to their intestinal bacterial flora and that populations could show strongly distinctive gut floras. The Dutch stock may, like that from the Isles of Scilly (McWhirter 1969), be resistant to those pathogens which cause mortality in English mainland stocks.

Table 16.3 gives the breeding data for spot-number. Parents did not include 0 spot females or 0 and 1 spot females. Offspring included all spot classes. An analysis of variance shows that the male and female offspring are not equal in variance (F = 1.29, d.f. 654 and 684, $P < 0.001$). This means that, strictly speaking, h^2 must be estimated separately for each sex from the regressions on single parent values (Falconer 1981). For comparison, Fig. 16.4 shows the regressions for each sex of mean offspring values on both parent and mid-parent values. The corresponding estimates of h^2 are given in Table 16.4. The brood sizes varied widely (83.75 ± 19.2, range 8–260). Therefore Table 16.4 includes estimates of h^2 obtained by weighting the mean offspring values according to the number of offspring in each family (Kempthorne & Tandon 1953, Reeve 1955). This procedure has little effect on the estimates. Those obtained for each sex using single (same sex) parent values may be higher due to sex dependent expression. Estimates for hindwing spot-characters obtained using mid-parent values are intermediate between those for the same sex, and the opposite sex parent values (Brakefield & Noordwijk in prep.). The estimates of h^2 do not differ significantly

Table 16.3. Comparisons of hindwing spot-number of offspring with parental values in *M. jurtina*.

a) Males

Brood number	Parents m–f	Mid-parent	Spot-number of male offspring											Total	Spot-average
			0	1	2	3	4	5	6	7	8	9	10		
2	4–4	4	—	—	1	—	6	1	6	—	—	—	—	14	4.79
3	6–4	5	—	—	1	—	41	—	11	—	—	—	—	53	4.38
4	6–10	8	—	—	—	—	3	—	4	—	9	—	4	20	7.40
5	6–6	6	—	—	—	—	2	1	12	—	13	—	1	29	6.86
6	6–2	4	—	—	—	—	4	—	—	1	—	—	—	5	4.60
7	6–2	4	—	—	—	—	9	—	5	—	—	—	—	14	4.71
13	10–4	7	—	—	—	—	4	1	17	—	17	—	9	48	7.27
14	10–6	8	—	—	—	—	3	—	41	—	82	—	5	131	7.36
19	6–6	6	—	—	—	—	1	—	4	—	—	—	—	5	5.60
20	8–8	8	—	—	—	1	15	1	16	—	5	—	2	40	5.60
21	10–6	8	—	—	—	—	—	—	3	—	5	—	1	9	7.56
22	4–4	4	2	—	2	—	18	1	3	—	—	—	—	26	3.81
24	8–6	7	—	—	—	—	29	8	37	2	54	—	—	130	6.34
27	6–6	6	—	—	—	—	28	6	40	—	16	—	—	90	5.67
33	8–6	7	—	—	—	—	13	1	25	1	14	—	—	54	6.04
39	4–8	6	—	—	—	—	10	—	5	—	2	—	—	17	5.06

b) Females

Brood number	Parents m–f	Mid-parent	Spot-number of female offspring											Total	Spot-average
			0	1	2	3	4	5	6	7	8	9	10		
2	4–4	4	1	—	3	1	9	—	—	—	—	—	—	14	3.21
3	6–4	5	1	1	39	3	20	—	2	—	—	—	—	66	2.73
4	6–10	8	—	—	1	—	7	—	11	—	9	—	2	30	6.27
5	6–6	6	—	—	5	—	14	—	15	1	11	—	—	46	5.46
6	6–2	4	—	—	2	—	1	—	—	—	—	—	—	3	2.67
7	6–2	4	1	—	5	—	4	1	—	—	—	—	—	11	2.82
13	10–4	7	—	—	4	1	8	1	12	—	7	—	2	35	5.60
14	10–6	8	—	—	3	—	27	1	44	2	46	—	1	124	6.25
19	6–6	6	—	—	—	—	2	—	10	—	—	—	—	12	5.67
20	8–8	8	—	—	—	—	2	—	20	1	9	—	—	32	6.47
21	10–6	8	—	—	—	—	1	—	5	—	7	—	1	14	7.14
22	4–4	4	3	—	7	1	10	—	2	—	—	—	—	23	3.00
24	8–6	7	—	—	9	1	32	1	50	—	12	—	—	105	5.24
27	6–6	6	—	—	1	—	31	2	37	—	20	—	—	91	5.69
33	8–6	7	—	—	4	1	16	—	10	—	2	—	—	33	4.58
39	4–8	6	—	—	1	—	4	1	9	—	1	—	—	16	5.31

N.B. Spot-values in tables of heritability are for both wings and so are double those given in the usual tables of flying populations.

from unity with the exception of that for male offspring on mid-parent values. The heritable nature of the character is evident in Fig. 16.5. There is no evidence for a difference between the sexes.

Jarvis and Høegh-Guldberg have made detailed investigations of the genetic relationships between two European lycaenids, *Aricia agestis* and *A. artaxerxes*. They demonstrated a genetic basis for a number of quantitative characters separating the species and the subspecies/races of each (Jarvis 1966, Høegh-Guldberg 1968, Høegh-Guldberg & Jarvis 1970). The characters studied in adults included underside spot variation (spot-number and size), upperside orange lunulation, wing size and ground colour. An intermediate or heterotic distribution of phenotypes relative to the parental stocks was evident for each trait in the crosses.

Early experiments on several species of Nymphalinae and Lycaenidae demonstrated that the adult phenotype could be altered by subjecting the pupae to extremes of temperature (Süffert 1924, Kühn 1926, Köhler & Feldotto 1935, Krodel 1904). Every pattern element was found to have its particular sensitive period. Lorković (1938, 1943) and Høegh-Guldberg (1971) produced effects on underside spot pattern in two lycaenids by prolonged pupal cooling. Høegh-Guldberg & Hansen (1977), working on *A. artaxerxes*, found that a lower spot-number was sometimes produced by subjecting insects just before or after pupation to one or more

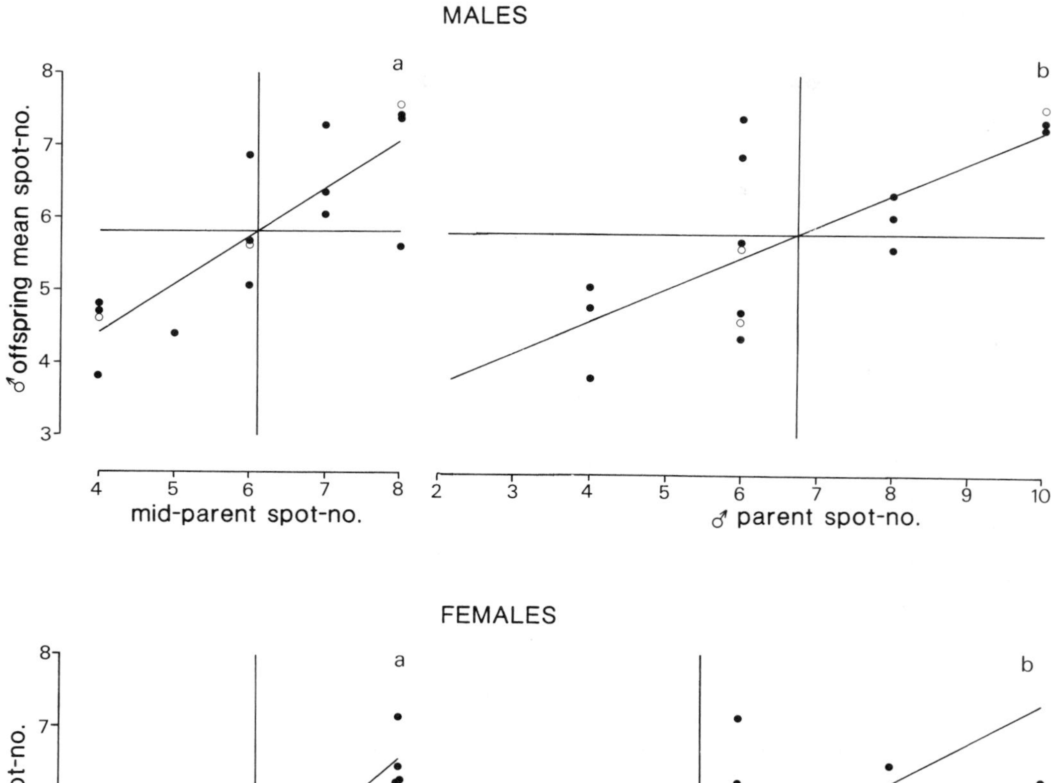

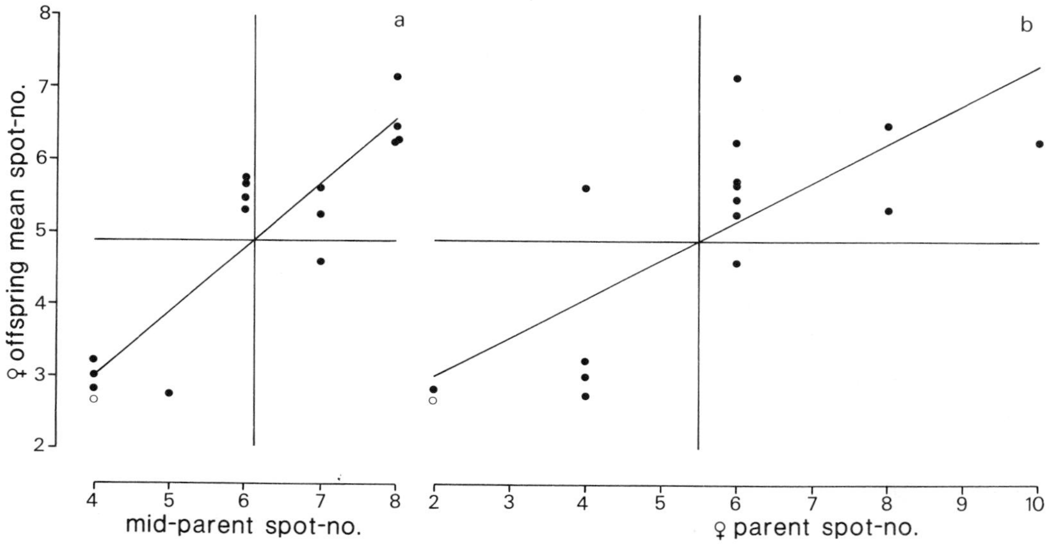

Fig. 16.4. Heritability of hindwing spot-number in *M. jurtina*. (Circles show mean offspring values in broods plotted against (a) mid-parent value and (b) parent value. Open circles indicate a brood size of <10. Note, in (b) slope of regression = $0.5h^2$.)

Table 16.4. Heritability (mean ± S.E.) of spot-number in *M. jurtina* (see text for details of method of analysis).

	Single (same sex) parent	Mid-parent
Male offspring:		
unweighted regression	0.88 ± 0.21	0.66 ± 0.11[1]
weighted regression	0.87 ± 0.22	—
Female offspring:		
unweighted regression	1.08 ± 0.25	0.89 ± 0.11[1]
weighted regression	1.05 ± 0.20	—

[1]These estimates are not significantly different (t = 1.43, d.f. = 28).

periods of cooling at +2–5°C for 9–12 hours. The experimental pupae produced some rare forms. The authors suggest that their results may be a possible explanation for such aberrations in nature. Similar experiments to those of Høegh-Guldberg & Hansen on material from a single brood of *M. jurtina* did not detect an environmental effect on spot phenotype. Relative differences in spot variation between populations of *M. jurtina* are maintained when samples of wild-collected larvae are reared under similar conditions in the laboratory (see below). This observation and the finding of a high heritability for spot-number indicates that differences between, or changes within populations in the frequency distribution of spot-number (= spot frequency) can more confidently be interpreted as due to the action

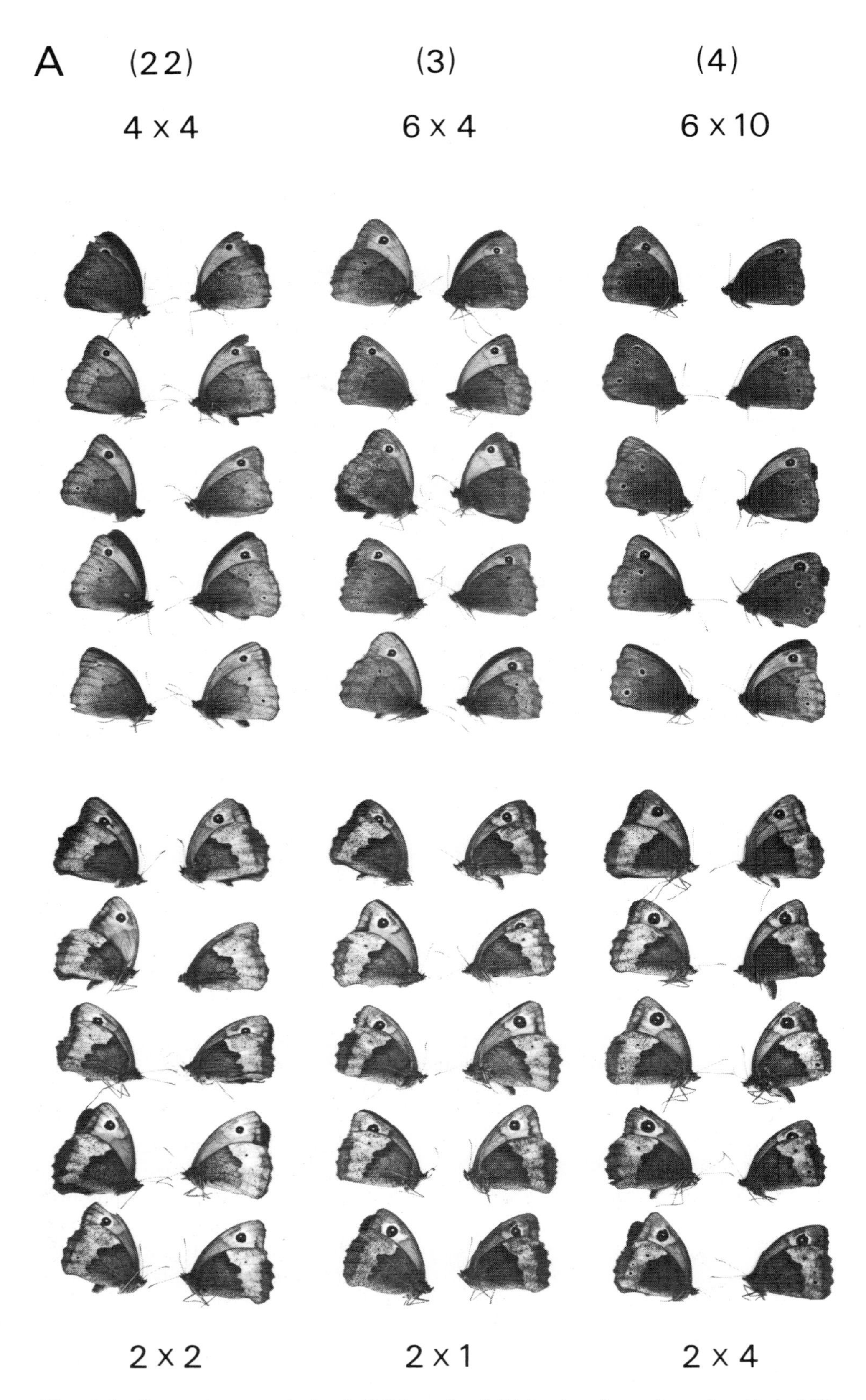

Fig. 16.5. Spot pattern variation in F1 broods of *M. jurtina*. In each section (A and B) a random sample of 10 males (above) and females (below) are shown in columns for each of three broods (reference numbers given in brackets, see Table 16.3). The hindwing spot-number and the forewing spot size index for the male × female parents are indicated above and below each brood respectively.

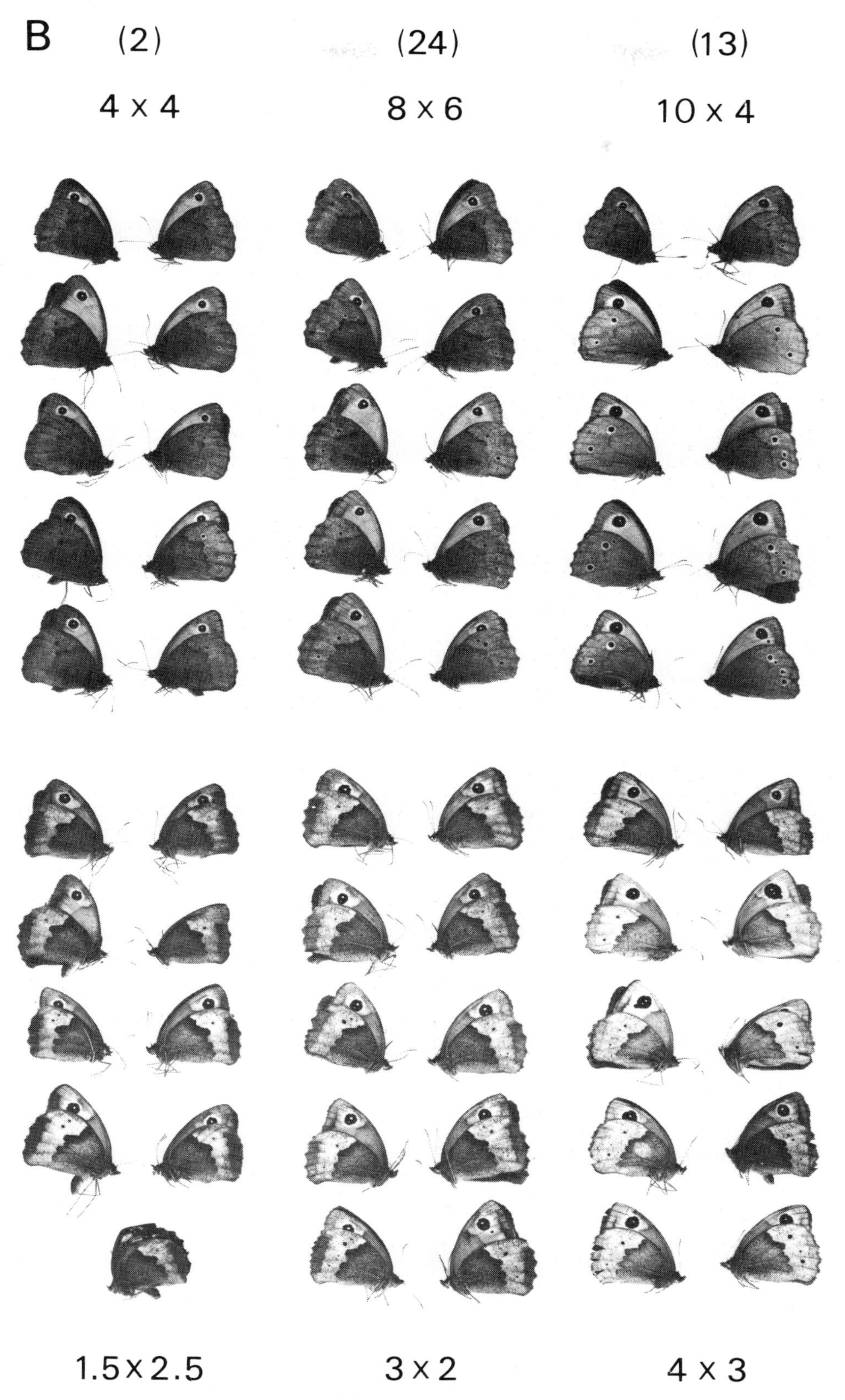

Within each section parents showed a trend of increasing spot expression from left to right. The heritable nature of the spot characters (see text) is then evident in the corresponding trend amongst the illustrated offspring. Brood 13 includes some butterflies in which the usually rare hindwing spot at position 4 is expressed.

of natural selection on genetic variation (see following section).

Examples from the breeding data for *M. jurtina* of the influence of the spot-placing of the parents of each sex on that of their offspring are shown in Figs 16.6 and 16.7. This relationship was investigated in greater detail by calculating the average spot position (Fig. 16.1) for parents and offspring of each sex. The regressions are shown in Fig. 16.8 and the estimates for h^2 are given in Table 16.5. They indicate a rather high heritability of about 0.6.

A preliminary analysis of the control of forewing spot size was made. Butterflies of each sex were scored by comparing with a set of size standards on a scale from 0 to 5. The regressions are shown in Fig. 16.9 and samples of butterflies from six broods in Fig. 16.5. The estimates of heritability are lower than those obtained for hindwing spot-number (Table 16.6). However, particularly when the higher expectation of measurement error is considered, they indicate that for this material there is a significant genetic influence on forewing spot size. This is also evident in the samples of F1 progeny shown in Fig. 16.5. The broods also reveal evidence for genetic

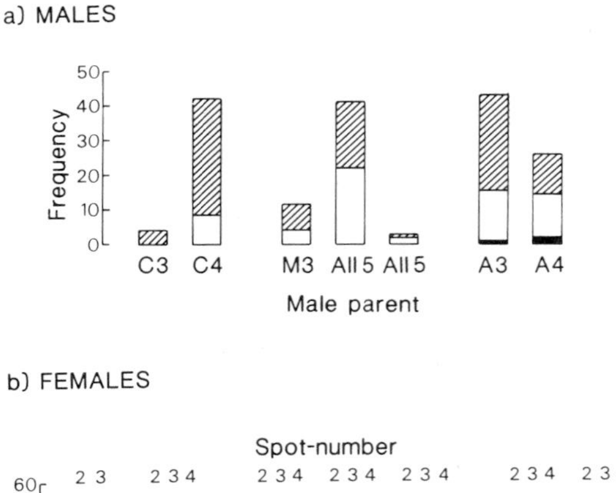

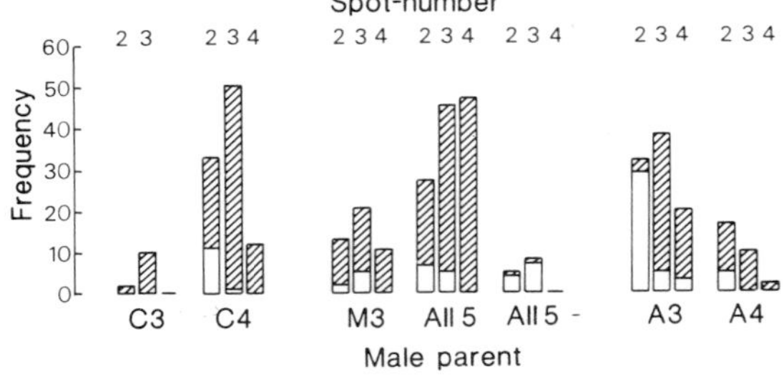

Fig. 16.6. The spot-placing in *M. jurtina* reared from pairings between costal 3 females and males of differing spot-placing bias. In (a) 3 spot male offspring are shown and in (b) 2–4 spot female offspring. Spot-placing bias: ▨, costal; □, neutral; ■, anal.

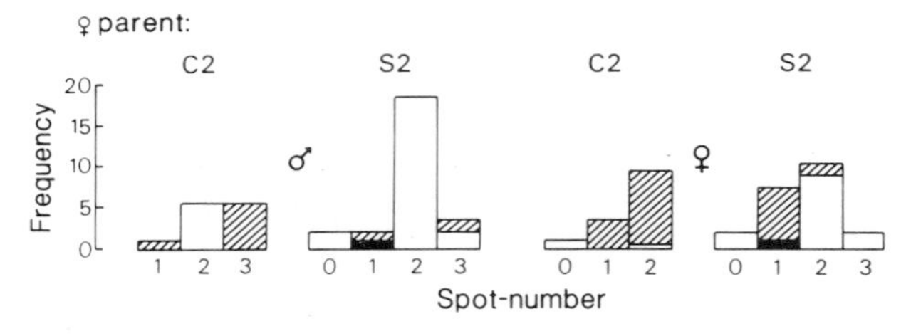

Fig. 16.7. The spot-placing of male (on left) and female *M. jurtina* reared from pairings between splay 2 males and a costal 2 or splay 2 female. Spot-placing bias: ▨, costal; □, neutral; ■, anal.

Table 16.5. Heritability (mean ± S.E.) of average spot position in *M. jurtina*.

	Single (same sex) parent	Mid-parent
Male offspring	0.57 ± 0.19	0.46 ± 0.12
Female offspring	0.80 ± 0.25	0.61 ± 0.20

An average spot position is calculated from the combined data for all spots. A consequence of this is that a weighted regression analysis would not be valid. Analyses using means for average spot position in individuals do not yield significantly different estimates.

Table 16.6. Heritability (mean ± S.E.) of underside forewing spot size in *M. jurtina*.

	Single (same sex) parent	Mid-parent
Male offspring:		
unweighted regression	0.40 ± 0.36	0.80 ± 0.21
weighted regression	0.42 ± 0.28	—
Female offspring:		
unweighted regression	0.66 ± 0.27	0.59 ± 0.20
weighted regression	0.56 ± 0.23	—

Table 16.7. The percentage frequency (mean ± S.E.) of bipupilled or 'double' underside forewing spots in F1 broods (*n* given) of *M. jurtina*.

	Parents of broods		
	both double (n=2)[1]	f. only double (n=7)	both single (n=7)
Male offspring	68.63 ± 3.25	16.79 ± 4.33	5.49 ± 2.24
Female offspring	98.57 ± 1.43	83.98 ± 7.57	68.59 ± 8.88

[1]Brood nos 13 (n=83) and 24 (n=235)

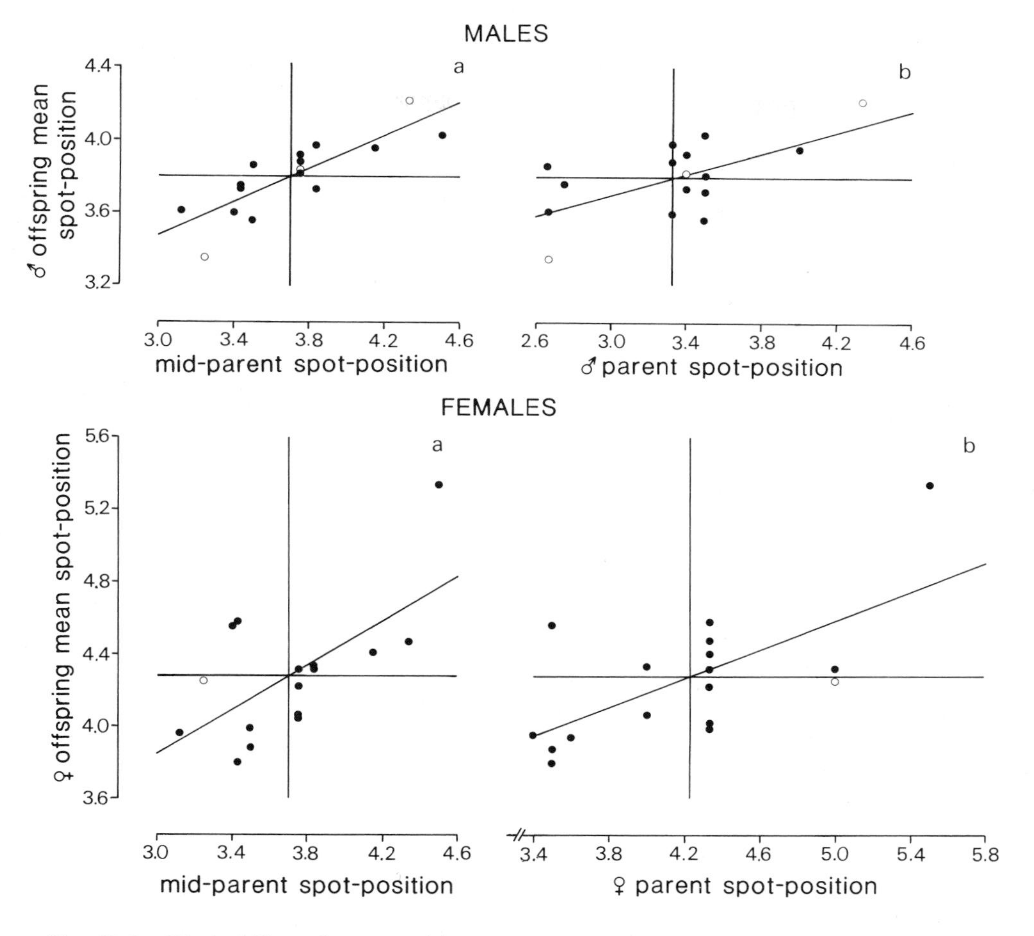

Fig. 16.8. Heritability of average hindwing spot-position in *M. jurtina*. (Circles show mean offspring values in broods plotted against (a) mid-parent values and (b) parent value. Open circles indicate a brood size of < 10.)

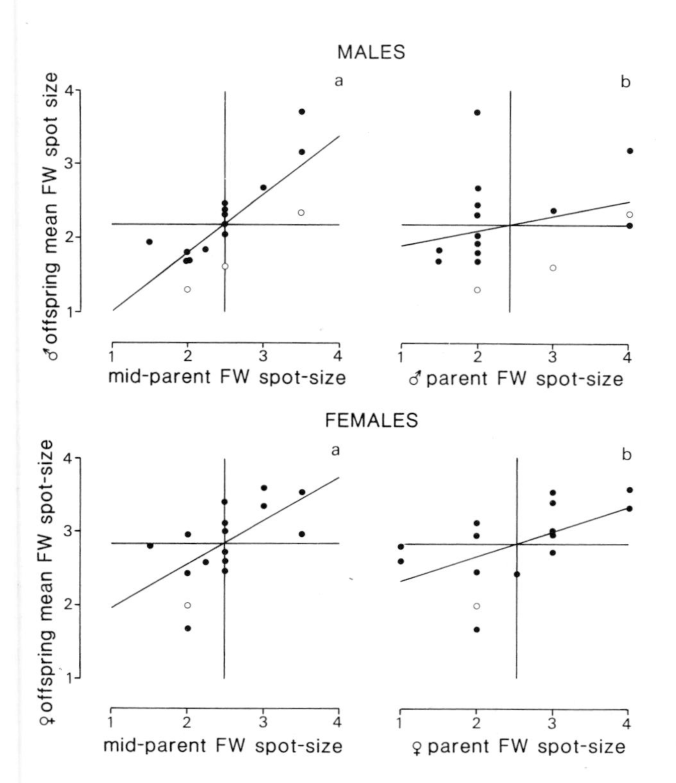

control of the bipupillation of forewing spot (Table 16.7). The complete breeding data for each character will be published elsewhere (Brakefield & Noordwijk in prep.).

Field Studies on Variation in Quantitative Characters

Most of the ecogenetical studies on *M. jurtina* have been concerned with quantifying spot-number variation of populations within a geographical area, usually over a number of years (generations). This approach embraces two possible means of demonstrating that selection influences specific morph or genotype frequencies: to search for consistent correlations between such frequencies and particular environmental factors, or to follow changes in the frequencies (in large populations) over many generations. Many aspects of the comparative studies

Fig. 16.9. Heritability of underside forewing spot size in *M. jurtina*. (Circles show mean offspring values in broods plotted against (a) mid-parent value and (b) parent value. Open circles indicate a brood size of < 10.)

of variation in *M. jurtina* have been described in detail by Ford (1975*a*). I shall therefore only briefly outline the results whilst discussing interpretations of them in more detail. The final subsection describes some studies of variation in other species.

Island Populations

The populations of *M. jurtina* within the Isles of Scilly (an archipelago situated off the southwest coast of England) show a male spot frequency that is unimodal at 2 spots, with little variation in spot average. Both sexes are characterized by high costality indices (McWhirter & Creed 1971). Most female populations on the three large islands (>275ha) show a 'flat-topped' spot frequency with similar numbers of 0, 1 and 2 spot classes. In contrast, those inhabiting the small islands (<16ha) show a variety of spot frequencies which tend to fall into three groups: unimodal at 0 spots, bimodal at 0 and 2 spots and unimodal at 2 spots (Dowdeswell *et al.* 1960, Creed *et al.* 1964). In an early study McWhirter (1957) suggested that these groups reflected three different types of habitat (see below).

Some authors have attributed these results to the workings of genetic drift and the founder principle (Wright 1948). Thus Waddington (1957) considered that the differences between small islands result from periods of 'intermittent drift' which correspond to bottle-necks in population size. Again, Dobzhansky & Pavlovsky (1957) suggested that the small island populations were derived from founder groups with differing gene frequencies from which relatively stable but different gene pools were developed. Ford and his co-workers disagree (and see MacArthur & Wilson 1967). Ford (1975*a*) discusses the hypothesis that those populations which occupy the large (and diverse) islands result from natural selection producing a gene complex simultaneously adapted to a wide range of environments. In contrast, populations occupying the small islands (and isolated, small areas on the large islands) tend to be dissimilar to each other, being closely adjusted to particular and different environments. In this context Ford argues that the following observations are of particular significance. On several occasions populations have been found to pass through an extreme bottle-neck in size with no subsequent change in spot frequency, although this may be different in the period of low numbers (Creed *et al.* 1964). On the other hand, some changes in spot frequency have not been associated with periods of low numbers but rather with a change in habitat, for example when grazing by cattle ceased on Tean (Dowdeswell & Ford 1955; Dowdeswell *et al.* 1957) and when exceptional drought occurred on St Martin's and Tresco (Dowdeswell *et al.* 1960). Similar observations of a coincidence of an unusually warm and dry summer with a change in spot frequency have been made by Bengtson (1978) working on several populations on two small islands in southern Sweden.

The Boundary Phenomenon

A more or less abrupt transition in female spot frequency in populations occurs between the west of Dorset and the east of Cornwall. This is the so-called 'boundary phenomenon'. Changes are also found in spot-placing variation (McWhirter & Creed 1971) and in allelic frequencies at two esterase loci (Handford 1973*a*) across the boundary region. When discovered in 1956 there was a particularly sharp discontinuity in female spot frequency from being unimodal at 0 to bimodal at 0 and 2 spots (Creed *et al.* 1959). Despite diverse ecological conditions, populations to either side of the boundary itself show a high degree of homogeneity of spot frequency. In subsequent years it was found that the boundary was sometimes less abrupt and that its geographical position could move considerable distances (up to 60km) east or west between generations (Creed *et al.* 1970). It is these shifts in position which is the most difficult feature of the boundary phenomenon to account for.

Creed *et al.* (1962) interpreted these observations as demonstrating the action of very powerful selection which differed on each side of the boundary. The nature of these forces remains unknown. Others, such as Handford (1973*a*), have elaborated on this interpretation: for example invoking a switch over between two co-adapted genetic systems at a critical point in an environmental gradient. 'Genes are said to be co-adapted if high fitness depends upon specific interactions between them' (Wallace 1968: 305; for a discussion in relation to clines see Endler 1977). McWhirter & Creed (1971) have emphasized the interactions of the spot-number and spot-placing systems within the populations of the boundary region. Ford (1975*a*) discusses the phenomenon in relation to sympatric evolution in which distinct races or local forms can arise without isolation, past or present. Clarke (1970) put forward an alternative hypothesis that the boundary region represents a zone of hybridization between two parapatric races and that individuals within it are particularly prone to developmental instability. Dennis (1977: 250–51) considers that the alternate gene complexes are more likely to have originated in an earlier geographical isolation of the two main population groups than from the processes of sympatric evolution. Dennis discusses some geological and historical features of southwest England. He suggests that for some 4500yr prior to the Sub Boreal period differences in vegetation cover and climate could have led to such a disjunctive distribution of *M. jurtina*. Handford (1973*a*) considers that Clarke's hypothesis cannot alone account for the orderly or abrupt changes in spot frequency which occur both in space and time. Ford (1975*a*) argues that developmental instability is not evident within the boundary region since a mosaic of groups of populations with differing spot frequencies is not found. In this context a detailed and local study of spot frequency change around the position of the discontinuity would be worthwhile.

These hypotheses could be examined or tested more objectively by an experiment designed to detect the existence of differing co-adapted gene complexes across the boundary. The experiment would involve making the appropriate crosses among stocks from populations across the boundary region together with a control cross with a population from an isolated area (e.g. Isles of Scilly), and a series of such crosses within a transect of similar dimensions to the boundary region in an area where spot frequencies are more or less stable and uniform (e.g. S. England). Similar methods could be utilized to those developed by Oliver (1972*a*,*b*, 1979) to study genetic differentiation in species of butterflies, including *Pararge megera*. A possible means of investigating developmental instability within populations is the analysis of the departure of a set of metrical characters from perfect bilateral symmetry (Soulé 1967). Soulé & Baker (1968) studied asymmetry in such characters (including spot measurements) in six populations of *Coenonympha tullia*. The frequency of asymmetry of spot-number in the broods of *M. jurtina* (Table 16.3) was 3.5%. Asymmetry of spot-size is more frequent. Mason *et al.* (1967) found that up to 25% of the size variation in pattern elements of *Euphydryas editha* was due to asymmetry.

Sheppard (1969) emphasized the interest of boundaries of the type described in *M. jurtina* to population and evolutionary geneticists because they incorporate some of the elements of disruptive selection and some of those of subspeciation. Laboratory experiments on the quantitative character of sternopleural chaeta number in *Drosophila melanogaster* have shown how disruptive selection, even in the presence of high gene-flow between selected lines, can lead to divergence and effective isolation (e.g. Thoday & Boam 1959, Millicent & Thoday 1961). The response of chaeta number to both disruptive and directional selection in such experiments is slower than occurs when populations within the boundary region switch-over between the characteristic spot frequencies. The heritability of chaeta number is, however, lower (being about 0.5) than that for spot-number. Clarke & Sheppard have investigated disruptive selection on a quantitative character in *Papilio dardanus* (see also Ch.14). The inheritance of tails in this butterfly is due to a single pair of alleles, autosomal but sex-controlled. The males are non-mimetic and tailed. In most of Africa many of the female forms are mimetic. The females, like the models for these forms, are tailless. In the Ethiopian race the majority of females are tailed and non-mimetic. A minority are mimetic but differ from similar forms elsewhere in having tails. The genetic and morphometrical analyses of Clarke & Sheppard (1960*a*,*b*, 1962*b*) have shown that in Ethiopia there is disruptive selection acting on the females and favouring the reduction in tail length in the mimetic forms but discouraging it in the non-mimetic females. Their results indicate the presence of modifier loci in this race which enhance the difference in mean tail length.

Two local discontinuities in spot frequency in *M. jurtina* have been found in the Isles of Scilly, one on Great Ganilly and the other on White Island (Dowdeswell *et al.* 1960, Creed *et al.* 1964, Ford 1975*a*). In the latter case the difference between the two areas of the island was only detected after these areas were isolated, at least partially, by storm damage. The areas differ in vegetation and exposure and the populations they support show differences in esterase variation (Handford 1973*b*). A further discontinuity of this type occurs along a 5km transect on the coast near St. Andrews in Scotland (Brakefield 1979*a*). The climate becomes more maritime along the transect but there is no obvious habitat change.

High Spotting and Aestivation Behaviour

Scali and co-workers have sampled many populations of *M. jurtina* in Tuscany, central Italy. High spot averages prevailed. In mainland populations females tend towards the 'flat-topped' spot frequency whilst males are unimodal at 2 spots with very few lower spotted individuals (Scali 1971*a*). In populations on two large offshore islands even higher spot averages were found (Scali 1972). The male populations did not exhibit their usual uniformity. Significant changes between generations occurred in two male populations with an estimated 72.7% elimination of high spotted phenotypes in one. This latter population showed some evidence of a reverse change in the following year. A climatic factor may have caused the initial changes since they coincided with those in several mainland populations. Scali pointed out their parallel in the widespread spotting shifts in populations in S. England and the Isles of Scilly which were associated with unusual climatic conditions in 1955–57 (Creed *et al.* 1959, 1962, Dowdeswell *et al.* 1960). Two earlier examples of intergeneration changes in male populations in mainland Tuscany coincided with habitat changes. Nearby colonies remained unaffected (Scali 1971*a*).

The reproductive biology of *M. jurtina* in Tuscany has been studied in detail (Scali 1971*b*, Masetti & Scali 1972, Scali & Masetti 1975). At lower altitudes adult emergence occurs over a short period of about three weeks from late May. Copulation takes place, after which the males die. Females then always undergo a long aestivation during the hottest season. In late August and September the eggs which have matured during aestivation are fertilized from stored sperm and laid. A difference in spot frequency has consistently been found in populations between females flying before and after aestivation. The spotting shifts always tend towards lower values with usually a change from a 'flat-topped' or unimodal at 2 spots distribution to one unimodal at 0 spots. The calculated selection against high spotted phenotypes (2–5 spots) amounts in many instances to 65–70%. In mountain populations aestivation does not occur and the adult flight period extends from late June until early September. A mixed strategy is found in a population at intermediate altitude with some butterflies emerging early, followed by aestivation of females, and others late with no subsequent aestivation. An investigation of the control of aestivation behaviour would be most interesting.

Spot Stabilizations

Dowdeswell & McWhirter (1967) examined museum samples of *M. jurtina* from throughout its distribution. They described a system of stabilization areas characterized by populations with particular spot frequencies. The largest of these, called the General European, extended from Britain (except the southwest) through much of continental Europe. Here the spot frequency is unimodal at 2 spots in males and at 0 spots in females. Dowdeswell & McWhirter considered that changes between stabilizations were sharp and resulted from prolonged and violent alterations in selection (cf. the boundary region).

The type of data analysed by Dowdeswell & McWhirter and the distribution of their samples suggests that distinctions between stabilizations are somewhat imprecise. This is supported by additional samples. In several stabilization areas enclaves of populations with distinctive spot frequencies are found. In Scotland, females in the Grampian Mountains tend towards a bimodal spot frequency whilst the spot variation of neighbouring populations is typical of the General European area (Table 16.8; Forman *et al.* 1959). Samples obtained from Ireland up to 1967 consistently showed very low spot averages (Dowdeswell & McWhirter 1967, Frazer & Willcox 1975) but much higher frequencies of spotted females were later found in two populations in a different region (Table 16.8). Unusually high spotted populations have sometimes been found in central and southern England and in coastal regions of the continental General European area (Frazer & Willcox 1975, Brakefield unpublished).

Populations of *M. jurtina* tend to show one of a limited number of types of spot frequencies. Thus whilst females are often unimodal at 0 or 2, or bimodal at 0 and 2 and may change from one to the other, they are very rarely unimodal at 1 spot. This feature has been attributed to the occurrence of co-adapted gene complexes (McWhirter & Creed 1971, Handford 1973*a*). However, the probability of a mode occurring at 1 spot may be less than at 0 or 2 spots. The set of spot frequencies which will result from selective processes will depend on the fitness relationships between spot genotypes and on the developmental relationship between genetic variation and spot phenotype. At the simplest level, the spot-number classes include differing numbers of spot types (Table 16.1). Within each type of spot frequency there is some variability in the height of the mode(s) in populations.

Sometimes a general change in spot variation has been detected in populations within part of a stabilization area (see above). The recent samples from Gairloch in northwest Scotland and from northwest England (Table 16.8) suggest that a change to high spotting has occurred in these areas in the last 25 years (early samples in Creed *et al.* 1959, 1962, Dowdeswell & McWhirter 1967; comparison—Gairloch $\chi_3^2 = 10.89$; West Kirby area $\chi_3^2 = 9.89$, $P < 0.05$ for each value). When considered together, the samples of *M. jurtina* indicate that the distribution map for stabilization areas given by Dowdeswell & McWhirter is an oversimplification and that the different types of spot frequencies may not represent such a discontinuous nature of variation as has been supposed.

Dowdeswell & McWhirter showed that a number of different stabilizations occur around the periphery of the species' distribution. They considered that populations in such areas are adjusted to specialized environments. I have analysed the change in spot-number variation between generations in samples from three areas collected over five-year periods (Brakefield 1979*b*). A greater constancy of female spot average between generations was found both within the ecologically more marginal populations of central-eastern Scotland and the geographically peripheral populations of the Isles of Scilly than within those more centrally located in southern England. The results were consistent with the hypothesis that adaptive specialization and selection favouring a relative homozygosity predominate in marginal populations of *M. jurtina*.

Variation in Other Species

Discontinuities involving characters supposed to have a polygenic basis have been described in some other butterflies. Owen & Chanter (1969) found differences between adjacent populations of the African nymphalid *Acraea encedon* in the underside hindwing

Table 16.8. Spot variation in some recent samples of female *M. jurtina* from the British Isles (full data given in Brakefield 1979*a*).

Locality	Year	Spot-number 0	1	2	3	4	5	Total	Spot average	Costality Index (%)
CE Scotland:										
Grampian Mts	1973–77	280	144	190	47	18	8	687	1.13	38.6
All others	1973–77	1597	600	309	122	24	2	2654	0.64	48.2
CW Ireland:										
Co. Limerick	1976	77	58	57	18	2	1	213	1.12	58.4
NW Scotland:										
Gairloch	1976–77	18	12	19	7	4	0	60	1.45	53.0
NW England:										
West Kirby	1975	43	24	24	10	1	0	102	1.04	62.3
Hightown	1976	95	78	131	78	17	1	400	1.62	69.6
Whitchurch	1976	28	20	24	12	1	1	86	1.31	66.7

'Grampian Mts' and 'All others' comprise 3 and 12 populations respectively. Each group of populations gave homogenous samples in each year.

pattern of 20 spots (but see also Ch.18). The smaller spots, which tended to be closer to the thorax, were the most frequently absent. A low rate of adult movement was detected between some populations. Lucas (1969) investigated clinal variation in wing pattern characters in populations of *Tisiphone abeona* along the southeast coast of Australia. Preliminary evidence was presented for a simple genetic control for some characters. Clines in several characters showed local steepenings which, with one exception, coincided in geographical position. Endler (1977: 76) discusses the possibility that this may be an effect of a partial barrier to gene-flow. The variation in underside spot-number and spot-size found in British populations of *Coenonympha tullia* represents a complex pattern of clines and races associated with isolated distributions (Dennis 1977, Porter 1980).

Ehrlich & Mason (1966) and Mason *et al.* (1968) in a morphometric analysis of spot pattern characters in two Jasper Ridge populations of *Euphydryas editha* found that changes in trend characters occurred in a uniform and concurrent manner over a period of eight generations (years). They suggested that such changes were caused by strong fluctuating selection, but the possibility that they were due to a complex penetrance system could not be excluded. The characters studied in *E. editha* affected the lightness of wing coloration (see also Le Gare & Hovanitz 1951, on *E. chalcedona*). Descimon & Renon (1975) showed that in France, *Melanargia galathea* becomes blacker in warm-dry areas. The interaction between temperature, insolation and pigment is complex (Papageorgis 1975, Turner 1977*a*, Watt 1968, 1969). The common occurrence of seasonal forms which differ in extent of melanin deposition is discussed by Shapiro (Ch.27).

Geographical and seasonal differences in wing size are found in many butterfly species (e.g. Baker 1972*b*, Ishii & Hidaka 1979). The influence of polygenic systems on wingspan has been demonstrated in crosses between subspecies of *Aricia* spp. (Høegh-Guldberg & Jarvis 1970) and of *Anthocharis cardamines* (Majerus 1979). The existence of genetic variability within populations of a species has not been rigorously examined. Population studies in species of moths have shown examples of both environmental and genetical influences on pupal or adult size (e.g. Danthanarayana 1976, Lorimer 1979, Richards & Myers 1980, Myers & Post 1981). Dempster *et al.* (1976) investigated size variation of *Papilio machaon* in England. A comparison of museum samples showed that insects from the isolated Wicken Fen had a longer wing relative to body size and a narrower thorax relative to body length than those from more continuous habitat in Norfolk. Examination of temporal changes showed that the butterflies from these areas were similar until 1880 when those from Wicken declined in size (wing shape remained similar). After 1920 the difference diminished, largely because Norfolk butterflies became smaller. Experiments indicated that a functional relationship might exist between body shape and flight speed. The changes at Wicken could be interpreted as due to selection against mobility associated with a reduced area of habitat. The later changes in Norfolk were less easy to explain because a change in habitat was gradual. An interesting parallel distribution of dwarf forms of *Hipparchia semele* and *Plebejus argus* occurs on a small isolated peninsula of carboniferous limestone in north Wales (Thompson 1944, Ford 1975*b*, Dennis 1972*b*,*c*, 1977). The forms also show differences in wing pattern characters and phenology to adjacent (for *H. semele*) or nearby populations. Dennis (1977:

253–56) suggests that a calcareous landscape and a warmer and drier microenvironment have been responsible for the evolution of these forms.

Geographical variation in the number of annual generations occurs in many species of butterflies in temperate zones (refs. to examples in Britain in Dennis 1977). Ecological aspects of differences between species are discussed by Gilbert & Singer (1975; see also Chs 2, 3) and Watt *et al.* (1979). Lees (1962*a*) showed that differences between stocks of *Coenonympha pamphilus* from two English populations with predominantly univoltine and bivoltine strategies respectively were maintained in identical laboratory conditions. The population origins differed in latitude and altitude. Lees (1965) was able to selectively increase the proportion of non-diapausing larvae in the normally univoltine stock from 23.7 to 46.3% over two years. The response to selection was not as rapid as that reported for the Gypsy moth (Hoy 1977). Some additional evidence has been obtained for genetic differentiation within butterfly species in response to the environmental factors determining diapause (e.g. Petersen 1949, Danilevski 1965, Jarvis 1966).

Selection on Spot Variation in *M. jurtina* during Development

The surveys of spot variation in *M. jurtina* in combination with the demonstration of high heritability provide strong evidence for selection, even though the specific factors involved remain unidentified. Examples of rapid changes in this variation have indicated that sometimes such selection can be very powerful. In this section I examine evidence that selection operates during pre-adult development.

Rearing Experiments

Dowdeswell (1961, 1962) reared samples of late instar larvae from two Hampshire populations collected by sweeping. The population at Middleton East was sampled in each of four years. Dowdeswell showed that sweeping probably samples larvae at random in relation to spot-number. Larvae collected before and after late May were treated as early and late samples respectively. Females emerging from the early samples consistently yielded a higher spot average than those from the late samples. Larval mortality was higher in the late samples mainly due to much heavier parasitism by *Apanteles tetricus* (about 25% parasitized). Dowdeswell noted that parasitism probably takes place during the first instar. He suggested that *A. tetricus* was the principal agency responsible for a selective elimination amounting to about 70% of 2–5 spot females in the late samples. However, if parasitism was not restricted to spotted genotypes (which seems unlikely) it must have caused a considerably higher mortality in the field than that detected in the laboratory to fully account for the differences between early and late samples. No causal relationship has been established between parasitism and spot variation. An experiment in which mid-instar larvae from a Scillonian population were introduced into experimental grass enclosures near Liverpool provided no evidence for selective elimination by birds or small mammals (Brakefield 1979*a*).

In three of the four years at Middleton East the combined reared females and the samples of flying adults showed significantly different spot frequencies. Table 16.9 shows that the relative fitness of the female spot classes declines with increasing spot-number. Although differences were not detected in individual years for males, the combined data do suggest that a similar relationship between fitness and spot-number occurs (Table 16.9). When the other Hampshire population is considered these conclusions are only supported for females (Dowdeswell 1962).

McWhirter (1965) reported that in the Isles of Scilly a powerful selective elimination of high spotted males occurs in the late-larval and pupal stages. McWhirter's records of the braconid parasites of *M. jurtina* in the Isles of Scilly do not include *A. tetricus*. I obtained comparable data to those of Dowdeswell for three local sites on St Martin's Island. Table 16.9 shows that a similar relationship between fitness and spot-number is found in this population as at Middleton East. The estimates for intensity of selection are similar for females in these populations but the intensity is much higher for males on St Martin's Island (Table 16.9). If selection takes a similar form in the populations it may have largely operated on males at Middleton East by the time of larval sampling. This is more likely when the faster development of males is considered.

Larvae have also been obtained from Cramond Island in eastern Scotland and Buckley in N. Wales (Brakefield 1979*a*; total sample size = 340). The larvae from each population were roughly comparable in development to early and late samples from Middleton East respectively. Only 1 of 116 deaths was due to a parasite. There was no evidence for selection in either population since the spot variation in the reared and flying samples was similar. The wide difference in spotting between the populations was maintained in the reared material.

Endocyclic Selection.

The results of the rearing experiments have been interpreted as due to a powerful selective elimination of high spotted genotypes in late

Table 16.9. The intensity of selection and the relative fitness of spot-number classes estimated from mid-late instar larvae to adults in two populations of *M. jurtina*.

Spot-number class	Females: Middleton East	Females: St Martin's Island	Males: Middleton East	Males: St Martin's Island
0	1.000	0.708	[	[
1	0.513	1.000	1.000	[
2	0.469	0.668	0.641	1.000
3	0.364	0.513	0.725	0.186
4	0.064	[0.079	0.590	0.189
5	—	[	0.212	0.032
Intensity of selection	0.294	0.360	0.252	0.697
Sample sizes:				
before selection:	339	229	240	216
after selection:[1]	929	69	1148	32

Data: Middleton East (1957-60) from Dowdeswell (1961): St Martin's Island (1976-77) from Brakefield (1979*a*).
Method of calculation given by O'Donald (1971). Maximum fitness is defined to be $w = 1$. Where the number before or after selection = 0, braces indicate some combining of spot classes. A dash is used when both numbers = 0. Intensity of selection, which equals − mean w, is calculated directly using estimates of relative fitness. For the combined 1959-60 female data from Middleton East, O'Donald showed that the estimates obtained by the direct method and using an unbiased quadratic model were similar.
[1]The data for St Martin's Island were kindly supplied by Professor K. G. McWhirter.

pre-adult development. To account for this it is necessary to predict an earlier counterbalancing selection favouring these genotypes (McWhirter 1967, Ford 1975*a*). A small sample of adults reared with heavy mortality from larvae collected on St Martin's Island in November before the winter period of slow growth suggested high spot averages (males = 3.69, n = 13; females = 2.47, n = 17; Brakefield 1979*a*). Therefore the initial period of directional selection may occur in early development.

One-generation cyclical selection has been called endocyclic by McWhirter (1967). The basic models of population genetics developed by Fisher, Haldane and Wright assume that the selective forces which act on individuals are constant. Sheppard (1953), Sheppard & Cook (1962), Kojima (1971) and others have pointed out that this assumption is invalid. Pasteur (1977) has suggested that endocyclic selection is widespread (also see Bishop 1969, Dowdeswell 1971). He indicates that as such selection helps to maintain the genetic variability and evolutionary potential of the species, it should be common in eukaryotes. Goux (1978) developed a simple model of a single diallelic locus which shows that an endocyclic pattern of selection can produce a stable polymorphism if the mean fitness of the heterozygotes over the whole life cycle is greater than that of either homozygote.

Some further examples of endocyclic selection in *M. jurtina* have been proposed. Masetti & Scali (1978) and Scali & Masetti (1979) found that adult allelic frequencies at the phosphoglucomutase locus remained homogeneous through two successive generations in two Italian populations, whilst the intervening larval populations showed different frequencies from the adults. Scali & Masetti (1973) showed that in certain Italian populations a sex ratio in favour of females (up to 5.4: 1) at the embryonic stage became sequentially reduced until adults showed a probable excess of males. The relevance of Scali & Masetti's results to other populations is uncertain. The broods of *M. jurtina* (Table 16.3) show no evidence for heterogeneity of sex ratio (χ^2_{15} = 21.86) or for an overall departure from a ratio of unity (χ^2_1 = 0.62).

Effects of Spot Genes on Development

Intraseasonal changes in spot frequency have been detected in adult populations (Creed *et al.* 1959, Dowdeswell 1962, Beaufoy *et al.* 1970, Scali 1971*a*, Brockie 1972, Scali & Masetti 1975, Brakefield 1979*a*). They do not occur in all populations (Tudor & Parkin 1979). When a shift has been detected it is nearly always towards lower spot averages in the later sample(s). These observations could be due to differential development rates (Scali 1971*a*).

The date of emergence in the broods of *M. jurtina* was recorded. Analysis of the male data provides evidence for genetically determined differences in the rate of development of the spot-number classes. Figure 16.10 shows that in four of the seven larger broods the early and later portions of the emergence differed in spot frequency. In each example the spot average is lower in the later portion. A similar difference is only found in one of the six larger broods of females (brood no. 13:χ^2_2 = 13.04, $P < 0.001$). The evidence in females is therefore inconclusive. Differences in development time between spot classes could readily affect their relative fitness. That major

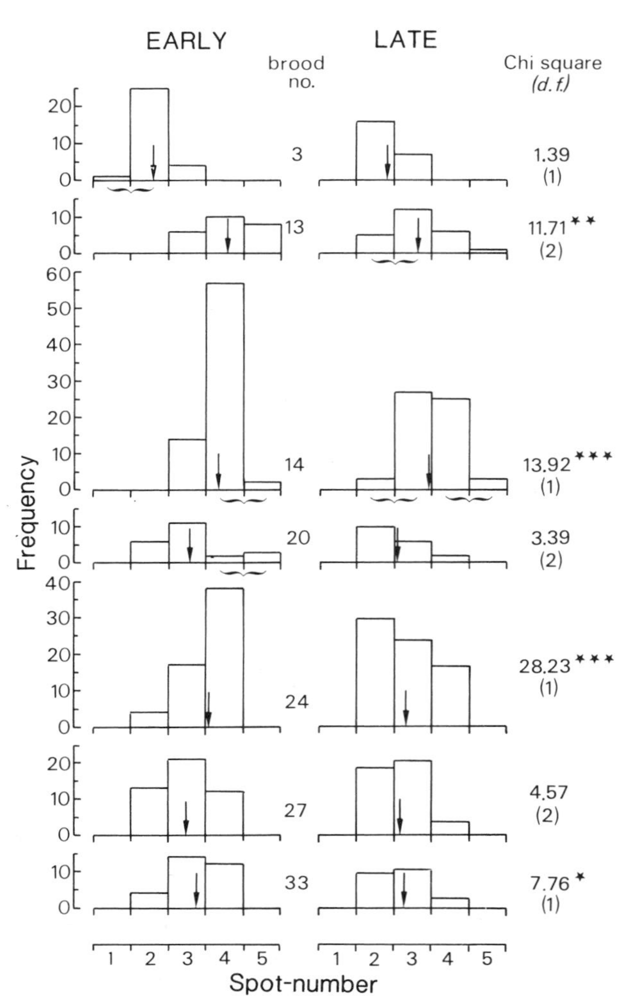

Fig. 16.10. Timing of emergence of male spot-number classes in broods of *M. jurtina*. (Each brood with $n > 35$ was divided as closely as possible into two with respect to date of emergence to give an early (first) and a late half. Arrows show mean spot-number. Spot frequencies are compared by χ^2 (chi-square) with braces indicating groupings where necessary: $*P < 0.05$; $**P < 0.01$; $***P < 0.001$.)

genes determining adult phenotypes in Lepidoptera can influence development time has been suggested by studies on *Colias* butterflies (Graham *et al.* 1980) and on several melanic moths (Bishop *et al.* 1978).

Differences in the timing of emergence of two male spot types were detectable in three broods (Fig. 16.11). Unfortunately the data could not distinguish between spot types within any of the spot-number classes. A difference of this form was evident in an earlier emergence of anal-3 than costal-3 males in two populations near Liverpool which were sampled daily during the emergence period in 1978 ($P = 0.04$ and $P < 0.001$; from Brakefield 1979*a*). A difference was not found in one of these populations in 1977. The spot-placing of female populations in Scotland becomes more anal with increasing altitude (Fig. 16.12). Males also show higher frequencies of anal-spot types (particularly anal-3) at higher

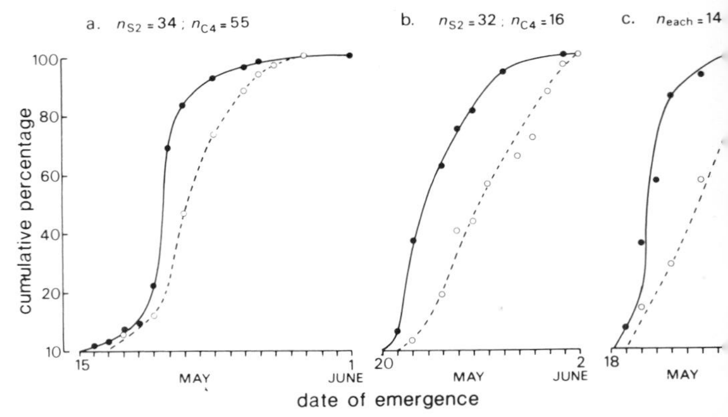

Fig. 16.11. Cumulative emergence curves for splay 2 (○) and costal 4 (●) males in broods of *M. jurtina* in which each spot type comprised >15% of all males.

altitudes. McWhirter & Creed (1971) have shown that the anality of populations increases northwards through Britain. These observations suggest the hypothesis that anal spot types are favoured at high altitudes and in northern latitudes because of a faster development rate. A relationship might also exist between spot variation and the variability which occurs in the timing and length of the flight period in British populations. This is sometimes associated with differences in soil type and habitat (Pollard 1979*a*; and see Thomson 1971, Ford 1975*a*, Brakefield 1979*b*).

It is unlikely that differential development rates could fully account for Dowdeswell's (1961, 1962) results, since the differences he found between early and late samples of larvae were not always paralleled by intraseasonal changes of spotting in the flying populations. Such changes have also not been recorded in the Isles of Scilly. The reared and flying samples of males from St Martin's Island differed particularly widely (spot average = 3.69 and 2.50).

Does Visual Selection Influence Spot Variation?

The hindwing spots of *M. jurtina* have been described as trivial or unimportant to the individual (Ford 1955: 220, 1973, Sheppard 1969, Scali 1971*b*). I suggest that this may be an invalid assumption. No reference has been made to the prominent and variable forewing eyespot in connection with the significance of the hindwing spot variation. In this section a model is developed to account for the phenotypic variation in spot pattern in terms of visual selection.

Functions of Spot Patterns

Beak-damage patterns suggest that small eyespot markings on butterfly wing margins direct the attacks of birds away from the vulnerable body (Marshall & Poulton 1902, Swynnerton 1926, Carpenter 1941,

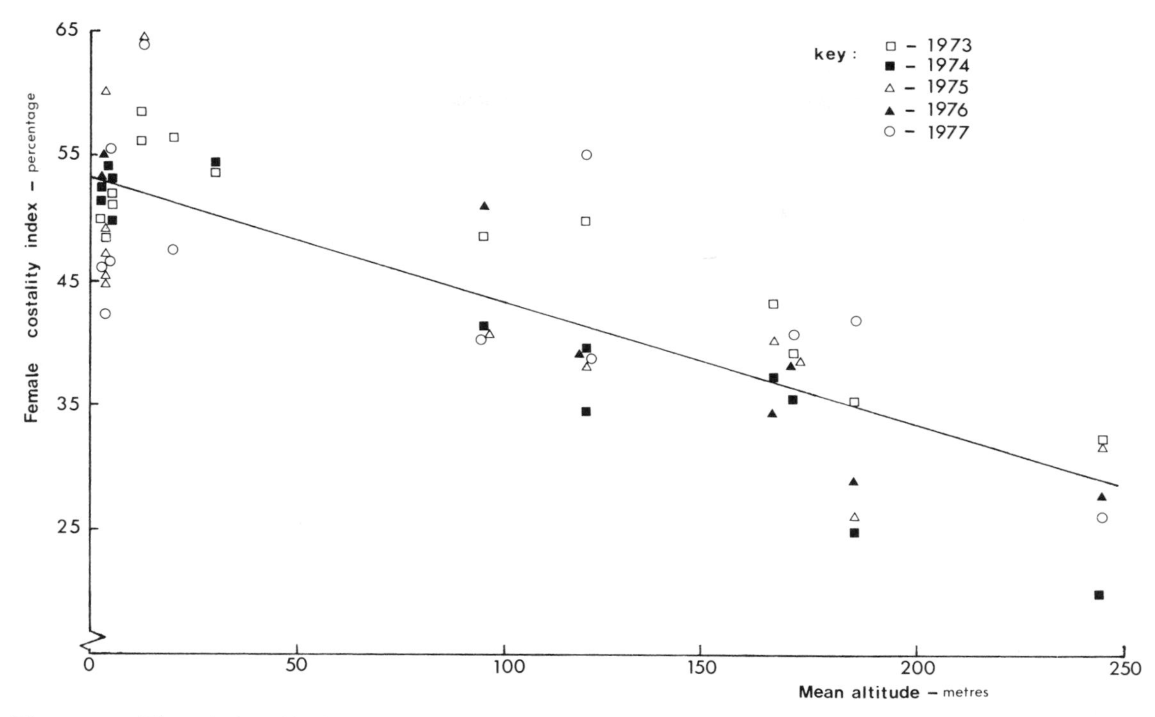

Fig. 16.12. The relationship between altitude and female spot-placing in populations of *M. jurtina* in central-eastern Scotland (from Brakefield 1979*a*). Fitted regression line: $g = 53.19\text{-}0.0987\,x$.

Blest 1957; also see Robbins 1980, 1981, Larsen 1982*b*). Similar observations have been made in relation to predation by lizards (Van Someren 1922, Brockie 1972, Ford 1975*b*). Experimental evidence was obtained by Blest (1957) working with painted mealworms and captive birds. Additional experiments supported the hypothesis that whilst a weak stimulus (e.g. small undifferentiated eyespots) evokes an approach from a bird, a strong stimulus (large solid-coloured eyespots) causes withdrawal (Schneirla 1965, Coppinger 1969, 1970). These responses have been attributed to anti-predator mechanisms of deflecting and startling respectively (Cott 1940, Edmunds 1974*a*). The effect of startling (or confusing) may be enhanced by a form of flash-coloration in which the stimulus is rapidly exposed in response to disturbance (Blest 1957).

Most organisms exhibit some form of cryptic coloration (Cott 1940, Edmunds 1974*a*). The evolution of spot patterns in butterflies must have been closely integrated with that of such coloration (see Schwanwitsch 1948). The adaptive significance of cryptic coloration has been demonstrated by numerous experimental studies (refs. in Endler 1978, 1980; examples involving butterfly pupae in Baker 1970, Wiklund 1975*a*). Endler (1978) provides a useful amplification of the term cryptic: 'a pattern is cryptic if it resembles a random sample of the background perceived by the predator at the time and age, and in the microhabitat where the prey is most vulnerable to visually hunting predators'. The matching of pattern and background must extend to features of grain, geometry, contrast and colour. The visibility of an organism's colour patterns is also influenced by predator vision and hunting tactics and by prey behaviour. The effects of these factors are interdependent and may vary from place to place. The optimum cryptic pattern is determined by their interaction (Endler 1978).

Adult Behaviour and Survivorship of *M. jurtina*

M. jurtina is a resident species which typically shows a low rate of movement from favourable areas of grassland habitat (Dowdeswell *et al.* 1949, Brakefield 1979*a*, 1982*a*, Tudor & Parkin 1979, Pollard 1981*b*). Males and females within a population show similar survivorship curves. The available estimates for English populations indicate that the adult life expectation is 5–12 days. I have estimated that adult mortality, on average, accounts for 50–60% of a female's potential egg supply of about 180 eggs. The dispersal rates for the sexes are similar (Brakefield 1979*a*, 1982*a*). However, males fly more often than females (Table 16.10; see also Baker 1978, Pollard 1981*b*). Males show a slow exploratory flight with frequent changes in direction when searching for mates. Females rest on vegetation for long periods. They fly to lay eggs and to feed. My field observations and Baker's tracking experiments show that in general males are more active and make more frequent changes in behaviour than females. In common with many species of butterflies, male *M. jurtina* are selected to maximize the number of

Table 16.10. The frequency of some behavioural activities of *M. jurtina* recorded immediately prior to capture in 1977.

Sex	Activity			Total	Percentage
	flying	resting	feeding		flying
(a) Hightown population					
males	239	132	27	398	60.1
females	33	44	9	86	38.4
(b) Scottish populations ($n = 16$)					
males	245	47	25	317	77.3
females	255	204	87	546	46.7

Observations were made when conditions were suitable, but not necessarily optimal, for activity (from Brakefield 1982*a*).

Table 16.11. Comparisons of dispersal distances of the individual and combined female spot types of *M. jurtina* at Hightown by use of the Mann-Whitney test (from Brakefield 1979*a*).

(a) Comparison of the most numerous spot types, 1976 (table gives values of Z, sample sizes shown in parentheses).

	costal 1	costal 2	splay 2	costal 3 (43)
nought (65)	0.19	1.10	3.19**	1.70†
costal 1 (39)		0.35	3.03**	1.62
costal 2 (37)			2.13*	0.50
splay 3 (32)				1.59

(b) Comparison of nought, costal 1 and costal 2 spot types combined (n_1) with all other spot types (n_2)

Year	Sample sizes		Z	U
	n_1	n_2		
1976	141	95	3.57***	
1977	18	22		331.5***
1978	43	18	2.36*	

2-tailed significance level: *$P < 0.05$; **$P < 0.01$; ***$P < 0.001$; †$0.1 > P > 0.05$.

matings whilst females mate only once (Scali 1971*b*), thus allowing maximum time for nectar feeding, egg maturation and locating breeding microhabitats (cf. Wiklund & Åhrberg 1978). The sexes in a population at Hightown near Liverpool showed different microdistributions over a grid of 7.5m squares (Brakefield 1979*a*, 1982*a*). This difference and temporal changes in microdistribution were correlated with the distribution of the adult resources of each sex. These observations confirm Handford's (1973*a*) suggestion that males and females occupy differing ecological niches.

Tudor & Parkin (1979), in a population study of *M. jurtina*, found changes between generations in the relative rate of recapture of different groups of spot classes. Some observed changes in spot frequency in males could have had a selective basis if differences in fecundity were associated with those in recapture rate. The results of applying Leslie's (1958) test and of a comparison of recapture rates with the Poisson distribution suggested that males within some populations do not have an equal probability of capture (Brakefield 1979*a*, 1982*b*). The survivorship curves of the most numerous spot types in each sex were similar. However, there is evidence that differences in dispersal behaviour occurred between some female spot types (Table 16.11). Figure 16.13 suggests that dispersal increased with increasing spot-number and that a similar but less marked relationship may be found in some male populations. There was also evidence for differences in microdistribution between some male spot types.

Additional intensive capture-recapture experiments are necessary to more firmly establish the relationship between dispersal and spotting. However, some support comes from another source. Bengtson (1981) investigated wing damage which resulted from attacks by birds, especially *Lanius collurio*, in five populations on two islands in Sweden. He found that spotted and unspotted females showed significant differences in beak-damage frequencies. This result can be accounted for if differences in behaviour influence the exposure and conspicuousness of these groups to avian predators. Such differences could also be a factor in the observed selective elimination of spotted females during aestivation in central Italy (e.g. Scali 1971*a*). Bengtson found that males and females showed different frequencies of beak-damage (13.3 and 8.5%). This could be due to various factors including differences in behaviour, age structure or

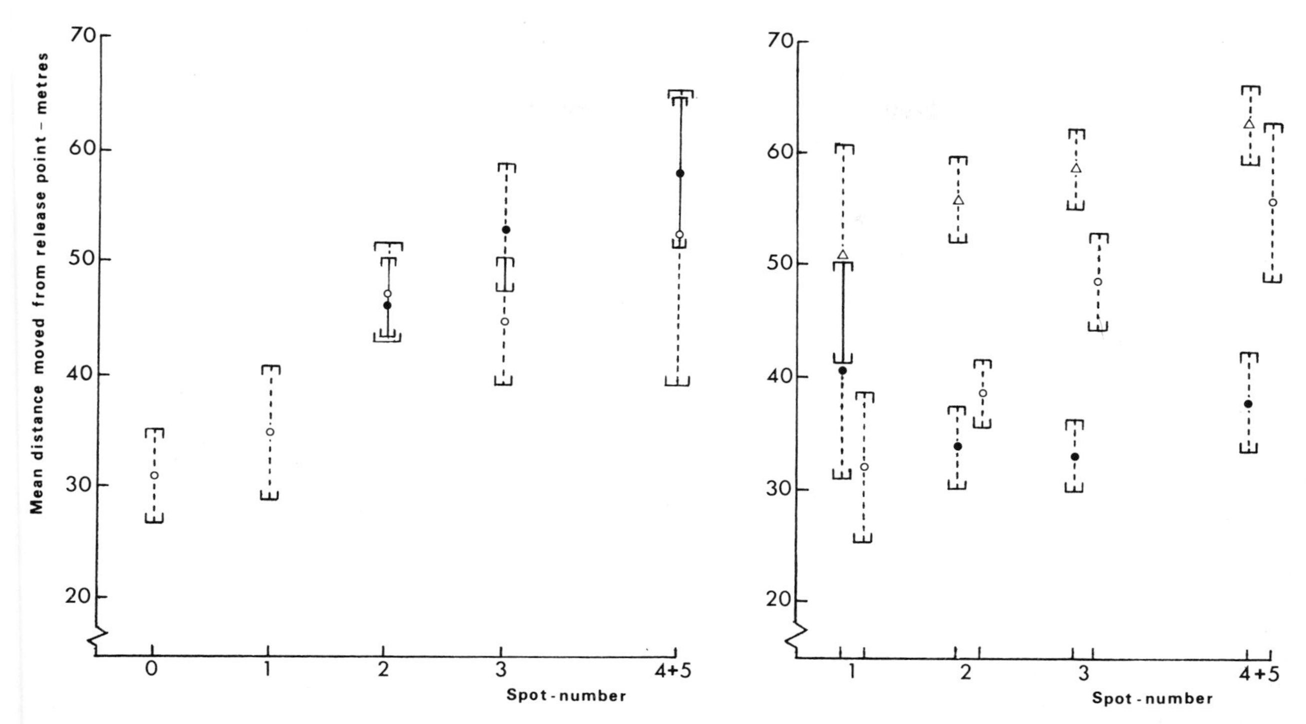

Fig. 16.13. Mean distance moved (± S.E.) by *M. jurtina* in relation to spot-number for the populations indicated (full details are given in Brakefield 1982*a*,*b*). a) Hightown 1976: ● males; ○ females. b) Hightown: ● males 1977; ○ females 1978. Hall Road: Δ males 1978.

palatability (see Edmunds 1974*b*). Lane (1957) presented adults to a captive Shama bird. In early trials it showed an apparent preference for males. Thereafter the sexes were taken indiscriminantly. Frazer & Rothschild have found more histamine-like substances in females (Rothschild pers. comm.). These observations merit further investigation. However, the sexes show a similar survivorship and Lane found that their taste is similar. Bowers & Wiernasz (1979) showed that the satyrine *Cercyonis pegala* is palatable. Samples from two populations showed similar overall frequencies of wing damage to those found by Bengtson for *M. jurtina*.

A Model for *M. jurtina*

The components of the model are outlined in Fig. 16.14. The behaviour of an adult *M. jurtina* can be related to four activity states (A–D). Details of behaviour, wing-positioning and reaction to predators differ between states. It is proposed that a spot pattern is both most likely to be effective in deflection (with subsequent escape) and most frequently elicited when a butterfly is changing its position or behaviour (state A, also B). Movement increases the likelihood of attracting a predator. At such times the butterfly usually has its wings closed above its body with the submarginal ring of spots displayed. Males are more likely to be encountered in these circumstances than females. It may then be more effective to have several spread out spots than the single large eyespot which is the effect of the female pattern (Fig. 16.3). The latter could increase the initial likelihood of a predator being attracted.

Females are more likely to be encountered whilst resting between bouts of activity (state C). The forewings are often withdrawn between the hindwings. Prominent hindwing spots may then reduce the effectiveness of crypsis (see below). If slightly disturbed, such a butterfly will frequently raise its forewing rapidly so exposing the eyespot (cf. Tinbergen 1958 and Ford's 1975*b* observations for the Grayling, *Hipparchia semele*). The trend in females towards a large eyespot with no alternative stimuli may then be advantageous and could be associated with both deflecting (Lane 1957) and startling anti-predator mechanisms. The effect of a single large stimulus is accentuated in females by a greater contrast between the eyespot and surrounding wing pigmentation, a more marked costal positioning of any hindwing spots, and a 35% larger white pupil area relative to eyespot area than in males (Fig. 16.3).

Whilst in an inactive state the forewings are withdrawn (state D). Therefore visual selection will only act on the hindwings and will then only favour cryptic properties. In uniform grassland habitats of predominantly linear backgrounds, prominent hindwing spots are probably disadvantageous (Cott 1940, Wickler 1968, Edmunds 1974*a*). This selection pressure will differ between males and females if they show differences in roosting behaviour.

I have outlined the model in relation to the sexual

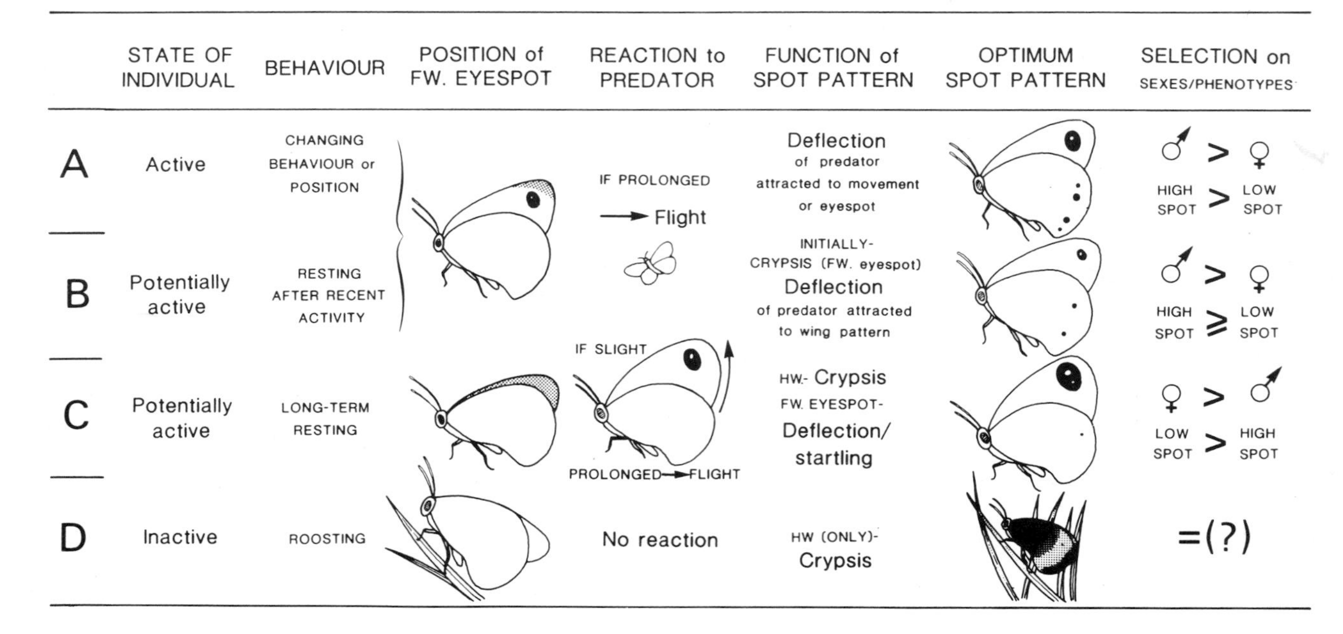

Fig. 16.14. A model to account for the phenotypic variation in spot pattern in *M. jurtina*. Four activity levels are labelled A–D on left. An outline of behaviour and of the proposed mechanism of selection follows for each state, from left to right. The final column indicates the predicted relative fitness of the sexes and hindwing spot phenotypes in relation to the model. (See text for full explanation.)

dimorphism in *M. jurtina*. It can also be applied to the spot variation within each sex. Males and females with larger and more numerous hindwing spots tend to show relatively large forewing eyespots (Fig. 16.3). This relationship is also found at the population level. The model proposes that the fitness of a given phenotype will depend on the relative probabilities of it being encountered by a predator whilst in each activity state (Fig. 16.14). This set of probabilities will vary between phenotypes if the positive relationship indicated between spot-number and dispersal rate is a real one. If it is assumed that dispersal is positively related to the level of activity, then the relationship between spotting and movement can be accounted for in the general prediction of the model that it is advantageous for a more active butterfly to show a heavier spot pattern. An anomaly is evident since in situations of long-term resting the model predicts that a large forewing eyespot and a lack of hindwing spots is favoured (state C). However, this disadvantage of hindwing spots could be outweighed by their advantage in complementing the forewing eyespot in other circumstances (especially A). A further complicating factor is that there may be some threshold of hindwing spot size (or contrast) below which visual selection is ineffectual.

Bengtson's (1978, 1981) studies of avian predation and spot variation in island populations of *M. jurtina* in Sweden were made over a five-year period. He found that a change in one year in the relative frequency of beak-damage in spotted and unspotted females coincided with changes in spot frequency, adult numbers and overall beak-damage frequency. These observations are consistent with a selective influence of avian predation on spot variation (Bengtson 1981).

The model in Fig. 16.14 predicts that relationships will occur between spot variation and habitat. In particular, it predicts that a high level of spotting will be found in types of habitat which favour high mobility and activity, or where the background is less uniformly linear and visual selection against large spots in relation to crypsis is weaker. These factors may coincide in mixed habitats of grass and scrub. Some authors have suggested that examples of differences in spot variation between populations of *M. jurtina* and of changes between generations were due to selective forces associated with differences in habitat (e.g. Dowdeswell & Ford 1955, Dowdeswell *et al.* 1957, 1960, Scali 1971*a*). McWhirter (1957) proposed that three types of female spot frequencies characteristic of populations in the Isles of Scilly might reflect differences in habitat. In particular, amongst the small islands areas of open, often exposed grasslands are usually associated with females unimodal at 0 spots whilst luxuriant habitats with patches of scrub tend to show high spot averages. This difference is therefore consistent with the prediction. It is well illustrated by the discontinuity found between the ends of White island (Ford 1975*a*). A similar association is found in the Grampian Mountains in Scotland. Here a high spotting on both hindwing and forewing (Table 16.8, Fig. 16.2) is found in three populations which are at low density, cover large areas of poor grassland and bracken, and probably show relatively high rates of long-distance movement (Brakefield 1979*a*, 1982*a*,*b*).

Although some examples of differences between populations can be interpreted with reference to the model, many components of it remain to be tested. Laboratory experiments could establish whether activity levels do vary between spot phenotypes. The effectiveness of the spot pattern, particularly in relation to its 'escape' functions, might be investigated by feeding trials (cf. Bowers 1980) and by a more detailed analysis of wing damage (cf. Sargent 1973, Bowers & Wiernasz 1979). The rates at which the sexes, spot phenotypes or artificially marked butterflies sustain wing damage from different types of predator attacks could be compared by using mark-release-recapture (MRR) techniques (Sheppard 1951, Robbins 1980, Silberglied *et al.* 1980). However, precise estimates would only be obtained of unsuccessful predation. The cryptic properties of wing patterns during periods of day-time roosting can be investigated by exposing dead insects placed in natural positions to predation (Kettlewell 1955, 1956).

Relevance to Other Species

Differences in type of grassland habitat are associated with the groups of populations or races of *Coenonympha tullia* in Britain which show striking differences in spot development (Brakefield 1979*a*, Porter 1980). The races of European and N. African species of this genus (including *tullia*) at higher altitudes and latitudes tend to show smaller spots (see Higgins & Riley 1975). Cooler environments are likely to result in less adult activity. The model (Fig. 16.14) would then predict more emphasis on cryptic properties and lower spotting. The effect would interact with variation in habitat.

The model can also be applied to the common incidence of seasonal forms in species of Satyrinae in the Old World tropics. Such forms frequently show striking differences in spot development (Owen 1971*a*) which are consistent with selection favouring crypsis in the dry season forms which aestivate and are relatively immobile and the active anti-predator functions of the spot pattern in the wet season forms (Brakefield & Larsen 1984).

Some examples of interspecific variation in spot pattern in the Satyrinae are consistent with a relationship between habitat and visual selection. Three species inhabit a meadow study site near Liverpool (Brakefield 1979*a*). These showed different microdistributions which could be summarized as follows: *Coenonympha pamphilus* in open, short vegetation; *Pyronia tithonus* along grass/scrub edges; and *M. jurtina* more uniformly within habitat subunits. *M. jurtina* was more mobile than the other species. The species graded by increasing conspicuousness of hindwing spotting are *C. pamphilus*, *M. jurtina* and *P. tithonus*. This trend could follow from differences between microhabitats in selection for crypsis. The larger forewing eyespot of *M. jurtina* may reflect an apparently more frequent concealing of this spot at rest and therefore a more important function in active escape.

The satyrine faunas of Europe (Higgins & Riley 1975), North America (Emmel 1975) and South America (pers. obs.) show a general trend of more conspicuous spotting in species from wooded or scrub habitats. Species of uniform grassland usually show a lack of hindwing spots and any forewing eyespots are often hidden in inactive insects. This distinction is particularly evident between Amazon rainforest species (e.g. *Euptychia*) and those of Andean grasslands (e.g. *Pedaliodes*). In the heterogeneous backgrounds characteristic of forest floor environments prominent spot patterns may enhance crypsis except at close predator-prey distances when they can function in deflecting or startling predators (see Barcant 1970, Stradling 1976).

The model for *M. jurtina* shows some features in common with the hypothesis developed by Young to account for interspecific differences in eyespot development in *Morpho* species. Young (1975, 1979*b*) studied the habit of some species in Neotropical forests of feeding on rotting fruit falls. He proposed that the prominent underside eyespot patterns in these butterflies afford some protection from opportunistic vertebrate predators in the vicinity of the food sources. Young (1980) classified species of *Morpho* into: (1) primarily low flying ground feeders with large eyespots and (2) partially (a) or occasionally (b) ground feeders, flying at intermediate to high levels, and having small eyespots. In museum samples a high proportion of the primarily ground feeders showed evidence of unsuccessful predator attacks whilst at rest than of the partially or occasionally ground feeders (χ^2_1 = 68.62; no difference between groups 2a and 2b). This supports the hypothesis that the large eyespots have evolved in response to the relatively high intensity and particular nature of predation at fruit falls. Selection may have operated in a similar way in some species of Satyrinae and Brassolinae (Young 1980). In some ground feeders (especially *M. peleides*) considerable intraspecific variation occurs in spot size (Young pers. comm.). The significance of this variability in relation to Endler's (1978) prediction that pattern diversity among morphs or species subject to predation on the same background should decrease with increased visual selection intensity merits further investigation.

Concluding Remarks

Although some trends in spot development between butterfly communities are consistent with an influence of visual selection, a detailed ecogenetical investigation is necessary to properly assess its role at the population, species or community level. It is my bias that the possibility of visual selection on even 'minor' pattern variation should not be discounted (e.g. Bowden 1979) without an adequate investigation (see Cain 1977). The model developed for *M. jurtina* is not proposed as a unifying explanation for the spot variation. Rather it should be considered in addition to the non-visual effects of the spot genes of which differential development rates may be one consequence. The model may indeed have an ability to explain by virtue of the input of many variables. However, it may be of value in the absence of alternative hypotheses providing a mechanistic connection between selective factors and the observed phenotypic variation. A more rigorous testing of the model's components may lead to more specific predictions. It is possible to predict the direction of changes in spot variation in response to general changes in habitat in natural (e.g. Tean) or experimentally manipulated populations. The apparent examples of associations between habitat differentiation and spot variation in *M. jurtina* have been found in peripheral or ecologically marginal areas. In centrally located areas populations usually show a similar spot variation. The hypothesis that co-adapted gene complexes are characteristic of these areas needs to be tested. Moreover, co-adaptation is an effect of differentiation not a cause or maintenance mechanism (see Endler 1977).

Ford & Ford's (1930) observations on phenotypic variation and population fluctuations in a colony of *Euphydryas aurinia* were consistent with a relationship between intensity of selection and population size. They found that a period of declining and then expanding population size was accompanied by a decrease and an increase in variation respectively. In view of the ultimate importance to ecological geneticists of understanding the control of population size in terms of genetic variation within a population, it is surprising that these observations were not followed by attempts to quantify similar phenomena using more refined techniques. Such a study could usefully be combined with surveys of enzyme variation. Various species of *Euphydryas*, for which some understanding of population dynamics has been gained, could provide study material (see Ch.2). Since these species lay egg masses, genetic analyses might be facilitated by using Lorimer's (1979) technique of comparing variance in quantitative characters between and among sibling groups.

Ecogenetical studies in quantitative characters in butterflies have been almost entirely limited to morphological traits in adults. Studies involving the pre-adult stages and also on fitness components such as fecundity and longevity would be valuable. The work on *M. jurtina* and some observations made on *E. editha* (Gilbert & Singer 1973, Singer 1971) suggest that the investigation of genetic variance for behavioural traits within and between populations, including dispersal, provides an exciting possibility for future research.

17. *Enzyme Variation Within the Danainae*

Ian J. Kitching

British Museum (Natural History), London

Electrophoretic studies on Lepidoptera enzymes have mainly addressed two problems. Many (e.g. McKechnie *et al.* 1975, Brittnacher *et al.* 1978, Menken *et al.* 1980) have used electrophoresis as a tool in population studies, together with mark-release-recapture (MRR) and other ecological techniques. Quite often (e.g. Brussard & Vawter 1975) attempts have been made to detect selection on enzyme loci and hence provide evidence in the selectionist-neutralist debate. The other main use of electrophoresis has been taxonomic, for the detection or identification of sibling species (e.g. Jelnes 1975, Hudson & Lefkovitch 1980, Suomalainen *et al.* 1981, Geiger 1981). In some cases, certain allozymes (*sensu* Prakash *et al.* 1969) have proved diagnostic.

The only published study on the application of electrophoresis at the suprageneric level in butterflies is that of Geiger (1981). However, he did not have a defined model for the interrelationships of the pierid genera under consideration and thus, although his results are very similar to intuitive estimates based on morphological data (R. I. Vane-Wright, pers. comm.), it is difficult in this case to assess the utility or validity of the method.

However, for the Danainae such a model is available. Ackery & Vane-Wright (in press *a*) have assessed the cladistic relationships of the genera, based on over 200 adult morphology characters. To test both this classification and a variety of classificatory techniques, new information has been collected on immature stage morphology and adult allozymes (Kitching 1983). This paper briefly summarizes some results from the latter, the starch gel electrophoresis of enzymes.

Initially, 33 enzymes were resolved, of which only the 17 most consistently typable, representing 22 zones of activity, were used routinely. Preliminary analysis was performed using the genetic distance function of Cavalli-Sforza & Edwards (1967), followed by a principal coordinate analysis. The most striking result is that the generic groupings *Tirumala*, *Danaus (Danaus)*, *D. (Salatura)* and *D. (Anosia)* form distinct clusters. The absence of overlap is due to several enzymes within each group being represented by a band or bands restricted to that group. For example, in *D. (Salatura)* (of which *D. genutia*, *D. melanippus*, *D. philene* and *D. affinis* have been studied) the enzyme aspartate aminotransferase-1 consists of three bands present in all species, but not found elsewhere. The general result has been corroborated by further analyses using four other genetic distance functions and weighted, pair-group hierarchical cluster analyses (Kitching 1983).

Tirumala is a complex of sibling species. Unfortunately, large samples were not obtained, and five of the six samples consisted of less than ten individuals. Despite this, the species cluster together. Disagreement only occurs when the relationship of *Tirumala* with other genera is considered. According to Ackery & Vane-Wright (in press *a*), *Tirumala* is the sister-group of *Danaus*, while the enzymes appear to favour a relationship with *Parantica*. However, this is largely a result of clustering by overall similarity. If a Wagner tree is constructed, in which the allozyme bands are coded as present or absent and the hypothetical ancestor is allocated state 0 for all bands, then the pattern of relationships is similar to that given by Ackery & Vane-Wright—although the root occurs in a different position. Work is in progress, using distance Wagner techniques (Farris 1970) to investigate this discrepancy.

D. plexippus is the only danaid that has been investigated previously. Eanes & Koehn (1979) were interested in population structure, and they concluded that in the eastern USA the Monarch was panmictic when considered on an annual basis. Importantly for this study, the band frequencies they obtained, using samples of 75 or more individuals, are very similar to those that I have obtained using eastern USA samples of 40 or less. Only phosphoglucomutase showed a significant disparity—due largely to a higher frequency in one of the three

The Biology of Butterflies
0-12-713750-5

bands. The frequency rank-order was the same. This similarity suggests that allozyme frequencies obtained for other species are reasonable estimates of the population values, despite a number of small samples.

D. plexippus from Brisbane, Australia, was found to have only a few differences from the eastern USA samples, with extra bands in four of the enzymes occurring at reasonable frequencies ($>10\%$). If the Californian population were screened, it could shed some light on the origin of the Australian Monarch, which only reached the continent about 150 years ago (Walker 1914).

Danaus (Anosia) contains four species, of which *D. gilippus* and *D. chrysippus* could be considered conspecific (Vane-Wright 1978). Unfortunately, the principal coordinate analysis is equivocal. Kenyan *chrysippus aegyptius* is as distant from *gilippus* as it is from Australian *chrysippus petilia*, while a sample of nominotypical *chrysippus* from Thailand and a Nigerian sample of *c. aegyptius* are much closer to the Kenyan sample than the Australian. The allozyme data would thus appear to suggest that if the American *D. gilippus* is given specific status, then *D. chrysippus petilia* should be also. This contradicts Pierre (1980), who considered *D. chrysippus* to be monotypic.

Average heterozygosity ($\overline{H}$) depends on both the number of allozymes (or alleles) and their frequencies. The $\overline{H}$ values for most of the 26 danaine samples investigated fell into the range 10–25%, similar to most lepidopterans so far studied. However, there was one exception. Three populations of the *D. philene/D. affinis* complex were sampled: *D. p. ferruginea* from Papua New Guinea, *D. a. affinis* from Australia and *D. a. malayana* from Malaysia. Their $\overline{H}$-values are 15% ($n=35$), 11% ($n=2$) and 4% ($n=52$) respectively.

The sample of *D. a. malayana* was collected from mangrove swamp near Kuala Selangor. Although small in area, the swamp supported a population of hundreds of adult butterflies. There does not appear to be a similar habitat in the vicinity (this requires confirmation). Twenty of the 22 enzymes were monomorphic (in other species the number is usually 10–15), and this even included phosphohexose isomerase, highly polymorphic in other danaines, including the closely related *D. genutia*. The two enzymes found to be polymorphic are NAD^+-dependent malate dehydrogenase (MDH) and an esterase (EST-A).

EST-A in *D. a. malayana* comprises two bands, with a distribution strongly suggestive of X-linkage. Females only possess one or other band, while in males, double-banded, putative heterozygotes also occur. This proved to be widespread throughout the Danainae studied (although one double-banded *Euploea tulliolus* was detected). MDH is more or less monomorphic throughout the Danainae sampled; variant bands detected occur at frequencies of less than 5%. However, in *D. a. malayana* there are two bands, the one occurring at a frequency of about 0.6 being the widespread allozyme. The other band is apparently unique.

Thus the Kuala Selangor population of *D. affinis* is probably isolated, it is reasonably large and has a very unusual enzyme profile. It may have passed through a population bottleneck (as might occur if the mangrove largely dried out at some period), or it may have (relatively recently?) been founded by one or a very few individuals. Whatever the reason, this population ought to be studied more thoroughly.

Enzymes can be seen to have potential in helping to elucidate both inter- and intra-specific relationships in butterflies. Investigation of further danaine populations could clarify some of the points raised above, but would no doubt raise an equal or even greater number of new questions.

18. *Mimicry, Migration and Speciation in* Acraea encedon *and* A. encedana

Ian J. Gordon

*Department of Zoology, University of Cape Coast, Cape Coast, Ghana**

Acraea encedon is an African butterfly notable for its colony founding habits and the consequent partial ecological and genetic isolation of its local populations, its polymorphism despite evident distastefulness, its all-female broods and distorted sex ratios, and its Müllerian co-mimicry with *Danaus chrysippus* (Owen 1971*a*). *A. encedana* is a recently described species (Pierre 1976*a*) previously included in *A. encedon*, and exhibits these same features. Resemblance between the two species is so close that in east Africa adults can often be distinguished only by their genitalia (Pierre 1976*b*). In Sierra Leone, caged males of *A. encedana* attempted to mate with female *A. encedon* but the pairs quickly broke up and the females failed to oviposit (Owen *et al.* 1973*a*, as forms of *encedon*). It thus appears that the difference in genitalia ensures reproductive isolation, although Pierre (1976*b*) does report the occurrence of rare intermediates in museum collections, while the status of some of the hilltop populations studied by Owen & Chanter (1969) in Uganda is questionable. The overall impression is one of recent speciation, with *A. encedon* and *A. encedana* forming a sister-species pair (see also Pierre 1983).

Although both species mimic *D. chrysippus* they do not do so to the same degree. All three major morphs of the danaine are mimicked by the acraeines, producing three mimicry rings: *chrysippus* f. *'aegyptius'—encedon* f. *'encedon'—encedana* f. *'encedana'*; *chrysippus* f. *'dorippus'—encedon* f. *'daira'—encedana* f. *'dairana'*; and *chrysippus* f. *'alcippus'—encedon* f. *'commixta'—encedana* f. *'alcippina'* (see illustrations in Owen 1971*a*). However, the resemblance of the *commixta* form to *alcippus* is feeble and in natural populations of *A. encedon* non-mimetic forms (especially *lycia*) often predominate (Owen 1970*a*). This species is thus only partially mimetic, unlike *A. encedana* which (with rare exceptions) is entirely so (Pierre 1976*b*). While the simultaneous occurrence of mimetic and non-mimetic forms is common in Batesian mimics and can be explained (Ford 1971, Turner 1977*a*), the same situation in a Müllerian mimic is both unusual and paradoxical. This has led Pierre (1976*b*) to question the reality of the mimetic relationship in *A. encedon*, in spite of his demonstration that all six mimetic forms of the two acraeines fall largely within the distributional ranges of the matching forms of *D. chrysippus*. As Pierre points out, the correlation in distribution is rather better for *A. encedana* than for *A. encedon*, but this arises in part because he does not distinguish between the non-mimetic grey-brown *infuscata* form (found in west Africa) and the orange-brown mimetic *encedon* form (found in the rest of Africa). Interestingly, the correlation in distribution is excellent for the west and central African mimicry ring of *alcippus—alcippina—commixta*, in spite of the poor mimicry of the last named form and doubts as to the efficacy of *alcippus* as a model (Owen 1970*a*, Rothschild *et al.* 1975*b*, Brower *et al.* 1975, Brower *et al.* 1978). These biogeographical facts suggest that the mimicry is actively maintained by predators and deepen the paradox surrounding the persistence of non-mimetic forms in *A. encedon*.

Evolution of Mimicry

In previous discussions of mimicry in *A. encedon* (Owen 1970*a*, 1971*a*) and *A. encedana* (Pierre 1976*b*) little consideration has been given to the contrasting patterns of distribution and abundance shown by the species involved. Both of the acraeines are colonial butterflies, patchily distributed in local populations (although much less so in *A. encedana*), while *D. chrysippus* is widely dispersed and extremely

*Department of Biology, Oxford Polytechnic, Headington, Oxford

The Biology of Butterflies
0-12-713750-5

abundant. In the following discussion I assume not only that this difference in dispersion patterns existed between the common ancestor of the acraeines and ancestral *D. chrysippus*, but also that it preceeded the evolution of mimicry. Although untestable, these assumptions are both parsimonious and avoid problems in explaining the origin of the non-mimetic *lycia* form of *A. encedon*. That this form is ancestral is suggested by its resemblance to the related *A. necoda* and the fact that it appears to be the universal recessive (Owen & Chanter 1971*a*).

The different dispersion patterns have important implications for an understanding of the nature and origin of mimicry. In established populations of the common ancestor of *A. encedon* and *A. encedana*, this species must have been more abundant than *D. chrysippus*, but in the vast majority of habitats at most times of the year the latter must have greatly outnumbered the former. This is certainly the case for the three species in Ghana today. Two things follow from this observation. First it is clear that mimicry must have arisen in the ancestral acraeine since a resemblance to it would only occasionally be of value to *D. chrysippus*. That the acraeines are the more dependent partners in the mimetic relationship is also suggested by a comparison of morph distributions in the three species (see maps in Pierre 1976*b*). This should be borne in mind when considering the possibility that *D. chrysippus* may be a Batesian mimic of *A. encedon* and *A. encedana* (Brower *et al.* 1975, Brower *et al.* 1978). Second, mimicry of *D. chrysippus* would be of most benefit to the ancestral acraeine during migration (using the term in the sense of movement between habitats, as adopted by Hassell & Southwood 1978). Migrating individuals would be likely to encounter predators with previous experience of the danaine but no previous experience of the acraeine. Thus between colonies there will be selection for mimicry. Within colonies however, there will be stabilizing selection for ancestral (non-mimetic) colour patterns to which local predators are already adequately educated.

It is possible that the resultant balance of selective forces (including the relative merits of 'going or staying') could provide the basis for a stable polymorphism in which mimetic and non-mimetic morphs co-exist. The probability of this happening will, among other things, depend upon the extent to which the dispersion pattern is contagious and upon the persistence of established colonies. If the dispersion pattern is only weakly contagious then the protection gained within colonies by non-mimetic morphs will be correspondingly small and a greater proportion of the population will be exposed to selection for mimicry. If populations are relatively unstable with a high degree of local extinction then many colonies will be founded by migrating mimetic forms, so that mimics may predominate even within

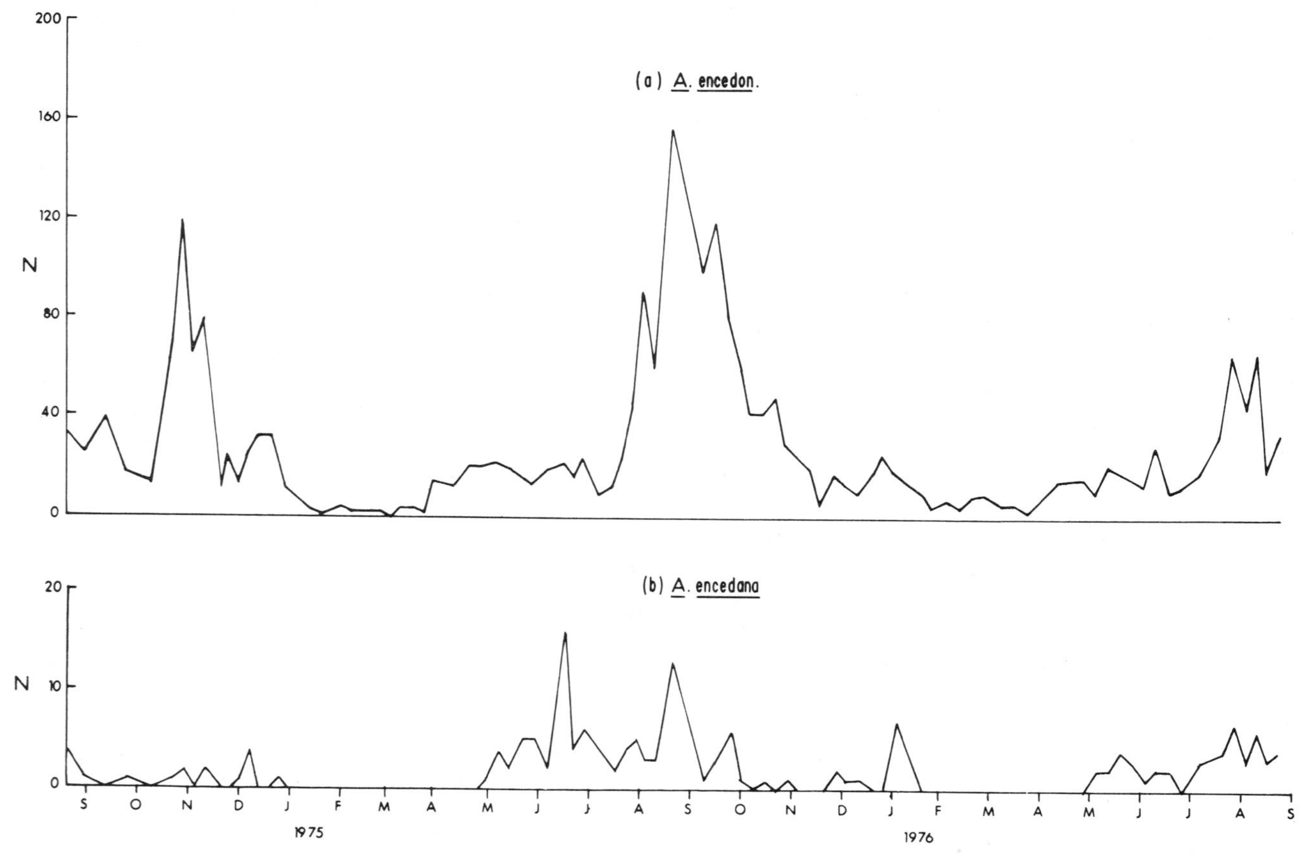

Fig. 18.1. Population changes in *A. encedon* and *A. encedana* at Brimsu Reservoir, Aug. 1974–Aug. 1976.

Table 18.1. Relative frequencies of male and female *A. encedon* and *A. encedana* between and within colonies.

	A. encedon			*A. encedana*		
	Male	Female	*n*	Male	Female	*n*
Within colony (Brimsu)	568	754	1322	54	81	135
Between colonies	1	32	33	1	47	48

χ^2, comparing relative frequencies of the two species = 168.98 ($P < 0.001$).
Note: data are combined for two periods, May–September 1975 and April–August 1976.

Table 18.2. Recapture rates of male and female *A. encedon* and *A. encedana* at Brimsu Reservoir (August 1974–August 1976).

	A. encedon		*A. encedana*	
	Male	Female	Male	Female
No. marked	1020	1198	56	95
Prop. recaptured	0.10	0.01	0.14	0.00

$\chi^2(P)$: 77.91 ($P < 0.001$): 11.62 ($P < 0.001$).

colonies and the ancestral colour pattern will be at a disadvantage in all habitats. A comparison in these respects between the wholly mimetic *A. encedana* and the partially mimetic *A. encedon* thus provides a test for the mimicry-migration hypothesis.

Figure 18.1 shows population data for the two species from a mixed colony at Brimsu Reservoir, near Cape Coast, Ghana. Throughout the two years of sampling, *A. encedon* was always more abundant than *A. encedana* and its maximum recorded colony size was greater by a factor of ten. Overall it comprised 93.5% of the total combined sample of the two species (n=2441). The same pattern of relative abundance was observed in five other mixed colonies in Ghana and in the populations studied by Pierre (1976*b*) in the Ivory Coast. Between colonies however, from April to July (during the first rains) the order of abundance is reversed (Table 18.1). These butterflies were collected on an opportunistic basis in a variety of habitats in which larval foodplants were absent or scarce, and were strays presumably in the act of migration. The difference in the relative frequencies of the two species in the migrating and Brimsu populations for the same periods is significant (χ^2 = 168.98, $P < 0.001$). The data in Table 18.1 also suggest that, in both species, migration is performed almost entirely by females. This conclusion is confirmed by an examination of recapture rates at Brimsu (Table 18.2) where males were recaptured significantly more often in both species. Owen & Chanter (1969) obtained an identical result in possibly mixed populations of the two species in Uganda. The ultimate explanation for this sexual difference in migratory behaviour is presumably that males which leave a resident breeding colony severely reduce their chances of encountering receptive females.

To return to the hypothesis under test, it is clear that *A. encedon* is much more contagiously dispersed than is *A. encedana*. Its colony sizes are considerably larger while its females appear to migrate much less readily than those of *A. encedana*. This is thus in accord with the mimicry-migration hypothesis. With regard to colony persistence the evidence is less certain. Both species were present at Brimsu in 1967 (Owen & Chanter 1971*b*), during the study period (1974–1976) and during recent visits in 1981. However, *A. encedana* disappears for three to four months during the dry season (Fig. 18.1) while *A. encedon* was recorded in every month during the two years of regular sampling. As it is possible that this absence was due to diapause or aestivation it cannot be concluded that colony extinction occurred. Nonetheless, the smaller colony size of *A. encedana* perhaps suggests a more fugitive existence. In any case the very considerable difference in dispersion pattern may be sufficient alone to explain the differential adoption of mimicry in the two species.

Speciation

With respect to speciation, analysis is greatly facilitated by its obvious recency and by clues provided by the mimetic relationship with *D. chrysippus*. Smith (1980) has argued convincingly (*pace* Pierre 1980) that the three major morphs of this species originated allopatrically during the Pleistocene, when montane and equatorial forest may have split it into three geographical races in Africa, namely *alcippus* to the west, *dorippus* to the north east, and *aegyptius* to the south. Forest is an effective barrier to migration in *A. encedon* and *A. encedana*, so the ancestral acraeines will probably have been similarly affected. Variations in the degree to which mimicry of the three *D. chrysippus* morphs has been perfected in the two acraeine species are thus relevant to the question of speciation. The poor mimicry in west and central Africa of the *commixta* form of *A. encedon* has already been noted. The pattern is right but the forewing colour is wrong, being dull grey-brown instead of bright orange-brown. In *A. encedana* however, mimicry is less good in east Africa where the *dairana* morph is a slightly less perfect mimic of *dorippus* than is the *daira* morph of *A. encedon*: according to Pierre (1976*b*) it tends to retain traces of the apical patterning of the *encedana* and *alcippina* forewings.

These comparisons suggest an allopatric origin for *A. encedana* during Pleistocene isolation in west Africa paralleling the origin of the *alcippus* race of *D. chrysippus*. A switch to a new foodplant and a less contagious dispersion pattern would have led to increased selection pressure for mimicry, the elimination of non-mimetic morphs, and the tracking of changes in the colour pattern of its model. With the subsequent expansion of open habitats following the end of the Pleistocene, sympatry would be established: *A. encedana* would spread to east and south Africa while *A. encedon* invaded the west. During this period the former would be under greater pressure to 'invent' or 'reinvent' mimetic forms than the latter. Thus while *A. encedon* has only partially succeeded in doing this in western Africa, *A. encedana* has been much more successful in eastern and southern Africa. This interpretation leads to the prediction (see also Ch.14) that the order of dominance in the latter species should be *dairana* > *encedana* > *alcippina*: an investigation of its genetics in eastern Africa where all three forms occur would be of much interest.

Vane-Wright (1981*b*) has suggested that 'the study of sympatric sister-species . . . involving a contrast between a mimetic and a non-mimetic species offers an exciting field for investigation'. I can only agree with this judgement and hope that field and laboratory research on the *A. encedon* complex in east Africa, dormant since the efforts of Denis Owen, will soon be resumed. The species involved are ideal subjects for the investigation of ecological genetics and evolution.

Acknowledgements

I am grateful to Dr Denis Owen for suggesting the work on *A. encedon*, to Lorna Depew, Kobina Yanney and Josephine Arthur for help in the field, to Professor Ray Kumar and Dr Peter Grubb for their criticism and advice, and to the University of Cape Coast for support.

19. Amplified Species Diversity: A Case Study of an Australian Lycaenid Butterfly and its Attendant Ants

Naomi E. Pierce

Museum of Comparative Zoology Laboratories, Harvard University, Cambridge, Massachusetts

The family Lycaenidae comprises perhaps 40% of all butterfly species (Vane-Wright 1978). This remarkable success may result from their frequent association, in the larval stages, with ants (Chs 3, 6). Downey (1962*b*) noted that of 833 documented life histories of lycaenid butterflies, 245 species had myrmecophilous larvae (Colour plate 3B, C) illustrates two Australian examples). Both Hinton (1951) and Malicky (1969) argued that ancestral lycaenids were myrmecophilous, and in his recent classification of the group, Eliot (1973) also suggested that symbiosis with ants was an early development in the evolution of the Lycaenidae. In this paper, I will use the interaction between an Australian lycaenid, *Jalmenus evagoras*, and its attendant ant, *Iridomyrmex* sp. 25 (ANIC) (*anceps* group) as a case study to illustrate two possible ways in which ant/larval associations may have contributed to diversification within the Lycaenidae.

Specializations of Lycaenid Larvae

All the lycaenid caterpillars examined to date possess at least one adaptation that appears to be specialized for associating with ants (although not all lycaenids do associate with formicids; Ch.6). Studded over their surfaces are small epidermal glands called 'pore cupolas' that are thought to exude ant attractants, or 'appeasement' substances (after Hölldobler 1970). Malicky (1969) described in detail the histology and distribution of these glands, and reported finding them on the larvae and pupae of all 52 lycaenid species examined, but not on the riodinid, *Hamearis lucina*. More recent examination of *H. lucina* with the scanning electron microscope has shown that the larvae do in fact possess pore cupolas (Roger Kitching, pers. comm.). Epidermal extracts of two species have been bioassayed and shown to secrete substances that are attractive to ants (Pierce 1983).

Many species also possess one or both of two other ant-associated structures. The Newcomer's organ (or dorsal organ) is located on the dorsum between the seventh and eighth abdominal segments. In several species it has been demonstrated to secrete a mixture of simple sugars and amino acids (Maschwitz *et al.* 1975, Pierce 1983). It is flanked on either side by a pair of eversible tentacles that may secrete attractants to ensure the company of ants while the larva travels (Claassens & Dickson 1977), or act as defensive structures if the dorsal organ is depleted or the caterpillar is alarmed (Downey 1962*b*).

Jalmenus evagoras and Ants

Together with colleagues at Griffith University, I have been studying populations of *Jalmenus evagoras* that occur in Mt Nebo, Queensland. The range of *J. evagoras* extends from Melbourne, Victoria as far north as Gladstone, Queensland, in both inland and coastal localities (Common & Waterhouse 1981). It is characterized by dense aggregations of caterpillars that feed on many species of *Acacia*, and are tended by several species of *Iridomyrmex* ants. Both the late instar larvae (Pierce 1983) and the pupae stridulate when disturbed (Downey 1966; Ch.6); the vibrations may serve to alert attending ants. Pupation occurs on the foodplant, and clusters of pupae are also vigorously tended by ants (Colour plate 3C).

How do larvae of *J. evagoras* benefit from their relationship with ants? Malicky (1970) emphasized that by producing ant-appeasement substances, lycaenid larvae and pupae escape from ant predation. To examine whether or not tending ants also protect

The Biology of Butterflies
0-12-713750-5

larvae and pupae, we performed ant exclusion experiments (c.f. Pierce & Mead 1981) and found that tending ants are extremely effective at defending larvae and pupae against certain parasitoids and predators (Colour plate 3D). For example, when ants were removed, pupae suffered 95% parasitism by the chalcid wasp, *Brachymeria reginia*, whereas tended pupae were untouched. Differences in survivorship of tended versus untended larvae and pupae due to predation by wasps (*Polistes (Polistella) variabilis*), jumper ants (*Myrmecia nigrocincta*) and a variety of small spiders was even more pronounced (Pierce 1983).

We also determined that ovipositing females of *J. evagoras* (Colour plate 3E), (Atsatt 1981*a*) like several other lycaenids lay eggs preferentially on plants with ants. We positioned two groups of foodplants in the field so that one contained larvae with ants whereas the other had an equal number of larvae without ants. After maintaining this arrangement for four days, the control and experimental trees were exchanged and monitored for another four days to exclude any possible 'tree' or position effect. In both treatments, females laid overwhelmingly on those trees containing ants as well as larvae. When the experiment was repeated using the membracid, *Sextius virescens* to attract ants, the females again oviposited significantly more often on trees with membracids and ants (Pierce 1983).

Ant Rewards

What kinds of reward do ants receive for their protective role? Chemical analysis of the 'honeydew' shows that it consists of fructose, glucose, and sucrose in concentrations ranging from 5–55% during the course of the day, and high amounts of the amino acid serine, estimated at 50mM (Pierce 1983). In addition, analysis of the secretions from the surface of the caterpillars provides strong evidence that epidermal glands secrete concentrated amounts of amino acids. Several lines of evidence were used in this determination. First we analysed washings of both larvae and pupae in a Dionex® amino acid analyser equipped with fluorescence detection for high sensitivity, and found large amounts of serine, smaller quantities of histidine, glutamic acid, lysine, and arginine, and traces of aspartic acid, threonine, glycine, alanine, valine, isoleucine, leucine, phenylalanine and tryptophane. We replicated these amino acid 'profiles' for many individuals, and took care to assure that our samples were not contaminated by defecation, regurgitation, or residues left after moulting. Radioactive tracer experiments showed that labelled serine consumed by the larvae was rapidly passed to their attendant ants. We then used the amino acid profiles to concoct a 'soup' containing similar combinations of amino acids secreted by the larvae and pupae. When we bioassayed this synthetic soup, it was found to be extremely attractive to workers of the attendant species, but of no interest to another species of ant (*Pheidole megacephala*) which does not favourably recognize the larvae and will attack them.

After these initial steps, an experiment was designed based on the observation that pupae of *J. evagoras* vary in their attractiveness to ants. We hypothesized that the attractiveness of pupae is directly related to the quantity of amino acids secreted by the pupae. To examine this possibility, an array of pupae was first monitored for ant attendance. These same pupae were then assayed for the amounts of amino acids secreted on their surfaces. There was a significant correlation between increasing attractiveness to ants and increasing amino acid concentration. Since the pupae do not possess a honeydew organ and can neither defecate nor regurgitate, we felt this was strong evidence that the amino acids in our samples were derived from epidermal glands only.

Finally, the chemical *o*-phthaldehyde was used to locate proteins and free amino acids on the caterpillars and pupae. The *o*-phthaldehyde, which we simply sprayed on the cuticle, combines with free amino acid groups and fluoresces under ultraviolet light. The dorsal organ, adjacent pore cupolas, and many of the modified setae located on the paired spines that run down the backs of the caterpillars, fluoresced with a purple hue that matched the colour emitted by serine treated with *o*-phthaldehyde (Pierce 1983).

Ants, Lycaenids and Nitrogen Rich Foodplants

We draw two main conclusions from these analyses. First, the ants tending *J. evagoras* may be receiving a significant nutritional reward in the form of amino acids as well as carbohydrates, the former being particularly important for the growth of ant larvae (e.g. Markin 1970, Brian 1956, 1973). The soup bioassays demonstrate that amino acids alone can serve as a strong phagostimulant for the ants, although it is quite likely that other compounds are secreted by the caterpillars and pupae to attract and appease ants. For those species of ants that are primarily nectar feeders and dependent on honeydew sources for their protein, caterpillars that secrete amino acids as well as carbohydrates could be an extremely important food resource (as suggested for extra-floral nectaries by Baker & Baker 1973*a*,*b*). Interestingly, the primary component of the lycaenid amino acid soup is serine, which is also a precursor of formic acid, and Gilmour (1965) remarks that it might therefore be of particular interest to formicine ants.

Second, these experiments indicate that while many

lycaenid caterpillars may benefit from ant attendance, they in turn must supply their associates with attractive rewards, and hence must feed on protein rich food sources. Indeed, the lycaenids are distinctive as a taxonomic group by their preference for nitrogen rich plant parts, such as flowers, seed pods, and terminal foliage (Mattson 1980). Many lycaenids feed on legumes (Ehrlich & Raven 1965), and a review of the species from Australia and South Africa shows a significant association between ant attendance and a preference for leguminous foodplants (Pierce 1983; but see also Ch.6). Moreover, some lycaenid larvae prey on homopterans, on ant larvae, and occasionally on each other (Hinton 1951).

Ant and Amplified Species Diversity Associations

Atsatt (1981*b*) discussed the adaptations that have evolved in the lycaenids as a consequence of associating with ants. He used the variables of ant abundance, ant predictability, and coincidence of ants with suitable foodplants to predict the nature of the likely interactions. What bearing might these different kinds of ant relationships have had on the evolution of diversity within the Lycaenidae? In the absence of a detailed fossil record providing unambiguous determinations of the relative ages of the butterfly families, I assume that the greater diversity of the Lycaenidae results from their relatively rapid speciation compared to other groups of butterflies rather than greater antiquity. I also assume that extinction rates of species do not differ greatly between the Lycaenidae and other butterfly families. Larval/ant associations may have influenced lycaenid speciation in at least two important ways.

Oviposition 'Mistakes'

The propensity of female lycaenids to use ants as ovipositional cues may have facilitated the process of host switching when the desired species of ant occurs on a novel foodplant (Atsatt 1981*a*,*b*). For example, during the course of a summer at Mt. Nebo, we observed *J. evagoras* switch onto four different species of *Acacia*, each of which was infested with membracids and *Iridomyrmex* sp. 25 before the switch occurred.

While few ovipositional 'mistakes' may have led to successful shifts, those changes would have been especially favoured in situations where the original foodplant was severely limited in occurrence within the range of the appropriate ant species (A 'mistake' refers to a case where a female oviposits on a plant that is not the usual larval foodplant and may or may not support larval growth; but see also Ch.7). In regions where ants are relatively scarce, there may have also been selection for ant generalists. However, I agree with Atsatt (1981*b*) who proposes that once adaptations to ensure ant associations have been achieved, it may be more difficult to switch ant hosts than to switch foodplants. Within the range of their tending ants, hostplant switching may have occurred more easily for lycaenids whose protection relies on a mobile ant guard than for those species of butterflies that are dependent on specific toxins for defence (e.g. Brower & Brower 1964).

Ant association may have thus amplified the speciation rate of lycaenids beyond the usual hostplant-based level. An increase in the number of ovipositional mistakes would lead to an increase in the numbers of opportunities for subsequent speciation. Lycaenid females may make novel foodplant choices more often than females of other butterfly families because they select for ants as well as for chemically suitable foodplants. Even if we assume that the probability of these mistakes resulting in successful shifts is the same for lycaenids as it is for other taxa, the actual number of successes would be higher simply because lycaenids make mistakes more often.

Once a successful shift is achieved, speciation could occur in a number of different ways. For example, several authors, most notably Endler (1977) have indicated that it is theoretically possible for speciation to occur in the absence of geographic isolation. For ant attended lycaenids, this situation could be pictured most readily when the new hostplant occupies a different range from the original host. Although there is limited empirical evidence (e.g. Huettel & Bush 1972, Bush 1975, Guttman *et al.* 1981, but see Jaenike 1981, Futuyma & Mayer 1980) that changes in hostplant selection alone could result in speciation, such a possibility seems unlikely for lycaenids given the constraints necessary to ensure isolation between the original and derived populations. Finally, population isolates could form concomitantly with hostplant shifts if gravid females are blown or disperse far away from their source populations, and this could then lead to divergence amongst geographically isolated populations.

Population Structure and Speciation

The rate of proliferation within the Lycaenidae may have also been strongly influenced by the population structure of the butterflies. Field observations of *J. evagoras* in Australia and of another lycaenid, *Glaucopsyche lygdamus* (which I have been studying in Colorado), suggest that these butterflies occur in small semi-isolated demes. Both species are patchily distributed in areas that are often widely separated from one another. Individuals of both species, like many other lycaenids (Scott 1974*b*, 1975*d*, Gilbert 1979) are non-vagile, and accordingly there may be little migrational interchange or gene flow between demes. This idea is supported by a mark-release-recapture study of *J. evagoras* in which we had extremely high recapture rates within an individual population (Pierce 1983). In addition, the size of these demes is often small. Both *G. lygdamus* and *J. evagoras* occur in quite limited areas, and in the case of *J. evagoras*, a deme is sometimes restricted to a single tree (Pierce 1983). Males of *J. evagoras* aggregate and compete for emerging females; clusters of as many as 15-20 males will surround a female pupa that is about to eclose (Colour plate 3F). Since not all males reproduce with equal success, the effective population size is thus further reduced.

Wright (1931, 1940) first showed that the structure and size of populations are important in determining the rate at which they evolve. Both he and subsequent workers (e.g. Mayr 1954, 1963, Lande 1976, Templeton 1980) have argued that the rate of evolution and speciation is much faster in species that have small and/or highly structured populations than in those that have large, panmictic populations. This argument has been used to explain the relatively rapid rate of speciation in placental mammals (Wilson *et al.* 1975, Bush *et al.* 1977), passerine birds (Baker, M. C. 1981), and herbaceous plants (Levin & Wilson 1976).

The patchy distribution and restricted size of populations of *J. evagoras* and *G. lygdamus* occur in spite of the availability of vast and continuous ranges of foodplants. I suspect that these range restrictions may result in part from selection for areas of foodplants that are both rich in nitrogen and coincident with tending ant species. Individual plants have been found to vary considerably in total nitrogen content, and in the 'quality' of nitrogen they produce (see Mattson 1980 for review). Many workers (e.g. Slansky & Feeny 1977, Morrow & Fox 1980, Auerbach & Strong 1981, Myers & Post 1981, Rauscher 1981*a*) have shown that such variation may have a strong impact on levels of herbivory. As previously mentioned, myrmecophilous Lycaenidae stand out because of their preference for protein rich foodplants and plant parts such as flowers and seed pods. This preference may carry over to discrimination between whole plants on the basis of their nitrogen content.

The predilection shown by lycaenids for nitrogen rich plants may be explained in part by the necessity of providing attendant ants with nutritional rewards in the form of amino acids. The distribution and size of lycaenid populations may have been restricted directly by the localized presence of potential attendant ants, and indirectly by the ants' requirements for nitrogenous rewards. By so doing, ant associations may have enhanced the rate of divergence of isolated or semi-isolated populations of butterflies, and hence their rate of speciation. In conclusion, the evidence gathered in studies of *J. evagoras* suggests two ways in which ants may have influenced the evolution of diversity within the Lycaenidae: first, by inducing a higher incidence of hostplant switching, and second, by modifying the population structure of the butterflies.

Acknowledgements

I thank S. Easteal, J. Coddington, and C. L. Remington for stimulating discussion about the evolutionary arguments presented in this paper. The comments of S. H. Bartz, M. A. Elgar, S. J. Gould, P. H. Harvey, B. Hölldobler, R. L. Kitching, R. C. Lewontin, E. Mayr, P. J. Rogers, D. P. A. Sands, R. E. Silberglied, M. F. J. Taylor and E. O. Wilson greatly improved the manuscript.

Colour Plate Section

Plate 1A. (Chapter 3). *Heliconius* species which are found around Sirena, Corcovado, Costa Rica. The left column from top to bottom includes *H. hecalesia, H. hewitsoni, H. erato, H. charitonia,* and *H. sara*. All of these are pupal mating species, and all breed in the Sirena area except *H. hecalesia* and *H. sara* which appear as rare immigrants. The right column from the top includes *H. hecale, H. pachinus, H. melpomene, H. ismenius,* and *Heliconius (Laparus) doris*. These appear to have more typical mating habits and all breed in the area.

Plate 1B. (Chapter 3). Interspecific hybridization within the genus *Heliconius*. This plate shows crosses between *H. cydno* and *H. melpomene* (upper left), *H. cydno* and *H. pachinus* (lower left), *H. pachinus* and *H. ismenius* (upper right), and *H. cydno* and *H. ismenius* (lower right). These crosses occurred under semi-natural greenhouse conditions at Austin and Irvine (J. Smiley) and will be discussed in detail elsewhere. These and other known hybrids (e.g. *H. hecale* × *H. melpomene*) show that species which differ in phenotype, host plant specialization, and habitat preference may nonetheless be genetically rather similar to one another.

Plate 1C. (Chapter 3). Some representative colour pattern groups found among Costa Rican mimetic butterflies. Note that a pericopid moth participates in each complex. See also Plate 1B.

Plate 1D. (Chapter 3). Pupal mating male *H. charitonia* on *H. cydno* pupa. The emerging *H. cydno* was killed by the activities of the *H. charitonia*.

Plate 1E. (Chapter 29). Above, *Precis octavia* dry-season form 'sesamus'; centre, *P. octavia* wet-season form 'natalensis'; below, *P. octavia* intermediate form.

Plate 1F. (Chapter 32). A recently emerged female of *Ornithoptera alexandrae* on its pupal exuvium.

Plate 2A. (Chapter 12). The presence of the same array of cardenolides in all life history stages of the Monarch butterfly reared on the Neotropical *Asclepias curassavica*. First instar larvae had limited access to foodplant and show few and small amounts of low rf cardenolides. Second through fifth instar larvae were collected after evacuating their guts prior to ecdysis, but still had some undigested plant material in their mid-guts which accounts for the green pigment and larger amounts of lower rf cardenolides compared to the eggs, 1st instar larvae, chrysalids, and adults.

Plate 2B. (Chapter 12). The uniform TLC patterns of cardenolides in the stomachs of 12 Black-headed Grosbeaks ingested from the overwintering Mexican Monarchs. The TLC channel on the far right is a composite of 158 wild butterflies from the same colony and is virtually identical to the bird stomachs. The next to last channel on the right is a composite of the contents of 5 Oriole stomachs and, although the cardenolide spots are weaker because the Orioles avoid ingesting most of the cardenolides in the butterflies, shows at least three of the same cardenolides. Note that the major cardenolide relative to digitoxin in both birds' stomachs and in the butterflies is very similar to that in the TLC pattern of Monarchs reared on *A. syriaca* shown in (D).

Plate 2C. (Chapter 12). The array of *A. curassavica* cardenolides found in the exoskeleton of the abdomen, thorax, and wings, and in the thoracic muscles and abdominal contents. Note the highly uniform distribution of the cardenolides throughout the body and exoskeleton of both sexes.

Plate 2D. (Chapter 12). The uniform array of cardenolides in Monarchs reared on *A. syriaca*. Note that the two TLC patterns are distinctive which allows Monarchs to be fingerprinted to their respective *Asclepias* foodplants.

The cleanup procedure for (A) is in Roeske *et al.* (1976), for B–D is in Brower *et al.* (1982). (A) was developed 2× in an ethylacetate:methanol (97:3) solvent system, (B–D) 4× in a chloroform:methanol: formamide (90:6:1) solvent system. In all channels, the total cardenolide spotted is given in digitoxin μg equivalents. In (A) all samples = 75μg. In (B) grosbeaks range from 30–75μg, orioles = 19μg, Monarchs = 75μg. In (C) plants = 100μg, butterfly material = 50μg. In (D), all butterfly material = 50μg. From Brower, Seiber, and Nelson *et al.*, unpublished data. For further discussion, see text.

A

D

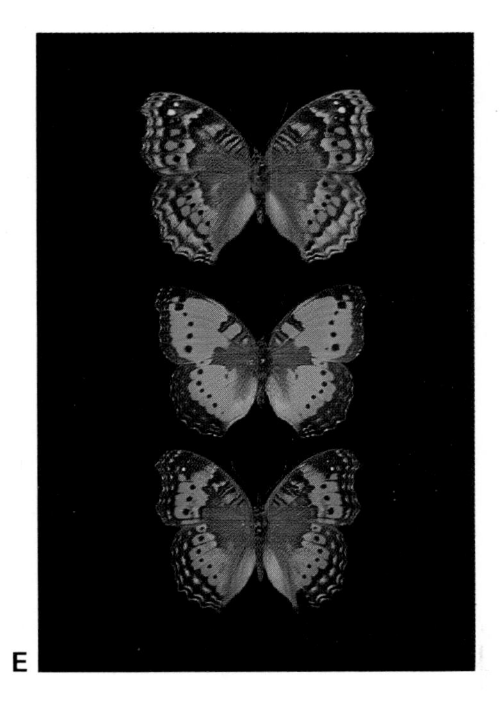

E

B

C

F

Plate 1

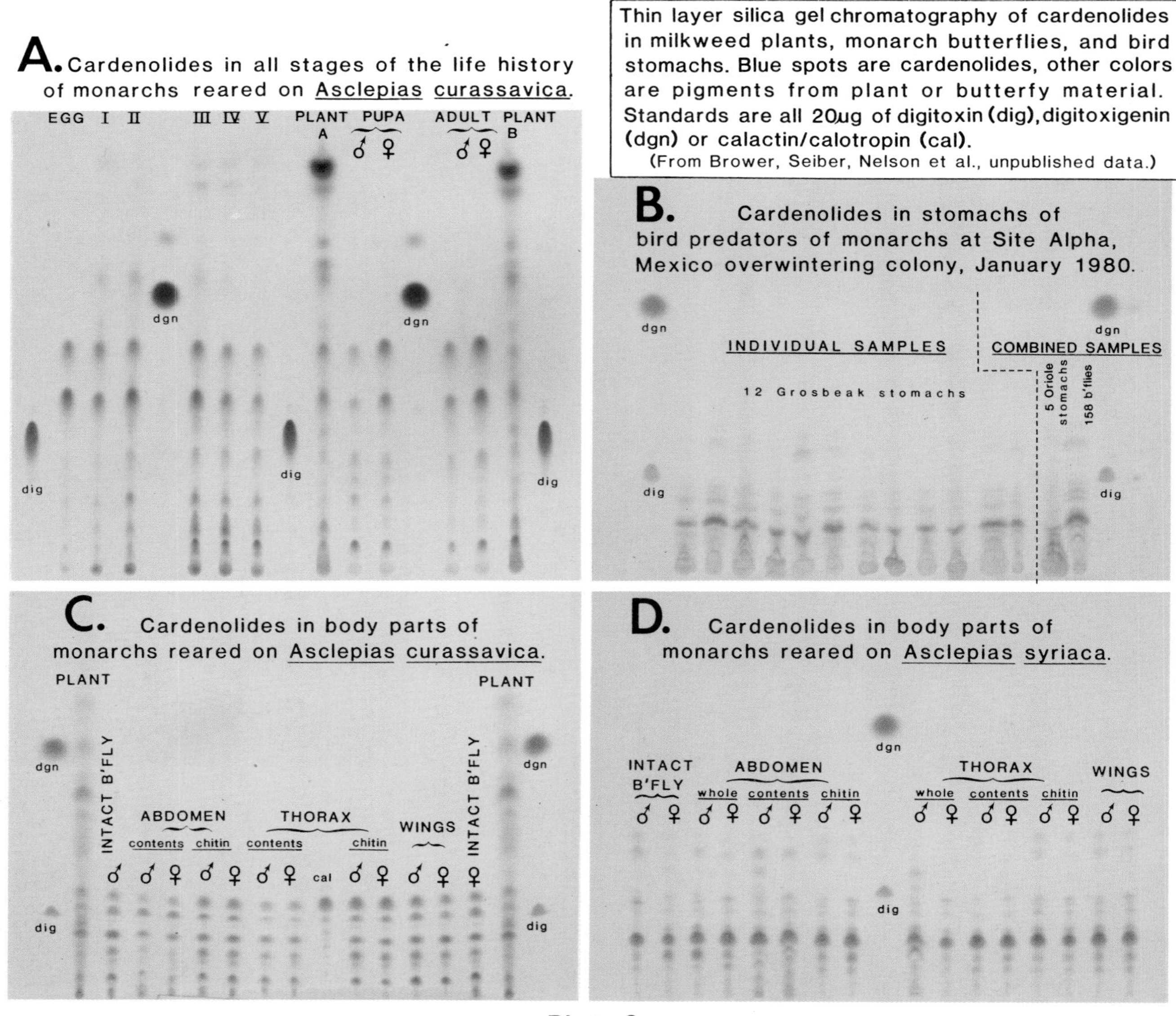
Thin layer silica gel chromatography of cardenolides in milkweed plants, monarch butterflies, and bird stomachs. Blue spots are cardenolides, other colors are pigments from plant or butterfy material. Standards are all 20µg of digitoxin (dig), digitoxigenin (dgn) or calactin/calotropin (cal).
(From Brower, Seiber, Nelson et al., unpublished data.)
A. Cardenolides in all stages of the life history of monarchs reared on Asclepias curassavica.
EGG I II III IV V PLANT A PUPA ♂ ♀ ADULT ♂ ♀ PLANT B
dgn
dig
B. Cardenolides in stomachs of bird predators of monarchs at Site Alpha, Mexico overwintering colony, January 1980.
INDIVIDUAL SAMPLES
COMBINED SAMPLES
12 Grosbeak stomachs
5 Oriole stomachs
158 b'flies
C. Cardenolides in body parts of monarchs reared on Asclepias curassavica.
PLANT
INTACT B'FLY
ABDOMEN contents chitin
THORAX contents chitin
WINGS
cal
D. Cardenolides in body parts of monarchs reared on Asclepias syriaca.
INTACT B'FLY
ABDOMEN whole contents chitin
THORAX whole contents chitin
WINGS

Plate 2

A

B

C

D

E

F

Plate 3

A

B

C

Plate 4

Plate 3A. (Chapter 19). Adult of *Hypochysops ignitus*—the ant association of this species is described by Common & Waterhouse (1981).

Plate 3B. (Chapter 19). Sugar ant (*Camponotus* spp.) guarding larvae of *Ogyris genoveva.*

Plate 3C. (Chapter 19). Workers of *Iridomyrmex* sp. 25 (ANIC) (*anceps* group) tending pupae of *Jalmenus evagoras.*

Plate 3D. (Chapter 19). Wasp (*Polistes variabilis*) attempting to attack a prepupa of *J. evagoras.* A worker of *I.* sp. 25 (ANIC) (*anceps* group), mandibles agape, is threatening the wasp.

Plate 3E. (Chapter 19). Female of *J. evagoras* laying eggs.

Plate 3F. (Chapter 19). Cluster of male *J. evagoras* surrounding a female pupa that is about to eclose.

Plate 4A. (Chapter 20). Males of *Colias eurytheme* (Pieridae) used in choice experiments. All have had their coloration altered by application of dyes using felt-tipped pens. The male of *C. eurytheme* normally has strong structural ultraviolet reflection from the dorsal wing surfaces. In the treated control male (upper left), yellow dye has been applied to the undersides of the wings, leaving the dorsal visible and ultraviolet reflection normal. The five experimental males have had yellow, red, green, blue and black dyes applied dorsally, changing the visible colour and suppressing the ultraviolet reflection. The control males mate successfully with conspecific females, but were rejected by female *C. philodice.* The experimental males were unacceptable to females of both species. The significant change in the experimental males is the loss of ultraviolet reflection, not the alteration of visible colour. See text, and Silberglied & Taylor (1978).

Plate 4B. (Chapter 20). Males of *C. philodice* used in the same experiments as those in Plate 4A. The yellow males of this species are ultraviolet-absorbing dorsally. Yellow dye was applied to the dorsal surface of the treated control male (upper left). Experimental males were dyed orange, red, green, blue, and black. All experimental males mated as successfully with conspecific females, as did the control males, but all were unacceptable to females of *C. eurytheme.* Females of *C. philodice* do not discriminate on the basis of either visible colour or ultraviolet reflection. See text, and Silberglied & Taylor (1978).

Plate 4C. (Chapter 20). Males of *Anartia amathea* (Nymphalidae) used in a mate-choice experiment. The treated control male (left) was modified with a clear-fluid felt-tipped marking pen; the experimental male (right) was dyed black. Experimental males mated as frequently as did treated control males. (Females painted black were not courted.) See text and Table 20.1.

Part VI
Sex and Communication

20. *Visual Communication and Sexual Selection Among Butterflies*

†Robert E. Silberglied

Smithsonian Tropical Research Institute, Balboa, Republic of Panama

The wings of butterflies are adorned with a wealth of pattern and a display of colour unrivalled in the living world. Some butterflies have uniformly-coloured, unpatterned wings of white; others reflect yellow, orange, red, green, blue, violet, brown or black. The colours are produced in various ways (Fox & Vevers 1960), including selective absorption by pigments (Ford 1945), tyndall scattering (Huxley 1976) and thin-film interference (Simon 1971). In some species, the wings bear simple patterns, such as contrasting colour along the veins. In others multicoloured patterns of anastomosing lines, ripple patterns or eyespots are found (Nijhout 1980*a*). These complex patterns are genetically determined and developmentally controlled (Nijhout 1978, 1980*b*, 1981). The coloration and pattern of a butterfly's dorsal and ventral wing-surfaces may be the same or different. The two sexes may be identical, slightly different, or in some cases so distinctive that they were originally described in different genera. Wing colour may be seasonally, geographical or locally variable or polymorphic. Furthermore, closely related species may appear extremely different, while unrelated species can display nearly identical colour patterns (Vane-Wright 1971, 1979*a*,*b*).

Attempts to understand this panoply of spectral diversity have resulted in major contributions to our knowledge of the evolutionary process. The study of butterfly colour and pattern has played an important role in the development of the theories of both natural (Darwin 1859) and sexual selection (Darwin 1874, Hingston 1933), mimicry (Bates 1862, Müller 1878*c*, 1879, Carpenter & Ford 1933, Wickler 1968) and genetic polymorphism (Sheppard 1961*a*, Robinson 1971). In return, information from ecological, behavioural, genetic, developmental and evolutionary studies has contributed greatly to our understanding the significance of butterfly colour and pattern.

Butterfly wing colours and patterns serve several important functions. Unlike some colours associated with metabolically important compounds (e.g. haemoglobin, chlorophyll, cytochrome), butterfly wing pigments and structural colours must serve biophysical functions because they are located outside the metabolic pool, in the dead cellular skeletons of the wing scales. Certain colours absorb or reflect radiation effectively (Watt 1968). A butterfly may regulate its body temperature by positioning itself and its wings with respect to the sun (Vielmetter 1954, Clench 1966, Watt 1968, Findlay *et al.* 1983), but only the basal regions of the wings play an important role in thermoregulation (Wasserthal 1975, Douglas 1979). We must look elsewhere if we are to understand the function of the coloration of the remainder.

The colour patterns of butterflies have usually been discussed in the context of one or both of two contrasting, communicative functions: protection from predators (Poulton 1908, Cott 1940), and social signals used during courtship (Silberglied 1977). Darwin (1874) devoted a chapter to the Lepidoptera in *The descent of Man and selection in relation to sex*. He provided numerous examples and observations to support his view that, 'although many serious objections may be urged, it seems probable that most of the brilliantly coloured species of Lepidoptera owe their colours to [inter-] sexual selection, excepting in certain cases . . . in which conspicuous colours have been gained through mimicry as a protection.' Sexual selection on pattern and colour may oppose other selective forces, such as predation (e.g. Carpenter & Ford 1933, Stride 1956, 1958*b*, Brower 1963, Turner 1977*a*, 1978, Vane-Wright 1971, 1975, 1976, 1979*a*, Ch.23). A fuller understanding of the nature and intensity of sexual selection for colour and pattern in butterflies is therefore of great importance in evaluating other evolutionary processes.

†Bob Silberglied died in a tragic accident on 13 January 1982, a few days after submitting his final draft of this paper.

The Biology of Butterflies
0-12-713750-5

Darwin's view that intersexual selection is of primary importance in the evolution of brilliant male butterfly coloration and sexual dimorphism has persisted with little change for more than a century. During this time a host of physiological and behavioural studies, both observational and experimental, have greatly augmented our knowledge of butterfly vision and its role in courtship. This paper explores the extent to which his view, which has been incorporated into numerous reviews, texts and theoretical discussions, is corroborated by our present knowledge of visual communication among butterflies.

Darwin's Views on Sexual Selection and Butterfly Coloration

Darwin (1874) was particularly impressed with sexual dimorphism as exhibited by many butterfly species, and by the disturbation of this phenomenon within genera. He was intrigued with the problem of why the sexes differ in some species, but not in close relatives. Were the colours and patterns attributable to direct effects of the environment in which the butterflies lived, were they largely protective, were they employed as attractive devices by the two sexes, or did they perhaps serve some unknown purpose? He pointed out that when the sexes differed, it generally affected the colour and pattern of the dorsal wing surfaces, those usually concealed when the butterfly is at rest. The male usually has the more beautiful, brilliant or striking dorsal coloration. According to Darwin, the female generally exhibits 'ancestral' features, from which the male pattern departs to greater or lesser degree. Thus females of closely related species tend to resemble one another more than do the males.

Darwin recognized that 'whenever colour has been modified for some special purpose, this has been, as far as we can judge, either for direct or indirect protection, or as an attraction between the sexes.' The similarity of the ventral wing surfaces in both sexes reflects the fact that they generally rest with the wings raised, with this surface exposed to view. The ventral surface is usually mottled or patterned in such a way as to resemble the background against which the butterfly rests, thus making it more difficult to detect.

That most nocturnal Lepidoptera were drab in coloration also impressed Darwin, as well as the fact that diurnal moths often have brilliant coloration. However, unlike butterflies, diurnal moths rarely exhibit strong sexual dimorphism. (The major exceptions involve sex-limited mimicry, e.g. *Callosamia promethea*, *Pericopis* spp.)

Darwin recognized that butterfly courtships are often prolonged. He knew that males are aggressive and competitive but more or less indiscriminate in their choice of mates, and that a female could refuse a courting male. The location of the colours, and the behaviour of the males, indicated to him an active function in display toward the opposite sex (Darwin 1880). Because butterflies exhibit colour preferences while feeding, and can be attracted to coloured decoys, he reasoned that they perceive and 'admire' bright colours. He believed that females were more excited by, and preferred, the more brilliant, beautiful males. By choosing mates on the basis of colour, females would select for males that departed from the duller ancestral pattern, resulting in the evolution of brilliant male coloration and striking cases of sexual dimorphism.

Though Darwin clearly perceived bright male coloration as being due to female sexual selection, he also discussed some evidence contrary to his view. He mentioned cases of matings to 'battered, faded or dingy males' (see also Ch.21) but dismissed these as being a result of protandry.

Wallace (1889) believed that the sexual dimorphism exhibited by many butterflies was often due to the acquisition of protective coloration by females, rather than to the development of brilliant colours in males by intersexual selection. However, the diversity of brilliant patterns of the males, and the similarity of closely-related females, argue strongly against this view, suggesting that protective coloration represents an ancestral character state that existed before the evolution of brilliant dorsal wing displays. Wallace felt that 'the varieties of colour and marking, forming the most obvious distinction between allied species, . . . have . . . in all probability been acquired in the process of differentiation for the purpose of checking the intercrossing of closely allied forms.' (It is difficult to understand how Wallace reconciled this "need for recognition" with his denial of female choice.) Both Darwin and Wallace argued that bright colours were not due to any special influence of tropical conditions.

In those few instances in which the females have the more brilliant patterns, Darwin suggested that the males prefer to mate with the more 'beautiful' females. When the sexes are alike, either the male has retained the ancestral colour pattern, or the male colours have been 'transferred' to the female (see Ch.23).

Darwin also recognized that mimicry affords a special explanation for the colours of many butterflies. When mimicry is limited to one sex, it is expressed in the female: the non-mimetic males generally retain the colour pattern typical of their phyletic group, from which the females depart. Darwin accepted Thomas Belt's explanation for this phenomenon, 'that the females had a choice of mates and preferred those that retained the primordial appearance of the group' (Belt 1874, see also Turner 1978).

Summarizing, Darwin (1874) said

> it is impossible to admit that the brilliant colours of butterflies, and of some few moths, have commonly been acquired for the sake of protection. We have seen that their colours and elegant patterns are arranged and

exhibited as if for display. Hence I am led to believe that the females prefer or are most excited by the more brilliant males; for on any other supposition the males would, as far as we can see, be ornamented to no purpose. With butterflies we have the best evidence, as the males sometimes take pains to display their beautiful colours; and we cannot believe that they would act thus, unless the display was of use to them in their courtship. Judging from what we know of the perceptive powers and affections of various insects, there is no antecedent improbability in sexual selection having come largely into play; but *we have as yet no direct evidence on this head, and some facts are opposed to the belief*. Nevertheless, when we see many males pursuing the same female, we can hardly believe that the pairing is left to blind chance—that the female exerts no choice, and is not affected by the gorgeous colours . . . with which the male is adorned' (my emphasis).

The Evidence

Several kinds of direct evidence bear on Darwin's conjecture. For instance, we need to ask of butterfly vision: Over what spectral range do they see? Can they discriminate colours from one another? Have they a well-developed ability to distinguish among patterns? What other factors (e.g. depth perception, motion, polarization) are important in their vision?

Which characteristics of butterfly colour patterns might be used for visual communication? How are the mechanisms of colour production related to the qualities of visual stimuli? Which features are constant, and which variable, within and between species? Do male and female butterflies behave during courtship in a manner consistent with the use of optical signals? Do males rely on visual stimuli to choose among females? Do females refuse, and so choose among, males? If so, do they respond to the males' colours?

Butterfly Vision

> When physiological methods uncover color receptors in the compound eye, their biological significance in a color vision system is usually assumed, even in the absence of behavioral data
>
> (Goldsmith & Bernard 1974)

Spectral sensitivity

The spectral range visible to butterflies extends from the ultraviolet (Lutz 1924) through the red (Eltringham 1919, Schlieper 1928), fully encompassing the visible spectrum of humans as well as that of other insects. It is the broadest visible spectrum known in the animal kingdom (Silberglied 1979).

Colour vision

It has long been supposed that butterflies are sensitive to colour (Eltringham 1919). Flower constancy exhibited while feeding, and responses of males to dead specimens and coloured dummies, strongly suggested that they could distinguish hues. Simple choice experiments (e.g. Lutz 1924; see McIndoo 1929) also appeared to reveal colour preferences, but none of the early experiments controlled for differences in stimulus intensity (brightness). It remained for Ilse (1928; Kühn & Ilse 1925) to demonstrate that true colour vision (hue discrimination) exists in butterflies. Additional behavioural studies (e.g. Ilse 1937, Tinbergen *et al.* 1942, Tinbergen 1958, Crane 1955, Ilse & Vaidya 1956, Magnus 1958*a*, Swihart 1969, 1970, 1972*b*, C. A. Swihart 1971, Swihart & Swihart 1970) reveal colour preferences that may change with age and reproductive state, context, or learning.

Neurophysiological studies (it is extremely frustrating to find that several neurophysiologists rarely mention the sexes of the individual animals they so painstakingly study) have confirmed and extended these behavioural findings. Electroretinograms, single-cell recordings, optomotor responses and other data reveal broad sensitivity from less than 300nm in the UV through 700nm in the red, and more than one type of colour receptor in both butterflies and skippers (e.g. Swihart 1963, 1964, 1967*a*, 1970, Post & Goldsmith 1969, Schümperli 1975, Bernard 1979). Swihart (1969) has modelled colour vision as a dichromatic system in a skipper, and 'at least a trichromatic system' in a *Papilio*, a *Heliconius* and a *Morpho* (Swihart 1970, 1972*a,b*), but since his experiments (and those of Schümperli 1975) do not include the UV portion of the spectrum, such conclusions and models must be considered premature and incomplete. It is probable that more than three colour receptors are present in some butterflies, as in some moths (e.g. four in *Spodoptera exempta*; Langer *et al.* 1979). The complexities of receptor structure (Bernard & Miller 1970), function (Swihart 1973) and integration (Swihart 1965, 1968, 1970, 1972*b*) are such that a full understanding of the neurophysiological basis of butterfly colour vision is still a long way off. (See also Goldsmith & Bernard 1974, Horridge 1975, Mazokhin-Porshnyakov 1969, Menzel 1975.)

Building upon the foundation of Crane's work with *Heliconius*, Swihart (1963, 1964, 1965) studied the integration of visual information in the central nervous system of several species (Swihart 1967*a*). He had found that 'the behavioral sensitivity of [*H. erato*] seems . . . to be related not to any modification of the [visual] receptors, but rather to the development of [neural] pathways which "selected" the output of those receptors which transduced information with special biological significance.' His findings generated the hypothesis that 'there was a selective advantage in developing neural mechanisms which demonstrate disproportionate sensitivity to the basic wing coloration, presumably because of the role played by such colors in releasing courtship behavior.' Studies on six species resulted in the discovery of a match between the spectral reflectance curve of the butterflies' wings, and the spectral efficiency curve for summated responses of high-order neurons in the medulla interna (Swihart 1967*a*). For example, the red and black *Heliconius erato* was found to have a peak response in the red, the green *Philaethria dido* and *Siproeta stelenes* a peak response in the green, and the iridescent blue *Morpho peleides* to have a maximal response to blue. He concluded that 'butterflies possess a neural mechanism which "selects" the output from various receptors in such a manner as to make the visual system respond maximally to stimulation with colors approximating the wing pigmentation.' It thus appears that colour preferences are effected by means of stimulus filtering in the butterfly brain. At the receptor level, contrast may be enhanced for biologically meaningful colours by selective back-reflection of narrow spectral regions from laminar tapetal interference filters (Bernard & Miller 1970, Bernhard *et al.* 1970, Miller & Bernard 1968, Ribi 1980).

Spatial resolution

The spatial resolution of a butterfly's eyes is related to the angular separation between ommatidia. However, the correspondence between interommatidial angles, and neurophysiological and behavioural measurements of resolution, is limited by complexities of ommatidial optics and receptor physiology (Goldsmith & Bernard 1974, Palka & Pinter 1975, Wehner 1975). In general, butterflies have a resolution of several degrees of arc, hundreds of times coarser than humans (0.5min of arc). From a behavioural standpoint, the amount of information obtained per solid angle of view is of greater interest than the simple angular resolution. Because the former is a squared function of the latter, the spatial information-gathering ability of butterfly vision may be tens of thousands of times lower than that of humans and other vertebrates. It is unlikely that pattern details of small objects, such as other butterflies, are resolved until they are extremely close (see Eltringham 1919, Mazokhin-Porshnyakov 1969, Yagi & Koyama 1963). Little is known about central processing of spatial information in butterflies (Swihart & Schümperli 1974, Schümperli 1975, Schümperli & Swihart 1978), but visual field organization and hierarchy generally resembling that of vertebrates has been found in some insects (Schümperli 1975, Wehner 1975).

The angle of view of butterfly eyes is very large, accounting in part for the difficulty of approaching alert species from any direction. One might think that the dorsal colours of a butterfly would not be visible to another that was below or in front of it (Wallace 1889). But the wide angle of view, the lowering of the wings to a nearly vertical position on the downbeat (Silberglied & Taylor 1978), and the forward projection of certain structural colours (Darwin 1880) make it likely that one butterfly can see the dorsal surface of another from most positions during flight. While there is little direct experimental evidence for depth perception in butterflies, a neurophysiological basis has been found in numerous protocerebral cells containing input from both eyes (Schümperli 1975).

The meagre behavioural data on pattern and spatial discrimination indicate that ovipositing female butterflies are capable of visually discriminating between leaves of different shapes (Rauscher 1978). They possibly detect eggs visually (Gilbert 1975; Ch.3) as well as chemically (Rothschild & Schoonhoven 1977, Ch.6). However, male *Hipparchia* butterflies do not discriminate among dummy butterflies of different shapes (Tinbergen *et al.* 1942, Tinbergen 1958), and pattern details play little role in species and sexual discrimination (Tinbergen *et al.* 1942, Tinbergen 1958, Crane 1955, Stride 1957, Magnus 1958*b*, Obara 1970, Rutowski 1977*a*) although there do appear to be some exceptions: *Papilio xuthus* (Hidaka & Yamashita 1975, 1976) and *Limenitis camilla* (Lederer 1960). In general, it does not appear that shape information is particularly important in the context of butterfly communication.

Temporal resolution and motion detection

The temporal resolution of the eye is generally measured in terms of the frequency at which successively presented images fuse. *Argynnis paphia* has been shown to have flicker fusion at about 150 images per second (A. Müller *in* Magnus 1958*a,b*), compared with about 40 per second in humans. It has been suggested that, for a moving insect or subject in view, the information gained from more frequent sampling of the visual environment may compensate to some extent for low spatial resolution.

A butterfly can detect a moving object if it differs in colour from the surrounding field, an ability that is greatly enhanced if a difference in brightness is also present (Kaiser 1975).

Polarization

Arthropod photoreceptors exhibit a general sensitivity to polarization, used at the behavioural level for orientation (von Frisch 1968). The iridescent reflection from the wings of some butterflies should be strongly polarized at certain angles. To my knowledge, this phenomenon has not actually been demonstrated, and its possible role in visual communication is unknown.

Non-imaging photoreceptors

Such receptors might also play a role in communication. Adult butterflies lack ocelli, but extra-ocular photoreceptors have recently been reported on the genitalia of both sexes in a small but diverse array of species (Arikawa *et al.* 1980). This 'anatomical provision for hindsight' (C. M. Williams) may serve some function during copulation.

Butterfly Colour Patterns

The colours and patterns of butterfly wings have been responsible for their great popularity among scientists, collectors and the general public. Attention has been given to the nature and synthesis of pigments, the mode of production of structural colours, the development and genetics of wing colour and pattern, and the role of visual signals in communication with vertebrate predators. However, the signal properties of butterfly colours and patterns, as related to communication, are as yet poorly understood. It is not intended to provide a comprehensive review of this formidable subject, which has been reviewed in general terms by Hailman (1977) and Hamilton (1973). Rather, a few features of butterfly coloration especially pertinent to the topic of this paper will be discussed. Readers are referred to the voluminous literature on butterfly coloration for reviews of other aspects (e.g. Crane 1954, Graham 1950, Nijhout 1978, 1981, Papageorgis 1975, Poulton 1908, Robinson 1971, Schwanwitsch 1924, Silberglied 1977, 1979, Turner 1977*a*, 1978, Vane-Wright 1975, 1976, 1979*a*).

Ultraviolet patterns

The human observer describes butterfly colours in terms of our system of colour nomenclature, which is inadequate for the task because butterfly vision includes an ultraviolet (UV) component. With some exceptions, diurnal terrestrial vertebrates do not see UV, so this spectral region may provide a 'private channel' for communication among insects (Silberglied 1979). A butterfly's wings may reflect little or much UV; as with 'visible' colours the reflection may be produced by scattering in the absence of UV-absorbing pigments, or by physical means. A great variety of UV patterns exists, even in 'visibly' concolorous groups such as pierids and satyrines, and striking differences in UV patterns between otherwise similarly coloured species occur in many genera (e.g. *Colias*, Silberglied & Taylor 1973, 1978; *Gonepteryx*, Mazokhin-Porshnyakov 1957, Nekrutenko 1968; *Phoebis*, Allyn & Downey 1977; *Pieris*, Bowden 1977; *Prepona*, Descimon *et al.* 1973–74) (cf Wynne-Edwards 1962: 34). Sexual dimorphism is more pronounced in the UV than in the 'visible' spectrum; local polymorphism, geographical variation and other colour phenomena also occur (Silberglied 1979). Since insect compound eyes are maximally sensitive in the UV, as much attention should be paid to UV patterns as potential sources of behaviourally meaningful signals, as to 'visible' reflection patterns. Furthermore, since vegetation generally absorbs UV, reflection in this region serves to maximize colour contrast in terms of insect vision.

Iridescence

Iridescence is usually defined as 'exhibiting a rainbow-like display of colours'. Most cases of iridescence in butterflies are attributable to reflection from multiple thin-film interference filters present in the outer layer of wing scales (Mason 1926–27, Anderson & Richards 1942, Simon 1971). In most such cases, including spectacular examples such as *Morpho* and many lycaenids, the range of reflected colour lies in the blue-to-violet end of the 'visible' spectrum; further investigation usually reveals intense UV reflection as well (Ghiradella *et al.* 1972, Ghiradella 1974). Due to the mechanism of colour production, such colours change in hue and intensity as the angle of the wing changes during flight (Crane 1954). At any particular angle, the reflection is of very high spectral purity and thus unlike other reflected light in the terrestrial environment. To an observer watching through a narrow spectral window (e.g. a television camera with a narrow-band filter; Eisner *et al.* 1969) or with species whose reflection barely enters the 'visible' spectrum (e.g. *Apatura* spp.), these butterflies are conspicuous at considerable distances as their wing reflection flashes 'on' or 'off'. The reflection may also be partially polarized at certain angles.

Such iridescent patterns are widespread among butterflies, especially in the three largest families, Lycaenidae, Nymphalidae and Pieridae. In pierids and some other groups, these spectacular patterns of iridescence are nearly always totally invisible to the human eye, as the reflection does not pass beyond 400nm into the violet. Iridescence, whether 'visible' or UV, is largely restricted to the dorsal wing surfaces, and is far more strongly developed in males.

Light colours

A large number of species are light in colour; nearly all of the Pieridae, for example. White and other light colours also provide contrast with vegetation, and the suggestion has been made at various times that white may be aposematic (Wallace 1889, Kettlewell 1965, T. Eisner, pers. comm., C. L. Remington, pers. comm.). In light coloured butterflies, the ventral wing surfaces are usually darker and bear cryptic patterns and colours. Where sexual dimorphism occurs among light-coloured species, the males usually have a uniform field of colour, while that of females is infuscated or obscured by other markings.

Orange and red

Orange and red patterns must be considered to have potential signal value, since butterflies, unlike other insects, can see these colours. Red has been demonstrated to have a communication function in *Heliconius* (Crane 1955).

Sexual dimorphism

When the sexes differ in coloration, females usually have a protective pattern, either more cryptic or more mimetic than males. The male pattern is often visible from a greater distance than that of the female, even in cases where the latter is mimetic of an aposematic unpalatable model. Female coloration also appears to be more variable, as has been shown quantitatively in moths (Fisher & Ford 1928).

Variation and polymorphism

In species with local polymorphism, either both sexes are polymorphic, or only the females. Male-limited polymorphism is virtually unknown in butterflies (Brower 1963, Turner 1978). Some species are polymorphic on a seasonal basis. The males of polymorphic species cannot use a single, narrow, visual searching-image for females. When both sexes are polymorphic, females likewise cannot use a single colour pattern to recognize conspecific males. Where geographic or seasonal variation or polymorphism occurs, there must be some compensation in the use of visual signals, by corresponding variation in the signal-receivers' behaviour (Burns 1967), a broad range of, or multiple, acceptable signals, constancy of the signal component (Rutowski 1981), or less reliance on visual signals for communication (Richards 1927, Brower 1963).

Mimicry

Mimetic species cannot use visual signals exclusively for communication, or confusion with similarly-coloured species would result. In sex-limited mimicry, the male cannot use visual signals alone for the recognition of conspecific females, but females could use them to recognize males. Brower (1963; see also Ch.25), in a brief but seminal discussion of sex-limited mimicry, suggested that in Müllerian mimicry where both sexes are mimetic, 'scent plays a more important role in courtship than sight.' It is alternatively possible that mimetic species use distinctive patterns of UV reflection for sex- and species-recogition (Remington 1973, Silberglied 1979).

A wide range of problems involving the use of visual signals by variable, polymorphic or mimetic species awaits investigation. For example, what visual signals are employed by male *Papilio dardanus*, and do these signals vary geographically with the female morphs? If red coloration is critically important in *Heliconius erato*, what signals are used by the yellow and blue Colombian *H. e. chestertonii*, the only subspecies that lacks red markings (see Turner 1975*a*)? How do similarly coloured sympatric congeners (e.g. *Adelpha* spp.) tell one another apart?

Butterfly Courtship

The courtship of butterflies is a complex ritual, involving the exchange of visual, chemical and tactile stimuli. While it differs in detail from one species to another, a general sequence of behaviour is followed by most species (Scott 1973*d*, Hidaka 1973, Silberglied 1977).

Courtship is usually initiated by the male. Males locate females by using one or both of two behaviour patterns: 'waiting' for females in locations where they are likely to be encountered, or 'seeking' females by active, persistent search of the habitat (Magnus 1963; these behaviours have also been named 'perching' and 'patrolling' by Scott 1973*d*, 1974*b*). Both types of behaviour may be found within a species (Dennis 1982), or in an individual at different times (Davies 1978).

Actual courtship begins in response to the sight of the female. It is easy to study the initial approach behaviour of males, by using 'dummies' of various sizes, shapes and colours (e.g. Crane 1955, Magnus 1958*b*, Petersen *et al.* 1952, Petersen & Tenow 1954, Silberglied & Taylor 1978, Stride 1957, 1958*a*, Tinbergen *et al.* 1942, Tinbergen 1958). The visual stimuli required to attract a male are surprisingly unspecific in many species. Movement, colour and size are usually important, while shape and detail of pattern are not (but cf. Lederer 1960, Hidaka & Yamashita 1975, 1976). Males of some species fly after anything that might be another butterfly, including falling leaves, birds, or a moving butterfly net. Further male behaviour appears to depend on the response of the chased object—other males may be chased away, birds avoided, and females courted.

In *Colias*, only a stationary visual stimulus is needed to release the entire sequence of male courtship behaviour, so that males may actually copulate with paper dummies (Silberglied & Taylor 1978). In other species (*Argynnis*; Magnus 1950, 1958*b*), the male requires motion and olfactory cues before continuing his courtship.

The later stages of courtship depend on the receptivity of the female. In those species carefully studied to date, olfactory signals released by the male are usually required for successful mating (e.g. Brower *et al.* 1965, Lundgren & Bergström 1975, Magnus 1950, Pliske & Eisner 1969, Rutowski

1977*b*, 1978*a*, 1980*a*, Stride 1958*a*, Tinbergen *et al.* 1942, Tinbergen 1958; cf. Pliske 1975*b*; for reviews see Scott 1973*d*, Silberglied 1977 and Ch.25). The more elaborate behaviour associated with courtship appears at this stage, when the male disseminates his odour(s) in the vicinity of, or directly upon, the female's antennae. Depending on the species, the male may dust her antennae during flight or while she is resting (*Danaus gilippus*; Brower *et al.* 1965), rub her antennae with one or both wings (Tinbergen *et al.* 1942, Temple 1953), or perform other specialized, species-specific behaviours (Crane 1955, 1957).

Most females encountered by males have already mated, and in most cases mated females respond negatively towards males. This negative response is characterized by 'rejection postures', 'ascending flights', and other behaviours which vary in detail from one species to another (Scott 1973*d*, Stride 1958*b*; see also Rutowski 1978*a*,*b*, Edmunds 1969), Chovet 1983). Males usually persist for a while in spite of the rejection posture, and a female that initially rejects a male may later accept him (Petersen & Tenow 1954).

In the final stage of courtship, the male positions himself alongside the female and probes laterally and ventrally along her wings with his abdominal tip, in an attempt to locate and couple with her genitalia. In most species, there is no visible indication that the female has accepted the male, other than absence of rejection posturing. In a few cases the female may be seen to adjust her position and lower her abdomen (Rutowski 1977*b*).

In some species, a receptive female may initiate courtship with a 'solicitation' flight towards a male (Crane 1955, Stride 1956, 1958*b*; Ch.21, Lederer 1960, Rutowski 1980*b*, 1981). Courtship may also be said to be initiated by the female in those few species where a sex-attractant pheromone is used (see also Ch.25). Such cases sometimes involve mating by the pharate or teneral female. Facultative mating with teneral females is common in many butterflies; in these circumstances the female may be unable to reject a male and prevent insemination (Taylor 1972).

In most species, the female becomes unreceptive after mating, and rejects subsequent matings for some time. Because the number of matings by a female may easily be determined by counting the number of spermatophores in the bursa copulatrix, substantial data exists on female mating frequency (Burns 1968; Ch.22). In general, females mate at most only a few times; large numbers of matings are very unusual (Ehrlich & Ehrlich 1978). Thus most mating attempts by males are not successful. Unsuccessful courtships may result from general female unreceptivity, female coyness or female preference (Ch.21).

While mating behaviour is usually initiated by males in response to visual stimuli, subsequent courtship behaviour is more complex. The male may require olfactory stimulation or he will not persist, and the female generally requires male olfactory stimuli before accepting (Ch.25). Tactile stimuli may be exchanged, especially in the later stages just prior to and during copulation. The role of visual stimulation during the later phases of courtship is poorly understood. In particular, the question of whether the female 'chooses' the male (or the male 'chooses' the female) on the basis of visual stimuli is not easily answered (see also Ch.21).

Visual Signals affecting Male Behaviour

The responses of male butterflies to visual stimuli have been studied in virtually every experimental investigation of butterfly courtship behaviour. Generally, dead specimens or 'dummies' have been used as stimuli, and the number of approaches by males counted as a measure of attractiveness. Such experiments are simple to perform, and the results have generally been consistent from one such study to another. One of their drawbacks is that little information is gained about the response of males to colour and pattern in later stages of courtship. Another problem with 'approach tests' is that the males' behaviour may easily be misinterpreted: males may be attracted to stimuli for reasons other than courtship (e.g. feeding, aggression, roosting, puddling, etc.). Some of these difficulties are circumvented by studying the behaviour of males toward living females in flight cages, or on tethers (e.g. Crane 1955, Brower 1959, Rutowski 1981, Silberglied & Taylor 1978).

Another method used to assess the importance of visual stimuli to males, compares mating frequencies of female colour phenotypes in polymorphic species relative to the frequencies of these morphs in the population (e.g. Smith 1973, Burns 1966, Pliske 1972; Chs 21, 22). Such data are obtained from mating pairs collected under field conditions, or from spermatophore counts in wild-caught females of the various morphs. While of great value in documenting assortative mating or sexual selection, this work does not necessarily identify the colour patterns as the signals used by male butterflies in mate choice. The correspondence between colour phenotypes and mating success may also be due to differences between female morphs in vigour, behaviour (coyness), pheromone production, longevity, or other properties correlated with or genetically linked to pattern loci.

In general, males are attracted to visual stimuli coloured most like the conspecific female (but see also Ch.23). The detailed pattern of the butterfly need not be duplicated in the more attractive dummies; only colour and UV reflection need resemble the female (e.g. Magnus 1963, Stride 1956, Obara 1970, Obara & Hidaka 1968, Silberglied & Taylor 1978). Often the ventral wing colour, exposed by the female when at rest, is most attractive. Both saturation and intensity of the colour make important contributions to a dummy's attractiveness. It has sometimes been possible to make artificial dummies that were more attractive than female specimens (Magnus 1958*b*, Silberglied & Taylor 1978).

In sexually dimorphic species, the colour of the male may be *deterrent* to other courting males, rather than neutral or attractive (but see also Ch.23). This was first discovered by Stride (1956, 1958*b*), who added the white of male *Hypolimnas misippus* wings to otherwise attractive female dummies. Males initially attracted to some of these dummies departed immediately, rather than persist in courtship. It is significant that the most pronounced deterrent effect was produced when he used white pieces of males' wings (rather than white areas from another species or an artificial material), which are intensely UV reflecting and iridescent (see also Ch.21 concerning this work). Rutowski (1977*a*) found that the UV flash of sitting, fluttering male *Eurema lisa* inhibited further courtship attempts by other males; courting males did not distinguish males lacking UV from females. In *Colias*, Silberglied & Taylor (1978) found that simply adding UV-reflection to an attractive dummy produced a 30- to 50-fold decrease in its attractiveness. They also found that mating males of *C. eurytheme* and *C. philodice* expose their hindwings when approached by another male; in *C. eurytheme* this action produces a sudden flash of intense UV-reflection that is surely repellent to the intruder.

In some species, male coloration elicits agonistic behaviour in other males. The deterrent or intimidating effect, of male colour on the behaviour of other males, is probably 'turned off' (or interpreted differently) during puddling behaviour, a predominately male activity (Arms *et al.* 1974).

Where studied, motion has been shown to significantly increase the attractiveness of a dummy (e.g. Tinbergen *et al.* 1942, Crane 1955, Magnus 1958*a*,*b*). In Magnus' classic experiments with *Argynnis paphia*, motion of the dummy was necessary to hold the attention of the male. He found that more and more effective super-normal stimuli could be created by increasing the degree of flicker in the dummy's apparent motion, until the flicker-fusion-frequency of the butterfly's eye was reached.

Size is another important stimulus. A graded response to size has usually been found, with low responses to dummies of the appropriate colour smaller than the female, good responses to dummies of the same size, and super-normal responses to giant dummies (Tinbergen *et al.* 1942, Tinbergen 1958, Silberglied & Taylor 1978). Males do not appear to discriminate against dummies shaped differently from the female, unless the differences assume extreme proportions (Tinbergen *et al.* 1942).

The details of wing-pattern usually do not contribute to the attractiveness of a dummy, except where the pattern affects the saturation or intensity of the entire wing. Thus the minor sexual differences of pattern found in genera like *Colias* (light spots in the dark borders of females), *Pieris*, etc. probably do not matter to males, at least in the initial attraction (Rutowski 1981).

Ethological studies of species with female polymorphism have sometimes demonstrated male preference for dummies resembling one of the morphs relative to others. Yellow females of *Pieris bryoniae* are less attractive than white ones (Petersen 1952). Magnus (1958*b*) found that the dark female form 'valesina' was less attractive to male *Argynnis paphia* than the normal phenotype. Extrapolating from the result that white is inhibitory to male *Hypolimnas misippus*, Stride (1956) tried to account for the rarity of its 'alcippoides' form (but cf. Edmunds 1969*b*, Smith 1976*a*, Ch.21). Burns (1966, 1967) used spermatophore counts from field-collected females to argue that male preferences maintain the non-mimetic female morph in *Papilio glaucus* (but cf. Prout 1967, Pliske 1972, Levin 1973, Barrett 1976; Ch.22). Male *Anartia fatima* are more attracted to dummies with white bands than to those with yellow bands; in this species the colours are not under genetic control, but change gradually from yellow to white with age (Emmel 1972, 1973*a*, cf. Taylor 1973*b*, Young & Stein 1976, Silberglied *et al.* 1979). The data and interpretation of several of these studies have been subject to dispute. I will discuss (below) only the relevance of dummy experiments to male mating behaviour under natural conditions.

Visual Signals affecting Female Behaviour

> Literally not one particle of evidence [exists] that the female is influenced by colour.
>
> (Wallace 1877)
>
> There is no proof to date for the assumption that any visual selection in favor of normally colored males takes place by butterfly females.
>
> (Magnus 1963)

In contrast with the numerous studies of the role of visual stimuli in male courtship behaviour, few attempts have been made to obtain similar information about females. There are many reasons for this. Male butterflies are aggressive in initiating courtship attempts, but females usually are initially passive or preoccupied with other activities. Any study of female courtship behaviour requires sexually active males. Second, 'determined' approaches can be used to bioassay attractive visual stimuli for males, but such tests can rarely be performed with females.

The coy responses of females are such that only actual copulation (or in rare cases where it is recognizable, solicitation flights or acceptance behaviour such as exposing the abdominal tip; Rutowski 1977*b*, 1981) can be used to compare the relative acceptability of various male phenotypes. Because females become unreceptive after mating, each can be used only once, so many virgin females are required. Third, many species have polymorphic females, and the experimenter may use the various female forms, dead specimens or even paper dummies as stimuli for courting males. But since males are rarely polymorphic, and because active males are required to test female responses, the investigator must modify living males with various experimental treatments to produce different colour or pattern phenotypes. Fourth, most experiments involving choice by females, between males from different groups, must be conducted in cages to keep the butterflies from dispersing, but many species will not behave normally in such an environment. Only those individuals that do not spend all their time trying to escape can be used, so large numbers must be reared to produce a cage-adapted population. Fifth, since mating behaviour changes with age, all experimental groups of each sex must be 'balanced'—they must have the same age distribution. Sixth, I have found that in nearly every experiment of this kind, untreated males mated more frequently than did treated control males of the same colour. This is probably due to minor injury of the latter during handling and treatment (see also Ch.4). A comparison between experimentally colour-modified males, and normally coloured, but *un*treated control males, is generally meaningless. Both untreated and treated control groups, as well as one or more experimental groups, should be included in the design of such experiments. For these and other reasons, few have attempted to perform them.

Crane (1955) and Rutowski (1981) used 'angling' techniques to assay for female responses to tethered males and artificial dummies. This method can succeed only if there is a visible bioassay for female acceptance behaviour.

Field data, on assortative mating in polymorphic species, are of limited value in determining the importance of male colour to females. Non-random mating and/or sexual selection are not necessarily due to visual discrimination and choice, even when there are coloration differences between phenotypes. While male phenotypes may differ in their relative representation in mated pairs, such a result may arise because of differences between morphs in thermal properties, vigour, courtship activity, competitiveness, pheromone production, longevity, etc. The rarity of male polymorphism is a serious practical limitation to the use of this technique, but it has been employed by Smith (1975*c*, 1980, 1981, Ch.21) in *Danaus chrysippus*. The method may also be used in species like *Chlosyne lacinia*, in which several colour forms occur sympatrically in both sexes. Field mating data may suggest that visual discrimination is occurring, but only experimental studies can test for the operation of visual signals.

Given these limitations, it is not surprising that data on female colour preferences are few, and needs to be augmented by additional studies of other species. Anyone with large cultures of living butterflies has the opportunity to test the ideas presented below. Experimental studies of the role of colour in mate selection in 'visibly' dimorphic species are particularly desirable.

Crane (1955) was the first to attempt the experimental study of the role of colour in female acceptance behaviour. She painted both sexes of *Heliconius erato* various colours, and observed the behaviour of other butterflies towards them. She claimed that 'the farther the altered color of the forewing from the normal scarlet in the spectrum, the less notice is taken of the butterfly, either by the opposite sex or as a general subject for social chases.' But she also remarked that 'positive courtship responses, sometimes including copulation, were obtained at least once for each color change effected in each sex', and that 'females were in general less influenced [in behaviour] by color change than were males.' Her discussion of this experiment includes no indication of sample size for each colour beyond 'at least one butterfly of each sex', and she did not include treated control individuals. She describes difficulties with variables of weather and the physiological state of the butterflies, and the reader is presented with a generalized summary of the results rather than quantitative data. It is not clear that social interactions were fully separable from sexual behaviour. Finally, most of her discussion of the role of colour deals with the responses of males.

Another series of experiments performed by Crane (1955) used cloth dummies angled from bamboo wands. The *single* female tested responded 'positively' to red and orange, but minimally or not at all to other colours. The 'female's responses were . . . gauged from the strength of her courting behavior, shown by the degree of abdomen elevation, extrusion of yellow gland, apposition of forewings and fluttering of hindwings.' But this is a description of female rejection behaviour (L. Gilbert pers. comm., J. Mallet pers. comm.). One may not conclude that female *H. erato* discriminate between males on the basis of colour.

Stride (1958*b*) studied the responses of female *Hypolimnas misippus* to two kinds of modified males. Two 'black' males had their white spots blackened chemically, a treatment that did not affect the UV-blue iridescence. Three 'colourless' males had most of their wing scales removed. Males were tested individually in a large flight cage containing six virgin females in each test. None of the 'colourless' males mated successfully (18 courtships), while both of the blackened males were successful (three courtships). All four untreated control males mated (four courtships). Unfortunately, Stride did not include a treated control group, so there is no way to evaluate the traumatic effect of scale removal. He did not control for odour (Brower 1963), so the results may have been affected by disruption of a chemical communication system involving wing-born pheromones (cf. Silberglied & Taylor 1978, Rutowski 1980*a*, Grula *et al.* 1980). His sample size is too small for statistical analysis. Nevertheless, the all-black males mated successfully, so one can probably agree with Stride's conclusion that 'the white spots on the wings of the males played little part in the courtship of the butterfly.' There was no female selection against males lacking the most conspicuous 'visible' pattern element.

Rutowski (1981) used his 'angling' technique to study solicitation flights of female *Pieris protodice*. By removing UV-absorbing pterins from the wings of males, he produced dummies that resembled females. He found that females use the UV-absorption of males as a sexual recognition signal. No colour modifications were performed in the 'visible' spectrum.

In *Danaus chrysippus*, a species in which both sexes are polymorphic, males heterozygous at a locus affecting colour pattern have a significantly higher mating success than either homozygote (Smith 1981, Ch.21). However, the minor colour differences between the heterozygote and one of the homozygotes are probably too small, in my opinion, to be acted upon by visual selection. Smith partly attributes the difference in mating success to inter-male competition, mediated through the interaction of climatic conditions and the thermoregulatory properties of these colour patterns.

Silberglied & Taylor (1978) reported an extensive series of experiments on two species of *Colias*. Their goal was to identify the basis of conspecific assortative mating by females. The experiments, performed in flight cages, involved large numbers of receptive females of both species, both untreated and treated control males as well as experimental groups, and balance for age distribution.

Colias eurytheme has orange, UV-reflecting males, while male *C. philodice* are yellow and UV-absorbing. Males of the yellow species were coloured orange by means of felt-tipped pens, and the males of the orange species yellow, by transplanting the discal area of the wings, of a yellow-winged, UV-reflecting strain of hybrid origin, onto the corresponding area of *C. eurytheme*. Females were still fully discriminatory: they mated conspecifically, and they *accepted* the peculiarly-coloured conspecific males as frequently as they did the normally-coloured treated control males of their own species. In another experiment, males *C. eurytheme* (Colour plate 4A) and *C. philodice* (Colour plate 4B) dyed orange, red, green, blue and black mated as frequently and as conspecifically as did treated control individuals!

The only visual component, the modification of which had any effect on female mating behaviour, was the ultraviolet reflection of *C. eurytheme*. Regardless of their 'visible' colour, males of this species whose UV reflection had been destroyed, suffered a significant drop in the number of successful conspecific matings. Yet they were not accepted by *C. philodice* females, even though the males of *C. philodice* are UV-absorbing, and even though *C. philodice* females accept their own males in any colour. In short, the females of *C. eurytheme* responded only to the UV component of wing coloration, and the females of *C. philodice* acted as though they were totally blind to the male colours or UV reflection!

The olfactory basis of mate selection has since been identified in these *Colias* species (Rutowski 1980*a*, Grula *et al.* 1980). So far as female discrimination is concerned, olfactory cues are far more important to the females of these species than are visual signals (Silberglied & Taylor 1978). It is significant that the females of these species are polymorphic for colour, while the males are not. This colour distribution agrees with the idea of stronger female than male sexual selection stabilizing coloration of the opposite sex. But the experimental results directly conflict this suggestion.

I have since carried out one additional experiment on the role of colour in female choice, this time with the brilliant red Neotropical species *Anartia amathea* (Colour plate 4C). It was performed in collaboration with Annette Aiello at the Smithsonian Tropical Research Institute, and is published here for the first time. The experiment meets all of the criteria discussed at the beginning of this section.

Freshly-eclosed but hardened *A. amathea* were used. Virgin females were mixed with equal numbers of males, divided into three groups: untreated control (red), treated control (red), and experimental (black). Treatment involved painting the dorsal wing-surfaces with clear (Ad Marker (R), Jacksonville, Florida: Warm Gray 1) or black (same brand, Super Black) felt-tip pens. The butterflies were released into a flight cage and subsequent matings recorded. Because there were three groups for each sex, there were nine mating combinations possible. The results are presented in Table 20.1.

Only 21 matings were obtained, so the number in each cell is very small. For both logical and statistical reasons (the total number of matings in each cell is the result of two phenomena: male attraction/courtship of females and female acceptance/rejection

Table 20.1. A mating experiment on sexual selection for colour in the brilliant red butterfly *Anartia amathea* (Nymphalidae)

	Females			
	Red (untreated control)	Red (treated control)	Black (experimental)	TOTAL
Males:				
Red (untreated control)	5	3	1	9
Red (treated control)	3	2	0	5
Black (experimental)	4	3	0	7
TOTAL	12	8	1	21

The experiment was performed in a 3m × 3m × 2.1m screened cage on Barro Colorado Island, Panamá, with stocks collected in the Darién region of eastern Panamá, in 1976.

behaviour, and the small sample size in each cell), the results in each of the nine cells cannot be meaningfully compared with one another. However, if we compare the number of matings by group within each sex, a distinct picture emerges.

Notice first the number of matings by females in each of the three groups. Black (experimental) females did not mate as frequently as did either of the red females (one-tailed binomial test of black vs. red treated control females, $P = 0.02$; Siegel 1956). It is highly unlikely that the acceptance/rejection behaviour of females is changed by painting them a different colour. The only logical conclusion is that the males spent little time courting them compared with the red females. This is entirely consistent with what we know about the role of colour in male behaviour.

On the other hand, notice that those females that mated did *not* discriminate between the red and black males (one-tailed binomial test of black vs. red treated control males, $P = 0.44$) (or if they did, they might have preferred the black!). Thus females did *not* choose on the basis of the red colour—the most conspicuous feature of the wings and the only colour present. A devil's advocate might argue that they prefer black, but this is hardly what we are concerned with when we speak of the brilliant colours that have been or could be so nicely explained by female sexual selection. This brilliant red butterfly sees red (Bernard 1979), and the males respond to red. Because the male has the more brilliant red coloration, it is only logical that it be the result of female selection. Yet it is clear even from this minor experiment that such is not the case.

Brower *et al.* (1971) have argued convincingly that *A. amathea* is an 'incipient mimic' of *Heliconius erato/melpomone*. This contention is supported by their experimental data which show that predators, having learned to discriminate *H. erato*, will also sight-reject *A. amathea*. Thus the red coloration might be 'explained' on that basis. But the male of *A. amathea* has a more brilliant red and a darker black, and would probably be a better mimic. Unless the red of the male serves some other role, we have a case in which the male is the better mimic than the female, which would be a unique situation so far as I am aware (Carpenter & Ford 1933).

Discussion

Is it possible, or wise, to attempt to make generalizations about the use of visual signals for communication among the 10 000–15 000 species of butterflies? I believe it is, because they all have to deal with the same problem: recognition of and communication with conspecific individuals. Their vision is constrained by the same solar spectrum. Their use of colour is limited by biosynthetic versatility, as well as by the activities of visually-oriented predators. Diverse solutions to the problem are found, including the discarding of visual communication for many functions. But a general picture of the use of visual signals by butterflies has emerged from the confusing wealth of species, colour patterns and behaviour.

Before proceeding, it is well to emphasize that exceptions will be found to such generalizations, and that the exceptional cases, when studied in depth, often serve to confirm, refute or qualify the hypotheses generated. Since so few intensive studies of butterfly communication have been carried out, I cannot hope to convince—only to raise doubt where it should exist, about the classical interpretation of the role of butterfly colour patterns in communication.

The two functions—protection from visually-oriented predators, and social signals used during courtship—may at first seem to be mutually exclusive, because the former often requires patterns that are difficult to detect (crypsis) or are confusing (mimicry), while the latter demands a high signal-to-noise ratio and uniqueness with respect to species and sex. But it should also be remembered that these functions involve different visual systems: the vertebrate lens-eye and the lepidopteran compound eye. It is possible that some patterns serve both functions by taking advantage of the differences between these visual systems (Silberglied 1979).

Vision and Pattern

Butterfly vision is characterized by excellent colour discrimination across an extremely broad visible spectrum, low spatial resolution and high sensitivity to motion or flicker. Colour contrast is increased in particular spectral regions by physical devices (e.g. tapetal reflection) and neurophysiological mechanisms (stimulus filtering). The ability of the insect's eye to detect a moving object is strongly enhanced by brightness- as well as colour-contrast. At the behavioural level, butterflies are capable of discriminating quite subtle differences of hue, intensity and saturation (e.g. the difference between greenish-yellow and yellowish-green on the ventral hindwings of male and female *Colias*; Silberglied & Taylor 1978), and of learning (C. A. Swihart 1971, Swihart & Swihart 1970).

Hence the extremely brilliant patterns that characterize the dorsal surfaces of so many male butterflies probably are not required for communication by any intrinsic limitations of the visual system. For iridescent colours, the change of hue with angle of the wing, narrow spectral reflectance at any given angle, abrupt intensity change with angle at any given wavelength, and strong ultraviolet component (and possible polarization) that contrasts strongly with vegetation, are features that would serve well as *signals for long-range communication*. *Morpho* can easily be seen from low-flying aircraft; the gleam of their wings, as they sail above the forest canopy, makes them appear as giant, blue, flashing beacons. Describing *M. rhetenor*, H. W. Bates (1864) reported that 'when it comes sailing along, it occasionally flaps its wings, and then the blue surface flashes in the sunlight, so that it is visible a quarter of a mile off.' Where iridescence is not involved, the flicker of the alternating dorsal and ventral surfaces exposed by beating the wing, often enhanced by colours that contrast well with the green of vegetation, might perform the same function. The patterns of females, being more variable and less conspicuous, ought not to be as apparent over the same visual distances as are those of the males.

The mode of colour production in butterflies allows for an overlay of transparent scales producing structural colour over a base of pigmented scales. As a result, these two types of coloration may evolve and be expressed independently of one another. We see a striking example of this in males of *Hypolimnas misippus* and related species, where a patch of intense ultraviolet reflection overlays both white and black. In terms of insect vision, this produces a white bull's-eye, surrounded by a ring of pure ultraviolet in a field of black. While the pigments have not been studied, only melanin or some other dark pigment would be required to achieve this effect, because white is produced by diffuse scattering, and the UV reflection by a transparent interference filter overlay. Thus, a remarkable range of coloration, with diverse optical properties, is made possible by the combination of pigmentary and structural colours. A glance through a collection reveals the extent to which these possibilities have been realized.

Thus we find two solutions to the problem of providing increased or decreased brightness- and colour-contrast. Reflectance over much of the spectrum may be increased (or decreased) (the pierid/satyrine solution). Alternatively, a particular colour may be increased (or decreased) in intensity beyond anything found in the habitat, by optical interference (the iridescence solution). The latter allows for the combination of intense reflection (or absorption) with *any* underlying visible coloration, and/or restriction of iridescence to the ultraviolet. The use of dark borders around a brilliant iridescence (as in most *Morpho*, some *Hypolimnas*, many *Colias*, etc.) provides a high-contrast edge to the intense colour field borne on the wings.

A visual signal may be effective in any spectral region that coincides with the spectral sensitivity range of the receiver. In butterflies we should expect to find important visual signals scattered from the UV to the red. Feeding and mate-finding behaviour are mediated in part by visual signals throughout the spectrum, and the butterfly's responses to differences of hue, saturation and intensity are most evident in these contexts. However, a surprising number of important signals involving sexual communication are effected through bright coloration at all wavelengths (white), UV-reflectance and iridescence. Many of these appear to be all-or-none signals, such as 'I am a male' or 'go away'. They differ from the full-spectrum group in that the signal's function is of great value to the sender as well as to the receiver.

Iridescent colours and/or ordinary high-intensity reflection, as well as large size, are characteristic of butterflies of open spaces, where long-range signalling may be important. The most spectacular examples of such colours are found in wide-ranging pierids and nymphalids that frequent the upper canopy, and species that use treefall gaps, shafts of sunlight and other openings in the forest. Such colour patterns are relatively less well developed in the satyrines, ithomiines and other groups that frequent deep forest or dense vegetation. While there are numerous exceptions, I believe most lepidopterists would agree with these generalities (cf. Hingston 1933, Papageorgis 1975).

Courtship and Intersexual Selection

Male and female behaviour during courtship agrees well with observations on other organisms, and with recent sexual selection theory (see Blum & Blum 1979, Maynard Smith 1978, O'Donald 1980, Parker 1974, Rutowski 1982, Thornhill 1976, 1979, 1980, Trivers 1972, West-Eberhard 1979, Williams 1975). Males are aggressive toward rivals, and persistent in locating and courting females, while females are coy and effective at rejecting males. Female mate-rejection behaviour has been described in numerous species (Scott 1973*d*). A male butterfly usually cannot rape an unreceptive, rejecting female unless she is teneral (Taylor 1972).

Female rejection behaviour is elicited by courting males in a variety of situations. Recently fertilized females are generally unreceptive. A receptive female may respond with rejection behaviour in the initial stages of courtship, or she may respond negatively if the courting male lacks appropriate signals (e.g. another species with different pheromones). While UV visual signals are employed in a few cases, pheromones are almost always involved; a male that has been experimentally deprived of the necessary pheromones is usually rejected (e.g. Tinbergen *et al.* 1942, Tinbergen 1958, Pliske & Eisner 1969; Ch.25).

Male butterflies depart immediately after copulation: there is no behavioural post-mating investment by males on behalf of their offspring. Males do not control access to resources. Besides gametes, the male's only contribution toward his offspring is the spermatophore and its contents, which in one species have been shown to be metabolized by the female and incorporated into the contents of eggs (Boggs & Gilbert 1979; but see also Chs 3, 6).

If a female were somehow to assess male quality during courtship, in terms of potential paternal investment, she would have to do so on the basis of the spermatophore he would produce if she accepted him. Older males that have previously mated produce smaller spermatophores, and also place the female at risk to predation due to prolonged copulation times (Rutowski 1979, L. E. Gilbert, cited in Thornhill 1976, Silberglied, unpublished). But because of her low visual resolution, it is unlikely that a female can discriminate between young and old males on the basis of colour pattern or wing wear. Most of a female's efforts in a generally short life are spent locating suitable hostplants for oviposition (Watt 1968; Ch.6). A female is unlikely to waste valuable time selecting among males on the basis of colour or pattern, features that are likely to be poor predictors of age and male quality (cf. Thornhill 1980). A female that mated with a male who produced a small spermatophore (or who was judged unsatisfactory on any other grounds) could always remate sooner than she otherwise would. (Because strong sperm precedence exists in butterflies, a male should produce large spermatophores to prevent this—Ehrlich & Ehrlich 1978; see also Rutowski 1978*a*, 1978*b*.) Another male adaptation that may perform a similar function is the antiaphrodisiac pheromone (Gilbert 1976). The female should minimize the time to copulate with a healthy male, so as to get on with oviposition—and the identification and health of a male may be better determined by chemical signals than by visual ones (Rutowski 1978*b*, 1981; Ch.21). Chemical signals may be more reliable at close range than visual ones, due to variation in lighting conditions and in the relative positions of sender and receiver. Initial rejection behaviour by receptive females may also be interpreted as a means of assessing male vigour or potency (Rutowski 1979).

Butterfly Coloration and Male Behaviour

Male butterflies definitely use colour as a signal to recognize females. This has been demonstrated in nearly every experimental study. But generally males do not encounter more than one female at a time, and they respond to anything that *might* be a female by approaching it. In natural conditions, males must decide whether to chase sequentially presented stimuli that might be conspecific females. A wild male rarely chooses between different stimuli presented simultaneously, as is usually the case in 'dummy' experiments. Males make inspection flights in response to a much broader range of visual stimuli than one might predict on the basis of results of approach tests.

Unlike females, males may mate many times; variance in the number of matings by males is probably quite high. Because receptive females are scarce and competition among males is high, one should not expect to find strong male discrimination against conspecific receptive females. In fact, males often persistently court, and sometimes mate with, females of other species and even on rare occasions with other males (Tilden 1980, Silberglied, unpublished).

It is unusual to find unmated older females (Ehrlich & Ehrlich 1978), even in species having female polymorphism with 'unattractive' forms such as 'valesina' of *Argynnis paphia* (Magnus 1958*b*) and 'turnus' of *Papilio glaucus* (Burns 1966, 1967; cf. Prout 1967, Pliske 1972, Levin 1973; Ch.22). Such females may or may not attract as many males, over as great a distance or as often, as 'normal' phenotypes, but it is unlikely that the availability of willing males is ever limiting, or that copulation by such females is significantly delayed. Females may also compensate for lower attractiveness to males by performing 'solicitation' flights.

For these reasons I do not believe that the colour preferences shown by males, under experimental conditions, necessarily translate into significant sexual selection on female colour in the field. There exists a graded response to sign stimuli, but it is in the male's best interest to mate at every opportunity, and to drive away or avoid competing males. A male's colour preferences serve him well in *locating* conspecific females, in shunning or thwarting rivals, and in wasting less time with members of other species. But because opportunities for choice between receptive females rarely occur, male colour preferences probably do not serve to stabilize female pattern to any great extent (Magnus 1963, Pliske 1972). The high variability of female coloration, and female polymorphism, are indirect reflections of this (cf. Richards 1927).

Butterfly Coloration and Female Behaviour

It is not surprising that we know so little of female preferences in butterflies. This situation is a reflection of the difficulties involved with such experiments, including the few female behaviours that can be used to measure the relative acceptability of males, the requirements that receptive (i.e. virgin) females be used and that a separate female be used for each data point (i.e. mating), and the fact that many, if not most, species do not mate readily in cages. There has also been a widespread assumption of colour discrimination by females, hence little interest in such experiments. Yet all experiments of this type point to the fact that colour is used less by females than by males, as a basis for discrimination, that females use colour little, if at all, and that when females use visual cues, these are primarily in the UV (Silberglied & Taylor 1978, Rutowski 1981).

However, I do not deny that females choose. They exhibit effective rejection behaviour, especially after mating and during other periods of unreceptivity. The evidence simply does not support choice on the basis of 'visible' colour characters. Olfactory and ultraviolet signals are used. Perhaps if we were as receptive and attentive to olfactory stimuli as we are to bright colours and complex patterns, we would appreciate the fragrant world of male butterflies as a sensory nirvana surpassing even their kaleidoscopic adornments.

A hypothesis cannot be proven; one can only attempt to refute it by means of experiment and observation. Darwin's hypothesis is that female choice is based on male colour. Experiments have been designed to try to disprove this hypothesis. Females have been offered choices among males bearing different colours. They exhibited little preference, if any, and at most far less preference for colour of their mates than do males. Hence the hypothesis is falsified, in spite of its logic, in spite of its apparent success at accounting for many colour phenomena in butterflies, and in spite of its attractiveness to evolutionary theorists.

The argument might be made that sexual selection is operating, and that females are choosing males on the basis of colours not changed in the experiments (e.g. the black of *Colias* spp. and *Anartia amathea*). However, the question is, to what extent are the *spectacular* patterns and colours of butterflies, as we see and interpret them, a product of sexual selection by females? Ultraviolet signals and dull 'background' colours are not at issue here; the former were unknown to Darwin, and the latter have never been suggested as products of intersexual selection.

Male Intrasexual Selection — an Alternative Hypothesis

The season of love is that of battle.

(Darwin 1874)

If female colour preferences are unimportant selective forces on male coloration, we are left with the problem of accounting for the peculiar sexual distribution of colour and pattern in butterflies. The problem is now more acute, for the old argument is now reversed: if colour is more important to the male than to the female, we should expect females to be less variable in colour than males.

I would like to suggest an *alternative hypothesis*, one that may explain the distribution of many butterfly colour patterns, and yet be in accord with the experimental and observational data on butterfly courtship. I believe that *intrasexual communication between males* (see also Ch.23), rather than inter-sexual or interspecific communication, *is the major selective agent responsible for brilliant male coloration*, low male colour variability, lack of male-limited polymorphism, and absence of male sex-limited mimicry.

Darwin (1874) recognized conflict between males as the other important component of sexual selection, but his discussion of characters produced by intra-sexual selection was concerned primarily with physical structures, such as the horns of beetles, narwhals and moose. He was aware of aerial battles between males in some butterfly species, but did not link such observations with sexual dimorphism for colour or pattern. Darwin appreciated that inter-male displays figure in the evolution of bird coloration, but he did not consider brilliant butterfly wing colours as important signals used for communication between males. He instead attributed them almost entirely to intersexual selection by females.

The alternative hypothesis, of male intrasexual selection, as an important factor in the evolution of butterfly sexual dimorphism, is not new. Wallace (1877, 1889), Hingston (1933) and Huxley (1938*a*) refuted intersexual selection in general, in favour of intrasexual selection among males. In particular, Hingston (1933) discussed the anomalous distribution of colour and pattern in butterflies. He presented a wide variety of observational evidence in support of his view that the brilliant dorsal colours of many male butterflies function in fighting. To Hingston, male butterflies engage in 'psychological warfare', 'a battle of bravado, gesticulation and threat'; 'their colours are . . . their weapons'.

Other recent authors have mentioned male interactions as a possible factor in the evolution of sexual dimorphism, but usually without further elaboration or supporting evidence (e.g. Turner 1978). Vane-Wright (1980), reviewing possible combinations of sexual interactions between butterflies, has also suggested that colour patterns might play a role as male-male signals (see also Ch.23). The hypotheses presented by Wallace, Hingston and Huxley are in accord with many of the observable facts of butterfly colour distribution. Inter- and intrasexual selection are not mutually exclusive agents. In many cases, they serve equally well to 'explain' the same observations. Until recently, it has not been possible to weigh the relative importance of intersexual versus intrasexual selection, due to lack of direct evidence. Now, having demonstrated experimentally that female selection on male colour is weak at best, we must evaluate the evidence regarding male intrasexual selection on colour and pattern. This evidence comes from diverse sources, including field observations and experiments, as well as from the physical nature of colour and its distribution among butterfly species.

The male is the active, mate-locating sex in butterfly courtship. If brilliant colour had evolved as a signal facilitating mate location, we would expect to find it better developed in the female. This is not the case. Colours detectable at a great distance are not necessary for female choice, because butterfly courtship usually takes place in a space of less than one cubic metre.

Fighting between males has been reported in numerous species (Baker 1972*a*, 1978, Davies 1978, L. E. Gilbert in Maynard Smith & Parker 1976, Joy 1902, Richards 1927, Hingston 1933). The behaviour can be observed at food sources as well as in other, but not all, contexts (cf. puddling, migration; Shapiro 1970). In a few stout-bodied species, the wings are modified as weapons (e.g. *Charaxes* spp.; Owen 1971*a*), and physical damage may be inflicted during flights (Darwin 1874). But direct physical contact should be avoided, for such action may result in a pyrrhic victory, in which the winner may be as badly damaged as the vanquished. Selection for effective alternatives, such as ritualized threat displays, would be expected. The signals need not be visual—loud snapping noises accompany inter-male fights in *Hamadryas* (Darwin 1839, Swihart 1967*b*).

Males of many butterfly species are 'territorial' (Baker 1972*a*, 1978; see also Silberglied 1977). Such behaviour is best developed in species exhibiting 'waiting' mate-location behaviour (cf. Scott 1973*d*). Territorial defence in butterflies probably serves to prevent access by rival males to sites at which receptive females fly. (Among birds, food resources, perches, and nests are also defended.) For this reason, butterfly 'territories' are not necessarily fixed in location from day to day, or even from one time of day to another (Baker 1972*a*, Owen 1971*a*, Davies 1978). These differences between butterfly and bird territoriality have served to complicate discussions with semantic problems and confound researchers looking for strict analogies (e.g. Ross 1966, Scott 1973*d*). In species that 'hilltop' males defend prominences to which receptive females fly in search of mates (Shields 1968, Scott 1974*b*). In territorial or aggressive species, brilliant male colour may serve to intimidate rivals for prime locations. Males that are victorious in such encounters would receive a disproportionate share of matings, resulting in selection for stronger male signals.

Male butterflies of many species make inspection flights whenever anything that might be a female appears (e.g. Tinbergen *et al.* 1942, Lederer 1960, Stride 1957). As a result, frequent male-male interactions occur, especially in species exhibiting 'seeking' mate-location behaviour (though also in 'waiting' species). Mating success in such species should be related to the efficiency with which an area is searched for females. It would be of advantage to a male to identify his sex to other 'seeking' males from a great distance; thus reducing time wasted in fruitless homocourtships. It would be of advantage to the signal receiver for the same reason, and also because his time may be better spent searching areas not recently explored by another cruising male (cf. Rutowski 1981). While such behaviour may result in greater dispersion of males throughout the habitat, its explanation is based on individual advantage rather than group selection (Wynne-Edwards 1962).

Experimental evidence also supports the hypothesis that males respond to one another's colours. Colour patterns are more important to males than to females. Not only are males attracted to visual stimuli resembling females in approach tests, they are *repelled* by visual stimuli resembling other males (Obara 1970, Stride 1958*a* (but see also Ch.21), Rutowski 1977*a*, 1981, Silberglied & Taylor 1978). White, UV reflectance and iridescent colours seem to be most important in this regard. Some species have special displays in which the male's dorsal colours are exhibited while sitting (e.g. 'flutter response'; Obara & Hidaka 1964, Rutowski 1977*b*, 1978*a*; see also David & Gardiner 1961). Another striking example is illustrated by the stereotyped rejection behaviour of mating male *Colias eurytheme*, in which the hindwings are flashed at approaching intruders (Silberglied & Taylor 1978). However, to my knowledge no one has performed an experiment to determine if colour may affect the outcome of intermale combats between butterflies.

Males of the Grayling butterfly studied by Tinbergen and his colleagues (Tinbergen *et al.* 1942, Tinbergen 1958) are unusual in discriminating colours while feeding but not in relation to social interactions. This lack of colour use is consistent with the lack of intense colour and absence of strong sexual dimorphism in this species.

Like birdsongs, brilliant butterfly colours may serve as agonistic devices, used for threat, intimidation and rejection. The attractive, alluring, seductive characteristics of male butterflies, required for successful courtship are their odours. Male pheromone glands are widespread in nocturnal as well as diurnal Lepidoptera, and their pheromones, unlike female sex-'attractants', are chemically diverse (Ford 1945, Silberglied 1977; see also Ch.25). Intersexual selection has probably played the dominant role in the evolution of these scent organs and their secretions—not in the evolution of brilliant male colours.

Darwin (1874) noted the close analogy between the secondary sexual characters of birds and insects, especially in terms of beautiful colours and the distribution and nature of sexual dimorphism. 'Whatever explanation applies to the one class probably applies to the other; and this explanation . . . is sexual selection.' However, Darwin treated male coloration in birds largely as adornment evolved under intense intersexual selection. This view has not been supported by recent studies. According to Rohwer *et al.* (1980), 'bright plumages are evolved strictly for aggressive signalling. We know of no good support for the alternative hypothesis that bright coloration serves in female attraction or as an isolating mechanism.' In contrast, Baker & Parker (1979) concluded that 'bird coloration has evolved almost entirely in response to predation-based selective pressures. Although plumage and coloration are involved in species and sex recognition systems, they have not evolved in response to sexual selection pressures.' Thus, while ornithologists do not fully agree on the selective agents responsible for bird coloration, intersexual selection no longer has their enthusiastic support (but cf. Burley 1981 and Ch.21; see also Hingston 1933). Butterflies also differ from birds in the clearer physical partitioning of communicative and protective functions onto the dorsal and ventral wing surfaces, exposed, respectively, during display or at rest (Darwin 1874).

Baker & Parker (1979) also support the hypothesis that 'bright colours may commonly be favoured when an individual is anyhow obvious (e.g., through activity), and where it represents an "unprofitable" prey for a predator', as was suggested for insects by Jones (1932; see also Young 1971, Gibson 1980). Flash coloration—the sudden appearance of colour during flight and its sudden disappearance at rest—may startle or confuse predators. These and other 'predation hypotheses' do not conflict with, and are not mutually exclusive of, the suggested role of intrasexual selection as the major selective factor producing brilliant male coloration. They do not explain why brilliant coloration should be so advantageous for the male, but less so for the female. These hypotheses should be supported to the extent that they are in agreement with experimental data, but are unlikely to account for the phenomena of male brilliance, low variation, etc. Colour may also play other roles (e.g. interspecific signalling in social aggregations, such as puddling) that do not conflict with these conclusions.

Another hypothesis that has been proposed to account for brilliant male coloration is the so-called 'handicap principle' (Zahavi 1975). This suggestion, that females prefer to mate with males that have survived *in spite of* the 'handicap' of brilliant coloration, has serious theoretical flaws (Maynard Smith 1976, 1978, O'Donald 1980; cf. West-Eberhard 1979, Thornhill 1980). But even if that was not the case, Zahavi's hypothesis would predict female choice based on male colour. This prediction is not supported by the experimental evidence.

The concept of intersexual selection on colour pattern has figured prominently in discussions of butterfly systematics, genetics, behaviour, mimicry and evolution (e.g. Carpenter & Ford 1933; Wickler 1968, Vane-Wright 1971, Scott 1973*d*, Silberglied 1977, Turner 1977*a*, 1978). Some theories developed using this concept do not depend on male-female communication as a selective force, and may easily be modified by the partial or complete substitution of intrasexual for intersexual selection. Others will require major revision. Male intrasexual selection cannot account for all interesting colour phenomena in butterflies, many of which are better understood in the contexts of thermoregulation, camouflage, mimicry, and in some cases (especially UV), intersexual selection. These agents are not mutually exclusive, but act in concert, sometimes one or another playing a more important role, producing infinite variations, concerti and symphonies on the themes of colour, pattern and vision.

Summary

Butterflies have a visual spectral sensitivity extending from the ultraviolet to the red, the widest known among animals. They have excellent colour vision, and exhibit both innate and learned colour preferences. Butterfly wings bear diverse colour patterns, and colour is used by butterflies during feeding, oviposition and sexual behaviour. Nevertheless, there is little evidence to support Darwin's argument that intersexual selection, in the form of male and female mating preferences, acts as a potent force in the evolution of 'visible' butterfly coloration.

For males of many species studied, the colour of the female provides an important visual stimulus for mate-location and recognition. However, male behaviour is probably not an important selective factor in determining female colour, and males do not often discriminate between potential mates on its basis. This is so because male colour preferences are relatively broad, females are usually encountered and courted one at a time, the number of receptive females encountered usually limits male reproductive success, and males can mate many times. Female variation and polymorphism, sex-limited mimicry and low pattern diversity among females of closely related species provide indirect evidence that males' preferences probably do not act as important stabilizing selective forces on female colour and pattern.

Female butterflies exhibit well-developed rejection behaviour. The bases for discrimination by females have been little studied, and there are very few data on colour preferences. Available data reveal that female colour preferences are weak or absent, or are at most even less precise than those of males. The only visual signals on the males' wings that have been shown to affect female behaviour lie in the ultraviolet. As determinants of female acceptance behaviour, olfactory stimuli appear to be far more important than visual.

The brilliant colour patterns of male butterflies have signal properties that would serve well for long-range communication. It is suggested that male-male interactions may be the major selective agent in the evolution of bright colours and iridescence in males, low variability of male coloration, lack of male sex-limited mimicry, and several other general colour phenomena in butterflies. Recognition of other males, and advertisement of his own sex, are advantageous for a male in the contexts of agonistic, territorial and mate-location behaviour. This suggestion is consistent with conclusions of recent reviews of the evolution and function of bird coloration.

Acknowledgements

I would like to thank A. Aiello, S. Foster, I. Rubinoff, D. Roubik, N. G. Smith, M. J. West-Eberhard, H. Wolda, and other colleagues at the Smithsonian Tropical Research Institute for stimulating discussion of some of the ideas presented and for helpful comments on an earlier draft. Other helpful information, suggestions or access to unpublished material were provided by M. Boppré, B. Hölldobler, K. Mikkola, R. Rutowski, D. Schneider, D. A. S. Smith, and R. I. Vane-Wright. A special debt is owed to T. Eisner, who first kindled my interest in problems of coloration, and with whom several of the ideas presented were discussed on various occasions. The colour photographs for Plate 4 were prepared with technical aid of N. Smythe and K. Dressler. The Smithsonian Institution, Scholarly Studies Program and Fluid Research Fund are gratefully acknowledged for financial support of some of the experiments discussed.

21. *Mate Selection in Butterflies: Competition, Coyness, Choice and Chauvinism*

David A. S. Smith

Department of Biology, Eton College, Windsor, Berkshire

Butterfly populations, in common with those of other sexually reproducing animals, are composed of individuals varying in fitness. One component of this variation is the ability of individuals to compete for or propensity in choice of mating partners. A male butterfly who outshines his competitors in acquiring matings transmits his genes, and hence probably some of his successful qualities, to larger numbers of offspring. Similarly, a female who chooses or attracts and accepts such a male to sire her offspring maximizes her own fitness by blending her genes with his. Clearly, if mate selection departs from random its evolutionary potential must be considerable, as Darwin (1859, 1871) originally postulated. Much therefore hangs on the following questions: how commonly in wild populations does mating behaviour deviate from random panmixia? If preferential mating occurs, which qualities in competing males are ingredients of success (see Chs 20, 25)? What is the manner and extent of female choice, if indeed it occurs? Is there congruence between mating success and other components of fitness? The answer to the last question must be in the affirmative if competition and choice in mating are to be regarded as adaptive.

The aim of this paper is to review the evidence for mate preference and non-randomness in butterfly mating. For insects in general (Blum & Blum 1979), birds (Selander 1972) and primates (Crook 1972), extensive new knowledge has accumulated, especially in the last decade, but a review of this aspect of butterfly biology has not been attempted hitherto. As published data are scanty, I will make extensive use of my own field data on *Danaus chrysippus* in Tanzania, much of which are published here for the first time, and also of unpublished work by the late J. A. Unamba, on *Hypolimnas misippus* in Sierra Leone which I have kindly been permitted to examine and analyse.

Sexual Selection

In 1859, in a few pages in *The origin of species*, Darwin first outlined his ideas on sexual selection, clearly distinguishing it from natural selection: 'This (sexual selection) depends, not on a struggle for existence, but on a struggle between males for the possession of females; the result is not death to the unsuccessful competitor, but few or no offspring.' In *The descent of man* (1871), Darwin used the theory to explain many examples of sexual dimorphism and further emphasized the distinction between sexual selection and natural selection. He defined the former as: 'the advantage certain individuals have over others of the same sex and species solely in respect of reproduction.'

Darwin thus distinguished two aspects of sexual selection. First, males fight with one another for the possession of females—'the law of battle'. Huxley (1938*a*) gave the name 'intrasexual selection', which has since been generally adopted, to this aspect of sexual selection. Second, males compete for the attention of females leading to the evolution of features such as sex-limited colour patterns, vocalizations, pheromones and courtship displays. The latter was called 'epigamic selection' by Huxley and 'intersexual selection' by many other authors. Wallace (1889) accepted intrasexual selection in males as a matter of common observation, although he regarded it merely as a component of natural selection. However, he rejected intersexual selection as requiring of females a highly developed aesthetic sense which he found impossible to accept. Subsequent writers, with the exception of Poulton (1908), have mainly followed Wallace, accepting intrasexual selection but altogether discarding intersexual selection for lack of evidence (e.g. Grant 1963, Lack 1968; but see also Ch.20).

The Biology of Butterflies
0-12-713750-5

The reinstatement of sexual selection in both its elements was due to Fisher (1930) who overcame the anthropocentric and teleological objections to female choice:

> If, instead of regarding the existence of sexual preference as a basic fact to be established only by direct observation, we consider that the tastes of organisms, like their organs and faculties, must be regarded as the products of evolutionary change, governed by the relative advantage which such tastes confer, it appears . . . that occasions may not be infrequent when a sexual preference of a particular kind may confer a selective advantage and therefore become established in the species.

Fisher also considered that the origin of sexual selection might be due to female preference for male characters which had gained their initial impetus, not from sexual selection, but from some aspect of natural selection such as sexual recognition or sexual isolation. So long as sons, having inherited a preferred character from their father, had any advantage over males lacking it, the genes determining a preferred character and a preference would increase their frequency in tandem, becoming associated in linkage disequilibrium. As selection of one selects for the other, Fisher stated that selection would increase in geometric progression with time and a 'runaway process' set in train which would proceed exponentially until checked by powerful adverse natural selection.

Despite Fisher's cogent analysis, his work appears not to have been widely understood at the time and sexual selection has remained *in limbo* for most of this century. Wallace (1889) had considered that sexual dimorphism resulted from the evolution, by natural selection, of cryptic coloration which would afford female birds protection while incubating and similar considerations could apply to the oviposition period in female butterflies. He denied the possibility of female choice in mating. Huxley (1938*a*), in a confusing but much quoted paper, also largely disavowed female choice.

Three works, all published in the last ten years, have entirely succeeded in reinstating sexual selection and, in particular, female choice (but see also Ch.20). The appearance of *Sexual selection and the descent of man* (Campbell 1972) provided for the first time essential observational and experimental evidence for sexual selection. Subsequently, *Sexual selection and reproductive competition in insects* (Blum & Blum 1979) supplied a wealth of new information for insects derived from behavioural studies in field and laboratory. Finally, O'Donald (1980) has provided a comprehensive genetical theory of sexual selection in his book *Genetic models of sexual selection.* Renewed interest in sexual selection, owing much no doubt to the rise of sociobiology (Trivers 1972, Wilson 1975, Dawkins 1976), has resulted in a flood of experimental and field work which has completed the reinstatement of the theory. Parker (1979) observed 'there has been over the last decade a considerable volume of evidence from field and laboratory studies that both aspects of sexual selection can operate'.

Assortative and Disassortative Mating

Widespread confusion exists concerning the distinction not only between sexual and natural selection (Mayr 1972) but also between sexual selection and non-random mating. Darwin seems to have considered all forms of preferential mating to constitute sexual selection and, more recently, Ehrman (1972) defined sexual selection as any deviation from random panmixia. However, O'Donald (1980) pointed out that preferential mating, based on male competition or female choice, may be either random or non-random with respect to the phenotypes pairing. Consequently, sexual selection and other forms of non-random mating must be distinguished. To this end, O'Donald introduced the new terms 'random preferential mating' for Darwinian sexual selection without assortative mating and 'assortative preferential mating' when both preferential and non-random mating occur within the same population. Non-random mating may therefore occur with or without sexual selection (*sensu stricto*). Here I will use the term sexual selection to mean 'random preferential mating' unless otherwise stated.

Mating may therefore be non-random but without reproductive advantage to some individuals over others of the same sex i.e. without sexual selection. The result may be an excess either of homotypic (like) matings, known as assortative mating (= positive assortative mating (O'Donald 1980) or sexual isolation (Parsons 1973)), or of heterotypic (unlike) matings which are disassortative (= negative assortative). Either assortative or disassortative mating may also occur with sexual selection, so that distinguishing the two effects may not be simple in practice.

Species in which two or more morphs, karyotypes, races or ecotypes are sympatric may show assortative or, much more rarely, disassortative mating (Halliday 1978, O'Donald 1980). Very few examples of either have been reported for Lepidoptera in which they would be readily observed only in species showing unimodal polymorphism (Vane-Wright 1975; Ch.20) of a visual kind. Disassortative mating apparently occurred in *Callimorpha (Panaxia) dominula* reared in the laboratory (Sheppard 1952) although this was not confirmed in the wild (Haldane 1954). Assortative mating occurs in *Danaus chrysippus* (present paper) and probably awaits detection in many other polymorphic species. Assortative mating is to be expected where there is incipient speciation such as between sub-species *arthemis* and *astynax* of *Limenitis arthemis* or where formerly allopatric races have come to overlap, as in *Heliconius melpomene* in Surinam.

The Detection of Non-Random Mating

Non-random mating within a population generally embraces only a portion of individuals. To detect it in field populations requires the collection of much data, preferably over a range of climatic and ecological conditions. Mating behaviour may vary with season (present paper), the time of day (Hovanitz 1948) and the frequencies of both morphs and sexes (Muggleton 1979). Therefore, a sampling period of less than a year (tropical species) or an entire breeding season (temperate species) may produce misleading results. As non-random mating is essentially a statistical phenomenon its detection is much dependent on sampling effort. Data are easy enough to collect provided that mating pairs can be observed in sufficient numbers. The pair composition is the only information required for analysis (if it is assumed that pair-composition has no effect on sampling probability; see also Ch.20).

Measurement of sexual selection requires the additional information necessary to calculate the proportions of each morph both within mated pairs and in the non-mating population. Heterogeneity is tested by χ^2. A problem likely to be encountered in many species is the difficulty of catching sufficient numbers of one sex or the other as netted samples are often heavily biased towards males even when the true sex-ratio is 1:1. Differences between sexes in behaviour, habitat, eclosion time and longevity are so commonplace that these difficulties cannot be under-emphasized. In some tropical species, such as many acraeines, netted samples are commonly over 90% male while extreme female bias occurs in *Acraea encedon* and *A. quirina* (Owen 1971*a*, 1974*a*; see also Chs 18, 24). These problems are stressed in order to explain why reliable information on sexual selection and assortative mating in butterflies is difficult to obtain and therefore scarce. Several published insect studies are open to the criticism that the authors have lumped together what may be heterogeneous data to increase sample sizes, resulting in the confusion of different effects and erroneous conclusions (Lusis 1961, Burns 1966, Creed 1975).

Sexual Selection in *Hypolimnas misippus*

The account which follows is based on data collected through 1966–68 in Sierra Leone by J. A. Unamba, who died in 1969 before his work could be published or fully analysed. Unamba was able to investigate a very large population of the nymphaline *H. misippus*, inhabiting farmland and orchards at Newton, near Freetown, Sierra Leone. The species is multivoltine with overlapping generations. Breeding activity is particularly intense in the wet season (May–October) and the bulk of the data was acquired in the two exceedingly wet months of July and August 1967. Although adult butterflies are known to survive the dry season, breeding was not observed in the driest months (December–April). A similar pattern of population growth and decline with a wet season maximum occurs in Ghana (Stride 1956, Edmunds 1969*a*) and Tanzania (Smith 1976*a*).

Unamba captured and marked 5345 females, including 129 *in copula*, the largest field sample of *H. misippus* ever assembled. 397 females were dissected and scored for spermatophores in the bursa copulatrix, a method which gives an accurate index of mating activity (Burns 1968, Pliske 1973, Ch.22).

The colour polymorphism in *H. misippus* is of the multiple-female dual type (Vane-Wright 1975) i.e. polymorphism is restricted to the female, the male being monomorphic and unlike any of the female morphs (Fig. 21.1.). Genetic control is by two unlinked loci, A and M, each with two alleles (Ford 1953, Unamba 1968).

The *aa* females have an orange hindwing whereas the dominant *A*-allele gives a white area of variable extent. The forewing is mainly orange in the *mm*-genotype but the apical half is black and white in *M*- butterflies. Interaction between the A- and M-loci produces an array of intermediate phenotypes (Edmunds 1969*a*, Smith 1976*a*) the genetic control of which is now fairly clear (Unamba 1968).

All the forewing variants other than 'misippus' (*M*-) are found to be of the *mm* ('inaria') genotype. The presence of the hindwing allele *A* seems to prevent the full development of the 'inaria' forewing pattern. As a result of rearing and analysing 40 broods obtained from wild females in Sierra Leone and Tanzania, and examining others from various parts of Africa preserved at Oxford, it is now possible to assign probable genotypes to the whole range of *H. misippus* female phenotypes.

The data for field matings (Table 21.1) are analysed in strictly orthogonal and independent 2 × 2 tables the results from which are summarized in Table 21.2. The analysis shows that, although $\chi^2_{(4)}$ for the complete 2 × 2 × 2 table falls just short of significance at the 5% level, the effect due to the A-locus is significant at the 2% level and neither of the others is close to significance. Females with white on the hindwing (*A*-) were found mating more frequently than expected from their proportion in the population. This result is confirmed and extended by the spermatophore counts (Table 21.3). The analysis (Table 21.4) supports the mating result, showing that *A*-females acquire more spermatophores ($\bar{x}$ = 1.170) than *aa* ($\bar{x}$ = 0.783). The value of

Fig. 21.1. The phenotypes (genotypes) of *Danaus chrysippus* and *Hypolimnas misippus* mentioned in the text. Left-hand side, from top: 1 *D. chrysippus* form 'aegyptius' (brown) (*B-cc*) female; 2 form 'aegyptius' (orange) (*bbcc*) male; 3 form 'dorippus' (brown) (*B-C-*) female; 4 form 'dorippus' (orange) (*bbC-*) female; 5 form 'transiens' (brown) (*BbCc*) male. Right-hand side, from top: 6 *Hypolimnas misippus* form 'alcippoides' (*M-A-*) female; 7 form 'misippus' (*M-aa*) female; 8 form 'inaria' (*mmaa*) female; 9 form 'inaria-alcippoides' (*mmA-*) female; 10 monomorphic male.

Table 21.1. Number of field matings recorded for four genotypes of *Hypolimnas misippus*, near Freetown, Sierra Leone (data from J. A. Unamba).

	Genotypes				
Condition	*M-A-*	*M-aa*	*mmA-*	*mmaa*	TOTALS
Captured not mating	988	3159	278	791	5216
Captured *in copula*	37	70	6	16	129
TOTALS	1025	3229	284	807	5345

Table 21.2. Analysis of χ^2 for Table 21.1.

Effect tested	Value of χ^2	Degrees of freedom
A-locus on chances of mating or not	5.590*	1
M-locus on chances of mating or not	0.917	1
Interaction of loci	1.760	1
Triple interaction (A × M × mating)	1.191	1
TOTAL	9.458	4

$^{*}P < 0.05$

Table 21.3. Number of spermatophores recorded in four genotypes of female *Hypolimnas misippus*, near Freetown, Sierra Leone.

	Genotypes				
	M-A-	*M-aa*	*mmA-*	*mmaa*	TOTAL
Number of spermatophores 0	34	59	6	11	110
Number of spermatophores 1	95	60	19	21	195
Number of spermatophores 2–3	56	27	7	2	92
TOTALS	185	146	32	34	397
Mean no. spermatophores/female	1.189	0.794	1.063	0.735	0.995
Per cent mated	81.62	59.59	81.25	67.65	72.29
Per cent multiply mated	30.27	18.49	21.88	5.88	23.17

Table 21.4. Analysis of χ^2 for Table 21.3.

Effect tested	Value of χ^2	Degrees of freedom
A-locus on chances of mating or not	20.553***	1
M-locus on chances of mating or not	0.150	1
A-locus on chances of mating once or more	2.531	1
M-locus on chances of mating once or more	5.193*	1
Interaction of loci on chances of mating or not	0.512	1
Interaction of loci on chances of mating once or more than once	0.776	1
Interaction of A- and M-loci	1.218	1
TOTAL	30.933***	7

$^{*}P < 0.05$
$^{***}P < 0.001$

partitioning χ^2 is again clear in showing that, of the overall heterogeneity, much the greater part is due to the advantage white hindwinged females have over orange ones in achieving the first mating. However, they have no such advantage in acquiring second or subsequent matings. On the other hand, the M-locus exercises no influence on the first mating but is significant for multiple matings. None of the interaction terms is significant.

The rank order of mating capacity taking both single and multiple matings into account is *M-A-* > *mmA-* > *M-aa* > *mmaa*, a ranking of decreasing likeness to the male judged in terms of quantity of white. Therefore, male choice for what remains of an ancestral colour pattern, originally shared by both sexes and subsequently modified in the female under pressure of mimicry for the various morphs of *Danaus chrysippus*, might explain the data.

Stride (1956, 1957) carried out experimental work in Ghana to test the response of wild males to dead and artificial test females. His results are illuminating but suffer from the small sample sizes involved in individual experiments, which preclude statistical analysis. Nevertheless, Stride has shown that sight plays a crucial role in the early stages of *Hypolimnas* courtship. The female wing pattern exercised a much more potent influence than the female body in evoking male courtship behaviour. Models fitted with male wings completely blocked the male courtship sequence. The female hindwing pattern was found to be much more effective than the forewing and models with a male forewing and female hindwing exercised an almost undiminished attraction for males. Orange hindwings were much more attractive than white ones (taken from *D. chrysippus* morph 'alcippus') and Stride concluded from this that 'alcippoides' females, which he mistakenly thought to be absent from Ghana (Edmunds 1969*a*,*b*; Ian Gordon, pers. comm.), would be at a disadvantage in mating. However, in experiment 35 (Stride 1956, p.65) he painted a white patch about 1cm in diameter in the centre of an orange hindwing, i.e. almost exactly of the modal dimensions and position of the white patch in *aa*-females, and found that the eight males tested all responded up to the final courtship stage 5 (catalepsis). Thus, his own results were not in accord with one of his main conclusions to the effect that a white patch on the hindwing, of similar size to that which actually occurs in the population he was investigating, inhibited male courtship.

Edmunds (1969*b*) studied the frequency in the field of an avoidance response, known as the 'ascending flight', through which recently mated females communicate to courting males their unwillingness to mate. Comparing the frequencies of ascending flights in orange and white hindwinged forms, and regarding these frequencies as an index of previous mating, Edmunds found no significant difference. (Both Stride and Edmunds found that females giving the ascending flight response invariably contained at least one spermatophore and the behaviour was never followed by copulation.) The problem with Edmunds' data is that, although they lead us to question the validity of Stride's conclusion on the inhibitory effect of white hindwings, they tell us nothing about possible variation in female receptivity due to genotype, age or number of previous matings.

If the female's response to courtship is unaffected by her previous mating history, the frequency distribution of spermatophores should be random; but this is not the case (Table 21.5). Females mate once comparatively readily but become increasingly refractory after each additional mating. Thus, longevity must affect their chances of mating more than once. There is some evidence on this point. Edmunds (1969*a*) and Smith (1976*a*) found that the orange forms survived better than the white as the population declined from the wet to the dry season, the selective disadvantage to the latter being 63–82% in Ghana and 71% in Tanzania. As the locations are some 3000 miles apart, these results may have general significance. The recatchability of orange forms over a two-month mark-recapture study in Sierra Leone (Unamba 1968) was significantly superior to that of white ($\chi^2_{(1)} = 11.110$; $P < 0.001$) and could indicate superior survival of the former although other explanations (e.g. differential emigration) are possible. At the M-locus, there is some evidence for the selective value of the two alleles alternating with the seasons with *M* selected for as the population declines (Edmunds 1969*a*, Smith 1976*a*). It follows that a reproductive advantage to the *A*-allele, conferred either by male preference or a lower threshold of response (coyness) in *A*-females, is balanced by an advantage to *a* and *M* due to superior survival. Sexual selection for *A* is, therefore, countered by natural selection for *M*, explaining the significant χ^2 for the effect of the M-locus on chances of mating once or more than once (Table 21.4).

Unamba's spermatophore data (Table 21.3) reveal an aspect of sexual selection not to my knowledge conclusively demonstrated before in butterflies, namely that sexual selection could operate through female choice, with female genotypes varying in coyness. One aspect of the results deserving attention is the remarkably high overall frequency (27.7%) of unmated females. Burns (1968) stated: 'I find that a male lepidopteran has almost always beaten me to any given female'. Yet, Unamba's net clearly stole a march on many an *H. misippus* male. Extensive data for several butterfly families (Burns 1968, Pliske 1973) show that virgin females are rare throughout the long list of species investigated, the only exceptions being hilltopping species (Shields 1968)—

Table 21.5. Goodness of fit to the Poisson Distribution for number of spermatophores per female in *H. misippus* (Data from J. A. Unamba).

Number of spermatophores	Females observed	Females expected		χ^2
		P	n	
0	110	0.370	146.890	9.265
1	195	0.368	146.096	16.370
2	76	0.183	72.651	0.154
3	16	0.079	31.363	7.525
TOTALS	397	1.000	397.000	33.314*** (2 degrees of freedom)

***$P < 0.001$

which *H. misippus* is certainly not. As Unamba found no evidence for departure from a 1:1 sex-ratio, either in the field or laboratory reared broods, the data may indicate a generally high level of coyness, especially in the *aa* females of which 38.8% were unmated. As the females taken for dissection were drawn from a population at very high density, their chances of encountering males must have been good. If females of the various genotypes differ in coyness, female choice alone could explain the spermatophore data. On the other hand, if searching males respond more readily at a distance to females showing white on the hindwing so that they are mated earlier, this could result in the relatively novel idea of sexual selection based on male choice (cf. Vane-Wright 1979*a*,*b*, 1981*b*, Ch.23).

Darwin's observation (1871) that the early breeders in monogamous birds had a sexual advantage from successfully rearing more offspring has recently been confirmed in the Arctic skua by O'Donald (1980 and references therein). The same explanation could apply to multivoltine and polygamous butterflies, given a short generation time (egg–adult development time in *H. misippus* is only 23 days in equatorial Africa), continuous breeding and overlapping generations. If the *aa*-butterflies suffer a delay compared to the *A*- genotypes of, say, 2–3 days before mating, their generation time could be increased by some 10% resulting in the relative loss of an entire generation a year. Magnus (1958*a*, 1963) found that the female morph 'valesina' of *Argynnis paphia* was less attractive to males than the commoner male-like form but concluded that, as all females are eventually inseminated, they have no disadvantage. Clearly, Magnus could be wrong (see also Vane-Wright 1981*b*) but Silberglied, Ch.20, discounts the delay argument.

Stride (1956) concluded that the complete blockage on male courtship imposed by a full white hindwing in the female must mitigate against the perfection of mimicry in Ghana where *D. chrysippus* (the model) is entirely form 'alcippus'. I suggest that this is precisely why the full development of hindwing white in *H. misippus* is so rare throughout Africa, at the same time emphasizing that the presence of white in lesser degree, the common condition, was found neither by Stride nor Unamba to inhibit male courtship: indeed, Unamba's results prove that it does not. The correlation of female morphs for two rankings, namely resemblance to the male pattern in the extent of white and number of spermatophores acquired, is 1:24 and statistically significant at the 5% level. Therefore, while we can safely conclude that the mating capacity of the female morphs of *H. misippus* differ significantly, a confident verdict on whether the variation is due to male choice, female coyness or a combination of the two must await further and better experiments designed to elucidate the courtship roles of both sexes.

Non-random Mating in *Danaus chrysippus*

Danaus chrysippus was described by Poulton as the commonest butterfly in the world. Among many absorbing aspects of the biology of this butterfly are its complex and varied patterns of mate selection (Smith 1973, 1975*c*, 1981). Unlike *H. misippus*, *D. chrysippus* in East Africa and many other regions is unimodally polymorphic, i.e. all morphs are present in both sexes. The situation is unusual for a species which is presumed to be distasteful, has Müllerian mimics and is a model for numerous Batesian mimics. It is nevertheless ideal for the study of non-random mating (see also Ch.20) as the phenotype of both sexes is visual and often betrays the genotype. In East Africa, four principal morphs (Fig. 21.1) and some heterozygous phenotypes occur with varying frequencies throughout the region (Owen & Chanter 1968, Pierre 1974, D. A. S. Smith 1980).

The account which follows is based on the two morphs 'aegyptius' and 'dorippus' in Tanzania where the other two, 'alcippus' and 'albinus', are rare and seen mating very infrequently. The population studied inhabited the campus of the University at Dar es Salaam, an area of 235ha of irregularly maintained grassland, with scattered trees, gardens, patches of thick bush (where the butterfly did not occur) and sporadic small areas of untidy and shifting cultivation. At any one time there were areas of short grass, where annual and perennial flowers abounded, and others with long grass. Larval asclepiad foodplants (*Calotropis gigantea* and other milkweeds), adult nectar sources, trees favoured for roosting (e.g. *Acacia* spp.) and sources of pyrrolizidine alkaloids (various Leguminosae and Boraginaceae), the latter essential to males for the elaboration of their aphrodisiac pheromone (Chs 9, 25), were all available throughout the year. The ecological conditions, entirely made and sustained by human activity, were particularly favourable to the support of a dense population of *D. chrysippus*.

Sampling was begun in February 1972 and continued without interruption to December 1975, a total of 47 months, covering approximately 52 overlapping generations. Samples were netted in the field on several days each week and each butterfly was marked before release to avoid repeated scoring. Each butterfly was scored for phenotype, and composition of mated pairs noted. Concurrently, extensive breeding was carried out (Smith 1975*a*, 1980). The genetical information gained paved the way for increasingly sophisticated field identification, especially the scoring of heterozygotes, as the work progressed. Data for analysis with respect both to mated pairs and the population frequencies of genotypes, were assembled monthly, an essential procedure as both sex and morph ratios vary continuously. The generation time is also approximately one month. As monthly samples were too small to be of much value, they were amalgamated on the basis of population and mating characteristics into two annual 'seasons', one being verdant and relatively cloudy, wet and cool (April–August) and the other hot and dry punctuated by a short and variable rainy season (September–March). In the wet season the frequency of 'aegyptius' rises rapidly to a peak in July or August while 'dorippus' predominates in the dry season.

For the purposes of this paper, the following genetical information is necessary. The B-locus has two alleles, *B* giving a brown ground-colour and *bb* orange on both wings. *B* is semi-dominant to dominant but *BB* and *Bb* butterflies cannot be reliably separated by phenotype. Forewing pattern is determined by the C-locus, *C* giving a wing uniformly coloured (orange or brown) except for a narrow black margin (form 'dorippus') and *cc* a large black apical area traversed by a row of, sometimes confluent, white spots (form 'aegyptius'). *Cc* heterozygotes are phenotypically close to *CC* butterflies but are often distinguishable by a row of 'aegyptius' sub-apical spots visible on the underside and occasionally the upperside of the forewing (form 'transiens'). The expressivity of *c* in bred heterozygotes is modified by the B-locus but not by sex; phenotypic 'transiens' have a frequency of 64.2 ± 2.4% ($n=408$) in brown (*B-Cc*) butterflies and 39.5 ± 3.7% ($n=172$) in *bbCc* (orange) (original data). The B- and C-loci are linked with a cross-over value of 1.9% (Smith 1975*a*, 1976*b*). As known recombinants all resulted from cross-over in males, achiasmatic oögenesis is suspected as in other Lepidoptera (Suomalainen *et al.* 1974, Turner & Sheppard 1975). Linkage disequilibrium is marked by an excess of repulsion over coupling genotypes (D′ of Lewontin 1964 = −0.78). Among 61 wild mated pairs used for breeding, double heterozygotes in repulsion (*Bc/bC*) significantly exceeded the Hardy-Weinberg expectation suggesting the possibility of heterozygous advantage dependent on epistatic interaction between the two loci (Smith 1980).

Non-randomness between morphs 'aegyptius' and 'dorippus' within the mating population, without scoring separately the 'dorippus'-like heterozygotes, was investigated for each season, four wet and three dry (Table 21.6). In the first wet season, mating was disassortative (Smith 1973), a situation which did not recur throughout the investigation. The mating pattern changed to strongly assortative in the first dry season. The remaining periods showed no significant departure from randomness (low expected numbers in two cases would invalidate χ^2). The data show several mating regimes are possible, changes in pattern being perhaps due to climate or fluctuations in the frequencies of genotypes or sexes.

For a more detailed analysis wet and dry season data were examined by a multidimensional analysis of χ^2 (Lewis 1962), a test which gives a separate value for each interaction operating and for heterogeneity (Table 21.7). Wet seasons show no overall departure from randomness but heterogeneity is very significant, presumably because disassortative mating occurred in 1972. Mating was assortative in the dry seasons but heterogeneity is very significant and mating behaviour was clearly subject to unidentified sources of variation. Morph frequencies varied throughout in one or both sexes and interaction between sex and morph ratios may have caused the heterogeneity. Comparison of the bulked data for wet and dry seasons shows an overall predominance of assortative mating ($\chi^2_{(1)} = 13.919$; $P < 0.001$) but heterogeneity is, not surprisingly, still present ($\chi^2_{(1)} = 4.524$; $P < 0.05$).

To investigate the effect of morph frequency on mating, all monthly samples were ranked in relation to the frequency of 'aegyptius', which ranged from 4.9–51.7%. The data were examined in frequency bands of 5% which were amalgamated if adjacent

Table 21.6. Mate selection of morphs 'dorippus' (d) and 'aegyptius' (a) of *Danaus chrysippus* at Dar es Salaam, April 1972–August 1975.

Period	Observed pairings				n	$\chi^2_{(1)}$
	d × d	d × a	a × d	a × a		
April–August 1972	6	15	17	5	43	8.379**
September 1972–March 1973	60	7	4	8	79	17.416***
April–August 1973	60	31	21	19	131	1.594
September 1973–March 1974	89	16	16	4	125	—
April–August 1974	60	28	12	11	111	1.408
September 1974–March 1975	127	8	1	1	137	—
April–August 1975	34	32	17	9	92	0.945
TOTALS	436	137	88	57	718	—

** $P < 0.01$
*** $P < 0.001$

Table 21.7. Analysis of χ^2 for Table 21.6.

Aspect tested	Value of χ^2	Degrees of freedom
Wet seasons (April–August)		
Assortative mating	0.001	1
Female morphs × years	2.834	3
Male morphs × years	13.967**	3
Heterogeneity	15.592**	3
TOTAL	32.394	10
Dry seasons (September–March)		
Assortative mating	21.564***	1
Female morphs × years	8.559*	2
Male morphs × years	18.514**	2
Heterogeneity	25.931***	2
TOTAL	74.569	7

* $P < 0.05$
** $P < 0.01$
*** $P < 0.001$

Table 21.8. Mating preference of morphs 'dorippus' (d) and 'aegyptius' (a) of *D. chrysippus* in relation to morph frequency at Dar es Salaam, April 1972–August 1975.

Per cent 'aegyptius' (*cc*)	Observed (and expected) pairs				n	$\chi^2_{(1)}$
	d × d	d × a	a × d	a × a		
<20.0	302 (295.4)	42 (48.6)	20 (26.6)	11 (4.4)	375	10.848***
20.1–45.0	144 (138.8)	79 (84.2)	52 (57.2)	40 (34.8)	315	1.470
>45.0	4 (6.7)	8 (5.3)	11 (8.3)	4 (6.7)	27	2.852[1]

1. See text
*** $P < 0.001$

Table 21.9. Observed (and expected) pairings among the phenotypes 'dorippus', 'transiens' and 'aegyptius' of *D. chrysippus* at Dar es Salaam, April 1974–August 1975.

Female phenotypes	Male phenotypes			
	'dorippus'	'transiens'	'aegyptius'	TOTALS
'dorippus'	84 (70.1)	58. (65.0)	17 (23.9)	159
'transiens'	36 (40.6)	43 (37.6)	13 (13.8)	92
'aegyptius'	30 (39.3)	38 (36.4)	21 (13.3)	89
TOTALS	150	139	51	340

Table 21.10. Analysis of χ^2 for data sumarized in Table 21.9.

Effect tested	Value of χ^2	Degrees of freedom
Assortative mating	13.436**	4
Female morphs × periods	75.595***	4
Male morphs × periods	53.763***	4
Heterogeneity	6.886	8
TOTAL	149.680	20

**$P < 0.01$
***$P < 0.001$

bands showed departures from expectation in the same direction (Table 21.8). This method reveals a strong influence of morph frequency on mate selection. When 'aegyptius' was scarce (<20%), mating was assortative but, as it increased, mating became at first random and then disassortative. Although the χ^2 for disassortative mating, when 'aegyptius' >45%, only approaches significance, the three months involved were May–July of the 1972 wet season when the mean frequency of 'aegyptius' over the whole April–August period was uniquely high, averaging 44%, and disassortative mating was clearly significant (Table 21.6; Smith 1973).

From April 1974–August 1975, all butterflies caught, either mating or otherwise, were scored for all three C-locus phenotypes and also for *B-* and *bb*. A more comprehensive analysis is therefore possible. As before the data were analysed by multidimensional χ^2 tests to distinguish the separate effects due to non-randomness, period (season), morph fluctuation and residual heterogeneity. The data for the last three sampling periods (Table 21.6) are summarized in Table 21.9 for the C-locus. The χ^2 analysis (Table 21.10) is for a 3 × 3 × 3 contingency table, three male morphs × three female morphs × three periods. The raw data show a surplus over expectation for all three types of like (homogamic) pairings and a corresponding deficiency for 5:6 unlike (heterogamic) ones. The analysis shows that this effect is independent of variations in morph frequency and free from significant heterogeneity. Thus, over a period of 17 consecutive months the population mated assortatively, especially for pairings between homozygous dominants and homozygous recessives and to some extent among heterozygotes.

The B-locus data for the same samples are given in Table 21.11 with the analysis of χ^2 in Table 21.12. The overall effect was again highly assortative with insignificant heterogeneity over the three periods. A further analysis of the possible effect of genotype frequency on mate selection at the B-locus proved negative although the overall effect was highly assortative ($\Sigma\chi/\sqrt{g}$ = d = 3.238; $0.002 > P > 0.001$). Therefore, unlike the C-locus, genotype frequency at the B-locus has no effect on mating behaviour.

The final analysis for non-randomness concerned the combined effect of the B- and C-loci on mate selection. Inspection of the data (Tables 21.13 and 21.14) suggests that like matings exceeded and unlike matings fell below expectation. As χ^2 for heterogeneity is not significant, further partitioning of χ^2_4 for assortative mating, in independent 2 × 2 contingency tables, is valid. Four χ^2s can be obtained, one for each degree of freedom in the overall term. (Note that the sum of the four independent χ^2s is very close to the overall value indicating that the method, although not exact, partitions the heterogeneity with sufficient accuracy). The analysis indicates that assortative mating occurs for both brown and orange ground colour (B-locus) and for 'dorippus' and 'aegyptius' pattern (C-locus) simultaneously. This explains how brown 'dorippus', most of which are double heterozygotes in repulsion

Table 21.11. Observed (and expected) pairings among B-locus phenotypes at Dar es Salaam, April 1974–September 1975.

Female phenotypes	Male phenotypes brown (*B*-)	orange (*bb*)	TOTALS
brown (*B*-)	133 (114)	66 (85)	199
orange (*bb*)	64 (83)	81 (62)	145
TOTALS	197	147	344

Table 21.12. Analysis of χ^2 for data summarized in Table 21.11.

Effect tested	Value of χ^2	Degrees of freedom
Assortative mating	17.656***	1
Female morphs × periods	43.179***	2
Male morphs × periods	19.551***	2
Heterogeneity	0.858	2
TOTAL	81.244	7

***$P < 0.001$

phase (*Bc/bC*), tend to mate preferentially with each other, as do the homozygotes. A further aspect of the analysis is the detection of a marked sex-difference in 'aegyptius' when offered a choice between brown morphs, like themselves, and orange: the males mate at random and the females assortatively, thus supporting the theory that it is the females who exercise the choice (Darwin 1871, Fisher 1930, O'Donald 1980).

Conclusions on Non-Random Mating in *D. chrysippus*

The following main conclusions can be drawn from the analysis of non-random mating in *D. chrysippus*: mate selection is normally assortative with respect both to colour and pattern. Selection for pattern is assortative when 'aegyptius' is at low frequency and either random or disassortative when it is not—thus, I was wrong to suggest (Smith 1975*c*) that a 'rare male effect' (Petit 1954, Petit & Ehrman 1969, Ehrman 1970) might be operating: the reverse is the case. Selection for ground colour is strongly assortative and independent of phenotype frequency.

Table 21.13. Observed (and expected) pairings among B- and C-locus phenotypes (genotypes) of *D. chrysippus* at Dar es Salaam, April 1974–August 1975.

Female phenotypes	Male phenotypes: orange 'dorippus' (*bbC*-)	brown 'dorippus' (*B-C*-)	brown 'aegyptius' (*B-cc*)	Totals
orange 'dorippus' (*bbC*-)	81 (61.4)	47 (60.1)	15 (21.5)	143
brown 'dorippus' (*B-C*-)	40 (46.4)	53 (45.4)	15 (16.2)	108
brown 'aegyptius' (*B-cc*)	25 (38.2)	43 (37.5)	21 (13.3)	89
TOTALS	146	143	51	340

Table 21.14. Analysis of χ^2 for data summarized in Table 21.13.

Effect tested	Value of χ^2		Degrees of freedom	
Assortative mating	23.077***		4	
Assortative mating at the B-locus within the 'dorippus' phenotype		8.934**		1
Assortative mating at the C-locus		6.986**		1
'aegyptius' male choice, brown *v.* orange within the 'dorippus' phenotype		0.676		1
'aegyptius' female choice, brown *v.* orange within the 'dorippus' phenotype		6.730**		1
TOTAL for assortative mating		23.236***		4
Female morphs × periods	60.588***		4	
Male morphs × periods	41.893***		4	
Heterogeneity	13.490		8	
TOTAL	139.048		20	

**$P < 0.01$
***$P < 0.001$

Assortative mating operates to favour homogamic pairings between heterozygotes as well as homozygotes. Within the limitations of my data, choice is exercised by females but not by males.

Sexual Selection in *Danaus chrysippus*

Sexual selection was investigated along with assortative mating using the same division into seasons (Table 21.6). From April 1972–March 1974, mated butterflies were scored only as 'aegyptius' or 'dorippus' whereas from April 1974, B-locus phenotypes and heterozygotes at the C-locus were scored in addition. The results for the two loci were analysed separately in two different ways; first by season and second, by frequency of the recessive phenotype, either *bb* or *cc*.

From April 1972–March 1974, there was no apparent sexual selection in either sex when the results were analysed by season and the data are omitted. I reported sexual selection for 'aegyptius' males from April–June or July in 1972 and 1973 (Smith 1975*c*): the results showed sexual selection for 'aegyptius' males in the early part of the wet season, followed by selection for 'dorippus' as the dry season set in. However, with hindsight this conclusion might be seen as a case of special pleading. A more accurate assessment of sexual selection in males emerges from scoring the *Cc* heterozygotes (see below).

Analysis of the results in relation to morph frequency indicates that male 'dorippus' are relatively successful when common ('aegyptius' < 20%) but lose their advantage at lower frequencies (Table 21.15A). The presentation in Table 21.15B shows that the proportion of male 'aegyptius' found mating is stable at around 16.4% over the whole of its wide frequency range whereas in 'dorippus' the proportion mating is 23.9% at high population frequency dropping to 15.6% at low frequency and the difference is highly significant. The female results (Table 21.16) also show a clear but different relationship with frequency. Form 'dorippus' was more successful at both extremes of its frequency range with no selection over a wide range of intermediate frequencies (50–90%). This apparently puzzling result may be ascribed to interaction between C- and B-loci, the latter being scored only over the last 17 months. Form 'dorippus' is predominantly orange (*bbC-*) when *aegyptius* is rare and brown (*B-C-*) when it is common. Thus, different 'dorippus' genotypes probably underpin its mating success at the two extremes of its frequency range. This interpretation is supported by the data in Table 21.16B, showing that the mainly brown phenotypes ('transiens' and 'aegyptius') had an advantage in the wet season and the predominantly orange 'dorippus' in the dry period.

Selection in relation to frequency in *D. chrysippus* is quite different from that reported in the polymorphic ladybird *Adalia bipunctata* (Muggleton 1979, O'Donald & Muggleton 1979) or the milkweed beetle *Tetraopes tetraophthalmus* (Eanes *et al.* 1977) in which each morph had a mating advantage only when rare. In contrast, the *D. chrysippus* morphs in both sexes have an advantage, if at all, only when common, with the exception of female 'dorippus' explained above. This suggests that mating success is not frequency-dependent but rather a reflection of overall fitness; both sexes of 'dorippus' show an advantage when common probably because of their superior adaptation to the hot, dry climatic conditions which prevail at that time.

Over a 17-month period (April 1974–August 1975), 340 pairings and all unmated individuals captured were scored, marked and released. For the first time,

Table 21.15. Mating success of the male morphs 'dorippus' and 'aegyptius' in relation in morph frequency at Dar es Salaam.

A.

Per cent male 'aegyptius'	'aegyptius' mated	'aegyptius' non-mated	'dorippus' mated	'dorippus' non-mated	n	$\chi^2_{(1)}$
≤20.0	32 (45.4)	166 (152.6)	351 (337.6)	1120 (1133.4)	1669	5.851**
≥20.1	108 (103.7)	546 (550.3)	231 (235.3)	1253 (1248.7)	2138	0.306

B.

Per cent male 'aegyptius'	Frequency of 'aegyptius' in population	Frequency of 'aegyptius' in mated pairs	Frequency of 'dorippus' in population	Frequency of 'dorippus' in mated pairs
≤20.0	0.119	0.084†	0.881	0.916***
≥20.1	0.306	0.318†	0.694	0.681***

†$\chi^2_{(1)} = 0.014$; $0.95 > P > 0.90$
***$\chi^2_{(1)} = 32.140$; $P < 0.001$.

Table 21.16. Mating success of the female morphs 'dorippus' and 'aegyptius' in relation to morph frequency at Dar es Salaam.

A.

Per cent female 'aegyptius'	'aegyptius'		'dorippus'			
	mated	non-mated	mated	non-mated	n	$\chi^2_{(1)}$
≤10.0	5 (13.3)	37 (28.7)	138 (129.7)	271 (279.3)	451	7.409**
10.1–50.0	158 (158)	420 (420)	363 (363)	965 (965)	1906	0.003
≥50.1	29 (38.2)	143 (133.8)	36 (26.8)	85 (94.2)	293	6.112*

B.

Period	Proportions mating			
	'dorippus'	'transiens'	'aegyptius'	$\chi^2_{(2)}$
April–August 1974	0.214	0.255	0.250	0.820
September 1974–March 1975	0.346	0.222	0.196	8.960*
April–August 1975	0.241	0.343	0.390	5.864

*$P < 0.05$
**$P < 0.01$

Table 21.17. Observed (and expected) numbers of C- and B-locus phenotypes of *D. chrysippus* mated and unmated at Dar es Salaam, April 1974–August 1975.

A. C-locus (full data for males in Smith 1981)

Sex	Condition	'dorippus'	'transiens'	'aegyptius'	TOTAL
Male	Mated	150 (165.5)	139 (101.7)	51 (72.8)	340
Male	Unmated	755 (739.5)	417 (454.3)	347 (325.2)	1519
				$\chi^2_{(2)}$ = 26.519***	
Female	Mated	159 (154.4)	92 (99.4)	89 (86.2)	340
Female	Unmated	391 (395.6)	262 (254.6)	218 (220.8)	871
				$\chi^2_{(2)}$ = 1.080	

B. B-locus

Sex	Condition	brown	orange	TOTAL	$\chi^2_{(1)}$
Male	Mated	194 (178.0)	146 (162.0)	340	3.676
Male	Unmated	795 (811.0)	754 (738.0)	1549	
Female	Mated	197 (175.5)	143 (164.5)	340	7.587**
Female	Unmated	428 (449.5)	443 (421.5)	871	

**$P < 0.01$
***$P < 0.001$

the 'transiens' phenotype was consistently recorded in pairs. In males, 'transiens' showed significantly higher mating success (Table 21.17A) than both homozygous phenotypes (Smith 1981) and the analysis (Table 21.8A) shows that this was independent of season and changes of morph-ratio. Residual heterogeneity was not significant. In contrast, there was no consistent sexual selection in females and heterogeneity was very significant.

As *CC* and *Cc* butterflies are not always recognizable phenotypically, male genotype numbers (Table 21.19) were estimated from the phenotypes observed by applying a correction for expressivity derived from breeding data and the corrected numbers used to calculate fitness. The fitness of *CC* was low (9–18%) in both wet seasons, rising to 79% in the dry season, whereas *cc* males had a fitness of 60–64% in the wet seasons falling to 15% in the dry season. It is therefore probable that the failure to score heterozygotes in 1972–73 led to selection in males remaining largely undetected. The reality of the sexual advantage shown by 'aegyptius' males during part of the April-August seasons of 1972 and 1973 (Smith 1975*c*) is supported by their high fitness,

Table 21.18. Analysis of χ^2 for data summarized in Table 21.17.

Aspect tested	Value of χ^2	Degrees of freedom
A. Sexual selection at C-locus in males		
Morphs × proportions mating	26.519***	2
Proportions mating × periods	14.669***	2
Morphs × periods	124.168***	4
Heterogeneity	6.723	4
TOTAL	172.080	12
B. Sexual selection at the C-locus in females		
Morphs × proportions mating	1.080	2
Proportions mating × periods	6.503*	2
Morphs × periods	130.946***	4
Heterogeneity	18.153**	4
TOTAL	156.682	12
C. Sexual selection at the B-locus in males		
Morphs × proportions mating	3.677	1
Proportions mating × periods	15.772***	2
Morphs × periods	93.475***	2
Heterogeneity	0.884	2
TOTAL	113.807	7
D. Sexual selection at the B-locus in females		
Morphs × proportions mating	7.587**	1
Proportions mating × periods	6.503*	2
Morphs × periods	134.441***	2
Heterogeneity	1.383	2
TOTAL	149.913	7

*$P < 0.05$
**$P < 0.01$
***$P < 0.001$

Table 21.19. Fitness of C-locus homozygotes relative to the heterozygote in male *D. chrysippus*.

Period	Condition	Genotypes			TOTAL
		CC	*Cc*	*cc*	
April–August 1974	Mated	3	85	23	111
	Unmated	149	307	155	611
	Fitness	*0.091*	*1*	*0.596*	
September 1974–March 1975	Mated	59	76	2	137
	Unmated	209	196	47	452
	Fitness	*0.788*	*1*	*0.146*	
April–August 1975	Mated	5	61	26	92
	Unmated	114	197	145	456
	Fitness	*0.178*	*1*	*0.643*	

Genotype numbers are based on corrections for expressivity.

relative at least to homozygous *dorippus*, for the corresponding periods in 1974 and 1975. Differential and seasonally alternating mating fitnesses of the magnitude observed must make a major contribution to the cycling in morph frequency observed in four successive years (1972–75).

Analysis of the same data with respect to the B-locus (Table 21.17B) leads to quite different and interesting conclusions (Table 21.18C,D). While acknowledging the possibly material fact that heterozygotes at this locus cannot be identified reliably, it is clear that *B-* females had a selective advantage over the whole period (heterogeneity is not significant), while in males sexual selection for *B*- was at the most slight and does not reach formal significance. The influence of the two loci also differed, as for assortative mating, in the absence at the B-locus of any relationship between selection and phenotype frequency in both sexes.

Sexual selection in the Dar es Salaam population of *D. chrysippus* is a powerful selective mechanism affecting both sexes but in quite different ways. Selection for heterozygous males through both wet and dry seasons must ensure the stability of the C-locus polymorphism, while the oscillating fitnesses of both homozygotes could be either cause or effect of the synchronized cycling in morph frequency observed throughout the study period. How such powerful selection for heterozygous males might operate is, unfortunately, not clear. Female preference has been demonstrated and must contribute to their success but it is unlikely to be the only factor as it implies synchronized seasonal switches of the secondary preference (for either *CC* or *cc* males). However, as the female morphs are themselves subject to seasonal cycling and tend to chauvinism in their choice, the overall bias of their choice could shift with the changing morph frequencies giving rise to assortative preferential mating.

Male competition must be considered as an additional or alternative possibility. *Cc* males (but not females) show heterosis for size compared to both homozygotes (Smith 1980) which could give them a competitive edge through faster flight or greater life expectancy (the latter is under investigation). The mean forewing length of mated males ($\bar{x}$ = 37.9mm, n = 161) did not differ significantly from those unmated ($\bar{x}$ = 38.0mm, n = 881) between October 1974 and September 1975; thus selection based on size alone is ruled out. Alternatively, the male pheromone, essential to successful courtship (Meinwald *et al.* 1974, Schneider *et al.* 1975, Boppré 1979, Ch.25) might differ qualitatively or quantitatively between genotypes. To test this hypothesis, 50 pairs of hairpencils and wing pouches from wild males of each genotype at Dar es Salaam were analysed for me by Professor J. Meinwald and his co-workers at Cornell University. No differences in pheromone components were detected. However, it is known that male *D. chrysippus* must spend several days collecting pheromone precursors (pyrrolizidine alkaloids) from plants; therefore, those males able to maintain the highest foraging rates are likely to achieve earlier matings and greater overall success. Observations in a flight cage suggest that old, worn males are usually the most successful in mating, probably because they have had more time to accumulate pheromone alkaloids. Therefore, male mating prowess could be age-dependent and superior survival would favour heterozygous males. Survival is however an aspect of natural selection, not sexual selection: illustrating the difficulty, especially acute in male competition, of distinguishing sexual and natural selection (Wallace 1889, Mayr 1972; Ch.20).

The major colour differences between genotypes could influence metabolic rate and hence activity levels. The amount of black and brown pigment is greatest in 'aegyptius' and least in 'dorippus', with most 'transiens' intermediate. The darker coloration of 'aegyptius', could allow it to maintain a higher body temperature, due to superior absorbance of radiant heat, during the cooler, cloudier months. It is interesting that 'aegyptius' is the only form in the cooler subtropical parts of the species' range at both its northern and southern limits (Smith 1981). Conversely, enhanced reflectance from its paler colouring could provide relative protection for *dorippus* during the hot, dry season, a suggestion supported by the fact that it is the only form in the arid regions of northern Kenya and Somalia (Smith 1980, Pierre 1980). The intermediate heterozygote could be better adapted to median radiation values and therefore better able to maintain optimum activity through the vicissitudes of a climate which can feature substantial wet or dry periods out of season, with successive years rarely having a consistent pattern (Smith 1981). If this argument is valid, natural selection taking the form of male competition will contribute substantially to the observed sexual selection in *D. chrysippus*. However, female choice between morphs certainly occurs and the evidence points to its being actuated by visual rather than olfactory cues.

Sexual selection for females (Vane-Wright 1980, 1981*b*, Ch.23) is not yet an established concept. Apart from *H. misippus* and *D. chrysippus* (this paper), the only reported cases in butterflies occur in two species with female-limited polymorphisms, *Papilio glaucus* (Burns 1966; but see also Ch.22) and *Argynnis paphia* (Magnus 1958*a*, 1963). Both authors assumed that selection for the female phenotypes resulted from visual choice by males and alternative possibilities were not explored. Yet, in *D. chrysippus*, male choice can probably be discounted: given a choice between orange and brown mates, females mate assortatively but males of the same genotype do not (Table 21.14). Variation in the threshold of response to courtship (coyness) between female genotypes is therefore a preferred hypothesis for *D. chrysippus* and could be tested by spermatophore counts. Coyness is hardly a suitable term to apply to *D. chrysippus* females which, like those of *D. gilippus* and *D. plexippus* (Burns 1968, Pliske 1973) are highly promiscuous, but variation for promiscuity would produce the same effect in terms of sexual selection. A further factor, probably rare in butterflies, is active female competition or assertiveness. Stride (1958*b*) and I have both observed that virgin females placed in a flight cage with inactive males, which have had no recent access to pheromone precursors, may eventually become so frustrated that they take the initiative in courtship by flying in front of or below passing males. They may even resort to persistent jostling of roosting males. In regions, such as East Africa, where at certain times of the year female *D. chrysippus* may outnumber males by as much as 4:1 (Owen & Chanter 1968, Smith 1975*b*), female competition probably becomes intense. Even where the sex-ratio is 1:1, males not only emerge from the pupa some 24 hours after their sisters but then require several days foraging for alkaloids before they can court effectively. Thus, three factors, all probably unusual in butterflies, might promote female competition—proclivity to promiscuity, low sex ratios and moderate protogyny.

Fisher (1930) supposed that mating preferences are the consequence of variation in female response to courtship with both preferred character and preference under genetic control. The sons of preferred males and females exercising the preference would inherit both preferred character and preference together and the two would increase in tandem either to fixation or to an equilibrium balanced by adverse natural selection. The two loci would tend to become associated in linkage disequilibrium (O'Donald 1980). This model is supported by the *D. chrysippus* data. The preferred male character is expressed in the 'transiens' (*Cc*) genotype for which there is sexual selection; the gene for the preference, indicated by its selection only in females, could be either gene *B* itself or a gene closely linked to it. Genes *B* and *C* are associated in linkage disequilibrium as predicted by Fisher, with double heterozygotes in repulsion phase predominant (Smith, D. A. S. 1980).

As mate selection by females is assortative at both loci, but more powerfully so at the B-locus, *Bc/bC* females should prefer males in the order *Bc/bC* > *Bc/Bc* > *bC/bC*. *Bc/Bc*-female preference may be the same or perhaps *Bc/Bc* > *Bc/bC* > *bC/bC* depending on the relative strength and constancy of the preference for *Cc* males conferred by *B*- in females and the opposed propensity to assortative mating at the C-locus. Powerful positive assortment for *B*- and *bb* in predominantly *B*- (brown) populations could produce the observed negative assortment at the C-locus (Table 21.6). Gene *b* in females may be either a 'null' allele, the *bb* genotype mating at random, or it could promote active choice for *bC/bC* males. The former is marginally preferred from the evidence in Table 21.13. Whether female preference is constant (total) or partial (O'Donald 1980), it will not give rise to frequency-dependent selection for male genotypes as the frequencies of female genotypes are always oscillating within wide limits. Thus the genes for preferred characters and preferences are both subject to broadly synchronized cycling. The frequencies of pairings predicted from the sexual selection and assortative mating data were largely confirmed through progeny tests using 61 wild mated pairs (Smith 1980).

The highly complex mating system in *D. chrysippus* is given a further twist by the markedly different sex-ratios found within genotypes in Tanzania. The sex-ratio is normally high in *CC*, low in *cc* and around 1:1 in *Cc*. Therefore male 'dorippus' are more frequent (82%) than females (57.4%) ($\chi^2_{(1)} = 7.6$; $P < 0.01$). It follows that, discounting sexual selection and assortative mating, pairing frequencies were still markedly different from expectation on a null hypothesis that the alleles have equal frequency in the two sexes (Table 21.20). The departure from expectation is largely due to the heterotypic pairings which make the largest contribution to χ^2. Although mating within this small sample is random, as there is a surplus of 'dorippus' males and 'aegyptius' females unable to find mates of their own kind, hybridization and thus gene exchange between morphs is an enforced consequence of the sex-ratio within each morph. The sex-ratio imbalance is possibly an adaptation ensuring a steady production of heterozygotes, countering chauvinism (assortative mating) between two recently allopatric races, hybridization between which has led to heterozygous advantage. It is clear from Table 21.21 that there is a genetic component to the sex-ratio imbalance and it is known that at least one of the controlling genes is linked to the B- and C-loci (Smith 1975*b*). In backcross broods at the C-locus the sex-ratio is significantly biased towards females if the female parent is *aegyptius* (*Bc/Bc*) but not in F_2 broods from *dorippus* or *transiens* (*Bc/bC*) females; thus the female parental genotype affects the sex-ratio (Smith 1975*b*, 1976*b*), the genetic control of which deserves further study. Comparing the two

Table 21.20. Mating frequencies at the C-locus tested against the null hypothesis that allele frequencies based on pairs used for progeny tests are identical in the two sexes.

Male genotype	*C-*	*C-*	*cc*	*cc*	TOTALS
Female genotype	*C-*	*cc*	*C-*	*cc*	—
Pairs observed	31	19	4	7	61
Pairs expected	29.6	12.9	12.9	5.6	61
Contributions to $\chi^2_{(3)}$	0.063	2.904	6.125	0.350	9.442**

*$P < 0.05$

Table 21.21. Comparison of the sex-ratios obtained from two types of cross.

Number of broods	Parental genotypes		offspring			
	male	female	male	female	n	$\chi^2_{(1)}$
10[1]	*Bc/bC*	*Bc/Bc*	39	72	111	9.811**
10[2]	*Bc/bC*	*Bc/bC*	99	65	164	7.049**

1. There is heterogeneity, four of the broods being either all-female or significantly biased to females.
2. There is heterogeneity, three of the broods being significantly biased to males and the others normal.
$^{**}P < 0.01$.

types of cross (Table 21.21), which together comprised 33% of wild mated pairs from which progeny were obtained, it is easy to see how a seasonal shift in the selective value of the two female genotypes will automatically alter the sex-ratio. Furthermore, such a shift could alter the balance and force of female choice and exert a decisive influence on the outcome of male competition.

The situation in *D. chrysippus* highlights a major problem confronting the student of sexual selection, namely distinguishing male competition from female choice. While, in purely statistical terms, the two lead to the same result, their ethology is quite different. If male competition is fierce, females may be undiscriminating: if it is not, female choice must largely determine the outcome. The two aspects are probably best regarded as opposite sides of the same coin: females can choose only among males which find them and are equipped with the appropriate courtship stimuli, a scenario which includes both male competition and female choice. The evidence for *D. chrysippus* suggests that both aspects operate in sexual selection.

Discussion

Some of the findings for *Drosophila* (e.g. Parsons 1973) probably have relevance to at least some butterfly populations, including those reviewed here.

Much experimental evidence is available, particularly for *D. pseudoobscura* and *D. persimilis*, regarding mating speed in relation to polymorphic chromosomal inversion sequences, which are effectively single gene complexes. Substantial mating speed differences were observed between homokaryotypes in both *D. pseudoobscura* (Speiss & Langer 1964*a*) and *D. persimilis* (Speiss & Langer 1964*b*). Male karyotype was more important in determining mating speed in the former but females had markedly more influence in the latter. Heterozygous advantage, due to the greater activity and persistence in courtship of heterokaryotype males or to their more ready acceptance by females, has also been frequently reported (e.g. Speiss *et al.* 1966). These results show several similarities to my own for *D. chrysippus*, namely the differences in mating frequency between homozygotes, heterozygous advantage restricted to males and variation in female receptivity. It is also worth noting that *D. chrysippus* is polymorphic for a complex of at least three linked genes, which could be enclosed in inversions as in the two *Drosophila* species.

The classical experiments of Bateman (1948) on *D. melanogaster* showed that the contribution of males to the next generation was much more variable than that of females. This is certainly also the case in *D. chrysippus* (Table 21.19). Clearly, many homozygous 'dorippus' males fail to mate in the April–August season while 'aegyptius' males are equally unsuccessful in the September–March period. On the other hand, virtually all wild females collected for genetical experiments laid fertile eggs. Thus, sexual selection will lead to 'undiscriminating eagerness' in males and 'discriminating passivity' in females (Bateman, 1948). Only when the sex-ratio is seriously disturbed, as for example in *D. chrysippus* (Owen & Chanter 1968, Smith 1975*b*), *Acraea encedon* (Owen & Chanter 1969, Owen 1970*b*, 1971*a*) and perhaps *Hypolimnas bolina* (Clarke *et al.* 1975; Ch.24), are females likely to become active in courtship. Indeed, in extreme cases of low sex-ratio such as occur in *A. encedon*, where many females remain unmated and contribute nothing to the next generation (Owen *et al.* 1973*b*), 'traditional' sex roles may be to some extent reversed.

Maynard Smith (1956) showed in *D. subobscura* that females were able to discriminate between outbred and inbred males on the basis that outbred males were more skilled in courtship. Females who preferred outbred males left four times as many viable offspring as those which mated at random. Female choice on similar grounds could easily explain the heterozygous advantage for form 'transiens' in *D. chrysippus* males.

Frequency-dependence of the 'rare male type' described by Petit & Ehrman (1969) in *D. melanogaster* is to be expected in Lepidoptera with unimodal polymorphisms. It has been claimed to occur in laboratory studies of the moth *Callimorpha dominula* (Sheppard 1952) but not so far in butterflies. The frequency effects described here for *D. chrysippus*

Table 21.22. Modes of sexual selection and non-random mating in Lepidoptera.

I. **Non-random mating without sexual selection**
  A. *Assortative mating (sexual isolation):* homotypic or homogamic pairings in excess (e.g. *L. arthemis*)
    1. Choice frequency-dependent
    2. Choice independent of frequency
  B. *Disassortative mating:* heterotypic or heterogamic pairings in excess (e.g. *C. dominula*)
    1. Choice frequency-dependent
    2. Choice independent of frequency

II. **Random preferential mating**
  A. *Intersexual (epigamic) selection*
    1. Based on choices among courting partners
      a. Choice frequency-dependent
      b. Choice independent of frequency
        i. Choice by females among actively competing males (e.g. *D. chrysippus*)
        ii. Choice by males between passive females at long range (e.g. *A. paphia*)
        iii. Choice by males between actively competing females at close range (e.g. *A. encedon*)
    2. Based on differences of breeding time
      a. Seasonal variation in readiness to breed (e.g. *Catopsilia florella*?)
      b. Diurnal variation in willingness to breed (e.g. *Colias* spp.?)
  B. *Intrasexual selection*
    1. Precopulatory competition
      a. Differential ability (usually male) in finding mates (e.g. *D. plexippus*?)
      b. Territorial exclusion (e.g. *P. aegeria*)
      c. Differential male ability to attract female attention based on
        i. Olfactory cues (pheromones) (e.g. Danainae)
        ii. Visual cues (e.g. *D. chrysippus*)
      d. Differential female ability to attract male attention based on
        i. Olfactory cues (e.g. *Lymantria dispar*?)
        ii. Visual cues (e.g. *H. misippus*?)
      e. Differential female coyness (threshold of response to courtship) (e.g. *H. misippus*)
    2. Postcopulatory competition
      a. Sperm displacement (e.g. *D. chrysippus*)
      b. Mating plugs (e.g. Acraeinae)
      c. Repellents (antiaphrodisiacs) (e.g. Heliconiinae)
      d. Prolonged copulation (e.g. Danainae)
      e. Mated pair leaves vicinity of competing suitors (e.g. *D. chrysippus*)

III. **Assortative preferential mating**: I and II operate simultaneously (e.g. *D. chrysippus*)

Based on E. O. Wilson (1975), extended to include non-random mating without sexual selection and adapted for relevance to Lepidoptera.

differ fundamentally and probably stem from natural selection rather than sexual selection.

Female influence on mating speed (Speiss & Langer 1964*b*) probably explains the variation in spermatophore counts between female genotypes in *H. misippus*. It is perhaps worth noting that the least coy, i.e. the fastest mating genotype is *M-A-*, which predominantly produced 2:1 segregations at both loci in breeding experiments (Ford 1953, Unamba 1978). This suggests that most 'alcippoides' are *MmAa* and thus may have heterozygous advantage.

Modes of sexual selection and non-random mating which either have been or are likely to be detected in butterflies are listed in Table 21.22. Some of the examples are speculative.

Assortative mating without sexual selection is probably common where geographical races meet and overlap, as for example between subspecies *arthemis* and *astyanax* of *Limenitis arthemis* in North America or in the several areas (e.g. the Guianas) of rampant polymorphism in *Heliconius melpomene* and *H. erato* (Turner 1971*b*). Species of Lepidoptera polymorphic in both sexes in which assortative mating might be expected include *Acraea encedon* and *A. encedana* (Pierre 1976*a,b*; Ch.18) in Africa and *Zygaena ephialtes* in Europe. It is especially predictable in *Hypolimnas dubius* in which both ecological and behavioural differences between morphs are known to exist. Assortative mating may often signify either incipient speciation or recent contact between previously isolated races and it is to the latter that it should be attributed in *D. chrysippus* (Smith 1980).

Disassortative mating seems always to be rare and I know of no record in butterflies except for *D. chrysippus* (Smith 1973) in which its significance is obscure. Female choice for males of different genotype from themselves will bring about dis-assortative mating.

Little is known of sexual selection in butterflies and its importance is probably underestimated. In some species, male precopulatory competition must play a significant part, as in pupal mating *Heliconius* (Ch.3). In this case, female choice must be discounted (see also Ch.19). The aerial takedown courtship of *Danaus plexippus* is perhaps the nearest thing to unbridled machismo found in butterflies although subsequent female rejection occurs commonly (Pliske 1975*c*) and indicates that the behaviour falls short of rape. Male choice in response to visual cues at a distance has been postulated but not, in my view, conclusively demonstrated in *H. misippus* by Stride (1956). Burns (1966) reported, on the basis of spermatophore counts, that male *Papilio glaucus* prefer the yellow to the black female morphs but Pliske (1972) found no selection of this kind and rightly pointed out that Burns had lumped together heterogeneous data (see also Ch.22). Magnus (1958*a*) demonstrated male choice favouring the male-like female form in *Argynnis paphia* against the dark female form 'valesina'. He believed that in dense populations all 'valesina' females mated successfully and were at no sexual disadvantage. This is the most convincing example of male choice but the conclusion may not be correct if the later mating females produce fewer offspring (O'Donald 1980); some disadvantage to 'valesina' would in any case be expected at low population densities.

Differential ability in locating and attracting mates probably occurs in male *D. chrysippus* but female choice also has importance and the two effects are virtually impossible to disentangle on present evidence. Territorial exclusion of more than a temporary nature is probably rare but it occurs, for example, in species which compete for lookout posts (*H. misippus*) or sunspots (*Pararge aegeria*; Davies 1978).

Post-copulatory competition is widespread in insects (Craig 1967, Gilbert 1976, Parker 1979) and occurs in butterflies, taking the form either of mating plugs (Acraeinae and Parnassiinae), antiaphrodisiac pheromones (Heliconiinae) or sperm displacement. Sperm displacement may be fairly general but seems to be prominent in the promiscuous Danainae in which a uniquely high spermatophore count is common (e.g. $\bar{x}$ = 4.02, max. = 15 for *D. gilippus*; $\bar{x}$ = 2.23, max. = 8 for *D. plexippus*, Pliske 1973). A wild *D. chrysippus* female, confined in a flight cage at Dar es Salaam, mated on five successive days with the same male: Pliske reported a similar case for *D. gilippus*. Clearly, the refractory period in female Danainae may be remarkably brief. Parker (1970, 1979) showed that sperm displacement is highly effective in the fly *Scathophaga stercoraria* and the same is the case in *D. chrysippus*. Reared broods from females captured *in copula* almost invariably segregated as expected if the male observed mating had fertilized the eggs. It is possible that female danaines utilize the contents of displaced spermatophores for general nourishment or towards the maturation of eggs. It is also clear that a female *D. chrysippus* or *D. gilippus* will, in the course of her long life, lay eggs fertilized by many different males. The promiscuity will result in a high rate of genetic recombination. In marked contrast, multiple mating was rare in *Heliconius erato* ($\bar{x}$ = 1.1, max. = 3 spermatophores, Pliske 1973) thus demonstrating the effectiveness of the male-derived antiaphrodisiac (Gilbert 1976).

Data on the duration of copulation in butterflies are scanty but, in this respect too, the danaines are probably exceptional. Pliske (1973) observed the mean duration of copulation (T^f) in *D. gilippus* as 7–8h (range 1.7–12h) which is similar to *D. chrysippus* ($\bar{x}$ = 4.2h, range 1.25–12h). In the latter species, mating activity is rare before midday and reaches a peak in the late afternoon. In the post-nuptial flight, the male carries the female to a secluded spot in long grass or a tree where the pair can remain unmolested until after dark. Thus a female is probably unable to mate twice in the same day. She will normally spend the morning following copulation laying eggs sired by the male with which she mated the previous afternoon. Sexual selection must operate powerfully in favour of those males who can displace the previous male's spermatophore at the earliest opportunity.

The role of female choice is troublesome to demonstrate but it is the most logical deduction from the analysis in Table 21.14. Moreover, the evidence that selection for the *B* gene occurs in female but not in male *D. chrysippus* suggests that this may itself be a gene determining female preference as predicted by Fisher (1930). Variation in the female threshold of response to courtship (coyness) probably explains the sexual selection found by Unamba in *H. misippus* although male choice cannot be ruled out entirely. Further investigation of this phenomenon is particularly desirable and *Papilio dardanus*, *P. memnon* or *P. polytes*, all of which have female-limited polymorphisms, would be particularly suitable subjects. The rather special case of the predominantly female populations of *A. encedon* should also be re-examined to establish how females work out what must be very intense competition for mates (Owen 1971*a*). In similar, though less extreme situations, *D. chrysippus* females solicit mates.

Our poor understanding of sexual selection stems from a failure to distinguish it clearly from sexual recognition and sexual isolation, the difficulty in separating male and female roles, and from frequently unresolved confusion between it and natural selection (particularly evident in male competition). Nevertheless, there is much evidence for the two species described in detail here that sexual selection is a factor of major importance in the maintenance of the genetic polymorphisms: indeed, in *D. chrysippus* heterozygous advantage acting through sexual selection could alone in theory maintain the polymorphism although, in practice, many aspects of natural selection are also involved. Balance between natural selection for mimicry and longevity, on the one hand, and sexual selection on the other, as found in *H. misippus*, may explain polymorphisms in many other species.

Given the theoretical and semantic difficulties involved in distinguishing some aspects of sexual selection from each other and from natural selection, it would perhaps be preferable, while continuing to strive for a deeper understanding of the behavioural, ecological and genetic mechanisms involved in mate selection, to abandon the search for rigid demarcations which can be restrictive and excessively pedantic.

Acknowledgements

I am much in debt to the late Josiah Unamba for his splendid data on *H. misippus*. His tragic early death in 1969 was a great loss to butterfly research. I thank the late Professor P. M. Sheppard, FRS, for many helpful comments, Dr Denis Owen, who supervised Unamba's research, for making his results available to me and the University of Sierra Leone for permission to publish it. Professor J. Meinwald kindly analysed pheromone samples for me. Dr P. O'Donald and Dr John Barrett have given valuable advice on statistical procedures.

22. *Absence of Differential Mate Selection in the North American Tiger Swallowtail* Papilio glaucus

Austin P. Platt, S. J. Harrison and Thomas F. Williams

Department of Biological Sciences, University of Maryland Baltimore County, Maryland

The North American tiger swallowtail, *Papilio glaucus*, ranges from central Canada to Florida and westwards to the Rocky Mountain foothills. Males are always of the tiger-striped (yellow) pattern (Edwards 1884, Scudder 1889, Clark & Clark 1951), but females can either be similar or dark. In eastern and southern areas they are dimorphic, but elsewhere only the yellow, tiger-striped females occur (Fig. 22.1). Where sex-limited dimorphism occurs, intrasexual competition for mates is likely to exist (Parker 1978).

The dark morph of *P. glaucus* mimics the unpalatable, widely sympatric blue swallowtail, *Battus philenor* (J. V. Z. Brower 1958*b*, Brower & Brower 1962, Platt & Brower 1968; Fig. 22.1). Inheritance studies of the yellow and dark (mimetic) female morphs of *P. glaucus* (Clarke & Sheppard 1955*b*, 1959, 1962*b*, Clarke & Willig 1977, Robinson 1971) show that both types usually yield female offspring like themselves. Broods with a few daughters unlike their mothers rarely occur, affecting both morphs. Since female Lepidoptera are the heterogametic sex, genetic control of 'dark' must presumably either be Y-linked or cytoplasmic—although the presence or absence of the heteropyknotic (or 'Smith') body affects its expression (Clarke *et al.* 1976). A number of gynandromorphs and other sexual mosaics, and intermediate 'speckled' females occur infrequently in wild populations (Clark & Clark 1951, 1983, Walsten 1977).

P. glaucus females are monomorphic 'yellow' in regions where *B. philenor* is either uncommon or absent. Burns (1966) postulated that dimorphism was maintained in other areas by counter-balancing forces of sexual selection and mimicry, yellow females being preferred by males, but the mimetic females having a survival advantage. This hypothesis was based on meagre and equivocal data (Burns 1967, Prout 1967), and has been questioned by more recent studies (Makielski 1972, Pliske 1972, Levin 1973) which failed to substantiate differences in mating frequency and fecundity. In the present study we likewise find no evidence of differential mate choice, and advance an alternative hypothesis to explain the persistence of female dimorphism.

Materials and Methods

Random, quantitative samples of dimorphic *P. glaucus* females were collected from three regions of Maryland, over an eight year period (Table 22.1). Dissection methods for spermatophores and eggs closely followed those of Burns (1966), Makielski (1972) and Pliske (1972). The age of each specimen was estimated on wing condition (Table 22.4). Three intermediate (speckled) specimens, excluded from our analyses, are discussed separately.

Data were pooled and analysed using χ^2 and *t*-tests (Table 22.2). The numbers of both types of female and the total number of females containing, respectively, one, two, three and four spermatophores, were fitted to a Poisson distribution (Table 22.3). Two linear regression analyses and accompanying *t*-tests were done to determine relationships between female age *versus* spermatophore number, and female age *versus* unlaid egg number (Table 22.4). Finally, examples of both *glaucus* females, together with a male and a specimen of *B. philenor*, were photographed under visible light and through a Wratten 18Å UV-transmitting filter, to compare their colour patterns under differing light conditions (Figs 22.1, 22.2).

Results

Preliminary analyses showed no temporal differences nor were there any spatial differences (Table 22.1) within Maryland samples so data were pooled for

The Biology of Butterflies
0-12-713750-5

Fig. 22.1. Morphs of *P. glaucus*. a: male; b: yellow disruptively-coloured female; c: dark, mimetic female; d: represents *Battus philenor*, the unpalatable model for dark female. (Photographed in visible light.)

Table 22.1. Wild-collected samples of dimorphic females of *P. glaucus* from Maryland. (Collected between 15 April–30 September, 1970–1977.)

Maryland Localities	Nos and frequencies of female morphs		
	Yellow	Dark	TOTALS
1. Northeastern (Cecil Co., nr. Conowingo Dam)	82 (55%)	66 (45%)	148
2. Central (Howard Co., Baltimore Co., Baltimore City)	21 (51%)	20 (49%)	41
3. Southern (Calvert Co., Anne Arundel Co., Charles Co., & St Mary's Co.)	87 (56%)	69 (44%)	156
TOTALS	190 (55%)	155 (45%)	345

final analysis. The samples consisted of 55% yellow and 45% dark females, reflecting apparently constant central Maryland frequencies.

The spermatophore counts (Tables 22.2, 22.3) indicate that 66% of all females contained either zero (two dark females only) or one spermatophore, 29% had two spermatophores, while only 5% contained either three or four (only a single yellow contained four). Among both morphs the observed and expected numbers in each category correspond

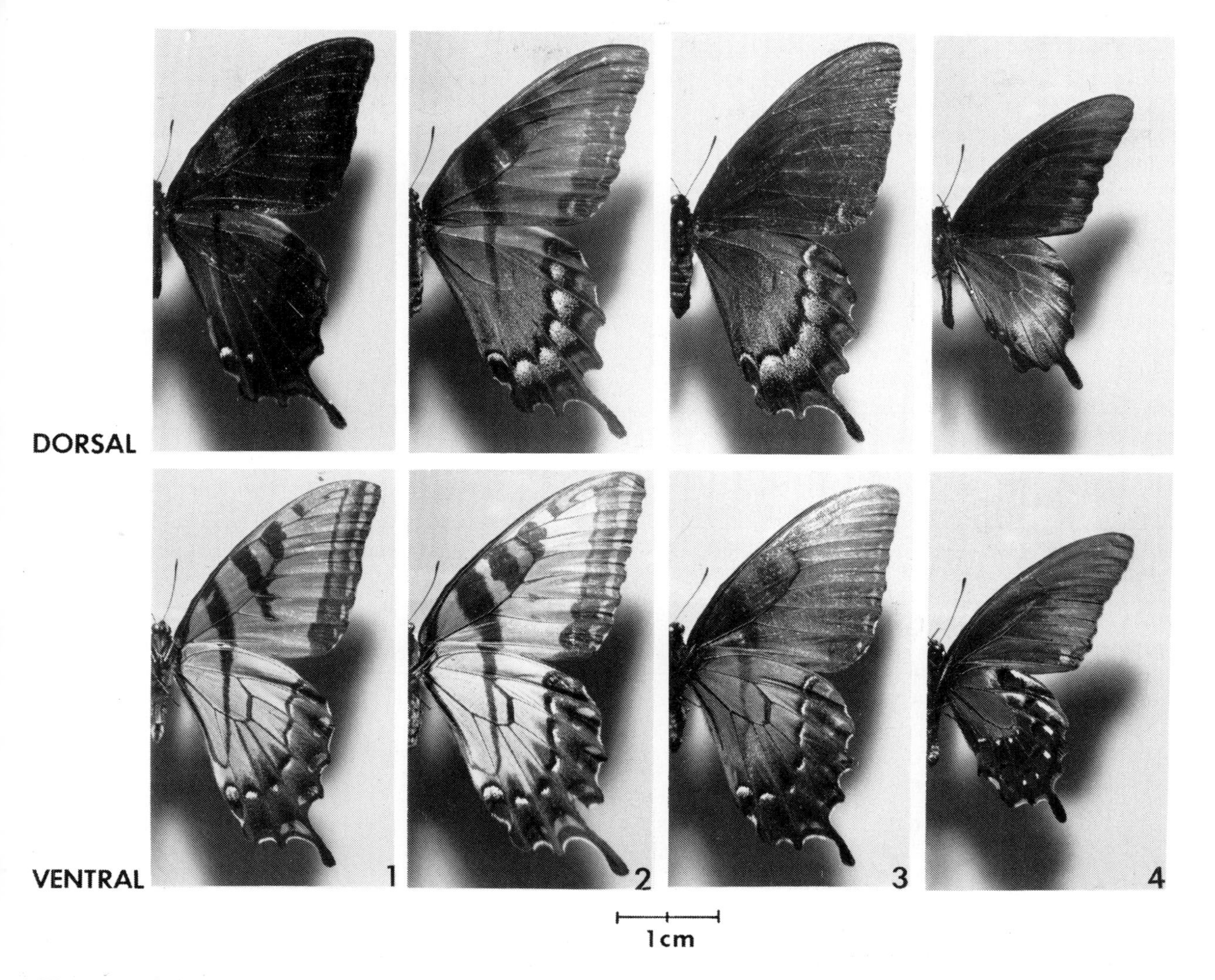

Fig. 22.2. Subjects as in Fig. 22.1, when photographed through a UV transmitting filter opaque to visible light.

Table 22.2. Analysis of spermatophores contained by dimorphic females of *P. glaucus*.

A. $\chi^2_{(2)}$ analysis of number of spermatophores per female

		No. of spermatophores per female				
		0–1	2	3–4	TOTALS	$\chi^2_{(2)}$
Yellow	Observed	125	56	9	190	0.08
	Expected	124.5	55.7	9.9		
Dark	Observed	101	45	9	155	0.12
	Expected	100.5	45.4	8.1		
Total nos		226	101	18	345	$\chi^2_{(2)} = 0.199$,
	Frequencies	(0.655)	(0.293)	(0.52)	(1.00)	$P > 0.90$

B. Total number and frequency (in parentheses) of spermatophores contributed by each sub-group. (Each spermatophore represents a separate mating.)

	No. of spermatophores per female					
	0	1	2	3	4	Totals and frequencies
Yellow	0	125	112	24	4	265 (0.55)
Dark	2	99	90	27	0	216 (0.45)
TOTALS	2	224	202	51	4	481 (1.00)

(a Z test on the total number of spermatophores contained by the yellow and dark females yields $Z = 0.03$, with $P > 0.97$)

Table 22.3. Analysis of *P. glaucus* spermatophore data for fit to the Poisson Distribution.

Observation and no. of spermatophores (parentheses)[1]	Theoretical frequencies $P(X)$	Observed and expected nos (parentheses) $f(X)$ Yellow females	Dark females	Both samples
0 (1)	0.677	125 (128.6)	101 (104.9)	226 (233.6)
1 (2)	0.264	56 (50.2)	45 (40.9)	101 (91.1)
2 (3)	0.052	8 (9.9)	9 (8.1)	17 (17.9)
3 (4)	0.007	1 (1.3)	0 (1.1)	1 (2.4)
TOTALS	1.000	190	155	345
$\chi^2_{(2)}$ =		1.20	1.76	2.19
P		>0.50	>0.25	>0.25

[1]Since there is no 'zero' spermatophore category in this analysis X has been defined as the number of times each female has been mated after the first mating, with $\lambda = 0.39$.

Table 22.4. Spermatophore no., egg no., and age of dimorphic females in *P. glaucus.*

	No. of spermatophores Yellow	Dark	Unlaid eggs Yellow	Dark	Age Yellow	Dark
x	1.395	1.394	23.6	19.8	1.90	1.99
± S.E.	±0.04	±0.05	±0.05	±1.3	±0.05	±0.06
t	0.02		2.10*		1.31	
d.f.	343		343		343	
P	>0.50		0.05 > P > 0.01		>0.10	

Results of linear regression analyses

	Age (X) *vs* *spermat. no. (Y)* *Yellow*	*Dark*	Age (X) *vs* *egg no. (Y)* *Yellow*	*Dark*
b ± S.E.	*0.27*	*0.21*	*−0.64*	*1.23*
	±0.06	*±0.07*	*±1.73*	*±1.91*
t	4.52**	2.94**	0.37	0.64
d.f.	188	153	188	153
P	<0.01	<0.01	NS	NS

Age based on wing wear (category 1, fresh (less than one week old), 2, worn (1–2 weeks), 3, tattered (more than two weeks).
*$P < 0.05$
**$P < 0.01$

closely, with the overall χ^2 value (d.f. = 2) being low ($P > 0.90$). Thus, the null hypothesis that the two female morphs on average contain equal numbers of spermatophores is accepted (Table 22.2A). In fact, the total numbers of spermatophores contained by the two morphs (Table 22.2B) are exactly proportional to their frequencies in our samples (Table 22.1). Also, the total numbers of females containing, respectively, one, two, three and four spermatophores approximate the expected Poisson distribution for 0, 1, 2 and 3 occurrences respectively (Table 22.3). On this evidence, insemination of Maryland *P. glaucus* appears to be random with respect to female colour morph.

Table 22.4 indicates that, while mean spermatophore number for both morphs is identical, yellow females contained a higher number of eggs ($P > 0.05$) than dark females. The dark females were slightly older than the yellow (estimated by wing wear), but the difference was not statistically significant. In the regression analyses (Table 22.4), the t-test on age *versus* spermatophore number, as might be expected, was clearly significant for both morphs ($P < 0.01$). However, the t-test on age *versus* egg number showed no significant difference ($P > 0.10$).

With respect to colour patterns, the tiger-striped (non-mimetic) insects closely resemble both the dark (mimetic) morph and the *B. philenor* model when viewed in UV light (Figs 22.1, 22.2). The similarity is somewhat greater on the dorsal surfaces, with both morphs being highly UV absorptive and the tiger-stripes tending to disappear. However, the tiger-striped pattern does remain faintly visible under UV, especially on the ventral surfaces (similar, dark stripes are also visible on the ventral surfaces of many mimetic individuals).

Discussion and Conclusions

The data suggest that courtship and mate selection in dimorphic *P. glaucus* populations are random with respect to female morph, the monomorphic males exerting no differential mate selection. These findings refute Burns' (1966) hypothesis that mate choice is capable of maintaining the female dimorphism. Only two unmated females were encountered amongst the 345 examined, and we are certain that all females are mated shortly after eclosion (i.e. before they have flown very far). As the numbers of females of both types containing multiple spermatophores (two, three or four) exhibit a Poisson distribution (Table 22.3), this suggests that additional matings also occur at random with respect to morph.

Re-examination of earlier spermatophore counts by Burns (1966) and Pliske (1972) reveals that their samples also fit Poisson distributions (Table 22.5). However, Makielski's (1972) Virginia sample departs ($P < 0.01$) from the Poisson values, having too many double-mated and two few single-mated females—apparently a result of sampling at the end of the flight season (August and September only). Nevertheless, her data, like our own, show that no significant differences exist between the mean number of spermatophores contained by dark and yellow females.

Table 22.5 presents *P. glaucus* samples in which the frequencies of the two female morphs vary between extreme yellow:dark ratios of 1:30 (North Georgia sample) to nearly 1:1 (Baltimore, Maryland samples). In no case is there any compelling evidence supporting differential mate selection. Thus, the argument that such selection may be frequency

Table 22.5. Frequency of yellow and dark female morphs in eight samples of *P. glaucus* from southeastern USA, with Poisson analyses of spermatophore counts in four of the populations. (From Burns 1966, with additional data.)

A. Samples and frequency of morphs

Localities	Inclusive collection dates	No.	Frequency of morphs: Yellow	Dark	Data source
1 Baltimore Co., Maryland	24–31 Aug. 1965	29	0.55	0.45	Burns 1966
2 Central Maryland (Baltimore, Cecil and Calvert Cos.)	14 Apr.–30 Sept. 1971–1977	345	0.55	0.45	Platt *et al.* this volume
3 Albermarl Co., Virginia	Aug.–Sept. 1966–1968	200	0.36	0.64	Makielski 1972
4 Mt Lake, Virginia and West Virginia	13–22 June 1965; 26 July–19 Aug. 1965	84	0.14	0.86	Burns 1966
5 Great Smoky Mts, N.C./Tenn.	11–21 Aug. 1954; 22 July 1959	15	0.07	0.93	Brower & Brower 1961
6 N. Georgia (Margret, Fannin, and Union Cos.)	15-16 Aug. 1959	33	0.03	0.97	Brower & Brower 1962
7 Highlands Co., Florida	1–4 Aug. 1972	220	0.69	0.31	Pliske 1972
8 Parker Isles, Florida (Highlands Co.)	13 June–6 Aug. 1959	517	0.92	0.08	Brower & Brower 1962
	6 Mar.–13 June 1956	501	0.94	0.06	

B. Results of fits of the frequency of multiple spermatophores to theoretical Poisson distributions in the above samples for which data are available

Samples	n	λ	$\chi^2_{(2)}$	P
1 Baltimore Co., Maryland	29	0.72	2.17	>0.25
3 Albermarl Co., Virginia	200	0.51	11.61**	<0.01
4 Mt Lake, Va./West Virginia	84	0.75	0.35	>0.75
7 Highlands Co., Florida	220	0.15	2.08	>0.25

**$P < 0.01$

dependent (Makielski 1972, Levin 1973) is also unsupported by the available data.

All data taken together lend support to the idea that both female morphs are being mated at random. The similarities between the morph frequencies in our samples, and the total number of spermatophores contained by each morph (Tables 22.1, 22.2) suggest that males may not even be able to distinguish the two types of females—a finding possibly explained by the UV observations (Lutz 1933, Remington 1973, Ferris 1972, Scott 1973*f*) (Fig. 22.2).

Two of our three 'speckled' females (excluded from data analyses above) contained two spermatophores, while the other contained one, suggesting that the speckled morphs have neither a mating advantage nor disadvantage. (Levin 1973 showed that a single insemination is sufficient to fertilize nearly all eggs that a female *glaucus* will produce during her life, and that both yellow and dark females have equal fecundities and progeny viabilities in laboratory cultures.)

Since preferential mating cannot be demonstrated to contribute towards the maintenance of female dimorphism in *P. glaucus*, another explanation must be sought. The dark morph is mimetic, and is apparently under selection for close resemblance to *B. philenor* (Poulton 1909, Brower 1958*b*, Brower & Brower 1962, Platt & Brower 1968). The tiger-striped morph, as in many similarly patterned *Papilio* species, apparently functions disruptively, breaking up the insect's outline in dappled sunlight. Mark-release-recapture experiments have been performed in which diurnally active male Promethea moths (*Callosamia promethea*) are released into the natural environment, some painted to resemble the yellow form of *P. glaucus* (Waldbauer & Sternberg 1975, Sternberg *et al.* 1977, Jeffords *et al.* 1979, Toliver *et al.* 1979, Sisson 1980). Differential recapture rates obtained between moths painted to resemble unpalatable models, mimics, and edible control species were attributed to avian predation. Indirect evidence for this is provided by the type of wing damage found amongst recaptured moths.

Since the frequencies of the dark and yellow morphs of *P. glaucus* appear to be relatively constant through time (i.e. from year to year) in any one locality, but vary between different localities in accordance with the prevalence of the model (Brower & Brower 1962; Table 22.5), we suggest that a balance exists between the relative strengths of the selective forces for mimetic and disruptive coloration. The major force maintaining this balance is almost certainly avian predation.

A similar explanation of disruptive coloration counter-balancing mimicry has already been suggested for the maintenance of banded and unbanded forms of the Banded Purple butterfly, *Limenitis arthemis-astyanax*, in intergrading (or blending) natural populations of these nymphaline butterflies (Platt & Brower 1968; Platt *et al.* 1971).

Acknowledgements

The authors wish to thank Mr J. H. Fales, Mr P. J. Kean, Mr E. Talley and the late Dr R. E. Silberglied for assistance with various aspects in the preparation of this paper.

23. *The Role of Pseudosexual Selection in the Evolution of Butterfly Colour Patterns*

Richard I. Vane-Wright

British Museum (Natural History), London

Butterflies, the dominant group of day-flying Lepidoptera, differ from almost all moths by their use of vision during the mate-selection stage of courtship (Chs 20, 25). They also differ in frequently exhibiting female-limited polymorphism and in lacking male-limited interspecific mimicry (Scudder 1877, Bernardi 1974, Vane-Wright 1975, Turner 1978). It is my contention that these phenomena in butterflies may be linked by an evolutionary process which I have called pseudosexual selection (Vane-Wright 1980).

Pattern Use and Signal Requirements

Colour patterns are used by butterflies for two principal purposes: defence against visually hunting predators, and for intraspecific signalling (Ch.20). Communication between the sexes may be intrasexual (male ↔ male; female ↔ female), or intersexual (female ↔ male; male ↔ female). Female butterflies tend to be solitary. Males, in contrast, interact vigorously, either defending a territory (Baker 1983; Ch.26), jostling in 'personal space' (many Pierinae), or apparently fighting over potential mates. At other times the males of the same species may form specific herds at salt or nitrogen sources ('mud-puddling' etc.). There thus appears to be a strong requirement for long-distance visual signalling amongst males, but not amongst females. With respect to intersexual signalling, the position seems less clear—although males usually locate adult females visually, males are promiscuous and so the only basic requirement for females needing to be mated is to fly in places where sexually active males occur. In addition, although females could theoretically generate epigamic selection affecting male visual signal pattern, there is little evidence that they do so—with the possible exception of UV colours (Ch.20—but see also Ch.21).

Are Males Narcissists?

Does female colour have no significant influence on males (cf. Ch.22)? If males are truly promiscuous, and even enter readily into interspecific unions, then it would seem unlikely that they would be influenced by variations in the colour patterns of potential mates. Indeed, males often appear insensitive even to clear mate-refusal signals, and may persist in trying to force themselves on unwilling females. Consistent with this, although relatively rare in males, polymorphism is widespread in female butterflies. Nonetheless, there is considerable evidence that males of many species, including some with polymorphic females (but see Ch.22), *are* sensitive to variations in female colour or pattern (Chs 20, 21). I believe that the only generalization that can be drawn from present studies is *not* that males respond most vigorously to the 'normal' colour of their females (Ch.20) but, like Narcissus, they show a preference for self-coloured, i.e. male-like or andromorph (Bernardi 1974) females (e.g. Magnus 1958*a*, 1963, on *Argynnis paphia*; Silberglied (Ch.20) on *Anartia amathea*; Vane-Wright (unpublished) on *Papilio ulysses*, *P. euchenor*, *P. aegeus*, *Cethosia* spp., *Vindula arsinoe*, and other spp.). A physical basis for this could lie in stimulus filtering by the butterfly brain, as suggested by the electrophysiological work of Swihart (see review in Ch.20).

Are Andromorph Females Transvestites?

The paradox of male preference and male promiscuity can be resolved if andromorph females evoke the same long-distance response as conspecific males: males would fly up to andromorphs in eager anticipation of a fight or joining a mud-puddle assemblage (Vane-Wright 1981*b*). On close approach, olfactory and other cues (Ch.25) would indicate their

The Biology of Butterflies
0-12-713750-5

error, and the males could then switch to courtship behaviour. Such a process, whereby one sex shows a preference for a particular morph of the opposite sex due to initial misidentification with a potential competitor or synergist of its own sex, I define as *pseudosexual selection.*

The process of pseudosexual selection could permit the maintenance of balanced female-limited polymorphisms. If andromorph or 'transvestite' females evoke a stronger long-distance response, they may get mated earlier and so potentially gain a longer post-mating life in which to lay eggs (Vane-Wright 1981*b*; Platt *et al.* (Ch.22) reject this notion in *Papilio glaucus*, however). In contrast, cryptic, disruptive or mimetic females may be less subject to attack by predators than are brightly-coloured andromorphs. Depending upon ecological circumstances at a particular time, one advantage or the other may be overwhelming, balancing or relatively non-existent. Monomorphism, simple sexual dimorphism and partial female-limited polymorphism are the most common categories of pattern morphism found in butterflies, and may all exist even within different populations or subspecies of a single species (e.g. *Appias nero, Papilio dardanus.*

Darwinian Transference and Wallacian Signals

Darwin (1874) considered, amongst other possibilities, that the best explanation of sexual dimorphism in butterflies involved the idea that male colours stimulate females, and so the males have become more brilliant through female (epigamic) selection. Features so evolved in males could later be *transferred* to the females during the course of evolution (decoupled morphs, Vane-Wright 1979*a*; see also Ch.20) but, in general, the females tended to retain their relatively dull, ancestral, protective colour patterns.

Wallace (1889), on the other hand, thought that brilliant colours (in either sex) were used to fulfil a need for specific recognition signals. He regarded the males as generally the more progressive sex with respect to signal coloration, while the females remained primitive or were separately modified (crypsis, mimicry, etc.) for protection against predators. Wallace dismissed both Darwinian transference and epigamic selection by females.

On the latter point, Wallace was probably correct. As shown by Silberglied (Ch.20), the evidence for widespread female epigamic selection of colour patterns in butterflies is unconvincing, and I agree with this despite Smith's (Ch.21) painstaking demonstration of some 'true' sexual selection affecting morphs of the singular *Danaus chrysippus*. However, I believe Wallace was wrong to dismiss Darwin's concept of transference.

If we reject female epigamic selection as a major force affecting butterfly colour patterns, but accept the notion of transference potential, the Darwin/Wallace theories can be combined in the form of a scenario: during speciation there may be a more or less dramatic, 'progressive' shift in male colour pattern for specific signalling purposes, not accompanied by a similar shift in female pattern, which remains 'primitive' (e.g. *Nessaea hewitsoni*, Vane-Wright 1979*b*). Subsequently, the new male pattern is stabilized for signal function (Turner 1978; Ch.20), while the female pattern, if advantageous, may become more or less modified for (greater) protection against predators. However, if the pattern adopted by the females is not particularly successful or, more likely, if predation drops to a lower level, pseudosexual selection may bring about the transference of male colour to the females—they become andromorphs through a mating advantage over alternatively patterned females (cf. Searcy 1982: 60, on mate choice and species identity). Such a process, bringing about secondary convergence of the pattern of one sex toward that of the other, would be an example of intraspecific mimicry (Wickler 1965, 1968, Vane-Wright 1976, 1981*b*), originally conceived by Wickler to be an important starting point for the evolution of many social signals.

We may thus envisage a three-step polymorphism pathway (Vane-Wright 1979*a*) leading to the evolution of a new species-specific colour pattern: speciation leading to simple sexual dimorphism, leading to partial female-limited polymorphism, leading to monomorphism, with the female pattern now secondarily convergent on the male (this may be manifest by small differences, females very often being the slightly less brightly-coloured sex in nearly monomorphic species).

What Can Pseudosexual Selection Explain?

The strong (visual) long-distance signalling requirement imposed on males by their own behaviour may help to explain why interspecific mimicry is never restricted to the male sex in butterflies. A notion of pseudosexual interactions may explain the anomalous observation that highly promiscuous males are nonetheless strongly affected by female colour. The idea of Darwinian transference, with pseudosexual selection as the process, may explain the widespread existence of partial female-limited polymorphism in butterflies (virtually unknown in other insects)—it may often represent no more than a 'frozen', perhaps only temporarily frozen, stage in the transference of a specialized male signal-pattern to the female.

Fighting is Not Enough

Here I must point out another paradox, which I think is answerable, and a fundamental problem—which remains unsolved. Hingston (1933) and Silberglied (Ch.20) have suggested that the principal function of bright male coloration in butterflies lies not in epigamic display (with which I agree), but for use in aggressive intrasexual interactions—with which I agree only in part. If male-male competition were the only factor, we might expect to find, following the arguments of Gadgil (1972), a number of butterfly species with dimorphic males: bright α 'fighters' and dull β 'sneaks', the latter group adopting a less energetic, non-fighting strategy to pick up females (this argument is *dependent* on females being unimpressed by male colour!). But male-limited visual polymorphism in butterflies is extremely rare (Bernardi 1974, Vane-Wright 1975; *Euthalia monina*, which has trimorphic males, is a possible candidate—Corbet & Pendlebury 1978). What other general factor, apart from the largely rejected epigamic selection, could stabilize male pattern? I believe we should seek the answer in the co-operative, non-aggressive interactions of male butterflies, such as mud-puddling, where the advantages of herding (Hamilton 1971), for example, are potentially very important to the day-to-day survival of such conspicuous organisms. The herding constraint alone may ensure that male deviants are nearly always at a nett disadvantage, and are so eliminated.

The fundamental problem is, simply, the origin of species. What happens during speciation such that the apparent stabilizing selection acting on male pattern is suddenly or miraculously released? The real origin of species, and its immediate consequences or prerequisites, remains the unsolved basic problem (Sibatani 1983).

Predictions and Tests

Platt *et al.* (Ch.22) suggest that any advantage of early mating to andromorph females is purely notional (cf. Magnus 1963), whereas Smith (Ch.21) implies that male-like females may have a real and measurable selective advantage. The clear demonstration that andromorphs do have a mating advantage, in at least some species or situations, is essential if pseudosexual selection is to be seriously regarded as a potential factor in colour pattern evolution. Before devising field experiments, however, some computer simulations should be undertaken, based on a model of butterfly reproductive strategy in which a number of factors can be varied. These would need to include variation for post-eclosion maturation, mode of egg production, degree of adult resource utilization, sex-ratio, mating frequency and average life-expectancy, in addition to separate mate-location times and survival values for andromorph and non-andromorph females. Other factors, such as degree of protandry, may be important. It does not seem to me that the outcome for particular combinations of values would necessarily be intuitively obvious.

With respect to field tests on experimentally altered males, effects on life-expectancy would be expected and, consequently, on overall reproductive success—but not in particular mating encounters through epigamic rejection (cf. Silberglied's (Ch.20) results on *Anartia amathea*). In making such experiments, males should not only be altered 'at random', but also to look like the males of other species in the study area. Of course, field or cage experiments with altered females offer the most direct test of the pseudosexual selection hypothesis, and these should be carried out with monomorphic, partial female-limited polymorphic, and sexually dimorphic species. The best tests might be based on polytypic species, whereby naturally more or less andromorphic females from different populations could be brought together and compared for speed of mating. However, perhaps the easiest initial experiments would be indirect, and involve mud-puddling males. Changed behaviour or survivorship of experimentals would be of considerable interest.

Acknowledgements

This brief paper would have been immeasurably improved had it been possible to continue discussions and undertake field work with Bob Silberglied. I would like to thank Georges Bernardi, Larry Gilbert and John Turner for discussion, at various times, of some of the ideas involved.

24. *Upsets in the Sex-Ratio of Some Lepidoptera*

Sir Cyril Clarke

Department of Genetics, University of Liverpool, Liverpool

Sex, as everyone knows, is inherited. The Almighty and Ronald Fisher between them decided that in general the ideal ratio between males and females should be 1:1. Here I outline two exceptions to this rule. One of them is in a moth, *Lymantria dispar*, where in a particular circumstance only males are produced, and the other in a butterfly, *Hypolimnas bolina*, where all-female broods are common. In both examples there should be a qualification—'all' should be 'almost all'.

All-Male Broods

In *L. dispar*, Goldschmidt (1934) stated that the most extreme form of intersexuality was found in the offspring of the cross between a 'very weak' female from the island of Hokkaido and a 'strong' male from the mainland of Japan. This produced only normal-looking males, and Goldschmidt postulated that half of these were completely transformed females. In recent studies using the same cross, E. B. Ford and I can confirm a large excess of males, but the sex-chromatin technique (see below) has shown that all the male insects tested either as larvae or adults were negative for the heteropyknotic body, i.e. were chromosomally male. The only positive insects were two normal looking females from one of the broods. We think it much more likely that both Goldschmidt's findings and the present ones result from the Haldane (1922) effect, and that the female embryos or tiny larvae died because of 'genic imbalance' (Clarke & Ford 1982).

Materials and methods

Egg-masses of *L. dispar* were obtained from Japan, Hokkaido and Aichi province. Breeding of the insects in England was carried out by standard methods.

The Smith (1945) S-status is tested (in both *L. dispar* and *H. bolina*) by removing, using sharp dissecting scissors, an abdominal proleg (Fig. 24.1). Enough tissue can be scraped from inside the proleg to make one good preparation; the material is teased out and spread as thinly as possible. After amputation each larva is kept separately and cotton wool applied to the wound. In the adult, gut or Malpighian tubule cells are used, from freshly killed insects.

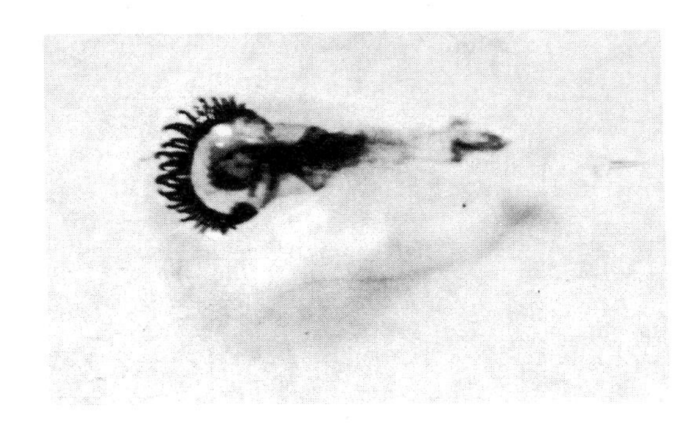

Fig. 24.1. Proleg tip of *H. bolina* larva showing tissue scraped out (as seen under dissecting microscope ×10).

The cells are not fixed before staining. Two drops of 2% orcein in 45% acetic acid are placed over the tissue and a coverslip added. After 10–15min the coverslip is firmly pressed to make a 'squash' preparation. The 'Smith' body, when present, can be clearly seen under a × 40 objective, as well as under a × 90 (oil immersion) (Figs 24.2, 24.3; Cross & Gill 1979).

Results

Table 24.1 gives the data in female Hokkaido × male Aichi cross; Table 24.2 gives the controls. In the cross it will be seen that, as judged from the S-testing, there is no suggestion of any sex reversal, all the males being S-negative (Table 24.1). In Table 24.2, for comparison, are recorded the findings in the pure races; here the overall sex ratio is near unity. We have, however, occasionally found discrepancies, the 'Smith' procedure not being entirely free from error.

The Biology of Butterflies
0-12-713750-5

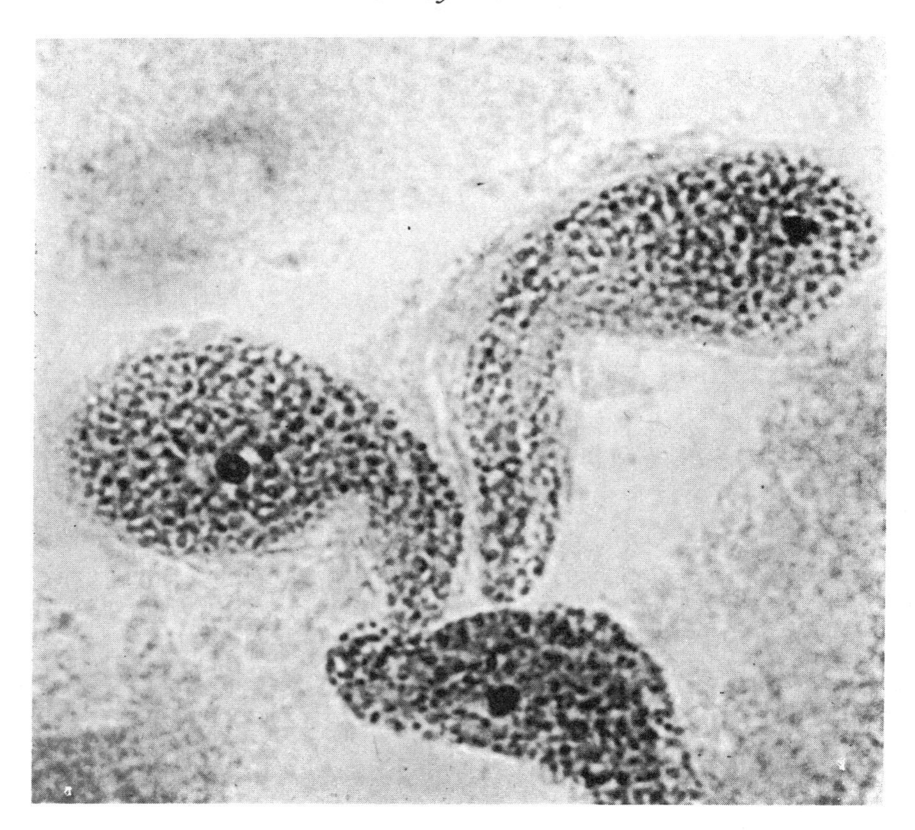

Fig. 24.2. Nuclei of tissue cells containing heteropyknotic body from a larva which developed into a female *H. bolina* butterfly (×90 oil immersion objective).

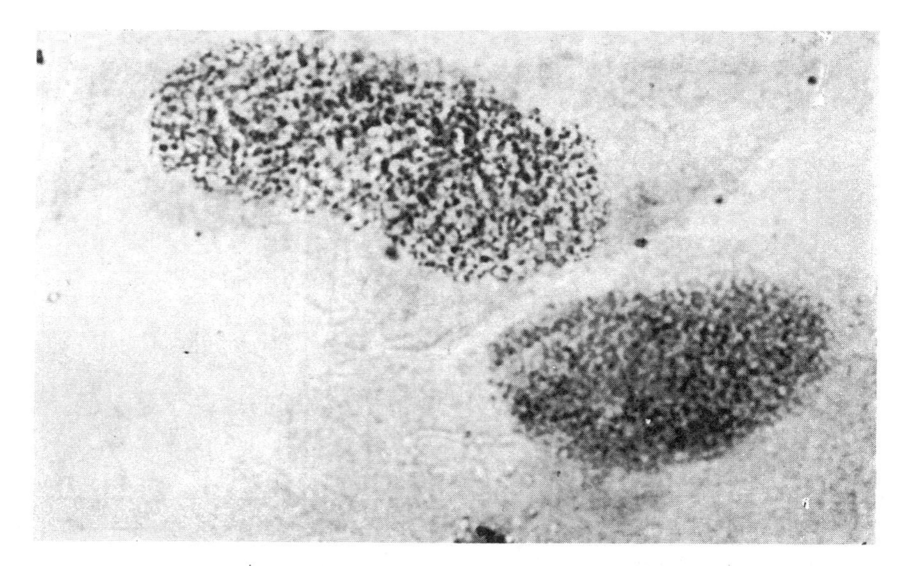

Fig. 24.3. Nuclei of tissue cells lacking heteropyknotic body from a larva which developed into a male *H. bolina* adult (×90 oil immersion objective).

Discussion

There is good evidence that partial sex-reversal in *L. dispar* is not associated with changes of chromosomal sex, for we showed that partially transformed females remained S-positive (Clarke & Ford 1980). If, as seems likely, the same is true for complete sex-reversal, we have very convincing evidence that the males are S-negative and that upset in the sex-ratio is due to other causes. A further argument against Goldschmidt's view is that our two females in brood 16510 appeared normal (see footnote to Table 24.1), whereas on the Goldschmidt hypothesis one would have expected at least some evidence of incomplete sex reversal.

Haldane (1922), in a well-known paper, stated in

Table 24.1. Details of broods of the *L. dispar* cross female Hokkaido ('very weak') × male Aichi ('strong').

Brood No.	Offspring male	Offspring female	No. of adults S-tested as moths	Results of S-testing of adults	No. of adults S-tested prospectively as larvae	Results of S-testing as larvae. (In dead larvae information about the gonads is given where known)
16508	45	0	25 males	25 S-neg.	4 males	4 S-neg.
16509	33	0	21 males	21 S-neg.	2 males	2 S-neg.
16510	42	2[1]	25 males 1 female	25 S-neg. 1 S-pos.	13 males 1 female	12 males S-neg., 1 male S-pos. as larva but neg. as adult in both gut and Malpighian tubules. The female was S-pos. as a larva. 5 larvae died, all S-neg., all with testes.
16513	11	0	7 males	7 S-neg.	0	2 larvae died, both S-neg.
16514	31	0	15 males	15 S-neg.	2 males	2 males S-neg.; 2 died as larvae both S-neg.; 1 died as larva, unscorable.[2]
16519	18	0	13 males	13 S-neg.	0	2 larvae died, 2 killed for dissection. Both latter had testes, all 4 S-neg.

[1]The two females (Brood 16510) appeared normal except that they were intermediate as regards colour between Aichi females (dark) and Hokkaido females (white). One female was mated to a sib, and the other was backcrossed to a Hokkaido male. Both females laid eggs but it is not yet known if these are fertile. We have also set up the reciprocal cross, female Aichi to male Hokkaido.
[2]With brood 16514, four additional small larvae were killed. All had testes and were S-negative (gut cells). See also Clarke & Ford (1983).

Table 24.2. Details of *L. dispar* broods of pure Hokkaido and pure Aichi stock

Brood no.	Offspring Male	Offspring Female	No. of adults S-tested prospectively as larvae (none tested as adults) Male	No. of adults S-tested prospectively as larvae (none tested as adults) Female	Results of S-testing as larvae. (In dead larvae information about the gonads is given where known)
Hokkaido 16494	2	4	2	4	2 males S-neg. 3 females S-pos. 1 female S-neg. 13 died as larvae: 9 S-neg. and 4 S-pos.; 1 S-neg. had testes and 1 S-pos. had ovaries.
Hokkaido 16537	4	3	2	1	2 males S-neg. 1 female S-neg. 1 died as larva, S-pos. and had ovaries.
Hokkaido 16538	12	7	0	1	1 female S-neg. 2 died as larvae, both S-neg. and had testes.
Hokkaido 16547	13	19	6	9	6 males S-neg. 9 females S-pos. 3 died as larvae, all S-pos. and one pupa was lost, S-neg.
Aichi 16522	2	10	2	3	2 males S-neg. 2 females S-pos. 1 female S-neg.

his summary that 'When in the F1 of a cross between two animal species or races one sex is absent, rare or sterile, that sex is always the heterozygous sex'. He gives many examples, some in the Lepidoptera and others in different orders, and his observation has subsequently been amply confirmed—though 'heterogametic' is the term now used (i.e. the female sex in the Lepidoptera). Haldane continued: 'As pointed out by Sturtevant, excess of homozygotes may be due to two distinct processes, a killing off of the heterozygotes or their transformation into members of the normally homozygous sex'.

The explanation of the Haldane effect was given by Clarke & Ford (1982) and is quoted here verbatim.

> Sex is controlled by a balance between predominantly male-determining and female-determining genes in the autosomes and sex chromosomes respectively. These are normally opposed so as to give a requisite excess of one set of sex-determinants or the other; yet they may take different *absolute* values in distinct races. Haldane therefore pointed out that in the homogametic sex all the chromosome types, including the X-chromosomes (XX) consist of one member from each race. However, in the heterogametic sex (XY) Y is dissimilar from X, containing fewer and different genes from the X. Therefore in female Lepidoptera the

male-determining genes in the X and the female-determining genes in the Y may no longer be respectively in effective excess in racial crosses and thus give rise to sexual abnormalities, but much more easily in the heterogametic sex than in the homogametic one . . . in racial crosses all the loci in the two X-chromosomes, being homologous, will on the whole produce an intermediate effect, though some of the characters they give rise to may be subject to genic interaction. In the heterogametic sex of the hybrids, the genes in the non-pairing region of the Y are non-allelic with any in the X, and to some extent they are selected to be different in the two races. It follows that the balance between the genes on the X from one race and the Y from another may be dissimilar in the F1 compared with the reciprocal cross, and this explains why the sex ratio may be very different in the two crosses—Goldschmidt (1934) for example found no excess of males in the cross female 'strong' Japan × male Hokkaido.

The evidence presented here in Tables 24.1 & 24.2 clearly indicates that the excess of males in Goldschmidt's classic cross is due to the Haldane effect of genic imbalance, and not to sex reversal. Brood 7207 of our earlier paper (Clarke & Ford 1980) was probably another instance of heterogametic lethality.

All-Female Broods

This phenomenon is well known in several species of butterfly (see Ch.21) but we have some detailed information on *Hypolimnas bolina* (Clarke *et al.* 1975). It has been known at least since 1924 (Poulton 1924*c*) that a proportion of females in the butterfly *H. bolina* produces only daughters, whereas others produce a 1:1 sex-ratio. Our breeding results amply confirmed this and we showed that the production of broods with the disturbed sex-ratio is inherited entirely through the female line. Only very occasionally are a few males produced. By sexing the embryos, larvae and a few pupae cytologically (see above), we obtained evidence that the deficiency of males was due to their very high mortality in the pre-adult stage. We suggested that the abnormal sex-ratio was due to an infective cytoplasmic factor (the presence of spirochaetes, as in *Drosophila*, having been ruled out), in contrast to *Acraea encedon* where meiotic drive of the Y chromosome had been postulated (Owen 1971*a*; see also Ch.21). We put forward the suggestion that the polymorphism in *H. bolina* is maintained by the 'infected' females being at a slight disadvantage and that their numbers are kept up by contagion from an unidentified reservoir species. However, we have never been able to infect 'bisexual' larvae by injecting tissue from 'unisexual' ones and such chromosomal studies as have been done do not indicate any abnormality—but further work is needed here. Field studies in Fiji in 1980 (Clarke *et al.* 1983) have shown that the ratio of abnormal to normal broods is much the same now as it was 60 years ago.

It seems unlikely that there are two different mechanisms in *H. bolina* and *A. encedon* and I think that sex chromatin studies as carried out in *H. bolina* should be done in *encedon*. In contrast to *L. dispar*, where the probable mechanism of the sex-ratio upset is known (see above), in the converse situation we have no idea as to the mechanism, but 'disturbed ground' (Owen 1971*a*) seems very unlikely in *H. bolina*, as unisexual broods have never been described in north Australia where much cultivation has been carried out.

25. *Chemically Mediated Interactions Between Butterflies*

Michael Boppré

Universität Regensburg, Zoologie, Federal Republic of Germany

Chemical communication in Lepidoptera has now been recognized for over 100 years, and even before this collectors took advantage of the long-range attraction of virgin female moths to obtain conspecific males (Fabre 1907; see also Mayer 1900). For the past 25 years or so, much attention has been given to these luring scents because of their potential for biological control in agriculture. Research has been mainly concentrated, not unnaturally, on moth pests, emphasizing chemical analyses of female secretions and tests of their potency as baits. In comparison, the study of courtship and close-range communication has received little attention.

Thus, although research on Lepidoptera pheromones could almost be described as 'fashionable', the butterflies appear much neglected. Few solid experimental data on their sexual behaviour are available—with regard to scents there are almost too few for a review paper. On the other hand, the information we do have—much of it more or less anecdotal—on butterfly 'scent organs', is sufficient to indicate a great diversity of problems and that butterflies are fascinating (although not necessarily convenient) organisms for basic studies on pheromone biology. This chapter aims to provide evidence for this allegation with respect to butterfly scents, by reviewing observations, data, suggestions and assumptions, and discussing them from a wide range of viewpoints. The intention is to give students of chemical communication mechanisms insight into the variety of phenomena involved in lepidopteran pheromone biology, partly by relating how studies on the subject developed and by summarizing both experimentally established facts and anecdotal evidence. To provide a basis for, and to stimulate renewed investigations on butterfly scents and sexual communication in the Lepidoptera, focal points of interest and ideas lacking substantiation are strongly emphasized throughout.

It is the common view that in butterfly courtship, the initial male search for females is guided by visual cues, and that at close-range, scents—particularly those of the males—come into play (e.g. Brower *et al.* 1965, Scott 1973*d*, Silberglied 1977, Ch.20). Although of great importance, chemical signals influence sexual communication only in part and, therefore, should not be seen in isolation from other cues. Nevertheless, concentration on the role of the (male) scents is justified by their adaptive complexity. Male pheromone systems of moths are considered in addition because of their similarity to those of butterflies.

Peculiar Androconial Organs—Scent Organs? Peculiar Scents—Pheromones?

Fritz Müller (1877*a*) recognized a musky odour originating from the alar fringes of a freshly emerged *Callidryas* male, a pierid butterfly. He thought that all patches, fringes, tufts, etc., restricted to the males of various Lepidoptera, were such 'odoriferous organs' even if they did not emit odours detectable by the human nose, and suggested that their evolution was the result of sexual selection (cf. Darwin 1871). It is Müller who deserves the credit for being the pioneer in exploration of male scents in the Lepidoptera.

Müller (1877*a*, 1878*a*) was also the first to discriminate between 'sexual odours' and 'protective odours' in butterflies (the 'epigamic and aposematic scents' of Dixey 1907). Odours thought to function in the defence of the insects (i.e. those not restricted to one sex, in general not associated with a special organ, and usually smelling unpleasant to humans) are not considered further in this paper.

Morphological structures, interpreted by Müller as organs that 'diffuse scents, which are undoubtedly agreeable to the females and arouse their sexual desires' (Müller *in* Longstaff 1912: 655), had long been known (e.g. DeGeer 1752), and conspicuous male organs were subsequently found in many

The Biology of Butterflies
0-12-713750-5

lepidopterans. The first record of microscopic male-limited structures was provided by Baillif (1825), who recognized fringed scales ('plumules') in *Pieris rapae*. Deschamps (1835) realized that these were male secondary sex characters and Watson (1865*a*) considered these plumules to be of taxonomic value—but the function of these peculiar scales was not seriously investigated. Baillif (1825) considered them deformities. Watson (1865*a*,*b*, 1868) assumed they would introduce air into the tracheae, and Wonfor (1869) stated 'They seem to me to have their analogies in the beard of man, the mane of the lion and the plumage of some birds'. Similarly, Scudder (1877) failed to appreciate that such scales might produce scents and only recognized a 'structural antigeny' (sex-limitation); he stated that the theory of sexual selection failed to explain the existence of these scale modifications, owing to their minute size.

Müller (1877*a*, 1878*a*) reported many more examples of *androconial* structures* which emit odours. His hypothesis of a sexually stimulating function was seldom doubted by subsequent investigators, especially since Weisman (1878) had indirectly supported this idea by demonstrating the presence of glandular cells beneath scales of wings—until then these were assumed to be composed of dead tissue. Numerous papers were published which not only reported on the phenomenology of androconial organs but also compared fine-structural details. Some of the most valuable reports, each of which deals with several species while also providing further references, include those by Aurivillius (1880), Dalla Torre (1885), Haase (1887*a*), Illig (1902), Swinton (1908), Dixey (1911, 1932), and Barth (1960) (see also articles by Müller in Longstaff 1912, and Dixey 1907, Clark 1927).

It is neither possible nor desirable to attempt here a complete survey of the morphology of androconial organs. Summarizing, androconia are modified scales with glandular bases. Individual scales may be variously modified to exhibit a fringed margin or an expanded surface, but most often they appear hair-like. Androconial scales may be scattered over the wings, but in general, they are arranged as discrete organs and often associated with veins, forming macroscopically visible patches, fringes, tufts, folds or pockets. Sometimes they occur as eversible structures, and they may also be found on other parts of the body. An enlarged surface, well-suited to increasing effective evaporation of volatile secretions, is characteristic of androconial organs. Various other forms of modified scales may protect these glandular areas, apparently to prevent uncontrolled evaporation of secretions.

Within the Rhopalocera, androconial organs are widely but irregularly distributed. They may occur in a given species or genus, while a close relative lacks them completely or perhaps has organs of very different construction. We are thus dealing with analogous structures, convergently evolved many times (Müller 1877*a*). Such structures are often of limited taxonomic value, although they certainly provide good characters in some groups (see e.g. Köhler 1900, Barth 1959, Boppré & Fecher 1977, Boppré in prep. 1984*a*).

Two especially peculiar features, the occurrence of 'binate organs' and the production of particles, are worth mentioning in some detail in order to show how elaborate androconial organs can be. Male danaine butterflies possess abdominal hairpencils extrudable by haemolymph pressure and containing up to five morphologically distinct hair types (Eltringham 1913, 1915, Boppré 1984*a*). Additionally, patches or pockets are often present on the hindwings, and a special behaviour pattern effects mechanical contacts between the two types of glandular organs (see below). Such binate androconial systems are also known from the nymphalid genus *Antirrhea*, where as many as six separate organs form three dual systems (Vane-Wright 1972*a*). The androconial organs of Danainae and of *Antirrhea* share the peculiarity of producing particles, a feature also found in Hesperiidae, some *Papilio* and many Heterocera (Boppré in prep. 1984*b*). At predisposed fracture points, androconial scales break off to form vast numbers of tiny particles (Müller 1878*b*, Eltringham 1937, Vane-Wright 1972*a*,*b*), termed *pheromone-transfer-particles* (Boppré 1976).

Although androconial organs can be extremely elaborate, it should be kept in mind that scents may also emanate from inconspicuous structures (e.g. Vetter & Rutowski 1978); also, scent-distributing organs may be separate from scent-secreting tissue. In a few cases, structures like typical androconial organs occur in both sexes (e.g. in the ithomiine genus *Methona*: Müller 1878*a*, Fox 1956 [both as *Thyridia*]).

An extensive body of detailed knowledge on androconial structures has been available for more than 50 years, and it is interesting to note that

*Instead of using the common term 'male scent organs', I call the structures in question 'androconial organs' to make clear that they do not necessarily emit a volatile chemical and thus to counter uncritical functional interpretations. The term 'androconia' was originally used by Scudder (1877) for the peculiar scales of male butterflies with reference to their masculine character 'since this is their single common peculiarity'. Often, androconial organs are also named after their outward appearance, e.g. 'hairpencils' in Danainae and 'fringes' in Ithomiinae. The term 'coremata' ('eversible sleeve-like bags, sometimes of great length and covered with hairs', Janse 1932; cf. Varley 1962) is nowadays mainly used in connection with Arctiidae, although it would also be applicable for many androconial organs in other families.

modern morphological investigations have not contributed any fundamental new ideas or discoveries. Nevertheless, the biological significance of male scents in Lepidoptera has largely remained a matter for speculation and has not been the subject of profound study, even when some field observations on the epigamic use of androconial organs have been reported (for reviews see Carpenter 1935, Brower *et al.* 1965).

Doherty (1891) described how 'male Euploeas often meet in great swarms, . . . most circling slowly around, many of them displaying their tufts, so that the air is noticeably permeated with their fragrance. Many different species meet on these occasions . . . I have often observed males flying around with expanded tufts . . .' and supposed that they were trying to attract the females from a distance. *Euploea core* males were occasionally noted to protrude their abdominal hairpencils in flight (Champion 1930, Fyson & Poulton 1930, Latter *in* Latter & Eltringham 1935). Latter once saw a male, sailing with expanded hairpencils, approached by a female flying upwind, while another flew to a male with expanded hairpencils held in his hand and hovered there for a few seconds. He concluded, like Doherty, that the anal brushes of *Euploea* distribute an aroma which 'actually causes the female to seek a male' (Latter *in* Latter & Eltringham 1935). Such a 'telegamic function' (Eltringham *in* Latter & Eltringham 1935) of androconial organs might also occur in *Troides amphrysus flavicollis*—males of this birdwing were reported to sometimes fly with the front wings only, while the hindwings drooped, thus opening the hindwing androconial fold which is otherwise closed in normal flight (Skertchly 1889).

Carpenter (*in* Carpenter & Poulton 1914) recorded the use of hairpencils by another danaine, *Amauris tartarea*—a male hovered about 10cm above a female that sat with her wings open. At intervals of a few seconds he rapidly extruded and withdrew his hairpencils. A male *Danaus chrysippus* has been observed using his hairpencils similarly by fluttering above and in front of a female, rapidly protruding and withdrawing the brushes and eventually coming down directly in front of her and protruding the brushes again (Carpenter & Poulton 1927). In both cases, however, this hairpencilling was not followed by copulation.

In addition to these observations on epigamic use of male danaine hairpencils, Lamborn noticed the hairpencils of two *Amauris* species extruded without a female being present. Males of *A. niavius* and *A. damocles* sat with open wings, passing the outer black hairs of the hairpencils to and fro over the hindwing patches (Lamborn *et al.* 1911, 1912, Lamborn & Poulton 1913, 1918). It was suggested that some secretion might be conveyed in this way from one organ to the other (Lamborn *et al.* 1911). Comparable contact behaviour was also recorded for *Parantica agleoides* (Lamborn 1921).

Male Pheromones in the Sexual Communication of Butterflies—Evidence and Implications from Experimental Studies

Although the anecdotal observations on the use of androconial organs did not give unqualified support to the Müller hypothesis, for several decades no attempt was made to establish the role of androconial organs or secretions by critical observation or experiment. Possibly, behavioural studies were not of interest because workers apparently felt certain

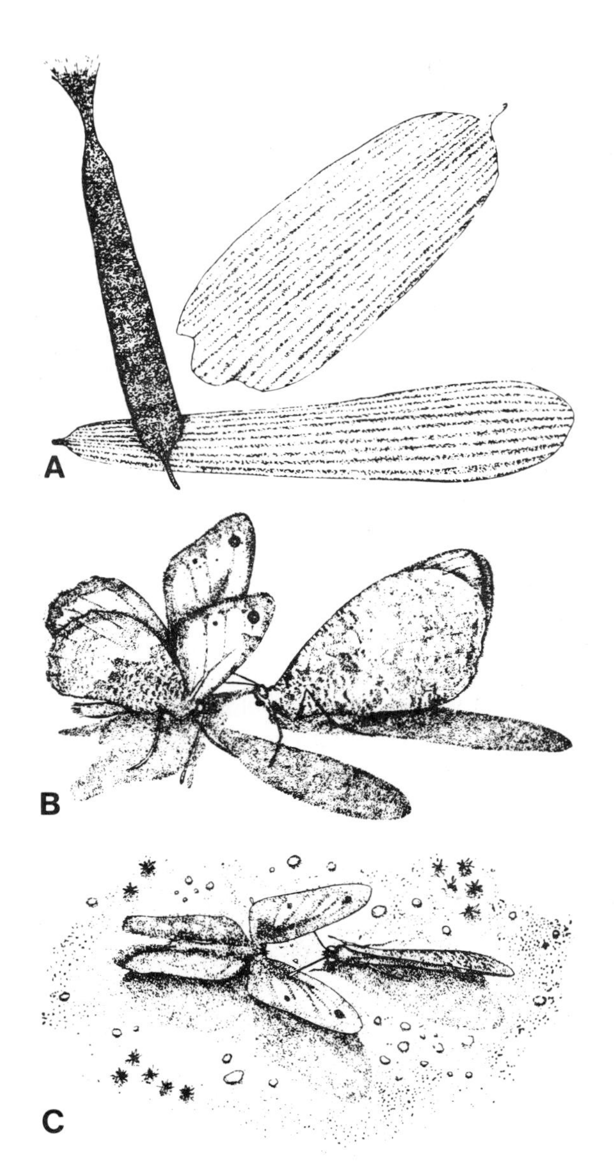

Fig. 25.1. Scales and courtship of *H. semele.* A: an androconial scale and two ordinary scales; B, C: bowing of male (left) during courtship. (From Tinbergen *et al.* 1942.)

of the adaptive significance of butterfly scents—and the Müller hypothesis well-fitted contemporary ideas on mate-selection (Darwin 1871). Thus, as late as 1941, Tinbergen and his collaborators were the first to conduct experimental investigations of butterfly courtship behaviour. They dealt mainly with visual signals, but included some simple experiments demonstrating the role of androconial organs in the Grayling, *Hipparchia semele* (Tinbergen 1941, Tinbergen *et al.* 1942). Courtship of *H. semele* was found to consist of several phases, two of which apparently directly involve the use of the forewing androconial patches (Fig. 25.1A; Sellier 1973*a*). After alighting and coming face to face with a settled female, the male Grayling flaps his wings in a distinct manner. The male then takes the female's antennae between his folded forewings (Fig. 25.1B,C), so bringing his androconial patches into contact with her sensory organs. This stimulates an unreceptive female to respond with fluttering (which repels the male), while a receptive female lifts her wings to permit genital contact. This distinctive behaviour suggests stimulation of the female by a secretion arising from androconial patches on the male forewings. In a series of experiments, comparing the courtship success of males with sealed androconial patches and sham-treated males, Tinbergen *et al.* established that male androconia provide necessary cues, probably of a chemical nature, for courtship success. However, this work, a classic study in the behavioural sciences, left many important questions unanswered, and the study has not been continued.

Courtship of the Silver-washed Fritillary, *Argynnis paphia*, was the subject of another detailed ethological investigation. Although concentrating on the role of visual signals, Magnus (1950, 1958*a*,*b*) described various behaviour patterns likely to involve chemical signalling by the male. Firstly, there is an aerial courtship phase in which the male accomplishes a characteristic dance during which he repeatedly 'undertakes' the flying female (Fig. 25.2C,D) and, secondly, during the ground phase, the male, while sitting face to face to the female, flaps his wings, taking the female's antennae between them (Fig. 25.2E). Behaviour in both phases suggests stimulation of the female by secretions from the male forewing folds (Fig. 25.2A,B—for morphology see Barth 1944, Sellier 1973*b*). However, proof of a signal function for these androconial structures is entirely lacking.

With their now famous studies on the courtship of the Queen butterfly, *Danaus gilippus berenice*, Brower *et al.* (1965) focused on the use of the male hairpencils. They released laboratory-reared females to wild males in their natural environment, and quantitatively analysed the complete courtship behaviour, which, as in the two species already described, is separable into various components and

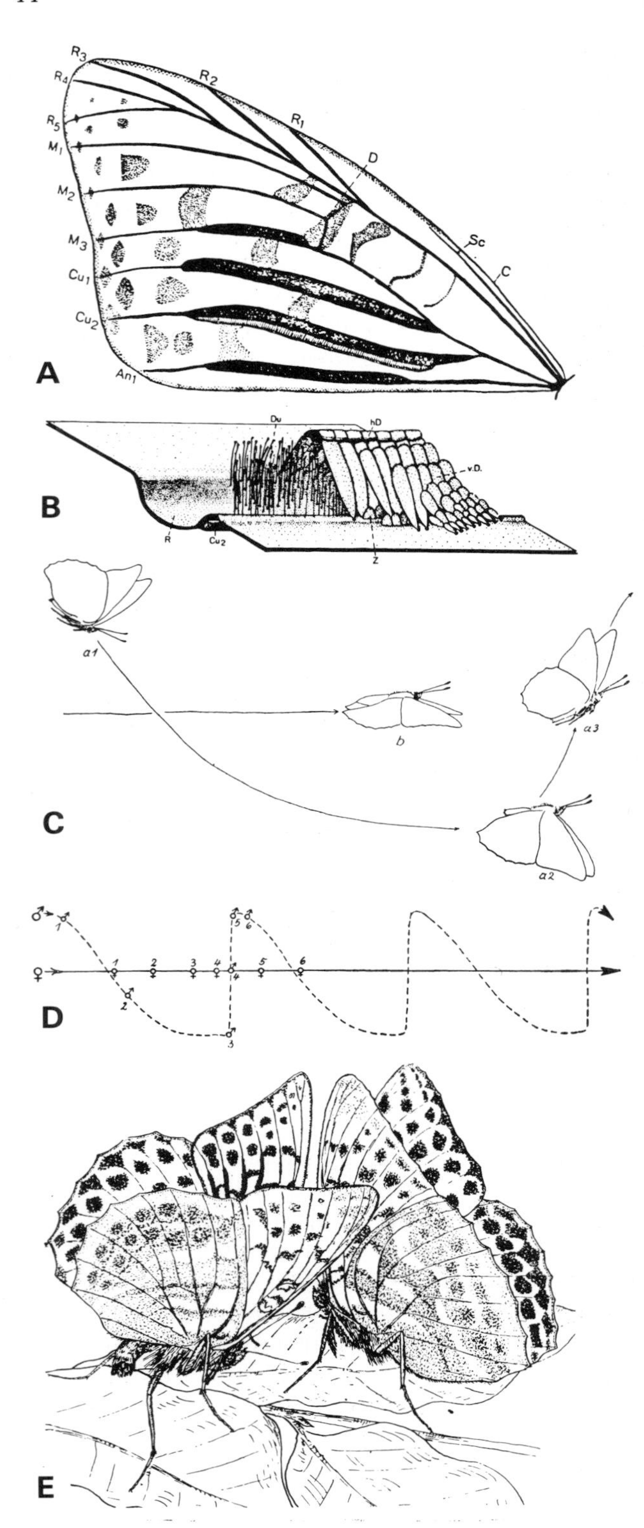

Fig. 25.2. Androconial organs and courtship of *A. paphia*. A: forewing of male showing location of androconia on M_3, Cu_1, Cu_2, and 2A ('An_1'). B: diagram of the most prominent swelling at Cu_1, Du = androconial scales, hD/vD = rear/front covering scales, Z = intermediate scales, R = groove. C, D: flight path and relative position of male and female during aerial courtship, a_{1-3} male, b female; E: bowing of male (left) during ground courtship. (A and B from Barth 1944; C, D and E from Magnus 1950.)

follows a stereotype stimulus-response reaction chain (Fig. 25.3). In two phases, the male abdominal hairpencils are protruded and expanded close to the female's head. Following 'aerial hairpencilling', a receptive female alights. This elicits 'ground hairpencilling' by the male; the female responds by folding her wings, thus allowing the male to alight and copulate. As all successful courtships were preceded by hairpencilling, the behaviour was interpreted to arrest the female's flight and also inhibit her from leaving once she had been induced to alight.

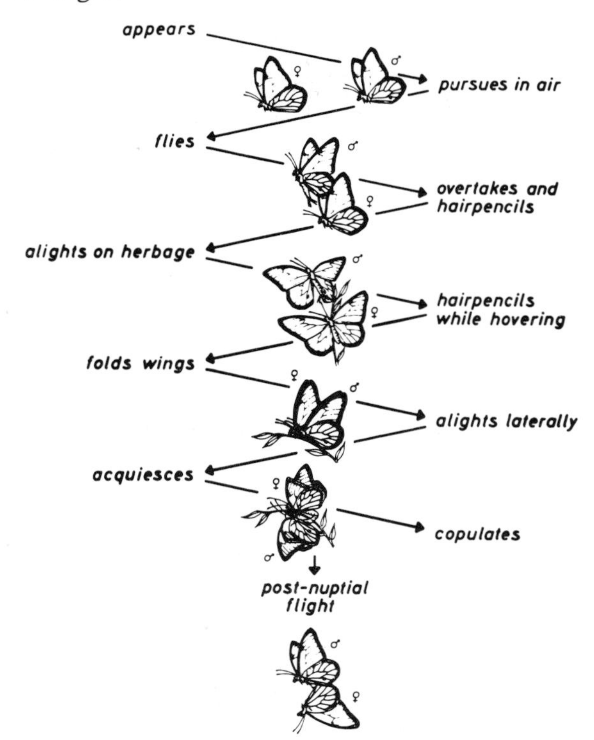

Fig. 25.3. Stimulus-response reaction chain in the courtship of *D. gilippus berenice*. Left: female behaviour, right: male behaviour. (From Brower *et al.* 1965.)

Removal of hairpencils in experimental males and sealing of antennal sensilla in females demonstrated that the androconial organs provide cues, perceived by the female antennal olfactory sensory receptors, which are essential for courtship success (Myers & Brower 1969). The structure and development of hairpencils and wing glands were investigated (Pliske 1968, Pliske & Salpeter 1971, see also Brower *et al.* 1965), and the major volatile component from the hairpencils chemically identified as a pyrrolizinone (Meinwald *et al.* 1969). Now named 'danaidone', it occurs in quantities up to 400μg per male (Boppré *et al.* 1978). Also, a terpenoid diol was identified (Meinwald *et al.* 1969). The perception of danaidone by certain antennal receptors ('short, thin-walled sensilla basiconica'; Myers 1968) was demonstrated by differential sealing of antennal zones in behavioural experiments (Myers & Brower 1969), and by employing electrophysiological techniques (Schneider & Seibt 1969, Schneider and Boppré unpublished). Finally, Pliske & Eisner (1969) showed that the absence of danaidone in indoor-raised males or in de-scented males reduces their courtship success—as does hairpencil extirpation. By applying danaidone and the terpenoid diol (or mineral oil) to the hairpencils, Pliske & Eisner (1969) could increase courtship success, demonstrating that danaidone provides a crucial signal, while the diol apparently serves as a fixative.

These mutually consistent studies with danaines provided the first evidence that androconial organs of Lepidoptera emit 'pheromones', defined by Karlson & Lüscher (1959) as 'substances, which are secreted to the outside of an individual and received by a second individual of the same species, in which they release a specific action, for example a definite behaviour or developmental process'.

Sexual communication of the Small Sulfur butterfly, *Eurema lisa*, was examined with respect to both visual and chemical cues (Rutowski 1977*a*,*b*). For the first time, a laboratory bioassay was developed to monitor the existence, function and source of a male butterfly pheromone. This test utilizes the fact that a receptive female extends her abdomen from between the hindwings in response to male courtship behaviour, and involves rubbing a female's antenna and thorax with a live male held in forceps (or with other stimuli such as different wing parts) and recording abdomen extension. By this means, Rutowski (1977*b*) clearly showed that male chemicals are required, in part, to elicit abdomen extension and are thus important pheromonal signals in close-range sexual behaviour. It was also demonstrated that the source of the pheromone is a paired, ventral androconial patch close to the base of the forewing, and that the scent is spread across wing surfaces by contact; thus, the entire wing serves as a disseminator of the androconial secretion.

Visual and chemical cues were investigated in a series of comparative studies on the sympatric and closely related *Colias eurytheme* and *C. philodice*, thus giving an insight into mechanisms of reproductive isolation. Elegant experiments by Silberglied & Taylor (1978) showed that while both *C. eurytheme* and *C. philodice* males rely on visual cues to locate and identify females, females of *C. eurytheme* utilize a visual signal as well as olfactory signals to distinguish males of their own species, whereas females of *C. philodice* rely solely on chemical cues for conspecific mate recognition. These two *Colias* are thus the first butterfly species for which an ethological isolation (strongly based on male scents) has been established. Initially, Taylor (1972) demonstrated reproductive isolation of *C. philodice* and *C. eurytheme* by either covering the eyes of virgin females with paint or by removing or covering their antennae with vaseline. It was found that mate selection can occur in the absence of visual stimuli and that the antennae play an important role, the perception of olfactory stimuli evidently maturing when most females were 20–30min old (Taylor 1973*a*). Chemical analyses (Grula & Taylor 1979, Grula *et al.* 1980) revealed the presence of a species-specific blend of wing volatiles in the males of both species: male *C. philodice* secrete three n-hexyl esters absent in *C. eurytheme*, male *C. eurytheme* produce 13-methylheptacosane lacking in *C. philodice*, and several straight-chain hydrocarbons occur on the wings of both species. The chemicals were biotested by recording electrical responses of antennal olfactory receptors (electroantennogram-technique) and by testing acceptance behaviour of females (extension of abdomen; cf. *E. lisa* above), as well as by checking courtship success of chemically modified field-caught males. Rutowski (1980*a*) confirmed the findings of Silberglied & Taylor (1978), and by biotests found the source of the male scent to be a patch of scales on the dorsal surface of the male hindwing near the wing base (cf. Illig 1902); it was also demonstrated that the forewings, by covering the patches when at rest or in flight, reduced evaporation of the pheromones. Finally, gas-chromatographic studies of male wing extracts of various genotypes derived from interspecific crosses of both *Colias* species, revealed that production of the major pheromone components is, as with UV-reflectance patterns (Ch.20), controlled by genes inherited as a linked complex on the X-chromosome (Grula & Taylor 1979).

Other studies, concerned with a variety of aspects and mostly investigated out of biological context, are summarized below. Because they only shed light on parts of complex systems, the findings—although interesting—should not be used to make hasty general conclusions, even if they support existing ideas. Nevertheless, further such studies should be encouraged, in order to obtain knowledge of as many aspects as possible, and so eventually provide a basis for experimentally testing overall hypotheses.

Danainae

The relevant literature on milkweed butterflies is reviewed by Ackery & Vane-Wright (in press *a*), and so I make only a few major points.

Chemical analyses of the androconial secretions of many species have supplemented the studies on *D. gilippus* (see above). Danaidone and/or related compounds (danaidal, hydroxy-danidal) were found to be major components of the hairpencils' scent in most investigated species of *Amauris*, *Danaus*, *Euploea*, *Lycorea*, and *Parantica*, and these pyrrolizines thus appear to be a basic feature of the subfamily. While providing essential courtship stimuli in *D. gilippus* (see above) and *D. chrysippus* (Seibt *et al.* 1972) and probably most or even all other danaine species, they are unlikely to be species recognition signals. However, the hairpencils of Danainae contain bouquets of volatiles (up to as many as 33 compounds, not yet identified; Boriack *in* Meinwald *et al.* 1974), and since these appear to have a species-specific composition (Schäfer, Boppré and Schneider, unpublished) it is hypothesized that androconial secretions may be important in reproductive isolation. So far, only some minor components of the hairpencil odours have been identified in *Amauris niavius*, *A. tartarea* (Meinwald *et al.* 1974) and *A. ochlea* (Petty *et al.* 1977). Interestingly, in *A. ochlea* they exhibit a topochemical distribution within the morphologically particularly complex androconial organs of this species (Petty *et al.* 1977, Broppé in prep. 1984*a*).

Danaidone was found to be absent in the hairpencils of indoor-raised *D. gilippus* and *D. chrysippus*, and such males have reduced courtship success (Pliske & Eisner 1969, Meinwald *et al.* 1971, 1974). This was later explained by the striking finding that Danainae as adults require secondary plant substances as precursors for biosynthesis of pyrrolizines (Ch.9; Edgar *et al.* 1973, Schneider *et al.* 1975).

From the respective position and shape of the glands, (Müller 1877*b* 1878*a*) had inferred the introduction of the abdominal hairpencils into the wing pockets in *Danaus*. Brower & Jones (1965), Seibt *et al.* (1972), Boppré (1977) and Boppré *et al.* (1978) confirmed this 'contact-behaviour'. The anecdotal field observations on contacts between parts of the abdominal hairpencils and the alar patches in *Amauris* were confirmed by greenhouse observations (Boppré 1977, in prep. 1984*a*), which also revealed that *Amauris* males exhibit contact-behaviour only in direct sunshine and only after a dark-light change, i.e. in the early morning and after artificial dark periods, respectively; this has yet to be explained.

Studies on *D. chrysippus* proved that mechanical contacts between the abdominal hairpencils and the hindwing pockets are also essential prerequisites for danaidone biosynthesis (Boppré *et al.* 1978). Possibly, the alar glands provide enzymes for conversion of the precursor into the pheromone component. In other species, contact-behaviour might also serve this function, but further experiments are needed, especially as *Lycorea* produces danaidone (Meinwald *et al.* 1966, Meinwald & Meinwald 1966) although it lacks wing glands, while in *Euploea* definite contact behaviour has yet to be demonstrated, despite repeated attempts to do so (Boppré, unpublished). In *Tirumala*, contact-behaviour serves, at least, to transfer particles from the alar pouches, where they are produced, into the abdominal hairpencils—which

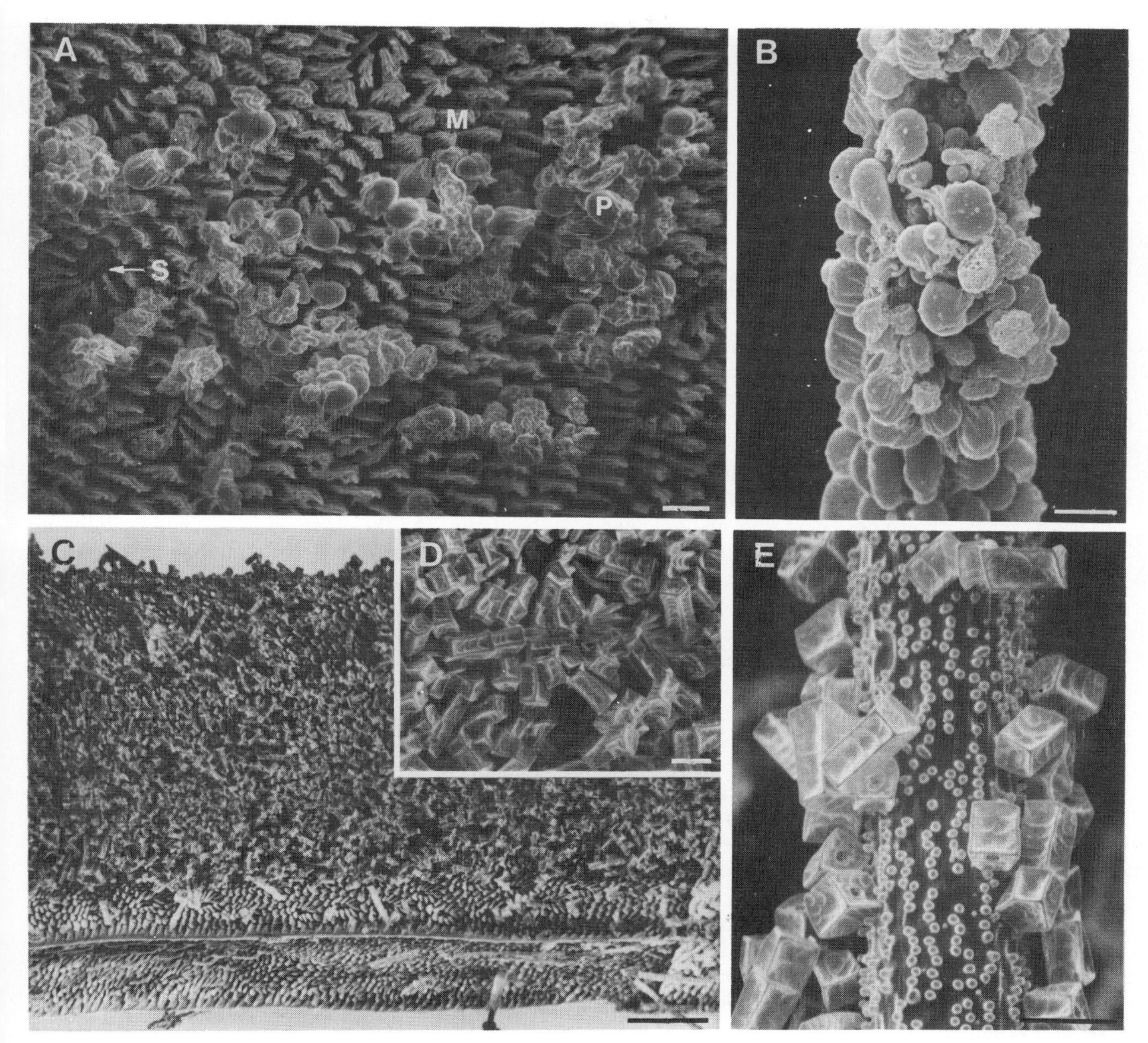

Fig. 25.4. Pheromone-transfer-particles on antennae of females *in coitu* of *Amauris ochlea* (A) and *A. niavius* (C, D; D = section of C) and on male hairpencil hairs of *A. ochlea* (B) and *A. niavius* (E). S = *sensillum basiconicum*; M = microtrichia; P = particles. Scale bars: A, B, D, E: 5μm, C: 50μm. (Original.)

later distribute the particles during courtship. Pliske & Eisner (1969), studying *Danaus gilippus*, obtained some indication that such particles were transferred from the male hairpencils onto a female's antenna during courtship hairpencilling; employment of scanning electron microscopy has since conclusively established this for a number of species (Fig. 25.4; Boppré 1979; in prep. 1984*a*,*b*).

For completeness, it must be mentioned that the best known danaine butterfly, the American Monarch (*Danaus plexippus*), is atypical of the subfamily with respect to sexual communication. Its courtship behaviour does not necessarily involve the hairpencils; instead, aerial takedowns, in which the males grab the females in mid air, predominate (Pliske 1975*c*). Both abdominal and alar androconial organs are morphologically reduced and do not contain danaidone, although both sexes take up and store pyrrolizidine alkaloids (Meinwald *et al.* 1968, Edgar *et al.* 1971, 1976*b*).

Ithomiinae

Androconial organs of Ithomiinae were initially recognized by Bates (1862). Müller (1878*a*), finding the long hairs at the costal edge of the hindwings to emit a distinct odour and noting the consistent occurrence of male tufts throughout the entire subfamily, thought there could 'hardly be any doubt that it [the tufts] has been acquired as a sexual

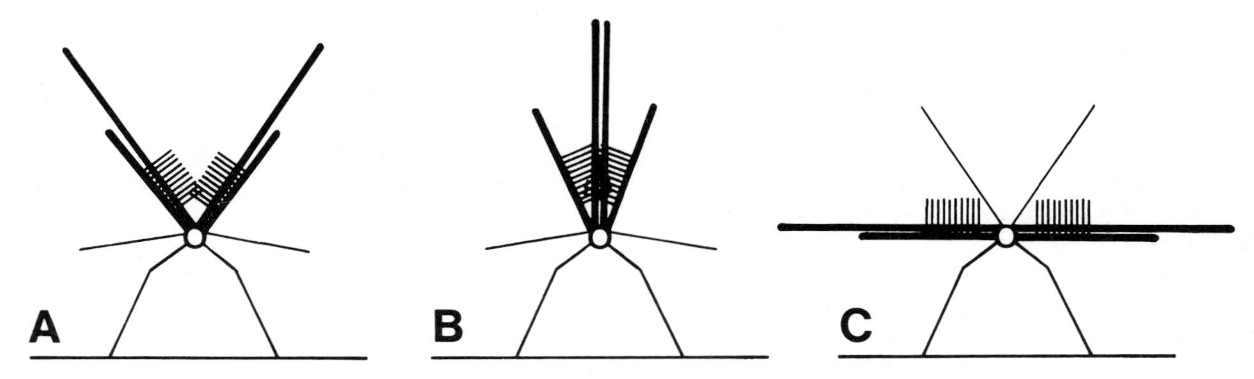

Fig. 25.5. Sketches of postures used by male ithomiines to expose scent scales (viewed from directly behind). A: scent scales exposed by shifting hindwings to rear. B: forewings held vertically and hindwings dropped slightly away from forewings. C: wings held horizontally, hindwings shifted slightly to the rear as in A. (After Haber 1978, with kind permission of the author.)

attraction by the males . . .'. Some evidence for this function was provided by Gilbert (1969), who observed male ithomiines perching in or near sunflecks along trails and exposing the scent scales of the hindwings (which are normally overlapped by the forewings). Like Müller, he also assumed that ithomiine males assembled females by this behaviour, which for ecological reasons he considered to be an optimum strategy (Gilbert 1969).

The function of the androconial organs was given more detailed attention by Haber (1978). He observed courtship and mating to take place in aggregations of several hundred butterflies comprising 20–30 species. The formation of aggregations (apparently serving as leks), which are often located hundreds of metres from hostplant concentrations, requires 3–7 days. Such aggregations tend to dissolve gradually after a few days, but may continue for as long as three months. An aggregation is independent of the individual butterflies which comprise it: many individuals enter and leave during its existence (Haber 1978).

Extensive observation and experiments clearly demonstrated that the androconial organs of the ithomiines stimulate the formation of these aggregations. As illustrated by Haber (1978), the scent scales are expanded in stereotyped postures (Fig. 25.5; cf. Gilbert 1969, Pliske 1975*b*) characteristic of species, but nonetheless varying considerably among conspecifics. Scent scale exposure occurs for 10–50s while sitting, followed by flying up and circling the area (males may also expose the hindwing fringes in flight); this scent-disseminating behaviour suggests the establishment of a broad odour plume made up by several individuals. The role of these scents in forming aggregations (which theoretically could also be mediated by visual signals) was proved by tests involving hidden presentation of excised androconial organs. Both sexes—including freshly mated females—were attracted in equal numbers, and almost 70% of the individuals attracted to the scent scales came to those of different species. This low specificity, and the observation that aggregations are made up of many species and that newly mated females are also attracted, renders the Müller/Gilbert hypothesis (that male scents function as sex-attractants) insufficient. Haber (1978) concluded that male scents in Ithomiinae act as pheromones to promote aggregating behaviour, which provides *two* advantages: aggregation of Müllerian mimics reinforces the protection conferred by mimicry, and the aggregations also serve as places for mating—which, moreover, can be located easily by the concentrated odour. Thus, not only mate-finding but (primarily?) predation would be the selective pressures for aggregating behaviour mediated by male scents.

The results of experiments on chemical communication by male Ithomiinae conducted by Pliske (1975*b*) appear to strongly contradict Haber's findings. Pliske considered a lactone (see below), released from the hairpencils of several species, to act as a male territorial-recognition pheromone, repelling not only conspecific males but also those of other lactone-producing species as well; the lactone would allow males to recognize one another and terminate male–male intra- and interspecific courtship pursuits. A second function of ithomiine hairpencils is seen by Pliske (1975*b*) in the dissemination of 'aphrodisiacs' during the later phases of courtship flight.

It seems unlikely that both the Haber and Pliske hypotheses can be correct. Perhaps the conflicting results are due to the fact that Haber studied butterflies at aggregation sites, while Pliske worked in different areas—apparently not even knowing of ithomiine aggregation behaviour—where the butterflies were gathering pyrrolizidine alkaloids (see below) and possibly in a very different motivational state. In any case, further studies are needed, and can take some advantage of the present knowledge concerning chemistry and production of scents.

Chemical analyses of androconial organs revealed the presence of a lactone in several species of several ithomiine genera (Edgar *et al.* 1976*b*). Its chemical structure is almost identical to the acid moiety of certain pyrrolizidine alkaloids (PAs), and it is most likely that male Ithomiinae cannot synthesize the lactone *de novo* but require appropriate PAs as precursors, in a manner closely similar to danaine pheromone production (see also Ch.9). Attraction to and feeding at withered plants containing PAs has long been recognized in Ithomiinae (e.g. Miles Moss 1947, Beebe 1955, Masters 1968, Pliske 1975*a*), and a few males of a lactone-disseminating species raised indoors were found to lack the lactone (Edgar *et al.* 1976*b*—see p.272 for main discussion of butterfly-PA relationships). According to Pliske (1975*a*) and Pliske *et al.* (1976), attraction to PA-containing plants is mediated olfactorally by esterifying acids liberated from alkaloids in rotting plant tissue; Pliske (1975*b*) also reports Ehrlich-positive androconial secretions, which are possibly danaidone-type chemicals (see Danainae).

Pieridae

Apart from the studies on *Eurema* and *Colias* reviewed above, little experimental work has been done on chemical communication in the Pieridae.

In the courtship of *Leptidea sinapis*, the females seem to rely on visual cues, although male scents may be involved; however, female odours do appear to be significant signals for the males (Wiklund 1977*c*).

The androconial secretions of four species of the genus *Pieris* have been analysed chemically. Bergström & Lundgren (1973) identified citral (neral:geranial—2:1) in *P. napi*, and tricosane in both *P. rapae* and *P. brassicae*. The same authors chromatographically detected additional minor components in *P. napi* and demonstrated that the wings of *P. rapae* and *P. brassicae* release a few (unidentified) volatiles, characteristic for each species. Chemical studies conducted by Hayashi *et al.* (1978) revealed the presence of seven monoterpenes (α-pinene, β-pinene, myrcene, *p*-cymene, limonene, neral, geranial), *n*-undecane and an unidentified component on the wings of both *P. melete* and *P. napi japonica*; the presence of linalool in addition in *P. melete* only and the ratio of neral and geranial ranging from 0.77–1.04 in *P. melete* but from 1.84–2.43 in *P. napi japonica* suggested species recognition signals provided by androconial secretions for both species. Kuwahara (1979), reinvestigating *P. melete*, failed to find α-pinene, β-pinene, mycrene, *p*-cymene or linalool, but did find neral and geranial, and nerol and geraniol in addition. His analyses considered the absolute amounts and the ratio of the four components in relation to the age of the butterflies, and the time of day their wings were extracted: the total amounts of volatiles increased from 12μg in 8h-old indoor-raised males to up to 183μg 50h-old individuals, and the ratio neral/geranial increased from 0 to 1.5; significant differences in the neral/geranial ratio were found in some individually analysed field-caught males, which contained a maximum of 448μg citral (for further details see Kuwahara 1979). Summarizing, the chemical studies are not consistent and the volatiles identified have never been biotested, despite the fact that very extensive morphological data are available on *Pieris* (refs. in Bergström & Lundgren 1973).

Lycaenidae

Behavioural studies considering scents in Lycaenidae are similarly scanty. Lundgren & Bergström (1975) reported on the role of odours emitted by both sexes of *Lycaeides argyrognomon*, and on the identification of three volatiles (nonanal, hexadecyl acetate, and a cyclic sesquiterpene alcohol, perhaps torreyol (δ-cadienol)) which, in addition to some minor components, are present on the male wings of this species. They suggest that the androconial secretion both inhibits mate-refusal in unmated females while eliciting mate-refusal in mated or insufficiently stimulated females. Douwes' (1976*b*) study on the courtship of *Heodes virgaureae* also indicates the existence of female scents, but did not reveal an effect of the androconia. Series of experiments on the courtship of *Zizeeria maha argia* conducted by Wago (1977, 1978*a*,*b*; see also Wago *et al.* 1976) showed, apart from further evidence for the presence of female scents, that at close-range male odours have a repellent effect on approaching conspecific males; he also demonstrated that the androconial secretion is mechanically spread over the entire surface of the wings.

Our present knowledge of butterfly androconial systems can now be briefly summarized. We have almost countless morphological descriptions of androconial organs together with some anecdotal observations on their use in behaviour, but only a few experimental studies—which nonetheless demonstrate that male chemical signals can be vital in courtship and may also be involved in species isolation and/or recognition. Several androconial secretions have been analysed chemically. Androconial organs have certainly evolved many times independently (as demonstrated by their haphazard distribution and morphological diversity) and they appear to be exposed to strong selection pressures.

Despite this valuable data and some insights, our knowledge is still too fragmentary to permit the formulation of an overall theory on the function of androconia. We do not properly understand the advantages of using male scents, their mode(s) of action nor the selective pressures causing their elaboration. Androconial secretions can influence female mate-choice or acceptance, but there are good reasons for believing that this is not their only function. Even in well-known groups like the Danainae, there remain puzzling questions. What is the advantage of using pheromone-transfer-particles? Do the many components of the secretions function in species recognition? Why is a bouquet of various volatiles secreted? What is the overall relevance of danaidone, the production of which requires so many peculiar adaptations and so much effort on the part of individual males, which is nonetheless a major component of several sympatric species? Why are there binate organs? What is the significance of morphologically complex hairpencils? In general, we may suggest that androconial systems seem, in many species, unnecessarily complex if they serve only as 'identity cards' to determine the species and sex of would-be suitors. We may not only have to assume —and consider in future experimental approaches— a variety of functions for androconial organs, but also that they can have multiple roles even in a single species.

Female Scents in the Sexual Communication of Butterflies—Exceptional or Typical?

The emphasis so far on male androconial systems should not imply that only male butterflies employ odours in the course of sexual communication. Temple (1953) asserts that female butterflies produce a 'directive scent' which guides males to them. However, no thorough experimental study on female pheromones has been published and the evidence for their existence in the Rhopalocera remains circumstantial.

The most obvious putative examples of female sex pheromones in the butterflies are found in *Ornithoptera priamus caelestis* (Borch & Schmid 1973), in various species of Heliconiinae (Edwards 1881, Mitchell & Zim 1964, Gilbert 1975, 1976, D. Thomas 1978, Brown 1981), and in the lycaenid *Jalmenus evagoras* (Colour plate 3). In these cases the female *pupae* are reported to be attractive to males, which check them for emerging butterflies or even wait for emergence in order to mate immediately. Interestingly, attraction of female heliconiine pupae is not always species-specific (Gilbert and Longino, pers. comm.). Although with such pupal matings chemicals are the most likely signals providing the relevant cues, chemical mate-location is not yet established and the situation is somewhat complicated. Gilbert and Longino (pers. comm.), experimenting with Heliconiinae, found the probable release of an attractive pheromone by female pupae (since males can distinguish between male and female pupae) but they also demonstrated essential involvement of visual cues, and even of learning: pupae placed in the field without host plant patches failed to lure males. Pupae which had not been removed from their pupation site did attract males, but only from among those individuals which were previously observed to search the areas where the pupae appeared. In *Jalmenus evagoras*, however, only volatile chemical cues appear to mediate attraction— if a female pupa that is about to eclose is crushed by hand, males will be attracted (Pierce, pers. comm.).

In a quite different context, female scents appear to come into play in the Silver-washed Fritillary, *Argynnis paphia*. Treusch (1967) reported that virgin females respond peculiarly to approaching males. They bend their abdomen, directing the tip (where two small glandular sacks become visible—for morphology see Urbahn 1913, Götz 1951) towards the male and even follow his movements, the abdominal tip always pointing towards him. This behaviour suggests that the female is signalling receptively to the male; it is reported to occur only if the female is sexually the more excited of the pair. Treusch (1967) also states that gland exposure is elicited by chemical cues from the male androconia while the oriented movements are triggered by visual stimuli. However, Treusch does not provide any detailed evidence for this.

Further circumstantial evidence for the existence of female scents influencing male behaviour comes from numerous records of perching males locating females that are either hidden from sight or have unexpanded wings—visual cues are then either absent or certainly atypical (Scott 1973*d* provides many references; this paper is a most valuable review although the use of scents in butterfly courtship is treated uncritically). In experimental work it has been noted that dummies may attract a male but do not elicit complete courtship behaviour (Magnus 1958*a*) and that freshly killed females are more effective stimuli to males than females de-scented by means of treatment with heat or chemical solvents (Lundgren & Bergström 1975, Douwes 1976*b*). Wiklund's (1977*c*) study on the courtship of *Leptidea sinapis* emphasizes the likelihood that males recognize conspecific females, at least partly, by means of olfactory cues, which also appear to signal willingness of the female to mate. The ability of males to distinguish olfactorally between sexes and the importance of female odours in initiating male mating behaviour were well-documented experimentally in *Zizeeria maha argia* by Wago (1978*a*). Without being conclusive, all these reports nevertheless corroborate the common experience of naturalists that after a visually mediated approach, male butterflies appear to discriminate not only between species and sexes but also between receptive and unreceptive females—

which may be due to visual and/or chemical cues. However, if male-repellent female odours can be demonstrated, it is apparent that they are not necessarily secreted by the females—the existence of scent transfer from males to females during mating has been shown to occur (Gilbert 1976).

Summarizing, a variety of observations and experimental results strongly suggest the occurrence of female pheromones in butterflies, and perhaps chemical signals for close-range communication, may be widespread or even typical. It is plausible that females should signal their physiological state to a pursuer, but unfortunately nothing is known of the origin and chemical nature of assumed female pheromones—and we can only speculate about their function(s). The lack of both conspicuous 'scent organs' and odours detectable to humans make the study of female chemical signals particularly difficult.

Pheromones of Butterflies and Moths—Significant Differences?

While chemical signals in butterflies usually appear to be used during close-range (courtship) communication, moth pheromones generally mediate long-range mate-attraction. Volatile chemicals, emitted by receptive females, lure the male moth to his mate. These sex-attractants are released from glands, located at the abdominal tip, usually secreting a species-specific blend of aliphatic compounds. Males perceive the female pheromones by means of specialized antennal chemoreceptors, activation of which induces an upwind flight towards the odour source. For details and references on this type of pheromone communication by moths see Priesner (1973), Kramer (1978), Schneider (1980) and Steinbrecht & Schneider (1980); see also various articles in Birch (1974), Shorey & McKelvey (1977) and Ritter (1979). [At close-range, other chemical cues (together with tactile and even visual signals) of the female may become relevant; these may all be paralleled by female butterflies.]

Female moths and butterflies employ signals of different modalities with different specificities: thus mate-location is fundamentally different in the two groups. The luring scents of female moths contain information on sex, species and readiness to mate, whereas the visual cues used by butterflies appear neither sex- nor species-specific and to be devoid of information on the female's physiological state. This disadvantage can be partly compensated for in some species by such behaviour as 'hilltopping' (Shields 1968, Scott 1974*b*). At close-range, female butterflies may also use chemical signals.

However, despite these fundamental differences between Rhopalocera and Heterocera in relation to mate-finding strategies, there are obvious similarities in their close-range communication, particularly with respect to male scents. As in butterflies, androconial organs of moths show great morphological variation and may occur on almost any part of the body. They consist of glandular, modified scales, often forming tufts, bristles, brushes, etc., that are protected at rest. Employment of very complex mechanical, hydraulic and pneumatic mechanisms ensures sudden expansion of the androconial organs. (Elaborate structures appear to be at least as common in moths as in butterflies, perhaps more so.) The occurrence of these organs is scattered throughout the moths and, as in the butterflies, they must have evolved independently many times and are similarly of limited taxonomic value (but see Birch 1972). For details on the morphology of androconial organs in moths see the reviews already noted together with Haase (1887*b*), Stobbe (1912), Varley (1962) and Birch (1979).

Experimental behavioural studies in various species have shown that, in general, the androconial organs are expanded briefly (at most, for a few seconds) near to the female and prior to mating, and that they emit volatile, often multi-component chemicals (reviewed by Weatherston & Percy 1977) which are essential stimuli for acceptance by the female. Thus, excision of the organs or removal/sealing of the female antennae usually eliminates female receptivity (but see p.270). Behavioural studies considering androconial organs in moths include those by Birch (1970*a*,*b*,*c*), Thibout (1972), Grant (1974, 1976*a*,*b*), Jacobson *et al.* (1976), Colwell *et al.* (1978), Baker & Cardé (1979), Conner *et al.* (1981) and Zagatti (1981); for review and further refs. see Birch (1974).

Since the androconial systems of butterflies and moths greatly resemble each other not only in morphology but also in function, the overall discussion of adaptive significance of male pheromones will consider Lepidoptera in general—studies on butterflies and moths appear to complement our understanding in various ways.

Male Pheromones—Aphrodisiacs?

We have seen that male pheromones can convey vital information about the species and sex of a suitor, and may conceivably be involved in female mate-choice, androconial systems having plausibly evolved by epigamic selection. But are male pheromone systems of Lepidoptera really this simple?

Androconial secretions of Lepidoptera are usually termed *aphrodisiacs*: 'scents employed by one or the other sex—most frequently by the male and often as only one part of a complex pattern of courtship behaviour—to stimulate the opposite sex to copulate' (Butler 1964, 1967). This vague definition, although often criticised, is still commonly used to describe a 'functionally distinct group of pheromones' and to differentiate between pheromones that attract the opposite sex over a distance, and those released after the insects have come together and, 'as far as is known, clearly facilitate courtship and/or copulation' (Birch 1974). Apart from the anthropomorphism of the term 'aphrodisiac', this definition inappropriately combines situational and functional aspects. Release of a scent after the sexes have come together does not necessarily imply 'facilitation' of courtship or copulation: observing expansion of androconial organs close to a female prior to mating, and then calling the supposed scent released an 'aphrodisiac pheromone', involves unwarranted assumptions and ignores the possibility of additional functions. The uncritical use of the term aphrodisiac should be avoided. Although the deployment of androconial organs during close-range sexual interactions provides circumstantial evidence that male pheromones affect female mate choice, we must also consider a variety of more subtle—or even wholly different—questions. Are we dealing with sexual selection, or merely non-random mating? (Ch.21). *How* do androconial secretions act? Which traits of the males do they reflect? Do male scents influence the female's behaviour, physiology or sex-appeal independent of/in addition to mate choice? Are male scents only involved in epigamic interactions? Such questions as these, and many others, need to be answered.

Apparent effects of male scents on females have been described by investigators with wordings like 'inhibit her escape response', 'inhibit her general activity', 'induce mating', 'induce her to take an acceptance posture', 'overcome her coyness', 'seduce her'. . . . These phrases would reflect a variety of underlying physiological mechanisms which, unfortunately, have never been studied. The question remains: how do androconial secretions act? And what does the male signal inform the female about? Do female Lepidoptera select their mates and if so, which are the decisive cues? Female mate-choice in the Lepidoptera has received little attention, and we urgently need experimental investigations of its occurrence and its mechanisms (cf. Ch.21; Rutowski 1982, has surveyed the literature and discussed the subject extensively). Male scents may be capable of informing the female if the suitor belongs to the appropriate species (to prevent hybridization), but they apparently need not (only) do so—many androconial secretions (or components of them) are not species-specific. Possibly, androconial secretions vary (in amount/composition) due to, for example, age and number of matings, so that a female could recognize the male's physiological state; do male scents reflect traits for fitness-recognition (which are appreciated by the females)? Do androconial secretions serve as 'fitness-cards'?

Wing-gland secretions emitted by male *Galleria mellonella* and *Achroia grisella* attract females from a distance (in *Achroia*, in combination with sound) by evoking anemotactic flight (Röller *et al.* 1968, Dahm *et al.* 1971, Leyrer & Monroe 1973). In *Eldana saccharina*, another pyralid in which the male stimulates the female at long-range, Zagatti (1981) experimentally demonstrated that two androconial organs are involved in its sexual interactions, each serving different functions. One secretion elicits female searching behaviour (in this case not anemotactic flight, but undirected searching for a calling male by climbing up plant stems, etc.), but is insufficient for mate acceptance, which is dependent on the subsequent deployment of male abdominal hairpencils to elicit female wing-fanning, ovipositor extension and the copulation-acceptance posture.

Mate choice by females on the basis of male scents has been demonstrated in an olethreutine moth, *Grapholita molesta*, the subject of detailed analyses by Baker & Cardé (1979; see also Cardé *et al.* 1975, Baker *et al.* 1981). The males are attracted by female sex-pheromone which, at close-range (1–2cm), also elicits male abdominal-hairpencil display—the male chemical (together with wind) stimuli may then attract the female to the male over the remaining short distance. Based on these results, Baker & Cardé speculate that female mate-choice has been a primary factor in the evolution of complex androconial organs and associated behaviour patterns. As yet, there are no data available indicating that male traits are mirrored by androconial secretions and thereby be potentially decisive for female mate-choice. However, some evidence is provided by studies on *Utetheisa* moths, in which the males secrete a pheromone derived from stored toxins obtained by the larvae from their foodplants (Conner *et al.* 1981; cf. Culvenor & Edgar 1972). Conner & Eisner (*in* Eisner 1980; cf. Conner *et al.* 1981) suggest that the chemical signal could be a measure of the male's (potentially inherent) toxin-sequestering ability and be open, therefore, to epigamic sexual selection (see also Ch.21).

Most of the detailed studies on scents in Lepidoptera have concentrated on the use of androconial organs only in the final phases of courtship—their use in other situations has been ignored, or regarded as exceptional. As already mentioned, males of *Troides*, *Euploea* and Ithomiinae have been seen to display their androconial organs in the absence of females, and it was suggested more than 100 years ago that male scents might attract females. Several corresponding observations, including some of those just mentioned, have been made with moths. For example, male Hepialidae form dancing swarms, of 100–2000 individuals, to which females appear to be attracted—perhaps by the odour originating from the elaborate tibial brushes, which have a pleasant aromatic smell to the human nose. However, as expansion of the brushes during courtship flights was observed, and matings take place within the swarms, the true function(s) of the odour remain in doubt (Robson 1887, 1892, Bertkau 1882, Barrett 1882, Degeener 1902, Carolsfeld-Kausé 1959, Cadbury *in* Kettlewell 1973: 297, Turner 1976). *Creatonotos* may expand their coremata when resting near artificial lights (Pagden 1957). Robinson (*in* Varley 1962) saw males assembled with their coremata extended (perhaps forming a lek), and in the laboratory individuals or groups of this arctiid moth will also display their coremata in the presence of luring females (Schneider & Boppré 1981, Schneider *et al.* 1982). Recently, Willis & Birch (1982) reported on aggregations of another arctiid, *Estigmene acrea*. Again, the male coremata were expanded and both females and further males were attracted upwind—the males joined the assembly while the females copulated with individual males within it—which may thus be considered a true lek. However, there are two distinct mating systems in *Estigmene*, since the females also release a luring pheromone to attract males—each method of mate-location is effective at different times of the same night (Willis & Birch 1982).

These mostly fortuitous observations of a variety of androconial displays not first involving close-range sexual communication suggest that such behaviour will be found to be widespread—a point much in need of attention. In particular, the Ithomiinae and *Estigmene* clearly indicate that all chemical signals emitted by male Lepidoptera are not exclusively directed to females—i.e., they can also have intrasexual effects. In the light of this, the presence in males of receptors for their own scents—a basic difference to female pheromone systems—becomes intelligible. Male competitive interactions mediated by odours are likely to be common: a clear example is given by heliconiines, in which the females, after mating, emit a distinct scent repellent to other males (Gilbert 1976). Experimental crosses made between races show that the odour originates in the male and is transferred to the female during copulation; this 'antiaphrodisiac' helps to enforce monogamy by females (Gilbert 1976).

A variety of other reports also indicate that male scents may directly or indirectly affect male competition. Virgin females of *Actias selene*, when kept isolated from males, laid their eggs later than virgin females kept in the presence of males; antennaless virgin females, kept with males, behaved like isolated females (Benz & Schmid 1968). Male scents of *Heliothis virescens* appear to inhibit sex-attractant release by females (Hendricks & Shaver 1975). Museum specimens indicate that male *Ocinara* can leave androconial organs inside the abdomen of a female, which Dierl (1977) suggests might repel other males. Chemical signals of male *Pseudaletia unipuncta* decrease the tendency of other males to approach calling females, or to exhibit copulatory behaviour if already near a female (Hirai *et al.* 1978). These observations are not conclusive but they do lend support to the idea that male scents (not necessarily emanating from androconial structures) are involved in promoting the reproductive success of individual males. They also demonstrate that this could be achieved in various ways, either directly by repelling competitors (cf. Ch.20) and/or indirectly—e.g. by stimulating egg production or termination of female calling behaviour.

To allow for this possibility of intrasexual competition, future investigations must take into account the problem of delayed male pheromone effects: measuring direct courtship success alone is insufficient. If this is so, then not only those cases where courtship without previous display of androconial organs have been observed (e.g. Thibout 1972, Grant & Brady 1975, Gothilf & Shorey 1976, Hirai 1977) need to be reconsidered, but also those where the secretion of 'aphrodisiacs' by androconia has already been apparently demonstrated—because additional effects may well be involved.

To summarize: androconial organs are employed in a great variety of situations, we have proof or indications of a variety of functions for their products (short- and long-range stimulation, inter- and intrasexual signalling), and male scents can even have multiple uses in a single species. To date, there are too few data to enable us to properly understand the roles and modes of action of male pheromones in interactions among Lepidoptera. Considering the diversity of morphological structures and their analogous evolution, together with the variety of behavioural contexts in which scents come into play, and the diversity of proven or suggested functions male scents appear to have for both sexes, we should abandon searching for *the* function of androconial systems and, instead, stimulate experimental approaches on all possible aspects of the subject.

Male Pheromones, Defence, and Mimicry —Coadapted?

Occasionally, the possible defensive functions of male scents have been considered (Birch 1970*c*). For example, several male danaines (species of *Euploea*, *Lycorea* and *Tirumala*) and male sphingid moths expand their androconial organs when being handled (however, there are no reports on the use of these organs during attacks by natural predators). A defensive scent may repel a predator directly, and/or act as a 'startler' to alert a predator of another defence mechanism, the odour thus acting as an aposematic signal (Ch.12). I favour the idea—if androconial organs are truly involved in defence at all—that their sudden appearance may alert predators visually rather than odoriferously—their frequent bright coloration appears to be circumstantial evidence for this.

It has also been suggested that male scents could protect mating pairs (Birch 1974). In addition to its function of attracting females (Röller *et al.* 1968, Dahm *et al.* 1971), the strong scent of male wax-moths (*Galleria mellonella*) was thought to deter honey bees from attack (Leyrer & Monroe 1973). 'However, it is difficult to explain the adaptive advantage of a defensive structure that occurs in males alone' (Birch 1979).

A different relationship between defence and male pheromones came to light when it was found that danaine butterflies require toxic secondary plant substances for biosynthesis of a major pheromone component (for review see Boppré 1978). While the abdominal hairpencils of field-caught danaines (e.g. *Danaus chrysippus*) contain up to several hundred micrograms of danaidone, males raised indoors, or freshly emerged, entirely lack this hairpencil volatile. However, the males are strongly attracted to withered or damaged plants of various species, to which they apply fluid by means of their proboscides and reimbibe it with dissolved secondary plant substances. In this way they ingest pyrrolizidine alkaloids (*PA*s), which have in their molecular structure a moeity similar to that of danaidone and have been shown to be essential precursors for biosynthesis of the pheromone component. Recent studies suggest the butterflies can also gather PAs with the nectar of certain plants (Boppré *et al.*, unpubl.).

The observation that secondary plant substances serve as precursors for pheromone production is striking because the *adults* obtain them from withered plants *independent* of energy assimilation. However, the finding has additional significance because PAs are toxic to vertebrates (Mattocks 1973, Schoenthal 1968, Bull *et al.* 1968, McLean 1970; see also Ch.12). Rothschild *et al.* (1970, 1979*b*) reported storage of PAs (ingested by larvae) by arctiid moths and suggest that it accounts for their well-known protection. Unconverted PAs were also found in the bodies of both sexes of Danainae (Edgar *et al.* 1976*a*, 1979, Edgar 1982), and Eisner, Conner and Hicks (unpublished, see Eisner 1980, Conner *et al.* 1981) have established the defensive potency of PAs for insects (see also Boppré 1984*c*). Thus, we can be quite certain that PAs in danaine butterflies serve a dual function both as male pheromone precursors and protective chemicals (see also Ch.12).

The utilization of secondary plant substances by danaines requires us to consider several novel factors when discussing insect-plant relationships. On emergence butterflies may lack pheromones essential for reproduction and have to make considerable efforts to obtain them during their adult life; butterflies may actively collect defensive substances as adults; and they may gather essential chemicals *independent* of energy assimilation (= 'feeding' in common usage).

The Ithomiinae, the sister-group to the Danainae, are similarly attracted to PA-containing plants. From their androconia, as already noted, several species were found to secrete a lactone probably derived from PAs. The ability of ithomiines to store unconverted PAs and employ them like danaines for defensive purposes has not been investigated. However, their elaborate aposematism and mimicry and similar relationship to PA-containing plants is suggestive of a double (pheromone/defence) function. Haber's (1978) finding that ithomiines assemble by means of androconial (PA-derived?) secretions to form aggregations partly to avoid predation adds another dimension to the complex evolution of pheromones and defence.

Utilization of PAs is not restricted to Danainae and Ithomiinae. Adults of various species of moths also ingest PAs from withered plants (Pliske 1975*a*, Boppré 1978, 1981, Goss 1979) and some have been shown to store these toxins (Boppré *et al.* unpubl.). Since many of the species in question possess androconial organs, it appears that sequestering PAs for defence *and* chemical communication is a recurrent phenomenon in the Lepidoptera. Not only adult Lepidoptera ingest PAs—in various species the larvae feed on PA-containing plants. *Utetheisa ornatrix* is protected from various predators by storage of PAs ingested by its larvae (Eisner, Conner & Hicks, unpublished). The males also derive their major pheromone component from PAs (Conner *et al.* 1981; see also Culvenor & Edgar 1972). In *Creatonotos gangis* and *C. transiens* PAs are not only stored and used as pheromone precursors, but also regulate corematal organogenesis (Schneider & Boppré 1981, Schneider *et al.* 1982). This novel morphogenetic phenomenon strikingly demonstrates how complex and elaborate pheromone biology may ultimately prove to be.

Thus, relationships between male scents and defence do occur in the Lepidoptera. Are there also correlations between mimicry and the evolution of androconial systems? L. P. Brower (1963) has suggested that the occurrence of mimicry in the males of Müllerian co-mimics is possible because scent plays a more important role than sight. Vane-Wright (1972*a*,*b*), however, demonstrated that no simple relationship exists between the presence or absence of androconial organs and the occurrence of sexual dimorphism, polymorphism or mimicry. Nevertheless, for danaine butterflies it seems plausible to assume 'coevolution' of mimicry and chemical communication, although not in the way postulated by Brower. Assuming the dual function of PAs as pheromone precursors and defensive chemicals, Boppré (1978) suggested that danaine butterflies may have adapted to PAs for defence by first feeding co-incidentally on PA-containing nectar, then specializing on it, and finally obtaining PAs from withered plants. As convergent Batesian mimics arose, to avoid courtship with mimics, female danaines might have accepted males only in the presence of PA-odours. (The assumption that withered PA-plants originally served as mating-sites, I no longer consider to be essential.) However, plant-derived pheromones could have evolved independently from mimicry. Conner & Eisner (*in* Eisner 1980; see also Conner *et al.* 1981) proposed that PA-derived pheromones may allow females to assess the alkaloid content of males and therefore their degree of protectedness.

Regardless of origin, the elaboration of danaine androconial systems to their present level of complexity is very probably related to the development of Müllerian mimicry within this subfamily. Assuming that it is of advantage to a distasteful butterfly species to keep constant all those characters which a predator may recognize, speciation in the Danainae might have mainly affected the androconial system(s), and not external, visual features (Boppré 1978). Comparative morphological and chemical findings seen in relation to a cladistic classification of the Danainae support this hypothesis (Boppré *et al.* in prep.).

Because the danaines are classical mimicry models, the relationship between Danainae and PA-containing plants has been given close attention (see also Ch.12). Until recently, their protection was thought to be solely due to cardenolides (= cardiac glycosides) sequestered from their larval foodplants, as demonstrated by extensive studies with *Danaus plexippus* (Ch.12). However, the likely universal storage of PAs by all species of Danainae suggests that they provide the *principal* chemical defence of the subfamily (Boppré 1978, Edgar *et al.* 1979; but see also Ch.12), while only species of the genus *Danaus* possess the *additional* defence mechanism based on storage of cardenolides (cf. Ackery & Vane-Wright, in press *a,b*). There are two hypotheses on the origin of PA-utilization. Edgar *et al.* (1974; see also Edgar 1982, Ch.9) assume the origin of PA-utilization by the Danainae and Ithomiinae to lie in ancestral larval foodplants containing both PAs and cardenolides. In contrast, Boppré (1978) and Boppré 1984*c* suggest independent evolutionary origins for the utilization of PAs and cardenolides, i.e. an adaptation to PA-containing plants independent of larval foodplants—a possibility highlighted by the existence of several groups of unrelated insects having associations with PA-plants.

Although disagreements remain, it is likely that chemical communication, chemical defence and mimicry have evolved together, at least in some groups of the Lepidoptera. Renewed experimental investigations, particularly in the Danainae, Ithomiinae and Arctiidae, should lead to some consensus of opinion. Comparative studies on communication within models, mimics and polymorphic species, and between models and mimics, should also be undertaken. The existence of a mimetic species may impose selective pressures on the communication system of the model *independent of the operator*, an aspect which has not been considered in the discussion so far (cf. Boppré 1978).

Conclusions

I have tried to review the development of research on scents in butterfly communication over more than 100 years, and to evaluate the subject critically, not only by demonstrating results but also by stressing lack of solid information, avoiding overemphasis on anecdotes, and pointing out areas urgently in need of investigation. Our generally poor understanding appears strongly biased towards considering male scents only in the context of specific recognition or female mate-choice. In the future, we should encourage unconventional hypotheses and experimental designs that break away from the mental straitjacket of tending only to consider male scents as 'aphrodisiacs'. With regard to scents of female butterflies, solid data are entirely lacking on distribution, glandular origin, chemistry and behavioural effects.

Having emphasized the diversity of androconial systems, and noted that male scents serve inter- and intrasexual functions and can even have multiple effects, it is worth contrasting the uses of chemical signals by female moths with male Lepidoptera (Table 25.1). Because of our incomplete knowledge, such a comparison must be made with caution. Nevertheless, there are some notable differences, suggesting that the chemical language of each sex has a different 'syntax'. The female pheromone systems appear very uniform throughout the Heterocera—pheromones are typical for females, the glands are homologous, the signals are similar chemically and only perceived by males, they are released in similar or identical situations and can be related to a common function: attraction of conspecific males. Male pheromone systems in moths, as in the butterflies, are diverse in occurrence, chemistry and perception. This diversity in itself suggests a multiplicity of functions for male scents, mirroring their apparently repeated independent evolution. In conclusion, it seems entirely unrealistic to search for a general function of androconial organs. Scudder's (1877) view, that the masculinity of 'androconia' is their single

Table 25.1. Generalized comparison of some features of pheromone systems in male and female moths, demonstrating the different 'syntax in the chemical languages employed.

Pheromone	Female moths	Male moths and Butterflies
Distribution	typical	frequent but irregularly distributed
Evolution	once, or only a few times	Many times
Glandular source	modifications of intersegmental membranes between abdominal segments 8 and 9	+/− complex organs made up by modified scales on any part of the body
Chemistry	straight, even-numbered (mono- or di-unsaturated with terminal oxygen function of species-specific type or blend; extractable amounts: 1ng–1μg*	variety of types of compounds sometimes (often?) common for several species; extractable amounts: up to several 100μg*
Release	at definite times of day, up to several hours	briefly (few secs) when close to a female; long (up to hours)/repeatedly independent of female presence
Perception	by males only	by both sexes
Information on	sex; species; readiness to mate; ???	sex; species; fitness (?); ???
Functions	in intersexual interactions only: inducing anemotactic flight from distance (= attracting males) and mating behaviour (at close range)	in both inter- and intrasexual interactions inhibit escape/arrest; overcome coyness; permit mate-choice; induce mating posture; calling males/females; (forming aggregations/leks); repelling competitors; inducing oogenesis/egg-laying; terminate female calling behaviour; ???

*The striking difference in the extractable amounts is likely to be due to continuous pheromone production by calling females and unique synthesis, followed by atrophy of glandular cells and pheromone 'storage', in males.

common peculiarity, seems to be corroborated by our present functional understanding.

Finally, it should again be stressed that in butterflies in particular, chemical signals interact with visual (and other) cues, and must not be taken in isolation when trying to understand sexual communication or selection. At present our knowledge of visual communication in butterflies is similarly insufficient and requires a fresh approach (Chs 20, 21, 23). A much better understanding of both chemical and visual communication is necessary before we can elucidate the nature of selection pressures acting on courtship behaviour in the Lepidoptera.

It is striking that our knowledge on scents in sexual communication is so paltry although the phenomenon was first recognized over a century ago. Since then, various pleas 'for a new look at Lepidoptera with special reference to the scent distributing organs' (Varley 1962; see also Dixey 1911, Ford 1945) have been made in the hope of stimulating extensive investigations. Possibly, the acceptance that 'scent organs' are merely courtship stimuli not requiring further study accounts for the relative disinterest. Also it seems that experimental work on scents and behaviour in Lepidoptera was considered to be difficult to conduct, and required elaborate technical facilities and advanced chemistry. If these were in fact obstacles, this chapter hopefully helps to overcome them. It demonstrates both the serious need for more solid data and the great variety of questions unanswered. The study of butterfly behaviour is fraught with difficulties—but the results are rewarding, as can be seen from the experimental work reviewed above. These studies, on butterflies, moths and other insects, have helped develop techniques which can now be applied to other species—and many demonstrate that important questions can be investigated by relatively simple methods. Also, more knowledge of the general biology of butterflies would indirectly be of great value. It must be stressed that chemical communication is basically a biological subject, not chemical. Of course, at a certain stage in experimental work, collaboration with chemists is appreciated or even indispensable but—as we have seen—many fundamental questions can be studied without knowledge of the structural formulae of the secreted chemicals; we lack an understanding of the biological significance of some androconial secretions even though they have been identified chemically. Finally, it is worth underlining that studies on chemically mediated interactions between butterflies is not only a field for professionals—amateurs are also able to contribute much to our understanding.

Addendum (notes added in proof)

The courtship of *Troides oblongomaculatus* involves the transfer of particles (cf. p.260; Parsons 1983).

Sevastopulo (1944) noted that *Euploea core* males can often be seen flying or sailing slowly with their anal brushes protruded, without a female being present; in addition, he observed a courtship attempt where a male hovered above a female by means of quick, short strokes of his wings, at the same time protruding and withdrawing his hairpencils (see p.261).

Chovet (1983) describes the courtship behaviour of *Pieris brassicae*, and demonstrates that the female recognise the males on a chemical basis (see p.267).

Greenfield & Coffelt (1983) describe observational and experimental analyses of the reproductive behaviour of *Achroia grisella*, covering eco-ethogical aspects extensively (see p.270).

Concerning the complex pheromonal systems of *Eldana saccharina*, see additional papers by Atkinson (1981, 1982) and Zagatti *et al.* (1981) (see p.270).

T. C. Baker (1983) provides further information on the significance of male scents in *Grapholita molesta* (se p. 270).

Pheromones extractable from male forewings of *Plodia interpunctella* elicit a 'turning response' in sexually receptive females, which react differentially over a narrow range of extract concentrations (McLaughlin 1982); this finding again suggests that females may choose males on the basis of pheromone quantity, which could indicate male fitness (cf. p.270).

New data on the sexual behaviour of *Hepialus humuli* (and a thorough literature survey on mating in *Hepialidae*) by Mallet (1984) make it clear that the females are attracted by the males, although it is not yet established if male scents and/or visual cues are the luring signal (see pp.270–271).

Lekking behaviour possibly also occurs in *Eldana saccharina* (cf. Atkinson 1981) (cf. p.271).

In certain circumstances, danaine butterflies damage fresh leaves of PA-plants by scratching, and thereby gain access to the chemicals otherwise sealed within the tissues (Boppré 1983) (see p.272).

Acknowledgements

I am most grateful to A. W. R. McCrae, R. I. Vane-Wright and P. R. Ackery for their efforts in revising the manuscript linguistically. They, as well as D. Schneider and W. Wickler provided stimulating criticism, which is acknowledged with many thanks. Having recently left the Max-Planck-Institut für Verhaltensphysiologie, I would like to take the opportunity to express my most sincere gratitude to Professor Dr Dietrich Schneider and to my friends and colleagues at Seewiesen for all the advice, assistance and encouragement they have given me for the past ten years, and to the Max-Planck-Gesellschaft for financial and technical support.

Part VII
Migration and Seasonal Variation

26. The Dilemma: When and How to Go or Stay

R. Robin Baker

Department of Zoology, University of Manchester, Manchester

In July 1981, while this paper was in preparation, C. B. Williams, pioneer in the study of butterfly migration, died at the age of 92. This paper is dedicated to his memory, for without his tireless work half a century earlier, it would probably never have been written.

Every butterfly begins life in a place determined for it by its mother. Sometime later it dies in a place determined largely by its own behaviour. Through time and space between these two events and places, the individual traces an invisible path, its *lifetime track*.

Continually, throughout its life, the individual has to make decisions over whether to go elsewhere (i.e. to migrate) or to stay where it is. The lifetime track is the cumulative product of all these decisions. Those decisions and movements it makes as a larva (Ch.7) are no less interesting than those made as an adult, but their contribution to its final lifetime track is usually minimal compared to the contribution of movements as a winged adult. Yet, even when adult, some butterflies seem continuously to opt to stay rather than to migrate, and die within only tens or hundreds of metres of the place where they were first deposited as an egg. Others, however, visit areas 2000–3000km from where they were born. Every gradation between these two extremes can be found and it is this variety of lifetime tracks found amongst butterflies that is the subject of this article.

The apparent length of the lifetime track depends on how it is studied. Insofar as butterflies are born, die and travel, it is axiomatic that each has a lifetime track, although this has never been measured through its entire length for any butterfly or any other animal. It would be a thankless task to try, even with current technology. Instead, we try to reconstruct a typical lifetime track for an individual of a particular species by sampling many individuals within that species. Even then, the best we can manage is to document an individual's position once, twice or, if we are lucky, a few times along its track. Alternatively, we can attempt to follow it continuously for a while, and then try (with due caution) to extrapolate forwards and backwards through time and space.

As with all methods based on sampling, such techniques are fraught with difficulties, probably few of which have even been identified, let alone resolved. Eager to press on with speculating and theorizing, we simply hope that our reconstructed lifetime tracks reflect butterfly behaviour, not just entomological technique. Unfortunately, present studies are not reassuring.

Ask a population geneticist (Chs 2, 3) how far a butterfly or moth travels in its lifetime and often you will be convinced that most spend their entire adult life within the area being studied, with only a few moving between habitats and maintaining gene flow. Ask an entomologist using radar on the other hand, and he will talk in terms of convection currents, jet streams, wind displacement and moths travelling 200km or more overnight. The feeling that each specialist develops is different.

Sit in almost any grassy field in Europe on a sunny day in summer and watch. Within a few minutes a white butterfly (usually a Small White, *Pieris rapae*, or Large White, *P. brassicae*, but occasionally a Green-Veined White, *P. napi*) is likely to appear, fly in a more or less straight line across the field, and disappear out of sight. Follow one of these butterflies and, if you are fit enough, it will lead you in a more or less straight line across country (Fig. 26.1). Only if it encounters an area such as a cabbage field or a patch of nectar-bearing flowers, will it stop flying in a straight line and fly around within the confines of the habitat. Stay with the individual for a few minutes, hours, or overnight and eventually it will set off once more across country, most often adopting its original compass direction. Such a study (Baker

The Biology of Butterflies
0-12-713750-5

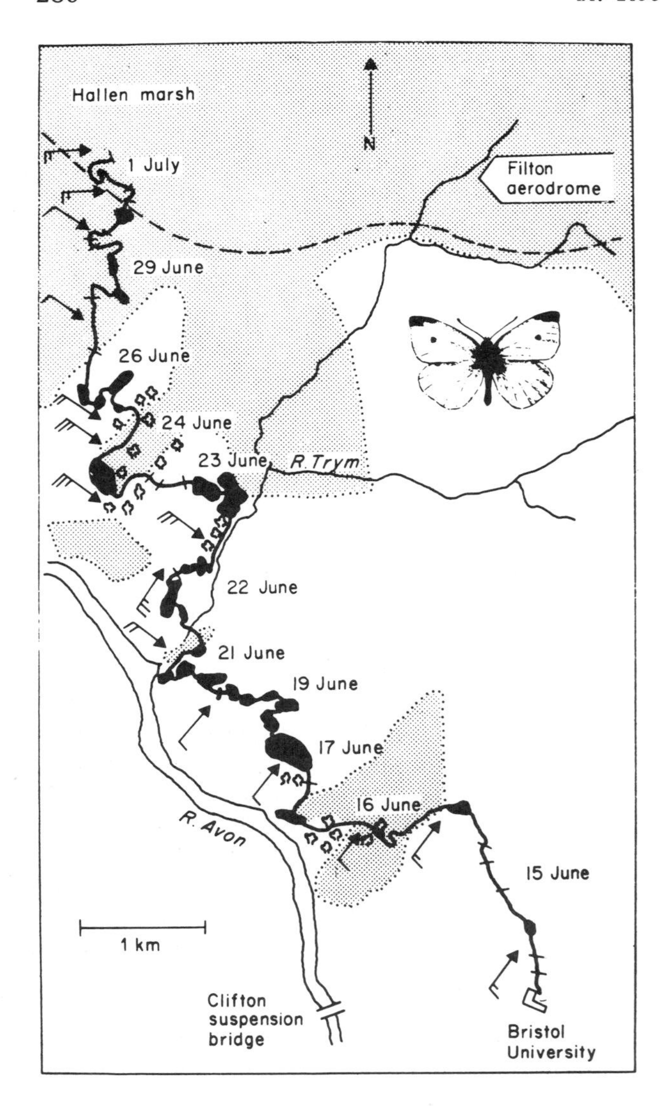

Fig. 26.1. Apparent cross-country travel of the Small White (*Pieris rapae*) observed by following individuals. Observation began outside Bristol University. Each individual was followed until it was lost from sight (bar), whereupon the observer waited until the next individual appeared flying in the same direction. That was then followed until lost. Solid black: areas such as gardens or waste ground in which the individual being followed flew to-and-fro and used resources by feeding, roosting, ovipositing, etc. Stippling: areas of relatively open grassland. (From Baker 1978.)

1978) suggests an average rate of cross-country movement for *P. rapae* of about 0.8km per hour of sunshine. Mark-release-recapture (MRR) experiments (Roer 1961*a*) on *P. brassicae* produced a slightly higher rate of migration 1.1 ± 0.3 (S.E.) km per hour of sunshine (Baker 1978). The marked individuals for which this rate was calculated were recaptured 11–95km from the point of release. Both of these studies, using different techniques, suggest a typical lifetime track for the Small and Large Whites of about 100–200km between its most distant points.

Recently, Jones *et al.* (1980), in a paper: Long-distance movement of *Pieris rapae*, concluded that 'Day to day movement is a random walk with a step length of (on fine days) about 450m per day. Consequently, a female who lives for 16 days will on average die 2km from her birthplace: our study covers a significant part of her total movement.' This study involved feeding larval *P. rapae* on an artificial diet containing dye so that when adult the females laid coloured eggs. The butterflies were released in a 1km square grid of cabbage plants and the distribution of eggs within that grid used to reconstruct the movements of the adults.

So which conclusion is correct: does the average Small White die 2km from its birthplace or 200km? We could try to avoid any conflict by suggesting that both conclusions were correct. We cannot, however, do this by resorting to sexual differences. Both Roer and I included females in our studies. Indeed, I have followed a female *P. rapae* on a cross-country journey of virtually 2km, one of the suggested distances covered by her lifetime track. The whole journey took less than an hour, and involved two stops while the butterfly fed. The butterfly flew on beyond 2km, but the entomologist stopped. Another possibility for conciliation relates to geographical area, Roer and I working in Europe, Jones *et al.* working in Australia. However, as Jones and her colleagues point out, it is difficult to reconcile their results with the rate at which *P. rapae* originally spread through Australia (Peters 1970). From an original appearance in Melbourne in 1939, the butterflies had reached the West Australian coast three years later, an absolute maximum of 25 generations. In the end, the authors prefer to believe that, despite a ban on the importation of brassicas to West Australia during the period, the spread of *P. rapae* in Australia occurred with human help. Perhaps they are right. I feel bound to point out, however, that the distance covered (about 4000km) is roughly consistent with lifetime tracks of 100–200km.

Can we explain these conflicting conclusions? It is well-known to students of winged vertebrates that the marking and recapturing stages of MRR studies should be carried out by different people, otherwise there is a real danger that the distances travelled will be grossly underestimated due to the tendency for more recaptures near to the release site (for other MRR problems, see Ch.4). To avoid this, students of bats (e.g. Leffler *et al.* 1979) often only analyse citizen recaptures. Ornithologists have long enjoyed the benefits of having their marked animals recovered by members of the public and some confidence can be felt in the migration patterns that have been reconstructed from the recoveries of ringed birds. Of course, ornithologists still agonize over how much can be deduced about the behaviour of live birds from the recovery of dead ones. To an entomologist, however, unsure whether a Small White travels 2 or 200km, the ornithologists' agonies seem trivial.

Notably, the butterfly studies of Roer (e.g. 1961*a*, 1968, 1969, 1970) on various European species and Urquhart (1960, 1976) on the American Monarch butterfly, *Danaus plexippus*, which have made much use of citizen recaptures, identified lifetime tracks hundreds or thousands of kilometres in length. The only MRR study not involving citizen recaptures that also identified long tracks (*ca* 100km) was Nielsen's (1961) classic study of *Ascia monuste* in Florida.

At the risk of being provocative, I confess a reluctance to believe any general statement concerning the lifetime track of a butterfly if it derives from an MRR study but is not based on citizen recaptures by an adequately alerted public. This does not mean that local studies tell us nothing about lifetime tracks. On the contrary, as shown in later sections, they tell us a great deal about the behaviour of butterflies that have opted to stay, if only temporarily, in an area of high suitability. But we must not extrapolate from such studies to general statements about entire lifetime tracks.

The Decision to Stay or Go

As already noted there are undoubtedly butterfly species which, generation after generation, produce individuals that die within only tens or hundreds of metres of their birthplace (see Ch.2). Yet others are undoubted travellers, staying nowhere for long, and ending their lives hundreds or thousands of kilometres from their birthplace. In view of the preceding discussion, this may seem a rash statement. However, the different predilections of different species to travel or stay can be experienced without recourse to the vagaries of MRR techniques. Stand in the middle of a bare or very-grazed field at intervals throughout summer. Some species, very common in the surrounding habitats, will never be seen to fly across. Others will be seen frequently. Yet others, the real travellers, will cross the field frequently but will not commonly be found anywhere nearby. The quotient U/S (where U is the number seen crossing an unsuitable habitat per unit time and S is the number present in a nearby suitable habitat) is a useful index of predilection to travel, though, of course, it tells us little about the distance between the furthermost points of the lifetime track.

Evidently individuals of some species are forever deciding to migrate elsewhere. Others continuously decide to stay where they are. Nor is such variation strictly interspecific. Over the course of three summers spent standing in fields in Britain, I saw nearly 100 transient Speckled Wood butterflies, *Pararge aegeria*. Not one was a male. Similarly, the British deme of *Papilio machaon* does not travel far from the East Anglian fens, but in continental Europe it seems to be much more of a traveller. Finally, even species that rarely travel across country may do so occasionally, given certain conditions. During the British drought of 1976, many Satyrinae (e.g. Ringlets, *Aphantopus hyperantus* and Marbled Whites, *Melanargia galathea*) that normally have a low U/S index were often to be seen travelling through areas they would normally avoid. Similarly, *Hipparchia semele*, a species with a low U/S index, occasionally migrates in some numbers (Feltwell 1976).

Southwood (1962, 1977) made the first modern attempt to explain this 'stay or go' variation among butterfly species. He pointed out that those that spend their lives near their birthplace are most often adapted to relatively permanent habitats, such as woodland and moorland. On the other hand, species that travel across country are most often adapted to temporary habitats, for example feeding as larvae on weeds. I expanded Southwood's model slightly (Baker 1969) to include factors such as inter-habitat distance and the spatial relationships of adult and larval requirements, and eventually formulated the 'initiation factor model' (Baker 1978) in which i, the probability of an individual initiating migration from its current habitat, is given by $h_q\ (1\ -\ m/S_d)$.

In this model, Southwood's original concept concerning temporary and permanent habitats is incorporated into the habitat quotient, h_q. Essentially h_q measures the probability that if an animal leaves its present habitat it will encounter a more suitable habitat. Obviously, species for which habitat quotients are highest are those adapted to those temporary habitats with a high rate of appearance of new, highly suitable, sites. S_d is a measure of that portion of an individual's reproductive potential that can still be influenced by the individual's actions. The inclusion of S_d in the initiation factor model allows variation with age in the decision to stay or migrate to be accommodated. Finally, m is a measure of the cost of migration between habitats, expressed in terms of by how much a migration reduces S_d. The factor m accommodates the energetic cost and survival risk associated with leaving the current habitat, as well as lost opportunities for feeding, ovipositing, copulating, etc. Clearly, m will tend to be least for species that experience the shortest inter-habitat distances, including species for which adult requirements (e.g. sites for feeding and roosting) are frequently encountered between perhaps more separated areas containing larval requirements.

Even in its embryonic form (Baker 1969), the initiation factor model accommodated in an encouraging way the apparent interspecific variation in predilection to travel among British butterflies. In its final form it can also accommodate variation

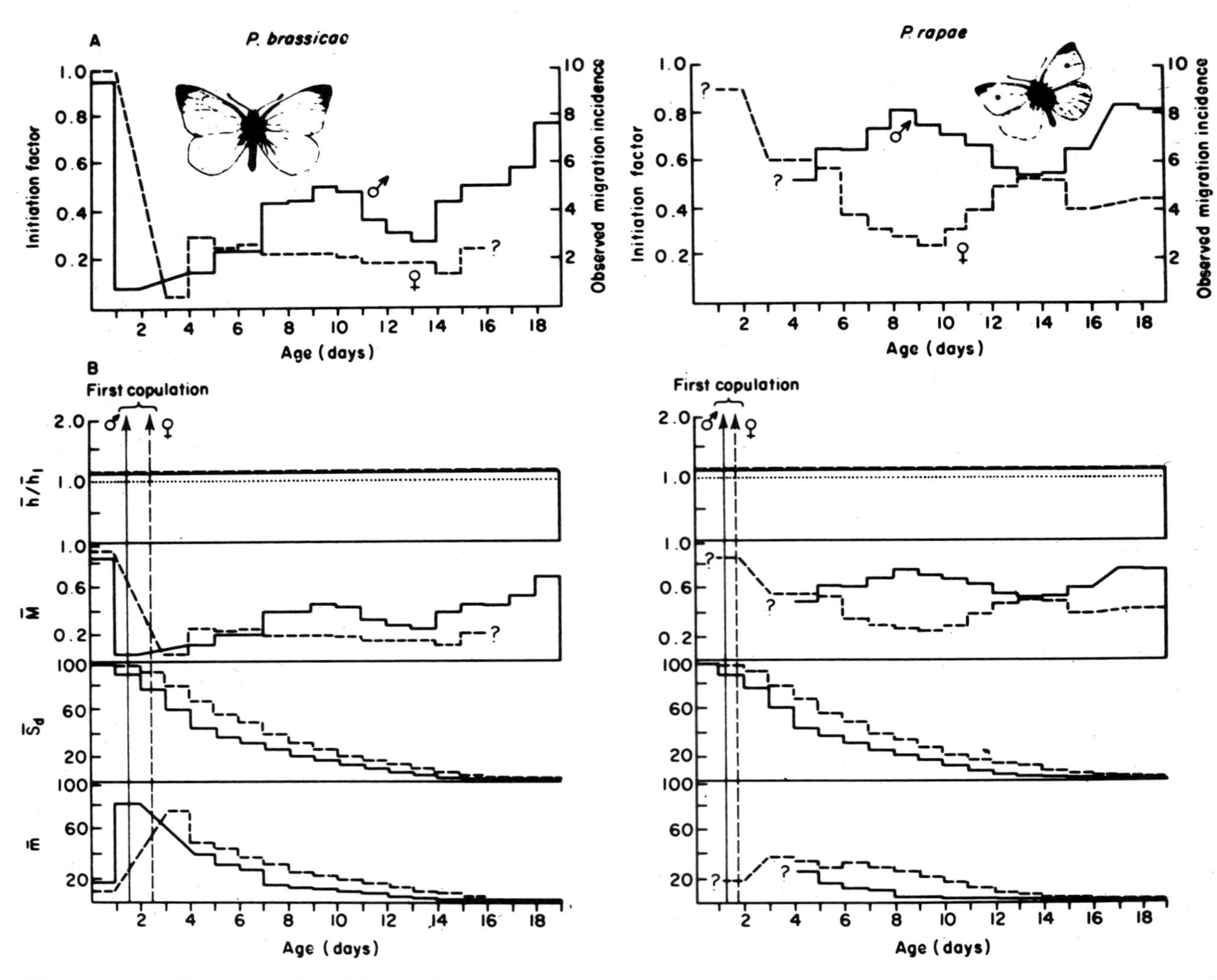

Fig. 26.2. Application of the initiation factor model to ontogenetic variation of migration incidence in the Small and Large Whites. *P. rapae* and *P. brassicae.* Observed migration incidence is derived from a U/S index (see text). Initiation factor (i) is matched arbitrarily. Habitat quotient ($h_q = \bar{h}/h_1$) is estimated. $\overline{M}$ is calculated from i and h_q by $\overline{M} = i/h_q$. Remaining action dependent reproductive potential ($\overline{S_d}$) is calculated for female *P. brassicae* from the data provided by David & Gardiner (1962) and is estimated for the others. Migration cost ($\overline{m}$) is then calculated given that $\overline{M} = i - \overline{m}/\overline{S_d}$. For further details see Baker (1978).

with age and between species, though a rigorous qualitative and predictive test of the model has yet to be attempted. The nearest to such a test so far is that for the Small and Large Whites shown in Fig. 26.2. Variation in U/S index for these species in Britain showed peaks in migration incidence for females of both species during the first two or three days of adult life. Such a peak was assumed by Jones *et al.* (1980) not to occur. Thus they released only 6- or 10-day-old females. Support for this peak, however, has come from studies of *P. rapae* in Japan by Ohsaki (1980). If we now use the U/S index as a base, make certain assumptions concerning habitat quotient for the two species, and calculate change in S_d with age for female *P. brassicae* from the data of David & Gardiner (1962), it is possible to use the initiation factor model to generate curves with age of migration cost. For female *P. brassicae*, this curve correlates well with the number of eggs to be laid the following day (data from David & Gardiner 1962). The curve for males suggests migration is most likely to occur when receptive females seem least likely to be abundant (Baker 1978).

The initiation factor model is a generalization that can be used to explain or predict all forms and extremes of variation in migration incidence in all animals. A specialized version of the initiation factor model was developed independently by Charnov (1976) and Parker & Stuart (1976). Known as the 'marginal value theorem' (after Charnov 1976), it can be applied to situations in which the only component of migration cost is search time (see also Ch.8) and there is a gradual decrease in rate of benefit from a habitat with time after arrival. In such situations, it allows the optimum stay-time in a habitat to be obtained graphically. As yet, the marginal value theorem seems not to have been applied to butterflies.

The initiation factor and marginal value models reflect the approach of behavioural ecologists to the decision by animals of whether and when to go or stay. Essentially, they are concerned with the decisions of individuals. Other approaches to variation in predilection to migrate have been made by physiologists and population ecologists.

Johnson (1969) proposed the existence of an 'oogenesis-flight syndrome' whereby mechanisms associated with oogenesis suppressed the urge of female insects to travel, most migration thus being pre-reproductive. The model has immediate appeal in interpreting the physiological mechanisms of those female insects that migrate only once in their lives, during the immediate post-teneral phase, and thereafter become resident in some suitable area. Overall, however, the 'oogenesis-flight syndrome' lacks generality (Baker 1978). Not only is it inapplicable to males, it does not really allow for variations in migration pattern with age or between species among females. It is particularly inapplicable to butterflies. True, both *Pieris* in Fig. 26.2 show a pre-reproductive peak for females and the known or presumed long-distance autumn migrations of Monarch, Painted Lady (*Cynthia cardui*) and Red Admiral (*Vanessa atalanta*) butterflies are performed by individuals in reproductive diapause. However, both *Pieris* continue to travel throughout life and the same Monarchs that migrated south in autumn are thought to migrate north in spring, depositing eggs along their migration track as they go. The male *Pieris* in Fig. 26.2 show, if anything, an increase in U/S index with age. Studies in Japan of the flight activity of tethered males of the skipper *Parnara g. guttata* (Ono & Nakasuji 1980) show a similar trend.

Population ecologists have tried to incorporate inter-specific and ontogenetic variation in migration incidence into current models of life-history strategy (e.g. Dingle 1972, 1980). It is, of course, axiomatic that the form of the lifetime track is part of life-history strategy. The question of interest is whether the various mathematical parameters in current models of life-history strategies can give useful insight into the form of the lifetime track. Dingle (1979) insists that I have 'missed the point' of the arguments of population ecologists—but without saying what that point is. As I understand it, there are two major suggestions. The first is an attempt to explain why many female insects show a pre-reproductive peak of predilection to migrate. The argument runs that migrants are colonizers and that, as a colonizer, upon arriving in a new habitat, the insect must be able to reproduce and leave descendants (Dingle 1972). Since pre-reproductive adults have their reproductive life ahead of them and at the same time have already survived the causes of juvenile mortality, their reproductive values are usually high relative to juveniles or older adults and thus have a high colonization potential. Evolution would be expected, therefore, to favour migration in most species just before reproduction. The second suggestion is that migratory species tend to be 'r-strategists' rather than 'K-strategists' and to show high intrinsic rates of increase (r) (reviewed by Dingle 1980).

Once again, the appeal of these suggestions is obvious if we think in terms of a female insect that migrates away from its natal area, settles permanently in a new area, and produces several generations of offspring filling the habitat before migration occurs once again. Maybe there are butterflies with this particular pattern. Indeed, Ohsaki (1980) has interpreted the results of local MRR data on *P. rapae* in these terms. Perhaps this is also the correct way to interpret the behaviour of *Ascia monuste*, as described by Nielsen (1961), again following non-citizen MRR studies. On the other hand, if the migration strategy of the more-mobile butterflies such as *Pieris rapae*, *P. brassicae* (Figs 26.1, 26.2) and northward migrating *Danaus plexippus* involves a steady progression across country scattering eggs or sperm along their lifetime track, producing offspring that will similarly travel inexorably across country, such ideas seem not so much wrong as irrelevant to the behaviour of the individual. Where lies the importance of colonization potential, reproductive value, or intrinsic rate of increase if neither the adult, its mate, nor its offspring is going to remain in one place long enough for these parameters to be relevant to that place? Somehow such concepts fail to touch the behaviour of the individual, no matter how powerfully the population arguments can be presented mathematically. Much more germane is the idea that scarcely has an individual arrived in an area than the area decreases in suitability to the point that somewhere else is better. In other words, the habitat quotient is always greater than 1 even when devalued by migration cost, except briefly just after the individual arrives. What can generate such a situation?

One possibility is that the individual's requirements continually change and that different activities are best carried out in different areas. First, for example, an individual may be motivated to feed. Then, having fed, to mate. Having mated, to roost, and so on. Another possibility is that the individual depletes the resources of an area, or having searched an area for a resource and failed to find it (e.g. a male for a female—Parker & Stuart 1976), the resource is most likely to be found by moving to a new area than persisting with the old. There are many possible reasons why an area should decrease in suitability to an individual with time after arrival. The most interesting, however, is seen when we take into account population density and the behaviour of other individuals. To appreciate the reasoning, we have to accept a number of hypotheses (developed independently and expressed differently by Parker & Stuart 1976 and Baker 1978): while in a currently suitable habitat an individual monitors features of the environment and only migrates when these exceed a certain threshold value; selection should set the migration threshold to these environmental features (e.g. food, shelter, wind strength) such that an individual migrates when the suitability of the current habitat just falls below $\overline{E}$, the mean expectation of migration (i.e. the average suitability of the likely destination devalued by the average migration cost); as an individual migrates and fails to find a suitable destination, $\overline{E}$ decreases and the individual's migration threshold gradually increases so that it accepts less and less suitable habitats; individuals differ in their migration threshold so that some individuals migrate before others; and when a habitat falls in suitability below the mean expectation of migration, just enough individuals leave to reduce competition and restore suitability of the habitat to the mean expectation of migration.

Field data in support of some of these points are presented in the references cited. There is no group selectionism in the arguments generating these hypotheses, each individual behaving in its own best interests. It can be shown theoretically that the situation stabilizes evolutionarily at that frequency distribution of migration thresholds within the population at which all individuals do equally well from their migration/residency strategy. Field evidence in support of that theory has been taken from G. A. Parker's classic study of a dung fly, *Scathophaga stercoraria* (Parker & Stuart 1976, Baker 1978).

So how does this theorizing help us understand the decisions of some butterflies of some species to travel forever across country while others stay forever in the area of their birth? Imagine a population of butterflies emerging at the start of a generation in habitats scattered throughout the species' range. Suppose the habitats are so permanent and so widely scattered that no habitat is below the mean expectation of migration ($\overline{E}$). Clearly all butterflies should opt to stay where they are and all should die nearby. Now suppose the suitability of some habitats is below $\overline{E}$, though for most it is not. Butterflies leave these habitats until their suitability is restored to $\overline{E}$, and search for habitats the suitability of which is greater than $\overline{E}$. Migration and redistribution ceases when all butterflies are in habitats more suitable than $\overline{E}$. Finally, imagine a situation in which, even once the butterflies have begun to migrate, there are so many that even once all habitats have equilibrated in suitability to $\overline{E}$, there are still butterflies between habitats searching for a site more suitable than $\overline{E}$. Every time one of these butterflies arrives in a territory it pushes its suitability below $\overline{E}$, a suitability that can only then be restored by an individual departing. If the individual that has just arrived has a higher migration threshold than those already there (third hypothesis above), the individual that departs will be one of the previous residents. In other words, a chain reaction is set in motion, individuals continually displacing and then being displaced by other individuals.

This analysis suggests that the nett result of all members of a population deciding to stay and go at times optimal for each individual should impart certain characteristics to demes and populations. Where the initiation factor is low, habitats never decreasing in suitability below $\overline{E}$, and migration is rare or non-existent, adjacent demes may fluctuate out of synchrony with one another, at the mercy of local vagaries in survival and reproduction. Such a situation seems to exist in demes of *Euphydryas editha* in California (Gilbert & Singer 1973, 1975; Ch.2). Where the initiation factor is higher and migration exceeds a certain level, individual decisions lead to the population redistributing itself in each generation. Inequalities between habitats are 'ironed-out', density finding its own level as if liquid, reverting at each site to a value determined primarily by the total population size, the number of habitats, and the mean expectation of migration. Where the physical suitability of a particular site remains stable over generations, the density in that site will relate almost entirely to the total number of butterflies of that species and the number of alternative sites. I can offer no evidence that such a situation exists in nature for butterflies, but I can for moths. Using light traps (which probably selectively capture migratory species —Baker & Sadovy 1978), Taylor & Woiwod (1980) have demonstrated that 263 moth species, caught over six years at 53 sites in Great Britain and adjacent mainland Europe, show a spatial and temporal stability at the different sites that is a power function of mean population density over an area and through time. They conclude that this stability must be an intrinsic property of the behaviour of the species. I suggest here that it is the unselected consequence of mobile individuals each making the best decision over when to stay and when to migrate.

How to Stay

Areas of marshland, chalkland, moorland and perhaps some types of woodland are occupied primarily by species with a low initiation factor and produce the familiar spectacle of large numbers of apparently resident individuals flitting around within often clearly demarcated boundaries. Such species have a low U/S index and, once marked, the same individual can be seen or recaptured in the same area perhaps days or even weeks later (but see also Ch.4). Such butterflies always seem to have opted to stay and exploit the area in which they were born or in which they first settled.

When any animal spends a long period of time in a restricted area, we might expect that it will exploit that area most efficiently if it learns which are the best places to do this or that. In others words we might expect it to establish an area of familiarity and from that area to crystallize a home range within which it will perform most of its activities (Baker 1978, 1982, Zalucki & Kitching 1982; Chs 2, 3).

Students of butterflies have been slow to study the home range behaviour of their subjects, some butterfly species presumably offering the same opportunities for such studies as do small mammals and birds. An indication of what may be possible comes from the pioneering work of Turner (1971*a*) and Gilbert (1975; Ch.3) on *Heliconius* species which seem to visit particular feeding and roosting sites routinely at specific times every day. The visits to feeding sites coincide with times of pollen release and nectar flow.

It is a relatively short step from becoming familiar with an area and learning how to exploit it efficiently to defending it from occupation by other individuals. There were many indications in the old entomological literature that some butterfly species may be territorial but it was not until the 1970s that work on their territoriality began in earnest. All studies so far seem to have been concerned with territories set up by males to give them access to females. There are no undisputed examples of a butterfly defending a larger area containing a variety of resources such as occurs with vertebrates (see also Ch.20). The nearest we can come to such an example is that of the stable home ranges of *Heliconius* observed over periods of months (see Ch.3). Such stability could indicate defence of a large area. On a smaller scale, however, the Peacock butterfly (*Inachis io*) establishes a territory at about midday every day in spring and early summer (Baker 1972*a*). Most of the time, successful defence is achieved simply by advertising its presence to incoming males by flying up to meet them (Fig. 26.3). Evidence that another male is already there is usually sufficient for the intruding male to continue on his way. Escalation occurs most often when an intruder arrives at a territory while the resident is temporarily absent, either with another male or with a female. When the resident returns both intruder and resident behave as if they were the occupier of the territory. In such cases the resident is still more likely to retain the territory though only during the first two interactions that follow. Success in these seems to depend on how long and how far the resident male leads the intruder away from the territory. Should the intruder return to the territory after the first two interactions, the decision as to which of the two stays is then determined by a contest of aerial manoeuverability in which the two adversaries spiral upwards (Fig. 26.3). A male that achieves and maintains the uppermost position during two successive spiral bouts retains the territory, the other male opting to leave.

In the Peacock the territory is positioned in a place favourable to intercept females. When one does pass through, the male follows her as she continues to travel across country. Copulation does not take place until the female goes to roost. In the Small Tortoiseshell (*Aglais urticae*) the territory is established at the oviposition site, a patch of nettles. A male defends an area within the nettle patch only until the arrival of a female. He then abandons defence of the area and concentrates on defence of the female (Baker 1978). This he does by positioning himself a centimetre or so behind, and tapping her at intervals with his antennae. Whenever another

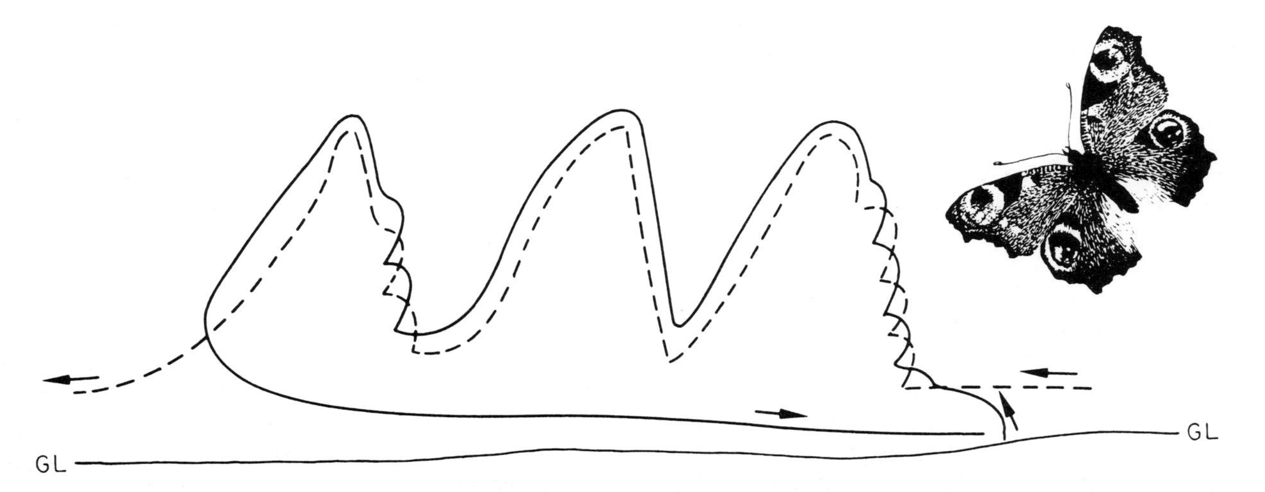

Fig. 26.3. Territorial interaction of resident (solid line) and intruding (dashed line) male Peacock butterflies (*Inachis io*) to show spiral, leap-frogging climbs and dives. GL = ground level. (From Baker 1972*a*).

male intrudes upon the pair the original male usually flies off before the intruder can settle, the latter giving chase. The pair of males may fly 100m or so from the female's position before the original male breaks loose from his pursuer and flies very fast, by a direct route, back to the female. He lands behind her, hits her with his antennae, and the pair fly immediately a few metres to a concealed position amongst the nettles before the intruding male also returns to the original site. Such interactions may take place many times during the course of an afternoon and the female is usually compliant with the defending male. Towards evening, however, when the female goes to roost, she is likely suddenly to drop into the nettles and the male has to follow very fast in order to stay with her. I have seen males successfully defend a female against six or more intrusions by other males, only to lose the female at this last moment. If he can stay with her during her initial scramble down into the nettles, she becomes quiescent and copulation takes place.

Other studies of butterfly territoriality have shown that although territorial locations may differ considerably (e.g. hilltops, for *Papilio zelicaon*, Gilbert *in* Maynard Smith & Parker 1976; sunspots for *Pararge aegeria*, Davies 1978), the behaviour shows broad similarity to that of the Peacock. In the Speckled Wood, however, Davies never saw an intruder win (but see Baker 1983).

There is one further difference between the territorial behaviour of Speckled Woods, on the one hand, and Peacocks and Small Tortoiseshells on the other. Whereas male Speckled Woods occupy the same small area of woodland for several days, perhaps their entire lives (Baker 1969), Peacocks and Small Tortoiseshells occupy their corner of a field or nettle patch, respectively, only for the afternoon, a temporary pause along their lifetime track. Next day they leave and travel across country until it is time to seek another area to defend, perhaps kilometres away, the following afternoon. Their behaviour, once they decide to go, could not be more different from the behaviour by which they stay.

How to Go: the Track

It may seem obvious that the behaviour shown once a butterfly leaves a suitable habitat will be different from the behaviour shown within that habitat, but the point is not always appreciated. The calculation that female *Pieris rapae* would die 2km or so from their birthplace was based on the assumption that behaviour shown over two days while within a 1km square cabbage field would persist for a further 14 days or so of life once the butterfly had left that field (Jones *et al.* 1980). The same study observed a high degree of directionality. Different individuals have different preferred directions but each individual flies with sufficient directionality to keep well within an angle of 45° over a distance of 1km. Within the field, however, individuals are likely to show different preferred directions on different days. Jones *et al.* conclude that directionality itself has no ecological relevance and is merely a mechanism for covering more ground (see Baker 1968*a*).

Within the cabbage field, this conclusion is almost certainly correct, and similar behaviour by a female Meadow Brown (*Maniola jurtina*), again within a habitat suitable for oviposition, is illustrated in Fig. 26.4. Jones *et al.* (1980) then remark, however, that their results exactly confirm the interpretation by Taylor *et al.* (1973) of orientation in the moth *Plusia gamma* flying *between* suitable habitats. Apart from the original interpretation for *P. gamma* being based on faulty analysis (see Baker 1978), this remark presupposes that orientation within habitats is subject to the same selective pressures as orientation between habitats. Yet, while within habitats, the need is for a flight path that covers ground but does not lead to premature departure. A succession of straight lines in different directions is ideal. Once the animal opts to leave the habitat, however, the requirements are different.

A butterfly that is searching for a habitat of a type that does not move and is stable over days or months suffers a disadvantage if it crosses and recrosses areas of open ground of low suitability. New habitats are unlikely to have appeared in such areas since they were last visited. An advantage lies, therefore, with individuals that maintain a fairly constant compass direction whenever they move between habitats (Baker 1968*a*, 1978). The main requirement, once a butterfly opts to search for a new habitat, is some means of straightening out its track. Moreover, there is an advantage in maintaining the same direction on successive occasions, each time it leaves one habitat in search of another.

The track of the female Meadow Brown (Fig. 26.4) hints at just such a pattern, departing from the field one day in more or less the same compass direction by which it arrived the previous day. Similar day to day constancy in the compass direction travelled by, for example, Small and Large Whites, and Small Tortoiseshells, has been observed many times (Baker 1978). Compass direction is of ecological relevance to the individual when travelling across country between suitable habitats.

How Far to Go

If, each time a butterfly sets off across country, it travels in a roughly constant compass direction, the distance it will travel from birth to death will depend in part on the distance between suitable habitats and in part on how often it leaves one such habitat for

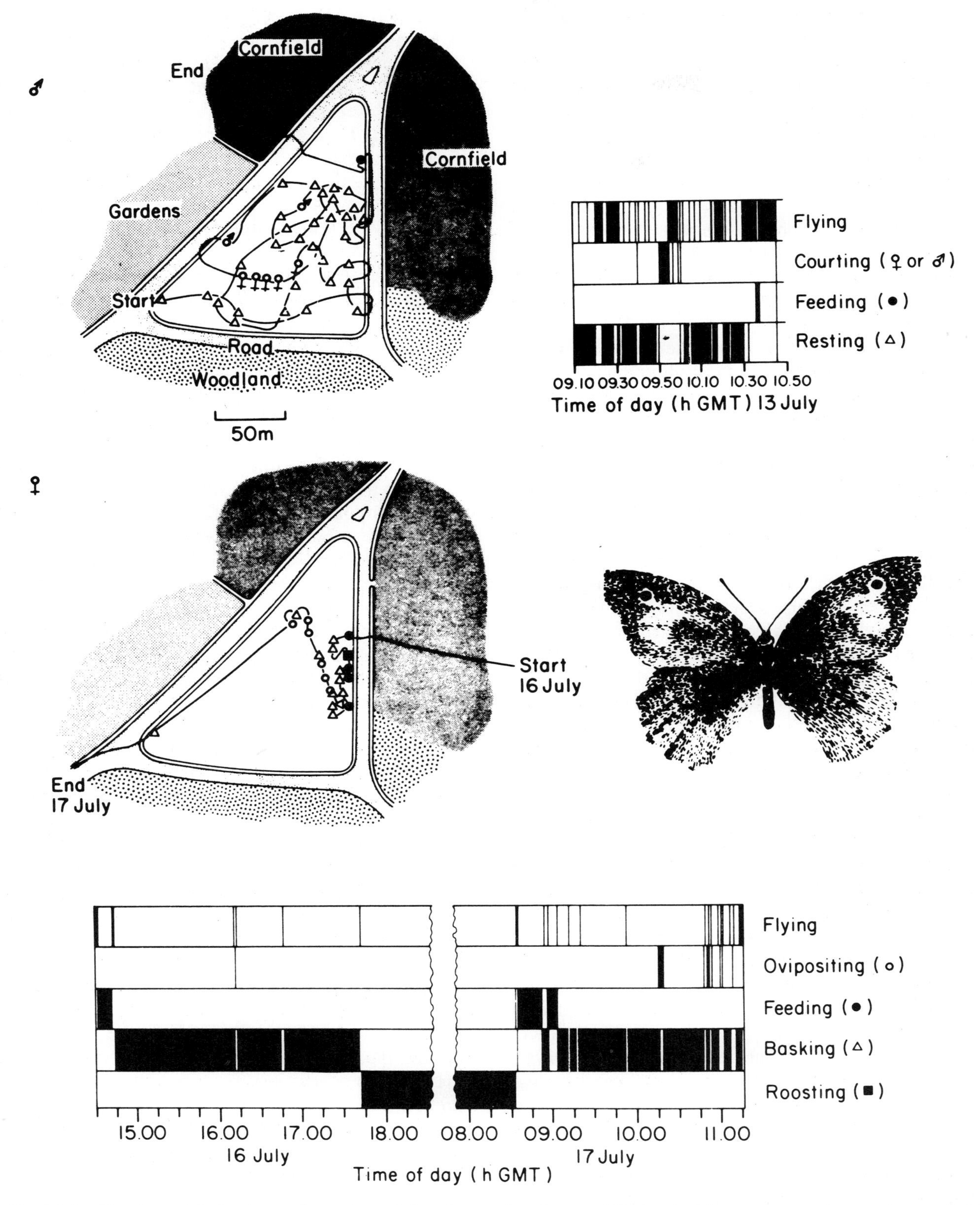

Fig. 26.4. Tracks and behaviour of a male and female meadow brown (*M. jurtina*) in a mown and grazed grass field near Bristol (from Baker 1978).

another. Butterflies that migrate across country only once in their lives, during redistribution after emergence, will have a lifetime track only as long as the distance between adjacent habitats. But a butterfly that is on the move across country throughout its life is likely to travel tens or hundreds of kilometres.

A butterfly that travels only metres or a few

kilometres from its birthplace during its lifetime experiences only local fluctuations in habitat suitability—as it travels up and down hills and along valleys, etc. Continuation in a particular compass direction for most such butterflies will have little consistent effect on the habitat suitability encountered. In contrast, if each generation of individuals of a species have, in the evolutionary past, regularly travelled distances of hundreds of kilometres, they may have experienced other selective forces. Over such distances, latitudinal or meteorological gradients in habitat suitability begin to emerge from the background noise caused by local fluctuations.

Mobile butterflies may equalize the suitability of habitats over large areas by their individual decisions of when to migrate, but the habitats themselves change in suitability with time of year. In temperate regions, with the first stutterings of spring, habitats are marginally suitable and provide food, shelter and oviposition sites for few individuals. As spring progresses into summer, suitability increases, reaches a peak, and begins to fade with autumn and the arrival of winter. The curve of suitability with season is reflected by the phenology of each species and, of course, the pattern described is only really true for species that are flying most of the year. For all temperate species, however, a wave of increased habitat suitability passes polewards in spring and summer and towards the equator in autumn and winter. No sooner has a mobile species redistributed itself and equalized the suitability of habitats than habitats in one geographical direction have increased in suitability above this level and those in another direction have decreased in suitability below it. The result is a continuous geographical gradient of habitat suitability for those species that have a lifetime track long enough to experience it. In arid tropical regions, similar gradients are established as rainfall zones move to and fro.

If the gradient of habitat suitability is steep enough and predictable enough, and if individuals of a species have lifetime tracks of appropriate length frequently enough, new selective pressures come into force. Individuals gain an advantage that hasten their travels up the gradient of habitat suitability, spending less time in the first habitats they encounter and concentrating their reproduction in the last habitats reached along their journey. Moreover, individuals with a physiology and behavioural repertoire such that they arrive at the zone of greatest suitability with the greatest economy of time and effort will also be at an advantage. The evolutionary result is a butterfly adapted to migrate with economy to a particular area, a specific distance away. For such species, the length of the lifetime track is already crudely fixed at birth.

Species that show such behaviour are probably scarce amongst butterflies. Autumn generation Monarchs in America and autumn generation Red Admiral and Painted Lady butterflies in America and Europe are the most likely temperate zone candidates (Williams 1958, Urquhart 1960, Baker 1978). In the tropics, *Belenois aurota*, *Andronymus neander*, *Catopsilia florella* and others (Williams 1958) may fit this pattern.

The Direction to Go

Butterflies that spend their entire lives travelling within an area a few hundred metres across are probably subjected to little by way of selection for direction of travel. When a butterfly abandons a habitat the primary requirement seems to be that it maintains a more or less straight line. This is best achieved by the individual having a preferred compass direction. Just which compass direction seems irrelevant as long as it remains more or less constant from one day to the next whenever it sets off across country. Butterflies with short lifetime tracks that do not experience latitudinal or meteorological gradients are thus subjected to directional selection only with respect to efficiency of search for the most suitable adjacent habitats. Yet, from each starting point, the direction of the most suitable adjacent habitat could be at any point of the compass. A parent, therefore, all of whose offspring are going to travel from their natal habitat, best avoids the dangers of an unnecessarily large proportion of its offspring passing through local habitat 'vacuums' if the offspring scatter in all directions. Parents gain, therefore, by manipulating (Alexander 1974, Trivers 1974) the preferred compass directions of their offspring such that equal numbers travel in all directions (i.e. a direction-ratio of 25:25:25:25—Baker 1969) when they depart from their natal habitat. Among British butterflies the species I have observed to produce direction-ratios nearest to 25:25:25:25 are *Anthocharis cardamines* (Fig. 26.5) and *Strymonidia w-album*. Another possible advantage to the parent in producing offspring with this direction ratio is that it will maximally reduce inbreeding.

Perhaps butterfly species show a direction-ratio of 25:25:25:25 more often than any other. When successive generations of a species regularly travel far enough to be exposed to geographical gradients of habitat suitability, however, we might expect the evolution of a different ratio. In such lineages some preferred compass directions may increase in the population at the expense of others with the direction-ratio showing a bias towards a particular compass direction.

It can be shown mathematically that if an environmental gradient favours migration in only one direction (e.g. north) throughout the entire year's

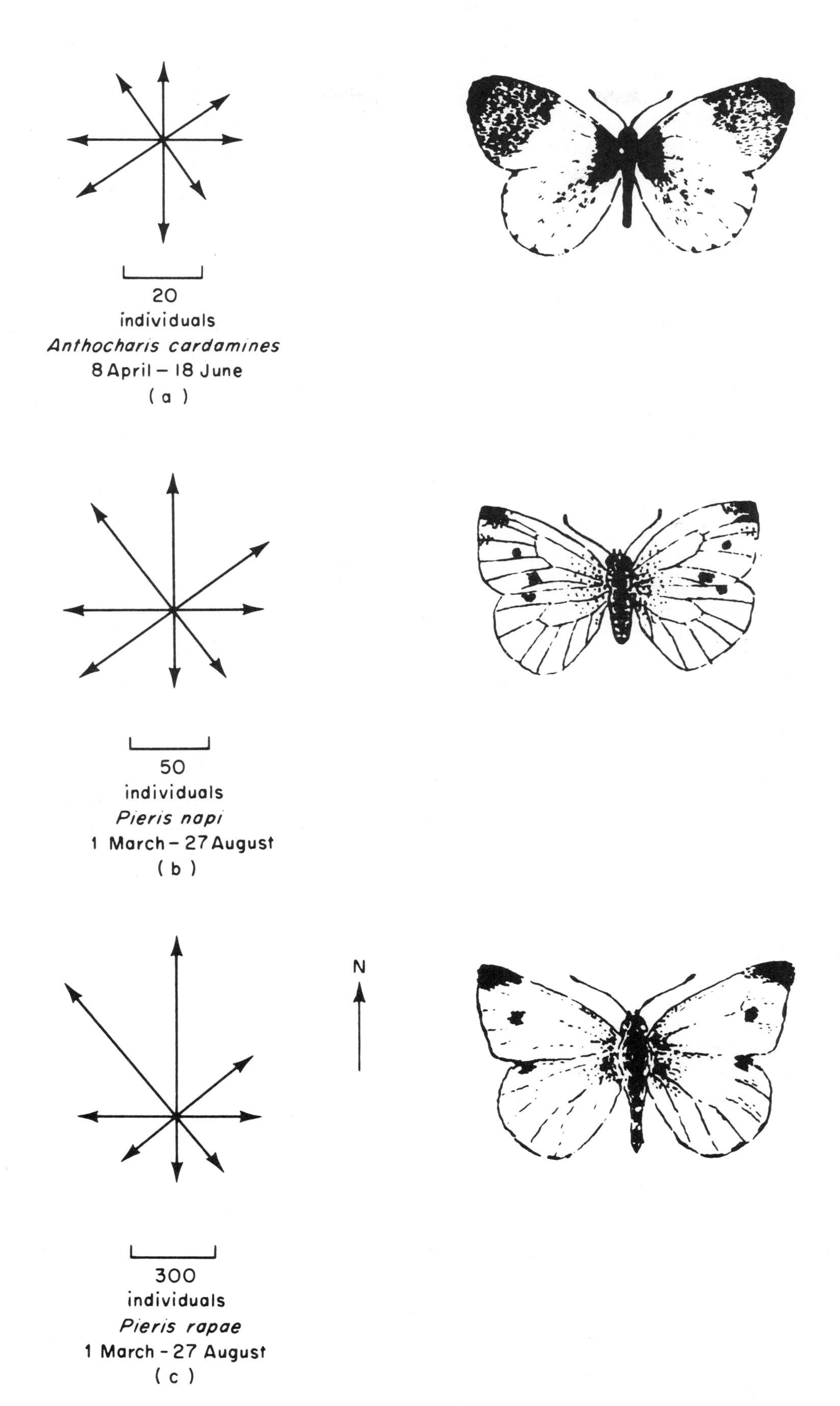

Fig. 26.5. Observed direction-ratios for three British pierids (from Baker 1978).

flight period of a particular species, selection cannot lead to an increase in the proportion of individuals showing that particular migration direction (Baker 1969). Only if the environmental gradient favours migration in one direction during one part of the flight period but in some other direction displaced more than 90 degrees from the first at some other time of year, can a bias for a particular direction evolve. For some bias to emerge, it is not necessary for the second direction to be 180 degrees different from the first. Nor is it necessary for the distance flown in the second direction to be equal to the distance flown in the first. However, if the two directions favoured by environmental gradients at different times of year are not exactly 180 degrees different and/or the distances flown in each direction are not equal, the result has to be a direction-ratio other than 100 per cent of the population flying in the two favoured directions. Lineages that do not conform to these rules disappear over the edge of the species' range and become extinct. Even before extinction, they suffer a disadvantage. After all, if all individuals fly up the gradient of increasing suitability, habitats in the opposite direction become empty and suitable for occupation by butterflies that do not fly in the main direction. The optimum direction-ratio for a mobile species is thus that which allows all individuals to travel across country but which reduces the suitability of all habitats to a common level (set by $\overline{E}$) and which keeps pace with the wave of increasing habitat suitability generated by the seasons, rainfall zones, etc.

In Britain *Pieris rapae* has a direction-ratio with a bias to the NNW throughout the spring and summer from May until mid- to late August. Thereafter, until its season ends in September or October, it flies more or less due South. Each year's season spans two, three or sometimes four generations, the generation that flies South in the autumn often being two generations removed from the butterflies that first began to fly NNW in spring. The direction-ratio for this species throughout the entire period is about 42:21:21:16 (Fig. 26.5). Different species seem to have different proportions of the population that fly in the main direction $\pm 45°$. For example, over 50% of British *Vanessa atalanta* fly in this sector of directions. Only when environmental gradients favour a second direction that is more or less precisely opposite to the first, and butterflies travel a more or less equal distance in each direction is it possible for a direction-ratio of 100:0:0:0 (all individuals flying in the main direction) to evolve (Baker 1969). The nearest that butterflies seem to have evolved to this situation, which we automatically think of as fairly standard for migrants such as birds, is a direction-ratio with 99% or so flying within a few degrees of the mean direction. The prime examples are tropical butterflies in arid areas (e.g. *Andronymus neander*, Williams 1976 and *Belenois aurota*, K. Adams in Baker 1978).

Such a high bias to the mean direction in tropical butterflies raises interesting, but still unsolved, questions. The extent of the bias should indicate a fairly precise return of a future generation to the place of origin of the initial generation. Yet many of the species are only ever seen to fly in one direction. *Belenois aurota* flies E through southern Africa in December and January (Williams 1958, Baker 1978) and *Andronymus neander* flies SSW through East Africa in March and April (Williams 1976). In neither case, however, has a corresponding movement in the opposite direction been reported. The most likely explanation is that successive generations of these species execute an annual migration circuit similar to those shown by demes of the Desert Locust, *Schistocerca gregaria* (Rainey 1963, Baker 1978) and various birds and large mammals that inhabit the same areas. On such a model, the eastward flight of *Belenois* through Zimbabwe and Botswana in January is just one leg of a circuit, the northerly, westerly and southerly legs being performed by successive generations bringing the population back more or less to its starting point each year. As yet, however, for butterflies such migration circuits remain in the realms of conjecture.

Dingle (1979) and Able (1980) have expressed misgivings over my aguments concerning the direction-ratio, condemning them as group-selectionist. This they are not. Consider the three major arguments: the advantage of a preferred compass direction to the individual is that it travels across country in a straight line; it gains no advantage from its particular preferred direction except in the sense that if it flew in some other direction it would disrupt a presumed stable ratio and suffer a disadvantage; and as a parent, the individual gains from producing offspring with the appropriate direction-ratio to fit the stable situation.

If this seems like group selection, then compare the arguments with current theories of the evolution of the sex-ratio (Fisher 1930, Hamilton 1964*b*, Trivers 1974): the advantage to the individual of being one gender or the other in a gonochoristic (separate sexes) species is that it can reproduce; it gains no advantage from being male or female except in the sense that if it were the other gender it would disrupt a presumed stable ratio and suffer a disadvantage; and as a parent, the individual gains from producing offspring with the appropriate sex-ratio to fit the stable situation. For many species the optimum effort to put into each sex is 50:50 (Ch.24). In some circumstances, however, a sex-ratio other than equality is optimum and individuals are favoured that accordingly put different effort to the production of offspring of the two sexes. Where lies group selection in the arguments for

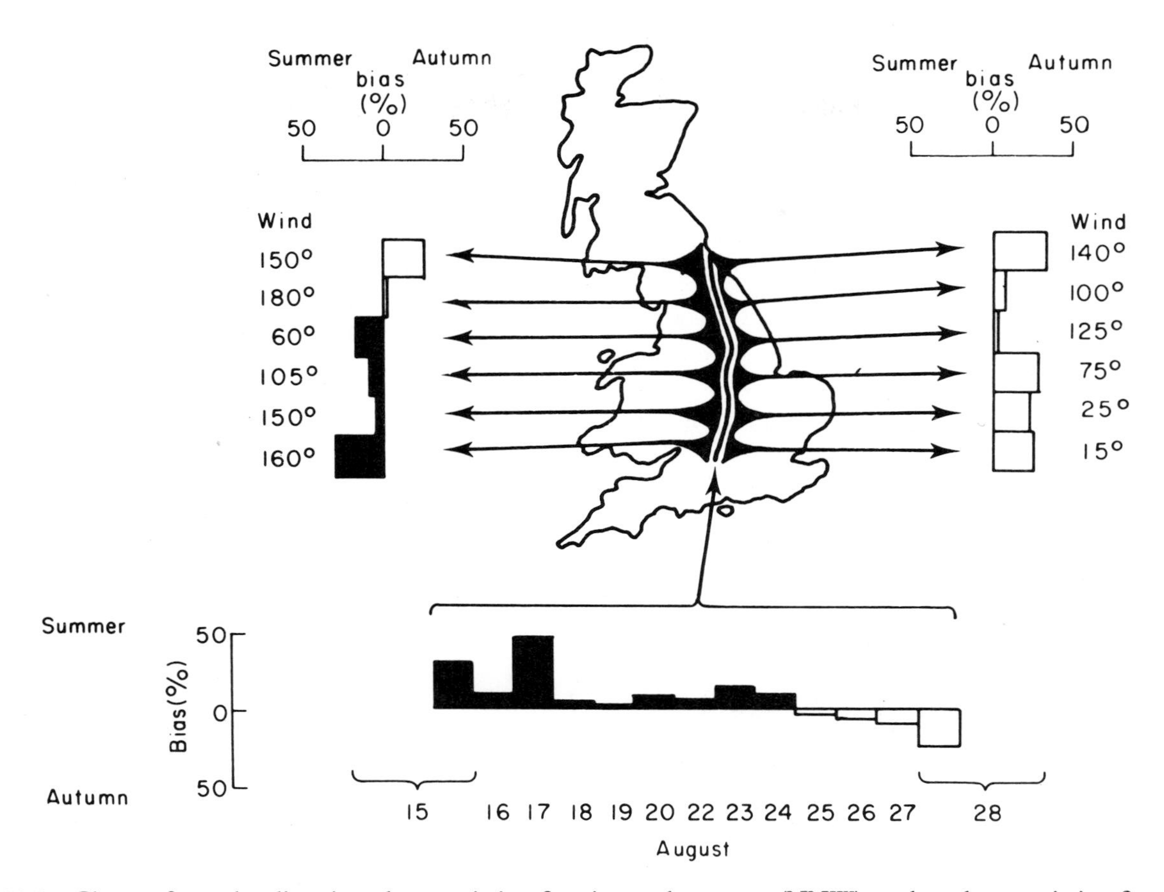

Fig. 26.6. Change from the direction characteristic of spring and summer (NNW) to that characteristic of autumn (S) during a single season by *P. rapae*. Histograms show bias toward summer (black) and autumn (open) directions during two transects between northeast and southern England and at a site in southern England. Wind shows the angle between downwind direction and the mean direction of the butterflies. For details see Baker (1978).

the evolution of either the sex- or direction-ratios?

The arguments for the evolution of the direction-ratio would predict that no matter what the preferred compass direction of an individual, it should produce offspring with the direction-ratio characteristic of the population. Rearing experiments with *Pieris rapae* support this prediction (Baker 1969).

The change from the spring and summer direction-ratio to the direction-ratio characteristic of autumn in the Small White cuts across the flight season of the autumn generation. This change in the preferred flight direction of individuals takes place within the space of a few days in any one place (Fig. 26.6) and has been shown to be a response to the length and temperature of night as experienced by the adult butterflies (Baker 1968*a*, 1978). The result is that a zone moves gradually southwards across Europe in August and September within which Small Whites change from their summer to autumn directions (Fig. 26.6; Baker 1969, 1978). Similar zones, moving south at different times and rates have been indicated by observations of *Colias croceus*, and various nymphalines (Baker 1969).

How to Go: Orientation

When an individual butterfly sets off across country between suitable habitats, it should adopt its particular preferred compass direction. We expect most such individuals as they migrate to be continuously on the look-out for the next suitable habitat. A few, members of the autumn generation of some temperate species or of a variety of arid zone tropical species, emerge as adults with a predisposition to migrate to a specific destination with the maximum economy of time and/or energy. Optimum orientation mechanisms should vary according to which of these categories the butterfly belongs.

Most butterflies fly when the sun shines, presumably for reasons of thermal efficiency. It has been shown for a variety of European butterfly species travelling across country (*P. rapae*, *P. brassicae*, *P. napi*, *Maniola jurtina*, *Aglais urticae*, *Inachis io*; Baker 1968*a*,*b*, 1969, 1980) that orientation is relative to the sun's azimuth. There seems to be no correction for the movement of the sun across the sky during the day (Fig. 26.7). For *P. rapae* and *Anthocharis*

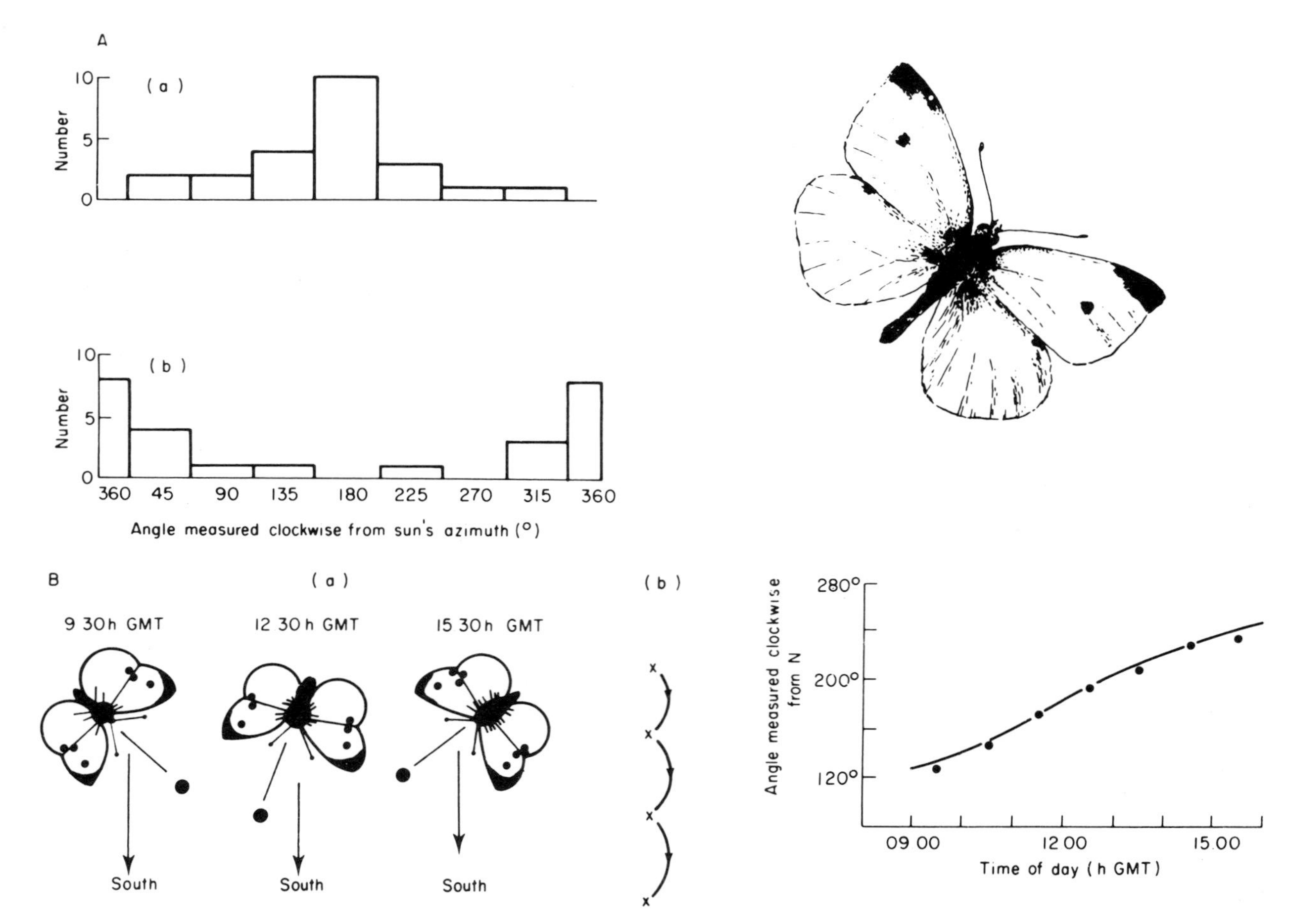

Fig. 26.7. Sun orientation without time compensation by *P. rapae.* Graph (lower right) shows the shift in sun's azimuth (line) with time of day and the mean compass direction of migrating Small Whites in autumn. A. individuals have a preferred angle of orientation to the sun's azimuth and when flying away from (a) or towards (b) the sun's azimuth, tended to fly in the same direction when released hours later or the next day in a different field; B: reconstruction of an individual's compass orientation and track. On this evidence, individuals change their compass orientation with time of day (a) as the sun (•) moves across the sky. The result over three days (b) is a curved track between overnight roosting sites (x) (from Baker 1978).

cardamines it has also been shown that each individual has its own preferred angle of orientation to the sun's azimuth and that this preference persists at least overnight (Baker 1969). The result is that each individual of these butterflies travels in slightly different directions at different times of day (e.g. North West in the morning, North East in the afternoon; Fig. 26.7).

Able (1980) has expressed doubts that sun orientation without time compensation of this type exists in butterflies. However, Kanz (1977) has demonstrated that Monarchs show a similar orientation mechanism and this also appears to be the case for a hermit crab (*Pagurus longicarpus*; Rebach 1978) that migrates a kilometre or so. A similar orientation technique, but with respect to the moon and stars, has been demonstrated for a moth, *Noctua pronuba* (Sotthibandhu & Baker 1979).

Most of the butterflies for which sun orientation without time compensation has been demonstrated are of the category that we should expect to be continuously vigilant for new habitats as they migrate. The gradual veering of track direction with time of day (Fig. 26.7) that comes from a lack of time compensation should not disadvantage a butterfly that needs only to avoid recrossing areas that it has already visited. Such a track should not be adequate, however, for a butterfly that is predisposed to migrate with the greatest economy to a particular area before beginning to search for suitable habitats. The best track for such a species should be the straightest route from place of emergence to the destination zone. Time compensation would seem to be an essential economy if sun orientation were to be used. Unfortunately, the critical data for those temperate and tropical species that may fit this category are not yet available.

The Monarch is perhaps the best-known of all butterfly travellers and its seasonal re-migrations are known in some detail (Fig. 26.8), thanks largely to

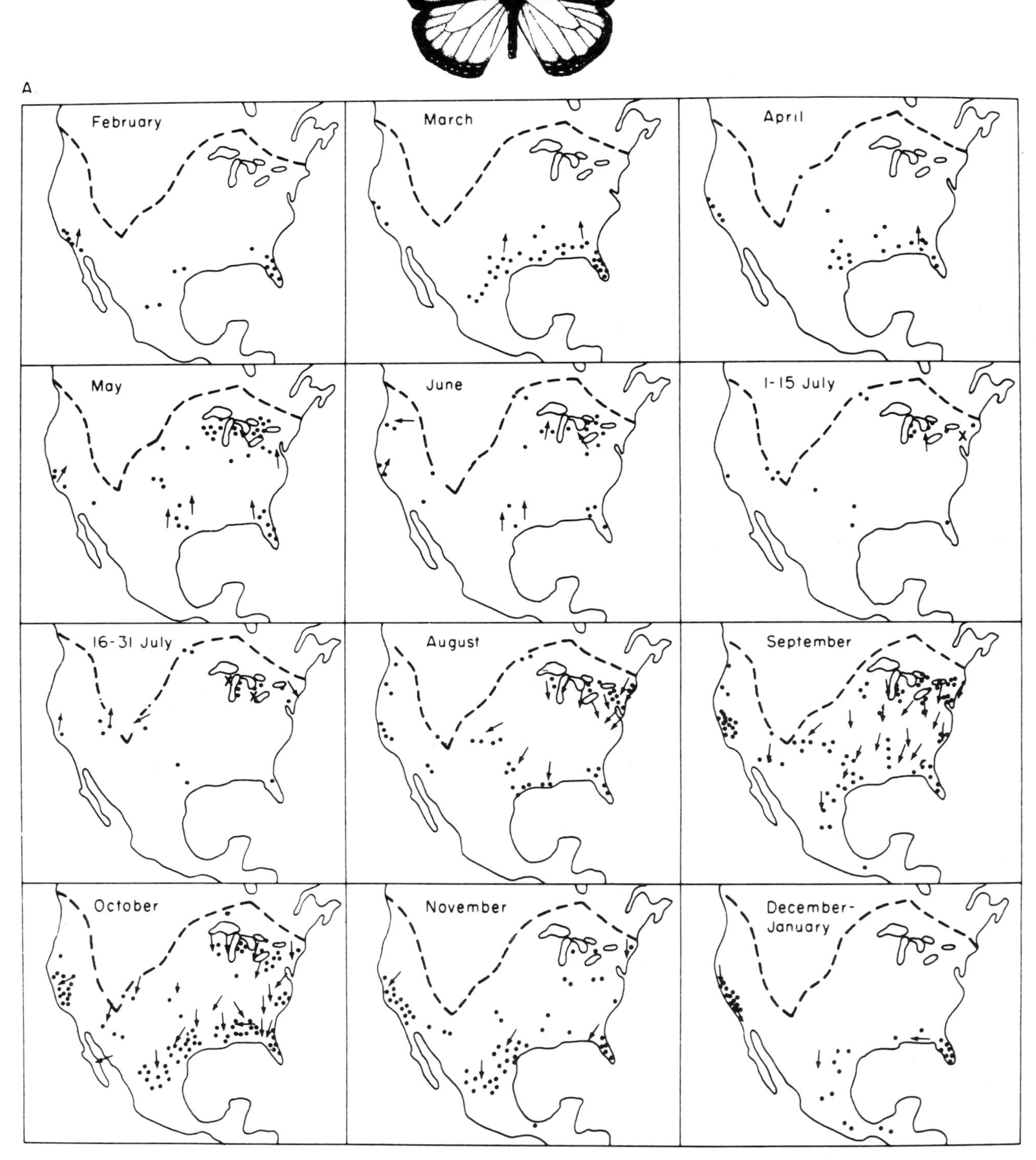

Fig. 26.8. Observations of occurrence (•) and flight directions (→) of North American *D. plexippus,* in each each month of the year. (From Baker 1978, compiled from data presented by Urquhart 1960.)

the tireless efforts of Urquhart (1960) and his helpers. My analysis of Urquhart's MRR data (Fig. 26.9) illustrates strikingly the way that fall Monarchs from the Great Lakes region set off at a fast rate in August and September and do not slow down until they have travelled 2000km or so and reached the zone that will in winter be south of the polar front. We should expect these 2000km at least to be travelled with the

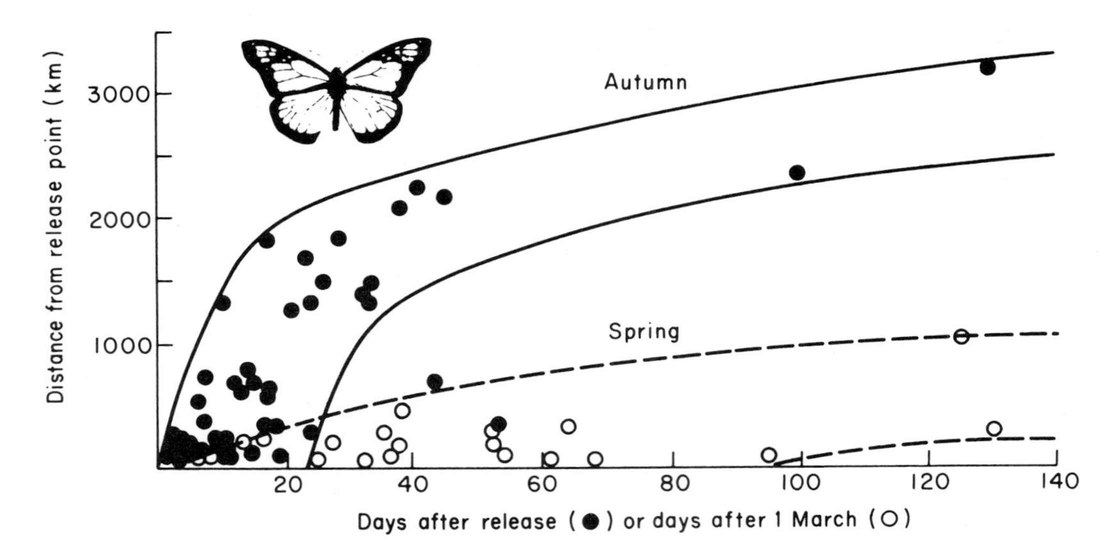

Fig. 26.9. Distance of recapture of North American *D. plexippus*, tagged in the Great Lakes Region (●) or California (○). (From Baker 1978, compiled from data presented by Urquhart 1960.)

greatest economy. Time compensated sun orientation seems an inevitable adaptation, as it does for arid tropical species migrating to a rainfall zone.

Schmidt-Koenig (1979), assuming that Monarchs from the eastern United States are migrating to overwinter in the Sierra Madre Occidentale in Mexico (after Urquhart 1976), has shown that even as far north as Ithaca, New York, migrating Monarchs orient to the southwest, roughly in accordance with the great circle route to the wintering area. He did not report on a lack of influence of time of day, though such a lack was implicit in his results. Such a lack has been confirmed for *Belenois aurota* in Botswana, by an observer primed carefully to look for the small shifts that would occur if sun orientation without time compensation were being used (K. Adams *in* Baker 1978). The butterflies flew precisely to the East throughout the day.

We do not know if arid tropical migrants orientate by the sun's azimuth but, if they do, they show time-compensation. We have some indication that Monarchs show sun orientation (Kanz 1977), but have conflict in that Kanz's results suggest the absence of time compensation, whereas Schmidt-Koenig's results suggest its presence. The conflict may not be intractable.

I have argued that temperate butterflies that overwinter as adults and produce an autumn generation that migrates long distances towards the equator will be polymorphic for migratory behaviour. A proportion should show non time-compensated sun orientation, migrate only tens or hundreds of kilometres, hibernate, and produce the earliest spring butterflies the following year (Baker 1972*b*, 1978). The remainder should show time-compensated sun orientation, migrate thousands of kilometres, remain mobile and feed during winter, and reproduce *en route* north the following spring. Red Admirals and Painted Ladies in Europe and Africa, together with the Monarch in America, could all fit into this category. A hint of polymorphic orientation behaviour in the European Red Admiral has been found (Baker 1978). Perhaps the same polymorphism could explain the apparent conflict in the results for Monarch orientation obtained by Kanz and Schmidt-Koenig.

Most animals that are known to have a sun compass are now also being found to have a magnetic compass that they can use at times when their celestial compass is not helpful because of cloud, or need to calibrate their sun compass (Schmidt-Koenig 1979, Able 1980, Baker 1981). We now know that moths use the moon for orientation on nights when the moon is visible, and almost certainly use stars on clear nights when the moon is not shining (Sotthibandhu & Baker 1979). On overcast nights they use a magnetic compass (Baker & Mather 1982). It can only be a matter of time before a magnetic compass is demonstrated in a butterfly (see Jones & Macfadden 1982), thus realizing one of C. B. Williams' earliest speculations on the orientation of migrating butterflies (Williams 1958; see also Moffat 1902).

How to Go: Economy

A butterfly flying across country searching for the next suitable habitat should place a premium on efficient scanning of the ground over which it is travelling. If it travels over the ground too fast it is likely to miss suitable habitats that pass at the limits of visual and/or olfactory range. On the other hand, if it travels too slowly it covers less ground than is possible per unit time. The trade-off between these two factors should be to travel across country, rather than through the air, at a relatively constant speed. Such a butterfly should compensate for the influence of wind, adjusting its orientation as well as its height and speed of flight to achieve a relatively constant cross-country motion. Evidence for such compensation was obtained by Nielsen (1961) in his work on *Ascia monuste* in Florida. By dying large numbers of butterflies red and then setting observers along the expected migration track, Nielsen demonstrated that *A. monuste* maintained a relatively constant speed across the ground of 10–15km/h over a wide range of wind speeds and directions. The butterflies flew low in the shelter of dunes and other features when the wind was strong and higher but often more slowly through the air when the wind was light. On quiet days most migrating butterflies in temperate regions fly relatively high above the ground, increasing the area perceived. On windy days, they fly low, taking maximum advantage of reduced wind speed near the ground but at the cost of a reduction in the area perceived.

Butterflies predisposed at emergence to cover a particular distance in a particular direction should experience different selective pressures. The premium is on behaviour that allows them to cover the distance to their destination with the maximum economy of time or energy. In migration, an economy in time is not necessarily an economy in energy and some trade-off is necessary. For a short-lived animal, such as a butterfly, the premium is likely to be shifted more towards economy of time, whereas for a long-lived animal, such as a bird or mammal, the premium may be shifted more towards economy of energy. Either way, the reaction to wind should be to travel at the height of the most favourable winds for travel to the destination zone. On days with a following wind this height will be high above the ground. On days with a head or side-wind, it will be low over the ground. Observations of fall Monarchs in the USA (Urquhart 1960, Schmidt-Koenig 1979) and various tropical butterflies (Williams 1958, 1976) flying low against headwinds and rising up into the air with tailwinds suggests that these butterflies are taking advantage of the wind consistent with economical travel to a particular destination zone. The behaviour contrasts with that of temperate 'searching' butterflies just described that adjust height in response to the speed, not the direction, of wind.

One other economy has been described for migrating Monarchs in fall (Gibo & Pallett 1979). On days with a tailwind, the butterflies used both lift, produced by winds blowing up slopes, and thermals to travel across country by soaring and gliding. When the weather is favourable, soaring is the main mode of migration, heights of at least 300m above the ground being achieved. The result is an enormous economy in distance travelled per unit of energy (and perhaps time).

Whether butterflies are continuously alert for their next suitable habitat or are economically migrating to a particular destination zone, they compensate for any tendency of the wind to displace them from their preferred track. C. B. Williams (1958, 1976) showed convincingly that butterflies maintain their preferred compass orientation with winds from all directions, and his observations have been confirmed visually many times (Nielsen 1961, Baker 1968*a*, 1969, 1978, Schmidt-Koenig 1979). Final confirmation of such behaviour of butterflies can be found in radar studies (Schaefer 1976). Walker and Riordan (1981) concluded that there were no data, for any butterfly, to support the hypothesis that upper air, or synoptic-scale wind systems regularly determine the direction of migration. The only investigation not to conclude that butterflies control their own track was the MRR study of Roer (1968, 1969) on *Algais urticae* and *Inachis io*. Although his results are consistent with downwind displacement, however, they are also consistent (Baker 1978) with the more abundant evidence that butterflies control their track and only migrate with the wind when downwind and preferred compass directions coincide. Even then, as already discussed, temperate migrants in the 'search' category resist being carried too fast downwind by reducing the height above ground at which they fly.

Discussion

Since the initial impetus given to the study of the lifetime track of butterflies by the pioneering work of C. B. Williams, the field seems to have entered a period of relative quiet. Paradoxically, Williams (*pers. comm.*) felt just before his death that the marvellous demonstration that some Canadian Monarchs over-winter in highland Mexico (Urquhart 1976, Urquhart & Urquhart 1978) had for many people removed much of the mystery and excitement from the study of butterfly migration, the extreme case having seemingly been proven. Yet there are so many questions about butterfly lifetime tracks left unanswered.

We still have little idea of the length and form of the lifetime track of most species, and it is essential that we determine soon which methods of study give reliable results. Surprisingly, studies of how butterflies stay, as in home-range and territorial behaviour, post-dated by decades studies of how butterflies migrate and there is still a need to apply to butterflies the techniques normally reserved for the ranging and defensive behaviour of mammals and birds. Despite their late initiation, such sedentary study is far easier than the study of cross-country movements. Modern telemetric techniques are still of little use for butterflies and the difficulties involved in following individuals seem enormous. MRR programmes that then rely primarily on citizen recapture are labour intensive and slow to produce results. The rewards, however, are high and Urquhart's lifetime work on the Monarch butterfly eventually produced the exciting conclusion for which we had all hoped. Is the North American Monarch unique or do other species of butterfly (or the Monarch elsewhere) show similarly long-distance movements? We might predict that comparable studies of species such as *Vanessa atalanta* and *Cynthia cardui* could be equally exciting and productive. Studies of *Danaus plexippus* outside of North America have begun (e.g. in Australia, Smithers 1977; see also Ch.17).

At the theoretical level, there is still much to be done on the evolution of direction-ratios and the consequences to populations of the behaviour of individuals. As far as orientation is concerned, we still await evidence for time-compensated sun, and magnetic, compass orientation during cross-country migration by butterflies. However, theoretical speculation and the study of orientation are relatively easy and the next decade must surely see progress in answering the remaining questions in these areas. The future looks less encouraging, however, for progress over the basic but vital questions concerning the true length of butterfly lifetime tracks. Half a century ago, there was the figure of C. B. Williams to prevail over sceptics and generate in others the fervour, excitement, and dedication necessary to establish (as did Nielsen, Roer and, of course, Urquhart) that long-distance travel by butterflies was a reality. Now, there is no comparable figure and the fervour seems to have dissipated. Let us hope both conditions are only temporary.

Acknowledgements

I thank Hodder & Stoughton for permission to use figures from my book (Baker 1978) and Josephine Martin for drawing the butterflies. Mrs Williams gave me inimitable hospitality during my last two meetings with C.B. to discuss butterfly migration. May Hurd typed the manuscript.

27. *Experimental Studies on the Evolution of Seasonal Polyphenism*

Arthur M. Shapiro

Department of Zoology, University of California, Davis, California

I suppose most of us, who are interested in Butterflies, are now so familiar with the phenomenon known as Seasonal Dimorphism, that we take it for granted and seldom pay much attention to it. . . .

N. D. Riley (1925)

This is the story of an attempt to work out the genetics of the epigenetic system controlling seasonal polyphenism in a group of butterflies. I view it as a noble failure: noble insofar as it was a difficult and important venture hitherto unattempted, a failure insofar as the resulting data gave no good match to theory, without underscoring inadequacies in the theory.

The experimental study of seasonal polyphenism in butterflies goes back to the mid-nineteenth century, when various British and European workers, shortly to be followed by Americans, began putting larvae and pupae in ice-chests. To the physiological mechanisms sketched out by 1900, photoperiod was added several decades later—necessitating a sweeping reassessment of prior work done under uncontrolled light regimes (Shapiro 1976*a*). In the past few years redundancies and 'fail-safe' systems have been found to be built in to the control of seasonal polyphenisms, and efforts have begun to construct a predictive theory of such control (Hoffmann 1978, Shapiro 1979*a*).

Throughout most of the neo-Darwinian era, great attention has been paid by evolutionists to polymorphisms of all kinds, while polyphenisms—and indeed, environmentally-influenced variation in general—languished in what has in American politics been called 'benign neglect'. This is not to say that such variation was completely ignored. It was of paramount importance to Waddington (1957, 1975) and in the models of the ecologist Richard Levins (1968), who was influenced by Waddington. It is hardly a secret that a reaction against the panselectionism of the neo-Darwinian synthesis is in motion; one of its beneficial consequences is a revival of interest in the role of epigenetic systems in the generation of evolutionary novelty. The atomistic models common in sociobiology have found a counterpoint in the revival of Goldschmidtian notions of saltatory evolution (Ch. 14) via changes in the regulation of embryonic development. Most of what has been published along these lines has been in the 19th Century tradition of bold theorizing and 'arm-waving'. Little in the way of experimental data has appeared. Might it be that the butterflies, which have contributed so much to organismal and population biology in general, and to the neo-Darwinian synthesis in particular, have something to contribute here as well?

Cyclic polymorphisms, such as the classic ladybird beetle (*Adalia bipunctata*) (Timofeef-Ressovsky 1940) and *Drosophila pseudoobscura* inversion (Dobzhansky *et al.* 1966), are a clumsy way to deal with an environment which is both seasonal and predictable. At any given moment a significant proportion of the population will be ill-adapted; there is a built-in 'genetic load'. How big a load? This depends basically upon the relationship between generation time and environmental periodicity. An organism with two generations a year which faces two seasonal selective regimes a year will be permanently out of phase: the genetic makeup of the 'summer' generation reflects selection in a 'winter' regime and *vice versa*—such an organism is like a military general always planning for the last war. The shorter the generation time relative to the environmental periodicity, the closer the organism may track its environment, and the less the load at any given time. However, polyphenism can reduce the load and the time lag of response to zero, if we assume that all individuals are equally competent to make correct developmental decisions and that the cue(s) they respond to in the environment is (are) trustworthy.

The Biology of Butterflies
0-12-713750-5

Polyphenisms, then, can be viewed as highly adaptive evolutionary responses to environmental seasonality, and we may wonder how they evolve. Bradshaw's (1973) verbal model of the evolution of dormancy in the dipteran *Chaoborus* rightly emphasized the reliability of the cue; it is on this basis that photoperiod rather then temperature tends to be the primary cue in both polyphenism and diapause in middle latitudes. In those environments where photoperiod fails as a reliable cue (montane habitats in the Colorado Rockies, Hoffmann 1978; coastal fog belts in California, Shapiro 1977) temperature generally replaces it. These observations, which are consistently in line with Darwinian expectations, further emphasize the adaptiveness of the developmental switch.

Shapiro (1976*a*) proposed that seasonal polyphenisms could evolve from pre-existing polymorphisms, probably via duplication of the relevant loci followed by selection of modifiers to turn the alleles on or off in appropriate environments. I also proposed that polyphenism might arise without a pre-existing polymorphism for a major gene, if hitherto non-adaptive developmental accidents (the 'morphoses' of Schmalhausen 1949) provided phenotypic raw material upon which selection could act in novel environments (the Baldwin effect or Waddington's 'genetic assimilation'). The well-known heat- and cold-shock phenotypes of *Nymphalis* and *Vanessa* were cited as examples.

These notions have been very difficult to test in the laboratory, using familiar Holarctic species. There seem to be two direct experimental approaches. One can try to convert a shock phenotype into a seasonal one by artificial selection, in effect duplicating with a lepidopteran the classic genetic assimilation experiments of Waddington, Bateman, and Milkman. In this, one is handicapped by the unsuitability of nymphalines as laboratory animals, and by the weakness of most of the extreme aberrants, making them unfit for breeding. After a frustrating attempt along these lines, I turned to the far less satisfying but more tractable approach of seeking differences in canalization of the normal phenotype among conspecific stocks from different climates (Shapiro 1981*a*, 1981*b*). However, this is only an indirect, correlational approach. The other direct experimental approach is to study the genetic basis of the ability to switch phenotype adaptively—in effect, the genetics of the epigenetic system. A pair of entities are needed, one neatly polyphenic and the other not, which are sufficiently closely related to be hybridized readily and carried through at the least the F_2 and various backcrosses without serious loss of fertility, vigour, or sexual balance, so the formal genetics can be done. The likelihood of finding such a pair of Holarctic pierids is meagre: the capacity to respond to environmental cues is very generally distributed within the major species-complexes, and where there is a marked contrast in polyphenic potential the level of genetic incompatibility may be high (cf. Shapiro 1980*a*). The opportunity to develop this approach materialized when I put into cultivation members of the Neantarctic-Andean *Tatochila sterodice* species-group, on which the rest of this paper is based.

The *Tatochila sterodice* Species-Group

> If it has taken 30 or 40 years to learn something of the caterpillar of *autodice*, which is the commonest species of the group, imagine when we will begin to know that which we lack, and which is so necessary to fill in the gaps in the data on *orthodice*, *pyrrhomma*, *argyrodice*, etc., etc.!
>
> E. Giacomelli (1915)

Tatochila is one of nine endemic Andean genera, comprising some 40 species, which with the Asiatic *Baltia* form a compact subtribe of the Pierini, near to *Pieris (sensu lato)* itself. The genus was monographed in 1959 by Herrera & Field. At that time the life-histories of all the species except *T. autodice* and *T. blanchardii* were either unknown or unpublished, and even they had not been figured. Herrera & Field recognized five species-groups, one of which corresponds to what I am calling the *sterodice* species-group (*sterodice* was incorrectly called *microdice* by Herrera & Field; Ackery 1975*a*). The group consists of *sterodice*, with four named subspecies (from north to south: *arctodice*, *macrodice*, *sterodice*, *fueguensis*), and two monotypic species, *vanvolxemii* and *mercedis*. *T. sterodice* in its various avatars has the widest latitudinal distribution of any South American butterfly, from central Colombia (3°N) to Tierra del Fuego (55°S), in and near the Andes; *T. vanvolxemii* occurs in grassland, scrub, and desert biomes in lowland Argentina while *T. mercedis* is virtually restricted to the Mediterranean-climate belt of central Chile. All are now associated with man, feeding on the introduced, weedy crucifers abundant in urban and agricultural environments. The life-histories of most are figured and described by Shapiro (1979*b*).

Tatochila s. sterodice and *T. s. macrodice* are multivoltine but monophenic. *T. vanvolxemii* is strongly seasonally diphenic in the male, but the female is invariant. *T. mercedis* shows continuous seasonal variation in both sexes. In these two species the polyphenisms are largely under temperature control, but there is a photoperiod component at least in the former. Studies of these polyphenisms were undertaken by Shapiro (1980*b*) in the hope of elucidating the phylogeny of the Andean pierines by differentiating between parallelism and convergence (*vis-à-vis* the Holarctic pierines) in their evolution. I concluded that the polyphenisms of these two

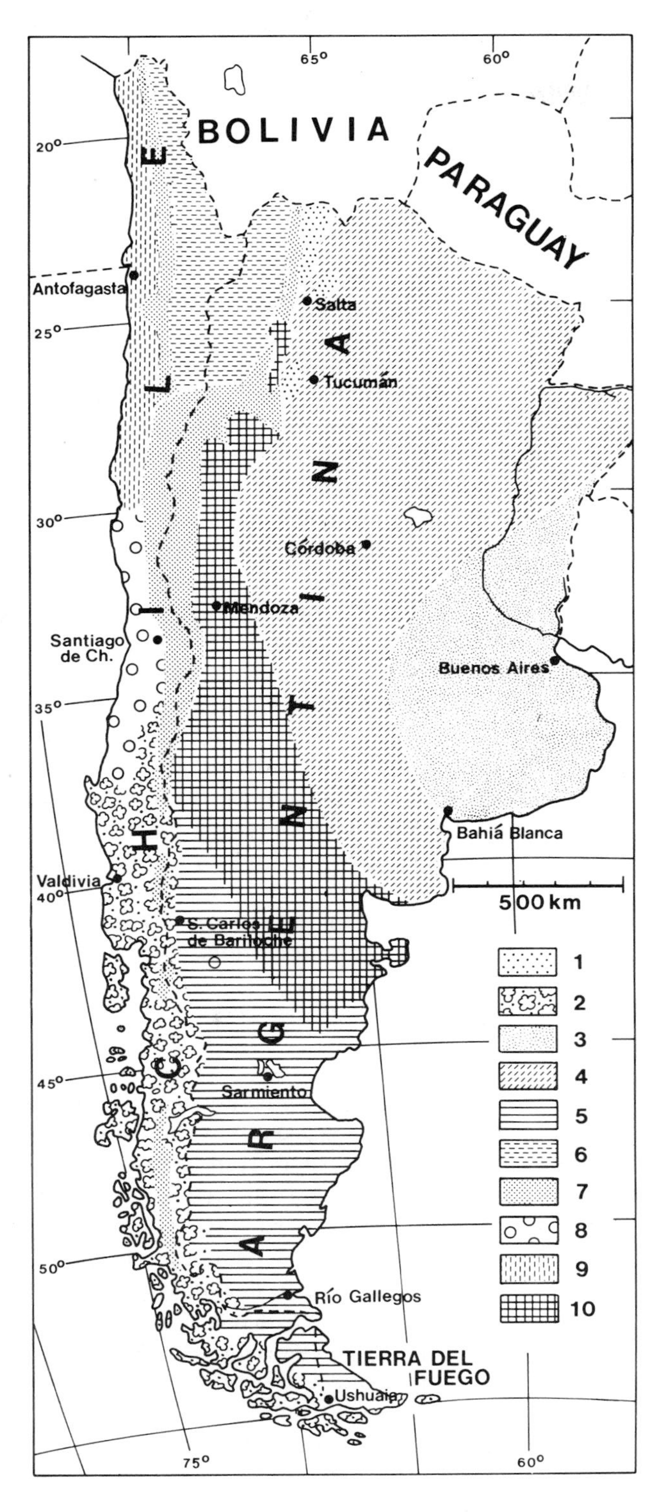

Fig. 27.1. Major biomes of Argentina and Chile, redrawn after Madsen *et al.* (1980). 1. Subtropical forest; 2. subantarctic (mostly *Nothofagus*) forest; 3. pampa; 4. xerophytic woodland and scrub (espinal); 5. Patagonian steppe; 6. puna; 7. Andean boreal and nival zones; 8. Chilean Mediterranean scrub (matorral) and valley grassland; 9. Atacama desert; 10. monte desert and arid montane (prepuna). Note how the hybrid zone shown in Fig. 27.2 nests at the convergence of biomes 2, 5, and 10, north of Bariloche.

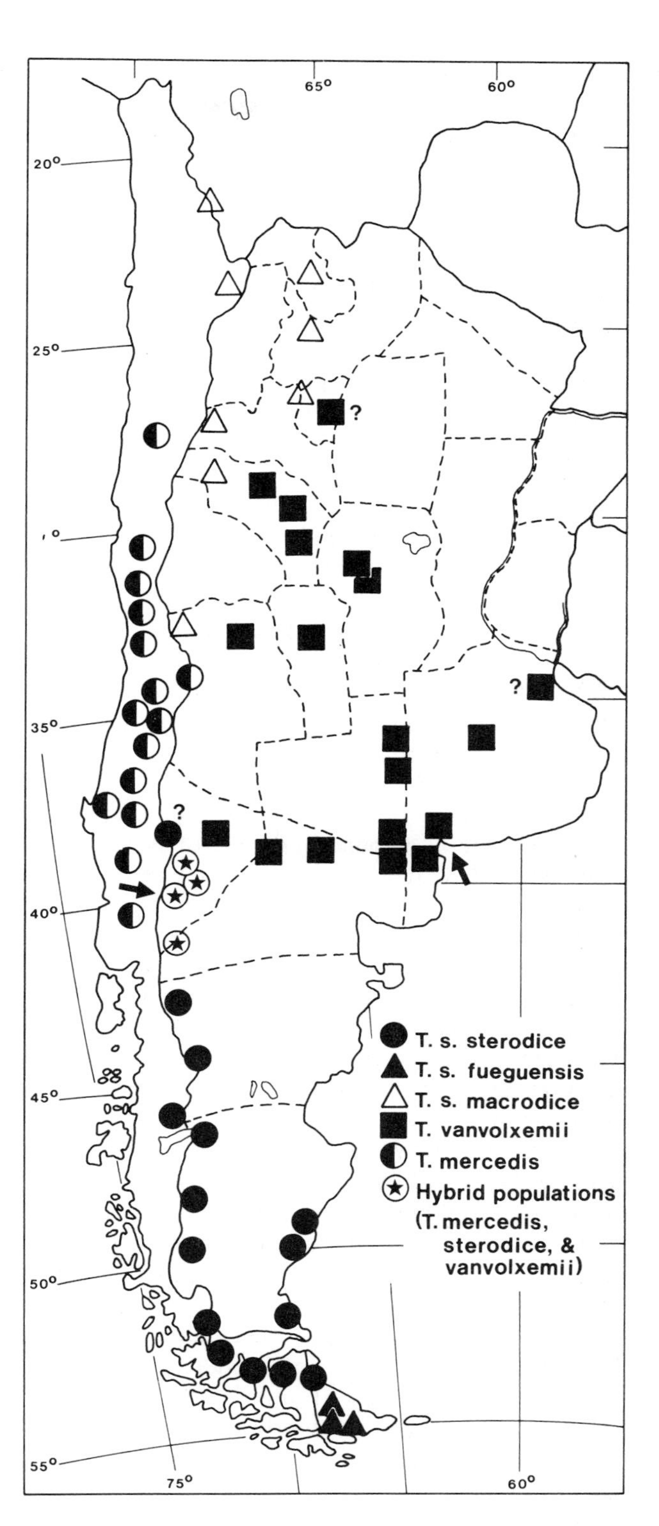

Fig. 27.2. Distribution of the *Tatochila sterodice* species-group in Argentina and Chile, incorporating 1981 collections. Arrows indicate the source populations for hybridization experiments described in text (*sterodice* from San Martín de los Andes, *vanvolxemii* from Bahía Blanca). Question marks indicate unconfirmed peripheral records.

species were recently and probably independently derived from a *sterodice*-like ancestor, and represented trends convergent to their Holarctic counterparts. The lightly-marked estival phenotypes were interpreted as derivative from the fully patterned vernal ones, which are equivalent to the invariant *sterodice* or *macrodice* patterns.

The possibility of natural hybridization within this group was recognized by Herrera & Field, who had limited distributional data and especially poor phenological information. What they took to be hybrids between *vanvolxemii* and *macrodice* were vernal phenotypes of the former; one such was taken hundreds of kilometres from the nearest possible contact zone. Hybridization, however, still seemed a real possibility. The genitalia of both sexes of all but *mercedis* are indistinguishable, and the assignment of rank (species or subspecies) to the taxa in this group was based on arbitrary and inexplicit notions of phenotypic distinctness. Shapiro (1980*b*) mapped the distributions of the taxa in the Southern Cone of South America, as then understood. This map showed all of them to be allopatric, or potentially parapatric in certain areas.

During the 1981 austral summer I travelled extensively in the 'Andean Frontier' region of northwestern Patagonia in the provinces of Río Negro and Neuquén, Argentina, where I thought there was a possibility of contact among two or more taxa. The reason for selecting this area for study is immediately apparent from Fig. 27.1—the convergence of three biomes north of Bariloche. The situation, which will be reported upon in detail elsewhere, bears so directly on the genetic studies to be presented herein that a brief summary seems necessary.

Figure 27.2 shows the revised distributions of the species-group with 1981 data incorporated. The zone of contact and intergradation discovered extends from Aluminé in the north to San Carlos de Bariloche in the south, in a series of localized urban-weedy populations within the mesic ecotone between the humid Andean forest and the Patagonian steppe. Large field samples were collected at Aluminé, San Martín de los Andes, Junín de los Andes, and Bariloche; in all these places (but least at Bariloche) there is widespread intergradation between *sterodice sterodice* and *mercedis*; both 'species' and intergrades may be reared in the progeny of single females. The hybrid zone corresponds latitudinally to the distribution of low passes in the lake district, where the otherwise Chilean *mercedis* can cross into Argentina; its southward limit in Chile is about opposite Bariloche, and that is where its phenotypic influence stops. To the north the zone is truncated by increasing aridity and the increasing height of the mountains. *T.s. sterodice* ends near Aluminé and *T.s. macrodice* commences farther north in the more arid climate in the province of Mendoza. The Sierras de Catan Lil, Espinazo del Zorro and Cerro Chachil provide a seemingly impenetrable barrier to the northeast; beyond them, at Zapala, where the desert begins, the population is pure *vanvolxemii*.

As Fig. 27.2 indicates, *vanvolxemii* extends in the hot desert east of the zone of intergradation. It seems to travel up the river bottoms into the zone, reaching San Martín de los Andes and even Bariloche. It is not clear how much gene flow results. Many hybrid and recombinant phenotypes are too nondescript to stand out in the complex, variable hybrid-zone populations, but one phenotype that is recognizable was picked out by Shapiro (1980*b*: Figs 3-3 and 4-3) as a putative hybrid, apparently correctly.

For reasons to be discussed elsewhere, I interpret this as a secondary parapatric hybrid zone (Woodruff 1973), facilitated by the conjunction of low passes, mesic (intermediate, ecotonal) climate, and the abundance of introduced weedy host plants. The correct taxonomic relationship among *Tatochila mercedis*, *vanvolxemii* and *sterodice* is problematical. A strict application of the biological species concept might reduce them to subspecies of a widespread polytypic species. For the sake of stability, and because the genetic relationships are far from settled, I will continue to refer to them as if I did not know how readily they cross. The important point, however, is that they *do* cross and that they in fact meet the criteria spelled out above for a laboratory system in which to study the genetics of polyphenism.

Materials and Methods

Butterflies of the *Tatochila sterodice* species-group are easily cultured and mate readily in cages. Lines used in this study were established from the progeny of single females of *T. s. sterodice* collected at San Martín de los Andes, Neuquén, 13 Jan. 81, and *T. vanvolxemii* at Bahía Blanca, province of Buenos Aires, 22 Jan. 81. San Martín is solidly within the hybrid zone. A long series taken there contains *sterodice*, *mercedis*, and recombinant phenotypes, and one apparent *vanvolxemii*. Of six broods reared from as many wild San Martín females, only the one used for hybridization consisted entirely of *sterodice*-phenotype animals; even here, some signs of introgression from *mercedis* were detected. Of course it would have been preferable for these experiments to use pure *sterodice* from south of the hybrid zone, but these were not available at the time. Bahía Blanca, on the coast some 900km from San Martín, supports a pure population of *T. vanvolxemii*. Although the population at Zapala, just northeast of the hybrid zone, seems equally pure, I elected to use a remote one to maximize the genomic differences between the lines. Rearing was by the techniques of Shapiro

(1980*b*), using *Brassica kaber* as the food plant for most broods (*Lepidium virginicum* was used occasionally in summer). Three environmental regimes were employed: continuous light (24L), 27°C (80.6°F); 10L-14D, 23.9/12.8°C (75/55°F); and 10L-14D, 21.1/4.4°C (70/40°F). The last was intended to induce diapause, so sibs of parents could be mated with later generations. At the time of writing, many of the resulting pupae have not yet eclosed, and data based on this regime are excluded from the present paper. It was initially intended to use all three regimes in all broods, but when it became clear that numbers would be statistically limiting (since only males could be scored for polyphenism, the females being invariant), a mixed strategy was adopted whereby some broods were more or less evenly divided among them while others were reared primarily in one regime, with a few token individuals in one or both of the others.

Various schemes for quantifying the phenotype of the ventral male hindwing were considered, only to be abandoned. No reflectance method was capable of discriminating among *pattern* differences when these were balanced by pattern intensity; methods involving scale counts at different places on the wing and combined in various ways were too cumbersome given the numbers involved. Ultimately the males were classified 'by eyeball' into six phenotypic classes (Fig. 27.3)—the tried-and-true method used by Castle & Wright for scoring coat colour and pattern in rodents. The breakdown of the males from all experimental broods completed to date is given in Table 27.1. It will be noted that the essentially monophenic *sterodice* produces mostly class V males, whilst vernal phenotypes of *vanvolxemii* are often of class VI (a few of class VI+, here included in VI, have been produced, mostly with pupal chilling).

Since I had no idea what sort of inheritance to expect, I decided to attempt all the possible matings through the F_2 and backcrosses, looking for Mendelian ratios but, failing there, applying the Castle-Wright-Mather methodology to estimate the number of loci or 'effective factors' involved in the trait. The resulting theoretical peregrinations are described in the next section. It was not possible to obtain the backcross to *sterodice* until July 1981, when I had F_1 individuals and ex-diapause *sterodice* at the same time, so this cross is omitted from the calculations and tables.

It should be stressed that the genetics of switching ability *per se*, of the epigenetic system, is approachable only indirectly through the genetics of the phenotype itself. The modifiers setting the thresholds at which the alternate phenotypes are produced cannot be quantified using only three rearing regimes, though some idea of their magnitude might be obtained by comparing the genetic analyses for phenotype under the three regimes. It should also be stressed that for purposes of this analysis only the ventral wing pattern as shown in Fig. 27.3 is being considered; all other characters, including the ground-colour of the ventral hindwing, will be ignored. Ground-colour, however, is also at least in part a polyphenic trait.

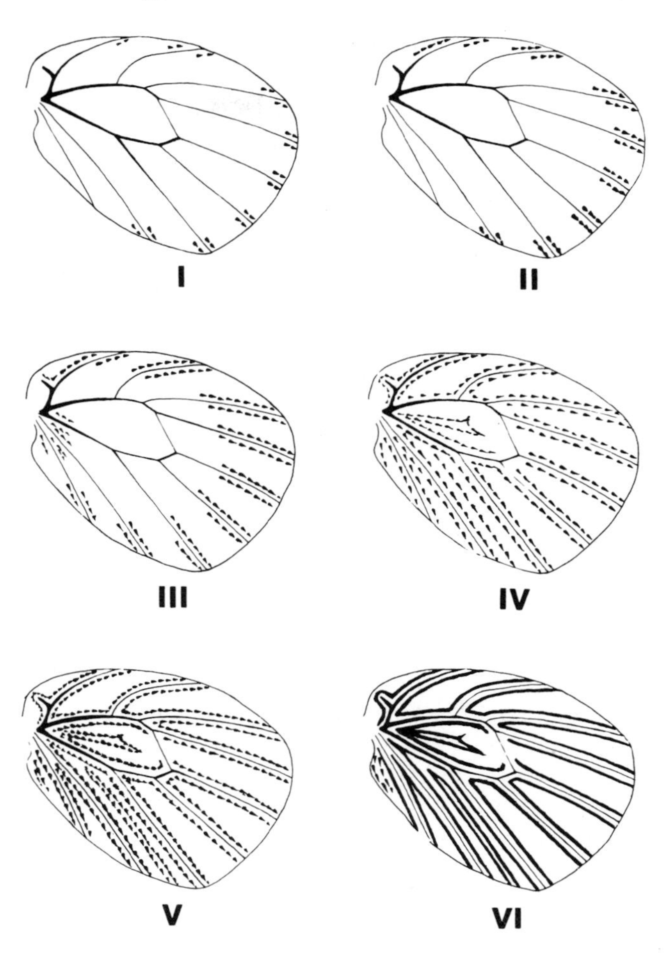

Fig. 27.3. Standards used for assigning male *Tatochila* to phenotypic classes.

Results and Non-Results

Two attempts were made to apply simple models to the phenotypic data. Neither was particularly successful. The most accessible phenotypic difference between *sterodice* and *vanvolxemii* is the extent of pattern in males reared under 24L,27°C. This regime produces light *vanvolxemii* and dark *sterodice* with essentially no overlap in phenotypes (Table 27.1). F_1 and F_2 individuals are phenotypically variable and exhibit essentially the entire range of parental phenotypes. As expected from essentially any model of genetic differentiation between *sterodice* and *vanvolxemii*, the F_2 exhibit greater phenotypic variation than the F_1. The recovery of parental extremes in the F_2 is clearly consistent with determination by a small number of loci. Their occurrence in the F_1 embarrassingly evokes classical

Table 27.1. Phenotypes of male *Tatochila* scored up to 16 July 1981.

Brood	Phenotypic classes I	II	III	IV	V	VI	TOTALS
valvolxemii 10L-14D 23.9/12.8°C	1	4	6	5	7	8	31
24L 27°C	18	16	1	0	0	0	35
sterodice 10L-14D 23.9/12.8°C	0	0	0	1	18	2	21
24L 27°C	0	0	0	2	11	2	15
F_1 no. 1 10L-14D 23.9/12.8°C	0	7	10	7	15	6	45
24L 27°C	2	3	8	4	4	0	21
F_1 no. 2 10L-14D 23.9/12.8°C	1	6	11	13	9	6	46
24L 27°C	1	0	1	0	0	0	2
F_1 no. 3 10L-14D 23.9/12.8°C	0	0	3	2	1	1	7
24L 27°C	1	12	38	11	2	0	64
F_1 no. 4 10L-14D 23.9/12.8°C	0	0	1	5	6	1	13
24L 27°C	0	0	2	0	0	0	2
F_1 no. 5 10L-14D 23.9/12.8°C	0	0	0	1	8	2	11
24L 27°C	0	0	1	2	3	0	6
F_2 no. 1 (male ex F_1 no. 5, female ex F_1 no. 3)							
10L-14D 23.9/12.8°C	1	0	0	5	9	6	21
24L 27°C	2	0	4	2	8	1	17
F_2 no. 2 (both ex F_1 no. 3)							
10L-14D 23.9/12.8°C	0	0	0	1	1	0	2
F_2 no. 3 (both ex F_1 no. 3)							
24L 27°C	1	0	0	2	0	0	3
F_2 no. 4 (male ex F_1 no. 5, female ex F_1 no. 2)							
10L-14D 23.9/12.8°C	0	2	0	0	0	1	3
Backcross (male V × female ex F_1 no. 1)							
10L-14D 23.9/12.8°C	1	3	4	8	1	0	17
24L 27°C	6	9	5	3	0	0	23
Backcross (male ex F_1 no. 5 × female V)							
10L-14D 23.9/12.8°C	0	0	0	0	3	3	6
24L 27°C	1	2	9	1	0	0	13

Notes: broods F_1 nos 1–4 were *sterodice* × *vanvolxemii*; no. 5 was the reciprocal. Three additional F_2 broods (nos 5, 6, and 7) produced only one adult each, all female. Although the *sterodice* line was from within the hybrid zone, there is no evidence of *mercedis* genes affecting polyphenic response in it. This was checked by crosses involving all three taxa (F_1 broods nos 6 and 7, and F_2 no. 8), not reported upon here.

nonexplanations involving large 'environmental effects' (not strictly applicable under laboratory regimes) and/or 'incomplete penetrance'. Both will, alas, be invoked below.

Given the variability of the F_1, a straightforward Mendelian analysis seemed precluded and resort was made to the methodology of quantitative genetics. The first approach used was to treat the phenotype distributions as a quantitative character and apply the classical procedure of Wright (Castle 1921, Wright 1952, 1968, Lande 1982) to estimate the minimum number of controlling loci. As a simplifying procedure, and to boost the numbers, replicate broods (including reciprocal crosses) were pooled. Table 27.2 shows the means and variances for the pooled F_1 and F_2 broods. A complete analysis was impossible due to the unavailability of the backcross to *sterodice*. As shown by Wright's original analysis and Lande's extension, a generally conservative estimate of the minimum number of loci underlying the difference in the parental lines can be obtained from

$$n_E = (\mu_{P_1} - \mu_{P_2})^2/8\sigma_S^2 \quad \text{(i)}$$

in which μ_{P_i} for i = 1,2 denotes the mean of parental strain i and σ_S^2 measures phenotypic variance due to

Fig. 27.4. Phenotypes of bred male *Tatochila*, ventral surfaces. Top row: left: *T. sterodice* (10L-14D, 23.9/12.8°C); centre: *T. vanvolxemii* (10L-14D, 23.9/12.8°C); right: *T. vanvolxemii* (10L-14D, 21.1/4.4°C). Second row: F_1 brood no. 1, all 10L-14D, 23.9/12.8°C. Third row: F_1 brood no. 2, all 10L-14D, 23.9/12.8°C. Fourth row: F_2 (male ex F_1 brood no. 3 × female ex F_1 brood no. 5): left and centre, 24L,27°C; right, 10L-14D, 23.9/12.8°C. Bottom row: backcrosses, male ex F_1 brood no. 1 × female *vanvolxemii*: left, 24L,27°C; centre, 10L-14D, 23.9/12.8°C; right, 10L-14D, 21.1/4.4°C. *T. sterodice* is seasonally invariant, while all the crosses shown were phenotypically flexible as in *T. vanvolxemii*.

Table 27.2. Pooled male phenotype data used for estimating the number of loci or 'effective factors', using the Castle-Wright method as modified by Lande. Long-day (24L, 27°C) regime only.

Grouping	Phenotypic classes						Total	Mean	Variance
	I	II	III	IV	V	VI			
vanvolxemii	18	16	1	0	0	0	35	1.51	0.32
sterodice	0	0	0	2	11	2	15	5.00	0.29
F_1 hybrids	4	15	50	17	9	0	95	3.13	0.88
F_2 hybrids	3	0	4	4	8	1	20	3.85	2.24

genetic differences rather than environmental effects. A standard formula for the latter is

$$\sigma_S^2 = \sigma_{F_2}^2 - \tfrac{1}{4}(2\sigma_{F_1}^2 + \sigma_{P_1}^2 + \sigma_{P_2}^2) \qquad \text{(ii)}$$

The fact, documented in Table 27.2, that the F_1 mean lies neatly between the parental means supports the additivity assumption that underlies formula (i). Applying (i) and (ii) to the data produces the estimate $\hat{n}_E = 0.92$. This is raised only to 1.1 by replacing (ii) with the less reliable estimator

$$\sigma_S^2 = \sigma_{F_2}^2 - \sigma_{F_1}^2 \qquad \text{(iii)}$$

which provides the largest estimate consistent with the data.

Unfortunately, no firm conclusion can be drawn from these small numbers. As emphasized by Wright (1968), these are *minimum* estimates and any estimated n_E value is consistent with *any* true number of controlling loci greater than or equal to the estimate. Nevertheless, the recovery of the extreme parental phenotypes in the relatively small number of surviving F_2 is suggestive of a relatively simple genetic basis. Their appearance in the F_1 can be ascribed to large 'environmental effects' (really decanalization) resulting from hybrid dysgenesis. To exploit these notions, a second analysis was attempted. Before proceeding to it, it should be noted that reciprocal differences apparently do exist in the female dorsal phenotypes in the F_1 and are suspected for the ventral hindwing of the males as well, so the pooling used in obtaining the estimates is perhaps on shaky ground. The n_E estimates for the 10L14D, 23.9/12.8°C regime were not calculated, but would surely be even smaller.

An alternative way to view the phenotypes in Fig. 27.3 and Fig. 27.4 is to differentiate only two types rather than six. Individuals in classes I–III are considered 'light'; those in IV–VI, 'dark'. Then a collapsed version of Table 27.2 can be approached cautiously with a simple genetic model for a dichotomous trait. Following the dubious example of human geneticists, a natural candidate is a one-locus model with intermediate penetrance. That is, we can suppose that there is a major gene governing pigmentation (at the light-temperature regime under consideration) at which *vanvolxemii* is monomorphic for a light allele A_1, and *sterodice* is monomorphic for a dark allele A_2. The appearance of both light and dark forms in the F_1 can be 'explained' by intermediate penetrance, reflecting the effects of various loci segregating in the background. Let h, the penetrance parameter, denote the probability that a heterozygote will be light. To test this simple model, we can use the F_1 and F_2 data to obtain separate estimates of h. The maximum likelihood estimates are 0.73 and 0.2 from the F_1 and F_2 data, respectively. This large discrepancy is extremely unlikely under the model proposed. Nevertheless, a one-locus interpretation of the phenotypic difference can be salvaged with a suitable story.

Naturally, *vanvolxemii* and *sterodice* are differentiated at many loci apart from the 'major gene' affecting ventral hindwing pigmentation in males (Fig. 27.5). The progressive decline in survivorship from the parental lines to the F_1 and F_2 (Table 27.3) demonstrates that several of these loci contribute to viability, since no pathogen appears to have been introduced during culture. Hence, a more careful accounting of the expected F_1 and F_2 genotype frequencies at the pigmentation locus must include not only potential pleiotropic effects on viability, but viability effects of other loci for which the species may differ genetically. In this way one can quickly generate many more viability parameters, describing hybrid breakdown, than organisms scored. In return, one obtains sufficient degrees of freedom to bring the observations into line with expectations. Although everything makes sense in the light of a sufficiently general model, the result is quite unilluminating.

Much of the trouble in handling these data comes from the peculiar phenotypic distribution in the good F_2 brood, from which the 'light' classes are almost absent. There is in fact reason to believe selective mortality did occur in this brood, albeit for a different phenotypic character which may or may not have been somehow coupled to ventral hindwing pattern: a substantial number of larvae and pupae were abnormally melanic, and all of these died. On the other hand, the entire major gene model is shaky at

Fig. 27.5. Dorsal and ventral surfaces of male *sterodice* × *vanvolxemii* bilateral mosaic reared in F_1 brood 2 (10L-14D, 23.9/12.8°C), presumably originating from a doubly fertilized binucleate egg. The two ventral hindwings are in classes IV and V; note the dissimilarity in virtually all pattern characters, as well as in size. This specimen clearly demonstrates the paramount importance of genetic (as against microenvironmental) determinants of pattern. Specimens resembling both sides were common in the brood.

Table 27.3. Survivorship of *Tatochila* broods in Table 27.1.

Brood		Ova laid	Hatched	Pupated	Eclosed	Survival (%)
vanvolxemii	10L	69	69	63	56	81.2
	24L	89	86	72	67	75.3
sterodice	10L	62	59	47	38	61.3
	24L	40	38	28	25	62.5
F_1 no. 1	10L	121	119	94	87	71.9
	24L	98	98	47	33	33.7
F_1 no. 2	10L	141	138	123	98	69.5
	24L	24	21	8	3	12.5
F_1 no. 3	10L	30	30	23	15	50.0
	24L	183	171	133	124	67.8
F_1 no. 4	10L	113	111	57	21	18.6
	24L	56	55	21	6	10.7
F_1 no. 5	10L	45	45	33	25	55.5
	24L	34	34	19	14	41.2
F_2 no. 1	10L	104	103	65	35	33.7
	24L	81	74	40	25	30.9
F_2 no. 2	10L	85	75	24	8	9.4
F_2 no. 3	24L	60	54	9	5	8.3
F_2 no. 4	10L	44	33	9	4	9.0
Backcross	10L	72	70	49	31	43.1
	24L	68	66	52	46	67.6
Backcross	10L	45	45	20	6	13.3
	24L	110	104	85	20	18.2

This is not a final tally, as it excludes diapausers, individuals eclosed since 16 July 1981, and individuals reared under other regimes. It includes both sexes since ova and larvae cannot be sexed.

best, especially given the propensity of *vanvolxemii* to produce a few dark males in large broods reared under estival-phenotype-inducing conditions (a propensity not expressed in the particular brood reported here, but often observed in other rearings).

Discussion

Falconer (1972) in his review of the Mather-Jinks book mentions the genetic architecture of metric traits. This seems to me to convey something. We might wonder, for instance, if a trait is determined by additive gene action, or whether there is complete dominance or overdominance or epistasis. I think this question is meaningful. We might ask if a metric trait in a population exhibits variability because of a few genetic factors of large effect or an unknown but presumably very large number of genetic factors with very small effects. If we could answer this definitely, we would have a useful result, which would suggest further activities of various sorts.

O. Kempthorne (1977)

The epigenetic system which directs the genotype/phenotype transition during development must itself be an epiphenomenon of the genome, and thus, according to Darwinian wisdom, a product of selection. Seasonal polyphenisms represent cases in which the developmental response to environment seems inescapably adaptive, even when we do not know how; there are two or more phenotypes, each canalized, and the epigenetic system somehow translates from environmental cues to a 'choice' among them. As already described, similar seasonal polyphenisms have originated at least twice in the Pierini—once in the Holarctic group to which *Pieris* itself belongs, and again in the *Tatochila sterodice* species-group in South America. In both cases the estival phenotype is the derivative one, and in both it has evolted farther (faster?) in males (in defiance of the conventional wisdom about the need for male pattern conservatism as a requirement for mate recognition, usually advanced in the context of female sex-limited mimicry—but see also Chs 20, 23). However, only in *Tatochila vanvolxemii* is it entirely restricted to males. Historically, such convergences have suggested inherent evolutionary predispositions to generation after generation of theorists. Recently, Ho & Saunders (1979) proposed that 'the intrinsic dynamical structure of the epigenetic system itself, in its interaction with the environment, is the source of non-random variations which *direct* evolutionary change . . .' (italics theirs; an autochthonous Baldwin effect). This role for the epigenetic system in the origin of evolutionary novelty was, as noted before, foreshadowed by Goldschmidt's ideas. In the Pierini one can see it extending beyond local seasonal adaptation and beyond the limits of genera and 'adaptive zones'. The pattern reduction seen in derived estival phenotypes of polyphenic mid-latitude pierines is mirrored in the invariant phenotypes of many low-latitude ones. Some are certainly primitively rather than derivatively reduced in pattern, but others give hints of being derived from higher-latitude ancestors with more complete patterns, perhaps by way of a polyphenic stage. Then we could visualize polyphenism as a bridge permitting the invasion of hot climates, after which the unneeded dark phenotype might be lost or effectively suppressed. Two candidates for such an origin come to mind. *Ascia monuste* crosses the Equator; it is strongly polyphenic in the female at its northern limit and weakly polyphenic in that sex in many other populations, but the male is seasonally constant. Its northwesternmost populations, the subspecies *raza*, have a complete ventral pattern only hinted at elsewhere. Perhaps more interesting is the odd *Theochila maenacte*, the only extra-Andean 'Andean pierine'. In this subtropical species the male is usually immaculate, but the female has a full *Tatochila* pattern. *T. maenacte* is closely related to, if not actually derived from, the *orthodice* species-group of *Tatochila*, which includes no polyphenic species. Male *maenacte*, however, may show quite a bit of the ventral hindwing pattern, especially in the southern Brazilian subspecies *itatiayae*; I have been unable to date to establish if this variability is seasonal, but it will not be surprising if it is.

If polyphenism can thus be connected to macroevolution and the invasion of new 'adaptive zones', it would be good to understand the underlying epigenetic system that controls it. Polikoff (1981) criticizes Waddington for failing to pursue his own ideas (canalization, genetic assimilation, epigenetic landscape) with sufficient rigour. Yet he did do the first assimilation experiments (Waddington 1953, 1961), and few have cared to follow. 'The theoretical basis of canalisation remains uncertain' and 'Waddington was never very clear on the subject of the mechanism involved in the canalisation of a novel developmental response; he appeared to suggest (1953) that it occurred via stabilising selection' (Ho & Saunders 1979). Whether and how stabilizing selection would be adequate to this task depends in part, at least, on the character of the genetic units to be selected, how 'selectable' they are. Experimental evidence bearing on this point is scanty. The classical assimilation experiments on *Drosophila* generally suggest that major genes are not responsible—the assimilated traits are polygenic (Waddington 1975). Such laboratory studies employ model systems intended to shed light on naturally-occurring, adaptive ones. Direct studies of the latter are difficult to come by. The genetics of diapause, which may be viewed either physiologically or as a type of developmental plasticity, are relevant but infrequently viewed in this context. The genetics of polyphenism has not been looked at before simply because one cannot do genetics in a genetically uniform population.

Even if disappointed, I am scarcely surprised that the Castle-Wright methodology did not work on *Tatochila*; it rarely does. Even though this approach is the most robust to the assumptions, where the equality assumption on the contributing loci is violated, it will tend to give numbers indistinguishable from 1 (R. W. Allard, pers. comm.), as it did here. (Given the small numbers, especially in the F_2 where 100 or more are desirable, and the inability to do a complete analysis including both backcrosses, Castle-Wright did surprisingly well.) To get at the true architecture of the epigenetic system controlling polyphenism will require several cycles of inbreeding followed by selection, as done by Allard & Harding (1963) and Wehrhahn & Allard (1965) for heading time in wheat, to successively expose and eliminate preponderant loci contributing to the character. It is much easier to do this with wheat than with butterflies, even such agreeable ones as *Tatochila*. Nothing of any generality is to be gained by further pursuing *ad hoc* viability models.

Even if the results of this exercise are dismayingly incomplete, it seems very likely that one locus is contributing a substantial share to the control of polyphenism, that several other loci could be exposed by peeling away the system layer by layer, and that the phenomenon is substantially less 'polygenic' than assimilated *Drosophila* phenocopies are. That says something: if there is a 'major gene' involved, one need only invoke ordinary Darwinian selection to evolve polyphenism, and neither genetic assimilation nor anything more arcane is necessarily required. Of course, what that selectable locus does in producing the final phenotype is still hidden from us inside a physiological-biochemical black box.

Acknowledgements

Collection of Argentine material was made possible by grants from the National Geographic Society (USA) and the Institute of Ecology, U. C. Davis (grant OPER-4806), and culture was supported in part by the Department of Zoology. Figures 27.1 and 27.2 were drawn by Virginia McDonald, and the photographs are by Samuel W. Woo. Most of the insight in matters genetical comes from conversations with Michael Turelli, Timothy Prout and Robert W. Allard, but any errors of interpretation, calculation, or omission are my own.

28. *Sunshine, Sex-ratio and Behaviour of* Euphydryas aurinia *Larvae*

Keith Porter

Department of Biology, Oxford Polytechnic, Oxford

Sunshine has a very strong influence on insect activity. Radiant energy can be utilized to raise body temperature (Digby 1955, Heinrich 1974, May 1976, 1979) and light is essential for visual communication and information (Callahan 1977, Kevan 1978, Mazokhin-Porshnyakov 1969, Silberglied 1979, Ch.20). Insects can detect and respond to wavelengths outside our own spectral sensitivity, notably in the ultraviolet (Eisner *et al.* 1969; Ch.20). At the same time, we cannot assume that 'black' insects absorb infrared: Heinrich (1972) investigated thoracic temperature in certain New Guinea butterflies thought to be in danger of over-heating in the sun. Measurements revealed that, in some species, the black scales actually reflected the near-infrared!

Most thermal studies on insects have only concerned adults (Clench 1966, Kammer & Bracchi 1973, Kevan & Shorthouse 1970, Rawlins 1980, Wasserthal 1975, Watt 1968), with relatively few on larval stages (Capinera *et al.* 1980, Casey 1976, Henson 1958, Shepherd 1958, Sherman & Watt 1973). However, larvae represent the growth phase of insects and, as such, are significantly influenced by their thermal environment.

Larval Behaviour of *Euphydryas aurinia*

In England, *Euphydryas aurinia* has six larval instars. Obligate overwintering diapause commences in September at the beginning of the fourth instar. The first three larval stages feed gregariously during July-late August within webs spun on *Succisa pratensis* (Dipsacaceae), and are not influenced by direct sunlight until mid third instar. The black post-diapause larvae, however, often bask in exposed positions. Fourth instars gain high heat-absorption rates by retaining the gregarious habit and basking in dense clusters. However, the number of larvae in these clusters decreases during February–April as the fourth instar progresses, until feeding and basking occur individually in the fifth and sixth instars (Table 28.1). This change in behaviour is almost certainly related to increasing individual food requirements: if gregarious feeding persisted throughout development, later instars would spend much potential feeding time moving on from defoliated plants. Table 28.2 summarizes the main features of larval behaviour, and also includes climatic data referred to below.

Table 28.1. Data for change in *E. aurinia* larvae (studied near Oxford, England) showing shift from gregarious behaviour in fourth instar to solitary habit by late fifth.

Date	Average number of larvae per web	Range	Instar
1 March 1980	87 ± 8.2	5–244	IV
13 March 1980	47 ±15.2	9–147	IV
1 April 1980	32 ± 10.3	15–24	IV–V
10 April 1980	3 ± 1.01	1–15	V
15 April 1980	1.00	1	V
20 April 1980	1.00	1	VI

Why Should Larvae Bask?

Basking would appear to expose the larvae to many dangers—what are the advantages? Using bead thermistors under a variety of conditions, I was able to establish (Porter 1981) that basking larvae can increase their body temperature substantially. In full sunshine larval body temperature regularly reached 35–37°C, while under dense cloud it equilibrated at ambient. Evidently, these jet-black larvae do absorb radiant energy from the sun.

The intensity of solar radiation in early spring is quite low, and body temperature of clustering fourth

The Biology of Butterflies
0-12-713750-5

Table 28.2. Basking behaviour of *E. aurinia* larvae. Climatic data are monthly averages for last 50 years obtained by the Radcliffe Observatory, Oxford.

Month	Stage	Colour	Web?	Greg?	Bask?	°C Temperature		Hours of sunshine
						Min	Max	
June	Egg					10.0	19.8	200.1
July	I	Pale brown	Yes	Yes	No	12.0	21.4	188.0
	II	Pale brown	Yes	Yes	No			
Aug.	III	Brown	Yes	Yes	No/Yes	11.7	20.9	175.5
	IV	Black		Yes	Yes			
Sept.						9.6	18.4	138.7
	IV	DIAPAUSE						
Feb.	IV	Black	Yes	Yes	Yes	1.3	7.3	68.6
March	V	Black	No	No	Yes	2.1	9.9	115.5
April	VI	Black	No	No	Yes	4.1	13.0	150.1
May	Pupa					7.0	16.7	191.1

instars never exceeded 37°C (Porter 1981). Later instars were never seen to bask during the hottest part of the day and maintained a temperature of 35–37°C by moving to less exposed sites if starting to overheat. Basking is part of a cycle larvae follow when feeding. First, an individual larva moves into the vegetation to feed—*Succisa* plants grow as low rosettes within the herb layer. In this phase the larval body temperature equilibrates with that of the vegetation. The larva next crawls out to bask in an exposed position (usually on a dead leaf), and so raises its body temperature. During this time the larva digests the food just eaten (indicated by frass production on the basking site). Once the gut is partly empty the larva completes the cycle by returning to feed. A relatively long time is spent basking as opposed to feeding, suggesting that digestion rate may be limiting—the mechanical process of ingestion is not likely to be as temperature sensitive as the biochemical process of digestion. Thus extensive basking may increase the overall assimilation rate by speeding up the digestion process. Many poikilotherms appear to have enzymes that work optimally at temperatures well above ambient (Heinrich 1977) and this appears to be the case for *E. aurinia*: laboratory experiments indicated that assimilation is maximal at 35°C (Porter 1982).

The Influence of Climate

This ability to use radiant energy from the sun enables larvae to grow rapidly in the cool northern spring, provided the weather is sunny. Table 28.2 shows such vernal conditions for Oxfordshire. Basking is probably not so important in summer, because air temperatures are much higher. It is also significant that the coldest days in spring are often associated with clear, sunny skies.

The Influence of Basking on Parasite Incidence

In southern England, *E. aurinia* has a specific braconid parasite, *Apanteles bignellii*. The parasitoid, which is trivoltine, attacks its univoltine host at three critical times: August, September and April–May. During the first two periods the host larvae cannot avoid attack and adult parasitoid-emergence is well-synchronized with host phenology. The attack on fifth–sixth instar larvae is, however, strongly influenced by weather conditions. If spring air temperatures are low but the skies are clear, host larvae develop much faster than parasitoid pupae (which are hidden in the vegetation). In such conditions the *Apanteles* emerge to find that most potential hosts have already pupated and are unsuitable for attack. This occurred near Oxford in 1979 (Table 28.3). During 1978–1980, the different conditions prevailed in each of the three years during the critical period for spring parasitoid attack: low air temperatures and short sunshine hours; low air temperatures and long sunshine hours; high air temperatures and long sunshine hours. The influence of each regime is predictable from the information given above—for example, in 1981 high air

Table 28.3. Timings of attack by *A. bignellii* on V and VI instar *E. aurinia* larvae, and relevant climatic data.

Year	% Parasites ex VI instar	Sample size	Date of 50% IV–V	Temp. range during period	Hours of sunshine
1978	18.06	72	10:4:78	−4.5–17°C	62.3
1979	7.7	129	25:4:79	−0.1–15°C	90.8
1980	74.56	169	5:4:80	0.8–25°C	104.0

The date for 50% IV–V instar moult is taken as the point at which parasitoid larvae emerge from pharate IV-instar hosts. The critical period occurs when parasitoid pupae are developing after emerging from the host.

temperatures maintained parasitoid synchrony, resulting in high (75%) parasitization of sixth instar larvae.

The Influence of Sunshine on *E. aurinia* Sex-Ratio

Apart from this direct influence of sunshine on parasitoid incidence, sunshine has an indirect affect on the butterfly's sex-ratio. Female *E. aurinia* larvae are on average larger than males, produce larger pupae, and pupate later. In years when *Apanteles* only catch the end of the larval period they therefore encounter a higher proportion of females. The sex-ratio of *E. aurinia* uninfluenced by parasitism is uncertain, but is expected to be close to 1:1. Results of extensive mark-release-recapture (MRR) work on the adult population suggest that there is an excess of males (Table 28.4)—but these daily sample estimates are almost bound to be male-biased due to higher male activity levels. A more accurate assessment can be obtained from the estimates of total adult population over the entire flight period, calculated separately for both sexes (i.e. the sexes are treated as different sub-populations for MRR purposes). These annual estimates do suggest real differences in sex-ratio that can be explained by the timing of parasitoid attack. When *Apanteles* synchrony is poor, as in 1979, the proportion of females in the butterfly population is lower than when synchrony is good (as in 1980). Poor synchrony causes a higher proportion of the slower-developing females to be attacked, whereas good synchrony leads to a more equal number of larvae of both sexes being killed (Porter 1983).

Table 28.4. *E. aurinia* sex-ratio expressed as percentage of males estimated from daily MRR adult samples, and estimated from total population size of each sex calculated separately from the MRR data.

	1978	1979	1980
Percentage males estimated from daily adult samples	71.4	74.4	66.6
Percentage males estimated from calculated total population size	66.9	70.1	61.1

Conclusions

Basking behaviour enables *E. aurinia* larvae to maintain some independence from air temperature. If spring temperatures are low, so long as the skies are relatively clear, the larvae can reach maturity by early May. Other species in Britain which adopt this larval basking behaviour and produce butterflies in June are *Melitaea cinxia* and *Mellicta athalia*. Larval basking by *E. aurinia* has an affect on synchrony with its parasite, *Apanteles bignellii*. The development rate of the parasitoid pupa is under the direct influence of ambient temperature, whilst the growth rate of *E. aurinia* larvae is influenced by the hours of sunshine. In turn, this affects the relative timing of parasitoid emergence in the spring and, because of a sexual difference in host larva maturation rate, can influence the adult sex-ratio of the butterfly.

29. *Seasonal Polyphenism in African* Precis *Butterflies*

Leonard McLeod

Quartier des Ecoles, St Pierre de Vassols, France

In many tropical and subtropical countries temperatures are sufficiently high to support successive generations of adult butterflies. However, there can be extreme seasonal changes, varying from severe drought to heavy rainfall. High temperatures with parched, stunted and brown foliage can change to cool conditions with luxuriant green vegetation. Thus a butterfly which is cryptic at one season may not be so at the next. Under such conditions the ability to produce only a single phenotype could be a disadvantage, and selection may result in the evolution of species having environmentally determined forms. The phenomenon of polyphenism is common in all climates with pronounced seasonality and apparently plays an important part in individual survival. Therefore, both the function and mechanisms of polyphenisms are an important area for study for butterfly biology. For temperate latitudes, it is known that photoperiod is the primary cue controlling adult phenotype of polyphenic butterflies, although temperature is also involved (Müller 1955, Shapiro 1968, 1971, 1973*b*, 1976*a*, Ch.27). In the tropics, where photoperiod remains almost constant throughout the year, this potential environmental cue can be discounted.

Polyphenism of *Precis*

In the tropical Pieridae there is hardly a species which does not exhibit seasonal polyphenism, but without doubt the phenomenon is most strikingly developed in the nymphaline genus *Precis*. Although polyphenism in *Precis* is often very remarkable, it is not universal, with sexually dimorphic species often lacking obvious seasonal variation (e.g. *P. westermanni*). In those *Precis* which do exhibit polyphenism, the dry-season forms tend to have more falcate forewings, tailed hindwings, and the underside ocelli reduced to mere points. These dry-season forms are particularly cryptic, and undoubtedly relate to the more sedentary habits of the butterflies during dry periods (see also Ch.16).

Some species, such as the Oriental *P. almana*, show seasonal variation in all the features mentioned, but not in the upperside wing colour. The species showing the most outstanding polyphenism of pattern and pigmentation of both wing surfaces is the African *P. octavia*. Although the wing shape variation of *octavia* is not as extreme as in several other species, the dramatic seasonal change in colour pattern makes it one of the curiosities of the insect world.

The Seasonal Variation of *P. octavia*

P. octavia was chosen as a subject for experimental investigation because of its suitability and availability. The two extreme forms normally encountered are 'natalensis', the wet-season form which is predominantly red with black markings on both surfaces, and the dry-season 'sesamus', a largely blue form with black and some red markings on the upperside, and almost totally black on the underside. Intermediates, although uncommon in nature, have been described, and may readily be produced under artificial conditions (see Colour plate 1E).

Behaviour and Predation

Differences between 'sesamus' and 'natalensis' are not confined to pattern and colour—there are also appreciable differences in their habits, as 'natalensis' shows pronounced hill-topping, frequenting the highest points in a neighbourhood, especially if they are open, while 'sesamus' is partial to shady spots and can be found in ravines, gorges and rocky, wooded slopes, and shows a marked affinity for disused mine shafts and cuttings. Both forms roost or aestivate, sometimes together, and may form compact groups of hundreds of butterflies.

The potential predators of *P. octavia* in these disparate situations are quite different. During the dry season, when 'sesamus' spends much of the day at rest in shady places, lizards are their main enemies

The Biology of Butterflies
0-12-713750-5

(see also Chs 2, 10, 12). While at rest a high degree of crypsis is advantageous—lizards carefully search out immobile prey at this time of year because of the sparsity of active insect life. During the luxuriant wet season, when 'natalensis' itself is very active, its main predators are themselves mostly cryptic. Hidden among flowers and larval foodplants, spiders, asilids, mantids and assassin bugs stealthily await the hapless butterflies. In these circumstances, crypsis is of little avail to *P. octavia.*

Between seasons, butterflies of the two extreme forms can often be seen roosting together and occasionally *in copula*. The latter observation poses interesting questions concerning the inter- and intrasexual significance of wing colour and markings in this butterfly. UV investigations could be important (Ch.20).

Control of Polyphenism in *P. octavia*

Although *P. octavia* was the focus of much attention at the beginning of this century, no definite conclusions were forthcoming concerning which environmental factor induces adult phenotype. It appears that the earliest studies were made by two Fellows of the Royal Entomological Society of London—Marshall and Poulton provided a great deal of information on the species but no conclusions. Their final recommendation (Marshall & Poulton 1902) was that 'in experiments in this most interesting of all known examples of seasonal change, it will be well to keep an open mind on all conceivable stimuli.' Later, Poulton (1906) reviewed information by K. St Aubyn Rogers concerning *P. octavia* on Mt Kilimanjaro. He stated 'By themselves they seem to suggest temperature and not moisture as the controlling factor.' Apparently the induction of seasonal polyphenism in the American *P. coenia* complex is still not fully understood (Klots 1951, Mather 1967) but current investigations might soon shed some light on this (A. Shapiro pers. comm.; D. Harvey, in prep.).

From preliminary investigations carried out in Kenya between 1964 and 1967, I had concluded that temperature was the only environmental factor involved in the control of adult seasonal phenotypes of *P. octavia* (McLeod 1968). However, in some of the experiments relative humidity was not adequately controlled. A few years later, D. F. Owen still maintained support for the old idea that rainfall and humidity were the pertinent factors: 'It is the relative amount of seasonal change rather than the actual rainfall that stimulates the production of seasonal forms' (Owen 1971*a*). Consequently I decided in 1972 to carry out a further experiment in England using modern, electronically-controlled constant environment cabinets.

In this experiment only two environmental factors were studied. These were temperature and relative humidity, both employed at two levels (30°C, 16°C; 90% RH, 30% RH). All insects were raised under a 12-h photoperiod and maintained under experimental conditions from early larval stages until adult emergence.

All insects were derived from stock captured at Byrne, near Richmond, Natal, South Africa. Adult phenotypes were to be scored from 0–6 according to the absence or presence of certain wing patterns and colours on the upperside, but only two scores were necessary: 'natalensis'–1, and 'sesamus'–6. All butterflies raised at a constant 30°C were 'natalensis' and those raised at 16°C were all 'sesamus'. These results indicate a clear correlation between adult phenotype and larval developmental temperature. There was no suggestion of a relationship between adult phenotype and relative humidity during development. Similar results have been obtained using other *Precis*, notably *P. archesia* (McLeod 1980).

Now that the possible influence of relative humidity can be discounted, we can return to my 1964–1967 work. The constant-temperature results obtained demonstrate a correlation between development temperature and adult phenotype. In addition, all the field data and other evidence reported by various authors point to the same conclusion (Marshall & Poulton 1902, Poulton 1906, Clark & Dickson 1957, Owen 1971*a*).

Loss of Polyphenism and Speciation

If a species with extreme seasonal polyphenism becomes restricted to a region where the factor inducing the polyphenism does not vary, could the ability to exhibit seasonal variation be lost? There is slight evidence for such an occurrence in Uganda. In the Entebbe region north of Lake Victoria, the mean maximum and minimum temperatures hardly vary from month to month (McLeod 1980). Both *P. octavia* and *P. archesia* occur there in only one form each, 'natalensis' and 'pelasgis' respectively. On nearby Mt Elgon, *P. octavia* can be found at high altitudes in both of its extreme forms, but this is not the case with *P. archesia*. Nowhere in Uganda or Sudan has *P. archesia ugandensis* ever been recorded in any form other than 'pelasgis' or 'semitypica' (very close to 'pelasgis').

I would like to suggest that instead of being the climax of evolution in the genus *Precis*, polyphenism is more likely to be an ancestral feature (cf. Ch.27). In isolation this ability to vary seasonally might be lost. Ultimately such populations might become reproductively isolated. Species such as *Precis coelestina*, *P. ceryne*, *P. terea*, *P. limnoria* and *P. frobeniusi* could have evolved in this way from an ancestral form similar to *P. archesia* f. 'pelasgis'. Until recently *P. frobeniusi* was classified as a form of *P. archesia*, and *P. limnoria* f. 'guruana' has frequently been mistaken for *P. archesia* f. 'pelasgis'. Although seasonal polyphenism does occur to a slight extent in these species, it is never extreme, and most of these species have a restricted distribution in comparison to *P. archesia*. Similarly, *P. antilope* and *P. cuama* could have evolved from an ancestral form similar to the red phenotypes of *P. octavia* and *P. actia*, and *P. artaxia* and *P. touhilimasa* from an ancestral form similar to the blue phenotype of *P. octavia*.

Larval Polymorphism and Polyphenism

A striking colour polymorphism occurs in the final larval instar of *P. octavia*. Two genetically controlled morphs have been described, 'purus', an almost immaculate form, and 'lineatus', having vertical stripes on each segment. (This polymorphism can also be seen, to a less marked extent, in the 4th instar.) Both larval morphs also exhibit polyphenism. This is probably the first recorded example of larvae being both polymorphic and showing seasonal variation.

The polyphenism of *P. octavia* larvae is readily understandable but the occurrence of polymorphism together with seasonal polyphenism is difficult to explain. The striped 'lineatus' larvae are probably aposematic and perhaps mimic (see also Ch.12) some distasteful Lepidoptera such as *Acraea zetes* and *A. terpsicore*, which have similarly coloured striped larvae and are widespread in eastern and southern Africa. The closely allied *P. archesia*, which also exhibits extreme adult polyphenism, has only one larval form throughout its range. These larvae are dark brown and are never cryptic on any of the larval foodplants, most of which are also foodplants of *P. octavia*. Further work on this interesting genus would obviously be worthwhile.

30. Seasonal Polyphenism in Four Japanese Pieris (Artogeia) *Species*

Osamu Yata, Toyohei Saigusa, Akinori Nakanishi, Hiroshi Shima

Biological Laboratory, Kyushu University, Fukuoka, Japan

Yoshito Suzuki

Department of Biophysics, Faculty of Science, Kyoto University, Kyoto, Japan

Akihiro Yoshida

Life Science Institute, Sophia University, Tokyo, Japan

All four *Pieris (Artogeia)* species (*rapae*, *canidia*, *melete* and *napi*) that occur in Japan, hibernate in the pupal stage. Diapause and related phenomena have been investigated experimentally in *rapae* (Kono 1970), but not in the others. We have now compared the seasonal polyphenism in both physiological and morphological characters of the four Japanese species, particularly in the pupal stage. We hope that such comparative studies will help to answer the question why only *rapae* amongst *P. (Artogeia)* species is a serious pest of cultivated crucifers (see also Ch.6). Here we give brief accounts of the photoperiodic response, diapause termination and pupal polymorphism associated with diapause. The reader is referred to Yata *et al.* (1979) for materials and methods.

Photoperiodic Response

Figure 30.1 shows the photoperiodic response curves of the four Japanese *P. (Artogeia)* species, from northern parts of Kyushu (33–34°N). In the progeny of spring-form females, the critical photoperiod was close to 11h in *rapae* and *canidia*. In *melete* and *napi*, on the other hand, a photoperiod of 12L-12D or shorter gave only diapause pupae, while that of 13L-11D or longer gave only non-diapause ones, the critical photoperiod being about 12.5h.

The results with summer-form mothers differed somewhat. Although a few pupae did not diapause even at 9L-15D the critical photoperiod was about 11h in *rapae*, about 12.5h in *canidia* and about 13h in both *melete* and *napi*. In *canidia*, *melete* and *napi* the critical photoperiod was thus longer in the progeny of the summer form than in that of the spring form. This finding of a 'maternal effect' is consistent with the result obtained by Hidaka & Takahashi (1967) in a nymphaline, *Polygonia c-aureum*.

The mean duration of the larval stage of each species is longer in diapause than in non-diapause groups. The difference between the two groups was larger in *melete* and *napi* than in *rapae* and *canidia*, being smallest in *rapae*. Furthermore, the larvae of *rapae* grew faster than those of the other three species in both the diapause and non-diapause groups.

Diapause Termination

Figure 30.2 shows the adult emergence from diapause pupae obtained in the foregoing experiments. The pupal diapause in *rapae* derived from Hokkaido stock was terminated by two or more months of chilling at −2 to 12°C, resulting in a distinct peak of adult emergence on incubation at a high temperature, whereas without exposure to cold, the adult emergence was more or less evenly distributed over a long period (Masaki 1955). We kept diapause pupae

The Biology of Butterflies
0-12-713750-5

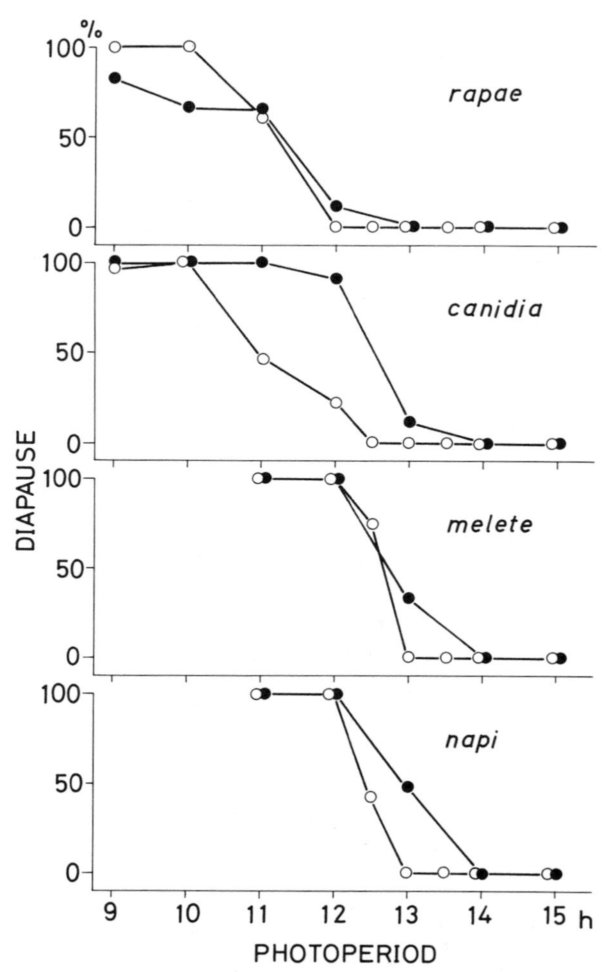

Fig. 30.1. Relation between the larval photoperiod and the percentage of diapause pupae in progenies derived from spring- (○) and summer-form (●) females in four Japanese *P. (Artogeia)* species (Yata *et al.* 1979).

at 20±1°C (15L-9D) after 60 days of chilling at 4±1°C (in continuous dark). In all species an emergence peak was observed after 10–15 days. In *napi* the adult emergence was initiated fast, highly synchronized and completed within 13 days after chilling. In *canidia*, in contrast, the emergence was apparently delayed and scattered over a long period of 52 days. In *rapae* there was only one major emergence peak (13 days after chilling) as in *melete* and *napi*, but some scattered emergences occurred over about one month.

Pupal Polymorphism

Pupal polymorphism was closely associated with the occurrence of diapause, but its expression differed among the species. First, the dorsolateral processes of the third abdominal segment were more prominent in non-diapause pupae than in diapause ones. This difference was distinct in *canidia*, *melete* and *napi*, but extremely slight in *rapae*. Johansson (1954) reported a similar dimorphism in *Pieris (P.) brassicae*. With respect to pupal colour, some workers have reported that the proportion of pale or green to brown tends to be higher in non-diapause than in diapause pupae (Bowden 1952, Johansson 1954). We found a colour difference between diapause and non-diapause pupae in all four species, but further investigations are required because the larvae pupated on various substrates, and the background colour and brightness were not controlled. The cuticle was somewhat more glossy in diapause than in non-diapause pupae, and this is attributed to the ultrastructural difference found in *rapae* by Kono (1973).

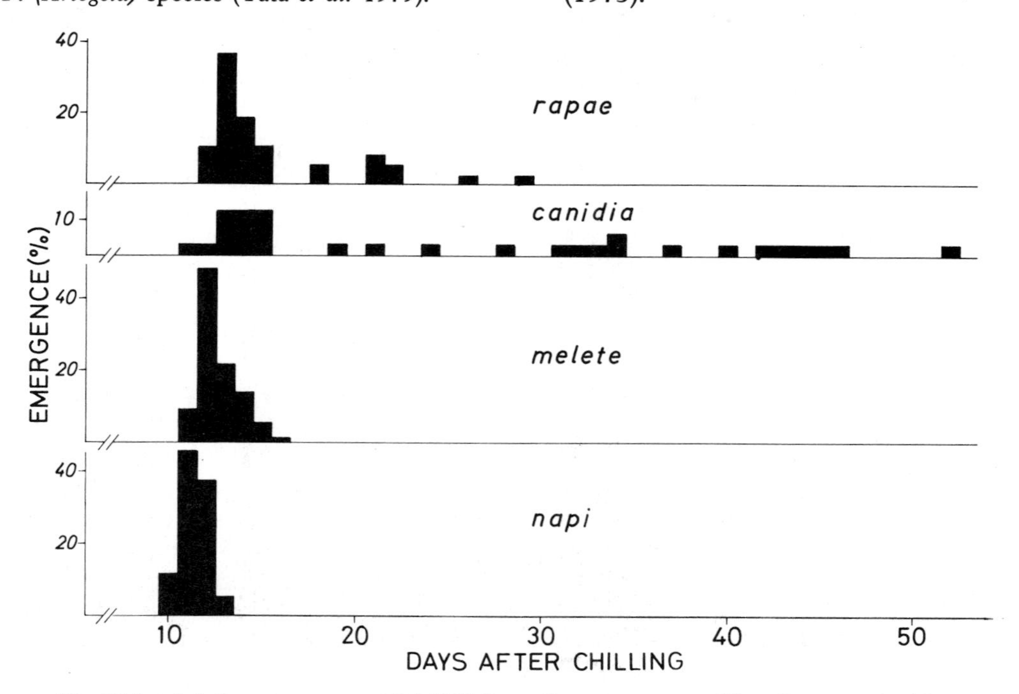

Fig. 30.2. Adult emergence at 20±1°C from diapause pupae of four Japanese *P. (Artogeia)* species after 60 days of chilling at 4±1°C (in continuous dark).

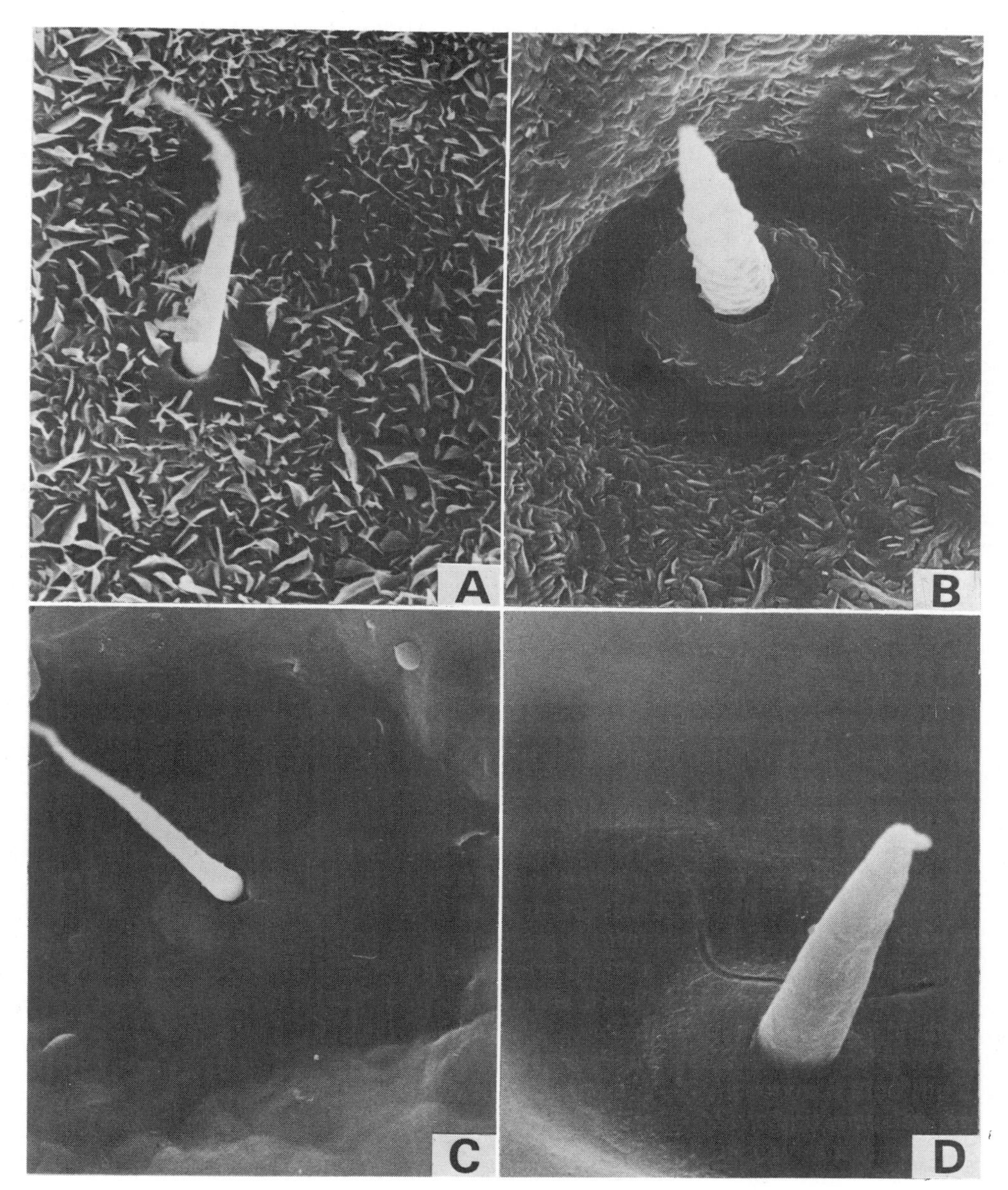

Fig. 30.3. Cuticular surfaces of non-diapause (A, B) and diapause pupae (C, D). A, C: *rapae*; B, D: *melete*. (× 3000)

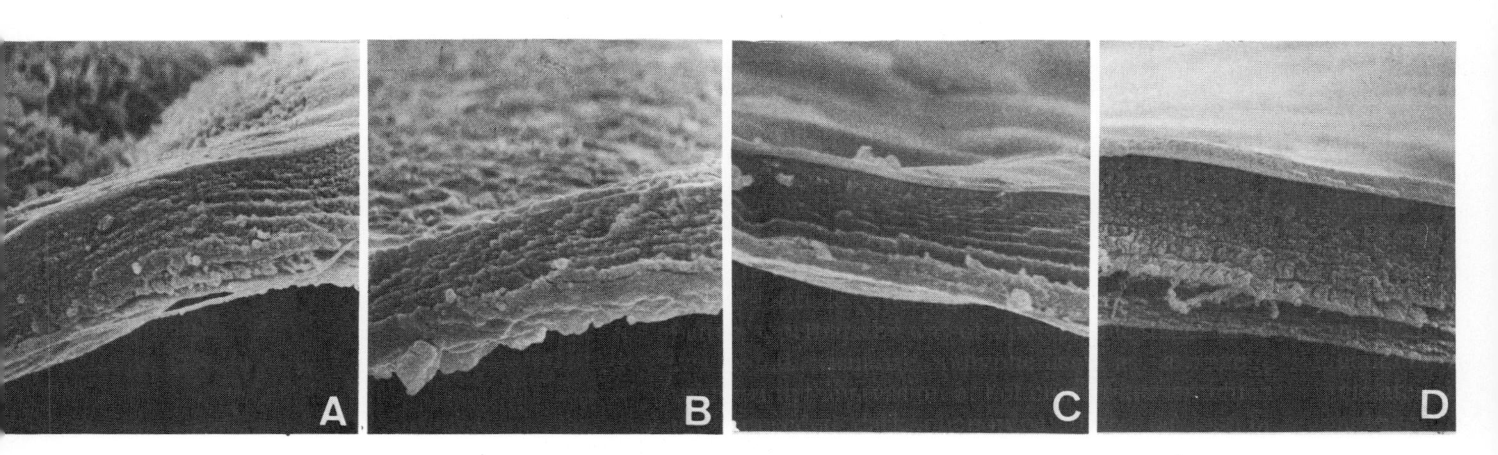

Fig. 30.4. Transverse section of non-diapause (A, B) and diapause pupal cuticles (C, D). A, C: *rapae*; B, D: *melete*. (× 3000)

As revealed by scanning electron microscopy (Fig. 30.3) the body surface of the nondiapause pupae of *rapae* and *canidia* is covered with numerous, irregularly arranged projections that look like scattered pieces of broken glass, and with irregular fine rugosities in *melete* and *napi*. In contrast, the cuticular surface of the diapause pupae is smooth (in all four species).

Figure 30.4 shows transverse sections of the pupal cuticle. In all the species the lamellate structure was observed in both diapause and non-diapause pupae. The cuticle of the non-diapause pupa is structurally almost the same in the four species, and seemed to be very fragile. The cuticle of the diapause pupa was covered with a thick, wax-like layer. This layer was 2–3 times thicker in *melete* and *napi* than in *rapae* and *canidia*. The cuticle of the diapause pupa is also thicker in *melete* and *napi* than in *rapae* and *canidia*. Generally, the cuticle of diapause pupae seemed to be harder than that of non-diapause pupae.

Conclusions

These observations suggest that *P.(A.) rapae* is not more specialized than the other *Artogeia* species, having the shortest critical photoperiod, and producing some non-diapause pupae even in a photoperiod of 9h. It showed the smallest difference in larval development time between the non-diapause and diapause groups, and only a very slight morphological difference between diapause and non-diapause pupae. In *rapae*, therefore, seasonal polyphenism, especially in the pupal stage, is less marked than in the other three *P. (Artogeia)* species. This perhaps surprising conclusion suggests that other factors must be responsible for its extensive distributional range and great ecological success in the temperate regions of the world.

Acknowledgements

We thank Professor Sinzo Masaki of Hirosaki University for his critical reading of the manuscript.

Part VIII
Conservation

31. *The Impact of Recent Vulcanism on Lepidoptera*

Robert M. Pyle

*Conservation Monitoring Centre, Cambridge, UK**

The eruptions of Mt St Helens in Washington State, USA, which started on 18 May 1980, have been the subject of enormous public and scientific interest, with a great deal of research being focussed on their biological implications. Among such investigations, entomological studies have considered pre- and post-eruption faunas both in the blast zone and at various points in the extensive ashfall zone. My home is near Mt St Helens and I carried out the last Lepidoptera sampling on the mountain prior to the eruptions. Thus I have observed at firsthand the effects of ashfall and established a faunistic 'baseline' against which blast zone recovery may be measured. In doing so, I have developed an interest in volcanic impact on Lepidoptera faunas elsewhere, and this chapter is intended as a brief introduction to the subject.

General Impacts on Invertebrates

Much of the large area in eastern Washington and Idaho included in the moderate to heavy ashfall zone from the 1980 eruptions of Mt St Helens (200 000km^2, Edwards & Schwartz 1981) consists of important agricultural land. Therefore, state, federal and university entomologists immediately began examining impacts of ash on both pest and beneficial insects in the region. Insect mortality was immediate and dramatic, leading to claims that the ash acted as a natural insecticide, some groups of insects being more susceptible than others. General characteristics of the main eruption are described by Rosenfeld (1980), the biological and physical properties of the ash by Fruchter *et al.* (1980), and the mechanisms for its impact on insects by Edwards & Schwartz (1981) and Brown & Hussain (1981). The ash, chemically similar to dacite, is relatively inert, but the particles are jagged and abrasive. Ash acts like diatomaceous earth (Fullers earth) as used against cockroaches: its effects are physical, rather than chemical. In the laboratory mortality from ash is caused by entrapment, abrasion of epicuticular waxes and cuticle causing water loss, and dessication enhanced by the absorbtive properties of the ash and excessive salivation during grooming. Brown & Hussain (1981) felt that ingestion or inhalation of ash were probably not lethal, though Edwards & Schwartz (1981) suggested both obstruction of spiracular valves and accumulation of ash boli in the gut as contributory factors. They found juvenile insects more susceptible (the major eruption took place in May), and the former authors considered active predaceous insects and parasitic species (especially Hymenoptera) particularly liable to ash kill.

The whole of Volume 37 of *Melanderia (Journal of the Washington State Entomological Society)* was devoted to reporting Mt St Helens entomological findings. Akre *et al.* (1981) reported the effects on social wasps and ants, Johansen *et al.* (1981) on bees, Smith (1981) on lotic Trichoptera, and Schanks & Chase (1981) on curculionid beetles.

Most of these authors are cautious about long-term impacts, and several seasons' monitoring will be necessary to assess these. Some species seem to be making a rapid comeback. However, at the time of writing (spring 1982) volcanic activity is continuing and additional ash falls cannot be ruled out. Gribbin (1982) discusses the possible climatic effects of major eruptions, concluding that dust-veil effects can alter regional weather and, conceivably, the general climate. This could have large-scale effects on insect populations. It is premature to judge the overall volcanic impact on the entomofauna, but as Edwards & Schwartz (1981) say: 'The effects of major insect mortality will have obvious consequences for their host plants, for pollination, and for insectivorous vertebrates that may have survived the initial ash fallout.'

*Permanent address Washington, USA

The Biology of Butterflies
0-12-713750-5

Specific Effects Upon Lepidoptera

Several authors have written on the effects of various eruptions on insects, including Kuwayama (1929, Mt Komagalake, Japan), Griggs (1919, Mt Katmai, Alaska), Segerstrom (1950, Paricutin, Mexico) and Wille) Fuetes (1975, Irazu, Costa Rica). However, none commented substantially on the Lepidoptera. Calvert *et al.* (1979) include vulcanism among the local catastrophes that could be responsible for the upper size limit of overwintering Monarch butterfly aggregations in Mexico. In 1951, Mt Lamington erupted periodically for six months above the Popondetta Plain in Papua New Guinea, and may have affected the endemic birdwing, *Ornithoptera alexandrae* (see Parsons 1980*a*,*b* and Ch.32).

The blast zone of Mt St Helens, radiating north and north-east from the crater, experienced near-total devastation almost certainly including the butterfly fauna. The area was generally neglected by lepidopterists, due to high rainfall and supposed sterility. In contrast to other, older Cascades volcanoes such as Mts Baker, Rainier and Hood, St Helens lacked well developed alpine meadows and associated Rhopalocera (Pyle 1974). Nevertheless, sampling under favourable conditions on 9 August 1979 demonstrated the existence of a richer butterfly fauna and associated flora than was previously thought (Pyle, unpublished). This fauna and flora occupied erosion gullies in old ash and pumice flows on the north-east above Timberline, at approximately 1600m elevation. The dominant plants included *Sorbus*, *Vaccinium*, *Castilleja*, *Lupinus* and *Spiraea.* Three nymphalids, three polyommatines and two diurnal moths were present, and two additional nymphalids were found at Spirit Lake, 5km to the north. For late summer in this part of Western Washington, this is a reasonably good sample. These sites were entirely destroyed by the May 18 eruption. Most of the species were first records for the entire Mt St Helens area, and these may now be locally extinct. Certainly the populations between the former Timberline Campground and Windy Pass, unknown prior to 1979, may be considered extinct due to the direct impact of vulcanism originating 2km away at the crater.

Several lepidopterists in the area had paid attention to butterfly populations within the ashfall zone. N. Curtis (pers. comm.) sampled a site in western Idaho where 1-2cm of ash were deposited. On three sample dates in July and August of 1977-78 (prior to eruption), 16-20 species of butterflies were recorded, with 67-104 specimens being taken. Similar dates at these sites in 1980 (post-eruption) yielded only six specimens of three species, all univoltine. However, Curtis found these sites to be almost normal in 1981. D. McCorkle (pers. comm.) found 1981 butterfly numbers low but species diversity near normal at the rich and long-studied Bear Canyon, Yakima County, Washington, where moderate amounts of ash fell the previous year. However, McCorkle also sampled at White Pass, farther south, where up to 0.5m of ash was still in evidence in midsummer, 1981. There he found virtually no *Speyeria* individuals in a habitat where several species of fritillaries are normally abundant at that time.

The hostplants of *Speyeria* (*Viola* spp.) have rough-textured leaves that are quick to acquire a coating of ash but slow to shed it, even in heavy rainfall. New leaves unfurling in an ashy environment become ash-coated. In contrast, grasses shed ash efficiently in the rain, and grass-feeding butterfly species do not show much depression in abundance. I found *Coenonympha tullia* in normal numbers in a field 25km north-west of Mt St Helens, two months after the major eruption. This field had a layer of ash 3-5cm deep with fresh grass growing through, uncoated. These satyrines would have been eggs or young larvae at the time of the eruption. J. P. Pelham (pers. comm.) found another satyrine, *Cercyonis pegala*, abundant at Moses Lake, Washington in 1981. This area suffered very heavy ashfall and still retained a blanket of ash 1-20cm in depth. The wood nymph population would have encountered ash during two larval generations. Three months following the first eruption, and again the following year, I found a grass-feeding skipper (*Ochlodes sylvanoides*) flying in the usual high numbers at a Western Washington site where 1-5mm of ash was still apparent. These skippers would have been larvae during and following the ashfall period.

Triphosa haesitata is a large geometrid moth that normally overwinters in large numbers in lava-tube caves on the south side of Mt St Helens. Systematic visits by biospeliologists (R. Crawford, pers. comm.) during the fall and winter of 1980-81 revealed a total absence of the moth from these caves. They and the external environment were well outside the blast zone, but they did receive heavy ashfall. The larval host of *T. haesitata* is *Acer circinatum* (vine maple)—a shrubby tree bearing deeply ridged, slightly tomentose, highly ash-retentive leaves. These anecdotal observations suggest the hypothesis that lepidopterans which eat rough-textured, ash-retentive foliage may suffer greater initial losses and take longer to recover from volcanic ashfall than those feeding on smooth-leaved hosts. It is also known that ash persists longer beneath the forest canopy, where *Speyeria* and *Triphosa* hostplants occur, in contrast to the open grasslands where hesperiines and satyrines fly.

Some concern was felt by local lepidopterists for several rare taxa occurring within the Mt St Helens impact area. *Speyeria zerene hippolyta*, the only federally designated threatened butterfly in

Washington and Oregon (Pyle 1976*a*) dwells in a small number of coastal localities where it was apparently not affected by the trace amounts of ashfall. A relictual population of *Boloria selene*, contained within the Moxee Bog Preserve belonging to The Nature Conservancy (Pyle 1976*a*), lay in the direct path of the southeasterly ash plume. Like much of Yakima County it received heavy fallout. While 1980 generations may have been depressed, J. Pelham and F. Krause (pers. comm.) found healthy generations of this trivoltine argynnid in 1981. (This seems to contradict the leaf-texture argument, since *Boloria* feeds on *Viola*. Voltinism and other phenological conditions surely play a role. A multivoltine species might be expected to recover faster than the univoltine *Speyeria*.) Concern for the only species of butterfly endemic to Washington State, the skipper *Polites mardon*, seems ungrounded. Its two major populations (in the highlands near Mt Adams and in the Puget Sound lowlands) appeared unaffected in 1981, except where damaged by other (anthropogenic) landscape changes (J. Hinchcliff, pers. comm.).

Recovery of Populations from Volcanic Effects

Recovery of Lepidoptera populations in vulcanized regions would seem to be a joint function of hostplant recovery and degree of survivorship *in situ*, or ease of recolonization from other sources. Little information on this has come to my attention. *Agrotis ipsilon* was the first species of Lepidoptera recorded on the new Icelandic island of Surtsey, following its volcanic origin (Odiyo 1975). This is a strongly migrant species, however, and its appearance soon after the eruption does not necessarily signal colonization. The butterflies present on Mt St Helens on 9 August 1979 represented a reasonably well developed fauna of host-specific species (Pyle, unpublished). These probably colonized the mountain since the previous major eruptions in the 1830s–50s (Pyle 1970). Their current extinction on Mt St Helens as a whole cannot be presumed—recent observations (R. Sugg, pers. comm.) show that a plant community including *Lupinus* survives on the west side of the volcano at 1500–2000m elevation. This community was probably similar to that on the northeast side where the last Lepidoptera sample was taken. Recovery, therefore, may involve progressive recolonization around the mountain in addition to (or rather than) from afar. Studies are planned to monitor butterfly recovery on Mt St Helens as far as Red Zone restrictions, red tape, and the mountain itself will allow.

Butterfly recovery presumes that community recovery is possible. The Valley of Ten Thousand Smokes in Katmai National Monument, Alaska, was scarified by the superheated gas eruption of Mt Katmai in 1911 (Griggs 1919). Tundra recovery in the valley itself was made impossible by the deposition of hundreds of metres of ash, and no butterflies whatever were seen there during my 1971 collecting trip. However, tundra on the slopes above the valley floor has recovered its basic pre-eruption flora and *Oeneis* butterflies are present. In the absence of other factors, local volcanic impacts are likely to be much heavier and last longer than are the widespread effects. The devastation of the immediate vicinity, and its chronic impact, would seem to be rapidly ameliorated outside the direct blast zone. Lava flows will affect communities in fundamentally different ways from explosive ash eruptions. Recovery from either, however, will reflect the degree of interruption of both ecological and biogeographical processes.

Long-Term Impacts

Recent vulcanism clearly seems to retard development of an arctic or arctic-alpine flora and fauna. This is shown by Katmai, the observations on Mt St Helens by Lawrence (1938), and by the Cascade Volcanoes themselves. Older peaks, such as Mt Baker, Glacier Peak and Mt Rainier in Washington, and Mt Hood in Oregon, have developed alpine faunas (albeit limited by their faunistic isolation), whereas younger members of the volcanic chain, notably pre-eruption Mt St Helens and Mt Lassen in California, lack such a fauna. This effect is exacerbated by distance from sources of recolonization, since volcanoes in a strong sense are islands. Isolation may further be exaggerated by human impact, which alters the landscape and diminishes dispersal and colonization opportunities.

Mt St Helens is surrounded by a vast expanse of national and private forest. It lacks a national park or wilderness area such as those that protect the forest environs of all other Cascade volcanoes. A regime of clear-felling, replacement of natural, diverse forest with monocultures of even-aged, single-species stands of conifers, and heavy herbicide application to retard competitive herbaceous and hardwood growth, along with dramatic erosion from logging roads built on steep grades in a high rainfall zone, must impede butterfly dispersal and survival. Furthermore, the political fate of Mt St Helens and its vicinity will drastically effect recolonization. Large areas have already been aerially sown with alien grass and legume seed that will favour adventive rather than specialized species of wildlife (A. B. Adams, pers. comm.). Management proposals range from the Mt St Helens Protective Association plan for a large national monument, to the timber industry's aim to salvage the maximum amount of fallen and standing timber, through the Governor's compromise measure

(Hooper 1982). Clearly, these kinds of precedents and sequelae to vulcanism can prejudice the long-term survival of regional Lepidoptera at least as much as the eruptive phase itself.

Discussion and Conclusion

Throughout much of the Pacific Rim, and certainly the Pacific Northwest, volcanic eruptions have been and continue to be major eco-geographic events. Lepidoptera distribution has surely been affected dramatically by large-scale lava flows in the Columbia Basin and elsewhere. On Hawaii, the localized distribution of the two endemic butterflies (*Vaga blackburni* and *Vanessa tameamea*) on kipukas—islands of surviving vegetation separated by recent lava flows—demonstrate the power of vulcanism over butterfly occurrence. Ashfall from volcanoes is seen to be potentially lethal to insects, and can depress butterfly and moth populations at least locally and temporarily. Recovery is linked to the two factors of plant recovery and dispersal potential. In fact, ecosystems in volcanic regions may include organisms pre-adapted for survival from and recovery after volcanic events. Investigating ants on Mt St Helens in 1981, R. Sugg (pers. comm.) concluded that certain ant species seem pre-adapted for survival under heavy ashfall. This seems also to be the case for certain adventitious plants, some of which serve as hosts or nectar sources for native Lepidoptera. It seems unlikely, therefore, that vulcanism would normally have profound consequences for Lepidoptera populations in the long-term, except where extreme endemism coincided with centres of intense eruptive activity. Nor is it reasonable to attribute the depauperate nature of the Western Washington butterfly fauna (Pyle 1976*a*) to periodic volcanic activity throughout the region.

However, recovery could be retarded and ultimate recolonization prevented by anthropogenic activities in conjunction with vulcanism: intensive forestry practices, reseeding with exotics, accelerated pesticide use and road building are examples of human forces that could interfere with normal recovery. Land-use designation following eruption can also have an inhibiting or abetting biological effect. Thus vulcanism and human disturbances may act together, even synergistically, to affect regional Lepidoptera populations. These seem to be able to absorb most volcanic effects without extreme adverse consequences. But pre- and post-eruption management of the volcanic landscape must be considered an important factor in the ultimate survival of butterflies and other organisms from volcanic events.

Postscript

Since this account was written, three studies have come to my attention that bear closely upon its content. All three deal with fritillary butterflies (*Speyeria*) in the Cascade Mountains of the American Pacific North-West.

Throughout the Pleistocene, the Cascades have been volcanically active. Crowe (1965) suggested that *Speyeria* populations were greatly reduced by ash- and pumice-fall from these eruptions; and that the resulting dry and barren deposits served as effective barriers, isolating surviving minor populations from one another and leading to modern variation among descendent demes. In particular, he accounts for the derivation of the distinctive, so-called 'Sand Creek-type' of *Speyeria egleis* in Oregon by this means.

Hammond (1983) also concluded that deposition of ash-pumice fields led to local extinctions, indeed to extermination of most life forms in their vicinity, from burial and suffocation. Mt St Helens, having deposited only 6–20cm of ash near Yakima, Washington, nonetheless depleted *Speyeria* severely as of 1981. Mt Mazama, in dramatic contrast, erupting some 6000 years BP, deposited up to 75m of ash, 'which must have virtually sterilized these areas of all pre-existing . . . *Speyeria* populations' and their host *Viola* plants. Yet Hammond has found that two species of violets and five of *Speyeria* have since colonized the Mazama ash-pumice fields. He concludes that the extremely xeric habitat afforded by such recent volcanic landscapes permits certain violets and fritillaries to colonize vigorously, while excluding others due to restrictive limitations in their physiology and ecology.

Indeed, the 'Sand Creek-type' of *S. egleis* has been found by Hammond & Dornfeld (1983) to correspond very closely in distribution to that of ash-pumice fields deposited by the eruptions of Mt Mazama, Mt Newberry and the volcanoes of the Three Sisters system, all in Oregon. They have named this population as a distinct subspecies which is virtually a volcanic endemic. Thus, while Cascadian eruptions have certainly had a negative impact on contemporary *Speyeria* butterflies, modern descendants have recovered, in some cases, to exploit the new volcanic habitats with great success, even to the point of generating new, autochthonous taxa.

Acknowledgements

I would like to thank the following individuals for helping me to obtain information reported in this paper: A. B. Adams, Elizabeth Betts, Carol Carver, Sharon Colman, Rod Crawford, Nelson Curtis, John Edwards, Robert Gara, John Hinchliff, Carl Johansen, David McCorkle, Michael Parsons, Jonathan Pelham, Merrill Peterson, Larry Schwartz, Rick Sugg, George Tamaki, Norman Tindale, O. Yata and Akemi Yoshikawa.

32. The Biology and Conservation of Ornithoptera alexandrae

Michael J. Parsons

*Lepidoptera Research Project, Insect Farming and Trading Agency, Bulolo, Morobe Province, Papua New Guinea**

Ornithoptera alexandrae is the world's largest butterfly. The wings of females of this spectacular insect (Colour plate 1F) may span over 25cm. *Ornithoptera* contains nine extremely dimorphic 'birdwing' species, all but three of which are indigenous to New Guinea. The adults and early stages are generally assumed to be aposematic—the larvae feed exclusively on *Aristolochia* vines, well known for their wide range of secondary defensive compounds (Ch.12). Because of their size, beauty and general unavailability, *Ornithoptera* are much sought after by insect collectors. *O. alexandrae* is wholly endemic to a small region of Papua New Guinea now much subject to ecological disturbance from the oil palm and logging industries. For these reasons, a detailed study of the biology and conservation requirements of Alexandra's Birdwing was begun by the Government of Papua New Guinea's Wildlife Division in April 1979 (Parsons 1980*a*,*b*,*c*, 1981).

Distribution and Habitat

Since its discovery in 1906 (Meek 1913), *O. alexandrae* has never been recorded outside a small coastal plain in Northern Province, Papua New Guinea, an area just over 1200km^2 in extent (Fig. 32.1). Roughly surrounding the town of Popondetta, the area is close to the Mt Lamington volcano which last erupted in 1951, causing extensive damage. The distributional survey of Straatman

Fig. 32.1. Distribution of *O. alexandrae* based on recent 10km distributional data. (Highland areas over 1000m are stippled.)

*Present address: Egham, Surrey, UK.

The Biology of Butterflies
0-12-713750-5

(1970) was updated by Parsons (1980*c*), who produced a (confidential) 10km^2 map. The map shows that *alexandrae* has only ever been found within ten 10km squares. Although *alexandrae* occurs further north-west than assumed by Straatman (1970), its distribution in the Popondetta area is almost entirely bordered by the Kumusi River to the west, the Embi Lakes to the east, the 400m contour of Mt Lamington to the south, and to within 10km of the coast to the north. Even within this area, *alexandrae* appears to be absent from many apparently suitable habitats (cf. Ch.2).

O. alexandrae occurs together with its larval foodplant, *Aristolochia schlechteri*, in primary and secondary lowland rainforest tracts up to 400m, on the well-drained volcanic ash soils of the Popondetta Plain, and in secondary hill forest on clay soils, from 550–800m. The vegetation on the Popondetta Plain is very species-rich, especially on the lower north-western slopes of Mt Lamington. These lowland tropical rainforests are more than 40m tall, with three distinct layers, plentiful lianas, epiphytes and buttress roots, leaving only a very sparse ground layer (Haatjens 1964). Both butterfly and vine appear to be absent from forests of more irregular structure, such as those found on the alluvial soils of the lower Mambare and Gira river floodplains to the north-east of the Popondetta Plain.

The two major disruptive influences on the vegetation pattern of the Popondetta region have been man and vulcanism. Widespread shifting cultivation on the lower slopes of Mt Lamington and the plain have reduced many areas of the original rainforest to secondary status, in various stages of regrowth. Locally, this practice has been aided by fire so that there are many areas of man-made grassland (now maintained for hunting wallabies). The greatest extent of grassland occurs on the very sandy soils of the north-eastern volcanic outwash fans, so that aerial photographs show a distinctive mosaic of linear tracts of forest and grassland.

The Hostplant, *Aristolochia schlechteri*

O. alexandrae is monophagous on *A. schlechteri*, a vine which occurs only in primary and advanced secondary forest. In primary sites the vine can be 40m tall, with all leaves in the upper canopy; such vines are certainly in excess of 20 years old. In secondary conditions the vine often spreads vegetatively, sending long side-shoots through the leaf litter to climb nearby trees. In general the vine sprawls rather than twines, but in primary forest the mainstems often go up the trunk of the supporting tree, without any attachment, until they reach the canopy.

Although *A. schlechteri* was first described from the Ramu River (400km to the north-east of Popondetta —Lauterbach 1914), it has otherwise only been recorded from localities either side of the Owen Stanley Range in south-east Papua New Guinea (Fig. 32.2). *A. schlechteri* is nevertheless far more widespread than *O. alexandrae* (Figs 32.1, 32.2). To the south, the vine is the foodplant of *Ornithoptera meridionalis*.

Straatman (1970) maintained that *schlechteri* seed is dispersed for short distances by rain, after the fruit has rotted. From my own observations of several half-eaten fruits I think it more likely that the seed is

Fig. 32.2. Distribution of *A. schlechteri* based on recent 10km distributional data. (Highland areas over 1000m are stippled.)

dispersed in the faeces of tree-dwelling marsupials, large ground-dwelling birds (such as cassowaries) or fruit-eating bats.

O. alexandrae Life-History

The life-history of *O. alexandrae* was described by Straatman (1971) and recent observations have been made by Parsons (1980*a*), who calculated that females can lay at least 240 eggs during an optimum lifespan. To sustain such egg production may require organic nitrogen obtained from pollen, as occurs in certain heliconiines (Pianka *et al.* 1977; Ch.3) and, perhaps, South American troidines (DeVries 1979). In secondary forest, as with most *Ornithoptera*, the eggs are laid singly, often on the bark of the supporting tree or on debris (such as sticks—see also Ch.7), close to the vine mainstem and frequently only about one metre or less from the ground. In primary forest, however, *alexandrae* oviposits high above the ground, usually on the underside of the vine leaves. Eggs take about 12 days to hatch. Larval development takes about 70 days, generally with five instars, although six occur sometimes. Pupation takes place on the underside of the leaves of nearby trees, up to 10m away. Pupal duration averages 40 days, giving a total egg-adult development time of about 122 days.

First instar larvae feed on the new, yellowish and tender leaves and shoots. Later instars feed on the older leaves and stems. Shortly before pupation, as in some other troidines, the larvae may move to the base of the vine and ringbark the mainstem, causing the upper parts to die. If the vine is young, the lower part is often eaten down to the ground.

Straatman (1971) found an interesting difference in the developmental rate of *O. alexandrae* larvae placed on *Aristolochia tagala* when compared with its development on *A. schlechteri*. These results, although based on limited data, clearly showed that on *A. tagala*, on which the females will not oviposit, larval development was faster by about 24 days. *A. tagala*, commonly utilized by *O. priamus* and *Troides oblongomaculatus*, is a much softer-leaved vine and far more widespread, occurring throughout New Guinea and much of Indo-Australia. The leaves are more easily masticated, and provide greater nutrition per volume of foliage than the sclerophyllous leaves of *A. schlechteri*.

Abundance

Straatman (1970, 1971) listed a number of mortality factors for *O. alexandrae*. These included bugs, ants and heavy rain affecting the eggs; wasps, ants, tachinids, tree rats, small marsupials and fungal diseases affecting the larvae; ants attacking pre-pupae and soft, new pupae; chalcids, tree rats and small marsupials affecting the hard pupae; and spiders trapping the adults. Parsons (1980*a*) adds Kingfishers (Kookaburras) and the Grey-breasted Brush Cuckoo, *Cacomantis variolosus*, to the list of foes attacking larvae, and suggests that the latter may be an important specialist larval predator. In general, like most birdwings, *O. alexandrae* adults appear to suffer little predation and may survive in the wild up to three months (Straatman 1971) and possibly longer. Often seen soaring above the canopy, the butterflies probably feed from the flowers of the rainforest trees. Around Popondetta, females can be seen at the Tulip Tree, *Hibiscus*, *Ixora*, *Poinciana* and *Poincettia*.

Little is known of adult movement. Straatman (1971) found that some marked males remained in a garden area for three months, suggesting that they may exhibit home range behaviour. If this is so, it may help to explain the very sporadic occurrence of the species throughout the area of apparently suitable habitat.

Adult abundance is difficult to assess: the high-flying habit and low density render mark-recapture methods unhelpful. My estimates are subjective, therefore, but field surveys and discussions with local landowners suggest that the species is rare, even in its known localities (see also D'Abrera 1975). When working in such areas, it is possible to find 25 larvae of *O. priamus* in the time it takes to find one or two *alexandrae* caterpillars. Although Straatman (1970) suggested that *O. alexandrae* could be abundant, he has since qualified this (pers. comm.) to mean that he estimated a total maximum population in some localities to be at a density of 25 individuals (all early stages) per acre. Most areas surveyed by Straatman yielded far fewer early stages or none at all.

What Limits *O. alexandrae*?

The reasons for the extremely limited distribution and low population density of *O. alexandrae* remain unclear. It appears that no single factor is responsible, but certain ideas are being tested. A major problem of obvious significance is the host-specialization of *alexandrae*, and the reasons for its monophagy (cf. Ch.9). Perhaps apparently suitable *A. schlechteri* vines from certain areas lack appropriate ovipositional cues; biochemical investigations are now being carried out.

Another likely factor is competition with *O. priamus* (Parsons 1980*a*). Larvae of both species can be found on the same vine, and *O. priamus* often occurs in parts of the Popondetta region where *alexandrae* is absent. Interspecific competition may not account for the present day distribution of *O. alexandrae*, but it may be the main reason for its monophagy on a foodplant which is apparently suboptimal with respect to development rate (see above). The 'better' foodplant, *A. tagala*, could have been its host in the past. Smiley (1978*a*), studying apparently analogous situations in *Heliconius*, termed such specializations 'ecologically induced monophagy'.

I have found the two common birdwings at Bulolo, Morobe Province (*O. priamus* and *T. oblongomaculatus*) to suffer 70% or higher mortality from two unidentified chalcid egg-parasitoids. The smaller of the two apparently also attacks *Pachliopta polydorus*, another common *A. tagala*-feeding troidine. As suggested to me by A. J. Pontin, if such a generalist egg-parasitoid occurs in the Popondetta region, by continuously breeding in the eggs of the common birdwings in the area, it may seriously deplete rare species such as *O. alexandrae*. New Guinea birdwings often lay their eggs on objects near their foodplants, rather than directly on them—a possible method for reducing the chances of egg predation or parasitization (Ch.7). Egg parasitoids may well be, therefore, an important factor in the ecology of *O. alexandrae*.

Conservation Problems

Concern for *O. alexandrae* was first evinced by its inclusion under the Fauna Protection ordinance of 1966, and the distribution surveys carried out by the Australian Government. At that time most of the birdwing's habitat remained intact, but much has now already been lost to the growing oil palm industry centred immediately on the range of *alexandrae* in the Popondetta Plain, and by logging. Equally important are local extinctions caused by slash and burn agriculture. Before the advent of large-scale exploitation, the effects of slash and burn were negligible, but now, aggravated by an increasing human population, it is an important element in the accelerating loss of remaining suitable habitat.

The Popondetta oil palm project, overseen by the Department of Primary Industry (DPI), is being developed in two ways (Anon. 1976). Small holdings are expected to cover about 5500 hectares (Fig. 32.3), while a private company is developing another 4000ha. An extensive area of habitat important for *O. alexandrae* (Straatman 1970; pers. obs.) was cleared in early 1979. Although a considerable part of the project affects existing grassland, or land formerly under cocoa, rubber and coffee cultivation, I estimate (based partly on information from D. Manton, DPI) that about 2700ha of forest that is actual or potential habitat for the butterfly is being converted to oil palm—with no long-term limit set to the ultimate area to be exploited.

Practical Measures

About 60 000ha to the west of Popondetta has been loosely defined as the Kumusi Timber Area

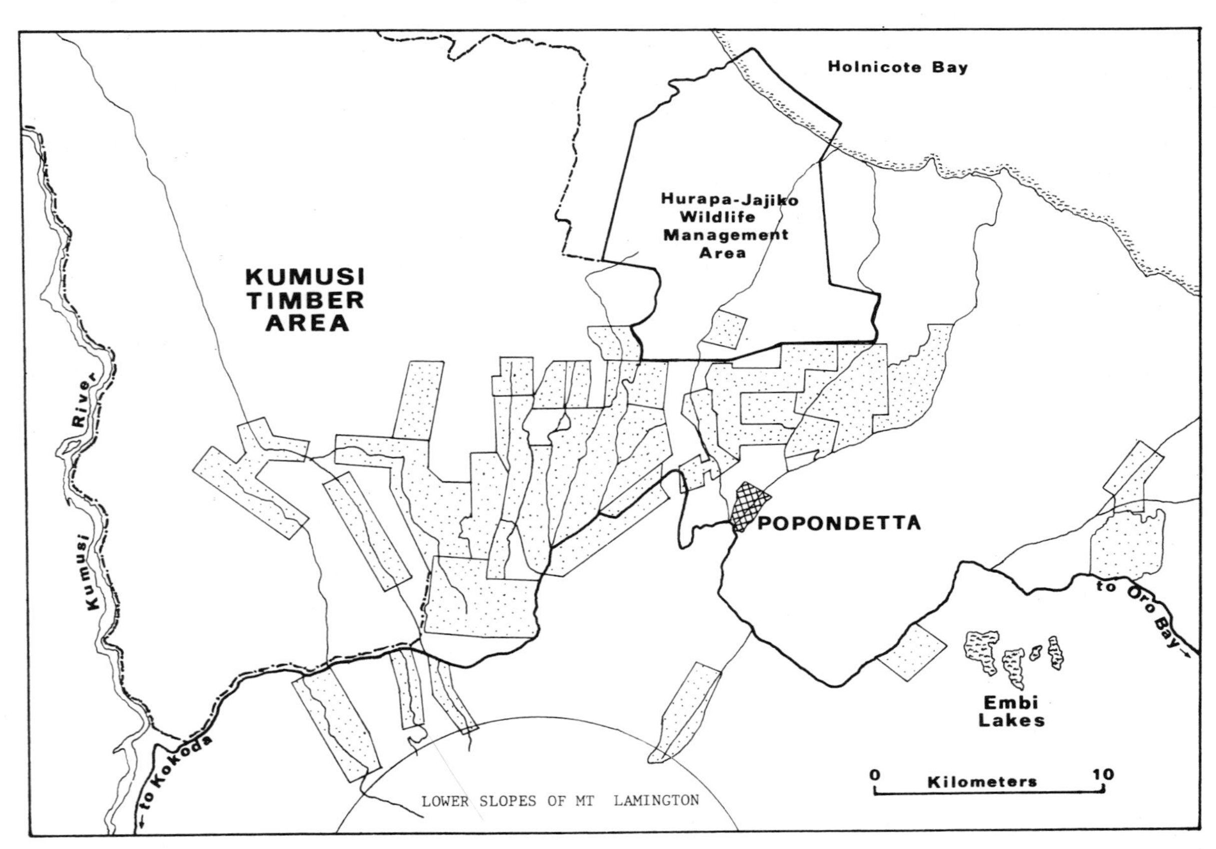

Fig. 32.3. The Popondetta region. The extent of the Popondetta Smallholder Oil Palm Development Project is shown as on the World Bank Map. (Main oil palm blocks are stippled.)

(Fig. 32.3). In September 1980, I and other members of the Wildlife Division surveyed this region, to establish an environmental plan detailing the land-use potential of various zones within it. As a result (Parsons 1980*c*, 1981), three areas where the butterfly occurred in favourable habitat are now being processed for designation as *O. alexandrae* reserves, to be exempted from logging and road access, and purposely managed for the benefit of the butterfly.

A few areas of government land have been rejected for use as oil palm plantations, and the Division of Wildlife has applied for the title deeds of these small areas. Being government-owned, they offer the best possible protection for the butterfly (once planted with *A. schlechteri*), because they can be fully protected by law.

Wildlife Management Areas (WMA) in Papua New Guinea can be set up at the request of traditional landowners, in discussion with the Wildlife Division. In Popondetta the Hurapa-Jajiko WMA of about 10 000ha has been established (Fig. 32.3) and certain areas within it are known to be used by *O. alexandrae*.

As a trial, two central rows of mature oil palms in a small experimental block at the Popondetta Agricultural College have been planted out with rooted cuttings of *A. schlechteri*. Each row contains 19 palms spaced at intervals of 5m. Approximately 7m tall, the palms provide ideal semi-shade with plenty of purchase up the trunks. The main aim is to observe the growth of the vine in these conditions and to see if *O. alexandrae* will eventually utilize them.

Production of *A. schlechteri* cuttings was begun in September 1980, with a long-term aim of creating new areas of potential habitat, especially where reserves can be established on government-owned land. It is hoped that a constant supply of rooted cuttings can be maintained at Lejo Agricultural Station, for use by all those interested in the conservation of *O. alexandrae*.

The Future

Because of the now intense agricultural use of land in the Popondetta region, the policy of the Wildlife Division has, of necessity, been to obtain as much land as possible for the conservation of *O. alexandrae*. Surveys to identify potential new reserve areas are continuing.

Education about the conservation problems of *O. alexandrae* has been extensively carried out by the Wildlife Division, in Northern Province. Response has been favourable from various companies and Higaturu Oil Palms Pty. Ltd., for example, have offered to assist by planting out forested areas on their estates with *A. schlechteri* cuttings.

In the long-term, once the future of *O. alexandrae* is assured, its economic importance cannot be overlooked. Fenner (1975) and Pyle & Hughes (1978) have discussed in detail the concept of satisfying the large worldwide demand of collectors for *Ornithoptera* by farming certain species. For the common, non-protected Papua New Guinea bird-wings, *T. oblongomaculatus* and *O. priamus*, this has already been done successfully. The unique ecological requirements of species such as *O. alexandrae* present many problems, but these are not insurmountable.

33. The Conservation of Butterflies in Temperate Countries: Past Efforts and Lessons for the Future

Jeremy A. Thomas

Institute of Terrestrial Ecology, Furzebrook Research Station, Wareham, Dorset, UK

Entomologists have long deplored the decline of certain butterflies (e.g. Grote 1876, Conquest 1897, Frohawk 1925, Ford 1945) but have lacked the resources or expertise to remedy this. Nor, except in the United Kingdom, have they significantly influenced the activities of general conservation bodies, whose main efforts have been to combat pollution, preserve whole biotopes and protect spectacular vertebrates. Naturally, it was hoped that insects would benefit from the first two categories, but the needs of butterflies *per se* have rarely been considered. This passive attitude is now changing: there is growing lay concern about a dearth of butterflies in the countryside (Thomas 1981*a*), and it is also clear that an unacceptable number of extinctions has occurred in nature reserves and national parks (Heath 1981*a*, Pyle *et al.* 1981, Thomas 1980*b*, Peachey 1982). Butterflies have consequently entered the agenda of most conservation bodies, and numerous committees and societies are being formed to consider the problem. These advise at world (Pyle 1981*a*,*b*), continental (Heath 1981*a*), national (Morris 1981*a*,*b*, Opler 1981, Pyle 1981*c*, Blab & Kudrna 1982), and local levels (Powell 1981, Weems 1981).

The aim of this review is to consider how well butterflies have been served by previous conservation measures, and to suggest how scarce resources might be used more effectively in future. First, it is necessary to describe the range of organizations on which these measures depend, and to define the problems that they face with butterflies. This has not been attempted in detail before, and thus much of this review describes changes in the status that are occurring in temperate countries, how these are being monitored, and suggests reasons for the changes so far as they are known. I have also tried to concentrate on the practical problems of conservation, and so devote little space to aspects of advisory work and legislation already covered by others (Pyle *et al.* 1981, Morris 1981*a*,*c*, Orsak 1981). They and other authors (e.g. Ehrlich & Ehrlich 1981*b*, Gilbert 1980; Chs 2, 3) have already made an eloquent case as to why butterflies should be conserved; I merely add a few points.

Most of this paper describes experiences in the UK. This was inevitable, for Britain's few species have been studied more comprehensively than those occurring elsewhere, and no other country has a history of insect conservation that is long enough to be judged on results (Morris 1981*c*, Pyle *et al.* 1981, Pyle 1981*c*), although current efforts in the USA and Holland now match those in the UK. If I dwell on certain failures, it is so that past mistakes should not be repeated, for the problems that have already been faced in the UK are similar to those being debated in many other highly populated, developed, temperate countries. The different problems and resources available for conserving tropical butterflies in third world countries are described by Gilbert (1980) and Parsons (Ch.32).

Organizations and Attitudes

United Kingdom

The practical conservation of butterflies in the UK depends almost entirely on the activities of generalist organizations whose remits cover all groups of wildlife. The most important are the Government's Nature Conservancy Council (NCC), the 42 voluntary local County Trusts, and the National Trust (NT). The main thrust of NCC and County Trusts

The Biology of Butterflies
0-12-713750-5

has been to establish nature reserves throughout the UK: the former's are usually chosen as prime examples of a biotope, are large, and have a permanent warden; the latter's tend to be smaller, without wardens, and are often selected to conserve a few rare species. These reserves are complemented by others established by the Royal Society for the Protection of Birds (RSPB; largely for birds), Local Councils (LNRs), the Forestry Commission (FNRs), and private individuals. The NT rarely selects properties because of a wildlife interest, but is committed to maintaining those fortuitously acquired.

NCC also has other duties, which include contracting research programmes and encouraging wildlife conservation on commercially managed land. This is mainly done through establishing Sites of Special Scientific Interest (SSSIs), over whose futures they have some influence but no control (Moore 1981).

Most monitoring and research has been carried out by the Institute of Terrestrial Ecology (ITE), partly under contract to NCC. ITE has pioneered a scientific approach to the conservation of butterflies (Hall 1981) and advises most organizations. The priorities of all conservation bodies have been strongly influenced by public opinion, the media, and specialist advice and pressure. Two bodies exist specifically to promote insect conservation: the Joint Committee for the Conservation of British Insects (JCCBI) and the British Butterfly Conservation Society (BBCS). The former is an advisory committee representing the major entomological societies in the UK (Morris 1981*b*). It has also undertaken small practical projects on its own, such as devising a code for collecting, and encouraged other groups to compile a Red Data Book (Heath 1981*b*) and to promote 'Butterfly Year' with the BBCS (Thomas 1981*a*). The BBCS, founded in 1968 and totalling 1800 members by 1981 (Tatham 1981), has become an increasingly important and vocal lobbyist, but is still too small to undertake major practical projects.

Other Countries

Practical conservation elsewhere depends on a similar structure of executive generalist bodies whose actions are influenced by pressure groups and public opinion, although fewer agencies exist than in the UK. In the USA the Xerxes society has been at the forefront of insect conservation for a decade, whilst Societas Europea Lepidopterologica is beginning to influence European policies. Strong outside pressure is also being exerted in weak areas by the IUCN-WWF, which now has a Lepidoptera Specialist Group (Pyle 1981*a*).

Assessing the Status of Butterflies

United Kingdom

Despite the presence of many entomologists in the UK, conservationists were gravely handicapped until recently by ignorance of species' status, the rates at which changes were occurring, and the location and boundaries of important sites. Thus, although the declines of a few spectacular species, such as *Maculinea arion*, were detected early enough for attempts to be made to save some colonies (Hunt 1965), others such as *Carterocephalus palaemon* reached the verge of extinction in England before anyone noticed (Farrell 1975).

Data-gathering has been transformed in recent years by four complementary schemes, described below. These now provide an outstanding service that is unmatched elsewhere. The main constraint on their effectiveness is that the results are not currently being pooled or seen by any single organization.

National butterfly mapping scheme

In the mid-1960s, ITE's Biological Records Centre (BRC) extended its botanical recording scheme to cover Lepidoptera and other animal taxa. The aim was to encourage the UK's many (mainly amateur) entomologists to pool past and present butterfly records. Technical details are given by Heath & Perring (1978), Hall (1981) and Heath & Harding (1981). To date, nearly 2000 individuals have participated. A 'definitive' atlas is now in preparation (e.g. Fig. 33.1). These data provide a vital foundation to most conservation projects.

Other European countries, and some American states, have also started mapping schemes, usually based on the techniques developed by BRC, but none has achieved the same cover (Leclercq & Gaspar 1971, Clench 1979). Modern developments in data handling mean that new schemes can rapidly produce

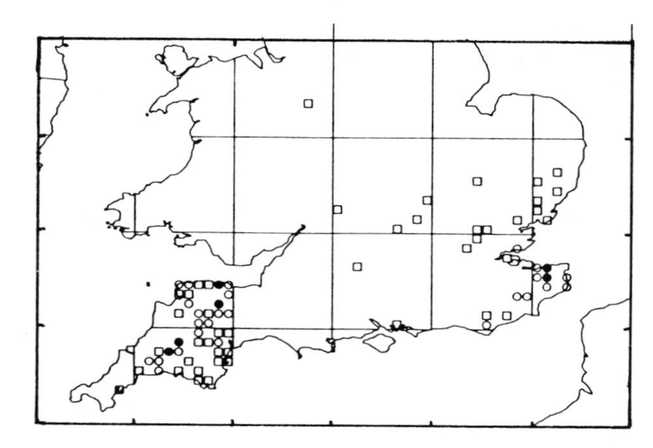

Fig. 33.1. The changing distribution of *M. athalia* in Britain, plotted by 10km squares. From Warren *et al.* 1981. □ Pre 1910; ○ 1910–1980; ● 1981.

print-outs of species lists, recording dates, named localities and grid references. These should be easily and cheaply obtainable if the data are to be fully used by conservationists.

Local mapping schemes

These resemble the BRC scheme, but each is restricted in range, usually to a county boundary (Smith & Brown 1979, Gower 1980). More detailed recording and more even coverage is obtainable, and it is possible to draw maps to a scale of tetrads or 1km squares. The data are particularly helpful in pinpointing outstanding sites which may need protection. Most local schemes are run by the County Trust or museum, and are financed by voluntary subscription. Many counties have yet to start a mapping centre and their encouragement (perhaps through the RSNC or BBCS) is a high priority.

Species surveys

Although serious changes in status may now be detected through mapping schemes, the data are rarely adequate for a decline to be quantified with precision. Moreover, conservationists want to know the location and boundaries of all colonies, and, if possible, their population sizes. To obtain such information it has been necessary to make an intensive survey of all recorded and potential sites for a species.

To date, national surveys have been made of nine butterflies in the UK (Hunt 1965, Buxton & Connolly 1973, Farrell 1975, Thomas 1974, 1976*a*,*b*, Simcox & Thomas 1980, Warren 1981, Warren *et al.* 1981, D. J. Simcox, C. D. Thomas pers. comm.). There is a pressing need for similar data on at least five other species.

Heath (1981*b*) has produced a similar list for Europe. Recent UK surveys have been organized by ITE or JCCBI. Details necessarily differ with the species, but a general pattern has evolved (Thomas 1983*b*). This is similar to that suggested by Clench (1979), but incorporates a simple method for quickly assessing the size of each population. Most surveyors also compare the conditions found on former sites with those that now support colonies of various sizes. A conservation report is then written in lay terms and distributed free to relevant landowners and conservationists. Most recommendations receive prompt action.

Butterfly monitoring scheme (BMS)

Pollard *et al.* (1975) and Pollard (1977, 1979*a*, 1981*a*, see also Ch.5) describe a simple method of transect recording being used to measure relative changes in adult numbers of every species of butterfly on 81 UK sites. Most sites are nature reserves, which, between them, support at least one population of all but three resident species. This national scheme began in 1976 and seems likely to become one of the most important tools in butterfly conservation. It reveals: (1) National and regional fluctuations in numbers for each species. It may now be possible to identify long-term declines before these are apparent (as extinctions) to the mapping schemes. (2) Local changes in numbers on or within part of a site. If a species is increasing or decreasing disproportionately to the regional trend, or to other parts of the site, it is probable that the habitat has changed locally and the reason may be obvious. If desired, the causes of any increase may be deliberately maintained or applied to other sites, or the causes of declines may be rectified. Pollard (Ch.5) gives examples of changes revealed by the BMS in populations of two species.

The present number of sites (81) whose data contribute to the national scheme is all that can be conveniently analysed. However, this quick method of recording provides such valuable information that many other wardens have started their own 'Pollard Walks' (see Hall 1980), and can compare the changes on their sites with the national and regional trends. This technique should be used on all reserves with important butterfly colonies. Another benefit is the enthusiasm for butterflies that it has kindled among many wardens.

The BBCS has also been running a scheme with the same aim of monitoring changes in adult numbers. However, the recording method is less objective and much less confidence can be placed on the results. The drive of the many BBCS participants might be better employed in transect recording. An important contribution could be made by concentrating on commercially-managed sites (the BMS is deliberately biased towards nature reserves).

The only comparable scheme elsewhere is the annual 'July 4th Count' in the USA, organized by the Xerxes Society (Powell & Sorenson 1980). Again the recording technique is not rigorous, and the results are best regarded as qualitative.

Results: recent changes in status of butterflies

This summary is based on the results of the four schemes described above, data from autecological studies of 14 species (24% of UK butterflies) and research described below.

There are now 55 UK-resident butterfly species and three immigrants that regularly breed. Among residents, 85% form closed sedentary populations; the rest range from locally mobile (e.g. *Pieris napi*) to semi-migratory species (e.g. *Pieris brassicae*). Many colonies fluctuate greatly in size, and several populations of most species still exist that attain very high numbers in 'good' years. However, most breeding sites are rapidly disappearing, and where this has not obviously occurred, the long-term trend (as distinct from annual fluctuations) of several butterflies has been downwards. This has resulted in numerous local extinctions among more sedentary species, whilst the rate at which new colonies have been formed by more mobile butterflies has rarely matched the increased extinction rate of other populations. Overall, it seems that during the last 25 years, probably three (*Thymelicus acteon*, *Leptidea sinapis*, *Pararge aegeria*) resident butterflies have increased their number of colonies in the UK, roughly eight (*Pieris brassicae*, *P. rapae*, *Gonepteryx rhamni*, *Celastrina argiolus*, *Aglais urticae*, *Inachis io*, *Polygonia c-album*, *Ladoga camilla*) species have held their own, whilst 44 have declined in at least a major part of their ranges. Some declines have been extremely rapid: the number of colonies of *Lysandra bellargus* has approximately halved every 12 years since the war and is probably now down to 70–80 sites (Thomas 1983*a*), whilst *Mellicta athalia* was found in only eight woodland blocks in 1980 and had become extinct in two of them by 1981 (Fig. 33.1). The presumed extinctions of *Carterocephalus palaemon* in England in 1976 (it still survives in Scotland), and of *Maculinea arion* in 1979 were the first species losses for at least 50 years, but more extinctions seem inevitable soon unless declines are deliberately checked.

Other Temperate Countries

Most other developed countries have apparently experienced similar declines to those in the UK, although they are less well documented. There have been numerous local extinctions and at least two endemics have already been lost in the USA, where several more species are threatened, especially in California (Pyle 1976*b*, Pyle *et al.* 1981, Orsak 1978, Anon. 1980*a*,*b*). In Europe, Holland has recently lost eight butterfly species and another 20 are considered to be in imminent danger (Heath 1981*a*). Similar losses have occurred throughout northern France, whilst Belgium, East Germany, Finland and Denmark have each had at least one recent extinction and expect several more. Heath (1981*a*) considers that 4% of all 362 European species are imminently endangered with extinction on the Continent, and that another 14% are 'vulnerable', i.e. expected to become endangered if the causes of their declines do not abate soon.

Identifying the Causes of Changes in Status

It is normally essential to discover why a species is in decline if it is to be successfully conserved. Examples are given later of theories that were used as the rationale for supposed remedies, only to be proved tragically wrong. An obvious cause of local extinctions has been the fundamental destruction of entire biotopes, such as the clearance of forests, 'improvement' of pasture or its conversion to arable land, the drainage of wetlands, and urbanization. This has caused extensive losses both in temperate countries (USA: Pyle 1976*a*,*b*, Pyle *et al.* 1981; UK: Ford 1945; Europe: Heath 1981*a*) and the tropics (Gilbert 1980, Pyle *et al.* 1981). Some of these processes have been occurring for thousands of years in Europe (Rackham 1981), but all have intensified dramatically during the present century.

It was often assumed that wildlife would be conserved merely by preventing such gross losses from occurring. However, many butterflies have become extinct on nature reserves throughout the world (Section 5) and numerous species have been lost (and a few colonizations have occurred) on commercially managed land that has superficially changed little, if at all (Ford 1945, Benham 1973, Muggleton 1973). Many hypotheses have been advanced to account for such changes. One suggestion is that biotopes may actually have altered inconspicuously, for example in structure (Frazer 1961, 1977, Spooner 1963, Hunt 1965, Lipscomb & Jackson 1964). But most ideas take the premise that the habitats have not changed and that other factors must be responsible. The ones that have most influenced UK conservationists have blamed the increased isolation of many colonies, bad weather or a change in the climate, insecticides and pollutants, and butterfly collectors. The first three of these ideas are particularly depressing as little can be done to remedy them; the conclusion has been to switch scarce resources to tractable problems.

Until recently, these various ideas were speculative or, at best, based on anecdotal observation. Little was known even about the most basic ecological requirements of most species. For example, to my own (unexceptional) present knowledge of larval foodplants, those listed in the UK's two recent textbooks (Howarth 1973*b*, Goodden 1978) are either unspecific (e.g. 'grasses') or missing a major species, or (in two cases) wrong for one-third of the listed butterflies. However, in the past 20 (mainly five) years, the autecology of 14 (25% of residential) UK species has been researched to varying depths. Studies of nine species (*Lycaena dispar*, *Strymonidia pruni*, *Thecla betulae*, *Maculinea arion*, *Pieris rapae*, *Papilio machaon*, *Leptidea sinapis*, *Ladoga camilla* and *Euphydryas aurinia*) included a compilation of lifetables, although none approaches the completeness of some research on moths (Dempster 1971*a*). A quicker approach was used with the remaining five species *Mellicta athalia*, *Melitaea cinxia*, *Lysandra bellargus*, *Thymelicus acteon* and *Hesperia comma*). Studies were made of the population structure and behaviour of all stages, of the conditions found on a range of sites that supported variously sized (or no) colonies, and of environmental changes that have preceded fluctuations in numbers on fixed sites over several years. Similar but less detailed empirical data are being obtained for many other species through the BMS (Pollard 1979*a*; Ch.5), whilst Peachey (1980, 1981) and K. Porter (pers. comm.) have studied the changes in adult distribution and numbers in a large commercially-managed woodland complex.

Great advances have also been made in understanding other aspects of the biology of butterflies (e.g. Baker 1969, Brown & Ehrlich 1980, Ehrlich *et al.* 1980, Gilbert & Singer 1973, 1975, Shields 1968, Wiklund 1977*a*, Chs 2, 3, 6, 7, 10, 16, 26). With an increased understanding of insect population ecology (e.g. Southwood 1968), it is now possible to reassess the causes of changes in butterfly populations more objectively than hitherto. In doing so, the hypotheses that have most influenced conservationists are separately considered. Most space is given to a review of how chance changes in land management can profoundly alter the carrying capacities of sites for butterflies: all species have proved extremely particular about the niches they occupy, and seemingly minor habitat alterations often result in poor survival of the young stages and/or reduced natality. This is the overwhelming cause of long-term changes in status among species that have been studied. Some local sedentary species are also hindered by an inability to colonize suitable habitat readily. Other factors, notably the weather, may cause short-term fluctuations, but have had little effect on the long-term survival of a population. The main examples are again taken from the UK.

General Restrictions on Distribution in the UK

The small number of species that occur in Great Britain, and the even smaller fauna of Ireland, is probably a consequence of isolation from mainland Europe, past and present climates, limited biotopes, and the poor mobility of many species. With the addition of edaphic factors, these also determine the range and local distribution of each species within the UK. Some, such as *Lysandra bellargus* and *L. coridon* are limited by their foodplant to particular geological formations, but the former is also restricted within such areas to sites that have unusually warm local climates (Muggleton 1973, Thomas 1983*a*). Similar restrictions can be listed for most species (Beirne 1947, Ford 1945, Dennis 1977).

Habitat Changes in the UK

General

The UK landscape (and most of Europe) has been drastically altered by man for at least 6000 years; for a millenium all but a small fraction has been deliberately managed to grow wood or food. The way in which these commodities have been produced has also varied, and has been transformed in almost every aspect in the present century (Dudley Stamp 1962, Nicholson 1970, Mellanby 1981, Rackham 1980, 1981). One side effect has been the creation and destruction of different habitats at various rates of change. Most butterflies have far more specialized requirements than was once supposed; apart from a few migratory species, the UK is evidently left with those butterflies that have been able to tolerate (and often flourish under) the conditions that man has fortuitously created over the centuries.

Woodland habitats

Following the last glacial maximum, virgin forest recolonized most of the UK by 5000 BC, but was reduced to 15% of the land surface in the next 6000 years. This diminished to 10% in the twelfth and thirteenth centuries and to the present 8% since 1800 (Rackham 1981). Since *c.* AD 1000, these fragments have changed more in structure than in size, for nearly all have been managed to produce various wood crops. Primeval temperate forest is generally considered to have had a patchy structure (Nicholson 1970). Then, for nearly 1000 years until this century, almost all woods were managed as coppices, often with standards. This created a continuum of very open conditions that were maintained in the early stages of succession. Standard trees were usually felled once the canopy reached 30% cover, and the shrub layer of coppice stools was often cut every three or four years over most of the UK, with very few cycles exceeding 10 years (Rackham 1971, 1981). Such regimes create ideal conditions for *Viola* feeding Fritillaries (Peachey 1980, Hall 1981), *Mellicta athalia* (Warren *et al.* 1981), *Hamearis lucina* (K. Porter pers. comm.), and *Leptidea sinapis* (Warren 1981). Thus, although these species lost most of their biotope through primeval clearances, they probably increased within the remaining fragments once these were actively managed. All were much commoner in Victorian coppices than in modern shadier woods. On the other hand, *Ladoga camilla* (Pollard 1979*b*) and *Pararge aegeria* (Pollard *et al.* 1975, Peachey 1980) need shady conditions and probably declined greatly both through the loss of biotope and the conversion of the remainder to open early successional habitats. It seems likely that other woodland species became extinct under these twin changes. One butterfly that perhaps only just survived is *Strymonidia pruni.* This forms small closed populations on sunny sheltered stands of *Prunus spinosa*, is extremely slow to colonize new habitat outside its immediate neighbourhood, and cannot inhabit woods that are frequently or drastically cleared (Thomas 1974, 1976*b*). It is unlikely that any colony could survive under a short coppice cycle, and, until recently, all populations were restricted to the east Midlands forest belt where alone there was an unbroken tradition for many centuries of long cycles spanning 20 years or more (Peterken & Harding 1975, Peterken 1976, Thomas 1974).

The demand for coppice products and large timbers slumped towards the late nineteenth century (Rackham 1981). Most wood is now grown as a crop that is planted in a large-scale operation, left (with occasional thinnings) to mature, and harvested after several decades, often by a single clear-felling. This leads to much more uniform conditions throughout each wood at any one time, and to deep shade during most of the cycle. This is exacerbated by the widespread replacement of native broadleaved trees by quick-growing exotic conifers. Few of the UK's butterflies feed on deciduous trees, but the shade cast by mature conifers is much deeper. Studies by Peachey (1980) and BMS data indicate that these changes have profoundly altered the capacity of modern woods to support most butterflies (notably the *Viola* feeding Fritillaries). This is certainly true of four woodland species that have been intensively researched: *Strymonidia pruni*, *Mellicta athalia*, *Leptidea sinapis* and *Ladoga camilla*.

S. pruni (Collier 1962, pers. comm., Thomas 1973, 1974, 1976*b*, 1980*b*) was well distributed in the east Midlands at the turn of the century, and at first survived well as coppices were abandoned. Odd colonies were lost because their woods were destroyed, but most sites were left to produce a 'natural' timber crop. However, local extinctions became common once these were harvested, and soon greatly outnumbered the establishment of fresh colonies. Scrub often grew rapidly in the replacement plantations, and became suitable for *S. pruni* after seven to ten years, but the poor mobility of this butterfly meant that it was only colonized in the rare cases when the discrete breeding stands had escaped clearance or, as is often the case nowadays, were deliberately spared. Several other colonies of *S. pruni* now exist only because of past re-introductions.

Although *S. pruni* has declined in the east Midlands and depends partly on conservationists for its survival there, the cessation of short-cycled coppicing elsewhere means that some woods in other regions may now contain suitable habitat for this butterfly. *S. pruni* was successfully introduced to the Weald in 1952, where it soon formed at least two colonies. Both were destroyed after nine years when the wood was cleared for agriculture, but in 1975 I found a colony in a nearby wood which still survives, 30 years after the original introduction. Indeed, it has supported the largest known British population of this species for at least six years.

M. athalia (Warren *et al.* 1981) also has limited powers of mobility in the UK and depends on the generation of a continuous supply of its habitat within fairly short distances of existing populations. It breeds only in woodland that is at an early stage of succession or regeneration, and is ill-adapted to most modern woods. Large populations may still develop in new plantations, but usually become extinct after about five years. This has already eliminated *M. athalia* from most sites in SW England (Fig. 33.1). The largest two (of only three) populations left there occur in three woods that were finally planted in 1979–80. As these will now be left to mature, their colonies will almost certainly become extinct soon, unless conservationists intervene. The other West Country colony occurs in an older plantation, but survives because the forester has a market for Christmas trees, so the site receives an annual thinning.

M. athalia survives in eastern Kent, in some of the last commercially coppiced woodland in England, but has declined there also as the area of coppice has diminished and been replaced by conifers. Its future is bleak because the local market for coppice products, a woodchip factory, is becoming uneconomic. If this closes, other forms of silviculture will be used in E. Kent and *M. athalia* will probably be eliminated.

L. sinapis (Warren 1981), also rather sedentary in the UK, declined considerably as coppices became overgrown in the first half of this century. However, new habitat has now been created fortuitously in Forestry Commission woodland, through their practice of maintaining wide rides for access, and ditches. Some abandoned railways have also developed suitable habitat. *L. sinapis* has spread quickly into these new sites, although in several cases the initial colonizers have been artificially introduced.

Ladoga camilla (Pollard 1979*b*) finds open woodland unsuitable and this butterfly was extremely restricted before the present century. However, the shadier conditions that developed in abandoned coppices enabled it to colonize much of the woodland in southern England. Although a mobile species, this spread did not occur widely until the late 1930s and 1940s after large populations had built up in its old sites during a series of unusually warm summers. *L. camilla* has maintained its extended range despite the generally cooler summers of the 1960s and 1970s, although often in lower numbers, but has again become extinct in some woods that became very densely shaded or which had few sunny rides.

Agricultural habitats

About 80% of the UK is managed for agriculture, and a similar percentage (73%; 40 spp.) of resident butterflies breed entirely or largely in this historically recent biotope. These range from 28 species that inhabit open grassland to those that breed mainly along hedgerows. These species were probably once limited to tracks, woodland pasture, temporary openings of the canopy and to localities where trees grew poorly, and presumably spread and increased

Table 33.1. The number of butterfly species that can be supported by different farmland habitats in the UK.

	Maximum per site	Total number of species
Unimproved patchy pasture	23	28
Unimproved tall grassland	21	23
Unimproved short pasture	17	19
Uncut sown grassland	3	3
Grass/clover ley	1	1
Arable	2 (Brassica crops)	2 (Brassica crops)
Hedge/verge	16	16
Rough corners/verge	10	10

Compiled from Anon. 1977, amended from Pollard 1981*a*, Thomas & Merrett 1980, Thomas 1982, 1983*b*, Frazer 1961, 1977, and other surveys.

on ancient agricultural land. In the past 40 years, modern agri-business has been so intensive in some regions (Cambridgeshire; most flatter parts of Central and northern Europe) that most traditional farmland species have largely been reduced again to colonies in woodland. The wide, permanently open, grassy rides of many modern woods are very different from the freshly cleared conditions needed by most 'woodland' species, but can support large populations of many 'grassland' butterflies (Peachey 1980, Pollard 1981*b*).

The modern agricultural revolution was mainly caused by the development of power machines, chemicals, and new breeds of crops and livestock. Southwood (1971), Anon. (1977), Smith (1980), and Mellanby (1981) review the side-effects on wildlife. On efficient farms, this has caused either the destruction of all breeding habitats or their alteration in character to the benefit of certain species and the detriment of others (Table 33.1).

Butterflies that need shrubs have lost 20% of these through recent hedge removal (Pollard *et al.* 1974). Much higher losses have occurred locally; the extinction of *Thecla betulae* in Huntingdonshire is one probable consequence (Thomas 1974). This species has almost certainly also been harmed by the *way* hedges are now managed, for mechanical cutters often cut deeper and more uniformly than did manual labour. Thomas (1974) estimated that nearly 50% of overwintering eggs on *Prunus* twigs were killed annually in the Weald because of this. Populations were unlikely to survive in regions that contain fewer wood edges to act as reservoirs or where hedges were cut more often. Other hedgerow butterflies have almost certainly been lost through the destruction of the ground flora by persistent herbicides, but there is no documentary evidence of this.

Much greater losses have occurred within fields. About one-third of the UK is arable land (Anon. 1977) and can only support *Pieris rapae* and *P. brassicae*, which still often reach pest densities. In contrast, one unit of pasture or rough grazing often supported whole colonies of up to 23 species, excluding those that are associated with shrubs (Table 33.1). Pasture still comprises 40% of the UK, but nearly all has been 'improved' by herbicides, fertilizers, drainage, drilling, ploughing, or seeding; native grasses and herbs have been replaced by a dense sward, mainly of cultivars and exotics. This eliminates all the known larval foodplants of UK butterflies, although invasive patches of *Urtica* spp. and *Carduus* spp. may support the mobile species, *Aglais urticae*, *Vanessa atalanta* and *V. cardui.* Otherwise, modern improved pasture is entirely bereft (Table 33.1).

Unimproved pasture is now rare in southern and lowland Britain where, for climatic reasons, most butterfly species occur. Only 0.4% of Huntingdonshire's clays remain in this state (Mellanby 1981), as does just 3% of the traditionally rich chalk outcrop of southern England (Smith 1980). The surviving fragments consist almost wholly of hills too steep for economic cultivation, or of non-agricultural land, such as military ranges. Some habitats for grassland butterflies are also provided by roadside verges and rough corners, but few species seem able to breed on such fragments (see 'isolation and area').

The diversity of habitats within surviving biotopes has also diminished. Traditionally, unimproved grassland was maintained in various seral stages ranging from close-cropped open turf, to tall and dense hay swards. Today, very little early successional grassland exists, because non-agricultural land is rarely stocked, and regular grazing is usually uneconomic on improved farmland. For example, only one out of 47 unimproved former *Maculinea arion* sites still had a short open turf when surveyed in 1972–76 (Thomas 1980*a*,*b*). This shift towards taller denser swards was greatly exacerbated with the loss of rabbits when myxomatosis was introduced in the early 1950s (Mellanby 1981).

Few grassland butterflies have been studied in detail, but the BMS returns (Pollard 1981*a*; Ch.5) and other research (Frazer 1961, 1977, Buxton & Connolly 1973, Simcox & Thomas 1980, Thomas

& Merrett 1980, Thomas 1981*b*, 1983*a*, *c*) leave little doubt that the carrying capacity of grassland varies greatly according to its structure and management, as is well known for other insects (Dempster 1971*a*, Morris 1971, 1979, Hutchinson & King 1980). The preponderance of overgrown swards has undoubtedly benefitted several grass-feeding Satyrinae and Hesperiidae, which may form large populations under these conditions (Lipscomb & Jackson 1964, Frazer 1961, Pollard 1981*b*, Ch.5), although local increases have been exceeded by the extinction of colonies on improved farmland. *Thymelicus acteon* alone has probably increased in absolute terms. In historical times, colonies have been largely restricted to warm hills in south-east Dorset, most of which have been abandoned rather than improved. The larval foodplant, *Brachypodium pinnatum*, has both spread on these sites (Wells 1976, Smith 1980) and grown into the tall clumps selected for oviposition (Frohawk 1934). This has allowed *T. acteon* to increase dramatically; one population increased 25-fold in four years compared with others in stable habitats. This butterfly is now one of the most widely distributed and abundant butterflies in S.E. Dorset (Thomas 1983*c*).

Local increases of Hesperiidae and Satyrinae have been offset by the decline of many species (especially Lycaenidae) whose low-growing foodplants are reduced or eliminated by invasive coarse grasses. Although many accounts are anecdotal, there is little doubt that *Aricia agestis*, *Lysandra coridon*, *Polyommatus icarus*, *Erynnis tages* and *Pyrgus malvae* have been lost or declined on numerous overgrown sites (Crane 1972, Davis *et al.* 1958, Lipscomb & Jackson 1964, Frazer 1961, 1977, Fearnehough 1972, Muggleton 1973, Wells 1976). In contrast, large populations of these species often survive in the same region on odd, well-cropped 'islands' (Thomas & Merrett 1980). Four other local species only use their foodplant if it is growing in a very short or sparse, open sward (*Hesperia comma*, *Lysandra bellargus*, *Maculinea arion*, *Eumenis semele* (calcareous grassland sites)). All disappeared from many localities before there was a discernible loss of foodplant. Two case-histories concern *Lysandra bellargus* and *Maculinea arion*.

Lysandra bellargus. About two thirds of the extinctions of *bellargus* have occurred on sites that still support large populations of its foodplant, *Hippocrepis comosa* (Frazer 1961, Lipscomb & Jackson 1964, Buxton & Connolly 1973, Muggleton 1973, Thomas 1983*a*). Colonies survive only on heavily grazed sites; by 1978 they had disappeared wherever *H. comosa* grew in turf that was over 5cm tall. Monitored populations have fluctuated under all grazing regimes in the short-term, but have been consistently larger (and less prone to extinction) under intensive cropping. They rapidly disappear when grazing is abandoned. This dependence on early seral grassland is probably explained by the fact that *L. bellargus* reaches the northern limit of its range in the UK and has always been restricted to unusually warm localities (south facing slopes in southern England). Tall or dense pasture may simply be too cold at the soil surface, where most of the immature period is spent; for example, a 5cm increase in turf height resulted in an 8°C drop in daytime ground temperature under *H. comosa* at one site. This may also be important indirectly through its effect on ants, which are generally more numerous in warm situations (Brian *et al.* 1976). *L. bellargus* is tended incessantly by ants from the second larval instar until adult emergence, and is presumably protected by them (cf. Chs 3, 19).

Maculinea arion. The obligate parasitic relationship of *M. arion* with *Myrmica* ants is well documented (Frohawk 1934). It has recently been discovered that only one species in this genus, *M. sabuleti*, is a suitable host in the wild, although odd individuals may survive in the nests of other *Myrmica* species (Thomas 1977, 1980*a*,*b*). Very large populations of *M. sabuleti* are needed to support a colony of *Maculinea arion* and, in the UK, adequate densities only occur on warm sites that have a very short sward. Changes in ant populations may be extremely rapid, with *Myrmica scabrinodis* (an unsuitable host) often replacing *M. sabuleti* in slightly taller turf, and both disappearing in even shadier conditions (Fig. 33.2). By contrast, *Thymus drucei*, on which the eggs of *Maculinea arion* are laid and the young larvae feed for a few weeks, is much more tolerant of shading, although it too disappears eventually on very overgrown sites.

M. arion was always extremely local in the UK and has been in periodic decline (since records began) until its presumed extinction in 1979. Roughly half its former sites have been destroyed by agricultural improvement or changes in land use, but large acreages survive that still contain far higher densities of *T. drucei* distributed over much greater areas than are needed to support a population of *M. arion* (Thomas 1976*a*, 1977). By the 1970s, none of these sites contained an adequate population of *Myrmica sabuleti* (it had entirely disappeared from several) although a few supported high densities of *Myrmica scabrinodis*. Losses of *Myrmica sabuleti* are attributable to a nearly universal cessation of domestic and rabbit grazing on the unproductive pasture of these sites, and with them has disappeared their parasitic butterfly. *M. sabuleti* populations can easily be manipulated by maintaining higher levels of domestic grazing than is economic (Fig. 33.2), whilst the recent return of high densities of rabbits to at least one former site has restored its densities of *M. sabuleti* to pre-extinction levels.

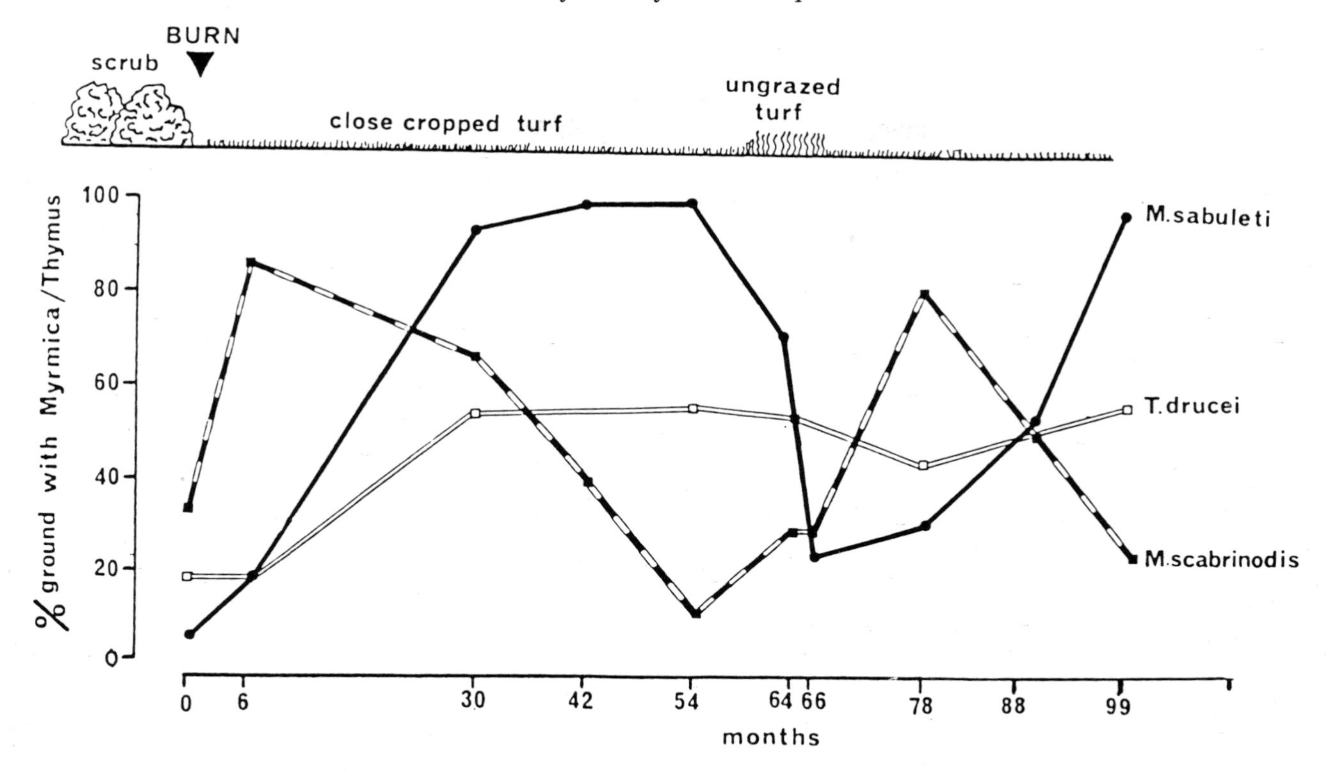

Fig. 33.2. Changes in status of *T. drucei* and the ants *M. sabuleti* and *M. scabrinodis*, under different management regimes.

Wetland habitats

Two prized butterflies were confined to the UK's wetlands: *Papilio machaon britannicus* and *Lycaena dispar dispar*. Both once bred in the E. Anglian Fens, but large reclamation projects started when drainage became technically possible in the seventeenth century. In 1851, the last great fenland mere of Whittlesea was drained, and *L. dispar dispar* disappeared in the same decade. The extinction of the last colony was probably hastened by collectors, but the national demise of the species was undoubtedly due to the destruction of its habitat (Duffey 1968).

A population of *P. machaon* survived in one fragment of fen at Wicken in Cambridgeshire, but became extinct in the early 1950s (Dempster *et al.* 1976). Frequent attempts were made to reintroduce it—but failed because this site has become too dry for British *machaon* in all but the wettest years. This is a side effect of the drainage of surrounding land for agriculture, which has caused the peat to shrink and oxidize, leaving Wicken Fen as an island of high ground (Dempster & Hall 1980). *P. machaon*'s foodplant, *Peucedanum palustre*, grows poorly under these drier conditions, and insufficient plants are large enough at Wicken to support a population of this butterfly, which requires prominent flowering plants for oviposition. There is now evidence that *Peucedanum* itself is declining there. Fortunately, a few populations of *P. machaon* still survive elsewhere in an ancient man-made biotope, the Norfolk Broads, although even there its habitat has become reduced through scrub encroachment or the erosion of the reed margins (Hall 1981).

Isolation and Area

Three different theories, discussed separately below, have been invoked to suggest that the absence or decline of butterflies on suitable-looking land can be blamed on increased isolation of the sites concerned. Each hypothesis is affected by the population structure and mobility of different species. Population structure is summarized for UK butterflies in Table 33.2, although for about 12% of species, the distinction between closed and open populations is poorly known, or may be indistinct or variable. For example, *Pieris napi* and *Anthocharis cardamines* (Courtney 1980) occur in more sedentary isolated populations in the north, but are listed as 'open' because adults are more mobile in the south, probably forming transient or dispersed colonies wherever their foodplants develop. In contrast, populations of the mobile butterflies *Thecla betulae* and *Apatura iris* breed at low densities over wide areas, usually incorporating many hedges or woods respectively within particular areas. Since the boundaries of these breeding areas are reasonably distinct, and the adults within them congregate to mate on one (possibly more) prominent point (cf. Shields 1968), these populations are considered to be predominantly closed.

The minimum area of biotope in the UK known

Table 33.2. The population structure of UK butterflies and minimum area from which a viable colony has been recorded.

Minimum Breeding Area (ha)	Closed Populations 0.5–1	1–2	2–5	5–10	10–50	> 50	Unknown area populations	Open or migratory populations
	T. lineola	*T. acteon*	*E. aurinia*	*B. selene*	*P. machaon*	*A. iris*	*C. palaemon*	*C. argiolus*
	T. sylvestris	*E. tages*	*M. cinxia*	*B. euphrosyne*	*T. betulae*		*A. artaxerxes*	*P. c-album*
	H. comma	*P. malvae*			*L. dispar*		*A. aglaja*	*A. cardamines*
	O. venata	*L. sinapis*			*L. camilla*		*A. adippe*	*P. napi*
	Q. quercus	*C. rubi*					*A. paphia*	*P. rapae*
	S. pruni	*L. phlaeas*					*P. aegeria*	*P. brassicae*
	S. w-album	*A. agestis*					*L. megera*	*A. urticae*
	C. minimus	*P. icarus*					*E. epiphron*	*N. io*
	P. argus	*L. bellargus*					*E. aethiops*	*N. polychloros*
	L. coridon	*M. arion*					*C. tullia*	*V. atalanta*
	H. lucina	*H. semele*					*P. tithonus*	*V. cardui*
	M. athalia						*A. hyperantus*	*C. croceus*
	C. pamphilus							
	M. jurtina							
	M. galathea							
Total	15	11	2	2	4	1	12	12

Compiled from various references including Dempster (1971*b*), Dennis (1977), Peachey (1980), Wiklund & Ahberg (1978), Courtney (1980), Sutton (1981), Paul (1977), Blackie (1951), Morton (in press), Dowdeswell (1981), Pollard (1981*a*,*b*), Turner (1963); K. Porter, M. Oates, C. D. Thomas pers. comm., and unpublished MRR data on 21 spp. gathered by author.

to have supported an isolated, medium-large colony of each series that forms a closed population (Table 33.2) is based on empirical data for all but the few butterflies that have been intensively studied. Several species can probably be supported by smaller areas, given a higher incidence of their habitat within this. Anecdotal data suggest that most of the 'unknown' species (especially Satyrinae and Hesperiidae) can be supported by the smaller categories of area listed.

Fragmentation (area)

Muggleton (1973), Muggleton & Benham (1975), and Morton (1979) suggested that *Maculinea arion* may have disappeared from sites that still looked suitable because, what had been assumed to be separate closed colonies in a region, were really components of a larger mobile population; individual sites were too small to support a colony alone once others in the neighbourhood had been destroyed. In fact, population studies later showed that this was not the cause of most (if any) extinctions: *M. arion* forms closed populations that can be (and often have been) supported in isolation by one tenth the area of many surviving sites. Instead, sites had changed inconspicuously and no longer contained enough (or any) nests of the obligate ant host (see 'habitat').

Few of the declines that had puzzled scientists now seem to be explicable by this hypothesis, in view of the population structure of most butterflies (Table 33.2). However, it probably explains some local losses of *Thecla betulae* (see 'habitat') and *Apatura iris*. Conservationists should also be aware that the scattered tiny fragments of land that they often advocate (wild patches in gardens, unploughed corners in fields etc) will probably benefit only the few, highly mobile species (Table 33.2).

Colonization

About 85% of UK butterflies form closed populations (Table 33.2). Apart from *Limenitis camilla* (Pollard 1979*b*), *Thecla betulae* (Thomas 1974), *Lycaena phlaeas* (Dempster 1971*b*), and perhaps *Melitaea cinxia* (Thomas & Simcox 1982), all that have been studied are reluctant to cross unsuitable habitat. Several are extremely sedentary (cf. Ehrlich 1961; Ch.2). The mobile species *Papilio machaon* rarely left an isolated fen to which it was introduced, even when few egglaying sites were available (Dempster & Hall 1980). At the other extreme *Strymonidia pruni* has sometimes failed to colonize new habitat within a few hundred metres in the same wood (Thomas 1974) and no mixing at all has been detected between large colonies of *Lysandra bellargus* that are divided by 100m of scrub (Thomas 1983*a*). Isolation of the latter species was due to the nature of the barrier rather than the distance.

Not only do fewer habitats now exist for most species in the UK, but most modern forms of land management are large-scale and drastic, greatly reducing the continuity within remaining sites (see 'habitat'). There is little doubt that many transient habitats are now under-exploited when they do occur, even if they are close to existing populations. This conclusion is supported by the success of introductions of several sedentary species (Table 33.3; this is probably a considerable underestimate

Table 33.3. Successful introductions of UK butterflies known to the author.

Species	Number of successful introductions	Survival of introductions (years) *still surviving	Sources
S. pruni	>4	60*, c60, several >20*, 30* (to Surrey)	Thomas (1974)
M. athalia	3	>40, >30, c14	Luckens (1980), Pratt (1981)
P. argus	1	31*	Dennis (1977)
M. galathea	1	c32*	K. J. Willmott (pers. comm.)
M. cinxia	1	23	Watson (1969)
E. aurinia	1	15*	
L. sinapis	6	13*, 8*, 7*, 7*, 5*, 2*	Warren (1981)
L. dispar	3	13, 13, 12	Duffey (1968)
H. comma	1	3*	B. Taylor (pers. comm.)
L. bellargus	1	2*	J. Bacon (pers. comm.).

of total successes). Most have been *re*-introductions to former sites that lost and later regained a particular habitat, but were not recolonized naturally. Many re-introduced populations are extremely large. All listed extinctions occurred because the habitat was later destroyed; however, another introduction of *Euphydryas aurinia* became so abundant in the absence of its parasite that it exhausted the food supply (Gilbert & Singer 1975).

Genetic deterioration

Most theorists state that isolated populations are unlikely to increase significantly in homozygosity (through inbreeding) unless they regularly go through bottlenecks of fewer than ten adults (Nei *et al.* 1975), or to lose genetic variabilty (through drift) unless effective breeding numbers remain under 50 adults for perhaps 20–30 generations (Franklin 1980, Soulé 1980). Species (like butterflies) with high reproductive rates capable of rapid increase should be much more robust than this, and both phenomena are negated by perhaps one immigrant every second generation (Nei *et al.* 1975, Hooper 1971). The few field data available for any species suggest that wild populations experiencing natural selection may be much more resistant to these changes than theory suggests; also, some species have an inherent resistance to inbreeding depression (Berry 1977, Gray & Ambrosen in press, Franklin 1980, Schwaegerle & Schaal 1979). Most authorities now agree with Berry's (1971, 1977) conclusion that any wild population that is regularly so small as to experience genetic problems is much more likely to have ecological difficulties related to the reason why the species became rare in the first place.

It is only possible to make a tentative assessment of whether populations of UK butterflies are threatened on theoretical genetic grounds. Many colonies of sedentary species are so isolated that it is hard to envisage that they are refreshed by an immigrant every second generation, if ever, but it is impossible to quantify this. Although isolation has increased, this is not a particularly new phenomenon: several flourishing populations have apparently been in this situation for decades. Thomas (1983*a*) found no correlation between the rate of extinction of *Lysandra bellargus* populations and the distance from their nearest neighbour.

Estimates made at no particular time in the short-term fluctuations of 1–75 colonies of 35 (66%) resident UK species give an enormous range of population size, from units to hundreds of thousands of adults (from sources in Table 33.2). But most closed colonies of sedentary species consist of about 50–200 adults in most years. Ehrlich (Ch.2) reports a similar situation in the USA. Populations of 15–50 individuals are not uncommon, but most probably represent temporary troughs; BMS data indicate that colonies of most species undergo considerable short-term fluctuations. With a few exceptions, it is unlikely therefore that UK butterfly populations are so consistently small as to risk losing genetic variability. This may not apply to *Cupido minimus*, which forms highly sedentary, isolated colonies that rarely seem to exceed 30 adults (unpublished ITE survey; A. C. Morton pers. comm.). But this has always been the case for *C. minimus*, which has not declined more than other butterflies in its biotopes.

No known population in the UK has experienced regular bottlenecks of under ten individuals. Three documented examples that once fell this low declined from much higher numbers in one or two generations (for ecological reasons, drought, or high density-dependent mortalities) and were extinct the next year: *Lycaena dispar* (Duffey 1977); *Papilio machaon* (Dempster & Hall 1980); *Maculinea arion* (Thomas 1980*a*,*b*). This is rather quick for inbreeding depression, and no indication of this, such as sterility or a failure to eclose, was found. Final extinctions were attributable to inverse density effects (uneven sex ratios and unsynchronized emergences) and chance deaths. The hypothesis that *M. arion* was experiencing local extinctions elsewhere in the UK through inbreeding (Muggleton & Benham 1975) has been criticized by Morton (1979) on theoretical grounds, and is untenable if empirical data are considered: most known populations had been isolated for many years, but became extinct over a 15-year period coinciding with the (inconspicuous) loss of the ant host on every site; most individual colonies declined in about five generations from genetically 'safe' normal levels of several hundred adults (Hunt 1965, Thomas 1976*a*, 1977, 1980*a*,*b*).

Some lay conservationists in the UK were undoubtedly influenced by theoretical geneticists a decade ago, when gloomier predictions than today's were fashionable: the argument that existing populations were already doomed was used (perhaps as an excuse) as the justification for not tackling difficult practical problems. This attitude has changed. However, known extinctions support Berry's (1971, 1977) conclusion that, if populations are allowed to become consistently so small as to incur possible genetic harm, they are very likely to be lost for other reasons.

Weather and Climate

Gilbert & Singer (1975; see also Chs 2, 3) list several ways in which variation in the weather can affect the survival and natality of individual butterflies. The effect of climatic factors on the size and distribution of populations is still unclear. Population analyses over three to eight generations indicate that variation in the weather was the key factor influencing changes in numbers in a colony of *Ladoga camilla* (Pollard 1979*b*), but not of *Lycaena dispar* (Duffey 1968) or *Thecla betulae* (Thomas 1974, unpublished). Short-term fluctuations of *Maculinea arion* (Thomas 1976*a*, 1980*a*, unpublished), *Strymonidia pruni* (over 17 generations on two sites; Thomas 1974*a*) and *Leptidea sinapis* (Warren 1981) were occasionally or largely attributable to the weather, but long-term trends in the size of colonies on all sites reflected changes in the habitat. The latter is also true for *Lysandra bellargus*, but no correlation was found between short-term variation in numbers and the weather over eight generations (1977–81) on seven sites; indeed, contrary to popular expectation, all colonies increased dramatically during a series of cool wet seasons, having declined (through drought) in 1976 (Thomas 1983*a*). Despite this confusing picture, BMS records show that the annual fluctuations of a particular species have usually been synchronized on most sites in a region, and the most plausible explanation is that this is somehow due to the weather (Pollard 1979*a*, 1981*a*, Ch.5). There is also little doubt that freak conditions can have an occasional catastrophic effect: one example is the crash of many species in 1976–77 after the severest drought ever recorded in north-west Europe (Dempster & Hall 1980, Pollard 1979*a*, Thomas 1980a, 1983*a*, Thomas & Merrett 1980).

It may prove to be an oversimplification to attribute most long-term changes in numbers to changes in habitats, and short-term fluctuations to variation in the weather. Nevertheless, it is interesting to compare the size range of these components, using such empirical data as are available: single species surveys made in the same generation show a range in population size between the smallest and largest viable colonies found, of from 15-fold (*Lysandra bellargus*; Thomas 1983*b*) to 1300-fold (*Mellicta athalia*; Warren *et al.* 1981); densities per unit area ranged from 27-fold (*Hesperia comma*; D. J. Simcox, C. D. Thomas pers. comm.) to 487-fold differences (*M. athalia*). Nearly all this variation was attributed to habitat differences. Short-term fluctuations, though dramatic, are of a lower order, as judged by variation in the mean regional indexes of 19 sedentary species recorded on several sites for the BMS (Pollard 1982): populations fluctuated in size by a factor of 1.3 (*Melanargia galathea*) to 22-fold (*Polyommatus icarus*) between the highest peak and lowest trough over six to eight years, and experienced a maximum change in consecutive years of from 1.1 (*M. galathea*) to 7-fold (*Lasiommata megera*).

The conclusion for the conservationist is that enough suitable habitat should be maintained on any site to support a large population that is unlikely ever to fall to so low a level during a short-term trough or catastrophe, that extinction is possible. The 1976 drought provided the coup-de-grace for one colony of *Papilio machaon* (Dempster & Hall 1980) and *Maculinea arion* (Thomas 1980*a*,*b*), and probably several of *Lysandra bellargus* (Thomas 1983*a*), but all were already very small due to inadequate habitats. To this extent, weather may affect distributions, for local extinctions may not be followed by natural recolonizations (see 'isolation'). Conversely, the weather probably contributed greatly to the spread of *Limenitis camilla* (Pollard 1979*b*), and perhaps later to that of *Pararge aegeria*, into increasingly shady woods, for colonizations only occurred after large numbers had temporarily developed in original sites as a result of a series of warm summers. Significantly, neither species has retreated from its expanded range during the predominantly cooler summers of the past 30 years. And there is little evidence to blame the decline of other species in the UK to a deterioration in the climate, as has been suggested for *Lysandra bellargus* in the Cotswolds (Muggleton 1973) and for *Maculinea arion* (Dover 1974). As with the putative genetic deterioration of colonies, the suggestion that changes in the climate now make large areas of the

UK unsuitable for certain species has sometimes been used as an excuse by conservationists for taking no action.

Insecticides and Air Pollution

Insecticides have often been blamed for declines of all wildlife, but there is little evidence that they have had more than occasional local effects on butterflies in Europe (Pyle *et al.* 1981), where most applications are restricted to arable land (Mellanby 1981). Spraying is more indiscriminate in the USA, where a higher proportion of species encounter insecticides. However, the long-term effect is not necessarily harmful, at least on mobile butterflies that form open populations. Dempster (1967, 1968, Ch.10) found that a high proportion of *Pieris rapae* larvae was killed when *Brassica* crops were sprayed, but a reinfestation soon occurred and survival was then higher than on unsprayed plots, because many of the more sedentary arthropod predators had also been killed; populations of these took longer to recover.

Most other butterflies that encounter insecticides in Europe breed on hedges, rough corners, etc. amid arable land. They receive lower doses than *Pieris*, which is deliberately sprayed. Typical applications of DDT and Dieldrin rarely kill *Aglais urticae* (Moriarty 1968, 1969), although occasional deaths of *Nymphalis io* larvae have been reported (P. I. Stanley pers. comm.). Moriarty did detect damaging sublethal effects among laboratory adult *A. urticae* that had been dosed as larvae, but wild individuals from an arable region contained under one hundredth of the harmful level of residues. Moreover, most species that breed on arable farms are mobile (like *Pieris*) and can presumably replenish local depletions and perhaps survive better there. Most other UK butterflies have local closed populations, but the doses they receive (if ever) from drift would be much lower and unlikely to be harmful. There is, however, considerable concern about the next generation of insecticides that is currently being tested, for these are much more toxic.

In the 1960s it was fashionable to blame insecticides for almost any decline of butterflies. Today, many Europeans blame air pollution (Heath 1981*a*). This is because of a reported dearth of certain species near some industrial areas, especially in Germany and Italy. BRC data show this around the English Midlands. No more positive evidence links the two phenomena, and research is needed to elucidate this matter.

Butterfly Collectors

Collectors have been blamed for mysterious declines in butterflies more often than any other factor (Bree 1852, Goss 1890, Dale 1903, Sheldon 1929). Even today, committees spend as long debating whether certain species should be protected by law as on most other aspects of conservation, and practical conservationists in the UK expend much energy and expense in wardening sites of local species during the adult flight period.

In fact, few temperate butterflies are spectacular enough to be taken indiscriminately for the popular market; most are caught selectively for collections. Most detached reviewers have concluded that collectors have had little if any harmful effect on UK butterfly populations, though all believe that heavy collecting might tip the balance against a small colony that was at a low ebb for other reasons (Ford 1945, Morris 1976). Thus, although isolated extinctions of *Lycaena dispar* (Duffey 1968) and *Mellicta athalia* Warren *et al.* 1981) were preceded by heavy bouts of collecting, numerous other local populations of *Maculinea arion* (Spooner 1963), *Strymonidia pruni* (Thomas 1976*b*), *Lysandra bellargus* (Thomas 1983*a*) and other species have sustained heavy collecting for decades without apparent harm. Conversely, nearly all extinctions have occurred on sites where collecting has not occurred.

This anecdotal evidence is supported by population studies. It is still not clear what level of collecting is safe; moreover, this will vary for any colony in any year. For example, the last (very small) population of *Maculinea arion* in the UK bred at high densities in a small area, but would have benefitted if collectors had escaped the wardens in 1973, and had taken up to half the population. This would have reduced the severe overcrowding of larvae in ants' nests that year, and the population in 1974 would have been higher, rather than reduced to one third its size through density-dependent mortalities (Thomas 1976*a*, 1980*a*,*b*). On the other hand, this population would probably have been harmed by lighter collecting in most others years.

Only species that form closed sedentary colonies are ever likely to be threatened by collectors, and then only small (<250 adults) populations of those species that fly often, and fairly weakly, in small areas of accessible terrain are at risk. *M. arion* fulfils all these criteria, but is still unlikely to be harmed unless several collectors work for about three weeks on a small site. Mark-recapture experiments of *M. arion* indicate that it is physically possible for one worker to take up to 85% of adults in one day. However, only about one-third of an entire emergence of this, and other species with similar lifespans, is alive on the peak day, and about 15% of adults on that day would be released by most collectors because they were worn (Thomas 1983*a*). Thus, the casual collector who coincided with the peak day of *M. arion* might, at worst, remove 25% of the population. Small colonies of other rare UK species, such as *Mellicta athalia* are also potentially at risk, but most butterflies are much less easily caught because they fly rapidly (*Papilio machaon*), or in inaccessible habitats (*Apatura iris*; woodland Theclinae), or rarely fly at all and are mostly overlooked (all Hesperiinae studied).

In retrospect, the UK code of collecting issued by the JCCBI strikes the right balance (Morris 1981*b*), and the behaviour and population structure of *Mellicta athalia* and *Maculinea arion* justify recent legal protection for those species, so long as it is recognized that collectors are not the cause of declines, but might be the last straw. Such an enlightened attitude has only been taken in legislature in the USA, where the Federal listing of any species must be accompanied by a recovery programme (Opler 1981). Other listed species in the UK do not warrant legal protection on present scientific knowledge, for example *Papilio machaon* and *Carterocephalus palaemon*. This applies to most species hurriedly being scheduled in other countries (Heath 1981*a*). At a practical level, the wardening of *M. arion* sites in the UK was justified, but the great efforts being spent to protect *Strymonidia pruni* and *Apatura iris* in certain woods are pointless since it is physically impossible to catch more than a very small proportion of their adults. Ironically, no such protection is being given to the few colonies that may be at risk: the smaller populations of *Mellicta athalia* and of other conspicuous sedentary species.

Practical Conservation

Here I consider whether butterflies have been adequately maintained by past conservation measures, given the biological conclusions reached above and the fact that most efforts have been made by generalist bodies that have rarely considered the particular needs of insects. Certain measures, such as the deterring of collectors, and laissez-faire attitudes prompted by theories about isolation or climatic change have already been discussed, and are not repeated. Nor is any account given of attempts to combat insecticides and pollutants, for these also seem largely irrelevant to butterfly needs. Instead, the role of nature reserves is examined, as are attempts both to conserve wildlife on other sites and to introduce butterflies to new localities.

Nature Reserves

The main effort of most conservationists in the UK, and increasingly elsewhere, has been to establish nature reserves. This coincides with the principal need of butterflies: the overwhelming threat to most is the loss of their particular habitats, whilst the main function of a reserve is to maintain habitats in general, with a special responsibility for those that are disappearing altogether. In the UK, perhaps 20% of butterfly species may soon be reduced largely or entirely to colonies that breed on reserves (Dempster *et al.* 1976, Duffey 1968, Thomas 1976*a,b*, 1983*a*, Thomas & Simcox 1982, Warren *et al.* 1981). This applies to a much higher percentage at some local levels. A similar situation exists in Europe (Heath 1981*a*), and a few indigenous species in the USA already exist only on reserves.

The structure of most temperate butterfly populations makes them well suited to reserve protection. Of UK species, 83% form closed, fairly sedentary colonies, of which all but one can be supported by under 50ha of suitable habitat (Table 33.2). Much smaller areas suffice for most butterflies, although, since it is difficult to maintain the habitats of a few species indefinitely on the same spot, extra space may sometimes be needed to provide continuity of a particular seral stage, such as freshly cleared woodland. It is also important that sufficient habitat should be present to support a population that is large enough to survive exceptional troughs caused, for example, by freak weather. Experience suggests that any site that is unable to carry a population of over 200–400 adults of most species in good years, is liable to suffer occasional extinctions during bad periods (Duffey 1977, Thomas 1980*a,b*). Fortunately, the size of most reserves in the UK greatly exceeds the area of habitat needed by a large population of most sedentary butterflies, and most reserves in other countries are even larger.

Nature Reserves are unlikely to support populations of (the few) species that form open or migratory colonies, although passing individuals may obviously breed on them. (Reserves to protect overwintering roosts of *Danaus plexippus* are a different, and important matter: Orsak 1981, Pyle 1981*b*, Brower 1977). Fortunately, most mobile species are widespread and common; only *Nymphalis polychloros* among mobile UK butterflies is endangered.

It is unlikely that nature reserves will ever occupy more than a very small proportion of any developed temperate country; those in the UK occupy 0.8% of the land, excluding NT land. The following questions must therefore be asked: How well are (especially local and rare) butterflies represented on existing reserves? Have habitats been maintained on these? Have their butterfly populations flourished?

Representation on nature reserves

Most formal UK nature reserves were obtained for their general richness in wildlife; most NT land for its physical beauty or as part of an estate. The number of butterflies on most properties is still poorly known, except for eleven rare species that have been surveyed (Table 33.4). However, reviews of statutory reserves (86% of which were NNRs) by

Table 33.4. The current representation of rare butterflies on UK nature reserves, and their history of survival over the past 20 years.

(a) Species	(b) Date	(c) No. of all known colonies at date in (b)	(d) Colonies on reserves at date in (b) Σ	% all colonies	NNR	SPNC/ County Trust	Other (LNR, FNR, RSPB, Private)	NT	(e) Extinctions on reserves 1960–1981	(f) % of colonies on reserves that became extinct 1960–1981	(g) Source
L. dispar	1982	1	1	100	1	—	—	—	1*	100*	Duffey (1968, 1977)
M. arion	1967–1979	4	4	100	—	2	—	1	4	100	Thomas (1976*a*, 1980*b*)
C. palaemon (England only)	1960–1975	16	4	25	3	—	1	—	4	100	Farrell (1975) updated
M. cinxia	1982	18	7	39	—	—	—	7	0	0	Simcox and Thomas (1980) updated
M. athalia	1982	29	7	24	1	—	5	1	2	100	Warren *et al.* (1981) pers. comm.
S. pruni	1982	30	11	25	2	4	5	—	0	0	Thomas (1976*b*) updated
H. comma	1982	52	33	63	7	5	2	19	9†	24	Simcox, C. D. Thomas, pers. comm.
L. bellargus	1982	*c.*75	19	25	3	6	2	8	4**	27	Thomas 1983*a* updated
T. acteon	1978	81	25	30	1	2	5	17	?	?	Thomas unpublished
L. sinapis	1979	87	10	12	3	4	3	—	?	?	Warren 1981
E. epiphron (England only)	1980	?	?	70	?	?	?	70% of all	?	?	Hemsley pers. comm.
Total		394	121	31	21	23	23	54	24	21	

*Reintroduced 1970.
**Excluding NT properties.
†Four reserves colonized during same period

Morris (1967) and of NNRs by Peachey (1982) indicate that butterflies are well represented on these sites. Recording cover was far from complete, but 33 (60% of) non-migratory species were reported from ten or more NNRs in 1980/81, and only three species were absent from any reserve (Peachey 1982). The latter all breed on other organizations' reserves.

Surveys of rarities suggest that the butterfly colonies on NNRs are matched by those on County Trust reserves, and again on sites owned by various other bodies (Table 33.4). All are dwarfed by the large number of (especially grassland) species found on NT properties. Considering these sites together, butterflies seem to have been surprisingly well served by the various criteria adopted for site selection. It should be noted, however, that some serious gaps existed until recently, and that 14% of the protected sites with rarities listed in Table 33.4 were obtained largely or wholly for that butterfly, generally after prolonged lobbying by entomologists. On the other hand, the owners (mainly NT) of 45% of reserves found to contain a rare species had been unaware of its presence.

Most surveyors of a local UK species have nevertheless urged that many more reserves should be obtained, because these often represent the only places where they are likely to survive. It is hard to disagree; the majority of colonies of rare species still occur on unprotected land (Table 33.4), and 65% of the populations of local species on NNRs are considered to be small (Peachey 1982). As has been described, small populations are apt to become extinct. Perhaps a national target of ten large conserved populations per species should be set. Representation is also very uneven on a local scale, especially on County Trust reserves: the chance of an important butterfly site being obtained for these has largely depended on whether a forceful entomologist happened to live there. There is an urgent need to discover the full representation of butterflies on all existing reserves. NCC (C. Peachey, pers. comm.) has started a review and the NT is making its own survey (J. Hemsley pers. comm.). Better cover for both might be achieved by involving the BBCS.

Information is hard to obtain for other countries, but Heath's (1981*a*) review of Europe indicates that butterflies are more poorly represented on reserves or National Parks than in the UK, except perhaps in Holland. The same seems to be true in much of the USA, although certain states such as California have made considerable advances. Mexican and USA conservationists, and IUCN/WWF are especially to be congratulated on protecting *Danaus plexippus* overwintering roosts (Pyle 1981*b*).

The survival of butterflies on reserves: management and causes of declines

No balanced assessment can be made yet of the history of butterflies on all UK reserves. However, the survival of eight species has been well documented, and good records exist for some individual reserves. In the following paragraphs case histories are presented for rarities on reserves, and then for particular sites.

Table 33.4 (columns e, f) shows that populations of rare UK butterflies have an appalling history of survival on nature reserves. Over a 20-year period, four species became extinct on every reserve on which they bred and two others were lost from a quarter of them. Losses of *H. comma* were partly offset by its recolonization of other reserves, and there was 100% survival of *S. pruni* and *M. cinxia*. Compared with populations on unprotected sites, *Mellicta athalia* fared worse on nature reserves than elsewhere; *Maculinea arion*, *Carterocephalus palaemon*, and *Melitaea cinxia* did much the same; and *S. pruni*, *H. comma*, and *Lysandra bellargus* have survived better. For *C. palaemon* and *Maculinea arion*, doing 'much the same' meant becoming extinct respectively in England and the UK.

All researched extinctions on nature reserves have been due to a failure by conservationists to maintain enough (or any) suitable breeding habitat, although the coup de grace has occasionally been administered by another factor. This has largely occurred through ignorance, although a lack of resources or interest have often contributed. *Strymonidia pruni* has been successfully conserved so far, because its scrub habitat remains suitable for breeding for perhaps 40–60 years, if protected from clearance. Protection has been given on all reserves, but attempts have been made on only one-third of these to create the new habitat that will eventually be needed, or to improve existing areas (Thomas 1980*b*).

The other listed species (Table 33.4, columns e, f) breed in earlier successional habitat that soon becomes unsuitable if left undisturbed. Natural regeneration (through annual cliff falls) occurs only for *Melitaea cinxia*, which no doubt explains why it has survived so well on its reserves, since even its presence was unknown until recently to the owners of most sites (Simcox & Thomas 1980). But neglect was nearly fatal for *Mellicta athalia*. Two of its finest sites were obtained as reserves, but both populations became extinct within a decade. This occurred because its need for early successional woodland was not appreciated, nor was manpower available for some years, and both sites became overgrown. Ironically, colonies flourished in commercial woods adjoining each reserve where forestry continued. Recent management based on Warren's (pers.

comm.) research has now led to the recolonization of both former reserves, and some of five other populations on new reserves have spread for the same reason.

Informed reserve management came just too late to save *Maculinea arion* in the UK, even though considerable resources were spent over many years in the attempt (Thomas 1980*a*,*b*). An early effort was made in the 1920s when one of the best sites was declared a nature reserve. At that time, extinctions were attributed to butterfly collectors, so the site was fenced off. Unfortunately, this also excluded herbivores, and the sward soon became too overgrown for *M. arion*, which promptly disappeared. Many different measures were taken in the next 50 years, directed since 1962 by a joint committee representing all interested conservation bodies (Howarth 1973*a*). Remaining colonies were found and protected from ploughing or collectors, the last four sites became reserves of various sorts, and great efforts were made to improve the habitat. Unfortunately, the underlying cause of decline (a reduction in grazing leading to a loss of the ant, *Myrmica sabuleti*) was unknown, and irrelevant or (occasionally) harmful measures were taken.

In retrospect, populations of *Maculinea arion* could almost certainly have been maintained with the available resources on its large unfarmed reserves, if its exact needs had been known earlier (Thomas 1976*a*, 1977, 1980*a*,*b*). Unfortunately, full-time research only began after the species had declined to 250–300 adults in the UK. Most of these bred on the only reserve that was still being adequately grazed, but the area of habitat was precariously small. By modifying the management, it later proved easy both to increase the density of *Myrmica sabuleti* nests in the original breeding area and to create the same conditions in other parts of the reserve and on a neighbouring hillside (Fig. 33.2; Thomas 1980*b*), probably sufficient to support a 'safe' population of over 1000 adult butterflies in 1981. But before this was accomplished, severe overcrowding in the original habitat followed by extreme drought reduced the small population to so low a level (two females, three males) that it was unable to recover (Thomas 1980*a*).

Lysandra bellargus and *Hesperia comma* have also been lost from several reserves that were left ungrazed (Table 33.4). However, fewer extinctions have occurred than elsewhere, because all reserves have been protected from ploughing or agricultural improvement, and several have been adequately managed. Indeed, both species now breed on one grazed reserve, which had been unsuitable in the years following myxomatosis (see below).

The UK's two fenland butterflies have a similar chequered history. *Lycaena dispar* became extinct in 1969 on its only reserve, Wood Walton Fen NNR, to which Dutch stock had been introduced in 1927 (Duffey 1968, 1977). This fen became drier and scrubbed over early in this century, but great efforts were made to recreate suitable breeding habitat through clearances, grazing (recently), peat digging, and planting the larval foodplant. Habitat of reasonable, but not ideal, quality was achieved, but Duffey (1977) concluded that too small an area of the 214ha reserve could be spared for this (30ha, increased from 1.8ha in 1954) to support a natural population of *L. dispar*. The colony was therefore nursed by protecting the young stages from predators, and was sometimes supplemented with captive stock, yet still fluctuated between only about 35 and 350 adults during the 1960s (Duffey 1968). This is a similar size to the last population of *Maxulinea arion*, and was likewise too small to withstand a catastrophe: in this case abnormal floods which submerged most foodplants during the egglaying period of 1968; only five adults (one male, four females) emerged the next year. A reintroduction was made in 1970–73, but its long-term survival will still depend on highly artificial aid (Duffey 1977).

Papilio machaon has a similar need for open fen, and became extinct on Wicken nature reserve in 1952 when the area of this habitat had declined from 120ha to 8ha, for reasons already described (Dempster & Hall 1980). Since then, 30ha of the reserve has been managed to recreate open conditions, which can be achieved on drying fen only by regular cutting. Unfortunately, this regime is unsuitable for *P. machaon*, as it destroys 25–30% of the butterfly population and causes the larval foodplant to grow weakly and eventually to decline. Dempster & Hall (1980) concluded that an isolated population of *P. machaon* could be supported at Wicken, but only if the management was radically changed so as to make the site wetter. This might be prohibitively expensive. Fortunately, recent management on a (wet) reserve in the Norfolk Broads has greatly improved its habitat for *P. machaon*, and the species should at least survive in the UK (Hall 1981).

The rarities may have survived less well on UK reserves than commoner species, which may need less specialized habitats. However, Peachey (1982) lists another 16 non-migratory species (out of 55 in the UK) that have been lost from at least one NNR. Moreover, despite the recent acquisition of many sites, 17 species were reported from *fewer* NNRs in 1981 than Morris (1967) listed on statutory reserves 15 years earlier. The two surveys are not strictly comparable, but there is little doubt that this difference reflects a severe decline by common and local species on some reserves. This is well documented on a few sites. On the other hand, most butterfly populations have at least survived on most nature reserves, and some have

Table 33.5. Changes in butterflies in Monks Wood compiled from Heath 1973, Thomas 1973, NCC records, E. Pollard and J. P. Dempster (pers. comm.).

Date	Losses	Gains
Nineteenth century	*M. arion**	
	A. crataegi	
	E. aurinia	
	B. selene	
	L. sinapis	
1914–25	*S. pruni*	*S. pruni*
	A. iris	(reintroduced)
1934		*P. c-album*
1939–46	*N. polychloros*	*C. palaemon*
		L. camilla
1953	Declared NNR	
1957	*H. lucina*	
late 1950s	*A. aglaja*	
1962	*A. adippe*	
1969	*A. agestis*	
1963	*B. euphrosyne*	
1966	*T. betulae*	
1969	*A. paphia*	
	E. tages	
early 1970s	*C. rubi*	
1971	*C. palaemon*	
1976	*M. galathea*	*L. sinapis**
1980	*S. w-album*	

*Probably not resident

even gained species. Examples of both cases are given.

Monks Wood has been famous for butterflies since 1828 (Thomas 1973). During the next 125 years, eight species were lost but four others colonized it or were re-introduced (Table 33.5). In 1954, the wood was bought as a nature reserve, and the object of its management became wildlife conservation. However, since then, another 12 species of non-migratory butterflies have disappeared; all of which are rare or local species (Table 33.5). The wood now supports just one national rarity, three local species, and another 19 butterflies that are common and widespread in the UK. Similar declines have occurred in that region at Castor Hanglands, another woodland reserve, once famed for its insects (Collier 1966, Peachey 1982), and also in all known commercially managed woods.

It is not wholly possible to explain the failure of these reserves, but some informed guesses may be made. *Strymonidia w-album* was lost when Monks Wood's elms were killed by Dutch Elm disease, and colonies of *Argynnis aglaja* and *Thecla betulae* (Thomas 1974) bred largely on neighbouring farmland, which is now unsuitable. This represents a failure to establish adequate boundaries for the 157ha reserve, if these species were considered important. Populations of the other butterflies were contained by the reserve. Most are species that breed in fairly recently cleared woodland. A major reason for their extinctions almost certainly lies in the present management of the site. Formerly, the entire wood was managed on a 20-year coppice cycle, with patches of 8ha being cleared annually. Then, in 1914–1918, over half the wood was clearfelled and the brush burnt back. This was left to regenerate over the next 40 years, mainly as a single-age stand. Since 1954, coppice has been re-introduced, but only to 9% of the wood (Steele & Schofield 1973) whilst the vast majority has continued to become increasingly shady. Extra light has been introduced by widening the rides, but these permanently-open, periodically-mown strips are really sheltered grassland, and support large populations of common grassland butterflies. It may well be that colonies of species that need freshly cut woodland could be supported by the small areas of coppice that now occur in Monks Wood, if the exact habitat requirements were known. In the absence of that knowledge, it is extraordinary that a wood that was obtained largely for species of the open habitats that predominated in it for perhaps a millenium should now be managed so that most is under heavy shade.

In contrast to Monks Wood, butterflies have increased in both diversity and numbers over the past decade at Old Winchester Hill. This isolated 80ha chalk grassland site had been ungrazed for several years in the late 1950s following myxomatosis, and at least one species (*Lysandra bellargus*), was lost. Grazing was reimposed when the site became a nature reserve, and the patchy rotational regime recommended by Morris (1971) for high insect diversity has recently been used. Most of the many resident species of butterfly that were present have increased in numbers relative to other sites (Pollard 1981*a*) and the new early seral habitat, which had been missing, has been colonized by *Hesperia comma*, presumably from a population 4km away (J. Bacon pers. comm.). No *L. bellargus* colony survived in the district, so this species was artificially introduced. It seems that its conditions have also been recreated, for the original introduction of 60 adults has already increased to over 5000 after three generations, outstripping the performance of other known colonies in the UK.

The survival of butterfly populations in nature reserves or parks outside the UK is badly documented, but there is reason to believe that the record elsewhere in Europe (Heath 1981*a*) and in the USA (Pyle *et al.* 1981) is equally poor.

Conservation on Unprotected Land

Many conservation bodies advise landowners to encourage wildlife on their properties. The relevance to butterflies of a few measures in the UK are considered below.

Sites of special scientific interest (SSSIs)

SSSIs are established by the NCC for their outstanding interest, but this does not prevent the owner altering or neglecting them. Even the fundamental destruction of a site is not prevented; the owner need merely inform NCC that this is planned. Traditionally, SSSI status did little to save butterflies from the harmful changes that were occurring throughout the countryside, but it did mean that colonies of rarities were not destroyed in ignorance. It sometimes also influenced local councils when planning permission was sought for a change in land use. More recently, there has been a trend towards entering into a management agreement with the owner, who may be compensated for loss of earnings. SSSIs are unlikely to contribute greatly to the conservation of butterflies unless this becomes commonplace.

Farmland and gardens

Farmers and householders are often urged to leave odd corners uncultivated (Anon. 1977), and even to plant *Urtica* species and nectar plants. As envisaged, these patches are so small and widely scattered that they are likely only to be used by the few butterflies that are migratory or have open populations; a recent survey revealed that gardens were almost bereft of Satyrinae, Hesperiidae, and Lycaenidae (Thomas 1982). To date, this practice is probably too rare an event to make any significant improvement to the populations of even mobile species, but clearly might do if it increased.

Road verges

In the past decade, there has been a reduction in spraying to reduce the herbage along roadsides, and cutting has become less frequent. This has occurred for economic reasons, but has probably greatly benefitted butterflies that need tall grassland. Some councils have sown new roadsides with wildflower seeds (Wells *et al.* 1981), sometimes biased towards larval foodplants and rich nectar sources, as a contribution to 'Butterfly Year'. Verges in Holland are often sown with wildflower seed, and there is great scope for research on the effect of this, of various mixes, and of different management regimes on butterfly populations.

Commercial woodland

There is considerable scope for improving the capacity of woods for butterflies (and other wildlife), with little or no loss of revenue. Peachey (1981) describes some of the important features of a very rich Forestry Commission wood, which the Forestry Commission are now prepared to maintain: broad flowery rides cut on different rotations; clearings to form glades where rides intersect; wide shrub and broadleaved tree borders to rides and edges; the preservation of blocks of scrub (for *Strymonidia pruni*). Similar schemes could be devised for many other sites; the BBCS has already started in a few areas.

Introductions

Many attempts have been made to introduce butterflies to sites in the UK. These range from the selling of pupae of mobile nymphalines for release in gardens, to carefully planned re-introductions of sedentary species to sites that have regained a habitat that was temporarily lost. The former merely fly away if no flowers are present, and rarely if ever breed, but several introductions of sedentary species have been extremely successful (Table 33.3), for reasons described above ('colonization'): perhaps one third to half the existing colonies of *Strymonidia pruni* in the UK stem from this practice (Thomas 1974).

Most introductions have been made by amateur entomologists who have sometimes succeeded using very small numbers; about 12 adults were used to form what, 30 years later, is now the largest colony of *Strymonidia pruni* in the country (A. E. Collier pers. comm.). Conservation bodies have taken an ambivalent attitude, although most now approve; NCC have maintained the introduced colony of *Lycaena dispar* for nearly 30 years and have recently re-introduced *Lysandra bellargus* to Old Winchester Hill (see 'nature reserves'). Positive leadership by conservationists would ensure that introductions were made only to sites containing suitable unexploited habitat; the most suitable source was used (strong colonies of the same race); adequate numbers were released; record centres were informed; and the results were monitored.

Conclusions and the Future

Individual successes, mistakes, and possible improvements have been described under each topic throughout this review. This section contains some general points as to where resources might be directed. For the forseeable future, the means to make a practical impact on the conservation of butterflies will be vested in generalist conservation bodies and sympathetic landowners. The growth of the former is greatly to be encouraged, but there is a more pressing need for a change in their attitudes. At present, the extent to which practical measures are relevant to the needs of butterflies depends mainly on chance, but is influenced by public opinion, media pressure, and the personal interests of individual employees.

Entomologists should work for two changes of attitude. One is to interest public opinion and the press in the plight of most butterflies. This has been achieved in the UK through 'Butterfly Year' (Thomas 1981*a*), which has both raised money to enlarge the conservation 'cake', and has strongly influenced conservation bodies that had previously ignored butterflies; most are now trying to improve existing reserves and to purchase outstanding sites. Several local authorities also intend to seed road verges and cuttings, fill refuse tips, etc., with the larval foodplants of butterflies. The other change should be to monitor the relevance of the entire wildlife conservation effort vis à vis butterflies and, where necessary, to advise or lobby to reduce anomalies. This could be achieved in the UK by an independent pressure group, such as the BBCS. That society may eventually become large enough to undertake major projects on its own, but would be more effective, meanwhile, if it concentrated on influencing others; at present, its local groups often have no contact with other conservationists.

Similar efforts should be made a high priority in other countries, as should be the establishment of popular pressure groups. The Xerxes Society in the USA leads the World in this respect. Growth in both fields is being pursued in Europe by the Societas Europea Lepidopterologica, but is still slow in many countries (notably France). Weak areas might be reinforced by efforts from IUCN/WWF. Ideally, the European Institute recommended by Heath (1981*a*) should be established.

It is essential to establish the extent of monitoring and surveys that has already been achieved in the UK if all problems are to be identified early enough for action to be taken. Other countries fall far short of this at present. In the UK, there is again no central organization that analyses the results of all schemes as a matter of course. To remedy this would cost little extra to the price of running existing schemes.

The threats that face most butterflies mean that the survival of many species will become increasingly dependent on nature reserves. The criteria used for site selection in the UK have worked rather well, although some gaps have only been filled after outside pressure from the JCCBI. Most populations have been entirely contained by reserve boundaries, and the few mistakes that have occurred could be prevented by consulting BBCS members, or other local entomologists, before any new reserve is obtained.

There is certainly a need for many more nature reserves in the UK, and a pressing one in other countries. However, it is much more important that conservationists should realize that obtaining a nature reserve is usually only the first step towards conserving its butterfly populations. Few butterfly habitats in temperate countries occur in climax vegetation, and site management is usually essential if earlier stages are to be maintained. Management must also occur on a grand enough scale to support a large population of every species of interest. At present this is rarely being done. Consequently, while most conservation bodies are stretching their resources to buy new reserves, they are simultaneously losing much of the interest on some current properties through inadequate (or no) management. This loss applies to most other wildlife. Insect populations react rapidly to changing conditions, and losses that are merely considered regrettable at present should be treated very seriously as indicators of a long-term change. For example, the lack of grazing that eventually leads to the demise of many low-growing perennial grassland plants is rapidly apparent in the decline of colonies of *Hesperia comma*, *Maculinea arion*, and *Lysandra bellargus* in N. Europe; drying Fens eventually lose *Peucedanum palustre*, but *Papilio machaon* soon disappears (Dempster & Hall 1980).

There are two other problems to be faced, even when the need for site management is appreciated. One is again organizational: the fate of individual reserves in the UK rests, in practice, almost entirely on local committees or the whim of the warden. To be fully effective, management of the suite of nature reserves that already exists in the UK should be coordinated so that certain sites have a special responsibility for maintaining or improving on certain habitats. Moreover, the overall emphasis of the suite should be biased towards those habitats that are no longer being created naturally on other land. This is not happening at present.

The second problem is that, once the aim of a reserve has been established, there is often great ignorance as to how this may be achieved. Knowledge about the exact requirements of butterflies is still woefully inadequate. This can only be provided through research by professional ecologists. Time and again in the UK, conservationists and ecologists have tried to guess the answers, but have nearly always been wrong. Ecological research is expensive, but costs less than the purchase and practical management of sites. At present in the UK, a species must reach the verge of extinction before funds are available (for example, to study *Maculinea arion* and *Mellicta athalia*). Solving the problem is by then much more difficult, takes longer, is more expensive, and remedies may be too late. There is still a need for several more population dynamics studies to be made of endangered butterflies, but short-cuts can probably now be made in most autecological studies, as with *Lysandra bellargus* in the UK (Thomas 1983*a*). And there is considerable scope for an extensive approach to study the effects of various management regimes on all the butterflies in different biotopes, without necessarily understanding why these effects have occurred.

The principles that apply to conservation on

reserves are also relevant to the maintenance of butterflies on other land. As the requirements of different species become known, there will almost certainly be greatly increased scope for maintaining populations in much smaller areas and with less effort than is now possible.

Another growth area lies in introductions. Occasional extinctions of sedentary species may always occur on nature reserves, and occasional replenishment may be needed. This will usually also be necessary if new habitats are created on isolated reserves. But the greatest potential is to utilize temporary habitats that are being created (often over large areas and at great expense) on commercially managed land, but which are not being colonized naturally. Previous introductions suggest that there is very great scope for this in the UK, and presumably elsewhere. As with all other measures, it is highly desirable that this should be planned and overseen by one responsible body.

Finally, the assumption has been made throughout this review that conservationists wish simply to maintain the traditional flora and fauna of their regions. Man's past activities probably eliminated many species before records began, and there is a case for also introducing foreign butterflies to the new habitats he is creating, rather than simply to maintain historically recent ones. Naturally, great care should be taken not to introduce a potential pest, but I would certainly welcome the enrichment of, for example, mature conifer plantations in the UK, if a harmless butterfly could be found that can live there.

Acknowledgements

I would like to thank E. Pollard, M. G. Morris, A. J. Gray, J. P. Dempster, R. M. Pyle, M. S. Warren, C. D. Thomas and J. Heath for valuable suggestions, and the Nature Conservancy Council for permission to quote some unpublished material.

Cumulative Bibliography

In compiling this section, we have adopted several conventions, some rather unusual. The titles of serials are largely in accordance with the *World List of Periodicals* (London, 1960); we have purposely omitted the quoted cities of publication for such well-known journals as *Science* [New York], *Nature* [London] and *Evolution* [Lancaster, Pa]. Volume numbers are given in bold face; separate parts are generally indicated in bold between parentheses, unless quoted together with a volume number. Where there is conflict between the actual and apparent year of publication, we have always given the actual year if this is known to us. For brevity, names of higher taxonomic groupings originally given in parentheses, together with authors' names for species and genera, have been excluded from titles. Books are mostly cited without pagination or publishers' names. By cross-referring articles appearing as parts of multi-author works to the volume titles under the editors' names, much unnecessary repetition has been avoided. We hope thus to have saved space without detracting from the clarity of citations.

Numbers in brackets following each reference indicate the chapters in this volume where the references are cited.

Abbott W. (1959). Local autecology and behavior in *Pieris protodice* with some comparisons to *Colias eurytheme*. *Wasmann J. Biol.* **17**: 279–298. [2]

Able K. P. (1980). Mechanisms of orientation, navigation and homing. Pp.284–373 in Gauthreaux (1980). [26]

Abo K. & Evans F. J. (1981*a*). Ingol esters, cytotoxic agents of the macrocyclic type. *J. Pharm. Pharmac.* **33**: 57. [13]

Abo K. & Evans F. J. (1981*b*). Macrocyclic diterpene esters of the cytotoxic fraction from *Euphorbia kamerunica*. *Phytochemistry* **20**: 2535–2537. [13]

Ackery P. R. (1975*a*). A new pierine genus and species with notes on the genus *Tatochila*. *Bull. Allyn Mus.* **(30)**: 9pp. [27]

Ackery P. R. (1975*b*). A guide to the genera and species of Parnassiinae. *Bull. Br. Mus. nat. Hist. (Ent.)* **31**: 73–105. [1]

Ackery P. R. & Vane-Wright R. I. (In press *a*). *Milkweed butterflies: their cladistics and biology*. London. [Intro, 1, 12, 25]

Ackery P. R. & Vane-Wright R. I. (In press *b*). Patterns of plant utilization by danaine butterflies. *Nota lepid* (Suppl.). [Intro, 25]

Adams M. J. & Bernard G. I. (1977). Pronophiline butterflies of the Sierra Nevada de Santa Marta, Colombia. *Syst. Ent.* **2**: 263–281. [3]

Adler P. H. (1982). Soil- and puddle-visiting habits of moths. *J. Lepid. Soc.* **36**: 161–173. [Intro]

Ahmad S. (ed.) (1983). *Herbivore insects: host-seeking behaviour and mechanisms*. New York.

Aiello A. & Silberglied R. E. (1979). Life history of *Dynastor darius* in Panama. *Psyche* **85**: 331–345. [6]

Akre R. D., Hansen L. D., Reed H. C. & Corpus L. D. (1981). Effects of ash from Mt St Helens on ants and yellowjackets. *Melanderia* **37**: 1–19. [31]

Alcock J. (1965). *The relative palatability of butterflies and the behavior of their avian predators*. Thesis, Amherst College, Mass., USA. [12]

Alcock J. (1969*a*). Observational learning by fork-tailed flycatchers (*Muscivora tyrannus*). *Anim. Behav.* **17**: 652–657. [12]

Alcock J. (1969*b*). Observational learning in three species of birds. *Ibis* **111**: 308–321. [12]

Alcock J. (1970). Punishment levels and the response of black-capped chickadees (*Parus atricapillus*) to three kinds of artificial seeds. *Anim. Behav.* **18**: 592–599. [12, 15]

Alcock J. (1973). The feeding response of hand-reared red-winged blackbirds (*Agelaius phoeniceus*) to a stinkbug (*Euschistus conspersus*). *Am. Midl. Nat.* **89**: 307–313. [12]

Aldrich J. K., Soderland D. M., Bowers W. S. & Feldaufer M. F. (1981). Ovarian sequestration of [14-C]-inulin by an insect: index of vitellogenesis. *J. Insect Physiol.* **27**: 379–382. [6]

Alexander A. J. (1961*a*). A study of the biology and behavior of the caterpillars, pupae, and emerging butterflies of the subfamily Heliconiinae in Trinidad, W.I. Part I. Some aspects of larval behavior. *Zoologica, N.Y.* **46**: 1–24. [12]

Alexander A. J. (1961*b*). A study . . . Part II. Molting,

and the behavior of pupae and emerging adults. *Zoologica, N.Y.* **46**: 105–124. [12]

Alexander R. D. (1974). The evolution of social behavior. *A. Rev. Ecol. Syst.* **5**: 325–383. [26]

Allard R. W. & Harding J. (1963). Early generation analysis and prediction of gain under selection in derivatives of a wheat hybrid. *Crop Sci.* **3**: 454–456. [27]

Allport N. L. (1944). *The chemistry and pharmacy of vegetable drugs.* Brooklyn, N.Y. [12]

Allyn A. C. & Downey J. C. (1977). Observations on male UV reflectance and scale ultrastructure in *Phoebis. Bull. Allyn. Mus.* **(42)**: 20pp. [20]

Anderson T. F. & Richards A. G. (1942). An electron microscope study of some structural colors of insects. *J. Appl. Phys.* **13**: 748–758. [20]

Andrewartha H. G. & Birch L. C. (1954). *The distribution and abundance of animals.* Chicago. [2]

Anon. (1950). *Webster's dictionary* (eds Neilson W. A., Knott T. A. & Carhart P. W.). Springfield, Mass. [12]

Anon. (1961). *Givaudan index* (2nd edn). New York. [12]

Anon. (1976). *Appraisal of the Popondetta Smallholder Oil Palm Development.* Papua New Guinea Govt Rep. (1160). [32]

Anon. (1977). *Nature conservation and agriculture.* NCC, London. [33]

Anon. (1980*a*). *The endangered San Francisco silverspot butterfly of California.* Xerxes Soc. Educ. Leaflet (5). [33]

Anon. (1980*b*). *The endangered San Bruno elfin butterfly of California.* Xerxes Soc. Educ. Leaflet (6). [33]

Aoki T., Yamaguchi S. & Uémura Y. (1982). Satyridae, Libytheidae. *Butterflies of the south-east Asian islands* **3**: 1–628 (ed. Tsukada E.). Tokyo. [1]

Aplin R. T. & Birch M. C. (1968). Pheromones from the abdominal brushes of male noctuid Lepidoptera. *Nature* **217**: 1167–1168. [25]

Aplin R. T. & Birch M. C. (1970). Identification of odorous compounds from male Lepidoptera. *Experientia* **26**: 1193–1194. [25]

Aplin R. T., Benn M. H. & Rothschild M. (1968). Poisonous alkaloids in the body tissues of the cinnabar moth (*Callimorpha jacobaeae*). *Nature* **219**: 747–748. [12]

Aplin R. T., d'Arcy Ward R. & Rothschild M. (1975). Examination of the large white and small white butterflies (*Pieris* spp.) for the presence of mustard oils and mustard oil glycosides. *J. Ent.* (A) **50**: 73–78. [12, 13]

Appel J. B. & Peterson N. J. (1965). Punishment: effects of shock intensity on response suppression. *Psychol. Rep.* **16**: 721–730. [12]

Applebaum S. W. & Birk, Y. (1979). Saponins. Pp. 539–566 in Rosenthal & Janzen (1979). [12]

Arikawa K., Eguchi E., Yoshida A. & Aoki K. (1980). Multiple extraocular photoreceptive areas on genitalia of butterfly *Papilio xuthus. Nature* **288**: 700–702. [20]

Arms K., Feeny P. & Lederhouse R. C. (1974). Sodium: stimulus for puddling behavior by tiger swallowtail butterflies, *Papilio glaucus. Science* **185**: 372–374. [2, 20]

Arnold E. N. (1981). Estimating phylogenies at low taxonomic levels. *Z. zool. Syst. EvolutForsch.* **19**: 1–35. [6]

Arnold G. W. & Hill J. L. (1972). Chemical factors affecting selection of food plants by ruminants. Pp.71–101 in Harborne (1972). [12]

Arnold R. A. (1980*a*). *The endangered Lepidoptera of San Bruno Mountain: their status, habitats, ecology and prospects for survival.* Unpublished MS. [2]

Arnold R. A. (1980*b*). *Great Basin silverspot butterfly study.* Bureau of Land Management. [2]

Arnold R. A. (1981). *Distribution, life history and status of three California Lepidoptera proposed as endangered or threatened species.* California Department of Fish and Game. [2]

Arnold R. A. (1983). Ecological studies of six endangered butterflies: island biogeography, patch dynamics and the design of nature preserves. *Univ. Calif. Publs Ent.* **(99)**: 1–161. [2]

Arnold R. A. & Fischer R. L. (1977). Operational mechanisms of copulation and oviposition in *Speyeria. Ann. ent. Soc. Am.* **70**: 455–468. [6]

Arnold S. J. (1978). The evolution of a special class of modifiable behaviors in relation to environmental pattern. *Am. Nat.* **112**: 415-427. [14]

Ashizawa H. & Muroya V. (1967). Notes on the early stages of *Calinaga buddha formosana. Spec. Bull. lepid. Soc. Japan* **3**: 79-85. [1]

Atkinson P. R. (1981). Mating behaviour and activity patterns of *Eldana saccharina. J. ent. Soc. sth. Afr.* **44**: 265-280. [25]

Atkinson P. R. (1982). Structure of the putative pheromone glands of *Eldana saccharina. J. ent. Soc. sth. Afr.* **45**: 93-104. [25]

Atsatt P. R. (1965). Angiosperm parasite and host: coordinated dispersal. *Science* **149**: 1398–1390. [2]

Atsatt P. R. (1970). The population biology of annual grassland hemiparasites. II. Reproductive patterns in *Orthocarpus. Evolution* **24**: 598–612. [2]

Atsatt P. R. (1981*a*). Ant-dependent food plant selection by the mistletoe butterfly *Ogyris amarillis. Oecologia* **48**: 60–63. [3, 6, 12, 19]

Atsatt P. R. (1981*b*). Lycaenid butterflies and ants: selection for enemy-free space. *Am. Nat.* **118**: 638–654. [Intro, 6, 12, 19]

Atsatt P. R. & O'Dowd D. J. (1976). Plant defense guilds. *Science* **193**: 24–29. [12]

Atsatt P. R. & Strong D. R. (1970). The population biology of annual grassland hemiparasites. I. The host environment. *Evolution* **24**: 278–291. [2]

Auerbach M. J. & Strong D. R. (1981). Nutritional ecology of *Heliconia* herbivores: experiments with plant fertilization and alternative hosts. *Ecol. Monogr.* **51**: 63–83. [19]

Aurivillius P. O. C. (1880). Über secundäre Geschlechtscharaktere norsicher Tagfalter. *Bih. K. svenska VetenskAkad. Handl.* **5**: 3–50. [25]

Avinoff A. & Sweadner W. R. (1951). The *Karanasa* butterflies, a study in evolution. *Ann. Carneg. Mus.* **32**: 1–251. [1]

Ayala F. J. & McDonald J. F. (1980). Continuous variation: possible roles of regulatory genes. *Genetica* **52/53**: 1–15. [16]

Baillif L. (1825). Examen microscopique de la Lupuline (matière active de Houblon). *J. Chemie Medicale Pharmacie et Toxicologie, Paris* **2**: 501-502 [referred to in Deschamps (1835); reference tentative, not seen]. [25]

Baker H. G. & Baker I. (1973*a*). Some anthecological aspects of the evolution of nectar-producing flowers, particularly amino acid production in nectar. Pp.243-264 in *Taxonomy and ecology* (ed. Heywood V. H.). London. [19]

Baker H. G. & Baker I. (1973*b*). Amino acids in nectar and their evolutionary significance. *Nature* **241**: 545. [19]

Baker M. C. (1981). Effective population size in a songbird: some possible implications. *Heredity* **46**: 209-218. [19]

Baker R. R. (1968*a*). A possible method of evolution of the migratory habit in butterflies. *Phil. Trans. R. Soc.* (B) **253**: 309-341. [2, 6, 26]

Baker R. R. (1968*b*). Sun orientation during migration in some British butterflies. *Proc. R. ent. Soc. Lond.* (A) **43**: 89-95. [2, 26]

Baker R. R. (1969). The evolution of the migratory habit in butterflies. *J. anim. Ecol.* **38**: 703-746. [2, 26, 33]

Baker R. R. (1970). Bird predation as a selective pressure on the immature stages of the cabbage butterflies, *Pieris rapae* and *P. brassicae*. *J. Zool., Lond.* **162**: 43-59. [2, 6, 10, 12]

Baker R. R. (1972*a*). Territorial behaviour of the nymphalid butterflies *Aglais urticae* and *Inachis io*. *J. anim. Ecol.* **41**: 453-469. [2, 20, 26, 33]

Baker R. R. (1972*b*). The geographical origin of the British spring individuals of the butterflies *Vanessa atalanta* and *V. cardui*. *J. Ent.* (A) **46**: 185-196. [16, 26]

Baker R. R. (1978). *The evolutionary ecology of animal migration.* New York & London. [2, 16, 20, 26]

Baker R. R. (ed.) (1980). *The mystery of migration.* London. [26]

Baker R. R. (1981). *Human navigation and the sixth sense.* London. [26]

Baker R. R. (1982). *Migration: paths through time and space.* London. [26]

Baker R. R. (1983). Insect territoriality. *Ann. Rev. Ent.* **28**: 65-89. [23, 26]

Baker R. R. & Mather J. G. (1982). Magnetic compass sense in the Large Yellow Underwing moth, *Noctua pronuba*. *Anim. Behav.* **30**: 543-548. [26]

Baker R. R. & Parker G. A. (1979). The evolution of bird coloration. *Phil. Trans. R. Soc. Lond.* (B) **287**: 63-130. [20]

Baker R. R. & Sadovy Y. (1978). The distance and nature of the light-trap response of moths. *Nature* **276**: 818-821. [26]

Baker T. C. (1983). Variations in male oriental fruit moth courtship patterns due to male competition. *Experientia* **39**: 112-114. [25]

Baker T. C. & Cardé R. T. (1979). Courtship behavior of the oriental fruit moth (*Grapholitha molesta*): experimental analysis and consideration of the role of sexual selection in the evolution of courtship pheromones in the Lepidoptera. *Ann. ent. Soc. Am.* **72**: 173-188. [25]

Baker T. C., Nishida R. & Roelofs W. L. (1981). Close-range attraction of female oriental fruit moths to herbal scent of male hairpencils. *Science* **214**: 1359-1360. [25]

Bang-Haas O. (1930). Zusammenstellung der faunistichen Literatur der paläarktischen Gross-schmetterlinge. *Novit. macrolepid.* **5**: 1-152. [1]

Barcant M. (1970). *The butterflies of Trinidad and Tobago.* London. [1, 16]

Barker J. F. & Herman W. S. (1973). On the neuroendocrine control of ovarian development in the Monarch butterfly. *J. exp. Zool.* **183**: 1-10. [6]

Barker L. M., Best M. R. & Domjan M. (eds) (1977). *Learning mechanisms in food selection.* Waco, Texas. [12]

Barrett C. G. (1882). Odour emitted by the male of *Hepialus hectus*. *Entomologist's mon. Mag.* **18**: 90. [25]

Barrett C. G. (1886). Singular habit of *Hepialus hectus*. *Entomologist's mon. Mag.* **23**: 110. [25]

Barrett J. A. (1976). The maintenance of non-mimetic forms in a dimorphic Batesian mimic species. *Evolution* **30**: 82-5. [20]

Barth R. (1944). Die männlichen Duftorgane einiger *Argynnis*-Arten. *Zool. Jb. (Anat.)* **68**: 331-362. [25]

Barth R. (1959). Phylogenetische Betrachtungen der Duftapparate einiger Nymphalinae. *An. Acad. bras. Cienc.* **31**: 557-565. [25]

Barth R. (1960). Órgãos odoríferos dos lepidópteros. *Bolm Parq. nac. Itatiáia* **(7)**: 159pp. [25]

Bartoshuk L. M., Lee C. & Scarpellino R. (1972). Sweet taste of water induced by artichoke (*Cynara scolymus*). *Science* **178**: 988-989. [12]

Bate-Smith E. C. (1972). Attractants and repellents in higher animals. Pp.45-56 in Harborne (1972). [12]

Bateman A. J. (1948). Intra-sexual selection in *Drosophila*. *Heredity* **2**: 349-368. [21]

Bates H. W. (1861). Contributions to an insect fauna of the Amazon valley. *J. Ent., Lond.* **1**: 218-245. [1]

Bates H. W. (1862). Contributions to an insect fauna of the Amazon valley. *Trans. Linn. Soc. Lond.* **23**: 495-566. [Intro, 1, 12, 20, 25]

Bates H. W. (1863). *The naturalist on the river Amazons.* London. [12]

Bates H. W. (1864). *The naturalist on the river Amazons* (2nd ed.). London. [20]

Bates H. W. (1864-5). Contributions to an insect fauna of the Amazon valley. *J. Ent., Lond.* **2**: 175-213, 311-346. [1]

Beattie J. R. (1976). *The Rhopalocera directory.* Berkeley, California. [1]

Beaufoy E. M., Beaufoy S., Dowdeswell H. W. & McWhirter K. G. (1970). Evolutionary studies on *Maniola jurtina*: the southern English stabilisation, 1961-68. *Heredity* **25**: 105-112. [16]

Beddard F. E. (1892). *Animal coloration.* London. [12]

Beebe W. (1955). Two little-known selective insect attractants. *Zoologica, N.Y.* **40**: 27-32. [25]

Beebe W., Crane J. & Fleming H. (1960). A comparison of eggs, larvae and pupae in 14 species of heliconiine butterflies from Trinidad, W.I. *Zoologica, N.Y.* **45**: 111-154. [12]

Begon M. (1979). *Investigating animal abundance.* Baltimore. [2]

Behan M. & Schoonhoven L. M. (1978). Chemoreception of an oviposition deterrent associated with eggs of *Pieris brassicae*. *Entomologia exp. appl.* **24**: 163-179. [6]

Beidler L. (1966). A physiological basis of taste sensation. *Fd Res.* **31**: 275-281. [12]

Beidler L. M. (ed.) (1971). Taste. *Handbook of sensory physiology* 4, *Chemical senses* (2). Berlin. [12]

Beidler L. M. (1976). Taste stimuli as possible messengers. Pp.483–488 in Müller-Schwarze & Mozell (1976). [12]

Beirne B. P. (1947). The origin and history of the British macrolepidoptera. *Trans. R. ent. Soc. Lond.* **98**: 273–372. [33]

Bell T. R. (1911). The common butterflies of the plains of India. *J. Bombay nat. Hist. Soc.* **20**: 1115–1136. [6]

Bell W. & Cardé R. (In press.) *Chemical ecology of insects.*

Bellow T. H. (1979). First reports of pellet ejection in 11 species. *Wilson Bull.* **91**: 626–628. [12]

Belt T. (1874). *The naturalist in Nicaragua.* London. [20]

Bengtson S. A. (1978). Spot-distribution in *Maniola jurtina* on small islands in southern Sweden. *Holarctic Ecol.* **1**: 54–61. [16]

Bengtson S. A. (1981). Does bird predation influence the spot-number variation in *Maniola jurtina*? *Biol. J. Linn. Soc.* **15**: 23–27. [16]

Benham B. R. (1973). The decline (and fall?) of the large blue butterfly. *Bull. amat. Ent. Soc.* **32**: 88–94. [33]

Benson J. M., Seiber J. N., Bagley C. V., Keeler R. F., Johnson A. E. & Young S. (1979). Effects on sheep of the milkweeds *Asclepias eriocarpa* and *A. labriformis* and of cardiac glycoside-containing derivative material. *Toxicon* **17**: 155–165. [12]

Benson J. M., Seiber J. N., Keeler R. F. & Johnson A. E. (1978). Studies on the toxic principle of *Asclepias eriocarpa* and *A. labriformis*. Pp.273-284 in Keeler *et al.* (1978). [12]

Benson W. W. (1971). Evidence for the evolution of unpalatability through kin selection in the Heliconiinae. *Am. Nat.* **105**: 213–226. [6, 12]

Benson W. W. (1972). Natural selection for Müllerian mimicry in *Heliconius erato* in Costa Rica. *Science* **176**: 936–939. [10, 14]

Benson W. W. (1977). On the supposed spectrum between Batesian and Müllerian mimicry. *Evolution* **31**: 454–455. [14]

Benson W. W. (1978). Resource partitioning in passion vine butterflies. *Evolution* **32**: 493–518. [3, 6]

Benson W. W. (1982). Alternative models for infrageneric diversification in the tropics: tests with passion vine butterflies. Pp.608–640 in Prance (1982). [14]

Benson W. W., Brown K. S. & Gilbert L. E. (1976). Coevolution of plants and herbivores: passion flower butterflies. *Evolution* **29**: 659–680. [Intro, 1, 3, 6, 12, 14]

Benson W. W. & Emmel T. C. (1972). Demography of gregariously roosting populations of the nymphaline butterfly *Marpesia berania* in Costa Rica. *Ecology* **54**: 326–335. [2]

Bent A. (1965). *Life histories of North American blackbirds, orioles, tanagers and allies.* New York. [12]

Benz G. & Schmid K. (1968). Stimulation der Eiablage unbegatteter Weibchen des Mondspinners *Actias selene* durch ein Männchen-Pheromon. *Experientia* **24**: 1279–1281. [25]

Berenbaum M. (1981). An oviposition "mistake" by *Papilio glaucus*. *J. Lepid. Soc.* **35**: 75. [6]

Berenbaum M. (1983). Coumarins and caterpillars: a case for coevolution. *Evolution* **37**: 163–179. [Intro]

Bergström G. (1978). Role of volatile chemicals in *Ophrys*-pollinator interactions. Pp.207–231 in Harborne (1978). [12]

Bergström G. & Lundgren L. (1973). Androconial secretion of three species of butterflies of the genus *Pieris*. *Zoon.* (Suppl.) **1**: 67–75. [12, 25]

Bernard G. D. (1979). Red-absorbing visual pigment of butterflies. *Science* **203**: 1125–1127. [20]

Bernard G. D. & Miller W. H. (1970). What does antenna engineering have to do with insect eyes? *I.E.E.E. Student J.* **1978**: 2–8. [20]

Bernardi G. (1947). Revision de la classification des espèces holarctiques des genres *Pieris* et *Pontia*. *Miscnea ent.* **44**: 65–80. [1]

Bernardi G. (1974). Polymorphisme et mimétisme chez les Lépidoptères Rhopalocères. *Mém. Soc. zool. Fr.* **(37)**: 129-165. [Intro, 23]

Bernays E. A. (1981). Plant tanins and insect herbivores: an appraisal. *Ecol. Ent.* **6**: 353–360. [12]

Bernays E. A. & Woodhead S. (1982). Plant phenols utilized as nutrients by a phytophagous insect. *Science* **216**: 201–203. [12]

Bernays E. A., Edgar J. A. & Rothschild M. (1977). Pyrrolizidine alkaloids sequestered and stored by the aposematic grasshopper, *Zonocerus variegatus*. *J. Zool., Lond.* **182**: 85–87. [12]

Bernhard C. G., Boëthius J., Gemne G. & Struwe G. (1970). Eye ultrastructure, colour reception and behaviour. *Nature* **226**: 865–866. [20]

Bernstein C. (1980). Density-dependent changes in sex ratio in *Colias lesbia*. *Ecol. Ent.* **5**: 105–110. [2]

Berger L. A. (1981). *Les papillons du Zaire.* Bruxelles. [1]

Berry R. J. (1971). Conservation and the genetical constitution of populations. Pp.117–206 in Duffey & Watt (1971). [33]

Berry R. J. (1977). *Inheritance and natural history.* London. [33]

Bertkau P. (1882). Über den Duftapparat von *Hepialus hecta*. *Arch. Naturgesch.* **48**: 363–370. [25]

Bertram B. C. R. (1978). Living in groups, predator and prey. Pp.64–96 in Krebs & Davies (1978). [12]

Berube D. E. (1970). Host-plant finding by odor in adult *Coryphista meadi*. *J. Lepid. Soc.* **24**: 220–224. [6]

Berube D. E. (1972). *Behavioral and physiological adaptations in the evolution of foodplant specificity in a species complex of* Colias *butterflies.* PhD thesis, Yale Univ., New Haven, USA. [6]

Best P. J., Best M. R. & Lindsey G. P. (1976). The role of cue additivity in salience in taste aversion conditioning. *Learning and motivation* **7**: 254-264. [12]

Best P. J., Best M. R. & Henggeler S. (1977). The contribution of environmental non-ingestive cues in conditioning with aversive internal consequences. Pp.371-393 in Barker *et al.* (1977). [12]

Bethune-Baker G. T. (1910). A revision of the African species of the *Lycaenesthes* group of the Lycaenidae. *Trans. ent. Soc. Lond.* **1910**: 1–84. [1]

Bethune-Baker G. T. (1925). A revision of the Liphyrinae together with a description of the structure of the puparium of *Liphyra brassolis* and of the pupae of *Aslauga vininga* and *A. lamborni*. *Trans. ent. Soc. Lond.* **1924**: 199-238. [1]

Bielewicz M. (1967). The causes of the vanishing butterfly *Papilio podalirius* in Poland. *Chrońmy Przyr. ojcz.* **23**: 21–29. [2]

Bink F. A. (1972). Het onderzoek naar de grote vuurvlinder (*Lycaena dispar batava*) in Nederland. *Ent. Ber. Amst.* **32**: 225–239. [2]

Birch M. C. (1970*a*). Pre-courtship use of abdominal brushes by a nocturnal moth, *Phlogophora meticulosa. Anim. Behav.* **18**: 310–316. [25]

Birch M. C. (1970*b*). Structure and function of the pheromone-producing brush-organs in males of *Phlogophora meticulosa. Trans. R. ent. Soc. Lond.* **122**: 277–292. [25]

Birch M. C. (1970*c*). Persuasive scents in moth sex life. *Nat. Hist., N.Y.* **79**: 34–39. [25]

Birch M. C. (1972). Male abdominal brush-organs in British noctuid moths and their value as a taxonomic character. *Entomologist* **105**: 185–205, 233–244. [25]

Birch M. C. (ed.) (1974). *Pheromones* (pp.115–134, Aphrodisiac pheromones in insects, by Birch M. C.). Amsterdam. [25]

Birch M. C. (1979). Eversible structures. Pp.9–18 in *The moths and butterflies of Great Britain and Ireland* **9** (eds Heath J. & Emmet A.M.). London. [25]

Birket-Smith J. (1960). Results from the Danish Expedition to French Cameroons (1949–1950). XXVII—Lepidoptera (Parts I–III). *Bull. Inst. fr. Afr. noire (A)* **22**: 521–554, 924–982, 1259–1284. [1]

Bishop J. A. (1969). Changes in the genetic constitution of a population of *Sphaeroma rugicauda* (Crustacea: Isopoda). *Evolution* **23**: 589–601. [16]

Bishop J. A., Cook L. M. & Muggleton J. (1978). The response of two species of moths to industrialization in northwest England. I. Polymorphisms for melanism. *Phil. Trans. R. Soc.* (B) **281**: 489–515. [16]

Bisping R., Benz U., Boxer P. & Longo N. (1974). Chemical transfer of learned colour discrimination in goldfish. *Nature* **249**: 771–773. [14]

Blab J. & Kudrna O. (1982). *Hilfsprogramm für Schmetterlinge.* Steinfurt. [33]

Blackie J. E. H. (1951). Range and distribution of *Agapetes galathea. Entomologist* **84**: 132–135. [33]

Blakley N. R. & Dingle H. (1978). Competition: butterflies eliminate milkweed bugs from a Caribbean island. *Oecologia* **37**: 133–136. [3]

Blau W. S. (1980). The effect of environmental disturbance on a tropical butterfly population. *Ecology* **61**: 1005–1012. [2, 10]

Blau W. S. (1981). Life history variation in the black swallowtail butterfly. *Oecologia* **48**: 116–122. [7]

Blest A. D. (1957). The function of eyespot patterns in the Lepidoptera. *Behaviour* **11**: 209–256. [12, 16]

Blower J. G., Cook L. M. & Bishop J. A. (1981). *Estimating the size of animal populations.* London. [Intro.]

Blum M. S. (1981). *Chemical defenses of arthropods.* New York. [12]

Blum M. S. & Blum N. A. (eds) (1979). *Sexual selection and reproductive competition in insects.* New York. [20, 21]

Bobisud L. E. & Potratz C. J. (1976). One-trial versus multi-trial learning for a predator encountering a model-mimic system. *Am. Nat.* **110**: 121–128. [12]

Boggs C. L. (1981*a*). Nutritional and life history determinants of resource allocation in holometabolous insects. *Am. Nat.* **117**: 692–709. [6]

Boggs C. L. (1981*b*). Selection pressures affecting male nutrient investment at mating in heliconiine butterflies. *Evolution* **35**: 931–940. [6]

Boggs C. L. & Gilbert L. E. (1979). Male contribution to egg production in butterflies: evidence for transfer of nutrients at mating. *Science* **206**: 83–84. [6, 20]

Boggs C. L. & Watt W. B. (1981). Population structure of pierid butterflies. IV. Genetic and physiological investment in offspring by male *Colias. Oecologia* **50**: 320–324. [6]

Boggs C. L., Smiley J. T. & Gilbert L. E. (1981). Patterns of pollen exploitation by *Heliconius* butterflies. *Oecologia* **48**: 284–289. [3, 6]

Bolles R. C. (1970). Species-specific defense reactions and avoidance learning. *Psychol. Rev.* **77**: 32–48. [12]

Bondar G. (1940). Insetos nocivos e molestias do coqueiro (*Cocos nucifera*) no Brasil. *Bolm Inst. cent. Fom. econ. Bahia* **(8)**: 160pp. [6]

Boppré M. (1976). Pheromon-Transfer-Partikel auf einem Duftpinselhaar eines Monarchfalters (*Danaus formosa*). *Naturw. Rdsch.* **29**(9): cover. [25]

Boppré M. (1977). Pheromonbiologie am Beispiel der Monarchfalter. *Biol. unserer Zeit.* **7**: 161–169. [12, 25]

Boppré M. (1978). Chemical communication, plant relationships, and mimicry in the evolution of danaid butterflies. *Entomologia exp. appl.* **24**: 264–277. [Intro, 9, 12, 15, 21, 25]

Boppré M. (1979). *Untersuchungen zur Pheromonbiologie bei Monarchfaltern.* Thesis, München Univ., GFR. (Also circulated privately.) [25]

Boppré M. (1981). Adult Lepidoptera "feeding" at withered *Heliotropium* plants (Boraginaceae) in East Africa. *Ecol. Ent.* **6**: 449–452. [12, 25]

Boppré M. (1983). Leaf-scratching—a specialized behaviour of danaine butterflies for gathering secondary plant substances. *Oecologia* **59**: 414–416. [25]

Boppré M. (1984*a*). In prep. Androconial organs of *Amauris ochlea*, *Danaus chrysippus*, and *Tirumala petiverana*. [25]

Boppré M. (1984*b*). In prep. Pheromone-transfer-particles in some Lepidoptera. [25]

Boppré M. (1984*c*). In prep. On the defensive potency of pyrrolizidine alkaloids stored by insects. [25]

Boppré M. & Fecher K. (1977). Structural complexity of scent organs of male danaid butterflies. *Scanning Electron Microsc.* **2**: 639–644. [25]

Boppré M., Petty R. L., Schneider D. & Meinwald J. (1978). Behaviourally mediated contacts between scent organs: another prerequisite for pheromone production in *Danaus chrysippus* males. *J. comp. Physiol.* **126**: 97–103. [25]

Boppré M., Vane-Wright R. I. & Ackery P. R. (In prep.). Comparative studies on the androconial organs of *Amauris* butterflies. [25]

Borch H. & Schmid F. (1973). On *Ornithoptera priamus caelestis*, *demophanes* and *boisduvali. J. Lepid. Soc.* **27**: 196–205. [25]

Borison H. L. & Wang S. C. (1953). Physiology and pharmacology of vomiting. *Pharmac. Rev.* **5**: 193–230. [12]

Bourgogne J. (1951). Ordre des Lépidoptères. *Traité de Zoologie* **10**(1). Paris. [Intro, 12]

Bourquin F. (1953). Notas sobre la metamorfosis de *Hamearis susanae*, con oruga mirmecofila. *Revta Soc. ent.* **16**: 83–87. [6]

Bovey P. (1941). Contribution à l'étude génétique et biogéographique de *Zygaena ephialtes*. *Revue suisse Zool.* **48**: 1–90. [14]

Bowden S. R. (1952). Pupal colour and diapause in *Pieris napi*. *Entomologist* **85**: 175–178. [30]

Bowden S. R. (1977). *Pieris*—the ultra-violet image. *Proc. Trans. Br. ent. nat. Hist. Soc.* **10**: 16–22. [20]

Bowden S. R. (1979). Subspecific variation in butterflies: adaptation and dissected polymorphism in *Pieris (Artogeia)*. *J. Lepid. Soc.* **33**: 77–111. [16]

Bowers M. D. (1979). *Unpalatability as a defense strategy of checkerspot butterflies with special reference to* Euphydryas phaeton. PhD thesis, Univ. of Massachusetts, USA. [12]

Bowers M. D. (1980). Unpalatability as a defense strategy of *Euphydryas phaeton*. *Evolution* **34**: 586-600. [6, 12, 16]

Bowers M. D. (1981). Unpalatability as a defense strategy of western checkerspot butterflies (*Euphydryas*). *Evolution* **35**: 367–375. [6, 12]

Bowers M. D. & Wiernasz D. C. (1979). Avian predation on the palatable butterfly, *Cercyonis pegala*. *Ecol. Ent.* **4**: 205–209. [12, 16]

Bowman W. C. & Rand M. J. (1980). *Textbook of pharmacology* (2nd ed.). Oxford. [12]

Boyden T. C. (1976). Butterfly palatability and mimicry: experiments with *Ameiva* lizards. *Evolution* **30**: 73–81. [12, 14]

Bradshaw W. E. (1973). Homeostasis and polymorphism in vernal development of *Chaoborus americanus*. *Ecology* **54**: 1247–1259. [27]

Brakefield P. (1979*a*). *An experimental study of the maintenance of variation in spot pattern in* Maniola jurtina. PhD thesis, Liverpool Univ., UK. [16]

Brakefield P. (1979*b*). Spot-number in *Maniola jurtina*—variation between generations and selection in marginal populations. *Heredity* **42**: 259–266. [16]

Brakefield P. M. (1982*a*). Ecological studies on the butterfly *Maniola jurtina* in Britain. I. Adult behaviour, microdistribution and dispersal. *J. anim. Ecol.* **51**: 713–726. [16]

Brakefield P. M. (1982*b*). Ecological studies on the butterfly *Maniola jurtina* in Britain. II. Population dynamics: the present position. *J. anim. Ecol.* **51**: 727–738. [16]

Brakefield P. M. & Larsen T. B. (1984). The evolutionary significance of wet and dry season forms in some tropical butterflies. *Biol. J. Linn. Soc.* **22**.

Braveman N. S. (1977). Visually guided avoidance of poisonous food in mammals. Pp.455–473 in Barker *et al.* (1977). [12]

Bree W. T. (1852). A list of butterflies in the neighbourhood of Polebrook, Northampton; with some remarks. *Zoologist* **10**: 3348–3352. [33]

Breedlove D. E. & Ehrlich P. R. (1968). Plant-herbivore coevolution: lupines and lycaenids. *Science* **162**: 672–673. [2, 6]

Breedlove D. E. & Ehrlich P. R. (1972). Coevolution: patterns of legume predation by a lycaenid butterfly. *Oecologia* **10**: 99–104. [2, 6]

Brett L. P., Hankins W. G. & Garcia J. (1976). Prey-lithium aversions. III. Buteo hawks. *Behav. Biol.* **17**: 87–98. [12]

Brian M. V. (1956). Studies of caste differentiation in *Myrmica rubra*. 4. Controlled larval nutrition. *Insectes soc.* **3**: 364–394. [19]

Brian M. V. (1973). Feeding and growth in the ant *Myrmica*. *J. anim. Ecol.* **42**: 37–53. [19]

Brian M. V., Mountford M. D., Abbott A. & Vincent S. (1976). The changes in ant species distribution during ten years post-fire regeneration of a heath. *J. anim. Ecol.* **45**: 115–133. [33]

Brittnacher J. G., Sims S. R. & Ayala F. J. (1978). Genetic differentiation between species of the genus *Speyeria*. *Evolution* **32**: 199–210. [17]

Brock J. P. (1971). A contribution towards an understanding of the morphology and phylogeny of the ditrysian Lepidoptera. *J. nat. Hist.* **5**: 29–102. [1]

Brockie R. E. (1972). Evolutionary studies on *Maniola jurtina* in Sicily. *Heredity* **28**: 337–345. [16]

Bronstein P. M. & Crockett D. P. (1979). Rat pup's food consumption as a function of preweaning tastes and odors. Pp.157–171 in Müller-Schwarze & Silverstein (1979). [12]

Brooks M. & Knight C. (1982). *A complete guide to British butterflies*. London. [1]

Brower J. V. Z. (1958*a*). Experimental studies of mimicry in some North American butterflies. I. The monarch *Danaus plexippus* and viceroy *Limenitis archippus*. *Evolution* **12**: 32–47. [12, 19]

Brower J. V. Z. (1958*b*). Experimental . . . II. *Battus philenor* and *Papilio troilus*, *P. polyxenes* and *P. glaucus*. *Evolution* **12**: 123–136. [12, 22]

Brower J. V. Z. (1958*c*). Experimental . . . III. *Danaus gilippus berenice* and *Limenitis archippus floridensis*. *Evolution* **12**: 273–285. [12]

Brower J. V. Z. (1963). Experimental studies and new evidence on the evolution of mimicry in butterflies. *Proc. Int. Congr. Zool.* (16) **4**: 151–161. [14]

Brower J. V. Z. & Brower L. P. (1961). Palatability of North American model and mimic butterflies to caged mice. *J. Lepid. Soc.* **15**: 23–24. [12]

Brower L. P. (1959). Speciation in butterflies of the *Papilio glaucus* group. II. Ecological relationships and interspecific sexual behaviour. *Evolution* **13**: 212–228. [20]

Brower L. P. (1961*a*). Studies on the migration of the monarch butterfly. I. Breeding populations of *Danaus plexippus* and *D. gilippus berenice* in south central Florida. *Ecology* **42**: 76–83. [2, 12]

Brower L. P. (1961*b*). Experimental analyses of egg cannibalism in the monarch and queen butterflies, *Danaus plexippus* and *D. gilippus berenice*. *Physiol. Zool.* **34**: 287–296. [12]

Brower L. P. (1962). Evidence for interspecific competition in natural populations of the monarch and queen butterflies, *Danaus plexippus* and *D. gilippus berenice* in south central Florida. *Ecology* **45**: 549–552. [2]

Brower L. P. (1963). The evolution of sex-limited mimicry in butterflies. *Proc. Int. Congr. Zool.* (16)**4**: 173–179. [Intro, 12, 20, 25]

Brower L. P. (1968) (motion picture film). *Patterns for*

survival: a study of mimicry and protective coloration in tropical insects. Amherst (LCCN Fi-A-68). [12]

Brower L. P. (1969). Ecological chemistry. *Scient. Am.* **220**: 22-29. [12, 25]

Brower L. P. (1970). Plant poisons in a terrestrial food chain and implications for mimicry theory. *Proc. Ann. Biol. Coll. Oregon State Univ.* **(29)**: 69-82. [12]

Brower L. P. (1971). Prey coloration and predator behavior. Pp.360-370 in *Topics in the study of life*, The Bio source book (6), *Animal behavior*. New York. [12]

Brower L. P. (1977). Monarch migration. *Nat. Hist., N.Y.* **86**: 40-53. [2, 12, 33]

Brower L. P. & Brower J. V. Z. (1962). The relative abundance of model and mimic butterflies in natural populations of the *Battus philenor* mimicry complex. *Ecology* **43**: 154-158. [22]

Brower L. P. & Brower J. V. Z. (1964). Birds, butterflies, and plant poisons: a study in ecological chemistry. *Zoologica, N.Y.* **49**: 137-159. [6, 12, 14, 19, 25]

Brower L. P. & Brower J. V. Z. (1972). Parallelism, convergence, divergence, and the new concept of advergence in the evolution of mimicry. *Trans. Conn. Acad. Arts Sci.* **44**: 57-67. [14]

Brower L. P. & Glazier S. C. (1975). Localization of heart poisons in the monarch butterfly. *Science* **188**: 19-25. [12, 14, 15]

Brower L. P. & Jones M. A. (1965). Precourtship interaction of wing and abdominal sex glands in male *Danaus* butterflies. *Proc. R. ent. Soc. Lond.* (A) **40**: 147-151. [25]

Brower L. P. & Moffitt C. M. (1974). Palatability dynamics of cardenolides in the monarch butterfly. *Nature* **249**: 280-283. [12]

Brower L. P., Brower J. V. Z. & Collins C. T. (1963). Experimental studies on mimicry. 7. Relative palatability and Müllerian mimicry among neotropical butterflies of the subfamily Heliconiinae. *Zoologica, N.Y.* **48**: 65-84. [12]

Brower L. P., Brower J. V. Z. & Cranston F. P. (1965). Courtship behavior of the queen butterfly, *Danaus gilippus berenice*. *Zoologica, N.Y.* **50**: 1-39. [12, 20, 25]

Brower L. P., Brower J. V. Z. & Corvino J. M. (1967*a*). Plant poisons in a terrestrial food chain. *Proc. natn Acad. Sci. U.S.A.* **57**: 893-898. [12, 15]

Brower L. P., Cook L. M. & Croze H. J. (1967*b*). Predator responses to artificial Batesian mimics released in a neotropical environment. *Evolution* **21**: 11-23. [10]

Brower L. P., Ryerson W. N., Coppinger L. L. & Glazier S. C. (1968). Ecological chemistry and the palatability spectrum. *Science* **161**: 1349-1351. [12]

Brower L. P., Pough F. H. & Meck H. R. (1970). Theoretical investigations of automimicry. I. Single trial learning. *Proc. natn Acad. Sci. U.S.A.* **66**: 1059-1066. [12]

Brower L. P., Alcock J. & Brower J. V. Z. (1971). Avian feeding behaviour and the selective advantage of incipient mimicry. Pp.261-274 in Creed (1971). [12, 20]

Brower L. P., McEvoy P. B., Williamson K. L. & Flannery M. A. (1972). Variation in cardiac glycoside content of monarch butterflies from natural populations in eastern North America. *Science* **177**: 426-429. [6, 12, 15]

Brower L. P., Edmunds M. & Moffitt C. M. (1975). Cardenolide content and palatability of a population of *Danaus chrysippus* butterflies from West Africa. *J. Ent.* (A) **49**: 183-196. [12, 15, 18]

Brower L. P., Calvert W. H., Hedrick L. E. & Christian J. (1977*a*). Biological observations on an overwintering colony of Monarch butterflies (*Danaus plexippus*) in Mexico. *J. Lepid. Soc.* **31**: 232-242. [2, 12]

Brower L. P., Huberth J. C. & Lurtsema R. J. (1977*b*) (motion picture film). *Strategy for survival: behavioral ecology of the Monarch butterfly*. Hagerstown, Maryland. [12]

Brower L. P., Gibson D. O., Moffitt C. M. & Panchen A. L. (1978). Cardenolide content of *Danaus chrysippus* butterflies from three areas of East Africa. *Biol. J. Linn. Soc.* **10**: 251-273. [12, 18]

Brower L. P., Seiber J. N., Nelson C. J., Tuskes P. & Lynch S. P. (1982). Plant-determined variation in the cardenolide content, thin layer chromatography profiles, and emetic potency of monarch butterflies, *Danaus plexippus*, reared on the milkweed, *Asclepias eriocarpa*, in California. *J. Chem. Ecol.* **8**: 579-633. [12]

Brown F. M. & Heineman B. (1972). *Jamaica and its butterflies*. London. [1]

Brown I. L. & Ehrlich P. R. (1980). Population biology of the checkerspot butterfly, *Euphydryas chalcedona*: structure of the Jasper Ridge colony. *Oecologia* **47**: 239-251. [2, 33]

Brown J. J. & Hussain Y. B. (1981). Physiological effects of volcanic ash upon selected insects in the laboratory. *Melanderia* **37**: 30-38. [31]

Brown K. S. (1970). Rediscovery of *Heliconius nattereri* in eastern Brazil. *Ent. News* **81**: 129-140. [2]

Brown K. S. (1979). *Ecologia geográfica e evolucão nas florestas Neotropicais. Padrões Geográficos de Evolucão em Lepidópteros Neotropicais* (6). Campinas. [1]

Brown K. S. (1981). The biology of *Heliconius* and related genera. *A. Rev. Ent.* **26**: 427-456. [1, 3, 14]

Brown K. S. (1982). Palaeoecology and regional patterns of evolution in neotropical forest butterflies. Pp.225-308 in Prance (1982). [14]

Brown K. S. & Benson W. W. (1974). Adaptive polymorphism associated with multiple Müllerian mimicry in *Heliconius numata*. *Biotropica* **6**: 205-228. [3]

Brown K. S. & Benson W. W. (1977). Evolution in modern Amazonian non-forest islands: *Heliconius hermathena*. *Biotropica* **9**: 95-117. [14]

Brown K. S. & Vasconcellos Neto J. (1976). Predation of aposematic ithomiine butterflies by tanagers (*Pipraeidea melanonota*). *Biotropica* **8**: 136-141. [2, 12]

Brown K. S., Sheppard P. M. & Turner J. R. G. (1974). Quaternary refugia in tropical America: evidence from race formation in *Heliconius* butterflies. *Proc. R. Soc.* (B) **187**: 369-378. [14]

Brown K. S., Damman A. J. & Feeny P. (1981). Troidine swallowtails in southeastern Brazil: natural history and foodplant relationships. *J. Res. Lepid.* **19**: 199-226. [12]

Bruch C. (1926). Orugas mirmecofilas de *Hamearis epulus signatus*. *Revta Soc. ent.* **1**: 1–9. [6]

Brues C. T. (1920). The selection of food-plants by insects, with special reference to lepidopterous larvae. *Am. Nat.* **54**: 213–232. [Intro.]

Brues C. T. (1924). The specificity of food-plants in the evolution of phytophagous insects. *Am. Nat.* **58**: 127–144. [Intro.]

Brues C. T. (1946). *Insect dietary: an account of the food habits of insects*. Cambridge, Mass. [12]

Brussard P. F. & Ehrlich P. R. (1970*a*). The population structure of *Erebia epipsodea*. *Ecology* **51**: 119–129. [2]

Brussard P. F. & Ehrlich P. R. (1970*b*). Adult behavior and population structure in *Erebia epipsodea*. *Ecology* **51**: 880–885. [2]

Brussard P. F. & Ehrlich P. R. (1970*c*). Contrasting population biology of two species of butterfly. *Nature* **227**: 91–92. [2]

Brussard P. F. & Vawter A. T. (1975). Population structure, gene flow and natural selection in populations of *Euphydryas phaeton*. *Heredity* **34**: 407–415. [2, 17]

Brussard P. F., Ehrlich P. R. & Singer M. C. (1974). Adult movements and population structure in *Euphydryas editha*. *Evolution* **28**: 408–415. [2]

Bryk F. (1923). Baroniidae, Teinopalpidae, Parnassiidae. *Lepid. Cat.* **24** (27): 1–247. [1]

Bryk F. (1929–30). Papilionidae I–III. *Lepid. Cat.* **24** (35, 37, 39): 1–676. [1]

Bryk F. (1934). Lepidoptera, Baroniidae, Teinopalpidae, Parnassiidae. Pars I. *Tierreich* **64**: xxiii + 131pp. [1]

Bryk F. (1935). Lepidoptera, Parnassiidae. Pars II (subfam Parnassiinae). *Tierreich* **65**: li + 790pp. [1]

Bryk F. (1937*a*). Danaidae I: subfamilia: Danainae. *Lepid. Cat.* **28** (78): 1–432. [1]

Bryk F. (1937*b*). Danaidae II: subfam: Ituninae, Tellervinae, Ithomiinae. *Lepid. Cat.* **28** (80): 433–702. [1]

Buckner C. H. (1971). Vertebrate predators. Pp.21–31 in *Toward integrated control* (Proc. 3rd Annual N.E. For. Ins. Wk Conf.). USDA For. Serv. Res. Pap. NE-194. [12]

Bull L. B., Culvenor C. C. J. & Dick A. T. (1968). *The pyrrolizidine alkaloids*. Amsterdam. [12, 25]

Bullini L., Sbordoni V. & Ragazzini P. (1969). Mimetismo mulleriano in popolazione italiane di *Zygaena ephialtes*. *Archo. zool. ital.* **44**: 181–214. [14]

Burghardt G. M., Wilcoxon H. C. & Czaplicki J. A. (1973). Conditioning in garter snakes: aversion to palatable prey induced by delayed illness. *Animal Learning and Behavior* **1**: 317–320. [12]

Burley N. (1981). Sex ratio manipulation and selection for attractiveness. *Science* **211**: 721–722. [20]

Burns J. M. (1964). Evolution in skipper butterflies of the genus *Erynnis*. *Univ. Calif. Publs Ent.* **(37)**: iv + 214pp. [1]

Burns J. M. (1966). Preferential mating versus mimicry: disruptive selection and sex-limited dimorphism in *Papilio glaucus*. *Science* **153**: 551–553. [20–22]

Burns J. M. (1967). Selective forces in *Papilio glaucus*. *Science* **156**: 534. [20, 22]

Burns J. M. (1968). Mating frequency in natural populations of skippers and butterflies as determined by spermatophore counts. *Proc. natn Acad. Sci. U.S.A.* **61**: 852–859. [20, 21]

Burton J. (1971). Somerset butterflies and their conservation. *A. Rep. Somerset Trust Nature Conserv.* **1971**: 13–18. [2]

Burtt E. H. (1981). The adaptiveness of animal colors. *Bioscience* **31**: 723–728. [12]

Bush G. L. (1975). Modes of speciation. *A. Rev. Ecol. Syst.* **6**: 339–364. [19]

Bush G. L., Case S. M., Wilson A. C. & Patton J. L. (1977). Rapid speciation and chromosomal evolution in mammals. *Proc. natn Acad. Sci. U.S.A.* **74**: 3942–3946. [19]

Butler C. G. (1964). Pheromones in sexual processes in insects. *Symp. R. ent. Soc. Lond.* **(2)**: 66–77. [25]

Butler C. G. (1967). Insect pheromones. *Biol. Rev.* **42**: 42–87. [25]

Buxton R. D. & Connolly M. L. (1973). *Survey of the adonis blue,* Lysandra bellargus. Unpublished report, Joint Committee for the Conservation of British Insects. London. [33]

Cain A. J. (1977). The efficacy of natural selection in wild populations. *Spec. Publs Acad. nat. Sci. Philad.* **(12)**: 113–133. [16]

Callaghan C. J. (1977). Studies on Restinga butterflies. 1. Life cycle and immature biology of *Menander felsina* (Riodinidae), a myrmecophilous metalmark. *J. lepid. Soc.* **31**: 173–182. [1]

Callahan P. S. (1977). *Tuning in to Nature*. London. [28]

Calvert W. H. (1974). The external morphology of foretarsal receptors involved with host discrimination by the nymphalid butterfly *Chlosyne lacinia*. *Ann. ent. Soc. Am.* **67**: 853–857. [6]

Calvert W. H. & Brower L. P. (1982). The importance of forest cover for the survival of overwintering monarch butterflies (*Danaus plexippus*). *J. Lepid. Soc.* **35**: 216–225. [12]

Calvert W. H. & Hanson F. E. (1983). The role of sensory structures and preoviposition behavior in oviposition by the patch butterfly, *Chlosyne lacinia*. *Entomologia exp. appl.* **33**: 179–187. [6]

Calvert W. H., Hedrick L. E. & Brower L. P. (1979). Mortality of the monarch butterfly (*Danaus plexippus*) due to avian predation at five overwintering sites in Mexico. *Science* **204**: 847–851. [12, 31]

Campbell B. (ed.) (1972). *Sexual selection and the descent of man 1871–1971*. London. [21]

Campbell S. (1980). Is reintroduction a realistic goal? Pp.263–269 in Soulé & Wilcox (1980). [33]

Capinera J. L., Wiener L. F. & Anamosa P. R. (1980). Behavioral thermoregulation by late-instar range caterpillar larvae, *Hemileuca oliviae*. *J. Kans. ent. Soc.* **53**: 631–636. [28]

Capretta P. J. (1977). Establishment of food preferences by exposure to ingestive stimuli early in life. Pp.99–121 in Barker *et al.* (1977). [12]

Carcasson R. H. (1961). The *Acraea* butterflies of East Africa. *Jl E. Africa nat. Hist. Soc.* (Suppl.) **(8)**: i + 47pp. [1]

Carcasson R. H. (1963). The milkweed butterflies of East Africa. *Jl E. Africa nat. Hist. Soc. & Coryndon Mus.* **24**: 19–32. [1]

Carcasson R. H. (1975). The swallowtail butterflies of East Africa. Faringdon, UK. (First published as *Jl E. Africa nat. Hist. Soc.* (Suppl.) **(6)**, 1960.) [1]

Carcasson R. H. (1981). *Collins' handguide to the butterflies of Africa*. London. [1]

Carcasson R. H., Ackery P. R., Arora R., Huggins C. F., North S. M., Smiles R. L. & Vane-Wright R. I. (In prep.). A synonymic catalogue of the butterflies of the Afrotropical region. [1]

Cardé R. T., Baker T. C. & Roelofs W. L. (1975). Ethological function of components of a chemical sex attractant system in the Oriental fruit moth, *Grapholitha molesta*. *J. chem. Ecol.* **1**: 475–491. [25]

Carolsfeld-Krausé A. G. (1959). The mating of Hepialidae. *Entomologist's Rec. J. Var.* **71**: 33–34. [25]

Carpenter G. D. H. (1921). Experiments on the relative edibility of insects, with special reference to their coloration. *Trans. ent. Soc. Lond.* **1921**: 1–105. [12]

Carpenter G. D. H. (1935). Courtship and allied problems in insects. *Trans. Soc. Br. Ent.* **2**: 115–135. [25]

Carpenter G. D. H. (1941). The relative frequency of beak-marks on butterflies of different edibility to birds. *Proc. zool. Soc. Lond.* (A) **111**: 223–231. [10, 12, 16]

Carpenter G. D. H. (1942). Observations and experiments in Africa by the late C. F. M. Swynnerton on wild birds eating butterflies and the preference shown. *Proc. Linn. Soc. Lond.* **154**: 10–46. [10, 12]

Carpenter G. D. H. & Ford E. B. (1933). *Mimicry*. London. [12, 20]

Carpenter G. D. H. & Poulton E. B. (1914). Dr G. D. H. Carpenter's observation on the epigamic use of its anal brushes by a male *Amauris psyttalea*. *Proc. ent. Soc. Lond.* **1914**: cxi–cxii. [25]

Carpenter G. D. H. & Poulton E. B. (1927). Dr G. D. H. Carpenter's observation on the epigamic use of its anal brushes by the male *Danaida chrysippus*, in E. Madi, Uganda. *Proc. ent. Soc. Lond.* **1927**: 44. [25]

Casarett L. J. (1975). Toxicological evaluation Pp.11–25 in Casarett & Doull (1975). [12]

Casarett L. J. & Doull J. (eds) (1975). *Toxicology: the basic science of poisons*. New York.

Casey T. M. (1976). Activity patterns, body temperature and thermal ecology of two desert caterpillars. *Ecology* **57**: 485–497. [28]

Caspari E. (1941). The morphology and development of the wing patterns of Lepidoptera. *Q. Rev. Biol.* **16**: 249–273. [16]

Castle W. F. (1921). An improved method of estimating the number of genetic factors concerned in cases of blending inheritance. *Science* **54**: 223. [27]

Cavalier-Smith T. (1978). Nuclear volume control by nucleoskeletal DNA, selection for cell volume and cell growth rate, and the solution of the DNA C-value paradox. *J. Cell Sci.* **34**: 247–278. [16]

Cavalli-Sforza L. L. & Edwards A. W. F. (1967). Phylogenetic analysis: models and estimation procedures. *Evolution* **21**: 550–570. [17]

Chadha G. K. & Roome R. E. (1980). Oviposition behaviour and the sensilla of the ovipositor of *Chilo partellus* and *Spodoptera littoralis*. *J. Zool., Lond.* **192**: 169–178. [6]

Champion H. G. (1930). The protrusion of anal scent-brushes during flight by the male *Euploea core*. *Proc. ent. Soc. Lond.* **5**: 14–15. [25]

Chaney S. G. & Kare M. R. (1966). Emesis in birds. *J. Am. vet. med. Ass.* **149**: 938–943. [12]

Chanter D. O. & Owen D. F. (1972). The inheritance and population genetics of sex ratio in the butterfly *Acraea encedon*. *J. Zool., Lond.* **166**: 363–383. [21]

Chapman R. F. & Blaney W. M. (1979). How animals perceive secondary compounds. Pp.161–270 in Rosenthal & Janzen (1979). [6, 12]

Charlesworth D. & Charlesworth B. (1975). Theoretical genetics of Batesian mimicry. I. Single locus models. *J. theor. Biol.* **55**: 283–303. [3]

Charlesworth D. & Charlesworth B. (1976). Theoretical . . . II. Evolution of supergenes. *J. theor. Biol.* **55**: 305–324. [14]

Charnov E. L. (1976). Optimal foraging: the marginal value theorem. *Theoret. Pop. Biol.* **9**: 129–136. [26]

Chermock R. L. (1950). A generic revision of the Limenitini of the World. *Am. Midl. Nat.* **43**: 513–569. [1]

Chew F. S. (1975). Coevolution of pierid butterflies and their cruciferous foodplants. I. The relative quality of available resources. *Oecologia* **20**: 117–127. [6]

Chew F. S. (1977). Coevolution . . . II. The distribution of eggs on potential foodplants. *Evolution* **31**: 568–579. [6–8]

Chew F. S. (1979). Community ecology and *Pieris*-crucifer coevolution. *Jl N.Y. ent. Soc.* **87**: 128–134. [12]

Chew F. S. (1980). Foodplant preferences of *Pieris* caterpillars. *Oecologia* **46**: 347–353. [6, 7]

Chew F. S. (1981). Coexistence and local extinction in two pierid butterflies. *Am. Nat.* **18**: 655–672. [2, 6]

Chew F. S. & Rodman J. E. (1979). Plant resources for chemical defense. Pp.271–306 in Rosenthal & Janzen (1979). [6]

Chovet G. (1983). Le comportement sexuel de *Pieris brassicae*. *Annls Soc. ent. Fr. (N.S.)* **18**: 461–474. [20, 25]

Claassens A. J. M. & Dickson C. G. C. (1977). A study of the myrmecophilous behaviour and the immature stages of *Aloeides thyra* with special reference to the function of the retractile tubercles and with additional notes on the general biology of the species. *Entomologist's Rec. J. Var.* **89**: 225–231. [19]

Claesson A. & Silverstein R. M. (1976). Chemical methodology in the study of mammalian communication. Pp.71–93 in Müller-Schwarze & Mozell (1976). [12]

Claridge M. F. & Wilson M. R. (1978). Oviposition behaviour as an ecological factor in woodland canopy leafhoppers. *Entomologia exp. Appl.* **24**: 101–109. [6]

Clark A. H. (1925). Carnivorous butterflies. *Rep. Smithson. Instn* **1925**: 439–508. [12]

Clark A. H. (1926). Notes on the odor of some New England butterflies. *Psyche* **33**: 1–5. [12]

Clark A. H. (1927). Fragrant butterflies. *Rep. Smithson. Instn* **1926**; 421–446. [12, 25]

Clark A. H. (1947). The interrelationships of the several groups within the butterfly superfamily Nymphaloidea. *Proc. ent. Soc. Wash.* **49**: 148–149, 192. [1]

Clark A. H. (1948). Classification of the butterflies, with the allocation of the genera occurring in North America north of Mexico. *Proc. biol. Soc. Wash.* **61**: 77–81. [1]

Clark A. H. & Clark L. H. (1951). The butterflies of Virginia. *Smithson. misc. Collns* **(116)**: 239pp. [22]

Clark G. C. & Dickson C. G. C. (1957). Life history of *Precis octavia*. *J. ent. Soc. sth. Afr.* **20**: 257–259. [29]

Clark G. C. & Dickson C. G. C. (1971). *Life histories of*

the South African lycaenid butterflies. Cape Town. [1, 6]

Clarke B. (1970). Festschrift in biology. *Science* **169**: 1192. [16]

Clarke C. A. & Clarke F. M. M. (1983). Abnormalities of wing pattern in the Eastern Tiger Swallowtail, *Papilio glaucus*. *Syst. Ent.* **8**: 25-28.

Clarke C. A. & Ford E. B. (1980). Intersexuality in *Lymantria dispar*. A reassessment. *Proc. R. Soc.* (B) **206**: 381-394. [24]

Clarke C. A. & Ford E. B. (1982). Intersexuality in *Lymantria dispar*. A further reassessment. *Proc. R. Soc.* (B) **214**: 285-288. [25]

Clarke C. A. & Ford E. B. (1983). *Lymantria dispar*. A third reassessment. *Proc. R. Soc.* (B) **218**: 365-370. [24]

Clarke C. A. & Sheppard P. M. (1955*a*). A preliminary report on the genetics of the *machaon* group of swallowtail butterflies. *Evolution* **9**: 182-201. [14]

Clarke C. A. & Sheppard P. M. (1955*b*). The breeding in captivity of the hybrids *Papilio rutulus* female × *Papilio glaucus* male. *Lepid. News* **9**: 46-48. [22]

Clarke C. A. & Sheppard P. M. (1959). The genetics of some mimetic forms of *Papilio dardanus* and *Papilio glaucus*. *J. Genetics* **56**: 236-260. [22]

Clarke C. A. & Sheppard P. M. (1960*a*). The evolution of dominance under disruptive selection. *Heredity* **14**: 73-87. [16]

Clarke C. A. & Sheppard P. M. (1960*b*). The evolution of mimicry in the butterfly *Papilio dardanus*. *Heredity* **14**: 163-173. [14, 16]

Clarke C. A. & Sheppard P. M. (1960*c*). The genetics of *Papilio dardanus*. III. Race *antinorii* from Abyssinia and race *meriones* from Madagascar. *Genetics* **45** : 683-698. [14]

Clarke C. A. & Sheppard P. M. (1960*d*). Supergenes and mimicry. *Heredity* **14**: 175-185. [14]

Clarke C. A. & Sheppard P. M. (1962*a*). The genetics of *Papilio dardanus*. IV. Data on race *ochracea*, race *flavicornis*, and further information on races *polytrophus* and *dardanus*. *Genetics* **47**: 909-920. [14]

Clarke C. A. & Sheppard P. M. (1962*b*). Disruptive selection and its effect on a metrical character in the butterfly *Papilio dardanus*. *Evolution* **16**: 214-226. [14, 16]

Clarke C. A. & Sheppard P. M. (1962*c*). The genetics of the mimetic butterfly *Papilio glaucus*. *Ecology* **43**: 159-161. [22]

Clarke C. A. & Sheppard P. M. (1963). Interactions between major genes and polygenes in the determination of the mimetic patterns of *Papilio dardanus*. *Evolution* **17**: 404-413. [14]

Clarke C. A. & Sheppard P. M. (1971). Further studies on the genetics of the mimetic butterfly *Papilio memnon*. *Phil. Trans. R. Soc.* (B) **263**: 35-70. [14]

Clarke C. A. & Sheppard P. M. (1972). The genetics of the mimetic butterfly *Papilio polytes*. *Phil. Trans. R. Soc.* (B) **263**: 431-458. [14]

Clarke C. A. & Sheppard P. M. (1975). The genetics of the mimetic butterfly *Hypolimnas bolina*. *Phil. Trans. R. Soc.* (B) **272**: 229-265. [12]

Clarke C. A. & Willig A. (1977). The use of α-ecdysone to break permanent diapause of female hybrids between *Papilio glaucus* female and *Papilio rutulus* male. *J. Res. Lepid.* **16**: 245-248. [22]

Clarke C. A., Sheppard P. M. & Thornton I. W. B. (1968). The genetics of the mimetic butterfly *Papilio memnon*. *Phil. Trans. R. Soc.* (B) **254**: 37-89. [14]

Clarke C. A., Sheppard P. M. & Scali V. (1975). All-female broods in the butterfly *Hypolimnas bolina*. *Proc. R. Soc.* (B) **189**: 29-37. [21, 24]

Clarke C. A., Mittwoch U. & Traut W. (1977). Linkage and cytogenetic studies in the swallowtail butterflies *Papilio polyxenes* and *Papilio machaon* and their hybrids. *Proc. R. Soc.* (B) **198**: 385-399. [24]

Clarke C. A., Sheppard P. M. & Mittwoch U. (1976). Heterochromatin polymorphism and colour pattern in the tiger swallowtail butterfly, *Papilio glaucus*. *Nature* **263**: 585-587. [22]

Clarke C. A., Johnston G. & Johnston B. (1983). All-female broods in *Hypolimnas bolina*. A re-survey of West Fiji after 60 years. *Biol. J. Linn. Soc.* **19**: 221-235. [24]

Clarke J. C., Westbrook R. F. & Irwin J. (1979). Potentiation instead of overshadowing in the pigeon. *Behav. Neural Biol.* **25**: 18-29. [12]

Clausen C. P. (1940). *Entomophagous insects*. New York. [12]

Clench H. K. (1955). Revised classification of the butterfly family Lycaenidae and its allies. *Ann. Carneg. Mus.* **33**: 261-274. [1]

Clench H. K. (1966). Behavioral thermoregulation in butterflies. *Ecology* **47**: 1021-1034. [20, 28]

Clench H. K. (1967). Temporal dissociation and population regulation in certain hesperiine butterflies. *Ecology* **48**: 1000-1006. [2]

Clench H. K. (1979). How to make regional lists of butterflies: some thoughts. *J. Lepid. Soc.* **33**: 216-231. [33]

Cloudsley-Thompson J. L. (1981). Comments on the nature of deception. *Biol. J. Linn. Soc.* **16**: 11-14. [12]

Cochran W. G. (1963). *Sampling techniques* (2nd edn). New York. [5]

Cody M. L. (1974). *Competition and the structure of bird communities*. Princeton. [3]

Cody M. L. & Diamond J. M. (eds) (1975). *Ecology and evolution of communities* (pp.1-14, Introduction, by Cody M. L. & Diamond J. M.). Cambridge, Mass. [3]

Cohen J. A. & Brower L. P. (1982). Oviposition and larval success of wild monarch butterflies in relation to host plant size and cardenolide concentration. *J. Kans. ent. Soc.* **55**: 343-348. [12]

Cohen J. A. & Brower L. P. (1983). Cardenolide sequestration in the dogbane tiger moth (*Cycia tenera*). *J. Chem. Ecol.* **9**: 521-532. [12]

Cole B. J. (1981). Overlap, regularity, and flowering phenologies. *Am. Nat.* **117**: 993-997. [3]

Collier A. E. (1962). Butterflies in the Cranleigh district, 1961. *Entomologist's Rec. J. Var.* **74**: 45-47. [33]

Collier R. V. (1966). Status of butterflies on Castor Hanglands NNR 1961-1965 inclusive. *J. Northampt. nat. Hist. Soc.* **36**: 451-456. [33]

Colwell A. E., Shorey H. H., Gaston L. K. & Van Vorhis Key S. E. (1978). Short-range precopulatory behavior of males of *Pectinophora gossypiella*. *Behav. Biol.* **22**: 323-335. [25]

Colwell R. K. & Winkler D. W. (1983). A null model for null models in biogeography. Pp.990-999 in *Ecological communities: conceptual issues and evidence* (Strong D. R., Simberloff D. S. & Abele L. G., eds). Princeton. [3]

Common I. F. B. (1970). Lepidoptera. Pp.765-866 in *The insects of Australia* (ed. Mackerras, I. M.). Melbourne. [Intro, 1]

Common I. F. B. (1974). Lepidoptera. Pp.98-107 in *The insects of Australia* (suppl.). Melbourne. [1]

Common I. F. B. (1975). Evolution and the classification of the Lepidoptera. *A. Rev. Ent.* **20**: 183-203. [1]

Common I. F. B. & Waterhouse D. F. (1981). *Butterflies of Australia* (2nd edn). Sydney. [1, 9, 19]

Comstock J. A. & Garcia L. V. (1961). Estudios de los ciclos biologicos en lepidopteros Mexicanos. *An. Inst. Biol. Univ. Méx.* **31**: 349-448. [1]

Comstock W. P. (1961). *Butterflies of the American tropics.* The genus *Anaea*: a study of the species heretofore included in the genera *Anaea, Coenophlebia, Hypna, Polygrapha, Protogonius, Siderone* and *Zaretis.* New York. [1]

Condamin A. (1973). Monographie du genre *Bicyclus. Mém. Inst. Fond. Afr. noire* **88**: 1-324. [1]

Conn E. E. (1969). Cyanogenic glycosides. *J. agric. Fd Chem.* **17**: 519-526. [12]

Conn E. E. (1979). Cyanide and cyanogenic glycosides. Pp.387-412 in Rosenthal & Janzen (1979). [12]

Conner W. E. (1979). *Chemical attraction and seduction: the courtship of* Utetheisa ornatrix. PhD thesis, Cornell Univ., USA. [9]

Conner W. E., Eisner T., van der Meer R. K., Guerrero A. & Meinwald J. (1981). Precopulatory sexual interaction in an arctiid moth (*Utetheisa ornatrix*): role of a pheromone derived from dietary alkaloids. *Behav. Ecol. Sociobiol.* **9**: 227-235. [25]

Conquest G. H. (1897). The probable causes of the decadence of British Rhopalocera. *Entomologist* **30**: 102-104. [33]

Cook L. M., Brower L. P. & Alcock J. (1969). An attempt to verify mimetic advantage in a neotropical environment. *Evolution* **23**: 339-345. [10]

Cook L. M., Frank K. & Brower L. P. (1971). Experiments on the demography of tropical butterflies. I. Survival rate and density in two species of *Parides. Biotropica* **3**: 17-20. [2]

Cook L. M., Thomason E. W. & Young A. M. (1976). Population structure, dynamics and dispersal of the tropical butterfly *Heliconius charitonius. J. anim. Ecol.* **45**: 851-863. [2]

Coombes S., Revusky S. & Lett B. T. (1980). Long-delay taste aversion learning in an unpoisoned rat: exposure to a poisoned rat as the unconditioned stimulus. *Learning and Motivation* **11**: 256-266. [12]

Copp N. H. & Davenport D. (1978*a*). *Agraulis* and *Passiflora.* I. Control of specificity. *Biol. Bull.* **155**: 98-112. [6]

Copp N. H. & Davenport D. (1978*b*). *Agraulis* and *Passiflora.* II. Behavior and sensory modalities. *Biol. Bull.* **155**: 113-124. [6]

Coppinger R. P. (1969). The effect of experience and novelty on avian feeding behaviour with reference to the evolution of warning coloration in butterflies. Part I: reactions of wild-caught adult blue jays to novel insects. *Behaviour* **35**: 45-60. [12, 14, 16]

Coppinger R. P. (1970). The effect . . . II. Reactions of naive birds to novel insects. *Am. Nat.* **104**: 323-335. [12, 14, 16]

Corbet A. S. & Pendlebury H. M. (1978). *The butterflies of the Malay Peninsula* (3rd edn, revised by Eliot J. N.). Kuala Lumpur. [1, 23]

Corner E. J. H. (1964). *The life of plants.* London. [12]

Cornes M. A., Riley J. & St Leger R. G. T. (1973). A checklist of the Nigerian Papilionoidea. *Occ. Publ. Ent. Soc. Nigeria* **(11)**: 18pp. [1]

Cott H. B. (1940). *Adaptive coloration in animals.* London. [12, 16, 20]

Cott H. B. (1954). Allaesthetic selection and its evolutionary aspects. Pp.47-70 in Huxley *et al.* (1954). [12]

Cottrell C. B. (1984). Aphytophagy in butterflies: its relationship to myrmecophily. *Zool. J. Linn. Soc.* **80**: 1-57. [Intro, 1, 19]

Courtney S. P. (1980). *Studies of the biology of the butterflies* Anthocharis cardamines *and* Pieris napi *in relation to speciation in Pierinae.* PhD thesis, Durham Univ., UK. [6, 10, 33]

Courtney S. P. (1981). Coevolution of pierid butterflies and their cruciferous foodplants. III. *Anthocharis cardamines* survival, development and oviposition on different hostplants. *Oecologia* **51**: 91-96. [6, 7]

Courtney S. P. (1982*a*). Coevolution . . . IV. Crucifer apparency and *Anthocharis cardamines* oviposition. Oecologia **52**: 258-265. [6, 7]

Courtney S. P. (1982*b*). Coevolution . . . V. Habitat selection, community structure and speciation. *Oecologia* **54**: 101-107. [7]

Courtney S. P. & Courtney S. (1982). The "edge-effect" in butterfly oviposition: causality in *Anthocharis cardamines* and related species. *Ecol. Ent.* **7**: 131-137. [6]

Courtney S. P. & Duggan A. E. (1983). The population biology of the orange tip butterfly *Anthocharis cardamines* in Britain. *Ecol. Ent.* **8**: 271-281. [6, 7]

Cowan C. F. (1966). Indo-Oriental Horagini. *Bull. Br. Mus. nat. Hist. (Ent.)* **18**: 103-141. [1]

Cowan C. F. (1967). The Indo-Oriental tribe Cheritrini. *Bull. Br. Mus. nat. Hist. (Ent.)* **20**: 77-103. [1]

Craig G. B. (1967). Mosquitoes, female monogamy induced by male accessory gland substance. *Science* **156**: 1499-1501. [21]

Crane J. (1954). Spectral reflectance characteristics of butterflies from Trinidad, B.W.I. *Zoologica, N.Y.* **39**: 85-115. [20]

Crane J. (1955). Imaginal behavior of a Trinidad butterfly, *Heliconius erato hydara*, with special reference to the social use of colour. *Zoologica, N.Y.* **40**: 167-196. [20]

Crane J. (1957). Imaginal behavior in butterflies of the family Heliconiidae: changing social patterns and irrelevant actions. *Zoologica, N.Y.* **42**: 135-145. [20]

Crane R. (1972). Millhill-Shoreham. *Newl. Sussex Trust Nat. Conserv.* **38**: 3-4. [33]

Creed E. R. (ed.) (1971). *Ecological genetics and evolution: essays in honour of E. B. Ford.* Oxford.

Creed E. R. (1975). Melanism in the two-spot ladybird: the nature and intensity of selection. *Proc. R. Soc.* (B) **190**: 135-148. [21]

Creed E. R., Dowdeswell W. H., Ford E. B. & McWhirter K. G. (1959). Evolutionary studies on *Maniola jurtina*: the English mainland 1956-57. *Heredity* **13**: 363-391. [16]

Creed E. R., Dowdeswell W. H., Ford E. B. & McWhirter K. G. (1962). Evolutionary studies on *Maniola jurtina*: the English mainland, 1958-60. *Heredity* **17**: 237-265. [16]

Creed E. R., Dowdeswell W. H., Ford E. B. & McWhirter K. G. (1970). Evolutionary studies on *Maniola jurtina*: the "boundary phenomenon" in southern England, 1961 to 1968. Pp.263-287 in *Essays in evolution and genetics in honor of Theodosius Dobzhansky* (eds Hecht M. K. & Steere W. C.). New York. [16]

Creed E. R., Ford E. B. & McWhirter K. G. (1964). Evolutionary studies on *Maniola jurtina*: the Isles of Scilly, 1958-59. *Heredity* **19**: 471-488. [16]

Cripps C. (1947). Scent perception in some African myrmecophilous Lycaenidae. *Proc. R. ent. Soc. Lond.* (A) **22**: 42-43. [6]

Cronin E. H., Ogden P., Young J. A. & Laycock W. (1978). The ecological niches of poisonous plants in range communities. *J. Range Mgmt* **31**: 328-334. [12]

Cromartie W. J. (1975*a*). Influence of habitat on colonization of collard plants by *Pieris rapae*. *Envir. Ent.* **4**: 783-784. [6]

Cromartie W. J. (1975*b*). The effect of stand size and vegetational background on the colonization of cruciferous plants by herbivorous insects. *J. appl. Ecol.* **12**: 517-533. [6]

Crook J. H. (1972). Sexual selection, dimorphism and social organisation in the primates. Pp.231-281 in Campbell (1972). [21]

Cross W. & Gill A. G. (1979). A new technique for the prospective survey of sex chromatin using the larvae of Lepidoptera. *J. Lepid. Soc.* **33**: 50-55. [24]

Crossley A. C. & Waterhouse D. F. (1969). The ultrastructure of the osmeterium and the nature of the secretion in *Papilio* larvae. *Tissue Cell* **1**: 525-554. [12]

Crowe C. R. (1965). *Some interpretations of the postpluvial distribution of* Speyeria egleis *in the Oregon Cascades.* Privately circulated by Oregon Entomological Society. [31]

Croze H. J. (1964). *Secondary reinforcement as a component of Müllerian mimicry.* Snr Hons thesis, Amherst College, Mass., USA. [12]

Cullenward M. J., Ehrlich P. R., White R. R. & Holdren C. E. (1979). The ecology and population genetics of an alpine checkerspot butterfly, *Euphydryas anicia*. *Oecologia* **38**: 1-12. [2]

Culvenor C. C. J. (1968). Tumor-inhibiting activity of pyrrolizidine alkaloids. *J. Am. pharm. Ass.* **57**: 1112-1117. [13]

Culvenor C. C. J. (1978). Pyrrolizidine alkaloids—occurrence and systematic importance in angiosperms. *Bot. Notiser* **131**: 473-486. [9]

Culvenor C. C. J. & Edgar J. A. (1972). Dihydropyrrolizine secretions associated with coremata of *Utetheisa* moths. *Experientia* **28**: 627-628. [25]

Curio E. (1976). *The ethology of predation.* Berlin & New York. [12]

Curio E., Ernst U. & Vieth W. (1978). Cultural transmission of enemy recognition: one function of mobbing. *Science* **202**: 899-901. [12]

Cutting W. C. (1962). *Handbook of pharmacology.* New York. [12]

D'Abrera B. L. (1975). The largest butterfly in the world. *Wild Life, London.* Dec. 1975: 560-563.

D'Abrera B. L. (1977). *Butterflies of the Australian region* (2nd edn). Melbourne. [1]

D'Abrera B. L. (1980). *Butterflies of the Afrotropical region.* Melbourne. [1]

D'Abrera B. L. (1981). *Butterflies of the Neotropical region.* Part 1. Papilionidae and Pieridae. Melbourne. [1]

D'Abrera B. L. (1982). *Butterflies of the Oriental region.* Part 1. Papilionidae, Pieridae and Danaidae. Ferry Creek, Australia. [1]

Dahm K. H., Meyer D., Finn W. E., Reinhold V. & Röller H. (1971). The olfactory and auditory mediated sex attraction in *Archroia grisella*. *Experientia* **58**: 265-266.

Dale C. W. (1903). Historical notes on *Lycaena arion* in Britain. *Entomologist's mon. Mag.* **39**: 4-5.

Dalla Torre K. W. von (1885). Die Duftapparate der Schmetterlinge. *Kosmos, Stuttg.* **2**: 354-364, 410-423. [25]

D'Almeida R. F. (1922). Etudes sur les Lépidoptères du Brésil. *Mélanges Lépidoptèrologiques* **1**: viii + 226pp. [1]

D'Almeida R. F. (1966). *Catálogo dos Papilionidae Americanos.* São Paulo. [1]

D'Almeida R. F. (1978). *Catálogo dos Ithomiidae Americanos.* Curitiba. [1]

Danilevski A. S. (1965). *Photoperiodism and seasonal development of insects.* Edinburgh. [16]

Danthanarayana W. (1976). Environmentally cued size variation in the light-brown apple moth, *Epiphyas postvittana*, and its adaptive value in dispersal. *Oecologia* **26**: 121-132. [16]

D'Araújo e Silva A. G., Conçalves C. R., Galvão D. M., Gonçalves A. J. L., Gomes J., do Nascimento Silva M. & de Simoni L. (1967-68). *Quarto catologo dos insectos que vivem nas plantas do Brasil.* Rio de Janeiro. [1, 6]

Darlington P. J. (1938). Experiments on mimicry in Cuba, with suggestions for further study. *Trans. R. ent. Soc. Lond.* **87**: 681-695. [12]

Darwin C. (1839). *Narrative of the surveying voyages of HMS 'Adventure' and 'Beagle'.* London. [20]

Darwin C. (1859). *On the origin of species by means of natural selection, or the preservation of favoured races in the struggle for life.* London. [20, 21]

Darwin C. (1871). *The descent of man and selection in relation to sex.* London. [12, 21, 25]

Darwin C. (1874). *The descent of man and selection in relation to sex* (2nd edn). London. [20, 23]

Darwin C. (1880). The sexual colours of certain butterflies. *Nature* **21**: 237. [20]

David W. A. L. & Gardiner B. O. C. (1961). The mating behaviour of *Pieris brassicae* in a laboratory culture. *Bull. ent. Res.* **52**: 263-280. [20]

David W. A. L. & Gardiner B. O. C. (1962). Oviposition and hatching of the eggs of *Pieris brassicae* in a laboratory culture. *Bull. ent. Res.* **53**: 91-109. [6, 26]

David W. A. L. & Gardiner B. O. C. (1966). The effect of sinigrin on the feeding of *Pieris brassicae* larvae transferred from various diets. *Entomologia exp. appl.* **9**: 95-98. [6]

Davies N. B. (1978). Territorial defence in the speckled wood butterfly (*Pararge aegeria*): the resident always wins. *Anim. Behav.* **26**: 138-147. [20, 26]

Davis G. A. N., Frazer J. F. D. & Tynam A. M. (1958). Population numbers in a colony of *Lysandra bellargus* during 1956. *Proc. R. ent. Soc. Lond.* (A) **33**: 31-36. [2, 33]

Dawkins R. (1976). *The selfish gene.* Oxford. [12, 21]

Dawkins R. & Krebs J. H. (1978). Animal signals: information or manipulation? Pp.282-309 in Krebs & Davies (1978). [12]

Dawkins R. & Krebs J. R. (1979). Arms races between and within species. *Proc. R. Soc.* (B) **205**: 489-511. [12, 14]

Deb D. C. & Chakrovorty S. (1981). Effect of a juvenoid on the growth and differentiation of the ovary of *Corcyra cephalonica. J. Insect Physiol.* **27**: 103-111. [6]

Degeener P. (1901). Duftorgan von *Hepialus hectus. Z. wiss Zool.* **71**: 276. [25]

DeGeer K. (1752). *Mémoires pour sever à l'histoire des insects.* Stockholm. [25]

De Jong R. (1978). Functional morphology of the genitalia of *Carcharodus boeticus stauderi. Rev. Neth. J. Zool.* **28**: 206-212. [6]

De Meijere J. C. H. (1923). *Ceratopogon*-Arten als ectoparasiten anderer Insekten. *Tijdschr. Ent.* **66**: 137-142. [11]

Dempster J. P. (1967). The control of *Pieris rapae* with DDT. I. The natural mortality of the young stages of *Pieris. J. appl. Ecol.* **4**: 485-500. [2, 10, 12, 33]

Dempster J. P. (1968). The control . . . II. Survival of the young stages of *Pieris* after spraying. *J. appl. Ecol.* **5**: 451-462. [2, 10, 33]

Dempster J. P. (1969). Some effects of weed control on the numbers of the small cabbage white (*Pieris rapae*) on brussels sprouts. *J. appl. Ecol.* **6**: 339-345. [10, 33]

Dempster J. P. (1971*a*). The population ecology of the cinnabar moth, *Tyria jacobaeae. Oecologia* **7**: 26-67. [10, 33]

Dempster J. P. (1971*b*). Some observations on a population of the small copper butterfly *Lycaena phlaeas. Entomologist's Gaz.* **22**: 199-204. [2, 10, 33]

Dempster J. P. (1971*c*). Some effects of grazing on the population ecology of the cinnabar moth (*Tyria jacobaeae*). Pp.517-526 in Duffey & Watt (1971). [10]

Dempster J. P. & Hall M. L. (1980). An attempt at re-establishing the swallowtail butterfly at Wicken Fen. *Ecol. Ent.* **5**: 327-334. [2, 10, 33]

Dempster J. P. & Lakhani K. H. (1979). A population model for cinnabar moth and its food plant, ragwort. *J. Anim. Ecol.* **48**: 143-163. [10]

Dempster J. P. & Pollard E. (1981). Fluctuations in resource availability and insect populations. *Oecologia* **50**: 412-416. [5]

Dempster J. P., King M. L. & Lakhani K. H. (1976). The status of the swallowtail butterfly in Britain. *Ecol. Ent.* **1**: 71-84. [2, 8, 10, 16, 33]

De Nicéville L. (1890). *The butterflies of India, Burma, and Ceylon* **3**. Calcutta. [6]

Den Otter C. J., Behan M. & Maes F. W. (1980). Single cell responses in female *Pieris brassicae* to plant volatiles and conspecific odours. *J. Insect Physiol.* **26**: 465-472. [6]

Dennis R. L. H. (1972*a*). A biometrical study of a Welsh colony of the large heath butterfly, *Coenonympha tullia. Entomologist* **105**: 313-326. [16]

Dennis R. L. H. (1972*b*). *Eumenis semele thyone.* A microgeographical race. *Entomologist's Rec. J. Var.* **84**: 1-11, 38-44. [16]

Dennis R. L. H. (1972*c*). *Plebejus argus caernensis.* A stenoecious geotype. *Entomologist's Rec. J. Var.* **84**: 100-108. [16]

Dennis R. L. H. (1977). *The British butterflies, their origin and establishment.* Farringdon, Oxon. [16, 33]

Dennis R. L. H. (1982). Mate location strategies in the wall brown butterfly, *Lasiommata megera*: wait or seek? *Entomologist's Rec. J. Var.* **94**: 209-214, **95**: 7-10. [2]

Denno R. F. & Donnelly M. A. (1981). Patterns of herbivory on *Passiflora* leaf tissues and species by generalized and specialized feeding insects. *Ecol. Ent.* **6**: 11-16. [3]

Denno R. F. & McClure M. S. (eds) (1982; actually published 1983). *Impact of variable host quality on herbivorous insects.* New York.

Deschamps B. (1835). Recherches microscopiques sur l'organisation des ailes des Lépidopterès. *Annls Sci. nat.* (2) **3**: 111-137. [25]

Descimon H. & Renon C. (1975). Mélanisme et facteurs climatiques: II. Corrélation entre melanisation et certains facteurs climatiques chez *Melanargia galathea* en France. *Archs Zool. exp. gén.* **116**: 438-468. [16]

Descimon H., Mast de Maeght J. & Stoffel J. R. (1973-4). Contribution à l'étude des nymphalides neotropicales. Description de trois nouveaux *Prepona* Mexicains. *Alexanor* **8**: 101-105, 155-159, 235-240. [20]

Dethier V. G. (1941). Chemical factors determining the choice of foodplants by *Papilio* larvae. *Am. Nat.* **75**: 61-73. [6]

Dethier V. G. (1959*a*). Egg-laying habits of Lepidoptera in relation to available food. *Can. Ent.* **91**: 554-561. [6, 7]

Dethier V. G. (1959*b*). Food-plant distribution and density and larval dispersal as factors affecting insect populations. *Can. Ent.* **91**: 581-596. [2, 7]

Dethier V. G. (1970). Chemical interactions between plants and insects. Pp.83-102 in Sondheimer & Simeone (1970). [6]

Dethier V. G. (1973). Electrophysiological studies of gustation in lepidopterous larvae. II. Taste spectra in relation to food-plant discrimination. *J. comp. Physiol.* **82**: 103-134. [6]

Dethier V. G. (1978). Other tastes, other worlds. *Science* **201**: 224-228. [6]

Dethier V. G. (1980*a*). Evolution to receptor sensitivity to secondary plant substances with special reference to deterrents. *Am. Nat.* **115**: 45-66. [6]

Dethier V. G. (1980*b*). Food-aversion learning in two polyphagous caterpillars *Diacrisia virginica* and *Estigmene congrua. Physiol. Ent.* **5**: 321-325. [7]

Dethier V. G. & MacArthur R. H. (1964). A field's capacity to support a butterfly population. *Nature* **201**: 728-729. [2]

DeVries P. J. (1976). Notes on the behavior of *Eumaeus minyas* in Costa Rica. *Brenesia* **8**: 103. [12]

DeVries P. J. (1977). *Eumaeus minyas*, an aposematic lycaenid butterfly. *Brenesia* **12**: 269–270. [6, 12]

DeVries P. J. (1979). Pollen-feeding rainforest *Parides* and *Battus* butterflies in Costa Rica. *Biotropica* **11**: 237–238. [32]

DeVries P. J. (1983). Butterflies of Costa Rica: systematics, ecology and field site checklists. In *Costa Rican Natural History*. (Janzen D. H. ed.). Chicago. [1]

DeVries P. J. (1984). Of crazy-ants and Curetinae: are *Curetis* butterflies tended by ants? *Zool. J. Linn. Soc.* **80**: 59–66. [1]

DeVries P. J. (In prep. *a*). Species stratification by rainforest butterflies in Costa Rica: canopy and understory habitats. [3]

DeVries P. J. (In prep. *b*). Butterflies of Costa Rica and their natural history. Part I. [1]

DeVries P. J. (In prep. *c*). Pollen and rotting fruits in the diet of tropical butterflies. [Intro]

DeVries P. J., Kitching I. J. & Vane-Wright R. I. (In press). The systematic position of *Antirrhea* and *Caerois* with comments on the classification of the Nymphalidae. *Syst. Ent.* [1]

Dierl W. (1977). Bisher unbekanntes Duftstoffüberträger-system bei Schmetterlingen. *Naturw. Rdsch.* **30**: 264–265. [25]

Digby P. S. B. (1955). Factors affecting the temperature excess of insects in the sun. *J. exp. Biol.* **32**: 278–298. [28]

Dill L. M. (1975). Calculated risk-taking by predators as a factor in Batesian mimicry. *Can. J. Zool.* **53**: 1614–1621. [12]

Dillon L. S. (1948). The tribe Catagrammini. Part I. The genus *Catagramma* and allies. *Scient. Publs Reading publ. Mus.* (**8**): vii + 113pp. [1]

Dingle H. (1972). Migration strategies of insects. *Science* **175**: 1327–1335. [26]

Dingle R. H. (1979). Migration. *Science* **204**: 1007. [26]

Dingle R. H. (1980). Ecology and evolution of migration. Pp.2–101 in Gauthreaux (1980). [26]

Distant W. L. (1877). The geographical distribution of *Danais archippus*. *Trans. ent. Soc. Lond.* **1877**: 93–104. [12]

Dixey F. A. (1907). On epigamic and aposematic scents in Rhopalocera. *Rep. Br. Ass. Advmt Sci.* (D) **1906**: 600. [25]

Dixey F. A. (1909). On Müllerian mimicry and diaposematism. *Trans. ent. Soc. Lond.* **1908**: 559–583. [14]

Dixey F. A. (1911). The scents of butterflies. *Nature* **87**: 164–168. [25]

Dixey F. A. (1932). The plume-scales of the Pierinae. *Trans. ent. Soc. Lond.* **80**: 57–75. [25]

Dixon C. A., Erickson J. M., Kellett D. N. & Rothschild M. (1978). Some adaptations between *Danaus plexippus* and its food plant, with notes on *Danaus chrysippus* and *Euploea core*. *J. Zool., Lond.* **185**: 437–467. [12, 13, 15]

Doane C. C. (1968). Changes in egg mass density, size, and amount of parasitism after chemical treatment of a heavy population of the gypsy moth. *J. econ. Ent.* **61**: 1288–1291. [6]

Dobzhansky T., Anderson W. W. & Pavlovsky O. (1966). Genetics of natural populations. XXXVIII. Continuity and change in populations of *Drosophila pseudoobscura* in western United States. *Evolution* **20**: 418–427. [27]

Dobzhansky T. & Pavlovsky O. (1957). An experimental study of the interactions between genetic drift and natural selection *Evolution* **11**: 311–319. [16]

Doherty W. (1891). A list of the butterflies of Engano, with some remarks on the Danaidae. *J. Asiat. Soc. Bengal* **60** (2): 4–32. [25]

Dolinger P. M., Ehrlich P. R., Fitch W. L. & Breedlove D. E. (1973). Alkaloid and predation patterns in Colorado lupine populations. *Oecologia* **13**: 191–204. [2, 6]

Domjan M. (1977). Attenuation and enhancement of neophobia for edible substances. Pp.151–179 in Barker *et al.* (1977). [12]

Döring E. (1955). *Zur Morphologie der Schmetterlingseier*. Berlin. [1]

Douglas M. M. (1979). Hot butterflies. *Nat. Hist., N.Y.* **88**(9): 56–65. [14, 20]

Doutt R. L. (1959). The biology of parasitic Hymenoptera. *A. Rev. Ent.* **4**: 161–182. [12]

Douwes P. (1968). Host selection and host finding in the egg-laying female *Cidaria albulata*. *Opusc. ent.* **33**: 233–279. [6]

Douwes P. (1970). Size of, gain to and loss from a population of adult *Heodes virgaureae*. *Entomologica scand.* **1**: 263–281. [2, 4]

Douwes P. (1975). Distribution of a population of the butterfly *Heodes virgaureae*. *Oikos* **26**: 332–340. [2]

Douwes P. (1976*a*). Activity in *Heodes virgaureae* in relation to air temperature, solar radiation, and time of day. *Oecologia* **22**: 287–298. [6, 14, 25]

Douwes P. (1976*b*). Mating behaviour in *Heodes virgaureae* with particular reference to the stimuli from the female. *Entomologia germ.* **2**: 232–241. [25]

Douwes P. (1978). Adult feeding in the scarce copper. *Heodes virgaureae*. *Ent. Tidskr.* **99**: 1–10. [2]

Dover C. (1974). The ephemeral butterfly. *Daily Telegraph* 19 Aug. [33]

Dowden P. B. (1961). The gypsy moth egg parasite *Ooencyrtus kuwanai*, in southern Connecticut in 1960. *J. econ. Ent.* **54**: 876–878. [6]

Dowdeswell W. H. (1961). Experimental studies on natural selection in the butterfly, *Maniola jurtina*. *Heredity* **16**: 39–52. [16]

Dowdeswell W. H. (1962). A further study of the butterfly *Maniola jurtina* in relation to natural selection by *Apanteles tetricus*. *Heredity* **17**: 513–523. [16]

Dowdeswell W. H. (1971). Ecological genetics and biology teaching. Pp.363–378 in Creed (1971). [16]

Dowdeswell W. H. (1981). *The life of the meadow brown*. London. [33]

Dowdeswell W. H. & Ford E. B. (1955). Ecological genetics of *Maniola jurtina* on the Isles of Scilly. *Heredity* **9**: 265–272. [16]

Dowdeswell W. H. & McWhirter K. G. (1967). Stability of spot-distribution in *Maniola jurtina* throughout its range. *Heredity* **22**: 187–210. [16]

Dowdeswell W. H., Fisher R. A. & Ford E. B. (1940). The quantitative study of populations in the Lepidoptera. I. *Polyommatus icarus*. *Ann. Eugen.* **10**: 123–136. [2]

Dowdeswell W. H., Fisher R. A. & Ford E. B. (1949). The quantitative . . . II. *Maniola jurtina. Heredity* **3**: 67-84. [2, 16]

Dowdeswell W. H., Ford E. B. & McWhirter K. G. (1957). Further studies on isolation in the butterfly *Maniola jurtina. Heredity* **11**: 51-65. [2, 16]

Dowdeswell W. H., Ford E. B. & McWhirter K. G. (1960). Further studies on the evolution of *Maniola jurtina* in the Isles of Scilly. *Heredity* **14**: 333-364. [16]

Downes J. A. (1978). Feeding and mating in the insectivorous Ceratopogonidae (Diptera). *Mem. ent. Soc. Can.* **(104)**: 62pp. [11]

Downey J. C. (1962*a*). Host-plant relations as data for butterfly classification. *Syst. Zool.* **11**: 150-159. [6, 19]

Downey J. C. (1962*b*). Myrmecophily in *Plebejus (Icaricia) icaroides. Ent. News* **73**: 57-66. [19]

Downey J. C. (1966). Sound production in pupae of Lycaenidae. *J. Lepid. Soc.* **20**: 129-155. [19]

Downey J. C. & Allyn A. C. (1973). Butterfly ultrastructure. I. Sound production and associated abdominal structures in pupae of Lycaenidae and Riodinidae. *Bull. Allyn Mus.* **(14)**: 48pp. [12]

Downey J. C. & Allyn A. C. (1980). Eggs of Riodinidae. *J. Lepid. Soc.* **34**: 133-145. [6]

Downey J. C. & Allyn A. C. (1981). Chorionic sculpturing of eggs of Lycaenidae. Part I. *Bull. Allyn Mus.* **(61)**: 29pp. [6]

Downey J. C. & Fuller W. C. (1961). Variation in *Plebejus icaroides*. I. Food plant specificity. *J. Lepid. Soc.* **15**: 34-42.

Draeseke J. (1936). Lycaenidae I. *Lepid. Cat.* **25** (72): 48pp. [1]

Dressler R. L. (1981). *The orchids. Natural history and classification.* Cambridge, Mass. [12]

Drummond B. A. (1976). *Comparative ecology and mimetic relationships of ithomiine butterflies in eastern Ecuador.* PhD thesis, Univ. Florida, Gainesville, USA. [1, 3, 9]

Drummond B. A. (1981). Ecological chemistry, animal behavior, and plant systematics. *Solanaceae Newsl.* **8**: 59-67. [12]

Drummond B. A., Bush G. L. & Emmel T. C. (1970). The biology and laboratory culture of *Chlosyne lacinia. J. Lepid. Soc.* **24**: 135-142. [6, 12]

Dryja A. (1959). *Badania nad polimorfizmem Krasnika Zmiennego* (Zygaena ephialtes). Warsaw. [14]

Dudley Stamp L. (1962). *The land of Britain. Its use and misuse.* London. [33]

Duffey E. (1968). Ecological studies on the large copper butterflies, *Lycaena dispar batavus*, at Woodwalton Fen NNR, Huntingdonshire. *J. appl. Ecol.* **5**: 69-96. [2, 10, 33]

Duffey E. (1971). The management of Woodwalton Fen: a multidisiplinary approach. Pp.581-597 in Duffey & Watt (1971). [33]

Duffey E. (1977). The re-establishment of the large copper butterfly *Lycaena dispar batavus* on Woodwalton Fen NNR, Cambridgeshire, England 1969-73. *Biol. Conserv.* **12**: 143-158. [2, 33]

Duffey E. & Watt A. S. (eds) (1971). *Scientific management of animal and plant communities for conservation.* Oxford.

Duffey S. S. (1970). Cardiac glycosides and distastefulness: some observations on the palatability spectrum of butterflies. *Science* **169**: 78-79. [12]

Duffey S. S. (1977). Arthropod allomones: chemical effronteries and antagonists. *Int. Congr. Ent.* **(15)**: 323-394. [12]

Duffey S. S. (1980). Sequestration of plant natural products by insects. *A. Rev. Ent.* **25**: 447-477. [12]

Duffey S. S., Blum M. S., Fales H. M., Evans S. L., Roncadori R. W., Tiemann D. L. & Nakagawa Y. (1977). Benzoyl cyanide and mandelonitrile benzoate in the defensive secretions of millipedes. *J. Chem. Ecol.* **3**: 101-113. [12]

Duffey S. S., Underhill E. W. & Towers G. H. N. (1974). Intermediates in the biosynthesis of HCN and benzaldehyde by a polydesmid millipede, *Harpaphe haydeniana. Comp. Biochem. Physiol.* **47B**: 753-766. [12]

Duncan C. J. & Sheppard P. M. (1965). Sensory discrimination and its role in the evolution of Batesian mimicry. *Behaviour* **24**: 269-282. [14]

Dunlap-Pianka H. L. (1979). Ovarian dynamics in *Heliconius* butterflies: correlations among daily oviposition rates, egg weights, and quantitative aspects of oogenesis. *J. Insect Physiol.* **25**: 741-749. [6]

Dunlap-Pianka H. L., Boggs C. L. & Gilbert L. E. (1977). Ovarian dynamics in heliconiine butterflies: programmed senescence versus eternal youth. *Science* **197**: 487-490. [3, 6]

Dunning D. (1968). Warning sounds of moths. *Z. Tierpsychol.* **25**: 129-138. [12]

Dunning D. C. & Roeder K. D. (1965). Moth sounds and the insect-catching behavior of bats. *Science* **147**: 173-174. [12]

Dyar H. G. (1890). The number of moults of lepidopterous larvae. *Psyche* **5**: 420-422. [6]

Eanes W. F., Gaffney P. M., Koehn R. K. & Simon C. M. (1977). A study of sexual selection in natural populations of the milkweed beetle *Tetraopes tetraopthalmus*. Pp.49-64 in *Measuring selection in natural populations* Lecture notes in Biomathematics 19 (eds Christiansen F. B. & Fenchel T. M.). Berlin. [21]

Eanes W. F. & Koehn R. K. (1979). An analysis of genetic structure in the monarch butterfly, *Danaus plexippus. Evolution* **32**: 784-797. [17]

Eberhard W. G. (1975). The ecology and behavior of a subsocial pentatomid bug and two scelionid wasps: strategy and counterstrategy in a host and its parasites. *Smithson. Contr. Zool.* **(205)**: 39pp. [6]

Eberhard W. G. (1977). Aggressive chemical mimicry by a bolas spider. *Science* **198**: 1173-1175. [12]

Ebinuma H. & Yoshitake N. (1982). The genetic system controlling recombination in the silkworm. *Genetics* **99**: 231-245. [14]

Edgar J. A. (1975). Danainae and 1,2-dehydropyrrolizidine alkaloid-containing plants—with reference to observations made in the New Hebrides. *Phil. Trans. R. Soc.* **272**: 467-476. [9, 12]

Edgar J. A. (1982). Pyrrolizidine alkaloids sequestered by Solomon Island danaine butterflies. The feeding preferences of the Danainae and Ithomiinae. *J. Zool., Lond.* **196**: 385-399. [6, 9, 25]

Edgar J. A. & Culvenor C. C. J. (1974). Pyrrolizidine ester alkaloid in danaid butterflies. *Nature* **248**: 614-616. [12]

Edgar J. A. & Culvenor C. C. J. (1975). Pyrrolizidine alkaloids in *Parsonsia* species (family Apocynaceae) which attract danaid butterflies. *Experientia* **31**: 393–394. [9]

Edgar J. A., Culvenor C. C. J. & Smith L. W. (1971). Dihydropyrrolizine derivatives in the hair pencil secretion of danaid butterflies. *Experientia* **27**: 761–762. [25]

Edgar J. A., Culvenor C. C. J. & Robinson G. S. (1973). Hairpencil dihydropyrrolizines of Danainae from the New Hebrides. *J. Austr. ent. Soc.* **12**: 144–150. [9, 25]

Edgar J. A., Culvenor C. C. J. & Pliske T. E. (1974). Coevolution of danaid butterflies with host plants. *Nature* **250**: 646–648. [6, 9, 12, 25]

Edgar J. A., Cockrum P. A. & Frahn J. L. (1976*a*). Pyrrolizidine alkaloids in *Danaus plexippus* and *Danaus chrysippus*. *Experientia* **32**: 1535–1537. [9, 12, 25]

Edgar J. A., Culvenor C. C. J. & Pliske T. E. (1976*b*). Isolation of a lactone, structurally related to the esterifying acids of the pyrrolizidine alkaloids, from the costal fringes of male Ithomiinae. *J. chem. Ecol.* **2**: 263–270. [9, 12, 25]

Edgar J. A., Boppré M. & Schneider D. (1979). Pyrrolizidine alkaloid storage in African and Australian danaid butterflies. *Experientia* **35**: 1447–1448. [9, 12, 25]

Edgar J. A., Eggers N. J., Jones A. J. & Russell G. B. (1980). Unusual macrocyclic pyrrolizidine alkaloids from *Parsonsia heterophylla* and *P. spiralis* (Apocynaceae). *Tetrahedron Lett.* **21**: 2657–2660. [9]

Edmunds M. E. (1969*a*). Polymorphism in the mimetic butterfly *Hypolimnas misippus* in Ghana. *Heredity* **24**: 281–302. [21]

Edmunds M. E. (1969*b*). Evidence for sexual selection in the mimetic butterfly *Hypolimnas misippus*. *Nature* **221**: 448. [20, 21]

Edmunds M. E. (1974*a*). *Defence in animals*. Harlow, Essex. [10, 12, 16]

Edmunds M. E. (1974*b*). Significance of beak marks on butterfly wings. *Oikos* **25**: 117–118. [12, 16]

Edmunds M. E. (1976). Larval mortality and population regulation in the butterfly *Danaus chrysippus* in Ghana. *Zool. J. Linn. Soc.* **58**: 129–145. [15]

Edwards E. D. (1973). Delayed ovarian development and aestivation in adult females of *Heteronympha merope merope*. *J. Aust. ent. Soc.* **12**: 92–98. [6]

Edwards F. W. (1923). New and old observations on ceratopogonine midges attacking other midges. *Ann. trop. Med. Parasit.* **17**: 19–29. [11]

Edwards F. W. (1925). A midge attacking moths in Switzerland. *Entomologist's mon. Mag.* **61**: 228–229. [11]

Edwards J. S. & Schwartz L. M. (1981). Mount St Helens ash: a natural insecticide. *Can. J. Zool.* **59**: 714–715. [31]

Edwards M. & Edwards S. H. (1980). *West Sussex wildlife recording group: butterfly report*. Midhurst, Sussex. [33]

Edwards W. H. (1870). Notes on *Graptas c-aureum* and *interrogationis*. *Trans. Am. ent. Soc.* **3**: 1–9. [6]

Edwards W. H. (1881). On certain habits of *Heliconius charitonia*, a species of butterfly found in Florida. *Papilio* **1**: 209–215. [25]

Edwards W. H. (1882). Description of the preparatory stages of *Graptas interrogationis*. *Can. Ent.* **14**: 201–207. [6]

Edwards W. H. (1884). *Butterflies of North America 2*. Boston, Mass. [12, 22]

Ehrlich A. H. & Ehrlich P. R. (1978). Reproductive strategies in the butterflies: I. Mating frequency, plugging, and egg number. *J. Kansas ent. Soc.* **51**: 666–697. [2, 6, 20]

Ehrlich P. R. (1958). The comparative morphology, phylogeny and higher classification of the butterflies. *Kans. Univ. Sci. Bull.* **39**: 305–370. [1]

Ehrlich P. R. (1961). Intrinsic barriers to dispersal in the checkerspot butterfly, *Euphydryas editha*. *Science* **134**: 108–109. [2, 33]

Ehrlich P. R. (1962). A biting midge ectoparasitic on Arizona lycaenids. *J. Lepid. Soc.* **16**: 20–22. [11]

Ehrlich P. R. (1965). The population biology of the butterfly, *Euphydryas editha*. II. The structure of the Jasper Ridge colony. *Evolution* **19**: 327–336. [2]

Ehrlich P. R. (1980). The strategy of conservation, 1980–2000. Pp.329–344 in Soulé & Wilcox (1980). [2]

Ehrlich P. R. & Birch L. C. (1967). The 'balance of nature' and population control. *Am. Nat.* **101**: 97–101. [2]

Ehrlich P. R. & Ehrlich A. H. (1961). *How to know the butterflies*. Dubuque, Iowa. [1, 2, 8]

Ehrlich P. R. & Ehrlich A. H. (1981*a*). *Extinction: the causes and consequences of the disappearance of species*. New York. [2]

Ehrlich P. R. & Ehrlich A. H. (1981*b*). If butterflies disappear—disaster. *New York Times*, 30 May. [33]

Ehrlich P. R. & Ehrlich A. H. (1982). Lizard predation on tropical butterflies. *J. Lepid. Soc.* **36**: 148–152. [2]

Ehrlich P. R. & Gilbert L. E. (1973). Population structure and dynamics of the tropical butterfly *Heliconius ethilla*. *Biotropica* **5**: 69–82. [2, 3]

Ehrlich P. R. & Mason L. G. (1966). The population biology of the butterfly, *Euphydryas editha*. III. Selection and the phenetics of the Jasper Ridge colony. *Evolution* **20**: 165–173. [16]

Ehrlich P. R. & Mooney H. A. (1983). Extinction, substitution, and impairment of ecosystem services. *Bioscience* **33**: 248–254. [2]

Ehrlich P. R. & Murphy D. D. (1981). The population biology of checkerspot butterflies (*Euphydryas*). *Biol. Zbl.* **100**: 613–629. [2]

Ehrlich P. R. & Murphy D. D. (1982). Butterfly nomenclature: a critique. *J. Res. Lepid.* **20**: 1–11. [2]

Ehrlich P. R. & Raven P. H. (1965). Butterflies and plants: a study in coevolution. *Evolution* **18**: 586–608. [Intro, 1–3, 6, 8, 9, 12, 19]

Ehrlich P. R. & Raven P. H. (1967). Butterflies and plants. *Scient. Am.* **216**(6): 104–113. [12]

Ehrlich P. R. & White R. R. (1980). Colorado checkerspot butterflies: isolation, neutrality, and the biospecies. *Am. Nat.* **115**: 328–334. [2, 33]

Ehrlich P. R., Breedlove D. E., Brussard P. F. & Sharp M. A. (1972). Weather and the "regulation" of subalpine populations. *Ecology* **53**: 243–247. [2]

Ehrlich P. R., Holm R. W. & Parnell D. R. (1974). *The process of evolution*. New York. [2]

Ehrlich P. R., White R. R., Singer M. C., McKechnie S. W. & Gilbert L. E. (1975). Checkerspot butterflies: a historical perspective. *Science* **188**: 221–228. [Intro, 2, 7]

Ehrlich P. R., Murphy D. D., Singer M. C., Sherwood C. B., White R. R. & Brown I. L. (1980). Extinction, reduction, stability and increase: the responses of

checkerspot butterfly (*Euphydryas*) populations to California drought. *Oecologia* **46**: 101–105. [2, 33]
Ehrlich P. R., Launer A. E. & Murphy D. D. (1984). Realized sex ratio in a population of checkerspot butterflies, *Euphydryas editha.* (In press). [2]
Ehrman L. (1970). The mating advantage of rare males in *Drosophila. Proc. natn. Acad. Sci. U.S.A.* **65**: 345–348. [21]
Ehrman L. (1972). Genetics and sexual selection. Pp.105–135 in Campbell (1972). [21]
Eisner H. E., Wood F. & Eisner T. (1975). Hydrogen cyanide production in North American and African polydesmid millipeds. *Psyche* **82**: 20–23. [12]
Eisner T. (1970). Chemical defense against predation in arthropods. Pp.157–217 in Sondheimer & Simeone (1970). [12]
Eisner T. (1980). Chemistry, defence and survival: case studies and selected topics. Pp.847–878 in Locke & Smith (1980). [12, 25]
Eisner T. & Grant R. P. (1981). Toxicity, odor aversion, and "olfactory aposematism". *Science* **213**: 476. [12]
Eisner T. & Halpern B. P. (1971). Taste distortion and plant palatability. *Science* **172**: 1362. [12]
Eisner T. & Meinwald Y. C. (1965). Defensive secretion of a caterpillar (*Papilio*). *Science* **150**: 1733–1735. [12]
Eisner T. & Meinwald J. (1966). Defensive secretions of arthropods. *Science* **153**: 1341–1350. [12]
Eisner T., Hurst J. J. & Meinwald J. (1963). Defense mechanisms of arthropods. XI. The structure, function and phenolic secretions of the glands of a chordeumoid millipede and a carabid beetle. *Psyche* **70**: 94–116. [12]
Eisner T., Silberglied R. E., Aneshansley D., Carrel J. E. & Howland H. C. (1969). Ultraviolet video-viewing: the television camera as an insect eye. *Science* **166**: 1172–1174. [20, 28]
Eisner T., Pliske T. E., Ikeda M., Owen D. F., Vasquez L., Perez H., Franclemont J. G. & Meinwald J. (1970). Defense mechanisms of arthropods. XXVII. Osmeterial secretions of papilionid caterpillars (*Baronia, Papilio, Eurytides*). *Ann. ent. Soc. Am.* **63**: 914–915. [12]
Eisner T., Kluge A. F., Ikeda M. I., Meinwald Y. C. & Meinwald J. (1971). Sesquiterpenes in the osmeterial secretion of a papilionid butterfly, *Battus polydamas. J. Insect Physiol.* **17**: 245–250. [12]
Eisner T., Johnessee J. S., Carrel J., Hendry L. B. & Meinwald J. (1974). Defensive use by an insect of a plant resin. *Science* **184**: 996–999. [12]
Eisner T., Wiemer D. F., Haynes L. W. & Meinwald J. (1978). Lucibufagins: defensive steroids from the fireflies *Photinus ignitus* and *P. marginellus. Proc. natn. Acad. Sci. U.S.A.* **75**: 905–980. [12]
Ekholm S. (1975). Fluctuations in butterfly frequency in central Nyland. *Notul. ent.* **55**: 65–80. [2]
Eliot J. N. (1969). An analysis of the Eurasian and Australian Neptini. *Bull. Br. Mus. nat. Hist. (Ent.)* (Suppl.) **(15)**: 155pp. [1]
Eliot J. N. (1973). The higher classification of the Lycaenidae: a tentative arrangement. *Bull. Br. Mus. nat. Hist. (Ent.)* **28**: 373–506. [1, 19]
Eliot J. N. (In prep.). A review of the Oriental Miletinae. [1]
Eliot J. N. & Kawazoé A. (1983). *Blue butterflies of the* Lycaenopsis *group*. London. [1]
Eltringham H. (1913). On the scent apparatus in the male of *Amauris niavius. Trans. ent. Soc. Lond.* **1913**: 399–406. [25]
Eltringham H. (1915). Further observations on the structure of the scent organs of certain male danaine butterflies. *Trans. ent. Soc. Lond.* **1915**: 152–176. [25]
Eltringham H. (1919). Butterfly vision. *Trans. ent. Soc. Lond.* **1919**: 1–49. [20]
Eltringham H. (1923). *Butterfly lore*. Oxford. [10]
Eltringham H. (1937). On some secondary sexual characters in the males of certain Indian moths. *Trans. R. ent. Soc. Lond.* **86**: 135–150. [25]
Eltringham H. & Jordan K. (1913). Nymphalidae: subfam. Acraeinae. *Lepid. Cat.* **31** (11): 1–65. [1]
Emets V. M. (1977). Rare and disappearing species of diurnal butterflies in the Usmanskii Forest and possible measures for their protection. *Zool. Zh.* **56**: 1889–1890. [2]
Emmel T. C. (1970). The population biology of the neotropical satyrid butterfly, *Euptychia hermes*. I. Inter-population movement, general ecology, and population sizes in lowland Costa Rica (dry season, 1966). *J. Res. Lepid.* **7**: 153–165. [2]
Emmel T. C. (1972). Mate selection and balanced polymorphism in the tropical nymphalid butterfly, *Anartia fatima. Evolution* **26**: 96–107. [20]
Emmel T. C. (1973*a*). On the nature of the polymorphism and mate selection phenomena in *Anartia fatima. Evolution* **27**: 164–165. [20]
Emmel T. C. (1973*b*). Dispersal in a cosmopolitan butterfly species (*Pieris rapae*) having open population structure. *J. Res. Lepid.* **11**: 95–98. [2]
Emmel T. C. (1975). Satyridae. Pp.79–111 in Howe (1975). [2, 16]
Emmel T. C. (1976). *Butterflies, their world, their life cycle, their behaviour*. London. [12]
Emmel T. C. & Emmel J. F. (1969). Selection and host plant overlap in two desert *Papilio* butterflies. *Ecology* **50**: 158–159. [6]
Emmel T. C. & Leck C. F. (1970). Seasonal changes in organization of tropical rain forest butterfly populations in Panama. *J. Res. Lepid.* **8**: 133–152. [2]
Emmel T. C., Kilduff T. S. & McFarland N. (1974). The chromosomes of a long-isolated monotypic butterfly genus: *Tellervo zoilus* in Australia. *J. Ent.* (A) **49**: 43–46. [9]
Emsley M. G. (1963). A morphological study of imagine Heliconiinae with a consideration of the evolutionary relationships within the group. *Zoologica, N.Y.* **48**: 85–130. [1]
Emsley M. G. (1965). The geographical distribution of the colour-pattern components of *Heliconius erato* and *H. melpomene* with genetical evidence for the systematic relationships between the species. *Zoologica, N.Y.* **49**: 245–286. [14]
Enders F. (1975). The influence of hunting manner on prey size, particularly in spiders with long attack distances. *Am. Nat.* **109**: 737–763. [12]
Endler J. A. (1977). *Geographic variation, speciation and clines*. Princeton. [16, 19]
Endler J. A. (1978). A predator's view of animal color patterns. *Evolut. Biol.* **11**: 319–364. [14, 16]
Endler J. A. (1980). Natural selection on color patterns in *Poecilia reticulata. Evolution* **34**: 76–91. [16]
Engelmann F. (1979). Insect vitellogenin: identification,

biosynthesis, and role in vitellogenesis. *Adv. Insect Physiol.* **14**: 49–108. [6]

Erickson J. M. & Feeny P. (1974). Sinigrin: a chemical barrier to the black swallowtail butterfly, *Papilio polyxenes*. *Ecology* **55**: 103–111. [6]

Esaki T. (1940). Biological notes *in* Tokunaga M., Ceratopogonidae and Chironomidae from the Micronesian islands. *Philipp. J. Sci.* **71**: 226–228. [11]

Esau K. (1965). *Plant anatomy* (2nd edn). New York. [6]

Eshel I. (1972). On the neighbour effect in the evolution of altruistic traits. *Theoret. Pop. Biol.* **3**: 258–277. [12]

Estabrook G. F. & Jesperson D. C. (1974). Strategy for a predator encountering a model-mimic system. *Am. Nat.* **108**: 443–457. [12]

Evans F. J. & Kinghorn A. D. (1977). A comparative phytochemical study of the diterpenes of some species of the genera *Euphorbia* and *Elaeophorbia*. *Bot. J. Linn. Soc.* **74**: 23–35. [13]

Evans F. J. & Soper C. J. (1978). The tigliane, daphnane and ingenane diterpenes, their chemistry, distribution and biological activities. A review. *Lloydia* **41**: 193–233. [13]

Evans H. E., Ingalls T. H. & Binns W. (1966). Teratogenesis of craniofacial malformation in animals. III. Natural and experimental cephalic deformities in sheep. *Archs envir. Hlth* **13**: 706–714. [12]

Evans W. H. (1937). *A catalogue of the African Hesperiidae, indicating the classification and nomenclature adopted in the British Museum*. London. (Additions and corrections, *Ann. Mag. nat. Hist.* (11) **13**: 641–647 (1947); *ibid.* (12) **8**: 881–885 (1955).) [1]

Evans W. H. (1949). *A catalogue of the Hesperiidae from Europe, Asia and Australia in the British Museum (Natural History)*. London. (Revisional notes, *Ann. Mag. nat. Hist.* (12) **9**: 749–752 (1957).) [1]

Evans W. H. (1951). *A catalogue of the American Hesperiidae indicating the classification and nomenclature adopted in the British Museum (Natural History)*. Part I: *Introduction and group A, Pyrrhopyginae*. London. [1]

Evans W. H. (1952). *A catalogue of the American Hesperiidae* . . . Part II: *Groups B, C, D, Pyrginae*. Section 1. London. [1]

Evans W. H. (1953). *A catalogue of the American Hesperiidae* . . . Part III: *Groups E, F, G, Pyrginae*. Section 2. London. [1]

Evans W. H. (1954). A revision of the genus *Curetis*. *Entomologist* **87**: 190–194, 212–216, 241–247. [1]

Evans W. H. (1955). *A catalogue of the American Hesperiidae* . . . Part IV: *Groups H–P, Hesperiinae and Megathyminae*. London. [1]

Evans W. H. (1957). A revision of the *Arhopala* group of Oriental Lycaenidae. *Bull. Br. Mus. nat. Hist. (Ent.)* **5**: 85–141. [1]

Fabre J. H. (1907). *Souvenirs entomologiques*. (Dixième série). Paris. [25]

Falconer D. S. (1960). *Introduction to quantitative genetics*. New York. [6]

Falconer D. S. (1972). Review of Biometrical Genetics. *Heredity* **28**: 415–417. [27]

Falconer D. S. (1981). *Introduction to quantitative genetics* (2nd edn). London. [16]

Farquharson C. O. (1922). Five years observations (1914–18) on the bionomics of southern Nigerian insects, chiefly directed to the investigation of lycaenid life-histories and to the relation Lycaenidae, Diptera, and other insects to ants (with an additional multi-author appendix). *Trans. ent. Soc. Lond.* **1921**: 319-531. [1]

Farrell L. (1975). A survey of the status of the chequered skipper butterfly (*Carterocephalus palaemon*) in Britain. *Entomologist's Gaz.* **26**: 148–149. [33]

Farris J. S. (1970). Methods for computing Wagner trees. *Syst. Zool.* **19**: 83–92. [17]

Farris J. S., Kluge A. G. & Eckardt M. J. (1970). A numerical approach to phylogenetic systematics. *Syst. Zool.* **19**: 172–191. [14]

Fearnehough T. D. (1972). The butterflies of the Isle of Wight. *Entomologist's Rec. J. Var.* **84**: 102–109. [33]

Feeny P. (1970). Oak tannins and caterpillars. *Ecology* **51**: 565–581. [12]

Feeny P. (1975). Biochemical coevolution between plants and their insect herbivores. Pp.3–19 in Gilbert & Raven (1975). [6]

Feeny P. (1976). Plant apparency and chemical defense. *Recent Advances in Phytochemistry* **10**: 1-40. [9, 12]

Feeny P., Rosenberry L. & Carter M. (1983). Chemical aspects of oviposition behavior in butterflies. Pp.27–76 in Ahmad (1983). [6]

Feltwell J. (1976). Migration of *Hipparchia semele*. *J. Res. Lepid.* **15**: 83–91. [26]

Feltwell J. (1982). *Large White Butterfly—the biology, biochemistry and physiology of* Pieris brassicae. The Hague. [1]

Feltwell J. & Rothschild M. (1974). Carotenoids in 38 species of Lepidoptera. *J. Zool., Lond.* **174**: 441–465. [12]

Fenner T. L. (1975). *Proposal for experimental farming of protected birdwing butterflies with particular reference to* Ornithoptera alexandrae. Internal rpt, DPI, Port Moresby. [32]

Ferm R. (1977). *A comparative study of cardiac glycoside sequestering by* Danaus plexippus *and* D. chrysippus. Snr Hons Thesis, Amherst College, Mass., USA. [12]

Ferris C. D. (1972). Ultraviolet photography as an adjunct to taxonomy. *J. Lepid. Soc.* **26**: 210–215. [22]

Field W. D. (1971). Butterflies of the genus *Vanessa* and of the resurrected genera *Bassaris* and *Cynthia*. *Smithson. Contr. Zool.* **(84)**: 105pp. [1]

Field W. D. & Herrera J. (1977). The pierid butterflies of the genera *Hypsochila*, *Phulia*, *Infraphulia* and *Piercolias*. *Smithson. Contr. Zool.* **(232)**: iii + 64pp. [1]

Field W. D., Dos Passos C. F. & Masters J. H. (1974). A bibliography of the catalogs, lists, faunal and other papers on the butterflies of North America north of Mexico, arranged by state and province. *Smithson. Contr. Zool.* **(157)**: 104pp. [1]

Findlay R., Young M. R. & Findlay J. A. (1983). Orientation behaviour in the grayling butterfly: thermoregulation or crypsis? *Ecol. Ent.* **8**: 145–153. [20]

Fink L. S. (1980). *Bird predation on overwintering monarch butterflies*. BA thesis, Amherst College, Mass., USA. [12]

Fink L. S. & Brower L. P. (1981). Birds can overcome the cardenolide defense of monarch butterflies in Mexico. *Nature* **291**: 67-70. [12]

Fink L. S., Brower L. P., Waide R. B. & Spitzer P. R. (In press). Overwintering monarch butterflies as food for insectivorous birds in Mexico. [12]

Finn F. (1895). Contributions to the theory of warning colours and mimicry. I. Experiments with a babbler (*Crateropus canorus*). *J. Asiat. Soc. Beng.* **64**: 344–356. [12]

Finn F. (1896). Contributions to the theory . . . II. Experiments with a lizard (*Calotes versicolor*). *J. Asiat. Soc. Beng.* **65**: 42–48. [12]

Finn F. (1897*a*). Contributions to the theory . . . III. Experiments with a tupaia and a frog. *J. Asiat. Soc. Beng.* **66**: 528–533. [12]

Finn F. (1897*b*). Contributions to the theory . . . IV. Experiments with various birds. *J. Asiat. Soc. Beng.* **66**: 613–668. [12]

Fischer R. (1971). Gustatory, behavioral, and pharmacological manifestations of chemoreception in man. Pp.187–238 in *Gustation and olfaction* (eds Ohloff G. & Thomas A. F.). New York. [12]

Fisher R. A. (1927). On some objections to mimicry theory: statistical and genetic. *Trans. ent. Soc. Lond.* **1927**: 269–278. [12, 14]

Fisher R. A. (1930). *The genetical theory of natural selection.* Oxford. [6, 12, 14, 21, 25, 26]

Fisher R. A. (1958). *The genetical theory of natural selection* (2nd edn). New York. [12]

Fisher R. A. & Ford E. B. (1928). The variability of species in the Lepidoptera, with reference to abundance and sex. *Trans. ent. Soc. Lond.* **79**: 367–379. [20]

Floch H. & Abonnenc E. (1950). Ceratopogonides nouveaux de Venezuela: *Culicoides lichyi* et *Lasiohelea danaisi*. *Boln Ent. Venez.* **8**: 69–75. [11]

Fontaine M. (1981*a*). Biotopes et premiers états des espèces de Nymphalidae-Eunicinae observées au Zaire. *Lambillionea* **81**: 50–57. [1]

Fontaine M. (1981*b*). Lép: Nymphalidae sous-fam: Eurytelinae. Premiers états des espèces observées à Isiro (ex. Paulis) et environs immédiats Zaire: region de Haut-Zaire, sous region du Haut-Uele. *Lambillionea* **80**: 65-67. [1]

Fontaine M. (1982). Genre *Cymothoe*. Note sur les premiers états. *Lambillionea* **82**: 63–64, 67–72. [1]

Forbes W. T. M. (1939). Revisional notes on the Danainae. *Entomologica am.* (NS) **19**: 101–140. [1]

Ford E. B. (1945). *Butterflies.* London. [Intro, 10, 12, 20, 25, 33]

Ford E. B. (1953). The genetics of polymorphism in the Lepidoptera. *Adv. Genet.* **5**: 43–87. [Intro, 14, 21]

Ford E. B. (1955). *Moths.* London. [Intro, 12, 16]

Ford E. B. (1964). *Ecological genetics.* London. [Intro, 14]

Ford E. B. (1971). *Ecological genetics* (3rd edn). London. [14, 18]

Ford E. B. (1973). *Evolution studied by observation and experiment.* London. [16]

Ford E. B. (1975*a*). *Ecological genetics* (4th edn). London. [Intro, 12, 16]

Ford E. B. (1975*b*). *Butterflies* (revised edn). Glasgow. [16]

Ford H. D. & Ford E. B. (1930). Fluctuations in numbers and its influence on variation in *Melitaea aurinia*. *Trans. ent. Soc. Lond.* **78**: 345–351. [2, 10, 16]

Ford R. L. E. (1976). The influence of the Microgasterini on the populations of British Rhopalocera. *Entomologist's Gaz.* **27**: 205–210. [10]

Forman B., Ford E. B. & McWhirter K. G. (1959). An evolutionary study of the butterfly *Maniola jurtina* in the north of Scotland. *Heredity* **13**: 353–361. [16]

Forster W. A. & Wohlfahrt T. A. (1955). Tagfalter diurna. *Die Schmett. Mitteleuropas* **2**: 126pp. [1]

Fosdick M. H. (1973). A population study of the neotropical nymphalid butterfly, *Anartia amalthea*, in Ecuador. *J. Res. Lepid.* **11**: 65–80. [2]

Fountaine M. E. (1913). Five months butterfly collecting in Costa Rica in the summer of 1911. *Entomologist* **46**: 189–194, 214–219. [1]

Fountaine M. E. (1925–6). Amongst the Rhopalocera of the Philippines. *Entomologist* **58**: 235–239, 263–265; **59**: 9–11, 31–34, 53–57. [1]

Fox L. R. & Morrow P. A. (1981). Specialization: species property or local phenomenon? *Science* **211**: 887–893. [3]

Fox M. M. & Vevers G. (1960). *The nature of animal colours.* London. [20]

Fox R. M. (1940). A generic review of the Ithomiinae. *Trans. Am. ent. Soc.* **66**: 161–207. [1]

Fox R. M. (1956). A monograph of the Ithomiinae. Part I *Bull. Am. Mus. nat. Hist.* **111**: 1-76. [1, 9]

Fox R. M. (1960). A monograph . . . Part II. The tribe Melinaeini Clark. *Trans. Am. ent. Soc.* **86**: 109–171. [1]

Fox R. M. (1966). Forelegs of butterflies. I. Introduction: chemoreception. *J. Res. Lepid.* **5**: 1–12. [6]

Fox R. M. (1967). A monograph . . . Part III. The tribe Mechanitini Fox. *Mem. Am. ent. Soc.* **(22)**: (v) + 190pp. [1, 6]

Fox R. M. & Real H. G. (1971). A monograph . . . Part IV. The tribe Napeogenini Fox. *Mem. Am. ent. Inst.* **(15)**: (v) + 368pp. [1]

Fox R. M., Lindsey A. W., Clench H. K. & Miller L. D. (1965). The butterflies of Liberia. *Mem. Am. ent. Soc.* **(19)**: ii + 438pp. [1]

Fracker S. B. (1915). The classification of lepidopterous larvae. *Illinois biol. Monogr.* **2**: 1–169. [1]

Fraenkel G. (1959). The raison d'être of secondary plant substances. *Science* **129**: 1466-1470. [6, 9]

Franklin I. R. (1980). Evolutionary change in small populations. Pp.135–149 in Soulé & Wilcox (1980). [33]

Frazer J. F. D. (1961). Butterfly populations on the North Downs. *Proc. Trans. Br. ent. nat. Hist. Soc.* **1960**: 98–109. [16, 33]

Frazer J. F. D. (1977). The chalkhill blue butterfly (*Lysandra coridon*) at Burham Down 1955–74. *Trans. Kent Fld Club* **6**: 71–74. [33]

Frazer J. F. D. & Rothschild M. (1960). Defence mechanisms in warningly-coloured moths and other insects. *Int. Congr. Ent.* (11) **3**: 249–256. [12]

Frazer J. F. D. & Willcox H. N. A. (1975). Variation in spotting among the close relatives of the butterfly, *Maniola jurtina*. *Heredity* **34**: 305–322. [16]

Freeland W. J. & Janzen D. H. (1974). Strategies in herbivory by mammals. *Am. Nat.* **108**: 269–289. [12]

Freeman H. A. (1969). Systematic review of the Megathymidae. *J. Lepid. Soc.* **23** (Suppl. 1): 58pp. [1]

Friedlander M. (1975). Nucleated and anucleated, two types of concomitant spermatozoa in inseminated

female moths. Pp.75-82 in *Gamete competition in plants and animals* (ed. Mulcahy D. H.). Amsterdam. [6]
Friedlander M. & Gitay H. (1972). The fate of the normal-anucleated spermatozoa in inseminated females of the silkworm, *Bombyx mori*. *J. Morph.* **138**: 121–130. [6]
Friedlander M., Jans P. & Benz G. (1981). Precocious reprogramming of eupyrene-apyrene spermatogenesis commitment induced by allatectomy of the penultimate larval instar of the moth *Actias selene*. *J. Insect Physiol.* **27**: 267–269. [6]
Frings H. & Frings M. (1949). The loci of chemoreceptors in insects. *Am. Midl. Nat.* **41**: 602–658. [6]
Frings H., Goldberg E. & Arentzen J. C. (1948). Antibacterial action of the blood of the large milkweed bug. *Science* **108**: 689–690. [12]
Frohawk F. W. (1925). The scarcity and disappearance of butterflies. *Entomologist* **58**: 145–147. [33]
Frohawk F. W. (1934). *British butterflies*. London. [33]
Fruchter J. S., Robertson D. E., Evans J. C. & 19 others. (1980). Mt St Helens ash from 18 May 1980 eruption: chemical, physical, mineralogical, and biological properties. *Science* **209**: 1116–1125. [31]
Fujioka T. (1975). *Butterflies of Japan*. (in Japanese) Tokyo. [1]
Futuyma D. J. (1983). Evolutionary interactions among herbivorous insects and plants. Pp.207–231 in Futuyma & Slatkin (1983). [Intro]
Futuyma D. J. & Mayer G. C. (1980). Non-allopatric speciation in animals. *Syst. Zool.* **29**: 254–271. [19]
Futuyma D. J. & Slatkin M. (eds.) (1983). *Coevolution*. Sunderland, Mass.
Futuyma D. J. & Wasserman S. S. (1981). Foodplant specialization and feeding efficiency in the tent caterpillars *Malacosoma disstria* and *M. americanum*. *Entomologia exp. appl.* **30**: 106–110. [7]
Fyson D. R. & Poulton E. B. (1930). Mrs D. R. Fyson's observations of the epigamic behaviour of the male danaine butterfly *Euploea core* in Madras. *Proc. ent. Soc. Lond.* **5**: 48–49. [25]
Gadgil M. (1972). Male dimorphism as a consequence of sexual selection. *Am. Nat.* **106**: 574–580. [23]
Gaede M. (1931). Satyridae I–III. *Lepid. Cat.* **29** (43, 46, 48): 759pp. [1]
Galef B. G. (1977). Mechanisms for the social transmission of acquired food preferences from adult to weanling rats. Pp.123–148 in Barker *et al.* (1977). [12]
Gall L. F. (1981). Re. commercial sales of *Boloria acrocnema*. *News Lepid. Soc.* **1981** (3): 40–41. [2]
Gamzu E. (1977). The multifaceted nature of taste-aversion-inducing agents: is there a single common factor? Pp.477–509 in Barker *et al.* (1977). [12]
Ganchrow J. R. (1982). Affective influences on behavioural responses to the quality and intensity of gustatory stimuli. Pp.137–148 in *Determination of behaviour by chemical stimuli* (eds Booth D., Kroeze J. H. A. & Maller O.). London. [12]
Gans C. (1964). Empathic learning and the mimicry of African snakes. *Evolution* **18**: 705. [12]
Garcia J., Ervin F. R. & Koelling R. (1966). Learning with prolonged delay in reinforcement. *Psychonomic Science* **5**: 121–122. [14]
Garcia J. & Hankins W. G. (1974). The evolution of bitter and the acquisition of toxiphobia. Pp.39–45 in *5th international symposium on olfaction and taste* (ed. Denton D. A.). New York. [12]
Garcia J. & Hankins W. G. (1977). On the origin of food aversion paradigms. Pp.3–19 in Barker *et al.* (1977). [12]
Garcia J. & Koelling R. A. (1966). Relation of cue to consequence in avoidance learning. *Psychonomic Science* **4**: 123–124. [12]
Garcia J. & Rusiniak K. W. (1979). What the nose learns from the mouth. Pp.141–156 in Müller-Schwarze & Silverstein (1979). [12]
Garcia J., McGowan B. K. & Green K. F. (1972). Biological constraints on conditioning. Pp.3–27 in *Classical conditioning II: current research and theory* (eds Black A. H. & Prokasy W. F.). New York. [14]
Garcia J., Hankins W. G. & Rusiniak K. W. (1974). Behavioral regulation of the milieu interne in man and rat. *Science* **185**: 824–831. [12]
Gardner R. J. (1978). Lipophilicity and bitter taste. *J. Pharm. Pharmac.* **30**: 531–532. [12]
Gauthreaux S. A. (ed.) (1980). *Animal migration, orientation and navigation*. New York.
Geiger H. (1981). Enzyme electrophoretic studies on the genetic relationships of pierid butterflies. I. European taxa. *J. Res. Lepid.* **19**: 181–195. [Intro, 17]
Geiger H. & Scholl A. (1982). *Pontia daplidice* in Südeuropa—eine Gruppe von zwei Arten. *Mitt. schweiz ent. Ges.* **55**: 107–114. [17]
Gentry A. H. (1981). Distributional patterns and an additional species of the *Passiflora vitifolia* complex: Amazonian species diversity due to edaphically differentiated communities. *Pl. Syst. Evol.* **137**: 95–105. [3]
Ghent A. (1960). A study of the group-feeding behavior of larvae of the jack pine sawfly, *Neodiprion pratti banksiannae*. *Behaviour* **16**: 110–148. [6]
Ghiradella H. (1974). Development of ultraviolet reflecting butterfly scales: how to make an interference filter. *J. Morph.* **142**: 395–409. [20]
Ghiradella H., Aneshansley D., Eisner T., Silberglied R. E. & Hinton H. (1972). Ultraviolet reflection of a male butterfly: interference color caused by thin-layer elaboration of wing scales. *Science* **178**: 1214–1217; **179**: 415. [20]
Giacomelli E. (1915). El género *Tatochila*: lo que sabemos y lo que ignoramos de él. *Anales Mus. Argentino Ciencias Nat. "Bernardino Rivadavia"* **26**: 403–415. [27]
Gibb J. A. (1954). Feeding ecology of tits, with notes on treecreeper and goldcrests. *Ibis* **96**: 513–543. [12]
Gibb J. A. (1958). Predation by tits and squirrels on the eucosmid, *Ernarmonia conicolana*. *J. anim. Ecol.* **27**: 375–396. [10]
Gibb J. A. (1960). Populations of tits and goldcrests and their food supply in pine plantations. *Ibis* **102**: 163–208. [12]
Gibb J. A. (1966). Tit predation and the abundance of *Ernarmonia conicolana* on Weeting Heath, Norfolk, 1962–63. *J. anim. Ecol.* **35**: 43–53. [10]
Gibo D. L. & Pallett M. J. (1979). Soaring flight of monarch butterflies, *Danaus plexippus*, during the late summer migration in southern Ontario. *Can. J. Zool.* **57**: 1393–1401. [26]
Gibson D. O. (1980). The role of escape in mimicry and

polymorphism: I. The response of captive birds to artificial prey. *Biol. J. Linn. Soc.* **14**: 201–214. [14, 20]

Gifford D. (1965). *A list of the butterflies of Malawi.* Blantyre, Malawi. [1]

Gilbert L. E. (1969). *Some aspects of the ecology and community structure of ithomid butterflies in Costa Rica.* Research Report Advanced Population Biology Course OTS, July–Aug. 1969. Ciudad Universitaria, Costa Rica. [25]

Gilbert L. E. (1971). Butterfly-plant coevolution: has *Passiflora adenopoda* won the selectional race with heliconiine butterflies? *Science* **172**: 585–586. [6]

Gilbert L. E. (1972). Pollen feeding and reproductive biology of *Heliconius* butterflies. *Proc. natn. Acad. Sci. U.S.A.* **69**: 1403–1407. [2]

Gilbert L. E. (1975). Ecological consequences of coevolved mutualism between butterflies and plants. Pp.210–240 in Gilbert & Raven (1975). [6, 20, 25, 26]

Gilbert L. E. (1976). Postmating female odor in *Heliconius* butterflies: a male-contributed anti-aphrodisiac? *Science* **193**: 419–420. [3, 12, 20, 21, 25]

Gilbert L. E. (1977). The role of insect-plant coevolution in the organization of ecosystems. Pp.399–413 in *Comportement des insects et milieu trophique* (ed. Labeyrie V.). Paris. [3]

Gilbert L. E. (1979). Development of theory in the analysis of insect-plant interactions. Pp.117–154 in *Analysis of ecological systems* (eds Horn D. J., Mitchell R. D. & Stairs G. R.). Columbus. [2, 3, 12, 19]

Gilbert L. E. (1980). Food web organization and the conservation of neotropical diversity. Pp.11–33 in Soulé & Wilcox (1980). [2, 33]

Gilbert L. E. (1982). The coevolution of a butterfly and a vine. *Scient. Am.* **247**: 110–121. [12]

Gilbert L. E. (1983). Coevolution and mimicry. Pp.263–281 in Futuyma & Slatkin (1983). [Intro, 3]

Gilbert L. E. & Ehrlich P. R. (1970). The affinities of the Ithomiinae and Satyrinae. *J. Lepid. Soc.* **24**: 297–300. [9]

Gilbert L. E. & Raven P. H. (eds) (1975). *Coevolution of animals and plants.* Austin, Texas.

Gilbert L. E. & Singer M. C. (1973). Dispersal and gene flow in a butterfly species. *Am. Nat.* **107**: 58–72. [2, 6, 16, 26, 33]

Gilbert L. E. & Singer M. C. (1975). Butterfly ecology. *A. Rev. Ecol. Syst.* **6**: 365–397. [Intro, 2, 3, 6, 12, 16, 26, 33]

Gilbert L. E. & Smiley J. T. (1978). Determination of local diversity in phytophagous insects: host specialists in tropical environments. *Symp. R. ent. Soc. Lond.* **(9)**: 89–104. [3]

Gilmour D. (1965). *The metabolism of insects.* London. [19]

Gittleman J. L. & Harvey P. H. (1980). Why are distasteful prey not cryptic? *Nature* **286**: 149–150. [12, 14]

Gittleman J. L., Harvey P. H. & Greenwood P. J. (1980). The evolution of conspicuous coloration: some experiments in bad taste. *Anim. Behav.* **28**: 897–899. [12, 14]

Godfrey E. J. (1930). A revised list of the butterflies of Siam, with notes on their geographical distribution. *J. Siam. Soc.* **7**: 203–387. [1]

Godman F. D. & Salvin O. (1879–1901). Insecta, Lepidoptera Rhopalocera. *Biologia cent. am. (Zool.)* **1**: xlvi + 487pp.; **2**: 782pp.; **3**: 112 pls. [1]

Goldschmidt R. B. (1934). *Lymantria. Biblphia genet.* **11**: 186. [24]

Goldschmidt R. B. (1945). Mimetic polymorphism, a controversial chapter of Darwinism. *Q. Rev. Biol.* **20**: 147–164, 205–230. [14]

Goldsmith T. H. & Bernard G. D. (1974). The visual system of insects. Pp.165–272 in *The physiology of the Insecta* **2** (2nd edn) (ed. Rockstein M.). New York. [20]

Goldstein A., Aronow L. & Kalman S. M. (1969). *Principles of drug action: the basis of pharmacology.* New York. [12]

Gollub L. R. & Brady J. V. (1965). Behavioral pharmacology. *A. Rev. Pharmac.* **5**: 235–262. [12]

Gonzalez R. C., Behrend E. R. & Bitterman M. E. (1967). Reversal learning and forgetting in birds and fish. *Science* **158**: 519–521. [12]

Goodale M. A. & Sneddon I. (1977). The effect of distastefulness of the model on the predation of artificial Batesian mimics. *Anim. Behav.* **25**: 660–665. [12, 14]

Goodden R. (1978). *British butterflies: a field guide.* Newton Abbot. [33]

Goodwin N. (1982) (movie film). *Animal imposters.* Boston. Mass. [12]

Goss G. J. (1979). The interaction between moths and plants containing pyrrolizidine alkaloids. *Envir. Ent.* **8**: 487–493. [25]

Goss H. (1890). *Lycaena arion* in the Cotswolds. *Entomologist's mon. Mag.* **26**: 24. [33]

Gossard T. W. & Jones R. E. (1977). The effects of age and weather on egg-laying in *Pieris rapae. J. appl. Ecol.* **14**: 65–71. [2, 6]

Gothilf S. & Shorey H. H. (1976). Sex pheromones of Lepidoptera: examination of the role of scent brushes in courtship behavior of *Trichoplusia ni. Envir. Ent.* **5**: 115–119. [25]

Götz B. (1951). Die Sexualduftstoffe an Lepidopteren. *Experientia* **7**: 406–418. [25]

Gould S. J. (1980). Is a new general theory of evolution emerging? *Paleobiology* **6**: 119–130. [14]

Gould S. J. & Lewontin R. C. (1979). The spandrels of San Marco and the Panglossian paradigm: a critique of the adaptationist programme. *Proc. R. Soc.* (B) **205**: 581–598. [2, 6, 7]

Gould S. J. & Vrba E. S. (1982). Exaptation—a missing term in the science of form. *Paleobiology* **8**: 4–15. [12]

Goux J-M. (1978). Sélection endocyclique et polymorphisme stable. *C. r. hebd. Séanc. Acad. Sci., Paris* (D) **286**: 125–127. [16]

Gowers S. (1980). *Butterfly distribution maps.* Dorchester, Dorset. [33]

Graham M. W. R. de V. (1950). Postural habits and colour pattern evolution in Lepidoptera. *Trans. Soc. Br. Ent.* **10**: 217–232. [20]

Graham S. M., Watt W. B. & Gall L. F. (1980). Metabolic resource allocation vs. mating attractiveness: adaptive pressures on the 'alba' polymorphism of *Colias* butterflies. *Proc. natn. Acad. Sci. U.S.A.* **77**: 3615–3619. [16]

Grant G. G. (1974). Male sex pheromone from the wing

glands of the Indian meal moth, *Plodia interpunctella*. *Experientia* **30**: 917–918. [25]

Grant G. G. (1976*a*). Courtship behavior of a phycitid moth, *Vitula edmandsae*. *Ann. ent. Soc. Am.* **69**: 445–449. [25]

Grant G. G. (1976*b*). Female coyness and receptivity during courtship in *Plodia interpunctella*. *Can. Ent.* **108**: 975–979. [25]

Grant G. G. & Brady U. E. (1975). Courtship behavior of phycitid moths. I. Comparison of *Plodia interpunctella* and *Cadra cautella* and role of male scent glands. *Can. J. Zool.* **53**: 813–826. [25]

Grant G. G., Smithwick E. B. & Brady U. E. (1975). Courtship behavior . . . II. Behavioral and pheromonal isolation of *Plodia interpunctella* and *Cadra cautella* in the laboratory. *Can. J. Zool.* **53**: 827–832. [25]

Grant V. (1963). *The origin of adaptations*. New York. [21]

Gray A. J. & Ambrosen H. (In press). Genetic aspects of population size in *Agrostis setacea* Curt. and their implications for nature conservation. In *Area and isolation* (ed. Hooper. M. D.). Abbots Ripton, Cambridge, UK. [33]

Greene H. W. & McDiarmid R. W. (1981). Coral snake mimicry: does it occur? *Science* **213**: 1207-1212. [12]

Greenfield M. D. (1981). Moth sex pheromones: an evolutionary perspective. *Fla. Ent.* **64**: 4-17. [25]

Greenfield M. D. & Coffelt J. A. (1983). Reproductive behaviour of the lesser waxmoth, *Achroia grisella*: signalling, pair formation, male interactions, and mate guarding. *Behaviour* **84**: 287-315. [25]

Gribbin J. (1982). Do volcanoes affect the climate? *New Scient.* **93**: 150–153. [31]

Griggs R. F. (1919). Scientific results of the Katmai expeditions of the National Geographic Society. IV. The character of the eruption as indicated by its effects on nearby vegetation. *Ohio J. Sci.* **19**: 173–209. [31]

Grote A. R. (1876). A colony of butterflies. *Am. Nat.* **10**: 129–132. [33]

Grula J. W. & Taylor O. R. (1979). The inheritance of pheromone production in the sulphur butterflies *Colias eurytheme* and *C. philodice*. *Heredity* **42**: 359–371. [25]

Grula J. W., McChesney J. D. & Taylor O. R. (1980). Aphrodisiac pheromones of the sulfur butterflies *Colias eurytheme* and *C. philodice*. *J. chem. Ecol.* **6**: 241–256. [20, 25]

Guppy L. (1894). Notes on some Trinidad butterflies. *J. Trin. Fld Nat. Club* **2**: 170–174. [12]

Gustavson C. R. (1977). Comparative and field aspects of learned food aversions. Pp.23–43 in Barker *et al.* (1977). [12]

Guttman S. I., Wood T. K. & Karlin A. A. (1981). Genetic differentiation along host plant lines in the sympatric *Euchenopa binotata* complex (Homoptera: Membracidae). *Evolution* **35**: 205–217. [19]

Haase E. V. (1887*a*). Duftapparate indo-australischer Schmetterlinge. I. Rhopalocera. *Dt. ent. Z.* **1**: 92–107. [25]

Haase E. V. (1887*b*). Duftapparate . . . II. Heterocera. *Dt. ent. Z.* **1**: 159–179. [25]

Haase E. V. (1896). *Mimicry in butterflies and moths*. Part II. (transl. Childs C. M.). Stuttgart. [12]

Haatjens H. A. (ed.) (1964). General report on the lands of the Buna—Kokoda area, Territory of Papua New Guinea. *Land Res. Ser. C.S.I.R.O. Aust.* **(10)**: 113pp. [32]

Haber W. A. (1978). *Evolutionary ecology of tropical mimetic butterflies*. PhD thesis, Minnesota Univ., USA. [1, 3, 25]

Haig E. F. G. (1936–38). Butterflies of Nigeria. *Niger. Fld* **5**: 12–28, 114–128; **6**: 71–76; **7**: 41–42, 61–80. [1]

Hailman J. P. (1977). *Optical signals: animal communication and light*. Bloomington, Indiana. [20]

Haldane J. B. S. (1922). Sex ratio and unisexual sterility in hybrid animals. *J. Genet.* **12**: 101–109. [24]

Haldane J. B. S. (1924). A mathematical theory of natural and artificial selection. Part I. *Trans. Camb. phil. Soc.* **23**: 19–41. [14]

Haldane J. B. S. (1954). An exact test for randomness in mating. *J. Genet.* **52**: 631–635. [21]

Hall A. (1940). Catalogue of the Lepidoptera Rhopalocera (butterflies) of British Guiana. *Ent. Bull. Dept. Agric. Br. Guiana* **3**: 1–88. [1]

Hall M. L. (1980). *Butterfly monitoring: instructions for independent recorders*. Abbots Ripton, Cambridge, UK. [33]

Hall M. L. (1981). *Butterfly research in ITE*. Cambridge, UK. [33]

Halliday T. R. (1978). Sexual selection and mate choice. Pp.180–213 in Krebs & Davies (1978). [21]

Hamilton W. D. (1963). The evolution of altruistic behavior. *Am. Nat.* **97**: 354–356. [6]

Hamilton W. D. (1964*a*). The genetical evolution of social behaviour. I. *J. theor. Biol.* **7**: 1–16. [12, 14]

Hamilton W. D. (1964*b*). The genetical . . . II. *J. theor. Biol.* **7**: 17–52. [12, 14, 26]

Hamilton W. D. (1971). Geometry for the selfish herd. *J. theor. Biol.* **31**: 295–311. [14, 23]

Hamilton W. J. (1973). *Life's color code*. New York. [20]

Hammond P. C. (1983). The colonization of violets and *Speyeria* butterflies on the ash-pumice fields deposited by Cascadian volcanoes. *J. Res. Lepid.* **20**: 179–191. [31]

Hammond P. C. & Dornfeld E. J. (1983). A new subspecies of *Speyeria egleis* from the pumice region of central Oregon. *J. Lepid. Soc.* **37**: 115–120. [31]

Hancock D. L. (1978). *Phylogeny and biogeography of Papilionidae*. MSc thesis, Queensland Univ., Australia. [1]

Hancock D. L. (1983). Classification of the Papilionidae: a phylogenetic approach. *Smithersia* **(2)**: 48pp. [1]

Handford P. T. (1973*a*). Patterns of variation in a number of genetic systems in *Maniola jurtina*: the boundary region. *Proc. R. Soc.* (B) **183**: 265–284. [16]

Handford P. T. (1973*b*). Patterns of variation in a number of genetic systems in *Maniola jurtina*: the Isles of Scilly. *Proc. R. Soc.* (B) **183**: 285–300. [16]

Hankins W. G., Garcia J. & Rusiniak K. W. (1973). Dissociation of odor and taste in baitshyness. *Behav. Biol.* **8**: 407–419. [12]

Hanson F. E. (1976). Comparative studies on the induction of food choice preferences in lepidopterous larvae. Pp.71–77 in *The host-plant in relation to insect behaviour and reproduction* (ed. Jermy T.). Budapest. [6]

Harborne J. B. (ed.) (1972). *Phytochemical ecology*. London.

Harborne J. B. (ed.) (1978). *Biochemical aspects of plant and animal coevolution*. New York.

Harcourt D. G. (1961). Spatial pattern of the imported cabbageworm, *Pieris rapae*, on cultivated Cruciferae. *Can. Ent.* **93**: 945-952. [6]

Harcourt D. G. (1966). Major factors in survival of the immature stages of *Pieris rapae*. *Can. Ent.* **98**: 653-662. [2, 10, 33]

Hardy G. H. (1951). The courtship of *Euploea corinna*. *Entomologist's mon. Mag.* **87**: 8-9. [25]

Harvey P. H. & Greenwood P. J. (1978). Anti-predator defence strategies: some evolutionary problems. Pp.129-151 in Krebs & Davies (1978). [12]

Harvey P. H. & Paxton P. J. (1981). The evolution of aposematic coloration. *Oikos* **37**: 391-393. [12, 14]

Harvey P. H., Bull J. J., Pemberton M. & Paxton R. J. (1982). The evolution of aposematic coloration in distasteful prey: a family model. *Am. Nat.* **119**: 710-719. [12, 14]

Harvey P. H. [Bull J. J. & Paxton R. J.] (1983). Why some insects look pretty nasty. *New Scientist* **97**: 26-27. [14]

Hassell M. P. (1966). Evaluation of parasite or predator responses. *J. anim. Ecol.* **35**: 65-75. [10]

Hassell M. P. & Southwood T. R. E. (1978). Foraging strategies of insects. *A. Rev. Ecol. Syst.* **9**: 75-98. [18]

Haugum J. & Low A. M. (1978-9). *A monography of the birdwing butterflies.* **1** (1-3): *the genus* Ornithoptera. Klampenborg, Denmark. [1]

Haugum J. & Low A. M. (1978-9). *A monograph of the birdwing butterflies.* **1** (1-3): *the genus* Ornithoptera. Klampenborg, Denmark. [1]

Haukioja E. & Neimalaa P. (1977). Retarded growth of a geometrid larva after mechanical damage to leaves of its host tree. *Annls zool. fenn.* **14**: 48-52. [7]

Hayashi N., Kuwahara Y. & Komae H. (1978). The scent scale substances of male *Pieris* butterflies (*P. melete* and *P. napi*). *Experientia* **34**: 684-685. [12, 25]

Hayes J. L. (1981). The population ecology of a natural population of the pierid butterfly *Colias alexandra*. *Oecologia* **49**: 188-200. [2, 10]

Hayward K. J. (1948-67). *Genera et species animalium Argentinorum. Insecta, Lepidoptera (Rhopalocera).* **1** *(Hesperiidae: Pyrrhopyginae and Pyrginae);* **2** *(Hesperiidae: Hesperiinae);* **3** *(Nymphalidae, Heliconiidae);* **4** *(Papilionidae, Satyridae).* Tucuman, Argentina. [1]

Hayward K. J. (1969). *Datos para el estudio de la ontogenia de lepidopteros Argentinos.* Tucuman, Argentina. [1]

Hayward K. J. (1973). Catalogo de los Ropaloceros Argentinos. *Op. lilloana* **(23)**: 318pp. [1]

Hazel W. N. & West D. A. (1979). Environmental control of pupal colour in swallowtail butterflies: *Battus philenor* and *Papilio polyxenes*. *Ecol. Ent.* **4**: 393-400. [12]

Heath J. (1973). Lepidoptera. Pp.151-166 in Steele & Welch (1973). [33]

Heath J. (ed.) (1976). *The moths and butterflies of Great Britain and Ireland* **1**. London.

Heath J. (1981*a*). *Threatened Rhopalocera in Europe.* Strasbourg. [33]

Heath J. (1981*b*). British red data book—insects. *Atala* **7**: 49. [33]

Heath J. & Harding P. T. (1981). The Biological Records Centre and insect recording. *Atala* **7**: 47-48. [33]

Heath J. & Perring F. (1978). *Institute of Terrestrial Ecology Booklet.* Biological Records Center, Monks Wood, Huntingdon. [33]

Heath J. & Skelton M. S. (1975). *Provisional distribution maps: Lepidoptera Rhopalocera.* Abbots Ripton, Cambridge, UK. [33]

Hecker E. (1970). Cocarcinogens from Euphorbiaceae and Thymeleaceae. Pp.147-165 in *Pharmacognosy and phytochemistry* (eds Wagner H. & Horhammer L.). Berlin. [13]

Hefley H. M. (1937). The relations of some native insects to introduced plants. *J. anim. Ecol.* **6**: 138-144. [6]

Hegnauer R. (1964-73). *Chemotaxonomie der Pflanzen* **2-6**. Basel & Stuttgart. [9, 12, 13]

Heim de Balsac H. (1928). Fragments de bromatologie ornithologique. *Revue fr. Orn. Scient. prat.* **12**: 54-66. [12]

Heinrich B. (1972). Thoracic temperatures of butterflies in the field near the equator. *Comp. Biochem. Physiol.* **43**A: 459-467. [28]

Heinrich B. (1974). Thermoregulation in endothermic insects. *Science* **185**: 747-756. [28]

Heinrich B. (1977). Why have some animals evolved to regulate a high body temperature? *Am. Nat.* **111**: 623-640. [28]

Hendricks D. E. & Shaver T. N. (1975). Tobacco budworm: male pheromone suppressed emission of sex pheromone by the female. *Envir. Ent.* **4**: 555-558. [25]

Hennessy D. F. & Owings D. H. (1978). Snake species discrimination and the role of olfactory cues in the snake-directed behaviour of the California ground squirrel. *Behaviour* **65**: 115-124. [12]

Henning S. F. (1980). *Chemical communication between lycaenid larvae and ants.* MSc thesis, Univ. Witwatersrand, RSA. [Intro]

Henning S. F. (1983). Biological groups within the Lycaenidae. *J. ent. Soc. sth Afr.* **46**: 65-85. [Intro]

Henriksen H. J. & Kreutzer I. (1982). *The butterflies of Scandinavia in nature.* Odense. [1]

Henry T. A. (1949). *The plant alkaloids.* London. [12]

Henson W. R. (1958). The effects of radiation on the habitat temperatures of some poplar inhabiting insects. *Can. J. Zool.* **36**: 463-478. [28]

Heppner J. B. (ed.) (in prep.). Atlas of Neotropical Lepidoptera. [1]

Hering M. (1926). *Biologie der Schmetterlinge.* Berlin. [Intro, 10, 25]

Herman W. S. (1975). Endocrine regulation of post-eclosion enlargement of the male and female reproductive glands in monarch butterflies. *General comp. Endocr.* **26**: 534-540. [6]

Herman W. S. (1981). Studies on the adult reproductive diapause of the monarch butterfly, *Danaus plexippus*. *Biol. Bull.* **160**: 89-106. [6]

Herman W. S. & Barker J. F. (1977). Effect of mating on monarch butterfly oogenesis. *Experientia* **33**: 688-689. [6]

Herman W. S. & Dallman S. H. (1981). Endocrine biology of the painted lady butterfly *Vanessa cardui*. *J. Insect Physiol.* **27**: 163-168. [6]

Herrera J. & Field W. D. (1959). A revision of the butterfly genera *Theochila* and *Tatochila*. *Proc. U.S. natn. Mus.* **108**: 467-514. [27]

Heslop I. R. P. (1955). *Apatura iris* imago attacked by birds: with some observations on the status and future of this species. *Entomologist's Gaz.* **6**: 175-177. [10]

Hidaka T. (1972). Biology of *Hyphantria cunea* in Japan. XIV. Mating behavior. *Appl. Ent. Zool.* **7**: 116–132. [25]

Hidaka T. (1973). Logic of mating behavior of Lepidoptera. *Ann. N.Y. Acad. Sci.* **223**: 70–76. [20]

Hidaka T. & Takahashi H. (1967). Temperature conditions and maternal effect as modifying factors in the photoperiodic control of the seasonal form in *Polygonia c-aureum*. *Annotnes zool. Jap.* **40**: 200–204. [30]

Hidaka T. & Yamashita K. (1975). Wing color pattern as the releaser of mating behavior in the swallowtail butterfly, *Papilio xuthus*. *Appl. Ent. Zool., Tokyo* **10**: 263-267. [20]

Hidaka T. & Yamashita K. (1976). Change of meaning of color in the Umwelt of butterflies. *Seiro-seitai* **17**: 15–21. [In Japanese.) [20]

Higgins L. G. (1975). *The classification of European butterflies*. London. [1, 6]

Higgins L. G. (1981). A revision of *Phyciodes* and related genera, with a review of the classification of the Melitaeinae. *Bull. Br. Mus. nat. Hist. (Ent.)* **43**: 77–243. [1]

Higgins L. G. & Riley N. D. (1970). *A field guide to the butterflies of Britain and Europe*. London. [1, 2]

Higgins L. G. & Riley N. D. (1975). *A field guide to the butterflies of Britain and Europe* (3rd edn). London. [8, 16]

Hill D. S., Johnston G. & Bascombe M. J. (1978). Annotated checklist of Hong Kong butterflies. *Mem. Hong Kong nat. Hist. Soc.* **(11)**: iv + 62pp. [1]

Hinde R. A. (1966). *Animal behaviour: a synthesis of ethology and comparative psychology*. New York. [12]

Hingston R. W. G. (1933). *The meaning of animal colouration and adornment*. London. [20, 23]

Hinton H. E. (1946). On the homology and nomenclature of the setae of lepidopterous larvae, with some notes on the phylogeny of the Lepidoptera. *Trans. R. ent. Soc. Lond.* **97**: 1–37. [1]

Hinton H. E. (1948). Sound production in lepidopterous pupae. *Entomologist* **81**: 254–269. [12]

Hinton H. E. (1951). Myrmecophilous Lycaenidae and other Lepidoptera—a summary. *Proc. Trans. S. Lond. ent. nat. Hist. Soc.* **1949–50**: 111–175. [12, 19]

Hinton H. E. (1955). Protective devices of endopterygote pupae. *Trans. Soc. Br. Ent.* **12**: 49–93. [12]

Hinton H. E. (1974). Lycaenid pupae that mimic anthropoid heads. *J. Ent.* (A) **49**: 65–69. [12]

Hinton H. E. (1977). Subsocial behaviour and biology of some Mexican membracid bugs. *Ecol. Ent.* **2**: 61–79. [14]

Hinton H. E. (1981). *The biology of insect eggs* (3 vols). Oxford. [6]

Hirai K. (1977). Observations on the function of male scent brushes and mating behavior in *Leucania separata* and *Mamestra brassicae*. *Appl. Ent. Zool., Tokyo* **12**: 347–351. [25]

Hirai K., Shorey H. H. & Gaston L. K. (1978). Competition among courting male moths: male-to-male inhibitory pheromone. *Science* **202**: 644–645. [25]

Hirose J., Suzuki Y., Takagi M., Hiehata K., Yamasaki M., Kimoto H., Yamanaka M., Iga M. & Yamaguchi K. (1980). Population dynamics of the citrus swallowtail *Papilio xuthus*: mechanisms stabilising its numbers. *Researches Popul. Ecol. Kyoto Univ.* **21**: 260–285. [10]

Hiura I. (1983). Systematics of *Hypolimnas* [all in Japanese]. *Rhopalocerist's Mag.* **6**: 3–27. [Intro.]

Ho M. W. & Saunders P. T. (1979). Beyond neo-Darwinism—an epigenetic approach to evolution. *J. theor. Biol.* **78**: 573–591. [27]

Høegh-Guldberg O. (1968). Evolutionary trends in the genus *Aricia*. *Natura jutl.* **14**: 3–76. [16]

Høegh-Guldberg O. (1971). *Polyommatus icarus* undersidevariabilitet. Kuldeforsøg med pupper. *Lepidoptera, KBH.* **2**: 2-5. [16]

Høegh-Guldberg O. & Hansen A. L. (1977). Phenotypic wing pattern modification by very brief periods of chilling of pupating *Aricia artaxerxes vandalica*. *J. Lepid. Soc.* **31**: 223–231. [16]

Høegh-Guldberg O. & Jarvis F. V. L. (1970). Central and north European Ariciae. *Natura jutl.* **15**: 1–119. [16]

Hoffman B. F. & Bigger J. J. (1980). Digitalis and allied cardiac glycosides. Pp.729-760 in *The pharmacological basis of therapeutics* (6th edn, eds Gilman A. G., Goodman L. S. & Gilman A.). New York. [12]

Hoffman F. (1931). *Euselasia eucerus*. *Ent. Rdsch.* **48**: 55–56. [6]

Hoffmann C. C. (1940–41). Catalogo sistematico y zoogeographico de los Lepidopteros Mexicanos. *An. Inst. Biol. Univ. Mex.* **11**: 639–739; **12**: 237–294. [1]

Hoffmann R. (1978). Environmental uncertainty and evolution of physiological adaptation in *Colias* butterflies. *Am. Nat.* **112**: 999–1015. [27]

Hoffmann W. E., Ng Y. C. & Tsang H. W. (1938). Life history studies in nine families of Kwangtung butterflies. *Lingnan Sci. J.* **17**: 227–246, 407–424, 515–532. [1]

Hogan J. A. (1965). An experimental study of conflict and fear: an analysis of behaviour of young chicks towards a mealworm. Part I. The behaviour of chicks which do not eat the mealworm. *Behaviour* **25**: 45–97. [12]

Hogan J. A. (1966). An experimental . . . Part II. The behaviour of chicks which eat the mealworm. *Behaviour* **27**: 273–289. [12]

Hogan J. A. (1977). The ontogeny of food preferences in chicks and other animals. Pp.71–97 in Barker *et al.* (1977). [12]

Holdren C. E. & Ehrlich P. R. (1981). Long-range dispersal in checkerspot butterflies: transplant experiments with *Euphydryas gillettii*. *Oecologia* **50**: 125–129. [2]

Holdren C. E. & Ehrlich P. R. (1982). Ecological determinants of food plant choice in the checkerspot butterfly *Euphydryas editha* in Colorado. *Oecologia* **52**: 417–423. [2, 7]

Hölldobler B. (1970). Zur physiologie der gast-wirt-beziehungen (Myrmecophilie) bei Ameisen. II. Das gastverhaltnis zu *Myrmica* und *Formica*. *Z. vergl. Physiol.* **66**: 215–250. [19]

Hölldobler B. (1971). Ants and their guests. *Scient. Am.* **224** (3): 86–93. [12]

Holling C. S. (1959). The components of predation as revealed by a study of small mammal predation of the European pine sawfly. *Can. Ent.* **91**: 293–320. [10]

Holling C. S. (1965). The functional response of predators to prey density and its role in mimicry and population

regulation. *Mem. ent. Soc. Can.* **45**: 1-60. [12, 15]
Hollis D. (ed.) (1980). *Animal identification, a reference guide.* **3.** *Insects.* London & Chichester. [1]
Holloway J. D. & Peters J. V. (1976). The butterflies of New Caledonia and the Loyalty Islands. *J. nat. Hist.* **10**: 273-318. [1]
Holt G. G. & North D. T. (1970). Effects of gamma irradiation on the mechanisms of sperm transfer in *Trichoplusia ni. J. Insect Physiol.* **16**: 2211-2222. [6]
Honda K. (1980). Odor of a papilionid butterfly. *J. Chem. Ecol.* **6**: 867-869. [12]
Honda K. (1983). Defensive potential of components of the larval osmeterial secretion of papilionid butterflies against ants. *Physiol. Ent.* **8**: 173-179. [12]
Hooper D. (1982). The spoils of St Helens. *Pacif. NW. Q.* **16**: 36-46. [31]
Hooper M. D. (1971). The size and surroundings of nature reserves. Pp.555-561 in Duffey & Watt (1971). [33]
Hooper P. T. (1978*a*). Pyrrolizidine alkaloid poisoning-pathology with particular reference to differences in animal and plant species. Pp.161-176 in Keeler *et al.* (1978). [12]
Hooper P. T. (1978*b*). Cycad poisoning in Australia. Pp.337-347 in Keeler *et al.* (1978). [12]
Hopkins A. D. (1917). A discussion of C. G. Hewett's paper on insect behaviour. *J. econ. Ent.* **10**: 92-93. [6]
Hopkins G. H. E. (1927). *Insects of Samoa.* Part III. *Lepidoptera* **(1)**: 64pp. London. [1]
Horridge G. A. (ed.) (1975). *The compound eye and the vision of insects.* Oxford. [20]
Hovanitz W. (1948). Differences in the field activity of two color phases of *Colias* butterflies at various times of day. *Contr. Lab. Vertebr. Biol. Univ. Mich.* **41**: 1-37. [21]
Hovanitz W. (1963). The relation of *Pieris virginiensis* to *P. napi.* Species formation in *Pieris*? *J. Res. Lepid.* **1**: 124-134. [8]
Hovanitz W. (1969). Inherited and/or conditioned changes in host plant preference in *Pieris. Entomologia exp. appl.* **12**: 729-735. [6]
Hovanitz W. & Chang V. C. S. (1963). Ovipositional preference tests with *Pieris. J. Res. Lepid.* **2**: 185-200. [6]
Hovanitz W. & Chang V. C. S. (1964). Adult oviposition responses in *Pieris rapae. J. Res. Lepid.* **3**: 159-172. [6]
Howard L. O. & Fiske W. F. (1911). The importation into the United States of the parasites of the gypsy moth and the brown-tail moth. *Bull. U.S. Bur. Ent.* **(91)**: 344pp. [6]
Howard R. W., McDaniel C. A. & Blomquist G. J. (1980). Chemical mimicry as an integrating mechanism: cuticular hydrocarbons of a termitophile and its host. *Science* **210**: 431-433. [12]
Howarth T. G. (1973*a*). The conservation of the large blue butterfly (*Maculinea arion*) in West Devon and Cornwall. *Proc. Trans. Br. ent. nat. Hist. Soc.* **5**: 121-126. [33]
Howarth T. G. (1973*b*). *South's British butterflies.* London. [1, 33]
Howe W. H. (ed.) (1975). *The butterflies of North America.* Garden City, New York. [1, 6]
Hoy M. A. (1977). Rapid responses to selection for a non-diapausing gypsy moth. *Science* **196**: 1462-1463. [16]
Hsiao C., Hsiao T. H. & Rothschild M. (1980). Characterization of a protein toxin from dried specimens of the garden tiger moth (*Arctia caja*). *Toxicon* **18**: 291-299. [13]
Hudson A. & Lefkovitch L. P. (1980). Two species of the *Amathes c-nigrum* complex distinguished by isozymes of adenylate kinase and by selected morphological characters. *Proc. ent. Soc. Wash.* **82**: 587-598. [17]
Huettel M. D. & Bush G. L. (1972). The genetics of host selection and its bearing on sympatric speciation in *Procecidochares* (Diptera: Tephritidae). *Entomologia exp. appl.* **15**: 465-480. [19]
Huheey J. E. (1961). Studies in warning coloration and mimicry. III. Evolution of Müllerian mimicry. *Evolution* **15**: 567-568. [12]
Huheey J. E. (1976). Studies . . . VII. Evolutionary consequences of a Batesian-Müllerian spectrum; a model for Müllerian mimicry. *Evolution* **30**: 86-93. [12, 14]
Huheey J. E. (1980*a*). Studies . . . VIII. Further evidence for a frequency-dependent model of predation. *J. Herpet.* **14**: 223-230. [12]
Huheey J. E. (1980*b*). Batesian and Müllerian mimicry: semantic and substantive differences of opinion. *Evolution* **34**: 1212-1215. [14]
Huheey J. E. (In press). Warning colouration and mimicry. In Bell & Cardé (in press) (now published, 1984). [12]
Hulstaert R. P. G. (1931). Lepidoptera Rhopalocera: fam. Danaididae: subfam. Danaidinae and Tellervinae. *Genera Insect.* **(193)**: 215pp. [1]
Humphries D. A. & Driver P. M. (1967). Erratic displays as a device against predators. *Science* **156**: 1767-1768. [12]
Humphries D. A. & Driver P. M. (1970). Protean defence by prey animals. *Oecologia* **5**: 285-302. [12]
Hunt O. D. (1965). Status and conservation of the large blue butterfly *Maculinea arion* L. Pp.35-44 in *The conservation of invertebrates* (eds Duffey E. & Morris M. G.). Monks Wood. [33]
Hutchinson G. E. & Ripley S. D. (1954). Gene dispersal and the ethology of the Rhinocerotidae. *Evolution* **8**: 178-179. [2]
Hutchinson K. J. & King K. L. (1980). The effects of sheep stocking level on invertebrate abundance, biomass and energy utilisation in a temperate sown grassland. *J. appl. Ecol.* **17**: 369-387. [33]
Huxley J. (1976). The coloration of *Papilio zalmoxis* and *P. antimachus* and the discovery of Tyndall blue in butterflies. *Proc. R. Soc.* (B) **193**: 44-53. [20]
Huxley J. S. (1938*a*). The present standing of the theory of sexual selection. Pp.11-42 in *Evolution* (ed. de Beer G. R.). Oxford. [Intro, 20, 21, 25]
Huxley J. S. (1938*b*). Threat and warning colouration in birds, with a general discussion of the biological functions of colour. *Proc. Int. orn. Congr.* **8**: 430-455. [12]
Huxley J. S., Hardy A. C. & Ford E. B. (eds) (1954). *Evolution as a process.* London.
Igarashi S. (1979). *Papilionidae and their early stages.* Tokyo. (In Japanese.) [1]
Illig K. G. (1902). Duftorgane männlicher Schmetterlinge. *Zoologica, Stuttg.* **38**: 1-34. [25]

Ilse D. (1928). Über den Farbensinn der Tagfalter. *Z. vergl. Physiol.* **8**: 658–692. [20]

Ilse D. (1937). New observations on responses to colours in egg-laying butterflies. *Nature* **140**: 544–545. [6, 20]

Ilse D. (1956). Behaviour of butterflies before ovipositing. *J. Bombay nat. Hist. Soc.* **53**: 486–488. [6]

Ilse D. & Vaidya V. (1956). Spontaneous feeding responses to colours in *Papilio demoleus*. *Proc. Indian. Acad. Sci.* **43**B: 23–31. [20]

Imms A. D. (1951). *Insect natural history* (3rd edn). London. [12]

Iriki S. (1941). The two sperm types in the silkworm and their functions. *Zool. Mag.* **53**: 123–124. [6]

Ishii M. & Hidaka T. (1979). Seasonal polymorphism of the adult rice-plant skipper, *Parnara guttata guttata* and its control. *Appl. Ent. Zool. Tokyo* **14**: 173-184. [16]

Itô Y., Miyashita K. & Gotoh A. (1960). Natural mortality of the common cabbage butterfly *Pieris rapae crucivora*, with considerations of the factors affecting it. *Jap. J. appl. Ent. Zool.* **4**: 1-10. [10]

Ives P. M. (1978). How discriminating are cabbage butterflies? *Aust. J. Ecol.* **3**: 261–276. [2, 6]

Iwase T. (1954). Synopsis of the known life-histories of Japanese butterflies. *Lepid. News* **8**: 95–100. [1]

Iwase T. (1964). Recent foodplant records of the Loochooan butterflies. *J. Lepid. Soc.* **18**: 105–109. [1]

Jacobs G. H. (1981). *Comparative color vision*. New York. [12]

Jacobson M., Adler V. E., Kishaba A. N. & Priesner E. (1976). 2-phenylethanol, a presumed sexual stimulant produced by the male cabbage looper moth, *Trichoplusia ni*. *Experientia* **32**: 964–966. [25]

Jaenike J. (1978). On optimal oviposition behavior in phytophagous insects. *Theoret. Pop. Biol.* **14**: 350–356. [6]

Jaenike J. (1980). A relativistic measure of variation in preference. *Ecology* **61**: 990–991. [6]

Jaenike J. (1981). Criteria for ascertaining the existence of host races. *Am. Nat.* **117**: 830–834. [19]

Jaenike J. (1982). Environmental modification of oviposition behavior in *Drosophila*. *Am. Nat.* **119**: 784–802. [6]

James L. F. (1978). Overview of poisonous plant problems in the United States. Pp.3-5 in Keeler *et al.* (1978). [12]

Janse A. J. T. (1932). *The moths of South Africa*. Durban. [25]

Janzen D. H. (1971). Seed predation by animals. *A. Rev. Ecol. Syst.* **2**: 465–492. [12]

Janzen D. H. (1978). The ecology and evolutionary biology of seed chemistry as relates to seed predation. Pp.163–206 in Harborne (1978). [12]

Janzen D. H. (1980). When is it coevolution? *Evolution* **34**: 611–612. [Intro.]

Järvi T., Sillen-Tullberg B. & Wiklund C. (1981*a*). The cost of being aposematic. An experimental study of the predation on larvae of *Papilio machaon* by the great tit *Parus major*. *Oikos* **36**: 267–272. [12, 14]

Järvi T., Sillen-Tullberg B. & Wiklund C. (1981*b*). Individual versus kin selection for aposematic coloration: a reply to Harvey & Paxton. *Oikos* **37**: 393–395. [14]

Jarvis F. V. L. (1966). The genus *Aricia* in Britain. *Proc. S. Lond. ent. nat. Hist. Soc.* **1966**: 37–60. [16]

Jeffords M. R., Sternburg J. G. & Waldbauer G. P. (1979). Batesian mimicry: field demonstration of the survival value of pipevine swallowtail and monarch colour patterns. *Evolution* **33**: 275–286. [22]

Jelnes J. E. (1975). Electrophoretic studies on two sibling species, *Thera variata* and *T. obeliscata*, with special reference to phosphoglucomutase and phosphoglucose isomerase. *Hereditas* **79**: 67–72. [17]

Jennersten O. (1980). Nectar source plant selection and distribution pattern in an autumn population of *Gonepteryx rhamni*. *Ent. Tidskr.* **101**: 109–114. [2]

Jermy T., Hanson F. E. & Dethier V. G. (1968). Induction of specific food preference in lepidopterous larvae. *Entomologica exp. appl.* **11**: 211–230. [7]

Johansen C., Eves J. D., Mayer D. F., Bach J. C., Nedrow M. E. & Kious C. W. (1981). Effects of ash from Mt St Helens on bees. *Melanderia* **37**: 20–29. [31]

Johansson A. S. (1951). The food plant preference of the larvae of *Pieris brassicae*. *Norsk ent. Tidskr.* **8**: 187–195. [6]

Johansson A. S. (1954). Diapause and pupal morphology and colour in *Pieris brassicae*. *Norsk ent. Tidskr.* **9**: 79–86. [30]

Johki Y. & Hidaka T. (1979). Function of the "warning coloration" in larvae of a diurnal moth, *Pryeria sinica*. *Appl. Ent. Zool.* **14**: 164–172. [12]

Johnson C. G. (1969). *Migration and dispersal of insects by flight*. London. [26]

Johnston G. & Johnston B. (1980). *This is Hong Kong butterflies*. Hong Kong. [1, 6]

Joicey J. J. & Talbot G. (1924–32). A catalogue of the Lepidoptera of Hainan. *Bull. Hill Mus. Witley* **1**: 514–538; **2**: 3–27, 183–191; **3**: 151–162; **4**: 257–262. [1]

Jones D. A., Keymer R. J. & Ellis W. M. (1978). Cyanogenesis in plants and animal feeding. Pp.21–34 in Harborne (1978). [12]

Jones D. L. & Miller J. H. (1959). Pathology of the dermatitis produced by the urticating caterpillar *Automeris io*. *Archs Derm.* **79**: 81–85. [12]

Jones D. S. & Macfadden B. J. (1982). Induced magnetization in the monarch butterfly, *Danaus plexippus*. *J. exp. Biol.* **96**: 1–9. [26]

Jones F. M. (1932). Insect coloration and the relative acceptability of insects to birds. *Trans. ent. Soc. Lond.* **80**: 345–385. [12, 20]

Jones F. M. (1934). Further experiments on colouration and relative acceptability of insects to birds. *Trans. R. ent. Soc. Lond.* **82**: 443–453. [12]

Jones F. M. (1937). Relative acceptability and poisonous foodplants. *Proc. R. ent. Soc. Lond.* **12**: 74–76. [12]

Jones R. E. (1977*a*). Search behaviour: a study of three caterpillar species. *Behaviour* **60**: 237–259. [2, 7]

Jones R. E. (1977*b*). Movement patterns and egg distribution in cabbage butterflies. *J. anim. Ecol.* **46**: 195–212. [2, 6]

Jones R. E. & Ives P. M. (1979). The adaptiveness of searching and host selection behaviour in *Pieris rapae*. *Aust. J. Ecol.* **4**: 75–86. [2, 6]

Jones R. E., Gilbert N., Guppy M. & Nealis V. (1980). Long-distance movement of *Pieris rapae*. *J. anim. Ecol.* **49**: 629–642. [2, 26]

Jones R. E., Hart J. R. & Bull G. D. (1982). Temperature, size, and egg production in the cabbage butterfly *Pieris rapae*. *Aust. J. Zool.* **30**: 223–232. [6]

Jones R. L., Burton R. L., Bowman M. C. & Beroza M. (1970). Chemical inducers of oviposition for the corn earworm, *Heliothis zea*. *Science* **186**: 856–857. [6]

Jordan C. T. (1981). *Population biology and host plant ecology of caper-feeding pierid butterflies in northeastern Mexico*. PhD thesis, Texas Univ., Austin, USA. [3, 6]

Joy N. H. (1902). (no title). *Proc. ent. Soc. Lond.* **1902**: xl–xli. [20]

Kafatos F. C. (1981). Structure, evolution and developmental expression of the silkmoth chorion multigene families. *Am. Zool.* **21**: 707–714. [6]

Kaiser W. (1975). The relationship between visual movement detection and colour vision in insects. Pp.359–377 in Horridge (1975). [20]

Kamil A. C. & Sargent T. D. (eds) (1981). *Foraging behavior: ecological, ethological and psychological approaches*. New York.

Kammer A. E. & Bracchi J. (1973). Role of the wings in the absorption of radiant energy by a butterfly. *Comp. Biochem. Physiol.* **45**A: 1057–1063. [28]

Kane S. (1982). Notes on the acoustic signals of a Neotropical satyrid butterfly. *J. Lepid. Soc.* **36**: 200–206. [Intro.]

Kanz J. E. (1977). The orientation of migrant and non-migrant butterflies, *Danaus plexippus*. *Psyche* **84**: 120–141. [26]

Karlson P. & Lüscher M. (1959). "Pheromone", ein Nomenklaturvorschlag für eine Wirkstoffklasse. *Naturwissenschaften* **46**: 63–64. (See also in translation, *Nature* **183**: 55–56.) [25]

Katsuno S. (1977). Studies on eupyrene and apyrene spermatozoa in the silkworm *Bombyx mori*. IV. The behavior of the spermatozoa in the internal reproductive organs of female moths. *Appl. Ent. Zool.* **12**: 352–359. [6]

Kawazoé A. & Wakabayashi M. (1977). *Coloured illustrations of the butterflies of Japan* (2nd edn). Osaka. (In Japanese.) [1]

Kear J. (1968). Plant poisons in the diet of wild birds. *Bull. Br. Orn. Club* **88**: 98–102. [12]

Keeler R. F. & Binns W. (1968). Teratogenic compounds of *Veratrum californicum*. V. Comparison of cyclopian effects of steroidal alkaloids from the plant and structurally related compounds from other sources. *Teratology* **1**: 5–10. [12]

Keeler R. F., van Kempen K. R. & James L. F. (eds) (1978). *Effects of poisonous plants on livestock*. New York.

Keller E. C., Mattoni R. H. T. & Seiger M. S. B. (1966). Preferential return of artificially displaced butterflies. *Anim. Behav.* **14**: 197–200. [2]

Kempthorne O. (1977). Status of quantitative genetic theory. Pp.719–760 in *Proc. Int. Conf. Quant. Genet.* **1976** (eds Pollak E., Kempthorne O. & Bailey T. B.). Ames, Iowa. [27]

Kempthorne O. & Tandon O. B. (1953). The estimation of heritability by regression of offspring on parent. *Biometrics* **9**: 90–100. [16]

Kendall R. O. (1976). Larval foodplants and life history notes for some metalmarks from Mexico and Texas. *Bull. Allyn Mus.* (**32**): 12pp. [6]

Kendall R. O. & Kendall C. A. (1971). Lepidoptera in the unpublished field notes of Howard George Lacey, naturalist (1856–1929). *J. Lepid. Soc.* **25**: 29–44. [2]

Kettlewell H. B. D. (1955). Selection experiments on industrial melanism in the Lepidoptera. *Heredity* **9**: 323–342. [16]

Kettlewell H. B. D. (1956). Further selection experiments on industrial melanism in the Lepidoptera. *Heredity* **10**: 287–301. [16]

Kettlewell H. B. D. (1959). Brazilian insect adaptations. *Endeavour* **18**: 200–210. [12]

Kettlewell H. B. D. (1965). Insect survival and selection for pattern. *Science* **148**: 1290–1296. [20]

Kettlewell H. B. D. (1973). *The evolution of melanism*. Oxford. [25]

Kevan P. G. (1978). Floral coloration, its colorimetric analysis and significance in anthecology. Pp.51–78 in *The pollination of flowers by insects* (ed. Richards A. J.). London. [28]

Kevan P. G. & Shorthouse J. D. (1970). Behavioural thermoregulation by high arctic butterflies. *Arctic* **23**: 268–279. [28]

Kim C-W. (1976). *Distribution atlas of insects of Korea* (1). *Rhopalocera*. Seoul. [1]

King C. E. (1971). Resource specialization and equilibrium population size in patchy environments. *Proc. natn. Acad. Sci. U.S.A.* **68**: 2634–2637. [6]

Kingsbury J. M. (1964). *Poisonous plants of the U.S. and Canada*. N.J., USA. [12]

Kingsbury J. M. (1975). Phytotoxicology. Pp.591–603 in Casarett & Doull (1975). [12]

Kingsbury J. M. (1978). Ecology of poisoning. Pp.81–91 in Keeler *et al.* (1978). [12]

Kingsolver J. G. (1983*a*). Thermoregulation and flight in *Colias* butterflies: elevational patterns and mechanistic limitations. *Ecology* **64**: 534–545. [3]

Kingsolver J. G. (1983*b*). Ecological significance of flight activity in *Colias* butterflies: implications for reproductive strategy and population structure. *Ecology* **64**: 546–551. [3]

Kirby W. F. (1871). *A synonymic catalogue of diurnal Lepidoptera*. London. [1]

Kirby W. F. (1877). *A synonymic catalogue of diurnal Lepidoptera. Supplement, March 1871–June 1877*. London. [1]

Kirchberg E. (1942). Genitalmorphologie und natürliche Verwandtschaft der Amathusiinae und ihre Beziehungen zur geographischen Verbreitung der Subfamilie. *Mitt. münch. ent. Ges.* **32**: 44–87. [1]

Kiritani K. & Dempster J. (1973). Different approaches to the quantitative evaluation of natural enemies. *J. appl. Ecol.* **10**: 323–330. [10]

Kirton L. G., Tan M.-W. & Kirton C. G. (1982). The life histories of *Euploea crameri bremeri* and *Idea hypermnestra linteata*. *Malay. Nat. J.* **36**: 29–43. [9]

Kitching I. J. (1983). *An analysis of danaine classifications using different character sets and methods*. PhD thesis, Univ. London, UK. [Intro, 17]

Kitching R. L. (1981). Egg clustering and the southern hemisphere lycaenids: comments on a paper by N. E. Stamp. *Am. Nat.* **118**: 423–425. [6, 12]

Klijnstra J. W. (1982). Perception of the oviposition deterrent pheromone in *Pieris brassicae*. *Proc. int. Symp. Insect-Plant Relationships* **(5)**: 145–151. [6]

Klopfer P. H. (1957). An experiment on empathic learning in ducks. *Am. Nat.* **91**: 61–63. [12]

Kloppers J. J. (1976). Butterflies—a word of warning. *Fauna Flora Pretoria* **34**: 8–9. [2]

Klots A. B. (1933). A generic revision of the Pieridae together with a study of the male genitalia. *Entomologica am.* **12**: 139–242. [1]

Klots A. B. (1951). *A field guide to the butterflies*. Boston, Mass. [29]

Kobayashi S. (1965). Influence of parental density on the distribution pattern of eggs in the common cabbage butterfly *Pieris rapae crucivora*. *Researches Popul. Ecol.* **7**: 109–117. [6]

Kobayashi S. (1966). Process generating the distribution pattern of the common cabbage butterfly *Pieris rapae crucivora*. *Researches Popul. Ecol.* **8**: 51–61. [6]

Kobayashi S. & Takano H. (1978). Survival rate and dispersal of adults of *Pieris rapae crucivora*. *Jap. J. appl. Ent. Zool.* **22**: 250–254. [2]

Köhler F. (1900). Die Duftschuppen der Gattung *Lycaena*, auf ihre Phylogenie hin untersucht. *Zool. Jb. (Syst.)* **13**: 105–124. [25]

Köhler W. & Feldotto W. (1935). Experimentelle untersuchungen über die modifikabilität der flügelzeichnung, ihrer systeme und elemente in den sensiblen perioden von *Vanessa urticae*, nebst einigen beobachtungen an *Vanessa io*. *Arch. Julius Klaus-Stift. Vererb-Forsch.* **10**: 315–453. [16]

Kojima K. (1971). Is there a constant fitness value for a given genotype? No! *Evolution* **25**: 281–285. [16]

Kono Y. (1970). Photoperiodic induction of diapause in *Pieris rapae crucivora*. *Appl. Ent. Zool., Tokyo* **5**: 213–224. [30]

Kono Y. (1973). Difference of cuticular surface between diapause and non-diapause pupae of *Pieris rapae crucivora*. *Appl. Ent. Zool.* **8**: 50–52. [30]

Korshunov Y. P. (1972). A catalogue of the Rhopalocera in the fauna of the U.S.S.R. *Ent. Obozr.* **51**: 136–154, 352–368. (In Russian; translation in *Ent. Rev. Wash.* **51**: 83–98, 212–223.) [1]

Kramer E. (1978). Insect pheromones. Pp.205–229 in *Taxis and behaviour: receptors and recognition* (B) **5** (ed. Hazelbauer G. L.). London. [25]

Krebs J. R. & Davies N. B. (eds) (1978). *Behavioural ecology: an evolutionary approach*. Oxford.

Kristensen N. P. (1976). Remarks on the family-level phylogeny of butterflies. *Z. Zool. Syst. EvolForsch.* **14**: 25–33. [1]

Krodel E. (1904). Durch einwirkung niederer temperaturen auf das puppenstadium der *Lycaena*-arten: *coridon* und *damon*. *Allg. Z. Ent.* **9**: 49–55, 103–109, 134–136. [16]

Kudrna O. (1977). *A revision of the genus* Hipparchia. Faringdon, Oxon. [1]

Kudrna O. (ed.) (in prep.). Butterflies of Europe. [1]

Kühn A. (1926). Über die änderung des Zeichnungsmusters von Schmetterlingen durch Temperaturreize und das Grundschema der Nymphalidenzeichnung. *Nachr. Ges. Wiss. Göttingen Math. Physik. Kl.* **1926**: 120–141. [16]

Kühn A. & Ilse D. (1925). Die Anlockung von Tagfaltern durch Pigmentfarben. *Biol. Zbl.* **45**: 144–149. [20]

Kurentsov A. I. (1970). *The butterflies of the far east U.S.S.R.* Leningrad. (In Russian). [1]

Kurihara K. (1971). Taste modifiers. Pp.363–378 in Beidler (1971). [12]

Kurihara K. & Beidler L. M. (1968). Taste-modifying protein from miracle fruit. *Science* **161**: 1241–1243. [12]

Kuwahara Y. (1979). Scent scale substances of male *Pieris melete*. *Appl. Ent. Zool. Tokyo* **14**: 350–355. [25]

Kuwayama S. (1929). Eruption of Mt Komagatake and insects. *Kontyû* **3**: 271–273. [31]

Labine P. A. (1964). The population biology of the butterfly, *Euphydryas editha*. I. Barriers to multiple inseminations. *Evolution* **18**: 335–336. [6]

Labine P. A. (1968). The population . . . VIII. Oviposition and its relation to patterns of oviposition in other butterflies. *Evolution* **22**: 799–805. [2, 6]

Lack D. (1968). *Ecological adaptations for breeding in birds*. London. [21]

Lamas G. (1974). Supuesta extinción de una mariposa en Lima, Perú. *Rev. Peru Ent.* **17**: 119–120. [2]

Lamas G. (1977). Bibliografia de catalogos y lista regionales de mariposas de America Latina. Publ. spec. Soc. mex. Lepid. **(2)**: 44pp. (Adiciones. . . . *Boln inf. Soc. mex. Lepid.* **4**(5): 8–14, 1978.) [1]

Lamas G. (1978). Mariposas y conservació de la Naturaleza en el Perú. *Bol. Colonia Suiza Perú* **1978**: 61–66. [2]

Lamas G. (1979). Los Dismorphiinae de Mexico, America Central y las Antillas. *Revta Soc. Mex. Lepid.* **5**: 3–37. [1]

Lamborn W. A. (1914). On the relationship between certain West African insects, especially ants, Lycaenidae and Homoptera [with an additional multi-author appendix]. *Trans. ent. Soc. Lond.* **1913**: 436–524. [1]

Lamborn W. A. (1921). An oriental danaine butterfly brushing the brands on its hindwing. *Proc. ent. Soc. Lond.* **1921**: xcv. [25]

Lamborn W. A. & Poulton E. B. (1913). *Amauris egialea* stroking the brands of the hindwings with its anal tufts again observed by W. A. Lamborn. *Proc. ent. Soc. Lond.* **1913**: lxxxiii–iv. [25]

Lamborn W. A. & Poulton E. B. (1918). The relation of the anal tufts to the brands of the hindwings observed and the scent perceived in a male danaine butterfly by W. A. Lamborn. *Proc. ent. Soc. Lond.* **1918**: clxxii–iv. [25]

Lamborn W. A., Longstaff G. B. & Poulton E. B. (1911). Instances of mimicry, protective resemblance, etc., from the Lagos district. *Proc. ent. Soc. Lond.* **1911**: xlvi–vii. [25]

Lamborn W. A., Dixey F. A. & Poulton E. B. (1912). *Amauris egialea* stroking the brands of the hindwings with its anal tufts. *Proc. ent. Soc. Lond.* **1912**: xxxiv–vii. [25]

Lande R. (1976). Natural selection and random genetic drift in phenotypic evolution. *Evolution* **30**: 314–334. [19]

Lande R. (1982). The minimum number of genes contributing to quantitative variation between and within populations. *Genetics* **99**: 541–553. [27]

Lane C. (1957). Preliminary note on insects eaten and rejected by a tame shama (*Kittacincla malabarica*) with the suggestion that in certain species of butterflies and moths females are less palatable than males. *Entomologist's mon. Mag.* **93**: 172-179. [12, 16]
Lane R. P. (1977). Ectoparasitic adaptations in *Forcipomyia* from butterflies, with two new African species. *Syst. Ent.* **2**: 305-312. [11]
Langer H., Hamann B. & Meinecke C. C. (1979). Tetrachromatic visual system in the moth *Spodoptera exempta*. Preliminary note. *J. comp. Physiol.* **129**: 235-239. [20]
Langer T. W. (1958). *Nordens Dagsommerfugle*. Copenhagen. [1]
Larsen T. B. (1974). *Butterflies of Lebanon*. Beirut. [1]
Larsen T. B. (1982*a*). The butterflies of the Yemen Arab Republic. *Biol. Skr.* **23**(3): 1-87. [1]
Larsen T. B. (1982*b*). False head butterflies: the case of *Oxylides faunas*. *J. Lepid. Soc.* **36**: 238-239. [16]
Larsen T. B. & Larsen K. (1980). *Butterflies of Oman*. Edinburgh. [1]
Latter O. H. & Eltringham H. (1935). The epigamic behaviour of *Euploea (Crastia) core asela* with a description of the structure of the scent organs. *Proc. R. Soc.* (b) **117**: 470-482. [25]
Lauterbach C. (1914). Die Aristolochiaceen papuasiens. *Engler. Bot. Jb.* **52**: 104-107. [32]
Lawrence D. B. (1938). Continued studies on flora of Mt St Helens. *Mazama* **20**: [31]
Lawton J. H. & Strong D. R. (1981). Community patterns and competition in folivorous insects. *Am. Nat.* **118**: 317-338. [3]
Laycock W. A. (1978). Coevolution of poisonous plants and large herbivores on rangeland. *J. Range Mgmt* **31**: 335-342. [12]
Leclercq J. & Gaspar G. (1971). Atlas provisoires, hors-series. Gembloux. [33]
Lederer G. (1960). Verhaltenswiesen der Imagines und der Entwicklungsstadien von *Limenitis camilla camilla*. *Z. Tierpsychol.* **17**: 521-546. [20]
Lederhouse R. C. (1982). Factors affecting equal catchability in the swallowtail butterflies, *Papilio polyxenes* and *P. glaucus*. *Ecol. Ent.* **7**: 379-383. [Intro, 2]
Lederhouse R. C., Finke M. D. & Scriber J. M. (1982). The contributions of larval growth and pupal duration to protandry in the black swallowtail butterfly, *Papilio polyxenes*. *Oecologia* **53**: 296-301. [7]
Lee S-M. (1982). *Butterflies of Korea*. Seoul. [1]
Leech J. H. (1892-4). *Butterflies from China, Japan and Corea*. London. [1]
Lees E. (1962*a*). On the voltinism of *Coenonympha pamphilus*. *Entomologist* **95**: 5-6. [16]
Lees E. (1962*b*). Factors determining the distribution of the speckled wood butterfly (*Pararge aegeria*) in Great Britain. *Entomologist's Gaz.* **13**: 101-103. [2]
Lees E. (1965). Further observations on the voltinism of *Coenonympha pamphilus*. *Entomologist* **98**: 43-45. [16]
Leffler J. W., Leffler L. T. & Hall J. S. (1979). Effects of familiar area on the homing ability of the little brown bat, *Myotis lucifugus*. *J. Mammal.* **60**: 201-204. [26]
Le Gare M. J. & Hovanitz W. (1951). Genetic and ecologic analyses of wild populations in Lepidoptera. II. Color pattern variation in *Melitaea chalcedona*. *Wasmann J. Biol.* **9**: 257-311. [16]
Le Moult E. (1950). Revision de la classification des Apaturinae de l'ancien monde suivie d'une monographie de plusiers genres. *Miscnea ent.* (Suppl.): 68pp. [1]
Le Moult E. & Real P. (1962). Les *Morpho* d'Amerique du sud et centrale: historique, morphologie, systématique. *Novit. ent.* (Suppl.): xiv + 296pp., 92 pls. [1]
Lenz F. (1929). Massenauftneten von *Melitaea aurelia var britomartis*. *Int. ent. Z.* **23**: 149-150. [2]
Leon M. (1974). Maternal pheromone. *Physiol. Behav.* **13**: 441-453. [12]
Leon M. (1979). Development of olfactory attraction by young Norway rats. Pp.193-209 in Müller-Schwarze & Silverstein (1979). [12]
Leslie D. H. (1958). Statistical appendix. *J. anim. Ecol.* **27**: 84-86. [16]
Levin D. A. (1976). The chemical defenses of plants to pathogens and herbivores. *A. Rev. Ecol. Syst.* **7**: 121-159. [12]
Levin D. A. & Berube D. E. (1972). *Phlox* and *Colias*: the efficiency of a pollination system. *Evolution* **26**: 242-250. [6]
Levin D. A. & Wilson A. C. (1976). Rates of evolution in seed plants: net increase in diversity of chromosome numbers and species numbers through time. *Proc. natn. Acad. Sci. U.S.A.* **73**: 2086-2090. [19]
Levin M. P. (1973). Preferential mating and the maintenance of the sex-limited dimorphism in *Papilio glaucus*—evidence from laboratory matings. *Evolution* **27**: 257-264. [20, 22]
Levins R. (1968). *Evolution in changing environments*. Princeton, N.J. [27]
Levins R. & MacArthur R. H. (1969). An hypothesis to explain the incidence of monophagy. *Ecology* **50**: 910-911. [6]
Lewis B. H. (1962). On the analysis of interaction in multi-dimensional contingency tables. *Appl. Statist.* **125**: 88-117. [21]
Lewis H. L. (1974). *Butterflies of the world*. London. [1]
Lewontin R. C. (1964). The interaction of selection and linkage. I. General considerations; heterotic models. *Genetics* **49**: 49-67. [21]
Lewontin R. C. (1965). Selection for colonizing ability. Pp.79-94 in *The genetics of colonizing species* (eds Baker H. G. & Stebbins G. L.). New York. [2]
Leyrer R. L. & Monroe R. E. (1973). Isolation and identification of the scent of the moth, *Galleria mellonella*, and a revaluation of its sex pheromone. *J. Insect Physiol.* **19**: 2267-2271. [25]
Lichy R. (1946). Documents pour servir a l'étude des Lepidoptères du Venezuela (3e. note): un cas de parasitisme sur les ailés des Lépidoptères. *Boln. Ent. Venez.* **5**: 1-4. [11]
Lieb C. C. & Mulinos M. G. (1934). Pigeon emesis and drug action. *J. Pharmac. exp. Ther.* **51**: 321-326. [12]
Liepelt von W. (1963). Zur Schutzwirkung des Stachelgiftes von Vienen und Wespen gegenuber Trauer-

fliegenschnapper und Garrtenrotschwanz. *Zool. Jb. (Physiol.)* **70**: 167–176. [12]

Lincoln P. E. (1980). Leaf resin flavonoids of *Diplacus aurantiacus*. *Biochem. Syst. Ecol.* **8**: 397–400. [2]

Lincoln D. E., Newton T. S., Ehrlich P. R. & Williams K. S. (1982). Coevolution of the checkerspot butterfly *Euphydryas chalcedona* and its larval food plant *Diplacus aurantiacus*: larval response to protein and leaf resin. *Oecologia* **52**: 216–223. [2, 7]

Lindsey A. W., Bell E. L. & Williams R. C. (1931). The Hesperioidea of North America. *J. scient. Labs Denison Univ.* **26**: 1–142. [1]

Linnaeus C. (1756). Odores medicamentorum. *Amoenitates Academicae* **3**: 183–201. [12]

Lipscomb C. G. & Jackson R. A. (1964). Some considerations of some present day conditions as they affect the continued existence of certain butterflies. *Entomologist's Rec. J. Var.* **76**: 63–68. [33]

Lloyd J. E. (1979). Mating behavior and natural selection. *Fla. Ent.* **62**: 17–23. [25]

Locke M. & Smith D. S. (eds) (1980). *Insect biology in the future*. New York.

LoLordo V. M. (1979). Constraints on learning. Pp.473–504 in *Animal learning: survey and analysis* (eds Bitterman M. E., LoLordo V. M., Overmier J. B. & Rashotte M. E.). New York. [14]

Lomnicki A. (1980). Regulation of population density due to individual differences and patchy environment. *Oikos* **35**: 185–193. [3]

Long D. B. (1953). Effects of population density on larvae of Lepidoptera. *Trans. R. ent. Soc. Lond.* **104**: 543–585. [6]

Long D. B. (1955). Observations on subsocial behaviour in two species of lepidopterous larvae, *Pieris brassicae* and *Plusia gamma*. *Trans. R. ent. Soc. Lond.* **106**: 421–437. [6]

Longstaff G. B. (1912). *Butterfly hunting in many lands*. London. [25]

Loop M. S. & Scoville S. A. (1972). Response of newborn *Eumeces inexpectatus* to prey-object extracts. *Herpetologica* **28**: 254–256. [12]

Lopez A. & Quesnel V. C. (1970). Defensive secretions of some papilionid caterpillars. *Caribb. J. Sci.* **10**: 5–7. [12]

Lorimer N. (1979). Patterns of variation in some quantitative characters of *Malacosoma disstria*. *Ann. ent. Soc. Am.* **72**: 275–280. [16]

Lorković Z. (1938). Studien über den speziesbegriff. II. Artberechtigung von *Everes argiades*, *E. alcetas* und *E. decolorata*. *Mitt. münch. ent. Ges.* **28**: 215–246. [16]

Lorković Z. (1943). Modifikationen und Rassen von *Everes argiades* und ihre Beziehungen zu den klimatischen Faktoren ihrer Verbreitungsgebiete. *Mitt. münch. ent. Ges.* **33**: 431–478. [16]

Lorković Z. (1957). Speziationsstuffen in der *Erebia tyndarus* Gruppe. *Biol. Glasn.* **10**: 61–110. [2]

Lucas A. M. (1969). Clinal variation in pattern and colour in coastal populations of the butterfly *Tisiphone abeona*. *Aust. J. Zool.* **17**: 37–48. [16]

Luckens C. J. (1980). The heath fritillary *Mellicta athalia* in Britain. *Entomologist's Rec. J. Var.* **93**: 229–234. [33]

Lundgren L. (1975). Natural plant chemicals acting as oviposition deterrents on cabbage butterflies (*Pieris brassicae*, *P. rapae* and *P. napi*). *Zoologica Scr.* **4**: 253–258. [6]

Lundgren L. & Bergström G. (1975). Wing scents and scent-released phases in the courtship behavior of *Lycaeides argyrognomon*. *J. Chem. Ecol.* **1**: 399–412. [12, 20, 25]

Lusis J. J. (1961). On the biological meaning of colour polymorphism of ladybeetle, *Adalia bipunctata*. *Latv. Ent.* **4**: 3–29. [21]

Lutz F. E. (1924). Apparently non-selective characters and combinations of characters, including a study of ultraviolet in relation to the flower-visiting habits of insects. *Ann. N.Y. Acad. Sci.* **29**: 181–283. [20]

Lutz F. E. (1933). "Invisible" colors of flowers and butterflies. *Nat. Hist., N.Y.* **33**; 565-576. [22]

Ma W. C. & Schoonhoven L. M. (1973). Tarsal chemosensory hairs of the large white butterfly *Pieris brassicae* and their possible role in oviposition behaviour. *Entomologia exp. appl.* **16**: 343–357. [6]

Mabille P. (1912). Hesperiidae: subfam. Pyrrhopyginae. *Lepid. Cat.* **21**(9): 1–18. [1]

MacArthur R. H. (1972). *Geographical ecology*. New York. [3]

MacArthur R. H. & Pianka E. R. (1966). On optimal use of a patchy environment. *Am. Nat.* **100**: 603–609. [8]

MacArthur R. H. & Wilson E. O. (1967). *The theory of island biogeography*. Princeton. [3, 16]

Macfie J. W. S. (1934). Fauna Sumatrensis: bijdrage 75, Ceratopogonidae. *Tijdschr. Ent.* **77**: 202–231. [11]

Macfie J. W. S. (1935). A new ceratopogonid from British Guiana. *Stylops* **4**: 265. [11]

Mackay D. A. (1982). *Searching behavior and host plant selection by ovipositing* Euphydryas editha *butterflies*. PhD thesis, Texas Univ., Austin, USA. [7]

Mackay D. A. & Singer M. C. (1982). The basis of an apparent preference for isolated hostplants by ovipositing *Euptychia libye* butterflies. *Ecol. Ent.* **7**: 299–303. [6, 7]

Madsen H. B., Schmidt Nielsen E. & Ødum S. (1980). The Danish scientific expedition to Patagonia and Tierra del Fuego 1978–79. *Geogr. Tidskr.* **80**: 1–28. [27]

Magnus D. B. E. (1950). Beobachtungen zur Balz und Eiablage des Kaisermantels *Argynnis paphia*. *Z. Tierpsychol.* **7**: 435–449. [20, 25]

Magnus D. B. E. (1958*a*). Experimentelle Untersuchungen zur Bionomie und Ethologie des Kaisermantels *Argynnis paphia*. I. Über optische Auslöser von Anfliegereaktionen und ihre Bedeutung für das Sichfinden der Geschlechter. *Z. Tierpsychol.* **15**: 397–426. [20, 21, 23, 25]

Magnus D. B. E. (1958*b*). Experimental analysis of some "overoptimal" sign-stimuli in the mating-behaviour of the fritillary butterfly *Argynnis paphia*. *Proc. Int. Congr. Ent.* (10)**2**: 405–418. [20, 25]

Magnus D. B. E. (1963). Sex-limited mimicry II—visual selection in the mate choice of butterflies. *Int. Congr. Zool.* (16)**4**: 179–183. [20, 21, 23]

Majerus M. E. N. (1979). The status of *Anthocharis cardamines hibernica*, with special reference to the Isle of Man. *Entomologist's Gaz.* **30**: 245–248. [16]

Makielski S. K. (1972). Polymorphism in *Papilio glaucus*—maintenance of the female ancestral form. *J. Lepid. Soc.* **26**: 109-111. [22]

Malcolm S. & Rothschild M. (1983). A danaid Müllerian mimic, *Euploea core amymone*, lacking cardenolides in the pupal and adult stages. *Biol. J. Linn. Soc.* **19**: 27-33. [13]

Malicky H. (1969). Versuch einer analyse der okologischen beziehungen zwischen lycaeniden und formiciden. *Tijdschr. Ent.* **112**: 213-298. [2, 12, 19]

Malicky H. (1970). New aspects on the association between lycaenid larvae and ants. *J. Lepid. Soc.* **24**: 190-202. [12, 19]

Mallet J. L. B. (1984). Sex roles in the ghost moth *Hepialus humuli* and a review of mating in the Hepailidae. *Zool. J. Linn. Soc.* **80**: 67-82. [25]

Mallet J. L. B. & Jackson D. A. (1980). The ecology and social behaviour of the neotropical butterfly *Heliconius xanthocles* in Colombia. *Zool. J. Linn. Soc.* **70**: 1-13. [6, 12]

Malo R. & Willis E. R. (1961). Life history and biological control of *Caligo eurilochus*, a pest of banana. *J. econ. Ent.* **54**: 530-536. [6]

Manders N. (1911). An investigation into the validity of Müllerian and other forms of mimicry, with special reference to the islands of Bourbon, Mauritius and Ceylon. *Proc. zool. Soc. Lond.* **1911**: 696-749. [12]

Manley W. B. L. & Allcard H. G. (1971). *A field guide to the butterflies and burnets of Spain.* Hampton, UK. [1]

Markin G. P. (1970). Food distribution within laboratory colonies of the argentine ant, *Iridomyrmex humilis. Insectes soc.* **17**: 127-158. [19]

Marsh N. A. & Rothschild M. (1974). Aposematic and cryptic Lepidoptera tested on the mouse. *J. Zool., Lond.* **174**: 89-122. [12, 13]

Marsh N. A. & Whaler B. C. (1980). The effect of honey bee (*Apis mellifera*) venom and two of its constituents, melittin and phospholipase A_2, on the cardiovascular system of the rat. *Toxicon* **18**: 427-435. [13]

Marsh N. A., Clarke C. A., Rothschild M. & Kellet D. N. (1977). *Hypolimnas bolina* a mimic of danaid butterflies, and its model *Euploea core* store cardio-active substances. *Nature* **268**: 726-728. [12]

Marsh N. A., Rothschild M., Scutt A., Evans F. & Tedstone A. (In prep.). A cytotoxin from the pupa of the spurge hawk-moth (*Hyles euphorbiae*) inhibiting the growth of malignant cells *in vivo* and *in vitro*. [13]

Marshall G. A. K. (1908). On diaposematism, with reference to some limitations of the Müllerian hypothesis of mimicry. *Trans. ent. Soc. Lond.* **1908**: 93-142. [14]

Marshall G. A. K. & Poulton E. B. (1902). Five years' observations and experiments (1896-1901) on the bionomics of South African insects, chiefly . . . mimicry and warning colours. *Trans. ent. Soc. Lond.* **1902**: 287-584. [12, 16, 29]

Masaki S. (1955). On the pupal diapause of *Pieris rapae*, with special reference to the effect of temperature on its elimination. *Jap. J. appl. Zool.* **20**: 98-104. [30]

Maschwitz U., Wüst M. & Schurian K. (1975). Bläulingsraupen als Zuckerlieferanten für Ameisen. *Oecologia* **18**: 17-21. [19]

Masetti M. & Scali V. (1972). Ecological adjustments of the reproductive biology in *Maniola jurtina* from Tuscany. *Atti Accad. naz. Lincei Rc.* **53**: 460-470. [16]

Masetti M. & Scali V. (1978). Phosphoglucomutase (PMG) genetic variability and selection in *Maniola jurtina. Atti Ass. genet. ital.* **22**: 207-209. [16]

Mason C. W. (1926-27). Structural colors in insects. *J. phys. Chem. Ithaca* **30**: 383-395; **31**: 321-354. [20]

Mason J. R. & Reidinger R. F. (1981). Effects of social facilitation and observational learning on feeding behavior of the red-winged blackbird (*Agelaius phoeniceus*). *Auk* **98**: 778-784. [12]

Mason J. R. & Reidinger R. F. (1982). Observational learning of food aversions in red-winged blackbirds (*Agelaius phoeniceus*). *Auk* **99**: 548-554. [12]

Mason L. G., Ehrlich P. R. & Emmel T. C. (1967). The population biology of the butterfly, *Euphydryas editha.* V. Character clusters and asymmetry. *Evolution* **21**: 85-91. [16]

Mason L. G., Ehrlich P. R. & Emmel T. C. (1968). The population . . . VI. Phenetics of the Jasper Ridge colony, 1965-66. *Evolution* **22**: 46-54. [16]

Masters J. H. (1968). Collecting Ithomiidae with heliotrope. *J. Lepid. Soc.* **22**: 108-109. [25]

Mather B. (1967). Variation in *Junonia coenia* in Mississippi. *J. Lepid. Soc.* **21**: 59-70. [29]

Mather K. & Jinks J. L. (1971). *Biometrical genetics* (2nd edn). London. [16]

Matile P. (1976). Vacuoles. Pp.189-224 in *Plant biochemistry* (eds Bonner J. & Varner J. E.). New York. [6]

Matthews E. G. (1977). Signal-based frequency-dependent defense strategies and the evolution of mimicry. *Am. Nat.* **111**: 213-222. [12]

Mattocks A. R. (1972). Toxicity and metabolism of *Senecio* alkaloids. Pp.179-199 in Harborne (1972). [12]

Mattocks A. R. (1973). Mechanisms of pyrrolizidine alkaloid toxicity. *Proc. Int. Congr. Pharmac.* (5) **2**: 114-123. [25]

Mattson W. J. (1980). Herbivory in relation to plant nitrogen content. *A. Rev. Ecol. Syst.* **11**: 119-161. [6, 19]

May M. L. (1976). Thermoregulation and adaptation to temperature in dragonflies. *Ecol. Monogr.* **46**: 1-32. [28]

May M. L. (1979). Insect thermoregulation. *A. Rev. Ent.* **24**: 313-349. [28]

May R. M. (1976). Patterns in multi-species communities. Pp.142-162 in *Theoretical ecology: principles and applications* (ed. May R. M.). Oxford. [3]

Mayer A. G. (1900). The mating instincts of moths. *Ann. Mag. nat. Hist.* (7)**5**: 183-190. [25]

Maynard Smith J. (1956). Fertility, mating behaviour and sexual selection in *Drosophila subobscura. J. Genet.* **54**: 261-279. [21]

Maynard Smith J. (1976). Sexual selection and the handicap principle. *J. theor. Biol.* **57**: 239-242. [20]

Maynard Smith J. (1978). *The evolution of sex.* Cambridge, UK. [Intro, 20]

Maynard Smith J. & Parker G. A. (1976). The logic of asymmetric contests. *Anim. Behav.* **24**: 159-175. [20, 26]

Mayr E. (1954). Change of genetic environment and evolution. Pp.157-180 in Huxley *et al.* (1954). [19]

Mayr E. (1963). *Animal species and evolution*. Cambridge, Mass. [6, 19]

Mayr E. (1972). Sexual selection and natural selection. Pp.87–104 in Campbell (1972). [21]

Mazokhin-Porshnyakov G. A. (1957). Reflecting properties of butterfly wings and the role of ultra-violet rays in the vision of insects. *Biophysics* **2**: 352–362 (English translation; original Russian in *Biofizika* **2**: 358–368). [20]

Mazokhin-Porshnyakov G. A. (1969). *Insect vision*. New York. [20, 28]

McCann C. (1952). Aposematic insects and their food plants. *J. Bombay nat. Hist. Soc.* **51**: 752–754. [12]

McDunnough J. H. (1912). Megathymidae. *Lepid. Cat.* **21**(9): 19–22. [1]

McIndoo N. E. (1929). Tropisms and sense organs of Lepidoptera. *Smithson misc. Collns* **81** (10): 59pp. [20]

McKechnie S. W., Ehrlich P. R. & White R. R. (1975). Population genetics of *Euphydryas* butterflies. I. Genetic variation and the neutrality hypothesis. *Genetics* **81**: 571–594. [2, 17]

McKey D. (1974). Adaptive patterns in alkaloid physiology. *Am. Nat.* **108**: 305–320. [12]

McKey D. (1975). The ecology of coevolved seed dispersal systems. Pp.159–191 in Gilbert & Raven (1975). [12]

McLaughlin J. R. (1982). Behavioral effect of a sex pheromone extracted from forewings of male *Plodia interpunctella*. *Envir. Ent.* **11**: 378-380. [25]

McLean E. K. (1970). The toxic actions of pyrrolizidine (*Senecio*) alkaloids. *Pharmac. Rev.* **22**: 429–483. [25]

McLeod L. (1968/1970). Controlled environment experiments with *Precis octavia*. *J. Res. Lepid.* **7**: 1–18; **8**: 53–54. [29]

McLeod L. (1980). *Precis archesia ugandensis*: a new subspecies. *Entomologist's Rec. J. Var.* **92**: 109–113. [29]

McNeill S. & Southwood T. R. E. (1978). The role of nitrogen in the development of insect/plant relationships. Pp.77–98 in Harborne (1978). [6]

McWhirter K. G. (1957). A further analysis of variability in *Maniola jurtina*. *Heredity* **11**: 359–371. [16]

McWhirter K. G. (1965). Intensive natural selection of spot genotypes in stable populations of *Maniola jurtina cassiteridum*. *Heredity* **20**: 160. [16]

McWhirter K. G. (1967). Quantum genetics of human blood-groups and phoneme-preferences. *Heredity* **22**: 162–163. [16]

McWhirter K. G. (1969). Heritability of spot-number in Scillonian strains of the meadow brown butterfly (*Maniola jurtina*). *Heredity* **24**: 314–318. [16]

McWhirter K. G. & Creed E. R. (1971). An analysis of spot placing in the meadow brown butterfly *Maniola jurtina*. Pp.275–289 in Creed (1971). [16]

McWhirter K. G. & Scali V. (1966). Ecological bacteriology of the meadow brown butterfly. *Heredity* **21**: 517–521. [16]

Meek A. S. (1913). *A naturalist in cannibal land*. London. [32]

Meinwald J. & Meinwald Y. C. (1966). Structure and synthesis of the major component in the hairpencil secretion of a male butterfly, *Lycorea ceres ceres*. *J. Am. chem. Soc.* **88**: 1305–1310. [25]

Meinwald J., Meinwald Y. C., Wheeler J. W., Eisner T. & Brower L. P. (1966). Major components in the exocrine secretion of a male butterfly (*Lycorea*). *Science* **151**: 583–585. [12, 25]

Meinwald J., Chalmers A. M., Pliske T. E. & Eisner T. (1968). Pheromones III. Identification of trans, trans-10-hydroxy-3, 7-dimethyl-2, 6-decadienoic acid as a major component in "hairpencil" secretions of the male monarch butterfly. *Tetrahedron Lett.* **1968** (47): 4893–4896. [25]

Meinwald J., Meinwald Y. C. & Mazzochi P. H. (1969). Sex pheromone in the queen butterfly: chemistry. *Science* **164**: 1174–1175. [12, 25]

Meinwald J., Thompson W. R., Eisner T. & Owen D. (1971). Pheromones. VII. African monarch: major components of the hairpencil secretion. *Tetrahedron Lett.* **38**: 3485–3488. [12, 25]

Meinwald J., Boriack C. J., Schneider D., Boppré M., Wood W. F. & Eisner T. (1974). Volatile ketones in the hairpencil secretion of danaid butterflies (*Amauris* and *Danaus*). *Experientia* **30**: 721–722. [12, 21, 25]

Melkert F. (1983). Aantekeningen over de biologie van *Issoria lathonia*. *Ent. Ber., Amst.* **43**: 97–98. [12]

Mellanby K. (1981). *Farming and wildlife*. London. [33]

Menken S. B. J., Webes J. T. & Herrebout W. M. (1980). Allozymes and the population structure of *Yponomeuta cagnagellus*. *Neth. J. Zool.* **30**: 228–242. [17]

Menzel R. (1975). Colour receptors in insects. Pp.121–153 in Horridge (1975). [20]

Meves F. (1903). Ueber oligopyrene und apyrene Spermien und uber ihre Enstehlung, nach Beobachtungen an *Palundina* und *Pygaera*. *Arch. mikrosk. Anat. EntwMech.* **61**: 1–84. [6]

Michelbacher A. E. & Smith R. F. (1943). Some natural factors limiting the abundance of the alfalfa butterfly. *Hilgardia* **15**: 369–397. [10]

Michener C. D. (1942). A generic revision of the Heliconiinae. *Am. Mus. Novit.* **(1197)**: 8pp. [1]

Mielke O. H. H. & Brown K. S. (1979). *Suplemento ao catálogo dos Ithomiidae Americanos de R. Ferreira d'Almeida*. Curitiba, Brazil. [1]

Miles Moss A. (1920). The Papilios of Pará. *Novit. zool.* **26**: 295–319. [1, 6]

Miles Moss A. (1933). Some generalizations on *Adelpha*, a neotropical genus of nymphalid butterflies of the group Limenitidi. *Novit. zool.* **39**: 12–20. [1]

Miles Moss A. (1947). Notes on the Syntomidae of Pará, with special reference to wasp mimicry and fedegoso, *Heliotropium indicum* (Boraginaceae), as an attractant. *Entomologist* **80**: 30–35. [25]

Miles Moss A. (1949). Biological notes of some Hesperiidae of Pará and the Amazon. *Acta zool. lilloana* **7**: 27-29. [1]

Miller J. R. & Strickler K. V. (In press). Plant herbivore relationships: locating and identifying the host. In Bell & Cardé (in press). [6]

Miller J. S. (1976). *The evolution of coloration in the larval and pupal stages of the monarch butterfly*. Hons thesis, Hampshire College, Mass., USA. [12]

Miller L. D. (1968). The higher classification, phylogeny and zoogeography of the Satyridae. *Mem. Am. ent. Soc.* **(24)**: 174pp. [1]

Miller L. D. & Martin Brown F. (1981). A catalogue/checklist of the butterflies of America north of Mexico. *Lepid. Soc. Mem.* **(2)**: vii + 280pp. [1]

Miller M. A., Drakontides A. B. & Leavell L. C. (1977). *Kimber-Gray-Stackpole's anatomy and physiology* (17th edn). New York. [12]

Miller W. H. & Bernard G. D. (1968). Butterfly glow. *J. Ultrastruct. Res.* **24**: 286–294. [20]

Millicent E. & Thoday J. M. (1961). Effects of disruptive selection. IV. Gene flow and divergence. *Heredity* **16**: 199–217. [16]

Minnich D. E. (1924). The olfactory sense of the cabbage butterfly, *Pieris rapae*, an experimental study. *J. exp. Zool.* **39**: 339–359. [6]

Misyalyunene I. S. (1978). Influence of temperature and sun radiation on the destruction intensity of cabbage butterfly caterpillars infected by entobacterin-3. *Liet. TSR Mokslu Akad. Darb.* (C) **3**: 67-72. [16]

Mitchell N. D. (1978). Differential host selection by *Pieris brassicae* (the large white) on *Brassica oleracea* ssp. *oleracea* (the wild cabbage). *Entomologia exp. appl.* **22**: 208–219. [6]

Mitchell R. T. & Zim H. S. (1964). *Butterflies and moths*. New York. [25]

Mitter C. & Brooks D. R. (1983). Phylogenetic aspects of coevolution. Pp.65–98 in Futuyma & Slatkin (1983). [6]

Moffat J. A. (1902). *Anosia archippus* does not hibernate. *Rep. ent. Soc. Ont.* **32**: 78-82. [26]

Moncrieff R. W. (1967). *The chemical senses* (3rd edn). London. [12]

Mooney H. A., Ehrlich P. R., Lincoln D. & Williams K. S. (1980). Environmental controls on the seasonality of a drought deciduous shrub, *Diplacus aurantiacus*, and its predator, the checkerspot butterfly, *Euphydryas chalcedona*. *Oecologia* **45**: 143–146. [2]

Mooney H. A., Williams K. S., Lincoln D. E. & Ehrlich P. R. (1981). Temporal and spatial variability in the interaction between the checkerspot butterfly, *Euphydryas chalcedona*, and its principal food source, the California shrub, *Diplacus aurantiacus*. *Oecologia* **50**: 195–198. [2]

Moore N. W. (1975). Butterfly transects in a linear habitat, 1964–73. *Entomologist's Gaz.* **26**: 71–78. [2, 4]

Moore N. W. (1981). Insect conservation in Britain: national nature reserves. *Atala* **6**: 26–27. [33]

Morgan C. L. (1896). *Habitat and instinct*. London. [12]

Moriarty F. (1968). The toxicity and sublethal effects of p,p'-DDT and dieldrin on *Aglais urticae* and *Chorthippus brunneus*. *Ann. appl. Biol.* **62**: 371–393. [33]

Moriarty F. (1969). Butterflies and insecticides. *Entomologist's Rec. J. Var.* **81**: 276–278. [33]

Morishita K. (1981). Danaidae. Pp.441–598 in *Butterflies of the south-east Asian islands* (ed. Tsukada E.) **2**. Tokyo. [1]

Morrell G. M. & Turner J. R. G. (1970). Experiments on mimicry. I. The response of wild birds to artificial prey. *Behaviour* **36**: 116–130. [14]

Morris M. G. (1967). The representation of butterflies on British statutory nature reserves. *Entomologist's Gaz.* **18**: 57–68. [33]

Morris M. G. (1971). The management of grassland for the conservation of invertebrate animals. Pp.527–552 in Duffey & Watt (1971). [33]

Morris M. G. (1976). Conservation and the collector. Pp.107–116 in Heath (1976). [33]

Morris M. G. (1979). Grassland management and invertebrate animals—a selective review. *Scient. Proc. R. Dubl. Soc.* (A) **6**: 247–257. [33]

Morris M. G. (1981*a*). Insect conservation in Britain: ecological background and voluntary effort. *Atala* **6**: 28–31. [33]

Morris M. G. (1981*b*). The Joint Committee for the Conservation of British Insects. *Atala* **7**: 44–46. [33]

Morris M. G. (1981*c*). Conservation of butterflies in the United Kingdom. *Beih. Veröff. Naturschutz Landschaftspflege Bad-Württ* **21**: 35-47. [33]

Morrow P. A. & Fox L. R. (1980). Effects of variation in *Eucalyptus* essential oil yield on insect growth and grazing damage. *Oecologia* **45**: 209–219. [19]

Morrow P. A., Bellas T. E. & Eisner T. (1976). Eucalyptus oil in the defensive oral discharge of Australian sawfly larvae. *Oecologia* **24**: 193–206. [12]

Morse D. H. (1971). The insectivorous bird as an adaptive strategy. *A. Rev. Ecol. Syst.* **2**: 177–200. [12]

Morton A. C. (1979). Isolation as a factor responsible for the decline of the large blue butterfly, *Maculinea arion*, in Great Britain. *Entomologist's mon. Mag.* **114**: 247–250. [33]

Morton A. C. (1982). The effects of marking and capture on recapture frequencies of butterflies. *Oecologia* **53**: 105–110. [Intro, 33]

Mosher E. (1915). A classification of the Lepidoptera based on the characters of the pupa. *Bull. Ill. St. Lab. nat. Hist.* **12**: 17–159. [1]

Moss J. E. (1933). The natural control of the cabbage caterpillars, *Pieris* spp. *J. anim. Ecol.* **2**: 210–231. [10]

Mostler G. (1934). Beobachtungen zur Frage der Wespen-Mimikry. *Z. Morph. Okol. Tiere* **29**: 381–455. [12]

Moulds M. S. (1977). *Bibliography of the Australian butterflies*. Greenwich, NSW. [1]

Muggleton J. (1973). Some aspects of the history and ecology of blue butterflies in the Cotswolds. *Proc. Trans. Br. ent. nat. Hist. Soc.* **6**: 77–84. [33]

Muggleton J. (1979). Non-random mating in wild populations of polymorphic *Adalia bipunctata*. *Heredity* **42**: 57–65. [21]

Muggleton J. & Benham B. R. (1975). Isolation and the decline of the large blue butterfly (*Maculinea arion*) in Great Britain. *Biol. Conserv.* **7**: 119–128. [2, 33]

Mühlmann H. (1934). Modellversuch künstlich erzeugte Mimikry und ihre Bedeutung für den "Nachahmer". *Z. Morph. Okol. Tiere* **28**: 259–296. [12]

Müller F. (1877*a*). Über Haarpinsel, Filzflecke und ähnliche Gebilde auf den Flügeln männlicher Schmetterlinge. *Jena. Z. Naturw.* **5**: 99–114. (English translation in Longstaff (1912): 604–615.) [12, 25]

Müller F. (1877*b*). As maculas sexuaes do individuos masculionos das especiès *Danais erippus* e *D. gilippus*. *Archos Mus. nac. Rio de J.* **2**: 25-29. (English translation in Longstaff (1912): 616-620) [12, 25]

Müller F. (1877*c*). Die Duftschuppen der männlichen Maracujáfalter. *Kosmos* **1**: 391–395. (English translation in Longstaff (1912): 655–659.) [12]

Müller F. (1878*a*). Notes on Brazilian entomology. *Trans. ent. Soc. Lond.* **1878**: 211–223. [12, 25]

Müller F. (1878*b*). A prega costal des Hesperideas. *Archos Mus. nac. Rio de J.* **3**: 41–50. (English translation in Longstaff (1912): 640–646.) [25]
Müller F. (1878*c*). Über die Vortheile der Mimicry bei Schmetterlingen. *Zool. Anz.* **1**: 54–55. [20]
Müller F. (1879). *Ituna* und *Thyridia*. *Kosmos* **5**: 100–108. (English translation, by R. Meldola, under title *Ituna* and *Thyridia*: a remarkable case of mimicry in butterflies, appears in *Proc. ent. Soc. Lond.* **1879**: xx–xxix.) [20]
Müller H. J. (1955). Die Saisonformenbidung von *Araschnia levana*—ein photoperiodisch gestewerter Diapause-Effekt. *Naturwissenschaften* **42**: 134–135. [29]
Müller W. (1886). Südamerikanische Nymphaliden-raupen: Versuch eines natürlichen Systems der Nymphaliden. *Zool. Jb.* **1**: 417–678. [1]
Müller-Schwarze D. & Mozell M. M. (eds) (1976). *Chemical signals in vertebrates*. New York. [12]
Müller-Schwarze D. & Silverstein R. M. (eds) (1979). *Chemical signals: vertebrates and aquatic invertebrates*. New York. [12]
Mummery R. S., Rothschild M. & Valadon L. R. G. (1975). Carotenoids in two silk moths *Saturnia pavonia* and *Actia luna*. *Comp. Biochem. Physiol.* **50**B: 23–28. [12]
Munroe E. (1948). *The geographical distribution of butterflies in the West Indies*. PhD thesis, Cornell Univ., USA. [3]
Munroe E. (1961). The classification of the Papilionidae. *Can. Ent.* (Suppl.) **(17)**: 51pp. [1]
Munroe E. & Ehrlich P. R. (1960). Harmonization of concepts of higher classification of the Papilionidae. *J. Lepid. Soc.* **14**: 169–175. [1, 12]
Muroya Y., Kubo K., Maeda K., Ashizawa H. & Ohtsuka K. (1967*a*). The early stages of Formosan butterflies. *Spec. Bull. lepid. Soc. Japan* **(3)**: 51–60. [1, 9]
Muroya Y., Ae S. A., Kubo K., Maeda K., Ashizawa H. & Ohtsuka K. (1967*b*). Miscellaneous notes on early stages of 28 species of Formosan butterflies. *Spec. Bull. lepid. Soc. Japan* **(3)**: 117–149. [1]
Murphy D. D. & Ehrlich P. R. (1980). Two California checkerspot butterfly subspecies: one new, one on the verge of extinction. *J. Lepid. Soc.* **34**: 316–320. [2]
Murphy D. D. & White R. R. (1983). Rainfall, resources and dispersal in southern populations of *Euphydryas editha*. *Pan-Pacif. Ent.* (in press). [2]
Murphy D. D., Launer A. E. & Ehrlich P. R. (1983). The role of adult feeding in egg production and population dynamics of the checkerspot butterfly *Euphydryas editha*. *Oecologia* **56**: 257–263. [2]
Muyshondt A., Muyshondt A. jr & Muyshondt P. (1976). Notas sobre la biologia de Lepidopteros de El Salvador. I. *Revta Soc. mex. Lepid.* **2**: 77–90. [9]
Myers J. (1968). The structure of the antennae of the Florida queen butterfly, *Danaus gilippus berenice*. *J. Morph.* **125**: 315–328. [25]
Myers J. (1969). Distribution of foodplant chemoreceptors in the female Florida queen butterfly, *Danaus gilippus berenice*. *J. Lepid. Soc.* **23**: 196–198. [6]
Myers J. (1972). Pheromones and courtship behavior in butterflies. *Am Zool.* **12**: 545–551. [12, 25]
Myers J. & Brower L. P. (1969). A behavioural analysis of the courtship pheromone receptors of the queen butterfly, *Danaus gilippus berenice*. *J. Insect Physiol.* **15**: 2117–2130. [12, 25]
Myers J. H. (1978). Selecting a measure of dispersion. *Envir. Ent.* **7**: 619–621. [6]
Myers J. H. & Post B. J. (1981). Plant nitrogen and fluctuations of insect populations: a test with the cinnabar moth—tansy ragwort system. *Oecologia* **48**: 151–156. [12, 19]
Nabokov V. (1945). Notes on neotropical Plebejinae. *Psyche* **52**: 1–61. [1]
Nahrstedt A. & Davis R. H. (1981). The occurrence of cyanoglucosides, linamarin and lotaustralin, in *Acraea* and *Heliconius* butterflies. *Comp. Biochem. Physiol.* **68**B: 575–577. [Intro, 12]
Nahrstedt A. & Davis R. H. (1983). Occurrence, variation and biosynthesis of the cyanogenic glucosides linamarin and lotaustralin in species of the Heliconiini. *Comp. Biochem. Physiol.* **75**B: 65–73. [Intro]
Nakamura I. (1976). Female anal hair tuft in *Nordmannia myratle*: egg-camouflaging function and taxonomic significance. *J. Lepid. Soc.* **30**: 305–309. [6]
Neck R. W. (1973). Foodplant ecology of the butterfly *Chlosyne lacinia*. I. Larval foodplants. *J. Lepid. Soc.* **27**: 22–23. [6]
Neck R. W. (1977). Foodplant . . . II. Additional larval foodplant data. *J. Res. Lepid.* **16**: 69–74. [6]
Nei M. T., Maruyama T. & Chakraborty R. (1975). The bottleneck effect and genetic variability in populations. *Evolution* **29**: 1–10. [33]
Nekrutenko Y. P. (1968). *Phylogeny and geographical distribution of the genus* Gonepteryx*: an attempt of study in historical zoogeography*. Kiev (in Russian, with English summary). [20]
Nelson C. J., Seiber J. N. & Brower L. P. (1981). Seasonal and intraplant variation of cardenolide content in the California milkweed, *Asclepias eriocarpa*, and implications for plant defense. *J. Chem. Ecol.* **7**: 981–1010. [12]
Neuhaus W. (1963). On the olfactory sense of birds. Pp.111–123 in *Olfaction and taste* (ed. Zotterman Y.). Oxford. [12]
Neustetter H. (1929). Nymphalidae: subfam. Heliconiinae. *Lepid. Cat.* **32** (36): 136pp. [1]
Nichol A. A. (1938). Experimental feeding of deer. *Tech. Bull. Ariz. agric. Exp. Stn* **85**: 1–39. [12]
Nicholson A. J. (1927). A new theory of mimicry in insects. *Aust. Zool.* **5**: 10–104. [Intro, 12, 14]
Nicholson M. (1970). *The environmental revolution*. London. [33]
Nicholson S. W. (1976). Hormonal control of diuresis in the cabbage white butterfly, *Pieris brassicae*. *J. exp. Biol.* **65**: 565–575. [6]
Nielsen E. T. (1961). On the habits of the migratory butterfly, *Ascia monuste*. *Biol. Meddr.* **23**: 1–81. [26]
Nielsen M. C. (1977). Invertebrate predators of Michigan Lepidoptera. *Gr. Lakes Ent.* **10**: 113–118. [12]
Nijhout H. F. (1978). Wing pattern formation in Lepidoptera: a model. *J. exp. Zool.* **206**: 119–136. [16, 20]

Nijhout H. F. (1980*a*). Pattern formation on lepidopteran wings: determination of an eyespot. *Devl Biol.* **80**: 267–274. [16, 20]

Nijhout H. F. (1980*b*). Ontogeny of the color pattern on the wings of *Precis coenia*. *Devl Biol.* **80**: 275–288. [20]

Nijhout H. F. (1981). The color patterns of butterflies and moths. *Scient. Am.* **245** (5): 140–151. [20]

Nixon G. E. J. (1974). A revision of the north-western European species of the *glommeratus*-group of *Apanteles* (Hymenoptera, Braconidae). *Bull. ent. Res.* **64**: 453–524. [10]

Norris M. J. (1936). The feeding-habits of the adult Lepidoptera Heteroneura. *Trans. R. ent. Soc. Lond.* **85**: 61–90. [Intro]

Nur U. (1970). Evolutionary rates of models and mimics in Batesian mimicry. *Am. Nat.* **104**: 477–486. [14]

Obara Y. (1970). Studies on the mating behavior of the white cabbage butterfly, *Pieris rapae crucivora*. III. Near-ultra-violet reflection as the signal of intraspecific communication. *Z. vergl. Physiol.* **69**: 99–116. [20]

Obara Y. & Hidaka T. (1964). Studies . . . I. The 'flutter response' of resting male to flying males. *Zool. Mag. Tokyo* **73**: 131–135. [20]

Obara Y. & Hidaka T. (1968). Recognition of the female by the male, on the basis of ultra-violet reflection, in the white cabbage butterfly, *Pieris rapae crucivora*. *Proc. Japan Acad.* **44**: 829–832. [20]

Odiyo P. O. (1975). Seasonal distribution and migrations of *Agrotis ypsilon*. *Trop. Pest Bull.* **4**: 1–26. [31]

O'Donald P. (1971). Natural selection for quantitative characters. *Heredity* **27**: 137–153. [16]

O'Donald P. (1980). *Genetic models of sexual selection*. Cambridge. [20, 21]

O'Donald P. & Muggleton J. (1979). Melanic polymorphism in ladybirds maintained by sexual selection. *Heredity* **43**: 143–148. [21]

Oehme F. W., Brown J. F. & Fowler M. E. (1975). Toxins of animal origin. Pp.570–590 in Casarett & Doull (1975). [12]

Ohsaki N. (1979). Comparative population studies of three *Pieris* butterflies, *P. rapae*, *P. melete* and *P. napi*, living in the same area. I. Ecological requirements for habitat resources in the adults. *Researches Popul. Ecol.* **20**: 278–296. [2, 3, 6, 8]

Ohsaki N. (1980). Comparative . . . II. Utilization of patchy habitats by adults through migratory and non-migratory movements. *Researches Popul. Ecol.* **22**: 163–183. [26]

Oliver C. G. (1972*a*). Genetic differentiation between English and French populations of the satyrid butterfly *Pararge megera*. *Heredity* **29**: 307–313. [16]

Oliver C. G. (1972*b*). Genetic and phenotypic differentiation and geographic distance in four species of Lepidoptera. *Evolution* **26**: 221–241. [16]

Oliver C. G. (1979). Genetic differentiation and hybrid viability between some Lepidoptera species. *Am. Nat.* **114**: 681–694. [16]

Olson S. L. (1981). The museum tradition in ornithology —a response to Ricklefs. *Auk* **98**: 193–195. [2]

Ono T. & Nakasuji F. (1980). Comparison of flight activity and oviposition characteristics of the seasonal forms of a migratory skipper butterfly, *Parnara guttata guttata*. *Kontyû* **48**: 226–233. [26]

Opler P. (1981). The federal endangered insect programme. *Atala* **6**: 60–62. [33]

Orians G. M. (1981). Foraging behavior and the evolution of discriminatory abilities. Pp.389–405 in Kamil & Sargent (1981). [12]

Orsak L. J. (1978). Endangered and extinct butterflies—why? *Xerxes Soc. Educ. Leaflt* (**3**). [33]

Orsak L. J. (1981). Introduction to the proceedings and an update on terrestrial arthropod conservation. *Atala* **6**: 1–18. [33]

Osada M. & Itô Y. (1974). Population dynamics of *Pieris rapae crucivora*, an introduced insect pest in Okinawa. I. Winter-spring generations. *Appl. Ent. Zool., Tokyo* **18**: 65–72. [2]

Oshima K., Honda H. & Yamamoto I. (1975). The osmeterium secretion of the caterpillars of swallowtail butterflies in Japan. *J. agric. Sci.* **20**: 117–120. [12]

Ourisson G., Rohmer M. & Anton R. (1979). From terpenes to sterols: macroevolution and microevolution. *Recent Adv. Phytochem.* **13**: 131–165. [13]

Owen D. F. (1959). Ecological segregation in butterflies in Britain. *Entomologist's Gaz.* **10**: 27–38. [6]

Owen D. F. (1970*a*) Mimetic polymorphism and the palatability spectrum. *Oikos* **21**: 333-336. [12, 18]

Owen D. F. (1970*b*). Inheritance of sex-ratio in the butterfly, *Acraea encedon*. *Nature* **225**: 662–663. [2, 21]

Owen D. F. (1971*a*). *Tropical butterflies*. Oxford. [Intro, 1, 2, 12, 16, 18, 20, 21, 24, 29]

Owen D. F. (1971*b*). Pupal color in *Papilio demodocus* in relation to the season of the year. *J. Lepid. Soc.* **25**: 271–274. [12]

Owen D. F. (1974*a*). Seasonal change in sex ratio in *Acraea quirina* and notes on the factors causing distortions of sex ratio in butterflies. *Entomologica scand.* **5**: 110–114. [21]

Owen D. F. (1974*b*). Exploring mimetic diversity in West African forest butterflies. *Oikos* **25**: 227-237. [14]

Owen D. F. (1975). Estimating the abundance and diversity of butterflies. *Biol. Conserv.* **8**: 173–183. [2]

Owen D. F. (1980). *Camouflage and mimicry*. Oxford. [12]

Owen D. F. & Chanter D. O. (1968). Population biology of tropical African butterflies. II. Sex ratio and polymorphism in *Danaus chrysippus*. *Revue Zool. Bot. afr.* **78**: 81–97. [21]

Owen D. F. & Chanter D. O. (1969). Population . . . Sex ratio and genetic variation in *Acraea encedon*. *J. Zool., Lond.* **157**: 345–375. [2, 16, 18, 21]

Owen D. F. & Chanter D. O. (1971*a*). Genetics of some polymorphic forms of the African butterfly *Acraea encedon*. *Entomologica scand.* **2**: 287–293. [18]

Owen D. F. & Chanter D. O. (1971*b*). Polymorphism in West African populations of the butterfly *Acraea encedon*. *J. Zool., Lond.* **163**: 481–488. [18]

Owen D. F. & Chanter D. O. (1972). Species diversity and seasonal abundance in *Charaxes* butterflies. *J. Ent.* (A) **46**: 135–143. [2]

Owen D. F. & Owen J. (1972). Systematics and bionomics of butterflies seen and collected in the forest region of Sierra Leone. Part I. Introduction, Papilionidae, Danaidae and Acraeidae. *Revue Zool. Bot. afr.* **85**: 287-308. [1]

Owen D. F. & Owen J. (1973). Systematics and bionomics ... Part II. Nymphalidae and Libytheidae. *Revue Zool. Bot. afr.* **87**: 585–613. [1]

Owen D. F., Owen J. & Chanter D. O. (1973*a*). Low mating frequency in predominantly female populations of the butterfly *Acraea encedon*. *Entomologica scand.* **4**: 155–160. [18]

Owen D. F., Owen J. & Chanter D. O. (1973*b*). Low mating frequencies in an African butterfly. *Nature* **244**: 116–117. [21]

Pagden H. T. (1957). The presence of coremata in *Creatonotus gangis*. *Proc. R. ent. Soc. Lond.* (A) **32**: 90–94. [25]

Pagenstecher A. (1901). Libytheidae. *Tierreich* **14**: ix + 17pp. [1]

Pagenstecher A. (1909). *Die geographische Verbreitung der Schmetterlinge. Jena.* [1]

Pagenstecher A. (1911). Libytheidae. *Lepid. Cat.* **23** (3): 1–12. [1]

Palka J. & Pinter R. B. (1975). Theoretical and experimental analysis of visual acuity in insects. Pp.321–337 in Horridge (1975). [20]

Pan M. L. & Wyatt G. R. (1971). Juvenile hormone induced vitellogenin synthesis in the monarch butterfly. *Science* **174**: 503–505. [6]

Pan M. L. & Wyatt G. R. (1976). Control of vitellogenin synthesis in the monarch butterfly by juvenile hormone. *Devl Biol.* **54**: 127–134. [6]

Pant G. D. & Chatterjee N. C. (1950). A list of described immature stages of Indian Lepidoptera. Part I. Rhopalocera. *Indian Forest Rec.* N.S. **7**: 213–255. [1]

Papageorgis C. (1975). Mimicry in neotropical butterflies. *Am. Scient.* **63**: 522–532. [3, 14, 20]

Parker F. D. (1970). Seasonal mortality and survival of *Pieris rapae* in Missouri and the effect of introducing an egg parasite, *Trichogramma evanescens*. *Ann. ent. Soc. Am.* **63**: 985–994. [2, 10]

Parker G. A. (1970). Sperm competition and its evolutionary consequences in insects. *Biol. Rev.* **45**: 525–567. [21]

Parker G. A. (1974). Courtship persistence and female guarding as male time investment strategies. *Behaviour* **48**: 157–184. [20]

Parker G. A. (1978). Evolution of competitive mate searching. *A. Rev. Ent.* **23**: 173–196. [22]

Parker G. A. (1979). Sexual selection and sexual conflict. Pp.123–166 in Blum & Blum (1979). [21]

Parker G. A. & Stuart R. A. (1976). Animal behavior as a strategy optimizer: evolution of resource assessment strategies and optimal emigration thresholds. *Am. Nat.* **110**: 1055–1076. [20]

Parr M. J., Gaskell T. J. & George B. J. (1968). Capture-recapture methods of estimating animal numbers. *J. biol. Educ.* **2**: 95–117. [2]

Parsons J. A. (1965). A digitalis-like toxin in the monarch butterfly, *Danaus plexippus*. *J. Physiol.* **178**: 290–304. [12]

Parsons M. J. (1980*a*). *A conservation study of* Ornithoptera alexandrae. First report, Feb. 1980. Internal report, Wildlife Division, Papua New Guinea. [31, 32]

Parsons M. J. (1980*b*). *A conservation* ... Second report, Aug. 1980. Int. rept, Wildlife Division, PNG. [32]

Parsons M. J. (1980*c*). *A conservation* ... Third report: the Kumusi Timber Area survey. Int. rept, Wildlife Division, PNG. [32]

Parsons M. J. (1981). *A conservation* ... Fourth report, Aug. 1981. Int. rept, Wildlife Division, PNG. [32]

Parsons M. J. (1983). Notes on the courtship of *Troides oblongomaculatus papuensis* in Papua New Guinea. *J. Lepid. Soc.* **38**: 83–85. [25]

Parsons M. J. (In prep.). Early stages of the butterflies of New Guinea in colour. [1]

Parsons P. A. (1973). *Behavioural and ecological genetics. A study in* Drosophila. Oxford. [21]

Pasteur G. (1972). *Le Mimetisme.* Paris. [12]

Pasteur G. (1977). Endocyclic selection in reptiles. *Am. Nat.* **111**: 1027–1030. [16]

Paul A. R. (1977). Some observations on the marbled white butterfly *Melanargia galathea*. *Entomologist's mon. Mag.* **112**: 127–130. [33]

Pavan M. & Dazzini M. V. (1976). Sostanze di difensa dei lepidotteri. *Pubbl. Ist. Ent. agr. Univ. Pavia* **3**: 1–23. [12]

Peachey C. (1980). The conservation of butterflies in Bernwood Forest. NCC Report, Newbury, Berks UK. [33]

Peachey C. (1981). The butterflies of Bernwood Forest, England. *Atala* **7**: 61–63. [33]

Peachey C. (1982). The representation of butterflies on National Nature Reserves. *Invertebrate Site Register Report* **10**(1). NCC, London. [33]

Pearse F. K. & Ehrlich P. R. (1979). B chromosome variation in *Euphydryas colon*. *Chromosoma* **73**: 263–274. [2]

Pennington K. M. (1978). *Butterflies of southern Africa* (eds Dickson C. G. C. & Kroon D. M.). Johannesburg. [1]

Perrins C. M. (1976). Possible effects of qualitative changes in the insect diet of avian predators. *Ibis* **118** 850–854. [12]

Peterken G. F. (1976). Long-term changes in the woodlands of Rockingham Forest and other areas. *J. Ecol.* **64**: 123–146. [33]

Peterken G. F. & Harding P. T. (1975). Woodland conservation in eastern England. Comparing the effects of changes in three study areas since 1946. *Biol. Conserv.* **8**: 279–298. [33]

Peters J. V. (1970). The cabbage white butterfly. *Aust. nat. Hist.* **16**: 300–305. [26]

Peters W. (1952). *A provisional check-list of the butterflies of the Ethiopian region.* Feltham, Middx, UK. [1]

Petersen B. (1949). On the evolution of *Pieris napi*. *Evolution* **3**: 269–278. [16]

Petersen B. (1952). The relations between *Pieris napi* and *P. bryoniae*. *Trans. Int. Congr. Ent.* (9) **1**: 83. [20]

Petersen B. (1954). Egg-laying and habitat selection in some *Pieris* species. *Ent. Tidskr.* **75**: 194–203. [6]

Petersen B. (1964). Monarch butterflies are eaten by birds. *J. Lepid. Soc.* **18**: 165–169. [12]

Petersen B. & Tenow O. (1954). Studien am Rapsweissling und Bergweissling (*Pieris napi* und *P. bryoniae*): Isolation und Paarungsbiologie. *Zool. Bidr.* **30**: 169–198. [20]

Petersen B., Törnblom O. & Bodin N.-O. (1952). Verhaltenstudien am Rapsweissling und Bergweissling (*Pieris napi* und *P. bryoniae*). *Behaviour* **4**: 67-84. [20]
Petit C. (1954). L'isolement sexuel chez *Drosophila melanogaster*. Etude du mutant white et de son allelomorphe sauvage. *Bull. biol. Fr. Belg.* **88**: 435-443. [21]
Petit C. & Ehrman L. (1969). Sexual selection in *Drosophila*. *Evolut. Biol.* **3**: 177-223. [21]
Pettit G. R., Hartwell J. L. & Wood H. B. (1968). Arthropod antineoplastic agents. *Cancer Res.* **28**: 2168-2169. [13]
Petty R. L., Boppré M., Schneider D. & Meinwald J. (1977). Identification and localization of volatile hairpencil components in male *Amauris ochlea* butterflies. *Experientia* **33**: 1324-1326. [12, 25]
Pianka H. D., Boggs C. L. & Gilbert L. E. (1977). Ovarian dynamics in heliconiine butterflies: programmed senescence versus eternal youth. *Science* **179**: 487-490. [32]
Pielou E. C. (1969). *An introduction to mathematical ecology*. New York. [3, 6]
Piepers M. C. & Snellen P. C. T. (1909-18). *The Rhopalocera of Java*. The Hague. [1]
Pierce N. E. (1983). *The ecology and evolution of lycaenid/ant symbiosis*. PhD thesis, Harvard Univ., Cambridge, Mass., USA. [19]
Pierce N. E. & Mead P. S. (1981). Parasitoids as selective agents in the symbiosis between butterfly larvae and ants. *Science* **211**: 1185-1187. [2, 3, 6, 12, 19, 33]
Pierre J. (1974). Polymorphisme et coupes infra-spécifiques africaines dans l'espèce *Danaus chrysippus*. *Bull. Mus. natn. Hist. nat. Paris* (3) **149**: 601-640. [21]
Pierre J. (1976*a*). Un nouveau cas d'espèce jumelles chez un papillon mimetique: *Acraea encedon*. *C. r. Seanc. Biol.* **282**: 731-734. [18, 21]
Pierre J. (1976*b*). Polymorphisme et mimétisme chez deux espèces jumelles, *Acraea encedon* et *A. encedana*. *Annls Soc. ent. Fr.* **12**: 621-638. [18, 21]
Pierre J. (1980). Variation géographique du polymorphisme et du mimetisme de *Danaus chrysippus* et d'*Hypolimnas misippus* en Afrique et en Asie. *C. r. somm. Séanc. Soc. Biogéogr.* **486**: 179-187. [17, 18, 21]
Pierre J. (1983). *Systématique evolutive, cladistique et mimétisme chez les lépidoptères du genre* Acraea. DSc thesis, Univ. Paris, France. [1, 18]
Pierre-Baltus C. (1978). Resultats d'élevages de *Neptis* à facies "melicerta" en Côte D'Ivoire; description de trois nouvelles espèces. *Lambillionea* **78**: 33-44. [1]
Pimentel D. (1961). Natural control of caterpillar populations on cole crops. *J. econ. Ent.* **54**: 889-992. [10]
Platt A. P. (1975). Monomorphic mimicry in nearctic *Limenitis* butterflies: experimental hybridization of the *L. arthemis-astyanax* complex with *L. archippus*. *Evolution* **29**: 120-141. [14]
Platt A. P. (1979). Oviposition site selection and behavior in *Limenitis* spp. *Md. Ent.* **1**: 9-10. [6]
Platt A. P. & Brower L. P. (1968). Mimetic versus disruptive coloration in intergrading populations of *Limenitis arthemis* and *astyanax* butterflies. *Evolution* **22**: 699-718. [22]
Platt A. P., Coppinger R. P. & Brower L. P. (1971). Demonstration of the selective advantage of mimetic *Limenitis* butterflies presented to caged avian predators. *Evolution* **25**: 692-701. [12]
Pliske T. E. (1968). *Chemical communication in the courtship of the queen butterfly*. PhD thesis, Cornell Univ., Ithaca, USA. [25]
Pliske T. E. (1972). Sexual selection and dimorphism in female tiger swallowtails, *Papilio glaucus*: a reappraisal. *Ann. ent. Soc. Am.* **65**: 1267-1270. [20-22]
Pliske T. E. (1973). Factors determining mating frequencies in some New World butterflies and skippers. *Ann. ent. Soc. Am.* **66**: 164-169. [21]
Pliske T. E. (1975*a*). Attraction of Lepidoptera to plants containing pyrrolizidine alkaloids. *Envir. Ent.* **4**: 455-473. [9, 12, 25]
Pliske T. E. (1975*b*). Courtship behavior and use of chemical communication by males of certain species of ithomiine butterflies. *Ann. ent. Soc. Am.* **68**: 935-942. [12, 25]
Pliske T. E. (1975*c*). Courtship behavior of the monarch butterfly, *Danaus plexippus*. *Ann. ent. Soc. Am.* **68**: 143-151. [20, 21, 25]
Pliske T. E. & Eisner T. (1969). Sex pheromone of the queen butterfly: biology. *Science* **164**: 1170-1172. [12, 20, 25]
Pliske T. E. & Salpeter M. M. (1971). The structure and development of the hairpencil glands in males of the queen butterfly, *Danaus gilippus berenice*. *J. Morph.* **134**: 215-241. [25]
Pliske T. E., Edgar J. A. & Culvenor C. C. J. (1976). The chemical basis of attraction of ithomiine butterflies to plants containing pyrrolizidine alkaloids. *J. Chem. Ecol.* **2**: 255-262. [12, 25]
Plowright R. C. & Owen R. E. (1980). The evolutionary significance of bumble bee color patterns: a mimetic interpretation. *Evolution* **34**: 622-637. [14]
Pocock R. I. (1911). On the palatability of some British insects with notes on the significance of mimetic resemblances (with notes by Poulton E. B.). *Proc. zool. Soc. Lond.* **1911**: 809-868. [12]
Polikoff D. (1981). C. H. Waddington and modern evolutionary theory. *Evolut. Theory* **5**: 143-168. [27]
Polis G. A. (1981). The evolution and dynamics of intraspecific predation. *A. Rev. Ecol. Syst.* **12**: 225-251. [12]
Pollard E. (1977). A method for assessing changes in abundance of butterflies. *Biol. Conserv.* **12**: 115-134. [2, 4, 5, 33]
Pollard E. (1979*a*). A national scheme for monitoring the abundance of butterflies: the first three years. *Proc. Trans. Br. ent. nat. Hist. Soc.* **12**: 77-90. [16, 33]
Pollard E. (1979*b*). Population ecology and changes in range of the white admiral butterfly *Ladoga camilla* in England. *Ecol. Ent.* **4**: 61-74. [2, 5, 10, 33]
Pollard E. (1981*a*). *Monitoring population changes in butterflies*. ITE Report, Abbots Ripton, Cambridge, UK. [33]
Pollard E. (1981*b*). Aspects of the ecology of the meadow brown butterfly, *Maniola jurtina*. *Entomologist's Gaz.* **32**: 67-74. [16, 33]
Pollard E. (1982). *Butterfly Monitoring Scheme 1981*. Abbots Ripton, U.K. [33]

Pollard E., Hooper M. & Moore N. W. (1974). *Hedges*. London. [33]

Pollard E., Elias D. O., Skelton M. J. & Thomas J. A. (1975). A method of assessing the abundance of butterflies in Monks Wood National Nature Reserve in 1973. *Entomologist's Gaz.* **26**: 79–88. [2, 33]

Porter K. (1980). A quantitative treatment of clinal variation in *Coenonympha tullia*. *Entomologist's mon. Mag.* **116**: 71–82. [16]

Porter K. (1981). *The population dynamics of small colonies of the butterfly* Euphydryas aurinia. D.Phil thesis, Oxford Univ., UK. [10, 28, 33]

Porter K. (1982). Basking behaviour in larvae of the butterfly *Euphydryas aurinia*. *Oikos* **38**: 308-312. [28]

Porter K. (1983). Multivoltinism in *Apanteles bigellii* and the influence of weather on synchronisation with its host *Euphydryas aurinia*. *Ent. exp. & appl.* **34**: 155-162. [28]

Post C. T. & Goldsmith T. H. (1969). Physiological evidence for color receptors in the eye of a butterfly. *Ann. ent. Soc. Am.* **62**: 1497–1498. [20]

Pough F. H. & Brower L. P. (1977). Predation by birds on great southern white butterflies as a function of palatability, sex and habitat. *Am. Midl. Nat.* **98**: 50–58. [2, 12]

Pough F. H., Brower L. P., Meck H. R. & Kessell S. R. (1973). Theoretical investigations of automimicry: multiple trial learning and the palatability spectrum. *Proc. natn Acad. Sci. U.S.A.* **70**: 2261–2265. [12]

Poulton E. B. (1887). The experimental proof of the protective value of colour and markings in insects in reference to their vertebrate enemies. *Proc. Zool. Soc. Lond.* **1887**: 191–274. [12]

Poulton E. B. (1906). [Report on exhibit of *Precis sesamus*]. *Proc. ent. Soc. Lond.* **1906**: lvii–lx. [29]

Poulton E. B. (1908). *Essays on evolution*. Oxford. [20, 21]

Poulton E. B. (1909). Mimicry in butterflies of North America. *Ann. ent. Soc. Am.* **2**: 203–242. [22]

Poulton E. B. (1924*a*). The terrifying appearance of *Laternaria* (Fulgoridae) founded on the most prominent features of the alligator. *Proc. ent. Soc. Lond.* **1924**: xliii–ix. [12]

Poulton E. B. (1924*b*). Modes of protection in the pupal stage of butterflies and moths. *SEast. Nat.* **29**: 72–77. [12]

Poulton E. B. (1924*c*). Mimicry in the butterflies of Fiji considered in relation to the Euploeine and Danaine invasions of Polynesia and to the female forms of *Hypolimnas bolina* in the Pacific. *Trans. ent. Soc. Lond.* **1923**: 564–691. [24]

Powell J. A. (1980). Evolution of larval food preferences in microlepidoptera. *A. Rev. Ent.* **25**: 133–159. [6]

Powell J. A. (1981). Endangered habitats for insects: California coastal sand dunes. *Atala* **6**: 41–55. [33]

Powell J. A. & Sorenson J. T. (1980). Butterfly Count results. *Atala* **7** (Suppl.): 1–10. [33]

Prakash S., Lewontin R. C. & Hubby J. L. (1969). A molecular approach to the study of genic heterozygosity in natural populations: IV. patterns of genic variation in central, marginal and isolated populations of *Drosophila pseudoobscura*. *Genetics* **61**: 841–858. [17]

Prance G. T. (ed.) (1982). *Biological diversification in the tropics*. New York. [14]

Pratt C. (1981). *A history of the butterflies and moths of Sussex*. Brighton. [33]

Price P. W. (1975). *Insect ecology*. New York. [3]

Price P. W., Bouton C. E., Gross P., McPheron B. A., Thompson J. N. & Weis A. E. (1980). Interactions among three trophic levels: influence of plants on interactions between insect herbivores and natural enemies. *A. Rev. Ecol. Syst.* **11**: 41–65. [3]

Priesner E. (1970). Über die Spezifität der Lepidopteren-Sexual-lockstoffe und ihre Rolle bei der Artbildung. *Verh. dt. zool. Ges.* **64**: 337-343. [25]

Priesner E. (1973). Artspezifität und Funktion einiger Insekten-pheromone. *Fortschr. Zool.* **22**: 49–135. [25]

Prosser C. L. & Brown F. A. (1961). *Comparative animal physiology*. Philadelphia. [12]

Prout T. (1967). Selective forces in *Papilio glaucus*. *Science* **156**: 534. [20, 22]

Punnett R. C. (1915). *Mimicry in butterflies*. Cambridge. [14]

Pyke G. H., Pulliam H. R. & Charnov E. L. (1977). Optimal foraging: a selective review of theory and tests. *Q. Rev. Biol.* **52**: 137–154. [6]

Pyle R. M. (1970). The history and landforms of Mt St Helens. Unpublished MS, deposited in Washington Univ., Seattle. [31]

Pyle R. M. (1974). *Watching Washington butterflies*. Seattle. [31]

Pyle R. M. (1976*a*). *The eco-geographic basis for Lepidoptera conservation*. PhD thesis, Yale Univ., New Haven, Conn., USA. [31, 33]

Pyle R. M. (1976*b*). Conservation of Lepidoptera in the United States. *Biol. Conserv.* **9**: 55–75. [2, 33]

Pyle R. M. (1981*a*). International problems for insect conservation. *Atala* **6**: 56–58. [33]

Pyle R. M. (1981*b*). The role of IUCN and WWF in Lepidoptera conservation. *Beih. Veröff. Naturschutz Landschaftspflege Bad-Württ* **21**: 15–18. [33]

Pyle R. M. (1981*c*). Lepidoptera conservation in Great Britain. *Atala* **7**: 34–43. [33]

Pyle R. M. (In prep.). The last Lepidoptera from Mt St Helens. [31]

Pyle R. M. & Hughes S. A. (1978). *Conservation and utilisation of the insect resources of Papua New Guinea*. Consultant report, Division of Wildlife, Papua New Guinea. [32]

Pyle R. M., Bentzien M. & Opler P. (1981). Insect conservation. *A. Rev. Ent.* **26**: 233–258. [2, 33]

Pyörnilä M. (1976). Parasitism in *Aglais urticae*. II. Parasitism of larval stages by tachinids. *Suom. hyönt. Aikak.* **42**: 133–139. [2]

Rabinowitch V. E. (1968). The role of experience in the development of food preferences in gull chicks. *Anim. Behav.* **16**: 425–428. [12]

Rackham O. (1971). Historical studies and woodland conservation. *Symp. Br. ecol. Soc.* **(11)**: 563–580. [33]

Rackham O. (1980). *Ancient woodland—its history, vegetation and uses in England*. London. [33]

Rackham O. (1981). Memories of our wildwood. *Natural World* **1**: 26–29. [33]

Rainey R. C. (1963). Meteorology and the migration of desert locusts. *Tech. Notes Wld. met. Org.* **(54)**: 115pp. [26]

Rand A. S. (1967). Predator–prey interactions and the evolution of aspect diversity. *Atas Simposio Biota Amazonica* **5**: 73–83. [12]

Rankin M. A. & Singer M. (In press). Insect movement: mechanisms and effects. In *Insect ecology* (eds Huffaker C. B. & Rabb R.). [2]

Rathcke B. J. & Poole R. W. (1975). Coevolutionary race continues: butterfly larval adaptation to plant trichomes. *Science* **187**: 175–176. [6]

Rauch P. K. (1977). *The defense strategy of the monarch larva,* Danaus plexippus: *an oral discharge examined.* Snr Hons thesis, Amherst College, Mass., USA. [12]

Rausher M. D. (1978). Search image for leaf shape in a butterfly. *Science* **200**: 1071–1073. [6, 8, 20]

Rausher M. D. (1979*a*). Egg recognition: its advantage to a butterfly. *Anim. Behav.* **27**: 1034–1040. [6, 7, 12]

Rausher M. D. (1979*b*). Larval habitat suitability and oviposition preference in three related butterflies. *Ecology* **60**: 503–511. [2, 6, 7]

Rausher M. D. (1980). Host abundance, juvenile survival, and oviposition preference in *Battus philenor*. *Evolution* **34**: 342–355. [2, 6, 12]

Rausher M. D. (1981*a*). Host plant selection by *Battus philenor* butterflies: the roles of predation, nutrition, and plant chemistry. *Ecol. Monogr.* **51**: 1–20. [6, 12, 19]

Rausher M. D. (1981*b*). The effect of native vegetation on the susceptibility of *Aristolochia reticulata* to herbivore attack. *Ecology* **62**: 1187–1195. [12]

Rausher M. D. (1982*a*). Ecology of host-selection behavior in phytophagous insects. Pp.223–257 in Denno & McClure (1982; actually published 1983). [6]

Rausher M. D. (1982*b*). Population differentiation in *Euphydryas editha* butterflies: larval adaptation to different hosts. *Evolution* **36**: 581–590. [7]

Rausher M. D. & Feeny P. (1980). Herbivory, plant density, and plant reproductive success: the effect of *Battus philenor* on *Aristolochia reticulata*. *Ecology* **61**: 905–917. [2]

Rausher M. D. & Papaj D. (1983). Demographic consequences of discrimination among conspecific host-plants by *Battus philenor* butterflies. *Ecology* **64**: 1402-1410.

Rausher M. D., Mackay D. A. & Singer M. C. (1981). Pre- and post- alighting host discrimination by *Euphydryas editha* butterflies: the behavioral mechanisms causing clumped distributions of egg clusters. *Anim. Behav.* **29**: 1220–1228. [6, 7]

Rawlins J. E. (1980). Thermoregulation by the black swallowtail butterfly, *Papilio polyxenes*. *Ecology* **61**: 345–357. [28]

Ray T. S. & Andrews C. C. (1980). Antbutterflies: butterflies that follow army ants to feed on antbird droppings. *Science* **210**: 1147–1148. [6]

Rebach S. (1978). The role of celestial cues in short range migrations of the hermit crab, *Pagurus longicarpus*. *Anim. Behav.* **26**: 835–842. [26]

Reeve E. C. R. (1955). The variance of the genetic correlation coefficient. *Biometrics* **11**: 357–374. [16]

Reichstein T. (1967). Cardenolide (herzwirksame Glykoside) als Abwehrstoffe bei Insekten. *Naturw. Rdsch.* **20**: 499–511. [12]

Reichstein T., Von Euw J., Parsons J. A. & Rothschild M. (1968). Heart poisons in the monarch butterfly. *Science* **161**: 861–866. [12]

Reiskind J. (1965). Behaviour of an avian predator in an experiment simulating Batesian mimicry. *Anim. Behav.* **13**: 466–469. [12]

Remington C. L. (1963). Historical backgrounds of mimicry. *Int. Congr. Zool.* (16) **4**: 145–151. [3]

Remington C. L. (1973). Ultraviolet reflectance in mimicry and sexual signals in the Lepidoptera. *Jl. N.Y. ent. Soc.* **81**: 124. [20, 22]

Renwick J. J. A. & Radke C. D. (1980). An oviposition deterrent associated with frass from feeding larvae of the cabbage looper, *Trichoplusia ni*. *Envir. Ent.* **9**: 318–320. [6]

Renwick J. J. A. & Radke C. D. (1981). Host plant constituents as oviposition deterrents for the cabbage looper, *Trichoplusia ni*. *Entomologia exp. appl.* **30**: 201–204. [6]

Rettenmeyer C. W. (1970). Insect mimicry. *A. Rev. Ent.* **15**: 43–74. [12, 14]

Reuter E. (1896). Über die Palpen der Rhopaloceren. Ein Beitrag zur Erkenntnis der verwandtschaftlichen Beziehungen unter den Tagfaltern. *Acta Soc. Scient. fenn.* **22**(1): xvi + 578pp. [1]

Rhoades D. F. (1979). Evolution of plant chemical defense against herbivores. Pp.3–54 in Rosenthal & Janzen (1979). [12]

Ribi W. A. (1980). The phenomenon of eye glow. *Endeavour* **5**: 2–7. [20]

Richards L. J. & Myers J. H. (1980). Maternal influences on size and emergence time of the cinnabar moth. *Can. J. Zool.* **58**: 1452–1457. [16]

Richards O. W. (1927). Sexual selection and allied problems in the insects. *Biol. Rev.* **2**: 298–360. [20]

Richards O. W. (1940). The biology of the small white butterfly (*Pieris rapae*) with special reference to the factors controlling its abundance. *J. anim. Ecol.* **9**: 243–288. [2, 10]

Richards O. W. (1949). The relation between measurements of the successive instars of insects. *Proc. R. ent. Soc. Lond.* (A) **24**: 8–10. [6]

Richards O. W. & Davies R. G. (1977). *Imms' general textbook of entomology* (10th edn). London. [Intro, 6]

Richter C. P. (1950). Taste and solubility of toxic compounds in poisoning of rats and man. *J. comp. physiol. Psychol.* **43**: 358–374. [12]

Richter C. P. (1953). Experimentally produced behavior reactions to food poisoning in wild and domesticated rats. *Ann. N.Y. Acad. Sci.* **56**: 225–239. [12]

Ricklefs R. E. (1973). *Ecology*. Newton, Mass. [2]

Ricklefs R. E. & O'Rourke K. (1975). Aspect diversity in moths: a temperate-tropical comparison. *Evolution* **29**: 313–324. [12]

Riddle O. & Burns F. H. (1931). Conditioned emetic reflex in pigeon. *Proc. Soc. exp. Biol. Med.* **28**: 979–981. [12]

Riemann J. G. & Gassner G. (1973). Ultrastructure of lepidopteran sperm within spermathecae. *Ann. ent. Soc. Am.* **66**: 154–159. [6]

Riley N. D. (1925). Annual address to the Members. *Proc. S. Lond. ent. nat. Hist. Soc.* **1924–25**: 63-81. [27]

Riley N. D. (1958). The genera of Holarctic Theclinae: a tentative revision. *Proc. Int. Congr. Ent.* (10) **1**: 281-288. [1]

Riley N. D. (1975). *Butterflies of the West Indies*. London. [1]

Ritter F. J. (ed.) (1979). *Chemical ecology: odour communication in animals*. Amsterdam. [25]

Rivers C. F. (1976). Diseases. Pp.57-70 in Heath (1976). [10]

Robbins R. J. (1978). Poison-based taste aversion in deer mice (*Peromyscus maniculatus bairdi*). *J. comp. physiol. Psychol.* **92**: 642-650. [12]

Robbins R. J. (1979). The effect of flower pre-exposure upon the acquisition and retention of poison-based taste aversions in deer mice: latent inhibition or partial reinforcement. *Behavl Neural Biol.* **25**: 387-397. [12]

Robbins R. K. (1980). The lycaenid "false head" hypothesis: historical review and quantitative analysis. *J. Lepid. Soc.* **34**: 194-208. [16]

Robbins R. K. (1981). The "false head" hypothesis: predation and wing pattern variation of lycaenid butterflies. *Am. Nat.* **118**: 770-775. [12, 16]

Robbins R. K. (1982). How many butterfly species? *News Lepid. Soc.* **1982**: 40-41. [1]

Robbins R. K. & Aiello A. (1982). Foodplant and oviposition records for Panamanian Lycaenidae and Riodinidae. *J. Lepid. Soc.* **36**: 65-75, **36**(3): [1] (Erratum). [1, 6]

Robbins R. K. & Small G. B. (1981). Wind dispersal of Panamanian hairstreak butterflies and its evolutionary significance. *Biotropica* **13**: 308-315. [Intro]

Robinson G. S. (1975). *Macrolepidoptera of Fiji and Rotuma: a taxonomic study*. Faringdon, U.K. [1]

Robinson M. H. (1969). Defenses against visually hunting predators. *Evolut. Biol.* **3**: 225-259. [12]

Robinson R. (1971). *Lepidoptera genetics*. Oxford. [16, 20, 22]

Robinson T. (1968). *The biochemistry of alkaloids*. New York. [12]

Robinson T. (1975). *The organic constituents of higher plants* (3rd edn). N. Amherst, Mass. [12]

Robinson T. (1979). The evolutionary ecology of alkaloids. Pp.413-448 in Rosenthal & Janzen (1979). [6, 12]

Robson J. E. (1887). On the flight and pairing of *Hepialus hectus* and *humuli*. *Entomologist's mon. Mag.* **23**: 186-187. [25]

Robson J. E. (1892). The genus *Hepialus*. *Entomologist's Rec. J. Var.* **3**: 52-56, 77-79, 100-101. [25]

Rodman J. E. & Chew F. S. (1980). Phytochemical correlates of herbivory in a community of native and naturalized Cruciferae. *Biochem. Syst. Ecol.* **8**: 43-50. [6, 7]

Rodriguez E. & Levin D. A. (1976). Biochemical parallelisms of repellents and attractants in higher plants and arthropods. *Recent Adv. Phytochem.* **10**: 214-270. [12]

Roeder K. D. (1967). Prey and predator. *Bull. Ecol. Soc. Am.* **13**: 6-9. [12]

Roelofs W. L. (1980). Pheromones and their chemistry. Pp.583-602 in Locke & Smith (1980). [12]

Roepke W. (1935-42). *Rhopalocera Javanica*. Wageningen. [1]

Roer H. (1959). Uber Flug und Wandergewohnheiten von *Pieris brassicae*. *Z. angew. Ent.* **44**: 272-309. [2]

Roer H. (1961*a*). Ergebnisse mehrjähriger Markierungsversuche zur Erforschung der Flug—und Wandergehwohnheiten europäischer Schmetterlinge. *Zool. Anz.* **167**: 456-463. [2, 26]

Roer H. (1961*b*). Zur Kenntnis der Populationsdynamik und des Migrationsverhaltens von *V. atalanta* in Paläarktischen Raum. *Beitr. Ent.* **11**: 594-613. [2]

Roer H. (1962). Experimentelle Untersuchungen zum Migrationsverhalten des Kleinen Fuchs (*Aglais urticae*). *Beitr. Ent.* **1**: 528-554. [2]

Roer H. (1965*a*). Zur Frage der Abhängigkeit des Wanderrichtung des Tagfalters *Aglais urticae* von der Luftströmung. *Proc. Int. Congr. Ent.* **(12)**: 415-416. [2]

Roer H. (1965*b*). Etiquetage des ailes pour l'etude des migrations des papillons. *Bull. Soc. ent. Fr.* **1965**: 41-44. [2]

Roer H. (1968). Weitere Untersuchungen über die Auswirkungen der Witterung auf Richtung und Distanz der Flüge des Kleinen Fuchses (*Aglais urticae*) im Rheinland. *Decheniana* **120**: 313-334. [2, 26]

Roer H. (1969). Zur Biologie des Tagpfauenauges, *Inachis io*, unter besonderer Berücksichtigung der Wanderungen im mitteleuropäischen Raum. *Zool. Anz.* **183**: 177-194. [2, 26]

Roer H. (1970). Untersuchungen zum Migrationsverhalten des Trauermantels (*Nymphalis antiopa*). *Z. angew. Ent.* **65**: 388-396. [2, 26]

Roeske C. N., Seiber J. S., Brower L. P. & Moffitt C. M. (1976). Milkweed cardenolides and their comparative processing by monarch butterflies (*Danaus plexippus*). *Recent. Adv. Phytochem.* **10**: 93-167. [12]

Rohan T. A. (1972). The chemistry of flavour. Pp.57-71 in Harborne (1972). [12]

Rohwer S., Fretwell S. D. & Niles D. M. (1980). Delayed maturation in passerine plumages and the deceptive acquisition of resources. *Am. Nat.* **115**: 400-437. [20]

Röller H., Biemann, K., Bjerke J. S., Norgard D. W. & McShan W. H. (1968). Sex pheromones of pyralid moths. I. Isolation and identification of the sex attractant of *Galleria mellonella*. *Acta ent. bohemoslovaca* **65**: 208-211. [25]

Rosenfeld C. L. (1980). Observations on the Mt St Helens eruption. *Am. Scient.* **68**: 494-509. [31]

Rosenthal G. A. & Janzen D. H. (eds) (1979). *Herbivores, their interaction with secondary plant metabolites*. New York.

Rosevear D. R. (1978). A few lepidopterous life-histories. *Niger. Fld.* **43**: 177-181. [1]

Ross G. N. (1963). Evidence for lack of territoriality in two species of *Hamadryas*. *J. Res. Lepid.* **2**: 241-246. [20]

Ross G. N. (1966). Life-history studies of Mexican butterflies. IV. The ecology and ethology of *Anatoli rossi*, a myrmecophilous metalmark. *Ann. ent. Soc. Am.* **59**: 985-1004. [6]

Ross G. N. (1975-77). An ecological study of the butterflies of the Sierra de Tuxtla in Veracruz, Mexico. *J. Res. Lepid.* **14**: 103-124, 169-188, 233-252; **15**: 41-60, 109-128, 185-200, 225-240; **16**: 87-130. [1]

Rothschild M. (1961). Defensive odours and Müllerian mimicry among insects. *Trans. R. ent. Soc. Lond.* **113**: 101–121. [12]

Rothschild M. (1963). Is the buff ermine (*Spilosoma lutea*) a mimic of the white ermine (*S. lubricipeda*)? *Proc. R. ent. Soc. Lond.* (A) **38**: 159–164. [12]

Rothschild M. (1964*a*). An extension of Dr Lincoln Brower's theory on bird predation and food specificity, together with some observations on bird memory in relation to aposematic colour patterns. *Entomologist* **97**: 73–78. [12]

Rothschild M. (1964*b*). A note on the evolution of defensive and repellent odours of insects. *Entomologist* **97**: 276-280. [12]

Rothschild M. (1966). Experiments with captive predators and the poisonous grasshopper *Poekilocerus bufonis*. *Proc. R. ent. Soc. Lond.* (C) **31**: 32. [12]

Rothschild M. (1967). Mimicry, the deceptive way of life. *Nat. Hist., N.Y.* **76**: 44–51. [12]

Rothschild M. (1971). Speculations about mimicry with Henry Ford. Pp.202–223 in Creed (1971). [12, 14]

Rothschild M. (1972*a*). Colour and poisons in insect protection. *New Scientist* **54**: 318–320. [14]

Rothschild M. (1972*b*). Secondary plant substances and warning colouration in insects. *Symp. R. ent. Soc. Lond.* **(6)**: 59–83. [12–14]

Rothschild M. (1972*c*). Some observations on the relationships between plants, toxic insects and birds. Pp.1–12 in Harborne (1972). [9, 12]

Rothschild M. (1976). [Exhibit]. *Proc. R. ent. Soc. Lond.* (C) **40**: 35. [14]

Rothschild M. (1978). Carotenoids in the evolution of signals: experiments with insects (1974–1976). Pp.259–276 in Harborne (1978). [12]

Rothschild M. (1979). Female butterfly guarding eggs. *Antenna, Lond.* **3**: 94. [6, 12]

Rothschild M. (1981*a*). The mimicrats must move with the times. *Biol. J. Linn. Soc.* **16**: 21–23. [12]

Rothschild M. (1981*b*). Mimicry, butterflies and plants. *Symb. bot. upsal.* **22**(4): 82–99. [14]

Rothschild M. & Edgar J. A. (1978). Pyrrolizidine alkaloids from *Senecio vulgaris* sequestered and stored by *Danaus plexippus*. *J. Zool., Lond.* **186**: 347–349. [9, 12]

Rothschild M. & Fairbairn J. W. (1980). Ovipositing butterfly (*Pieris brassicae*) distinguishes between aqueous extracts of two strains of *Cannabis sativa* and THC and CBD. *Nature* **286**: 56–59. [6]

Rothschild M. & Ford B. (1968). Warning signals from a starling, *Sturnus vulgaris*, observing a bird rejecting unpalatable prey. *Ibis* **110**: 104–105. [12]

Rothschild M. & Ford B. (1970). Heart poisons and the monarch. *Nat. Hist., N.Y.* **79**: 36–37. [12]

Rothschild M. & Kellett D. (1972). Reactions of various predators to insects storing heart poisons in their body tissues. *J. Ent.* (A) **46**: 103–110. [12]

Rothschild M. & Lane C. (1960). Warning and alarm signals by birds seizing aposematic insects. *Ibis* **102**: 328–330. [12]

Rothschild M. & Lane C. (1964). [No title]. *Proc. R. ent. Soc. Lond.* (C) **29**: 26–28. [12]

Rothschild M. & Marsh N. A. (1978). Some peculiar aspects of danaid/plant relationships. *Entomologia exp. appl.* **24**: 437–450. [12]

Rothschild M. & Schoonhoven L. M. (1977). Assessment of egg load by *Pieris brassicae*. *Nature* **266**: 352–355. [6, 12, 20]

Rothschild M., Reichstein R. T., Von Euw J., Aplin R. T. & Harman R. R. M. (1970). Toxic Lepidoptera. *Toxicon* **8**: 293–299. [12, 25]

Rothschild M., Von Euw J. & Reichstein T. (1972*a*). Aristolochic acids stored by *Zerynthia polyxena*. *Insect Biochem.* **2**: 334–343. [12]

Rothschild M., Von Euw J. & Reichstein T. (1972*b*). Some problems connected with warningly coloured insects and toxic defense mechanisms. *Mitteilungen der Basler Afrika Bibliographien, Basel* **4–6**: 135–158. [12, 14]

Rothschild M., Von Euw J. & Reichstein T. (1973). Cardiac glycosides (heart poisons) in the polka-dot moth *Syntomeida epilais* with some observations on the toxic qualities of *Amata phegea*. *Proc. R. Soc.* (B) **183**: 227–247. [12, 14]

Rothschild M., Gardiner B. O. C., Valadon G. & Mummery R. (1975*a*). The large white butterfly: oviposition cues, carotenoids, and changes of colour. *Proc. R. ent. Soc. Lond.* (C) **40**: 13. [6]

Rothschild M., Von Euw J., Reichstein T., Smith D. A. S. & Pierre J. (1975*b*). Cardenolide storage in *Danaus chrysippus* with additional notes on *D. plexippus*. *Proc. R. Soc.* (B) **190**: 1–31. [12, 18]

Rothschild M., Rowan M. G. & Fairbairn J. W. (1977*a*). Storage of cannabinoids by *Arctia caja* and *Zonocerus elegans* fed on chemically distinct strains of *Cannabis sativa*. *Nature* **266**: 650–651. [12]

Rothschild M., Valadon G. & Mummery R. (1977*b*). Carotenoids of the pupae of the large white butterfly (*Pieris brassicae*) and the small white butterfly *Pieris rapae*). *J. Zool., Lond.* **181**: 323–339. [12]

Rothschild M., Gardiner B. O. C. & Mummery R. (1978*a*). The role of carotenoids in the "golden glance" of danaid pupae. *J. Zool., Lond.* **186**: 351–358. [12]

Rothschild M., Marsh N. A. & Gardiner B. O. C. (1978*b*). Cardioactive substances in the monarch butterfly and *Euploea core* reared on leaf-free artificial diet. *Nature* **275**: 649–650. [12, 13]

Rothschild M., Aplin R. T., Baker J. & Marsh N. A. (1979*a*). Toxicity induced in the tobacco horn-worm (*Manduca sexta*). *Nature* **280**: 487–488. [12]

Rothschild M., Aplin R. T., Cockrum P. A., Edgar J. A., Fairweather P. & Lees R. (1979*b*). Pyrrolizidine alkaloids in arctiid moths with a discussion on host plant relationships and the role of these secondary plant substances in the Arctiidae. *Biol. J. Linn. Soc.* **12**: 305–326. [13, 25]

Rothschild M., Keutmann H., Lane N. J., Parsons J., Prince W. & Swales L. S. (1979*c*). A study of the mode of action and composition of a toxin from the female abdomen and eggs of *Arctia caja*: an electrophysiological, ultrastructural and biochemical analysis. *Toxicon* **17**: 285–306. [13]

Rothschild W. (1895). A revision of the papilios of the eastern hemisphere, exclusive of Africa. *Novit. zool.* **2**: 167–463. [1]

Rothschild W. & Jordan K. (1898–1900). A monograph of *Charaxes* and the allied prionopterous genera. *Novit. zool.* **5**: 545–601; **6**: 220–286; **7**: 281–524. [1]

Rothschild W. & Jordan K. (1903). A revision of the lepidopterous family Sphingidae. *Novit. zool.* **9**: 815–972. [13]

Rothschild W. & Jordan K. (1906). A revision of the American papilios. *Novit. zool.* **13**: 411–753. [1]

Roughgarden J. (1979). *Theory of population genetics and evolutionary ecology: an introduction.* New York. [12]

Rozin P. (1976). The selection of foods by rats, humans, and other animals. *Adv. Study Behav.* **6**: 21–76. [12]

Rozin P. (1977). The significance of learning mechanisms in food selection: some biology, psychology, and sociology of science. Pp.557–589 in Barker *et al.* (1977). [12]

Rubinoff I. & Kropach C. (1970). Differential reactions of Atlantic and Pacific predators to sea snakes. *Nature* **228**: 1288–1290. [12]

Rusiniak K. W., Hankins W. G., Garcia J. & Brett L. P. (1979). Flavor-illness aversions: potentiation of odor by taste in rats. *Behavl Neural Biol.* **25**: 1–17. [12]

Russell F. E. (1971). Pp.255–384 in *Marine toxins and venomous and poisonous marine animals* (ed. Russell F. E.). New York. [12]

Rutowski R. L. (1977*a*). The use of visual cues in sexual and species discrimination by males of the small sulfur butterfly *Eurema lisa*. *J. comp. Physiol.* **115**: 61–74. [20, 25]

Rutowski R. L. (1977*b*). Chemical communication in the courtship of the small sulfur butterfly, *Eurema lisa*. *J. comp. Physiol.* **115**: 75–85. [20, 25]

Rutowski R. L. (1978*a*). The courtship behavior of the small sulphur butterfly, *Eurema lisa*. *Anim. Behav.* **26**: 892–903. [20]

Rutowski R. L. (1978*b*). The form and function of ascending flights in *Colias* butterflies. *Behav. Ecol. Sociobiol.* **3**: 163–172. [20]

Rutowski R. L. (1979). The butterfly as an honest salesman. *Anim. Behav.* **27**: 1269–1270. [20]

Rutowski R. L. (1980*a*). Male scent-producing structures in *Colias* butterflies: function, localization and adaptive features. *J. Chem. Ecol.* **6**: 13–26. [20, 25]

Rutowski R. L. (1980*b*). Courtship solicitation by females of the checkered white butterfly (*Pieris protodice*). *Behav. Ecol. Sociobiol.* **7**: 113–117. [20]

Rutowski R. L. (1981). Sexual discrimination using visual cues in the checkered white butterfly (*Pieris protodice*). *Z. Tierpsychol.* **55**: 325–334. [20]

Rutowski R. L. (1982). Mate choice and lepidopteran mating behavior. *Fla Ent.* **65**: 72–82. [20, 25]

Rydon A. H. B. (1971). The systematics of the Charaxidae. *Entomologist's Rec. J. Var.* **83**: 219–233, 283–287, 310–316, 336–341, 384–388. [1]

Sakai S. (1981). *Butterflies of Afghanistan.* Tokyo. [1]

Sankowsky G. (1975). Some new food plants for various Queensland butterflies. *Aust. ent. Mag.* **2**: 55–56. [9]

Sargent T. D. (1973). Studies on the *Catocala* of southern New England. IV. A preliminary analysis of beak-damaged specimens, with discussion of anomaly as a potential anti-predator function of hindwing diversity. *J. Lepid. Soc.* **27**: 175–192. [12, 16]

Sargent T. D. (1981). Antipredator adaptations of underwing moths. Pp.259–284 in Kamil & Sargent (1981). [12]

Saxena K. N. & Goyal S. (1978). Host-plant relations of the citrus butterfly *Papilio demoleus*: orientational and ovipositional responses. *Entomologia expl. appl.* **24**: 1–10. [6]

Sbordoni V. & Bullini L. (1971). Further observations on mimicry in *Zygaena ephialtes*. *Fragm. ent.* **8**: 49–56. [14]

Sbordoni V., Bullini L., Scarpelli G., Forestiero S. & Rampini M. (1979). Mimicry in the burnet moth *Zygaena ephialtes*: population studies and evidence of a Batesian-Müllerian situation. *Ecol. Ent.* **4**: 83–93. [14]

Scali V. (1971*a*). Spot distribution in *Maniola jurtina*: Tuscan mainland 1967–1969. *Monitore zool. ital.* **5**: 147–163. [16]

Scali V. (1971*b*). Imaginal diapause and gonadal maturation of *Maniola jurtina* from Tuscany. *J. anim. Ecol.* **40**: 467–472. [16]

Scali V. (1972). Spot-distribution in *Maniola jurtina*: Tuscan archipelago, 1968–1970. *Heredity* **29**: 25–36. [16]

Scali V. & Masetti M. (1973). The population structure of *Maniola jurtina*: the sex-ratio control. *J. anim. Ecol.* **42**: 773–778. [16]

Scali V. & Masetti M. (1975). Variazioni intrastagionali dello spotting e selezione in *Maniola jurtina*. *Atti Accad. naz. Lincei Rc.* (8) **58**: 244–257. [16]

Scali V. & Masetti M. (1979). Zygotic and fertility selection for phosphoglucomutase variants in natural populations of *Maniola jurtina*. *Atti Accad. naz. Lincei Rc.* (8) **67**: 137–144. [16]

Schaefer G. W. (1976). Radar observations of insect flight. *Symp. R. ent. Soc. Lond.* **(7)**: 157–197. [26]

Schanks C. H. & Chase D. L. (1981). Effect of volcanic ash on adult *Otiorhynchus* (Col.: Curculionidae). *Melanderia* **37**: 63–66. [31]

Scharloo W. (1970). Stabilizing and disruptive selection on a mutant character in *Drosophila*. III. Polymorphism caused by a developmental switch mechanism. *Genetics* **65**: 693–705. [14]

Scharloo W., Hoogmoed M. S. & ter Kuile A. (1967). Stabilizing . . . I. The phenotypic variance and its components. *Genetics* **56**: 709–726. [14]

Scharloo W., Zweep A., Schuitema K. A. & Wijnstra J. G. (1972). Stabilizing . . . IV. Selection on sensitivity to temperature. *Genetics* **71**: 551–566. [14]

Schlieper C. (1928). Über die Helligskeitverteilung im Spektrum bei verschiedenen Insekten. *Z. vergl. Physiol.* **8**: 281–288. [20]

Schmalhausen I. I. (1949). *Factors of evolution.* Philadelphia. [27]

Schmidt-Koenig K. (1979). *Avian orientation and navigation.* London. [26]

Schneider D. (1974). The sex-attractant receptor of moths. *Scient. Am.* **231**(1): 28–35. [25]

Schneider D. (1975). Pheromone communication in moths and butterflies. Pp.173–193 in *Sensory physiology and behavior* (eds Galum R., Hillman P., Parnas I. & Werman R.). New York & London. [12]

Schneider D. (1980). Pheromone von Insekten: Produktion—Reception—Inaktivierung. *Nova Acta Leopoldina* **51** (237): 249–278. [25]

Schneider D. & Boppré M. (1981). Pyrrolizidin-Alkaloide als Vorstufen für die Duftstoff-Biosynthese und als Regulatoren der Duftorgan-Morphogenese bei *Creatonotos*. *Verh. dt. zool. Ges.* **1981**: 269. [25]

Schneider D. & Seibt U. (1969). Sex-pheromone of the queen butterfly: electroantennogram responses. *Science* **164**: 1173–1174. [12, 25]

Schneider D., Boppré M., Schneider H., Thompson W. R., Boriack C. J., Petty R. L. & Meinwald J. (1975). A pheromone precursor and its uptake in male *Danaus* butterflies. *J. comp. Physiol.* **97**: 245–256. [21, 25]

Schneider D., Boppré M., Zweig J., Horsley S. B., Bell T. W., Meinwald J., Hansen K. & Diehl E. W. (1982). Scent organ development in *Creatonotos* moths: regulation by pyrrolizidine alkaloids. *Science* **215**: 1264–1265. [25]

Schneirla T. C. (1965). Aspects of stimulation and organisation in approach/withdrawal processes underlying vertebrate behavioral development. *Adv. Study Behav.* **1**: 1–74. [16]

Schoenthal R. (1968). Toxicology and carcinogenic action of pyrrolizidine alkaloids. *Cancer Res.* **28**: 2237–2246. [25]

Schoonhoven L. M. (1977). On the individuality of insect feeding behavior. *Proc. K. ned. Akad. Wet.* (C) **80**: 341–350. [6]

Schoonhoven L. M., Sparnaay T., van Wissen W. & Meerman J. (1981). Seven-week persistence of an oviposition-deterrent pheromone. *J. chem. Ecol.* **7**: 583–588. [6]

Schremmer F. (1978). Zur Bionomie und Morphologie der myrmekophilen Raupe und Puppe der neotropischen Tagfalter-Art *Hamearis erostratus*. *Entomologica germ.* **4**: 113–121. [12]

Schrier R. D., Cullenward M. J., Ehrlich P. R. & White R. R. (1976). The structure and genetics of a montane population of the checkerspot butterfly, *Chlosyne palla*. *Oecologia* **27**: 279–289. [2]

Schuler W. (1974). Die Schutzwirkung künstlicher Batesscher Mimikry abhängig von Modellähnlichkeit und Beuteangebot. *Z. Tierpsychol.* **36**: 71–127. [12, 14]

Schuler W. (1980). Factors influencing learning to avoid unpalatable prey in birds: re-learning, new alternative prey, and similarity of appearance of alternative prey. *Z. Tierpsychol.* **54**: 105–143. [12]

Schuler W. (1982). Zur Function von Warnfarben: Die Reaktion junger stare auf wespenähnlich schwarz-gelbe Attrappen. *Z. Tierpsychol.* **58**: 66–78. [12, 14]

Schümperli R. A. (1975). Monocular and binocular visual fields of butterfly interneurons in responses to white- and coloured-light stimulation. *J. comp. Physiol.* **103**: 273–289. [20]

Schümperli R. A. & Swihart S. (1978). Spatial properties of dark- and light-adapted visual fields of butterfly interneurons. *J. Insect Physiol.* **24**: 777–784. [20]

Schwaegerle K. E. & Schaal B. A. (1979). Genetic variability and founder effect in the pitcher plant, *Sarracenia purpurea*. *Evolution* **33**: 1210–1218. [33]

Schwanwitsch B. N. (1924). On the ground plan of wing-pattern in Nymphalidae and certain other families of the rhopalocerous Lepidoptera. *Proc. zool. Soc. Lond.* **1924**: 509–528. [16, 20]

Schwanwitsch B. N. (1948). Evolution of the wing-pattern in Palaearctic Satyridae. IV. Polymorphic radiation and parallelism. *Acta. zool., Stockh.* **29**: 1–61. [16]

Scott J. A. (1970). Hilltopping as a mating mechanism to aid the survival of low density species. *J. Res. Lepid.* **7**: 191–204. [2]

Scott J. A. (1973*a*). Convergence of population biology and adult behaviour in two sympatric butterflies, *Neominois ridingsii* and *Amblyscirtes simius*. *J. anim. Ecol.* **42**: 663–672. [2]

Scott J. A. (1973*b*). Down-valley flight of adult Theclini in search of nourishment. *J. Lepid. Soc.* **27**: 283–287. [2]

Scott J. A. (1973*c*). Population biology and adult behaviour of the circumpolar butterfly, *Parnassius phoebus*. *Entomologica scand.* **4**: 161–168. [2]

Scott J. A. (1973*d*). Mating of butterflies. *J. Res. Lepid.* **11**: 99–127. [2, 20, 25]

Scott J. A. (1973*e*). Adult behavior and population biology of two skippers mating in contrasting topographic sites. *J. Res. Lepid.* **12**: 181–196. [2]

Scott J. A. (1973*f*). Survey of the ultraviolet reflectance of Nearctic butterflies. *J. Res. Lepid.* **12**: 151–160. [22]

Scott J. A. (1974*a*). The interaction of behavior, population biology, and environment in *Hypaurotis crysalus*. *Am. Midl. Nat.* **91**: 383–394. [2]

Scott J. A. (1974*b*). Mate-locating behavior of butterflies. *Am. Midl. Nat.* **91**: 103–117. [2, 19, 20, 25]

Scott J. A. (1974*c*). Population biology and adult behavior of *Lycaena arota*. *J. Lepid. Soc.* **28**: 64–72. [2]

Scott J. A. (1974*d*). Adult behavior and population biology of *Poladryas minuta*, and the relationship of the Texas and Colorado populations. *Pan-Pacif. Ent.* **50**: 9–22. [2]

Scott J. A. (1975*a*). Flight patterns among eleven species of diurnal Lepidoptera. *Ecology* **56**: 1367–1377. [2]

Scott J. A. (1975*b*). Movements of *Precis coenia*, a 'pseudoterritorial' submigrant. *J. anim. Ecol.* **44**: 843–850. [2]

Scott J. A. (1975*c*). Movements of *Euchloe ausonides*. *J. Lepid. Soc.* **29**: 24–31. [2]

Scott J. A. (1975*d*). Mate-locating behavior of western North American butterflies. *J. Res. Lepid.* **14**: 1–10. [19]

Scott J. A. (1977). Competitive exclusion due to mate searching behaviour, male-female emergence lags and fluctuation in number of progeny in model invertebrate populations. *J. anim. Ecol.* **46**: 909–924. [2]

Scott J. A. (1981). Hibernal diapause of North American Papilionoidea and Hesperioidea. *J. Res. Lepid.* **18**: 171–200. [2]

Scott J. A. & Opler P. A. (1975). Population biology and adult behavior of *Lycaena xanthoides*. *J. Lepid. Soc.* **29**: 63–66. [2]

Scriber J. M. (1973). Latitudinal gradients in larval feeding specialization of the world Papilionidae. *Psyche* **80**: 355–373. [8]

Scriber J. M. & Slansky F. (1981). The nutritional ecology of immature insects. *A. Rev. Ent.* **26**: 183–211. [6, 7]

Scudder S. H. (1877). Antigeny, or sexual dimorphism in butterflies. *Proc. Am. Acad. Arts Sci.* **12**: 150–158. [23, 25]

Scudder S. H. (1889). *The butterflies of the eastern United States and Canada 1.* Cambridge, Mass. [6, 22]

Searcy W. A. (1982). The evolutionary effects of mate selection. *Ann. Rev. Ecol. Syst.* **13**: 57–85. [Intro, 23]

Segerstrom K. (1950). Erosion studies at Paricutín, state of Michoacán, Mexico. *Bull. U.S. Geol. Serv.* **965**: 1–151. [31]

Seiber J. N. & Lee M. S. (1982). Cardiac glycosides (cardenolides) in species of *Asclepias*. In *Encyclopedic handbook of natural toxins.* 1. *Plant toxins* (eds Tu A. T. & Keeler R. F.). New York. [12]

Seiber J. N., Tuskes P. M., Brower L. P. & Nelson C. J. (1980). Pharmacodynamics of some individual milkweed cardenolides fed to the larvae of the monarch butterfly (*Danaus plexippus*). *J. Chem. Ecol.* **6**: 321–339. [12]

Seibt U., Schneider D. & Eisner T. (1972). Duftpinsel, Flugelfaschen und Balz des Tagfalters *Danaus chrysippus. Z. Tierpsychol.* **31**: 513–530. [12, 25]

Seitz A. (ed.) (1907–09). *The Palaearctic Region. Gross-Schmetterl. Erde* (1) *Palaearktische Fauna* **1**. (Supplement publ. 1929–32.) Stuttgart (in original German, and English and French edns). [1]

Seitz A. (ed.) (1907–24). *The American Rhopalocera. Gross-Schmetterl. Erde* (2) *Exotische Fauna* **5**. Stuttgart (in orig. German, and English and French edns). [1, 12]

Seitz A. (ed.) (1908–25). *The African Rhopalocera. Gross-Schmetterl. Erde* (2) *Exotische Fauna* **13**. Stuttgart (in orig. German, and English and French edns). [1]

Seitz A. (ed.) (1908–28). *The Rhopalocera of the Indo-Australian faunal region. Gross-Schmetterl. Erde* (2) *Exotische Fauna* **9**. Stuttgart (in orig. German, and English and French edns). [1]

Selander R. K. (1972). Sexual selection and dimorphism in birds. Pp.180-230 in Campbell (1972). [21]

Seligman I. M. & Doy F. A. (1972). B-hydroxy-n-butyric acid in the defense secretion of *Papilio aegeus. Comp. Biochem. Physiol.* **41**: 341–342. [12]

Seligman I. M. & Doy F. A. (1973). Biosynthesis of defensive secretions in *Papilio aegeus. Insect Biochem.* **3**: 205–215. [12]

Seligman M. E. P. (1970). On the generality of the laws of learning. *Psychol. Rev.* **77**: 406–418. [12]

Sellier R. (1973*a*). Contribution à l'étude de l'ultrastructure et du mode de fonctionnement de l'appareil androconial alaire chez les satyrides. *Alexanor* **8**: 65–70. [25]

Sellier R. (1973*b*). Recherches en microscopie électronique par balayage, sur l'ultrastructure de l'appareil androconial alaire dans le genre *Argynnis* et dans les genres voisins. *Annls Soc. ent. Fr.* **9**: 703–728. [25]

Semper G. (1886–1892). *Die Schmetterlinge der Philippinischen Inseln.* **5**. *Rhopalocera.* Wiesbaden. [1]

Seppänen R. (1981). Differences in spotting pattern between populations of *Aphantopus hyperantus* in southern Finland. *Annls zool. fenn.* **18**: 1–36. [16]

Sevastopulo D. G. (1944). Note on the courtship of *Euploea core core. Proc. R. ent. Soc. Lond.* (A) **19**: 138–139. [25]

Sevastopulo D. G. (1964). Lepidoptera ovipositing on plants toxic to their larvae. *J. Lepid. Soc.* **18**: 104. [6]

Sevastopulo D. G. (1973). The foodplants of Indian Rhopalocera. *J. Bombay nat. Hist. Soc.* **70**: 156–183. [1]

Sevastopulo D. G. (1975). A list of the food-plants of East African macrolepidoptera. I. Butterflies. *Bull. amat. Ent. Soc.* **34**: 84–92, 124–132. [1]

Sexton O. J. (1960). Experimental studies of artificial Batesian mimics. *Behaviour* **15**: 244–252. [12]

Shallenberger R. S. & Acree T. E. (1971). Chemical structure of compounds and their sweet and bitter taste. Pp.221–277 in Beidler (1971). [12]

Shapiro A. M. (1968). Photoperiodic induction of vernal phenotype in *Pieris protodice. Wasmann J. Biol.* **26**: 137–149. [29]

Shapiro A. M. (1970). The role of sexual behaviour in density-related dispersal of pierid butterflies. *Am. Nat.* **104**: 367–372. [2, 20]

Shapiro A. M. (1971). Occurrence of a latent polyphenism in *Pieris virginiensis. Ent. News* **82**: 13–16. [29]

Shapiro A. M. (1973*a*). Altitudinal migration of butterflies in the central Sierra Nevada. *J. Res. Lepid.* **12**: 231–235. [2]

Shapiro A. M. (1973*b*). Photoperiodic control of seasonal polyphenism in *Pieris occidentalis. Wasmann J. Biol.* **31**: 291–299. [29]

Shapiro A. M. (1974). Beak-mark frequency as an index of seasonal predation intensity on common butterflies. *Am. Nat.* **108**: 229–232. [12]

Shapiro A. M. (1975*a*). Ecological and behavioral aspects of coexistence in six crucifer-feeding pierid butterflies in the central Sierra Nevada. *Am. Midl. Nat.* **93**: 424–433. [3]

Shapiro A. M. (1975*b*). The temporal component of butterfly species diversity. Pp.181–195 in Cody & Diamond (1975). [2, 3]

Shapiro A. M. (1976*a*). Seasonal polyphenism. *Evolut. Biol.* **9**: 259–333. [27, 29]

Shapiro A. M. (1976*b*). The role of watercress, *Nasturtium officinale*, as a host of native and introduced pierid butterflies in California. *J. Res. Lepid.* **14**: 158–168. [2]

Shapiro A. M. (1977). Phenotypic induction in *Pieris napi*: role of temperature and photoperiod in a coastal Californian population. *Ecol. Ent.* **2**: 219–224. [27]

Shapiro A. M. (1979*a*). The evolutionary significance of redundancy and variability in phenotypic-induction mechanisms of pierid butterflies. *Psyche* **85**: 275–283. [27]

Shapiro A. M. (1979*b*). The life histories of the *autodice* and *sterodice* species groups of *Tatochila. Jl N.Y. ent. Soc.* **87**: 236–255. [27]

Shapiro A. M. (1979*c*). The phenology of *Pieris napi microstriata* during and after the 1975–77 California drought, and its evolutionary significance. *Psyche* **86**: 1–10. [2]

Shapiro A. M. (1979*d*). Weather and the lability of breeding populations of the checkered white butterfly, *Pieris protodice. J. Res. Lepid.* **17**: 1–23. [2]

Shapiro A. M. (1980*a*). Genetic incompatibility between *Pieris callidice* and *P. occidentalis nelsoni*: differentiation within a periglacial relict complex. *Can. Ent.* **112**: 463–468. [27]

Shapiro A. M. (1980*b*). Convergence in pierine polyphenisms. *J. nat. Hist.* **14**: 781–802. [27]

Shapiro A. M. (1980*c*). Egg load assessment and carryover diapause in *Anthocharis*. *J. Lepid. Soc.* **34**: 307–315. [3, 6]

Shapiro A. M. (1981*a*). Phenotypic plasticity in temperate and subarctic *Nymphalis antiopa*: evidence for adaptive canalization. *J. Lepid. Soc.* **35**: 124–131. [27]

Shapiro A. M. (1981*b*). Canalization of the phenotype of *Nymphalis antiopa* from subarctic and montane climates. *J. Res. Lepid.* **19**: 82–87. [27]

Shapiro A. M. (1981*c*). The pierid red-egg syndrome. *Am. Nat.* **117**: 276–294. [3, 6]

Shapiro A. M. (1981*d*). Egg-mimics of *Streptanthus* (Cruciferae) deter oviposition by *Pieris sisymbrii*. *Oecologia* **48**: 142–143. [3, 6]

Shapiro A. M. & Cardé R. T. (1970). Habitat selection and competition among sibling species of satyrid butterflies. *Evolution* **24**: 48–54. [6]

Shapiro A. M. & Masuda K. K. (1980). The opportunistic origin of a new citrus pest. *Calif. Agric.* **34**: 4–5. [2]

Shapiro I. D. (1977). Interaction of population biology and mating behavior of the fiery skipper, *Hylephila phylaeus*. *Am. Midl. Nat.* **98**: 85–94. [2]

Sharp M. A. & Parks D. R. (1973). Habitat selection and population structure in *Plebejus saepiolus*. *J. Lepid. Soc.* **27**: 17–22. [2]

Sharp M. A., Parks D. R. & Ehrlich P. R. (1974). Plant resources and butterfly habitat selection. *Ecology* **55**: 870–875. [2]

Shaw M. R. & Askew R. R. (1976). Parasites. Pp.24–56 in Heath (1976). [10]

Sheldon W. G. (1929). *Lycaena arion* in South Devon: an enquiry as to its present existence. *Entomologist* **62**: 100. [33]

Shepard H. H. (1931–36). Hesperiidae: subfamilia Pyrginae. *Lepid. Cat.* **22** (47, 64, 69, 74): 677pp. [1]

Shepard H. H. (1933). Hesperiidae: subfamilia Ismeninae. *Lepid. Cat.* **21** (57): 55pp. [1]

Shepard H. H. (1936). Hesperiidae: subfamiliae: Euschemoninae et Trapezitinae. *Lepid. Cat.* **22** (77): 35pp. [1]

Shepard H. H. (1937–39). Hesperiidae: subfamilia: Hesperiinae I, II. *Lepid. Cat.* **21** (83, 90): 206pp. [1]

Shepherd J. G. (1974*a*). Activation of saturniid moth sperm by a secretion of the male reproductive tract. *J. Insect Physiol.* **20**: 2107–2122. [6]

Shepherd J. G. (1974*b*). Sperm activation in saturniid moths: some aspects of the mechanisms of activation. *J. Insect Physiol.* **20**: 2321–2328. [6]

Shepherd R. F. (1958). Factors controlling the internal temperatures of spruce budworm larvae, *Choristoneura fumiferana*. *Can. J. Zool.* **36**: 779–786. [28]

Sheppard P. M. (1951). A quantitative study of two populations of the moth *Panaxia dominula*. *Heredity* **5**: 349-378. [16]

Sheppard P. M. (1952). A note on non-random mating in the moth *Panaxia dominula*. *Heredity* **6**: 239–241. [21]

Sheppard P. M. (1953). Polymorphism and population studies. *Symp. Soc. exp. Biol.* **7**: 274–289. [16]

Sheppard P. M. (1959). The evolution of mimicry: a problem in ecology and genetics. *Cold. Spring Harb. Symp. Quant. Biol.* **24**: 131–140. [12]

Sheppard P. M. (1961*a*). Some contributions to population genetics resulting from the study of Lepidoptera. *Adv. Genet.* **10**: 165–216. [20]

Sheppard P. M. (1961*b*). Recent genetical work on polymorphic mimetic Papilios. *Symp. R. ent. Soc. Lond.* **(1)**: 20–29. [14]

Sheppard P. M. (1962). Some aspects of the geography, genetics, and taxonomy of a butterfly. *Publs Syst. Ass.* **(4)**: 135–152. [14]

Sheppard P. M. (1963). Some genetic studies on Müllerian mimics in butterflies of the genus *Heliconius*. *Zoologica, N.Y.* **48**: 145–154. [14]

Sheppard P. M. (1969). Evolutionary genetics of animal populations: the study of natural populations. *Proc. Int. Congr. Genet.* (12) **3**: 261-279. [16]

Sheppard P. M. (1975). *Natural selection and heredity* (4th edn). London. [14]

Sheppard P. M. & Cook L. M. (1962). The manifold effects of the medionigra gene in the moth *Panaxia dominula* and the maintenance of the polymorphism. *Heredity* **17**: 415–426. [16]

Sheppard P. M. & Turner J. R. G. (1977). The existence of Müllerian mimicry. *Evolution* **31**: 452–453. [14]

Sheppard P. M., Turner J. R. G., Brown K. S., Benson W. W. & Singer M. C. (1984). Genetics and the evolution of Muellerian mimicry in *Heliconius* butterflies. *Phil. Trans. R. Soc.* (B). (in press). [14]

Sherman P. W. & Watt W. B. (1973). The thermal ecology of some *Colias* butterfly larvae. *J. comp. Physiol.* **83**: 24–40. [28]

Shettleworth S. J. (1972*a*). The role of novelty in learned avoidance of unpalatable prey by domestic chicks (*Gallus gallus*). *Anim. Behav.* **20**: 29–35. [12, 14]

Shettleworth S. J. (1972*b*). Constraints on learning. *Adv. Study Behav.* **4**: 1–68. [14]

Shields O. (1968). Hilltopping. *J. Res. Lepid.* **6**: 69–78. [2, 20, 21, 33]

Shirôzu T. (1960). *Butterflies of Formosa in colour*. Osaka (in Japanese). [1]

Shirôzu T. & Hara A. (1960–2). *Early stages of Japanese butterflies in colour*. 2 vols. Osaka (in Japanese). [1, 6]

Shirôzu T. & Yamamoto H. (1956). A generic revision and the phylogeny of the tribe Theclini. *Sieboldia* **1**: 329–421. [1]

Shorey H. H. & McKelvey J. J. (eds) (1977). *Chemical control of insect behavior: theory and application*. New York. [25]

Shreeve T. G. & Mason C. F. (1980). The number of butterfly species in woodlands. *Oecologia* **45**: 414–418. [3]

Sibatani A. (1983). Kinji Imanishi and species identity. *Riv. biol.* **76**: 25–42. [23]

Sibatani A. & Grund R. B. (1978). A revision of the *Theclinesthes onycha* complex. *Trans. Lep. Soc. Japan* **29**: 1–34. [1]

Sidhu G. S. (1975). Gene-for-gene relationships in plant parasitic systems. *Sci. Progr., Lond.* (1906) **62**: 467–485. [14]

Siegel S. (1956). *Nonparametric statistics for the behavioral sciences*. New York. [20]

Silberglied R. E. (1977). Communication in the Lepidoptera. Pp.362-402 in *How animals communicate* (ed. Sebeok T. A.). Inc ana. [12, 20, 25]

Silberglied R. E. (1979). Communication in the ultraviolet. *A. Rev. Ecol. Syst.* **10**: 373-398. [20, 28]

Silberglied R. E. & Taylor O. R. (1973). Ultraviolet differences between the sulphur butterflies, *Colias eurytheme* and *C. philodice*, and a possible isolating mechanism. *Nature* **241**: 406-408. [20]

Silberglied R. E. & Taylor O. R. (1978). Ultraviolet reflection and its behavioral role in the courtship of the sulfur butterflies, *Colias eurytheme* and *C. philodice*. *Behav. Ecol. Sociobiol.* **3**: 203-243. [14, 20, 25]

Silberglied R. E., Aiello A. & Lamas G. (1979). Neotropical butterflies of the genus *Anartia*: systematics, life histories and general biology. *Psyche* **86**: 219-260. [20]

Silberglied R. E., Aiello A. & Windsor D. M. (1980). Disruptive coloration in butterflies: lack of support in *Anartia fatima*. *Science* **209**: 617-619. [16]

Simberloff D. (1982). The status of competition theory in ecology. *Ann. Zool. Fennici* **19**: 241-253. [Intro]

Simcox D. J. & Thomas J. A. (1980). *The Glanville fritillary—survey 1979*. JCCBI, London. [33]

Simon H. (1971). *The splendor of iridescence: structural colors in the animal world.* New York. [20]

Simpson G. G. (1953). *The major features of evolution.* New York. [12]

Sims S. R. (1980). Diapause dynamics and host plant suitability of *Papilio zelicaon*. *Am. Midl. Nat.* **103**: 375-384. [Intro]

Sims S. R. (1983). The genetic and environmental basis of pupal colour dimorphism in *Papilio zelicaon*. *Heredity* **50** 159-168. [Intro]

Singer M. C. (1971). Evolution of foodplant preferences in the butterfly *Euphydryas editha*. *Evolution* **25**: 383-389. [6, 7, 16]

Singer M. C. (1972). Complex components of habitat suitability in a butterfly species. *Science* **176**: 75-77. [2, 7]

Singer M. C. (1982*a*). Quantification of host preferences by manipulation of oviposition behavior in the butterfly *Euphydryas editha*. *Oecologia* **52**: 224-229. [2, 3, 6, 7]

Singer M. C. (1982*b*). Sexual selection for small size in male butterflies. *Am. Nat.* **119**: 440-443. [2, 7]

Singer M. C. & Ehrlich P. R. (1979). Population dynamics of the checkerspot butterfly *Euphydryas editha*. *Fortschr. Zool.* **25**: 53-60. [2]

Singer M. C. & Ehrlich P. R. (in press). Diversity and specificity among tropical grass-feeding butterflies. *J. Res. Lepid.* [2, 3]

Singer M. C. & Gilbert L. E. (1978). Ecology of butterflies in the urbs and suburbs. Pp.1-11 in *Perspectives in urban entomology* (eds Frankie G. W. & Koehler C. S.). New York. [3]

Singer M. C. & Mandracchia J. (1982). On the failure of two butterfly species to respond to the presence of conspecific eggs prior to oviposition. *Ecol. Ent.* **7**: 327-330. [6, 7]

Singer M. C. & Wedlake P. (1981). Capture does affect probability of recapture in a butterfly species. *Ecol. Ent.* **6**: 215-216. [Intro, 2, 4]

Singer M. C., Ehrlich P. R. & Gilbert L. E. (1971). Butterfly feeding on lycopsid. *Science* **172**: 1341-1342. [6]

Sisson R. F. (1980). Deception: formula for survival. *Natn geogr. Mag.* **157**: 394-415. [22]

Skertchly S. B. J. (1889). On the habits of certain Bornean butterflies. *Ann. Mag. nat. Hist.* (6) **4**: 209-215. [25]

Slansky F. (1974). Relationship of larval food-plants and voltinism patterns in temperate butterflies. *Psyche* **81**: 243-253. [2]

Slansky F. & Feeny P. (1977). Stabilization of the rate of nitrogen accumulation by larvae of the cabbage butterfly on wild and cultivated food plants. *Ecol. Monogr.* **47**: 209-228. [19]

Slater J. W. (1877). On the food of gaily-coloured caterpillars. *Trans. ent. Soc. Lond.* **1877**: 205-209. [12]

Smart P. (1976). *The illustrated encyclopedia of the butterfly world.* London. [1]

Smiles R. L. (1982). The taxonomy and phylogeny of the genus *Polyura*. *Bull. Br. Mus. nat. Hist. (Ent.)* **44**: 115-237. [1]

Smiley J. T. (1978*a*). Plant chemistry and the evolution of host specificity: new evidence from *Heliconius* and *Passiflora*. *Science* **201**: 745-747. [3, 6, 7, 32]

Smiley J. T. (1978*b*). *Host plant ecology of* Heliconius *butterflies in north-eastern Costa Rica*. PhD thesis, Texas Univ., Austin, USA. [3]

Smiley J. T. (In prep.). Exploitation of host plant by a *Heliconius* butterfly community in two adjacent habitats. [3]

Smith C. J. (1980). *Ecology of the English chalk*. London. [33]

Smith D. A. S. (1973). Negative non-random mating in the polymorphic butterfly *Danaus chrysippus* in Tanzania. *Nature* **242**: 131-132. [20, 21]

Smith D. A. S. (1975*a*). Genetics of some polymorphic forms of the African butterfly *Danaus chrysippus*. *Entomologica scand.* **6**: 134-144. [21]

Smith D. A. S. (1975*b*). All-female broods in the polymorphic butterfly *Danaus chrysippus* and their ecological significance. *Heredity* **34**: 363-371. [21]

Smith D. A. S. (1975*c*). Sexual selection in a wild population of the butterfly *Danaus chrysippus*. *Science* **187**: 664-665. [20, 21]

Smith D. A. S. (1976*a*). Phenotypic diversity, mimicry and natural selection in the African butterfly *Hypolimnas misippus*. *Biol. J. Linn. Soc.* **8**: 183-204. [20, 21]

Smith D. A. S. (1976*b*). Evidence for autosomal meiotic drive in the butterfly *Danaus chrysippus*. *Heredity* **36**: 139-142. [21]

Smith D. A. S. (1978). The effect of cardiac glycoside storage on growth rate and adult size in the butterfly *Danaus chrysippus*. *Experientia* **34**: 845-846. [15]

Smith D. A. S. (1979). The significance of beak marks on the wings of an aposematic, distasteful and polymorphic butterfly. *Nature* **281**: 215-216. [12]

Smith D. A. S. (1980). Heterosis, epistasis and linkage disequilibrium in a wild population of the polymorphic butterfly *Danaus chrysippus*. *Zool. J. Linn. Soc.* **69**: 87-109. [18, 20, 21]

Smith D. A. S. (1981). Heterozygous advantage expressed

through sexual selection in a polymorphic African butterfly. *Nature* **289**: 174–175. [20, 21]
Smith R. & Brown D. (1979). *The Lepidoptera of Warwickshire: a provisional list. Part 1.* BRC, Warwick. [33]
Smith S. D. (1981). Preliminary report: effects of Mt St Helens ashfall on lotic Trichoptera. *Melanderia* **37**: 56–62. [31]
Smith S. G. (1945). The diagnosis of sex by means of heteropycnosis. *Scient. Agric.* **25**: 566–571. [24]
Smith S. M. (1975). Innate recognition of coral snake pattern by a possible avian predator. *Science* **187**: 759–760. [12]
Smith S. M. (1977). Coral snake pattern: recognition and stimulus generalisation by naïve great kiskadees (Aves: Tyrannidae). *Nature* **265**: 535–536. [12]
Smith S. M. (1978). Predatory behaviour of young great kiskadees (*Pitangus sulphuratus*). *Anim. Behav.* **26**: 988–995. [12]
Smithers C. N. (1977). Seasonal distribution and breeding status of *Danaus plexippus* in Australia. *J. Aust. ent. Soc.* **16**: 175–184. [26]
Snow D. W. (1971). Evolutionary aspects of fruit-eating by birds. *Ibis* **113**: 194–202. [12]
Sondheimer E. & Simeone J. B. (eds) (1970). *Chemical ecology.* New York.
Sondhi K. C. (1963). The biological foundation of animal patterns. *Q. Rev. Biol.* **38**: 289–327. [16]
Sotthibandhu S. & Baker R. R. (1979). Celestial orientation by the large yellow underwing moth, *Noctua pronuba. Anim. Behav.* **27**: 786–800. [26]
Soulé M. (1967). Phenetics of natural populations. II. Asymmetry and evolution in a lizard. *Am. Nat.* **101**: 141–160. [16]
Soulé M. (1980). Thresholds for survival: maintaining fitness and evolutionary potential. Pp.151–169 in Soulé & Wilcox (1980). [33]
Soulé M. & Baker B. (1968). Phenetics . . . IV. The population asymmetry parameter in the butterfly *Coenonympha tullia. Heredity* **23**: 611–614. [16]
Soulé M. & Wilcox B. A. (eds) (1980). *Conservation biology: an evolutionary-ecological perspective.* Sunderland, Mass.
South R. (1941). *The butterflies of the British Isles* (3rd edn). London. [7]
Southwood T. R. E. (1962). Migration of terrestrial arthropods in relation to habitat. *Biol. Rev.* **37**: 171–214. [26]
Southwood T. R. E. (ed.) (1968). Insect abundance. *Symp. R. ent. Soc. Lond.* **(4)**. Oxford. [33]
Southwood T. R. E. (1971). Farm management in Britain and its effect on animal population. *Proc. tall Timb. Conf. ecol. Anim. Control* **1971**: 29–51. [33]
Southwood T. R. E. (1977). Habitat, the templet for ecological strategies? *J. anim. Ecol.* **46**: 337–365. [26]
Speiss E. B. & Langer B. (1964*a*). Mating speed control by gene arrangements in *Drosophila pseudoobscura* homokaryotypes. *Proc. natn Acad. Sci. U.S.A.* **51**: 1015–1019. [21]
Speiss E. B. & Langer B. (1964b). Mating speed control by gene arrangement carriers in *Drosophila persimilis. Evolution* **18**: 430–444. [21]
Speiss E. B., Langer B. & Speiss L. D. (1966). Mating control by gene arrangements in *Drosophila pseudoobscura. Genetics* **54**: 1139–1149. [21]
Spooner G. M. (1963). On causes of the decline of *Maculinea arion* in Britain. *Entomologist* **96**: 199–210. [33]
Stamp N. E. (1980). Egg deposition patterns in butterflies: why do some species cluster their eggs rather than deposit them singly? *Am. Nat.* **115**: 367-380. [6, 12]
Stamp N. E. (1981*a*). Effect of group size on parasitism in a natural population of the Baltimore checkerspot *Euphydryas phaeton. Oecologia* **49**: 201–206. [6]
Stamp N. E. (1981*b*). Parasitism of single and multiple egg clusters of *Euphydryas phaeton. Jl N. Y. ent. Soc.* **89**: 89–97. [6]
Stanley S. M. (1979). *Macroevolution: pattern and progress.* San Francisco. [14]
Stanton M. L. (1979). The role of chemotactile stimuli in the oviposition preferences of *Colias* butterflies. *Oecologia* **39**: 79–91. [6]
Stanton M. L. (1980). *The dynamics of search: foodplant selection by* Colias *butterflies.* PhD thesis, Harvard Univ., Cambridge, Mass., USA. [6]
Stanton M. L. (1982*a*). Searching in a patchy environment: foodplant selection of *Colias philodice* butterflies. *Ecology* **63**: 839-853. [6]
Stanton M. L. (1983). Spatial patterns in the plant community and their effects upon insect search. In Ahmad (1983). [6]
Steele R. C. & Schofield J. M. (1973). Conservation and management. Pp.296–337 in Steele & Welch (1973). [33]
Steele R. C. & Welch R. C. (eds) (1973). *Monks Wood: a nature reserve record.* Huntingdon, UK.
Steinbrecht R. A. & Schneider D. (1980). Pheromone communication in moths sensory physiology and behaviour. Pp.685–703 in Locke & Smith (1980). [25]
Stempffer H. (1967). The genera of the African Lycaenidae. *Bull. Br. Mus. nat. Hist. (Ent.)* (Suppl.) **(10)**: 322pp. [1, 6]
Stempffer H. & Bennett N. H. (1953). A revision of the genus *Teriomima. Bull. Br. Mus. nat. Hist. (Ent.)* **3**: 77-104. [1]
Stempffer H. & Bennett N. H. (1958). Révision des genres appartenant au groupe des *Iolaus. Bull. Inst. fr. Afr. noire* **20**(A): 1243–1347. [1]
Stern V. M. & Smith R. F. (1960). Factors affecting egg production and oviposition in populations of *Colias eurytheme. Hilgardia* **29**: 411–454. [6]
Sternberg J. G., Waldbauer G. P. & Jeffords M. R. (1977). Batesian mimicry: selective advantage of color pattern. *Science* **195**: 681–683. [22]
Stichel H. (1909). Brassolidae. *Tierreich* **25**: xiv + 244pp. [1]
Stichel H. (1912). Amathusiidae. *Tierreich* **34**: xv + 248pp. [1]
Stichel H. (1928). Nemeobiinae. *Tierreich* **51**: xxx + 330pp. [1]
Stichel H. (1930–31). Riodinidae I–IV. *Lepid. Cat.* **26** (38, 40, 41, 44): 795pp. [1]
Stichel H. (1932). Brassolidae. *Lepid. Cat.* **27** (51): 115pp. [1]

Stichel H. (1933). Amathusiidae. *Lepid. Cat.* **27** (54): 171pp. [1]

Stichel H. (1938). Nymphalidae I: subfam.: Dioninae, Anetiinae, Apaturinae. *Lepid. Cat.* **30** (86): 374pp. [1]

Stichel H. (1939). Nymphalidae II, III: subfam.: Charaxidinae I, II. *Lepid. Cat.* **30** (91, 93): 375–794. [1]

Stichel H. & Riffarth H. (1905). Heliconiidae. *Tierreich* **22**: xv + 290pp. [1]

Stiles F. G. (1964). *Comparative bird behavior and the evolution of Müllerian mimicry.* Snr Hons thesis, Amherst College, Mass., USA. [12]

Stobbe R. H. (1912). Die abdominalen Duftorgane der männlichen Sphingiden und Noctuiden. *Zool. Jb.* **32**: 493–532. [25]

Stoddart D. M. (1976). Two hypotheses supporting the social function of odorous secretions of some old world rodents. Pp.333–355 in Müller-Schwarze & Mozell (1976). [12]

Stoddart D. M. (1979). Some responses of a free-living community of rodents to the odors of predators. Pp.1–10 in Müller-Schwarze & Silverstein (1979). [12]

Straatman R. (1962). Notes on certain Lepidoptera ovipositing on plants which are toxic to their larvae. *J. Lepid. Soc.* **16**: 99–103. [6]

Straatman R. (1970). *Summary of survey on ecology of* Ornithoptera alexandrae. Consultant rept. to Dept. Agric. Stock Fish., Papua New Guinea. [32]

Straatman R. (1971). The life history of *Ornithoptera alexandrae. J. Lepid. Soc.* **25**: 58–64. [32]

Straatman R. & Nieuwenhuis E. J. (1961). Biology of certain Sumatran species of *Atrophaneura, Trogonoptera,* and *Troides. Tijdschr. Ent.* **104**: 31–43. [6]

Stradling D. J. (1976). The nature of the mimetic patterns of the brassolid genera *Caligo* and *Eryphanis. Ecol. Ent.* **1**: 135–138. [16]

Stride G. O. (1956). On the courtship behaviour of *Hypolimnas misippus,* with notes on the mimetic association with *Danaus chrysippus. Br. J. Anim. Behav.* **4**: 52–68. [20, 21]

Stride G. O. (1957). Investigations into the courtship behaviour of the male of *Hypolimnas misippus,* with special reference to the role of visual stimuli. *Br. J. Anim. Behav.* **5**: 153–167. [20, 21]

Stride G. O. (1958*a*). On the courtship behaviour of a tropical mimetic butterfly, *Hypolimnas misippus. Proc. Int. Congr. Ent.* (10) **2**: 419–424. [20]

Stride G. O. (1958*b*). Further studies on the courtship behaviour of African mimetic butterflies. *Anim. Behav.* **6**: 224–230. [20, 21]

Strong D. R. (1978). Biogeographic dynamics of insect–host plant communities. *A. Rev. Ent.* **24**: 89–119. [2]

Strong D. R., Szyska L. A. & Simberloff D. S. (1979). Tests of community-wide character displacement against null hypotheses. *Evolution* **33**: 897–913. [3]

Stubbs A. E. (1981). Invertebrate conservation in the Nature Conservancy Council, Great Britain. *Atala* **7**: 58–60. [33]

Sturkie P. D. (1965). *Avian physiology* (2nd edn). Ithaca, N.Y. [12]

Süffert F. (1924). Bestimmungsfaktoren des zeichnungsmusters beim saison dimorphismus von *Araschnia levana-prorsa. Biol. Zbl.* **44**: 173–188. [16]

Süffert F. (1927). Zur vergleichenden Analyse der Schmetterlingszeichnung. *Biol. Zbl.* **47**: 385–413. [16]

Süffert F. (1929). Morphologische Erscheinungsgouppen in der Flügelzeichnung der Schmetterlinge, insbesondere die Querbindenzeichnung. *Wilhelm Roux Arch. EntwMech. Org.* **120**: 299–383. [16]

Suomalainen E., Cook L. M. & Turner J. R. G. (1974). Achiasmatic oogenesis in heliconiine butterflies. *Hereditas* **74**: 302–304. [21]

Suomalainen E., Lokki J. & Saura A. (1981). Genetic polymorphism and evolution in parthenogenetic animals. X. *Solenobia* species. *Hereditas* **95**: 31–35. [17]

Sutton R. A. (1981). The ecology and conservation of the silver studded blue butterfly (*Plebejus argus*). *Br. Butterfly Conserv. Soc. News* **(26)**: 37–39. [33]

Suzuki T., Uehara J. & Sugawara R. (1979). Osmeterial secretions of papilionid butterflies. I. Identification of myrcene from *Luehdorfia japonica* and *L.puziloi inexpectata. Appl. Ent. Zool.* **14**: 346–349. [12]

Swain T. (1976*a*). Secondary compounds: primary products. In *Secondary metabolism and coevolution* (eds Luckner M., Mothes K. & Nover L.). Leipzig. [12]

Swain T. (1976*b*). Angiosperm-reptile co-evolution. Pp.107–122 in *Morphology and biology of reptiles* (eds Bellairs A. d' & Cox C. B.). London. [12]

Swain T. (1977). Secondary compounds as protective agents. *A. Rev. Pl. Physiol.* **28**: 479–501. [12]

Swain T. (1978). Plant-animal coevolution: a synoptic view of the paleozoic and mesozoic. Pp.3–19 in Harborne (1978). [12]

Swain T. (1979). Tannins and lignins. Pp.657–682 in Rosenthal & Janzen (1979). [12]

Swihart C. A. (1971). Colour discrimination by the butterfly *Heliconius charitonius. Anim. Behav.* **19**: 156–164. [20]

Swihart C. A. & Swihart S. L. (1970). Colour selection and learned feeding preferences in the butterfly, *Heliconius charitonius. Anim. Behav.* **18**: 60–64. [20]

Swihart S. L. (1963).The electroretinogram of *Heliconius erato* and its possible relation to established behavior patterns. *Zoologica, N.Y.* **48**: 155–165. [20]

Swihart S. L. (1964). The nature of the electroretinogram of a tropical butterfly. *J. Insect Physiol.* **10**: 547–562. [20]

Swihart S. L. (1965). Evoked potentials in the visual pathway of *Heliconius erato. Zoologica, N.Y.* **50**: 55–62. [20]

Swihart S. L. (1967*a*). Neural adaptations in the visual pathway of certain heliconiine butterflies, and related forms, to variations in wing coloration. *Zoologica, N.Y.* **52**: 1–14. [20]

Swihart S. L. (1967*b*). Hearing in butterflies (*Heliconius, Ageronia*). *J. Insect Physiol.* **13**: 469–476. [Intro, 20]

Swihart S. L. (1968). Single unit activity in the visual pathway of the butterfly *Heliconius erato. J. Insect Physiol.* **14**: 1589–1601. [20]

Swihart S. L. (1969). Colour vision and the physiology of the superposition eye of a butterfly. *J. Insect Physiol.* **15**: 1347–1365. [20]

Swihart S. L. (1970). The neural basis of colour vision in the butterfly, *Papilio troilus*. *J. Insect Physiol.* **16**: 1623–1636. [20]

Swihart S. L. (1972*a*). The neural basis of colour vision in the butterfly, *Heliconius erato*. *J. Insect Physiol.* **18**: 1015–1025. [20]

Swihart S. L. (1972*b*). Modelling the butterfly visual pathway. *J. Insect Physiol.* **18**: 1915–1928. [20]

Swihart S. L. (1973). Retinal duality in the butterfly *Heliconius charitonius*. *J. Insect Physiol.* **19**: 2035–2051. [20]

Swihart S. L. & Schümperli R. (1974). Visual fields of butterfly interneurons. *J. Insect Physiol.* **20**: 1529–1536. [20]

Swinton A. H. (1908). The family tree of moths and butterflies traced on their organs of sense. *Societas ent.* **23**: 99–101, 114–116, 124–126, 131–132, 140–141, 148–150, 156–158, 162–165. [25]

Swynnerton C. F. M. (1915*a*). Experiments on some carnivorous insects, especially the driver ant *Dorylus*, and with butterflies' eggs as prey. *Trans. ent. Soc. Lond.* **1915**: 317–350. [6]

Swynnerton C. F. M. (1915*b*). Further notes on the eggs of butterflies. *Trans. ent. Soc. Lond.* **1915**: 428–430. [6]

Swynnerton C. F. M. (1915*c*). Concluding discussion. In: G. D. H. Carpenter (1942). Observations and experiments in Africa by the late C. F. M. Swynnerton on wild birds eating butterflies and the preferences shown. *Proc. Linn. Soc. Lond.* **154**: 10–46. [12]

Swynnerton C. F. M. (1915*d*). Birds in relation to their prey. Experiments on wood-hoopoes, small hornbills, and a babbler. *Jl S. Afr. Orn. Un.* **1915**: 22–108. [12]

Swynnerton C. F. M. (1919). Experiments and observations bearing on the explanation of form and colouring, 1908–1913, Africa. *Zool. J. Linn. Soc.* **33**: 203–385. [12]

Swynnerton C. F. M. (1926). An investigation into the defences of butterflies of the genus *Charaxes*. *Proc. Int. Congr. Ent.* (3) **2**: 478–506. [16]

Szentesi A., Toth M. & Dobrovolszky A. (1975). Evidence and preliminary investigations on a male aphrodisiac and a female sex pheromone in *Mamestra brassicae*. *Acta phytopath.* **10**: 425–429. [25]

Tabashnik B. E. (1980). Population structure of pierid butterflies. III. Pest populations of *Colias philodice eriphyle*. *Oecologia* **47**: 175–183. [2]

Tabashnik B. E. (1981). *Evolution into a pest niche:* Colias *butterflies and alfalfa*. PhD thesis, Stanford Univ., California, USA. [6]

Tabashnik B. E. (1982). Responses of pest and non-pest *Colias* butterfly larvae to intraspecific variation in leaf nitrogen and water content. *Oecologia* **55**: 389–394. [2]

Tabashnik B. E. (1983). Host range evolution: the shift from native legume hosts to alphalpha by the butterfly, *Colias philodice eriphyle*. *Evolution* **37**: 150–162. [2]

Tabashnik B. E. (1983). Host range evolution: the shift from native legume hosts to alfalfa by the butterfly, *Colias philodice eriphyle*. *Evolution* **37**: 150–162. [2]

Takata N. (1961*a*). Studies on the host preference of the common cabbage butterfly *Pieris rapae crucivora*. XI. Continued studies on the oviposition preference of adult butterflies. *Jap. J. Ecol.* **11**: 124–133. [6]

Takata N. (1961*b*). Studies . . . XII. Successive rearing of the cabbage butterfly with certain foodplants and its effect on the oviposition preference of the adult. *Jap. J. Ecol.* **11**: 147–154. [6]

Takata N. & Isheda H. (1957). Studies . . . II. Preference between cabbage and radish for oviposition. *Jap. J. Ecol.* **7**: 56–58. [2]

Talbot G. (1928–37). *A monograph of the pierine genus* Delias. Witley; completed London. [1]

Talbot G. (1932–35). Pieridae I–III. *Lepid. Cat.* **23** (53, 60, 66): 697pp. [1]

Talbot G. (1939). Butterflies I. *Fauna Br. India* **1939**: 600pp. [1]

Talbot G. (1949). Butterflies II. *Fauna India* **1949**: 506pp. [1]

Tanada Y. (1965). Factors affecting the susceptibility of insects to viruses. *Entomophaga* **10**: 139–150. [10]

Tarpy R. M. & Mayer R. E. (1978). *Foundations of learning and memory*. Glenview, Illinois. [12]

Tatham J. T. (1981). The British Butterfly Conservation Society. *Atala* **7**: 53–54. [33]

Taylor L. R. & Taylor R. A. J. (1977). Aggregation, migration and population mechanics. *Nature* **265**: 415–421. [2]

Taylor L. R. & Taylor R. A. J. (1979). A behavioural model for the evolution of spatial dynamics. Pp.1–27 in *Population dynamics* (eds Anderson R. M., Turner B. D. & Taylor L. R.). Oxford. [2]

Taylor L. R. & Woiwod I. P. (1980). Temporal stability as a density-dependent species characteristic. *J. anim. Ecol.* **49**: 209–224. [26]

Taylor L. R., French R. A. & Macaulay E. D. M. (1973). Low-altitude migration and diurnal flight periodicity: the importance of *Plusia gamma*. *J. anim. Ecol.* **42**: 751–760. [26]

Taylor O. R. (1972). Random vs. non-random mating in the sulfur butterflies, *Colias eurytheme* and *C. philodice*. *Evolution* **26**: 344–356. [20, 25]

Taylor O. R. (1973*a*). Reproductive isolation in *Colias eurytheme* and *C. philodice*: use of olfaction in mate selection. *Ann. ent. Soc. Am.* **66**: 621–626. [25]

Taylor O. R. (1973*b*). A non-genetic "polymorphism" in *Anartia fatima*. *Evolution* **27**: 161–164. [20]

Telfer W. H. (1965). The mechanism and control of yolk formation. *A. Rev. Ent.* **10**: 161–184. [6]

Telfer W. H. & Rutberg L. D. (1960). The effects of blood protein depletion on the growth of the oocytes in the cecropia moth. *Biol. Bull.* **118**: 352–366. [6]

Temple V. (1953). Some notes on the courtship of butterflies in Britain. *Entomologist's Gaz.* **4**: 141–161. [25]

Templeton A. R. (1980). Modes of speciation and inferences based on genetic distances. *Evolution* **34**: 719–729. [19]

Thibout E. (1972). Etude du comportement de pariade d'*Acrolepia assectella*. *Revue Comporte. Anim.* **6**: 157–164. [25]

Thoday J. M. & Boam T. B. (1959). Effects of disruptive selection. II. Polymorphism and divergence without isolation. *Heredity* **13**: 205–218. [16]

Thomas D. (1978). Short lived phenomenon. *News Lepid. Soc.* **1978** (3[4]): 7. [25]

Thomas J. A. (1973). The hairstreaks of Monks Wood. Pp.153–158 in Steele & Welch (1973). [33]

Thomas J. A. (1974). *Factors influencing the numbers and distribution of the brown hairstreak,* Thecla betulae, *and the black hairstreak,* Strymonidia pruni. PhD thesis, Leicester Univ., UK. [2, 33]

Thomas J. A. (1976*a*). *The ecology and conservation of the large blue butterfly.* Internal report, ITE, UK. [33]

Thomas J. A. (1976*b*). *The black hairstreak butterfly.* Internal rpt., ITE, UK. [33]

Thomas J. A. (1977). The ecology of the large blue butterfly. *Rep. Inst. terr. Ecol.* **1976**: 25–27. [33]

Thomas J. A. (1980*a*). Why did the large blue become extinct in Britain? *Oryx* **15**: 243–247. [2, 33]

Thomas J. A. (1980*b*). The extinction of the large blue and the conservation of the black hairstreak butterflies (a contrast of failure and success). *Rep. Inst. terr. Ecol.* **1979**: 19–23. [2, 33]

Thomas J. A. (1981*a*). Butterfly Year 1981–82. *Atala* **7**: 52–54. [33]

Thomas J. A. (1981*b*). Insect conservation in Britain: some case histories. *Atala* **6**: 31–36. [33]

Thomas J. A. (1982). Butterfly Countdown. *Natural World* **5**: 30–31. [33]

Thomas J. A. (1983*a*). The ecology and conservation of *Lysandra bellargus* in Britain. *J. app. Ecol.* **20**: 59–83. [33]

Thomas J. A. (1983*b*). A quick method for estimating butterfly numbers during surveys. *Biol. Conserv.* **27**: 195–211. [33]

Thomas J. A. (1983*c*). The ecology and status of *Thymelicus acteon* in Britain. *Ecol. Ent.* **8**: 427-435. [33]

Thomas J. A. & Merrett P. (1980). Observations of butterflies in the Purbeck Hills in 1976 and 1977. *Proc. Dorset nat. Hist. archaeol. Soc.* **99**: 112–119. [33]

Thomas J. A. & Simcox D. J. (1982). A quick method for estimating larval populations of *Melitaea cinxia. Biol. Conserv.* **22**: 315–322. [33]

Thomashow P. M. (1975). *The paradox of the cryptic chrysalid.* Snr Hons thesis, Amherst College, Mass., USA. [12]

Thompson J. A. (1944). A new subspecies of *Eumenis semele. Entomologist's Rec. J. Var.* **56**: 65. [16]

Thomson G. (1971). The possible existence of temporal sub-speciation in *Maniola jurtina. Entomologist's Rec. J. Var.* **83**: 87–90. [16]

Thomson G. (1973). Geographical variation of *Maniola jurtina. Tijdschr. Ent.* **116**: 185–226. [16]

Thornhill R. (1976). Sexual selection and paternal investment in insects. *Am. Nat.* **110**: 153–163. [20]

Thornhill R. (1979). Male and female sexual selection and the evolution of mating strategies in insects. Pp.81–121 in Blum & Blum (1979). [20]

Thornhill R. (1980). Competitive, charming males and choosy females: was Darwin correct? *Fla Entom.* **63**: 5–30. [20, 25]

Thorpe W. H. (1956). *Learning and instinct in animals.* London. [12]

Thorsteinson A. J. (1953). The chemotactic responses that determine host specificity in an oligophagous insect (*Plutella maculipennis*). *Can. J. Zool.* **31**: 52–72. [6]

Thorsteinson A. J. (1960). Host selection in phytophagous insects. *A. Rev. Ent.* **5**: 193–218. [6]

Tietz H. M. (1972). *An index to the described life histories, early stages and hosts of the macrolepidoptera of the continental United States and Canada* (2 vols). Sarasota, Fla. [1]

Tilden J. W. (1980). Attempted mating between male monarchs. *J. Res. Lepid.* **18**: 2. [20]

Timofeef-Ressovsky N. W. (1940). Zur analyse des Polymorphismus bei *Adalia bipunctata. Biol. Zbl.* **60**: 130–137. [27]

Tinbergen L. (1960). The natural control of insects in pinewoods. I. Factors influencing the intensity of predation by songbirds. *Archs néerl. Zool.* **13**: 265–343. [10]

Tinbergen N. (1941). Ethologische Beobachtungen am Samtfalter, *Satyrus semele. J. Orn.* **89** (3): 132–144. [25]

Tinbergen N. (1958). *Curious naturalists.* London. [16, 20]

Tinbergen N., Meeuse B. J. D., Boerema L. K. & Varossieau W. W. (1942). Die Balz des Samtfalters, *Eumenis* (= *Satyrus*) *semele. Z. Tierpsychol.* **5**: 182–226. (English translation: Pp.197–249 in *The animal in its world: explorations of an ethologist, 1932–1972,* 1, *Field studies.* Cambridge, Mass., 1972.) [20, 25]

Tite G. E. (1963*a*). A revision of the genus *Candalides* and allied genera. *Bull. Br. Mus. nat. Hist. (Ent.)* **14**: 197–259. [1]

Tite G. E. (1963*b*). A synonymic list of the genus *Nacaduba* and allied genera. *Bull. Br. Mus. nat. Hist. (Ent.)* **13**: 67–116. [1]

Tite G. E. & Dickson C. G. C. (1973). The genus *Aloeides* and allied genera. *Bull. Br. Mus. nat. Hist. (Ent.)* **29**: 225–280. [1]

Tojo S., Kiguchi K. & Kimura S. (1981). Hormonal control of storage protein synthesis and uptake by the fat body in the silkworm, *Bombyx mori. J. Insect Physiol.* **27**: 491–497. [6]

Tokanaga M. (1940). Ceratopogonidae and Chironomidae from the Micronesian Islands. *Philipp. J. Sci.* **71**: 205–230. [11]

Tokanaga M. (1960). Notes on biting midges. *I. Akitu* **9**: 72–76. [11]

Toliver M. E., Sternberg J. G. & Waldbauer G. P. (1979). Early flight periods of male *Callosamia promethea. J. Lepid. Soc.* **33**: 232–238. [22]

Traut W. & Mosbacher G. C. (1968). Geschlects-chromatin bei Lepidopteren. *Chromosoma* **25**: 343–356. [24]

Traynier R. M. M. (1979). Long-term changes in the oviposition behavior of the cabbage butterfly, *Pieris rapae,* induced by contact with plants. *Physiol. Ent.* **4**: 87–96. [6]

Treusch H. W. (1967). Bisher unbekanntes gezieltes Duftanbieten paarungsbereiter *Argynnis paphia*-Weibchen. *Naturwissenschaften* **54**: 592. [25]

Trimen R. (1869). On some remarkable mimetic analogies among African butterflies. *Trans. Linn. Soc. Lond.* **26**: 497–522. [12]

Trimen R. (1887). *South African butterflies* **1**. London. [12]

Trivers R. L. (1972). Parental investment and sexual selection. Pp.136-179 in Campbell (1972). [20, 21]

Trivers R. L. (1974). Parent-offspring conflict. *Am. Zool.* **14**: 249-264. [26]

Tsubaki Y. (1973). The natural mortality and its factors of the immature stages in the population of the swallow-tail butterfly *Papilio xuthus*. *Jap. J. Ecol.* **23**: 210-217. [2, 10]

Tsubaki N. (1977). A bioeconomic study on the natural population of the swallowtail butterfly, *Papilio xuthus*. *Ebino Biol. Lab. Kyushu Univ.* **2**: 37-54. [2]

Tsukada E. & Nishiyama Y. (1980). *Papilionidae. Butterflies of the south-east Asian islands* **1** (ed. Tsukada E.). Tokyo. [1]

Tudor O. & Parkin D. T. (1979). Studies on phenotypic variation in *Maniola jurtina* in the Wyre Forest, England. *Heredity* **42**: 91-104. [16]

Turcek F. J. (1963). Color preference in fruit- and seed-eating birds. *Proc. Int. Orn. Congr.* **(13)**: 285-292. [12]

Turner E. R. A. (1965). Social feeding in birds. *Behaviour* **24**: 1-46. [12]

Turner J. R. G. (1963). A quantitative study of a Welsh colony of the large heath butterfly, *Coenonympha tullia*. *Proc. R. ent. Soc. Lond.* (A) **38**: 101-112. [2, 16, 33]

Turner J. R. G. (1965). Evolution of complex polymorphism and mimicry in distasteful South American butterflies. *Proc. Int. Congr. Ent.* **(12)**: 267. [14]

Turner J. R. G. (1971*a*). Experiments on the demography of tropical butterflies. II. Longevity and home-range behavior in *Heliconius erato*. *Biotropica* **3**: 21-31. [2, 12, 26]

Turner J. R. G. (1971*b*). Studies of Müllerian mimicry and its evolution in burnet moths and heliconid butterflies. Pp.224-260 in Creed (1971). [12, 21]

Turner J. R. G. (1972). The genetics of some polymorphic forms of the butterflies *Heliconius melpomene* and *H. erato*. II. The hybridisation of subspecies of *H. melpomene* from Suriname and Trinidad. *Zoologica, N.Y.* **56**: 125-157. [14]

Turner J. R. G. (1975*a*). A tale of two butterflies. *Nat. Hist., N.Y.* **84**(2): 28-37. [12, 14, 20]

Turner J. R. G. (1975*b*). Communal roosting in relation to warning colour in two heliconiine butterflies. *J. Lepid. Soc.* **29**: 221-226. [12, 14]

Turner J. R. G. (1976). Sexual behaviour: female swift moth is not the aggressive partner. *Anim. Behav.* **24**: 188-189. [25]

Turner J. R. G. (1977*a*). Butterfly mimicry: the genetical evolution of an adaptation. *Evolut. Biol.* **10**: 163-206. [12, 14, 16, 18, 20]

Turner J. R. G. (1977*b*). Forest refuges as ecological islands: disorderly extinction and the adaptive radiation of Muellerian mimics. Pp.98-117 in *Biogeographie et evolution en Amerique tropicale* (ed. Descimon H.). Paris. [14]

Turner J. R. G. (1978). Why male butterflies are non-mimetic: natural selection, sexual selection, group selection, modification and sieving. *Biol. J. Linn. Soc.* **10**: 385-432. [14, 20, 23]

Turner J. R. G. (1979). Genetic control of recombination in the silkworm. I. Multigenic control of chromosome 2. *Heredity* **43**: 273-293. [14]

Turner J. R. G. (1981). Adaptation and evolution in *Heliconius*: a defense of neoDarwinism. *A. Rev. Ecol. Syst.* **12**: 99-121. [12, 14]

Turner J. R. G. (1982). How do refuges produce biological diversity? Allopatry and parapatry, extinction and gene flow in mimetic butterflies. Pp.309-335 in Prance (1982). [14]

Turner J. R. G. (1983*a*). Mimetic butterflies and punctuated equilbria: some old light on a new paradigm. *Biol. J. Linn. Soc.* **20**: 277-300. [14]

Turner J. R. G. (1983*b*). "The hypothesis that explains mimetic resemblance explains evolution"; the gradualist-saltationist schism. Pp.129-169 in *Dimensions of Darwinism* (ed. Grene M.). Cambridge & Paris. [14]

Turner J. R. G. & Sheppard P. M. (1975). Absence of crossing-over in female butterflies (*Heliconius*). *Heredity* **34**: 265-269. [14, 21]

Tutt J. W. (1899). *A natural history of the British Lepidoptera* **1**. London. [6]

Unamba J. A. (1968). *Ecological genetics of the butterfly* Hypolimnas misippus. PhD thesis, Univ. Sierra Leone, Freetown. [21]

Urbahn E. (1913). Abdominale Duftorgane bei weiblichen Schmetterlingen. *Jena Z. Naturw.* **50**: 277-358. [25]

Ureta R., E. (1941). Lepidopteros Ropaloceros de Bolivia. *Boln Mus. nac. Hist. nat. Chile* **19**: 31-41. [1]

Ureta R., E. (1963). Catologo de Lepidopteros de Chile. *Boln Mus. nac. Hist. nat. Chile* **28**: 53-149. [1]

Urquhart F. A. (1960). *The monarch butterfly*. Toronto. [6, 12, 26]

Urquhart F. A. (1976). Found at last: the monarch's winter home. *Natn Geogr. mag.* **150**: 161-173. [26]

Urquhart F. A. & Urquhart N. R. (1978). Autumnal migration routes of the eastern population of the monarch butterfly (*Danaus p. plexippus*) in North America to the over-wintering site in the Neovolcanic Plateau of Mexico. *Can. J. Zool.* **56**: 1754-1764. [26]

Vaidya V. G. (1969*a*). Form perception in *Papilio demoleus*. *Behaviour* **33**: 212-221. [6]

Vaidya V. G. (1969*b*). Investigations of the role of visual stimuli in egg-laying and resting behavior of *Papilio demoleus*. *Anim. Behav.* **17**: 350-355. [6]

Valle J. R., Picarelli Z. P. & Prado J. L. (1954). Histamine content and pharmacological properties of crude extracts from setae of urticating caterpillars. *Arch int. Pharmacodyn. Thér.* **98**: 324-334. [12]

Van den Berg M. A. (1971). Studies on the egg parasites of the mopani emperor moth, *Nudaurelia belina*. *Agric. Sci. S. Afr.* **3**: 33-36. [6]

Van der Pijl L. (1966). Ecological aspects of fruit evolution. *Proc. K. ned. Akad. Wet.* (C) **69**: 597-640. [12]

Van der Pijl L. (1969). *Principles of dispersal in higher plants*. Berlin. [12]

Van der Wall S. B. & Balda R. P. (1981). Ecology and evolution of food-storage behavior in conifer-seed-caching corvids. *Z. Tierpsychol.* **56**: 217-242. [12]

Van Emden H. F. (1978). Insects and secondary plant substances—an alternative viewpoint with special reference to aphids. *Phytochem. Soc. Eur. Symp.* **(15)** 309–323. [6]

Van Etten C. H. & Tookey H. L. (1979). Chemistry and biological effects of glucosinolates. Pp.471–500 in Rosenthal & Janzen (1979). [12]

Van Lenteren J. C., Bakker K. & van Alphen J. M. (1977). How to analyse host discrimination. *Ecol. Ent.* **3**: 71–75. [6]

Van Someren V. G. L. (1922). Notes on certain colour patterns in Lycaenidae. *Jl E. Afr. Uganda nat. Hist. Soc.* **17**: 18–21. [16]

Van Someren V. G. L. (1935–39). *Butterflies of Kenya and Uganda*. Supplement to vol. 1, and vol. 2. Nairobi. [1]

Van Someren V. G. L. (1974). List of foodplants of some East African Rhopalocera, with notes on the early stages of some Lycaenidae. *J. Lepid. Soc.* **28**: 315–331. [1, 6]

Van Someren V. G. L. (1975). Revisional notes on *Charaxes, Palla* and *Euxanthe*. Part X. *Bull Br. Mus. nat. Hist. (Ent.)* **32**: 65–136. [1]

Van Someren V. G. L. & Rogers K. St A. (1925–31). *Butterflies of Kenya and Uganda* [1]. Nairobi. [1]

Van Son G. (1949–79). *The butterflies of southern Africa*. Pretoria. Vol. **1**, *Papilionidae and Pieridae*, (1949); **2**, *Nymphalidae: Danainae and Satyrinae*, (1955); **3**, *Nymphalidae: Acraeinae*, (1963); **4**, *Nymphalidae: Nymphalinae* (ed. Vari L.), (1979). [1, 6, 12]

Vane-Wright R. I. (1971). The systematics of *Drusillopsis* and the supposed amathusiid *Bigaena*, with some observations on Batesian mimicry. *Trans. R. ent. Soc. Lond.* **123**: 97–123. [14, 20]

Vane-Wright R. I. (1972*a*). Pre-courtship activity and a new scent organ in butterflies. *Nature* **239**: 338–339. [1, 25]

[Vane-Wright R. I.] (1972*b*). Scent organs of male butterflies. *Rep. Br. Mus. nat. Hist.* **1969–1971**: 31–35. [25]

Vane-Wright R. I. (1975). An integrated classification for polymorphism and sexual dimorphism in butterflies. *J. Zool., Lond.* **177**: 329–337. [Intro, 20, 21, 23]

Vane-Wright R. I. (1976). A unified classification of mimetic resemblances. *Biol. J. Linn. Soc.* **8**: 25–56. [12, 14, 20, 23]

Vane-Wright R. I. (1978). Ecological and behavioural observations on Batesian mimicry. *Trans. R. ent. Soc. Lond.* **123**: 97-123. [20]

Vane-Wright R. I. (1979*a*). Towards a theory of the evolution of butterfly colour patterns under directional and disruptive selection. *Biol. J. Linn. Soc.* **11**: 141–152. [20, 23]

Vane-Wright R. I. (1979*b*). The coloration, identification and phylogeny of *Nessaea* butterflies. *Bull. Br. Mus. nat. Hist. (Ent.)* **38**: 27–56. [20, 23]

Vane-Wright R. I. (1980). A classification of sexual interactions, and the evolution of species-specific coloration in butterflies. *Nota lepid.* **3**: 91-93. [14, 20, 21, 23]

Vane-Wright R. I. (1981*a*). Only connect. *Biol. J. Linn. Soc.* **16**: 33–40. [12]

Vane-Wright R. I. (1981*b*). Mimicry and its unknown ecological consequences. Pp.157–168 in *Chance, change and challenge: the evolving biosphere* (eds Greenwood P. H. & Forey P. L.). London. [Intro, 3, 18, 23]

Vane-Wright R. I., Ackery P. R. & Smiles R. L. (1977). The polymorphism, mimicry, and host plant relationships of *Hypolimnas* butterflies. *Biol. J. Linn. Soc.* **9**: 285–297. [Intro]

Varley G. C. (1962). A plea for a new look at Lepidoptera with special reference to the scent distributing organs of male moths. *Trans. Soc. Br. Ent.* **15**: 29–40. [25]

Varley G. C. & Gradwell G. R. (1960). Key factors in population studies. *J. anim. Ecol.* **29**: 399–401. [2, 5]

Varley G. C., Gradwell G. R. & Hassell M. P. (1973). *Insect population ecology: an analytical approach*. Oxford. [15]

Varshney R. K. (1977). Index Rhopalocera Indica. An index of the local-lists of butterflies from India and neighbouring countries. *Rec. zool. Surv. India* **73**: 159–178. [1]

Vasconcellos Neto J. (1980). *Dinâmica de populacões* de Ithomiinae. PhD thesis, Univ. Estadual de Campinas, Brazil. [2]

Vaughan P. (1982). The palm king. *Nat. malay.* **7**: 22–25. [12]

Vazquez L. G. & Perez H. R. (1961). Observaciones sobre la biologia de *Baronia brevicornis*. *An. Inst. Biol. Univ. Mex.* **32**: 295–311. [1]

Verschaffelt E. (1911). The cause determining the selection of food in some herbivorous insects. *Proc. K. ned. Akad. Wet.* **13**: 536–542. [6]

Vetter R. S. & Rutowski R. L. (1978). External sex brand morphology of three sulfur butterflies. *Psyche* **85**: 383–393. [25]

Vickery M. L. & Vickery B. (1981). *Secondary plant metabolism*. Baltimore. [12]

Vielmetter W. (1954). Die Temperaturregulation des Kaisermantels in der Sonnenstrahlung. *Naturwissenschaften* **41**: 535–536. [20]

Vielmetter W. (1958). Physiologie der Verhaltens zur Sonnenstrahlung bei dem Tagfalter *Argynnis paphia*. I. Untersuchungen im Freiland. *J. Insect Physiol.* **2**: 13–37. [14]

Viette P. (1950). Lépidoptères Rhopalocères de l'Oceanie francaise. *Faune de l'Empire franc.* **13**: 1–101. [1]

Vince M. A. (1964). Use of the feet in feeding by the great tit, *Parus major*. *Ibis* **106**: 508–529. [12]

Von Euw J., Fishelson L., Parsons J. A., Reichstein T. & Rothschild M. (1967). Cardenolides in a grasshopper feeding on milkweeds. *Nature* **214**: 35–39. [12]

Von Euw J., Reichstein T. & Rothschild M. (1968). Aristolochic acid in the swallowtail butterfly *Pachlioptera aristolochiae*. *Israel J. chem.* **6**: 659–670. [12]

Von Frisch K. (1968). *The dance language and orientation of bees*. Cambridge, Mass. [20]

Waage J. K., Smiley J. T. & Gilbert L. E. (1981). The *Passiflora* problem in Hawaii: prospects and problems of controlling the forest weed *P. mollissima* with heliconiine butterflies. *Entomophaga* **26**: 275–284. [3]

Waddington C. H. (1953). Genetic assimilation of an acquired character. *Evolution* **7**: 118–126. [27]

Waddington C. H. (1957). *The strategy of the genes*. London. [16, 27]

Waddington C. H. (1961). Genetic assimilation. *Adv. Genet.* **10**: 257–294. [27]

Waddington C. H. (1975). *The evolution of an evolutionist.* Edinburgh. [27]

Wago H. (1977). Studies on the mating behavior of the pale grass blue, *Zizeeria maha argia.* II. Recognition of the proper mate by the male. *Kontyû* **45**: 92–96. [25]

Wago H. (1978*a*). Studies . . . III. Olfactory cues in sexual discrimination by males. *Appl. Ent. Zool. Tokyo* **13**: 283–289. [25]

Wago H. (1978*b*). Studies . . . IV. Experimental analyses of the role of the male odor in male–male interactions. *Zool. Mag. Tokyo* **87**: 240–246. [25]

Wago H., Unno K. & Suzuki Y. (1976). Studies . . . I. Recognition of conspecific individuals by flying males. *Appl. Ent. Zool. Tokyo* **11**: 302–311. [25]

Waldbauer G. P. (1968). The consumption and utilization of food by insects. *Adv. Insect Physiol.* **3**: 229–282. [6]

Waldbauer G. P. (1977). Mimicry in animals. *Int. Encycl. Psych. Psychol. Psychoan. Neurology* **7**: 231–235. [12]

Waldbauer G. P. & Sternberg J. G. (1975). Saturniid moths as mimics: an alternative explanation of attempts to demonstrate mimetic advantage in nature. *Evolution* **29**: 650–658. [22]

Walker J. J. (1914). The geographical distribution of *Danaida plexippus*, with especial reference to its recent migrations. *Entomologist's mon. Mag.* **50**: 181–193, 233–237. [17]

Walker T.J. & Riordan A. J. (1981). Butterfly migration: are synoptic-scale wind systems important? *Ecol. Ent.* **6**: 433–440.

Wallace A. R. (1867*a*). [Discussion on brightly coloured larvae]. *Proc. ent. Soc. Lond.* **1867**: lxxx–i. [12]

Wallace A. R. (1867*b*). On the Pieridae of the Indian and Australian regions. *Trans. ent. Soc. Lond.* **1867**: 301–416. [12]

Wallace A. R. (1877). The colours of animals and plants. *Macmillan's Magazine* **36**: 384–408, 464–471. [20]

Wallace A. R. (1889). *Darwinism* (1st edn). London. [Intro, 20, 21, 23]

Wallace B. (1968). *Topics in population genetics.* New York. [16]

Walsten D. M. (1977). Tigers without their stripes. *Bull. Field Mus. nat. Hist.* **48**: 12–13. [22]

Wang S. C. (1965). Emetic and antiemetic drugs. Pp.255–328 in *Physiological pharmacology* (eds Root W. S. & Hoffmann F. G.). New York. [12]

Warren B. C. S. (1926). Monograph of the tribe Hesperiidi (European species) with a revised classification of the subfamily Hesperiinae (Palaearctic species) based on the genital armature of the males. *Trans. ent. Soc. Lond.* **74**: 1–170. [1]

Warren B. C. S. (1936). *Monograph of the genus* Erebia. London. [1]

Warren B. C. S. (1944). Review of the classification of the Argynnidi, with a systematic revision of the genus *Boloria. Trans. R. ent. Soc. Lond.* **94**: 1–102. [1]

Warren B. C. S. (1981). *Supplement to monograph of the genus* Erebia. Faringdon, Oxon. [1]

Warren M. S. (1980). The ecology of the wood white butterfly. *Rep. Inst. terr. Ecol.* **1979**: 50. [2]

Warren M. S. (1981). *The ecology of the wood white butterfly,* Leptidea sinapis. PhD thesis. Univ. Cambridge, UK. [2, 5, 33]

Warren M. S., Thomas C. D. & Thomas J. A. (1981). *The heath fritillary: survey and conservation report.* JCCBI, London. [33]

Wasserman S. S. & Futuyma D. J. (1981). Evolution of hostplant utilization in laboratory populations of the southern cowpea weevil, *Callosobruchus maculatus. Evolution* **35**: 605–617. [6]

Wasserthal L. T. (1975). The role of butterfly wings in regulation of body temperature. *J. Insect Physiol.* **21**: 1921–1930. [14, 20, 28]

Watanabe M. (1976). A preliminary study on population dynamics of the swallow-tail butterfly, *Papilio xuthus*, in a deforested area. *Popul. Ecol. Kyoto Univ.* **17**: 200–210. [2, 10]

Watanabe M. (1978). Adult movements and resident ratios of the black-veined white, *Aporia crataegi*, in a hilly region. *Jap. J. Ecol.* **28**: 101–109. [2]

Watanabe M. (1979*a*). Population dynamics of a pioneer tree, *Zanthoxylum ailanthoides*, a host plant of the swallowtail butterfly, *Papilio xuthus. Researches Popul. Ecol. Kyoto Univ.* **20**: 265–277. [2]

Watanabe M. (1979*b*). Natural mortalities of the swallowtail butterfly, *Papilio xuthus*, at patchy habitats along the flyways in a hilly region. *Jap. J. Ecol.* **29**: 85–93. [2, 10]

Watanabe M. (1979*c*). Population sizes and resident ratios of the swallowtail butterfly, *Papilio polytes*, at a secondary bush community in Dharan, Nepal. *Kontyû* **47**: 291–297. [2]

Watanabe M. & Omata K. (1978). On the mortality of the lycaenid butterfly *Artopoetes pryeri. Jap. J. Ecol.* **28**: 367–370. [2, 10]

Watson A. & Whalley P. E. S. (1975). *The dictionary of butterflies and moths in color.* New York. [1]

Watson J. (1865*a*). On the microscopical examination of plumules. *Entomologist's mon. Mag.* **2**: 1-2. [25]

Watson J. (1865*b*). On certain scales of some diurnal Lepidoptera. *Mem. lit. phil. Soc. Manchr* (3) **2**: 63–70. [25]

Watson J. (1868). On the plumules or battledore-scales of Lycaenidae. *Mem. lit. phil. Soc. Manchr* (3) **3**: 128–133. [25]

Watson R. W. (1969). Notes on *Melitaea cinxia*, 1945–1968. *Entomologist's Rec. J. Var.* **81**: 18–20. [33]

Watt W. B. (1968). Adaptive significance of pigment polymorphisms in *Colias* butterflies. I. Variation of melanin pigment in relation to thermoregulation. *Evolution* **22**: 437–458. [16, 20, 28]

Watt W. B. (1969). Adaptive . . . II. Thermoregulation and photoperiodically controlled melanin variation in *Colias eurytheme. Proc. natn Acad. Sci. U.S.A.* **63**: 767–774. [16]

Watt W. B. (1983). Adaptation at specific loci. II. Demographic and biochemical elements in the maintenance of *Colias* PGI polymorphism. *Genetics* **103**: 691–724. [2]

Watt W. B., Hoch P. & Mills S. G. (1974). Nectar resources use by *Colias* butterflies: chemical and visual aspects. *Oecologia* **14**: 353–374. [6]

Watt W. B., Chew F. S., Snyder L. R. G., Watt A. G. & Rothschild D. E. (1977). Population structure of pierid butterflies. I. Numbers and movements of some montane *Colias* species. *Oecologia* **27**: 1-22. [2]

Watt W. B., Han D. & Tabashnik B. E. (1979). Population . . . II. A "native" population of *Colias philodice eriphyle* in Colorado. *Oecologia* **44**: 44-52. [2, 16]

Weatherston J. & Percy J. E. (1977). Pheromones of male Lepidoptera. Pp.295-307 in *Advances in invertebrate reproduction* **1** (eds Adiyodi K. G. & Adiyodi R. G.). Karivellur, India. [25]

Weems H. (1981). The Florida state endangered insect programme. *Atala* **6**: 37-40. [33]

Wehner R. (1975). Pattern recognition in insects. Pp.75-113 in Horridge (1975). [20]

Wehrhahn C. & Allard R. W. (1965). The detection and measurement of the effects of individual genes involved in the inheritance of a quantitative character in wheat. *Genetics* **51**: 109-119. [27]

Weismann A. (1878). Über Duftschuppen. *Zool. Anz.* **1**: 98-99. [25]

Welch H. E. (1963). Nematode infection. Pp.363-392 in *Insect pathology: an advanced treatise* **2** (ed. Steinhaus E. A.). New York, London. [10]

Wells T. C. E. (1976). *A report on an ecological survey of the chalk and limestone grasslands on the Royal Armoured Corps gunnery ranges, Lulworth, Dorset.* Abbots Ripton, Cambridge, UK. [33]

Wells T., Bell S. & Frost A. (1981). *Creating attractive grasslands using natural plant species.* NCC, Shrewsbury. [33]

Welty J. C. (1975). *The life of birds.* Philadelphia. [12]

West D. A. & Hazel W. N. (1982). An experimental test of natural selection for pupation site in swallowtail butterflies. *Evolution* **36**: 152-159. [12]

West-Eberhard M. J. (1979). Sexual selection, social competition and evolution. *Proc. Am. phil. Soc.* **123**: 222-234. [20]

Wheeler L. R. (1939). Deaths among butterflies. *Proc. Linn. Soc. Lond.* **152**: 79-88. [10]

Wheelwright N. T. & Orians G. H. (1982). Seed dispersal by animals: contrasts with pollen dispersal, problems of terminology, and constraints on coevolution. *Am. Nat.* **119**: 402-413. [12]

White R. R. (1974). Foodplant defoliation and larval starvation of *Euphydryas editha. Oecologia* **14**: 307-315. [2, 6]

White R. R. (1980). Inter-peak dispersal in alpine checkerspot butterflies. *J. Lepid. Soc.* **34**: 353-362. [2]

White R. R. & Levin M. P. (1981). Temporal variation in vagility: implications for evolutionary studies. *Am. Midl. Nat.* **105**: 348-357. [2]

White R. R. & Singer M. C. (1974). Geographical distribution of hostplant choice in *Euphydryas editha. J. Lepid. Soc.* **28**: 103-107. [2, 6]

Whittaker R. H. (1970). The biochemical ecology of higher plants. Pp.43-70 in Sondheimer & Simeone (1970). [12]

Whittaker R. M. & Feeny P. (1971). Allelochemics: chemical interactions between species. *Science* **171**: 757-770. [12]

Wickler W. (1965). Mimicry and the evolution of animal communication. *Nature* **208**: 519-521. [23]

Wickler W. (1968). *Mimicry in plants and animals.* London (translated from original German by Martin R. D.). [12, 14, 16, 20, 23]

Wiens J. A. (1981). Single-sample surveys of communities: are the revealed patterns real? *Am. Nat.* **117**: 90-98. [3]

Wigglesworth V. B. (1964). *The life of insects.* London. [12]

Wigglesworth V. B. (1972). *The principles of insect physiology* (7th edn). London. [12]

Wiklund C. (1974*a*). The concept of oligophagy and the natural habitats and hostplants of *Papilio machaon. Entomologica scand.* **5**: 151-160. [8]

Wiklund C. (1974*b*). Oviposition preferences in *Papilio machaon* in relation to the hostplants of the larvae. *Entomologia exp. appl.* **17**: 189-198. [6]

Wiklund C. (1975*a*). Pupal colour polymorphism in *Papilio machaon* and the survival in the field of cryptic versus non-cryptic pupae. *Trans. R. ent. Soc. Lond.* **127**: 73-84. [12]

Wiklund C. (1975*b*). The evolutionary relationship between adult oviposition preferences and larval host plant range in *Papilio machaon. Oecologia* **18**: 185-197. [6-8]

Wiklund C. (1977*a*). Oviposition, feeding and spatial separation of breeding and foraging habitats in a population of *Leptidea sinapis. Oikos* **28**: 56-68. [2, 6, 33]

Wiklund C. (1977*b*). Observationer over agglaggning, fodosok och vila hos Donzels blavinge, *Aricia nicia scandicus. Ent. Tidskr.* **98**: 1-4. [2, 6]

Wiklund C. (1977*c*). Courtship behaviour in relation to female monogamy in *Leptidea sinapis. Oikos* **29**: 275-283. [25]

Wiklund C. (1981). Generalist *vs* specialist oviposition behaviour in *Papilio machaon* and functional aspects on the hierarchy of oviposition preferences. *Oikos* **36**: 163-170. [2, 6]

Wiklund C. (1982). Generalist versus specialist utilization of host plants among butterflies. *Proc. int. Symp. Insect-Plant Relationships* (**5**): 181-191. [7]

Wiklund C. & Åhrberg C. (1978). Hostplants, nectar source plants and habitat selection of males and females of *Anthocharis cardamines. Oikos* **31**: 169-183. [2, 6, 8, 16, 33]

Wiklund C. & Fagerström T. (1977). Why do males emerge before females? A hypothesis to explain the incidence of protandry in butterflies. *Oecologia* **31**: 153-158. [2]

Wiklund C. & Persson A. (1983). Fecundity, and the relation of egg weight variation to offspring fitness in the speckled wood butterfly, *Pararge aegeria*, or why don't butterfly females lay more eggs? *Oikos* **40**: 53-63. [6]

Wiklund C., Persson A. & Wickman P. O. (1983) Larval aestivation and direct development as alternative strategies in the speckled wood butterfly, *Pararge aegeria*, in Sweden. *Ecol. Ent.* **8**: 233-238. [Intro]

Wilcoxon H. C. (1977). Long-delay learning of ingestive aversions in quail. Pp.419-453 in Barker *et al.* (1977). [12]

Wilcoxon H. C., Dragoin W. B. & Kral P. A. (1971). Illness-induced aversions in rat and quail: relative salience of visual and gustatory cues. *Science* **171**: 826-828. [12]

Wille A. & Fuetes G. (1975). Efecto de la ceniza del Volcán Irazú (Costa Rica) en algunos insectos. *Revta Biol. trop.* **23**: 165-175. [31]

Williams C. B. (1930). *The migration of butterflies.* Edinburgh. [2]

Williams C. B. (1958). *Insect migration.* London. [26]

Williams C. B. (1970). The migration of the painted lady butterfly, *Vanessa cardui*, with special reference to North America. *J. Lepid. Soc.* **24**: 157-175. [2]

Williams C. B. (1976). The migrations of the hesperid butterfly, *Andronymus neander*, in Africa. *Ecol. Ent.* **1**: 213-220. [26]

Williams E. (1981). Thermal influences on oviposition in the montane butterfly *Euphydryas gillettii. Oecologia* **50**: 342-346. [6]

Williams E. (In press). Movement and dispersal in a montane butterfly. *Oecologia.* [2]

Williams E., Holdren C. E. & Ehrlich P. R. (In press). The life history and ecology of *Euphydryas gillettii. J. Lepid. Soc.* **38**. [2]

Williams G. C. (1966). *Adaptation and natural selection.* Princeton. [12]

Williams G. C. (1975). *Sex and evolution.* Princeton. [20]

Williams K. S. (1983). The coevolution of *Euphydryas chalcedona* butterflies and their larval hostplants. III. Oviposition behaviour and host plant quality. *Oecologia* **56**: 336-340. [2]

Williams K. S. & Gilbert L. E. (1981). Insects as selective agents on plant vegetative morphology: egg mimicry reduces egg laying by butterflies. *Science* **212**: 467-469. [3, 6, 12]

Williams K. S., Lincoln D. E. & Ehrlich P. R. (1983*a*). The coevolution of *Euphydryas chalcedona* . . . II. Maternal and host plant effects on larval growth, development, and food-use efficiency. *Oecologia* **56**: 330-335. [2]

Williams K. S., Lincoln D. E. & Ehrlich P. R. (1983*b*). The coevolution of *Euphydryas chalcedona* . . . I. Larval feeding behaviour and host plant chemistry. *Oecologia* **56**: 323-329. [2]

Willis M. A. & Birch M. C. (1982). Male lekking and female calling in the same population of the arctiid moth, *Estigmene acrea. Science* **218**: 168-170. [25]

Wilson A. C., Bush G. L., Case S. M. & King M. C. (1975). Social structuring of mammalian populations and rate of chromosomal evolution. *Proc. natn Acad. Sci. U.S.A.* **72**: 5061-5065. [19]

Wilson E. O. (1971). *The insect societies.* Cambridge, Mass. [12]

Wilson E. O. (1975). *Sociobiology: the new synthesis.* Cambridge, Mass. [12, 21]

Wilson E. O., Bossert W. H. & Regnier F. E. (1969). A general method for estimating threshold concentrations of odorant molecules. *J. Insect Physiol.* **15**: 597-610. [12]

Windecker W. (1939). *Euchelia jacobaeae* und das Schutztrachtenproblem. *Z. Morph. Okol. Tiere* **35**: 84-139. [12]

Wirth W. W. (1956). New species and records of biting midges ectoparasitic on insects. *Ann. ent. Soc. Am.* **49**: 356-364. [11]

Wirth W. W. (1972). Midges sucking blood of caterpillars. *J. Lepid. Soc.* **26**: 65. [11]

Wolfson J. L. (1980). Oviposition response of *Pieris rapae* to environmentally induced variation in *Brassica nigra. Entomologia exp. appl.* **27**: 223-232. [6]

Wonfor T. W. (1869). On certain butterfly scales, characteristic of sex. *Q. Jl microsc. Sci.* (N.S.) **8**: 80-83. [25]

Woodhouse L. G. O. (1952). *The butterfly fauna of Ceylon.* Colombo, Ceylon. [1]

Woodruff D. S. (1973). Natural hybridization and hybrid zones. *Syst. Zool.* **22**: 213-218. [27]

Wright D. M. (1981). Historical and biological observations of Lepidoptera captured by ambush bugs. *J. Lepid. Soc.* **35**: 120-123. [12]

Wright S. (1931). Evolution in Mendelian populations. *Genetics* **10**: 97-159. [19]

Wright S. (1940). Breeding structure of populations in relation to speciation. *Am. Nat.* **74**: 232-248. [19]

Wright S. (1945). A critical review. *Ecology* **26**: 415-419. [12]

Wright S. (1948). On the roles of directed and random changes in gene frequency in the genetics of populations. *Evolution* **2**: 279-294. [16]

Wright S. (1952). The genetics of quantitative variability. Pp.5-41 in *Quantitative inheritance* (eds Reeve E. C. R. & Waddington C. H.). London. [27]

Wright S. (1967). "Surfaces" of selective value. *Proc. natn Acad. Sci. U.S.A.* **58**: 165-172. [6]

Wright S. (1968). *Evolution and the genetics of populations: genetic and biometric foundations.* Chicago. [27]

Wright S. & Clarke K. U. (1981). Photoperiod and larval body size: integrated factors controlling onset of the moulting cycle in *Heliconius melpomene. J. Zool., Lond.* **194**: 143-163. [6]

Wynne-Edwards V. C. (1962). *Animal dispersion in relation to social behaviour.* Edinburgh. [20]

Yagi N. & Koyama N. (1963). *The compound eye of Lepidoptera: approach from organic evolution.* Tokyo. [20]

Yamamoto M. (1981). Comparison of population dynamics of two pierid butterflies, *Pieris rapae crucivora* and *P. napi nesis*, living in the same area and feeding on the same plant in Sapporo, northern Japan. *J. Fac. Sci. Hokkaido Univ.* (6) *Zool.* **22**: 202-249. [2]

Yamamoto M. & Ohtani T. (1979). Number of eggs laid by *Pieris rapae crucivora*, compared with *P. napi nesis*, in Sapporo. *Kontyû* **47**: 530-539. [2, 6]

Yamanaka M., Suzuki Y. & Takagi M. (1978). Estimation of egg number oviposited per day by a female of *Papilio xuthus. Kontyû* **46**: 329-334. [6]

Yata O. (1981). Pieridae. Pp.205-438 in *Butterflies of the south-east Asian islands* **2** (1) (ed. Tsukada E.). Tokyo. [1]

Yata O., Shima H., Saigusa T., Nakanishi A., Suzuki Y. & Yoshida A. (1979). Photoperiodic response of four Japanese species of the genus *Pieris. Kontyû* **47**: 185-190. [30]

Young A. M. (1971). Wing coloration and reflectance in *Morpho* butterflies as related to reproductive behaviour and escape from avian predators. *Oecologia* **7**: 209-222. [20]

Young A. M. (1972). Breeding success and survivorship in some tropical butterflies. *Oikos* **23**: 318-326. [6]

Young A. M. (1973). Studies on comparative ecology and ethology in adult populations of several species of *Morpho* butterflies. *Stud. neotrop. Fauna* **8**: 17-50. [6]

Young A. M. (1975). Feeding behaviour of *Morpho* butterflies in a seasonal tropical environment. *Revta Biol. trop.* **23**: 101-123. [16]

Young A. M. (1977). Studies on the biology of *Parides iphidamas* in Costa Rica. *J. Lepid. Soc.* **31**: 100-108. [12]

Young A. M. (1978*a*). The biology of the butterfly *Aeria eurimeda agna* in Costa Rica. *J. Kans. ent. Soc.* **51**: 1-10. [9]

Young A. M. (1978*b*). "Disappearances" of eggs and larvae of *Heliconius* butterflies in northeastern Costa Rica. *Ent. News* **89**: 81-87. [2]

Young A. M. (1979*a*). Oviposition of the butterfly *Battus belus varus*. *J. Lepid. Soc.* **33**: 56-57. [12]

Young A. M. (1979*b*). The evolution of eyespots in tropical butterflies in response to feeding on rotting fruit: an hypothesis. *Jl N. Y. ent. Soc.* **87**: 66-77. [16]

Young A. M. (1980). The interaction of predators and "eyespot butterflies" feeding on rotting fruits and soupy fungi in tropical forests. *Entomologist's Rec. J. Var.* **90**: 63-69. [16]

Young A. M. & Moffett M. W. (1979). Studies on the population biology of the tropical butterfly *Mechanitis isthmia* in Costa Rica. *Am. Midl. Nat.* **101**: 309-319. [10]

Young A. M. & Stein D. (1976). Studies on the evolutionary biology of the neotropical butterfly *Anartia fatima* in Costa Rica. *Contr. Biol. Geol. Milwaukee publ. Mus.* **(8)**: 29pp. [20]

Young A. M. & Thomason J. H. (1974). The demography of a confined population of the butterfly *Morpho peleides* during a tropical dry season. *Stud. neotrop. Fauna* **9**: 1-34. [2]

Young L. C. H. (1907). Common butterflies of the plains of India. *J. Bombay nat. Hist. Soc.* **17**: 921-927. [6]

Zagatti P. (1981). Comportement sexuel de la pyrale de la canne à sucre, *Eldana saccharina*, lié à deux phéromones émises par le male. *Behaviour* **78**: 81-98. [25]

Zagatti P., Kunesch G. & Morin N. (1981). La vanilline, constituant majoritaire de la sécrétion aphrodisiaque émisse par les androconies du mâle de la Pyrale de la Canne à sucre: *Eldana saccharina*. *C.R. Acad. Sci. Paris* (3) **292**: 633-635. [25]

Zahavi A. (1975). Mate selection—a selection for a handicap. *J. theor. Biol.* **53**: 205-214. [20]

Zahn W. (1935). Über den Geruchssinn einiger Vögel. *Z. vergl. Physiol.* **19**: 785-796. [12]

Zahorik D. M. (1977). Associative and non-associative factors in learned food preferences. Pp.181-199 in Barker *et al.* (1977). [12]

Zahorik D. M. & Houpt K. A. (1977). The concept of nutritional wisdom: applicability of laboratory learning models to large herbivores. Pp.45-67 in Barker *et al.* (1977). [12]

Zahorik D. M. & Houpt K. A. (1981). Species differences in feeding strategies, food hazards, and the ability to learn food aversions. Pp.289-310 in Kamil & Sargent (1981). [12]

Zalucki M. P. & Kitching R. L. (1982). The analysis and description of movement in adult *Danaus plexippus*. *Behaviour* **80**: 174-198. [26]

Zukowski R. (1959). Extinction and decrease of the butterfly *Parnassius apollo* in Polish territories. *Sylwan* **103**: 15-30. [2]

Systematic Index

During the preparation of this index, opportunity has been taken to indicate authors' names for all Lepidoptera species and, where appropriate, for lepidopterous genera—authors' names for all taxa having been intentionally omitted from the main text. Many author names have been abbreviated; these are listed below. In addition, moth families are indicated (in parentheses), while butterfly subfamilies, in accordance with Fig. 1.1 (p.10), are denoted by superscript numbers (see list below). Thus the entry *Maniola jurtina* L.[26] indicates that this familiar brown butterfly was named by Linnaeus, and is treated by us as belonging to the Nymphalidae: Satyrinae. Likewise, Apaturini[30] indicates that the purple-emperor tribe of butterflies belongs to the Nymphalidae: Nymphalinae (in the system we have adopted—see Ch. 1).

Scientific nomenclature is, for both good and bad reasons, permanently unstable. Where names have recently changed for purely technical reasons, some of these have (for convenience) been left unaltered in the main text, but the more 'correct' form is indicated here. Thus the index entry *Phasis* (corr.[ectly] *Trimenia*) *argyroplaga* Dickson means that this African lycaenid is referred to the genus *Trimenia* in the most recent taxonomic literature, but is retained in the text under the older name *Phasis*. Cross-references are included. Another reason for change is less easy to deal with: disagreements (sometimes violent!) over rank (see Ehrlich & Murphy 1982). We have tried, where practical, to follow a conservative course—in particular, by judicious use of subgenerea. Thus the european White Admiral is treated by some authors as belonging to a separate genus, *Ladoga*, while others retain it in the older and more inclusive genus, *Limenitis*. We have allowed both in the text, but in the index we treat *Ladoga* as a subgenus of *Limenitis*, and cross-refer it in that way.

Abbreviations of Authors' Names

B., Boisduval; **B. & L.**, Boisduval & Leconte; **C.**, Cramer; **D.**, Doubleday; **D. & S.**, Denis & Schiffermüller; **E.**, Edwards; **F.**, Fabricius; **F. & F.**, Felder & Felder; **G.**, Godart; **G. & S.**, Godman & Salvin; **H.** Hübner; **Hew.**, Hewitson; **L.**, Linnaeus; **M.**, Moore; **Mén.**, Ménétriés; **O.**, Ochsenheimer; **R.**, Rottemburg; **Roth.**, Rothschild; **S.**, Staudinger; **W.**, Westwood.

Ciphers for Butterfly Subfamilies

HESPERIIDAE: **1**, Pyrrhopyginae; **2**, Pyrginae; **3**, Trapezitinae; **4**, Hesperiinae; **5**, Megathyminae; **6**, Coeliadinae. PAPILIONIDAE: **7**, Baroniinae; **8**, Parnassiinae; **9**, Papilioninae. PIERIDAE: **10**, Pseudopontiinae; **11**, Dismorphiinae; **12**, Pierinae; **13**, Coliadinae. LYCAENIDAE: **14**, Lipteninae; **15**, Poritiinae; **16**, Liphyrinae; **17**, Miletinae; **18**, Curetinae; **19**, Theclinae; **20**, Lycaeninae; **21**, Polyommatinae; **22**, Riodininae; **23**, Styginae. NYMPHALIDAE: **24**, Brassolinae; **25**, Amathusiinae; **26**, Satyrinae; **27**, Morphinae; **28**, Calinaginae; **29**, Charaxinae; **30**, Nymphalinae; **31**, Heliconiinae; **32**, Acraeinae; **33**, Danainae; **34**, Ithomiinae; **35**, Tellervinae; **36**, Libytheinae.

A

B

Subject Index

M

N

O

P